AF876399

Lecture Notes on Data Engineering and Communications Technologies

Volume 113

Series Editor

Fatos Xhafa, Technical University of Catalonia, Barcelona, Spain

The aim of the book series is to present cutting edge engineering approaches to data technologies and communications. It will publish latest advances on the engineering task of building and deploying distributed, scalable and reliable data infrastructures and communication systems.

The series will have a prominent applied focus on data technologies and communications with aim to promote the bridging from fundamental research on data science and networking to data engineering and communications that lead to industry products, business knowledge and standardisation.

Indexed by SCOPUS, INSPEC, EI Compendex.

All books published in the series are submitted for consideration in Web of Science.

More information about this series at https://link.springer.com/bookseries/15362

Aboul Ella Hassanien · Rawya Y. Rizk ·
Václav Snášel · Rehab F. Abdel-Kader
Editors

The 8th International Conference on Advanced Machine Learning and Technologies and Applications (AMLTA2022)

Springer

Editors
Aboul Ella Hassanien
Faculty of Computer and AI
Cairo University
Giza, Egypt

Václav Snášel
Department of Computer Science
VŠB-TUO
Ostrava-Poruba, Czech Republic

Rawya Y. Rizk
Port Said University
Port Fouad, Egypt

Rehab F. Abdel-Kader
Faculty of Engineering
Port Said University
Port Fouad, Egypt

ISSN 2367-4512 ISSN 2367-4520 (electronic)
Lecture Notes on Data Engineering and Communications Technologies
ISBN 978-3-031-03917-1 ISBN 978-3-031-03918-8 (eBook)
https://doi.org/10.1007/978-3-031-03918-8

© The Editor(s) (if applicable) and The Author(s), under exclusive license
to Springer Nature Switzerland AG 2022
This work is subject to copyright. All rights are solely and exclusively licensed by the Publisher, whether the whole or part of the material is concerned, specifically the rights of translation, reprinting, reuse of illustrations, recitation, broadcasting, reproduction on microfilms or in any other physical way, and transmission or information storage and retrieval, electronic adaptation, computer software, or by similar or dissimilar methodology now known or hereafter developed.
The use of general descriptive names, registered names, trademarks, service marks, etc. in this publication does not imply, even in the absence of a specific statement, that such names are exempt from the relevant protective laws and regulations and therefore free for general use.
The publisher, the authors and the editors are safe to assume that the advice and information in this book are believed to be true and accurate at the date of publication. Neither the publisher nor the authors or the editors give a warranty, expressed or implied, with respect to the material contained herein or for any errors or omissions that may have been made. The publisher remains neutral with regard to jurisdictional claims in published maps and institutional affiliations.

This Springer imprint is published by the registered company Springer Nature Switzerland AG
The registered company address is: Gewerbestrasse 11, 6330 Cham, Switzerland

Preface

This volume constitutes the refereed proceedings of the 8th International Conference on Advanced Machine Learning Technologies and Applications, AMLTA2022, held in Port Said University, Port Fouad, Egypt, during May 5–7, 2022. The 8th edition of AMLTA will be organized by the Scientific Research Group in Egypt (SRGE), Egypt, collaborating with Port Said University, Egypt, and VSB-Technical University of Ostrava, Czech Republic. AMLTA series aims to become the premier international conference for an in-depth discussion on the most up-to-date and innovative ideas, research projects, and practices in the field of machine learning technologies and their applications. The accepted papers cover current research on advanced machine learning technology, including deep learning technology, sentiment analysis, cyber-physical system, IoT, and smart cities informatics and AI against COVID-19, data mining, power and control systems, business intelligence, social media, and digital transformation, and smart systems.

We want to emphasize that the success of AMLTA2022 would not have been possible without the support of many committed volunteers who generously contributed their time, expertise, and resources toward making the conference an unqualified success. We express our sincere thanks to the plenary and tutorial speakers, workshop special session chairs, and international program committee members for helping us to formulate a rich technical program. We want to extend our sincere appreciation for the outstanding work contributed over many months by the Organizing Committee: local organization chair and publicity chair. We also wish to express our appreciation to the SRGE members for their assistance. Finally, thanks to the Springer team for their support in all stages of the proceedings' production.

Organization

Honorary Chair

Ayman Mohamed	President of Port Said University, Egypt

General Chairs

Rawya Y. Rizk	Port Said University, Egypt
Vaclav Snasel	Czech Republic

Co-chairs

Rehab F. Abdel-Kader	Port Said University, Egypt
Ashraf Darwish	Scientific Research Group in Egypt

International Advisory Board

Mohamed Abdel-Azim Mohamed, Egypt
Sherif Massoud El-Badwy, Egypt
Norimichi Tsumura, Japan
Mohamed F. Tolab, Egypt
Saad Darwish, Egypt
Mincong Tang, China
Tai-hoon Kim, China
Nagwa Badr, Egypt
Vaclav Snasel, Czech Republic
Janusz Kacprzyk, Poland
Siddhartha Bhattacharyya, India
Hesham El-Deeb, Egypt
Khaled Shaalan, UAE

Magdy Zakariya, Egypt
Diego Oliva, Mexico
Fatos Xhafa, Spain
Mohammad Alia, Jordan

Publication Chair

Aboul Ella Hassanien	Scientific Research Group in Egypt, Egypt

Program Chairs

Rehab F. Abdel-Kader	Port Said University, Egypt
Said Salloum	United Arab Emirates

Publicity Chairs

Mohamed AbdelFatah, Egypt
Mourad Rafat, Egypt

Technical Program Committee

Said El Kafhali	Morocco
Jatinderkumar R. Saini	India
Shakir Ullah	University of Louisiana Monroe, USA
Kiet Van Nguyen	University of Information Technology, VNU-HCM, Vietnam
Ehlem Zigh	INTTIC, Oran, Algeria
Gandhiya Vendhan	India
Randa Atta	Port Said University, Egypt
Sherif M. Abuelenin	Port Said University, Egypt
Heba Nashaat	Port Said University, Egypt
Heba Y. M. Soliman	Port Said University, Egypt
Heba M. Abdel-Atty	Port Said University, Egypt
Mohamed F. Abdelkader	Port Said University, Egypt
Anjali Awasthi	Concordia University, Canada
Abdelkrim Haqiq	Hassan 1st University, Morocco
A. V. Senthil Kumar	Hindusthan College of Arts and Science, India
Marius Balas	Aurel Vlaicu University of Arad, Romania
Mario Pavone	University of Catania, Italy
Mohamed Khalgui	University of Carthage, Tunisia
Muaz A. Niazi	COMSATS Institute of Information Technology, Pakistan
Nickolas S. Sapidis	University of Western Macedonia, Greece
Nilanjan Dey	Techno India College of Technology, India

Nizar Banu P. K.	B.S. Abdur Rahman University, India
Nizar Rokbani	University of Sousse, Tunisia
Peter Géczy	National Institute of Advanced Industrial Science and Technology (AIST), Japan
Philip Moore	University College Falmouth, UK
Valentina Balas	Aurel Vlaicu University of Arad, Romania
Viet-Thanh Pham	Hanoi University of Science and Technology, Vietnam
Brian Galli	Long Island University, USA
Camelia Pintea	TU Cluj-Napoca, Romania
Chakib Bennjima	University of Sousse, Tunisia
Christos Volos	Aristotle University of Thessaloniki, Greece
Arezki Fekik	Algeria
Ashraf Darwish	Egypt
Dabiah Alboaneen	Saudi Arabia
Abdelrahman Sayed Sayed	Egypt
Tien-Wen Sung	China
Mazen Juma	Saudi Arabia
Tarek Abd El-Hafeez	Egypt
Mourad Raafat	Egypt
Mohamed Torky	Egypt
Roheet Bhatnagar	India
Omar Reyad	Egypt
Walid Abdelmoez	Egypt
Pranay Yadav	India
Cai Dai	China
Nashwa Abdelbaki	Egypt
Mohammed A.-M. Salem	Egypt
Ghada Hamed	Egypt
Jacob Howe	UK
Ghazala Bilquise	United Arab Emirates
Ayman Haggag	Egypt
Amr Abdel Fatah Ahmed	Egypt
Fu-Hsiang Chang	China
Dimitris Ampeliotis	Greece
Fernando Serrano	Saudi Arabia
Manash Sarkar	India
Mohammed Habes	USA
Hsiao Chuan Wang	China
Saad Darwish	Alexandria University, Egypt
Lamiaa Hassan	Egypt
Nashwa Ahmad Kamal	Egypt
Faisal Talib	Aligarh Muslim University, India
Hajar Mousannif	Cadi Ayyad University, Morocco
Irene Mavrommati	Hellenic Open University, Greece

Jaouad Boumhidi	Sidi Mohammed Ben Abdellah University (USMBA), Morocco
Jesus Manuel Munoz-Pacheco	Autonomous University of Puebla, Mexico
Jihene Malek	Higher Institute of Applied Sciences and Technology Sousse, Tunisia
Jin Xu	Behavior Matrix LL, USA
Kemal Polat	Abant Izzet Baysal University, Turkey
Kusuma Mohanchandra	Dayananda Sagar College of Engineering, India
Laura Romero	University of Seville, Spain
Mehmet Cunkas	Selcuk University, Turkey
Mahdi Bastan	University of Eyvanekey, Iran
Mariem Ben Abdallah	Monastir University, Tunisia

Local Arrangement Chairs

Mohamed Abd Elfattah, Egypt
Heba Aboul Ella, Egypt

Contents

Intelligent Optimization

Deep Learning and Applications

Plant Leaf Diseases Detection and Identification Using Deep Learning Model

Dang Huu Chau[1], Duong Chan Tran[1], Hao Nhat Vo[1], Tai Thanh Do[1], Trong Huu Nguyen[1], Bao Quoc Nguyen[1], Narayan C. Debnath[2], and Vinh Dinh Nguyen[1(✉)]

[1] FPT University, Can Tho 94000, Vietnam
{dangchce140529,duongtcce140484,haovnce140475,trongnhce140372, taidtce140136,baonqce140454,vinhnd18}@fpt.edu.vn

[2] School of Computing and Information Technology, Eastern International University, Thu Dau Mot, Vietnam
narayan.debnath@eiu.edu.vn

Abstract. In agriculture, leaf diseases often appear in changing weather conditions. Changing weather conditions can be very rainy or very hot or very humid. These factors make plants susceptible to bacterial, fungal, and viral infections. This research investigates three common diseases on apple trees, such as black rot, fish scale disease, snow apple rust. We use the deep learning method-based Yolo-v5 model along with the proposed stable information based on auto-encoder to train a Plant-Village dataset with 5740 images; The proposed system uses Google Colab for training phases. The data set is divided into three parts: training, validation, and test. We used 70% dataset (4018 images) for training, used 20% dataset (1148) for performing the validation step, and 10% dataset (574 images) for a testing phase. After training, we obtained the result of 81.28% in terms of the detection rate and 91.93% in terms of the classification rate by using the Plant-Village dataset.

Keywords: Leaf diseases detection · Deep learning · Leaf diseases classification

1 Introduction

The establishment of The World Trade Organization (WTO) in 1995 and the Sanitary and Phytosanitary Measures Agreement (SPS Agreement)[1] facilitates the liberalization of commercialization generally and the commercialization of agricultural products particularly. Those have boosted trade, reduced trade barriers, and created energetic conditions for development. However, in Agricultural Countries, Newly Industrializing Countries (NICs), the SPS Agreement seems to have been consigned to oblivion. As the consequence, it causes some unfavorable

Supported by FPT University, Can Tho, Vietnam.

© The Author(s), under exclusive license to Springer Nature Switzerland AG 2022
A. E. Hassanien et al. (Eds.): AMLTA 2022, LNDECT 113, pp. 3–10, 2022.
https://doi.org/10.1007/978-3-031-03918-8_1

conditions for commercialization and leads not being too invested and focused production process. The difficulties of finding data sets on plant diseases; with using machine learning methods to recognize, classify, predict, analyze risks and provide solutions become not as easy as anything.

There are a vast majority of diseases and pests that agricultural products have to deal with currently. In apple trees, there are some typical diseases such as white root-rot, fire blight, black rot, fish scale disease, snow apple rust, bitter rot disease, blue mold disease, mosaic disease, etc. Those diseases and pests are alarming and threatening since they always tend to evolve to adapt to and resist human pests' control measures. A new disease or a new fungus could drastically reduce the quality of a crop, or even destroy the whole crop if an immediate solution is not found. Beyond the cause agriculture has not been focused on development, another reason is almost farmers might only rely on experience to diagnose disease; therefore, they do not have a method of prevention or treatment methodically, leading to reducing the quality of agricultural products. Furthermore, the difference is pretty enormous using pesticide residues compared to the standards of the consuming countries is also another cause. To sum up, Those are challenges for developing countries in consuming agricultural products, because of the fastidious and strict requirements of the "big brothers" in the economy on the quality of imported agricultural products [2,16].

In Vietnam, agriculture plays a very important role that is an activity for millions of farmers. Thus, it is necessary to investigate methods that can help a farmer to deal with many available plant diseases in agriculture. There are many methods that have been investigated to detect and classify plant diseases as discussed in Table 1. Deep learning has been applied to solve various problem domains from texture classification to autonomous cars. However, it is just a few studies of deep learning to improve the performance of an agriculture-based application. This research aims to investigate an efficient deep learning model to effectively detect and classify plant leaf diseases to help the farmer easy to recognize the plant diseases and find their corresponding meditation. The proposed system aims to use the deep learning-based Yolo-v5 method [13] and the proposed stable feature-base auto-encoder to detect and classify plant leaf diseases. Experimental results proved that our proposed system provided a good accuracy when detecting and classifying plant leaf disease with 81.28% in terms of the detection rate and 91.93% in terms of the classification rate. We organized our research paper as follows: the existing research in the field are presented in Sect. 2. Section 3 describes the proposed method in detail. We verify the results of our proposed system in Sect. 4. Discussion and limitations are discussed in Sect. 5.

2 Related Works

When the plant is infected it is really difficult for farmers to find out which disease in order to select an appropriate method for curing the plant. To help the farmer sort out current problems, many plant disease detection and classification algorithms have been proposed as shown in Table 1. Sardogan et al. used vector quantization and convolutional neural network (CNN) techniques to detect and classify the disease tomato leaf [3]. Gadade et al. introduced a segmentation method for detecting segmentation of infected regions [4]. Recently, Luna et al. introduce a new algorithm, name Diamante Max to detect and classify the tomato leaf disease [5]. Their method is designed to detect various kinds of leaf disease, such as phoma rot, leaf miner, and target spot. More recently, Le et al. found that ResNet with 18 layers is suitable for detecting and classifying apple leaf disease [6]. After investigating the existing methods of detecting and classifying plant diseases, we found that the existing plant leaf-detection-based approach is still not satisfactory for the industrial application in terms of processing time and accuracy. Therefore, in this research, we will investigate Yolo-v5 [13] to develop a robust method for detecting and classifying three kinds of plant diseases: apple-scab disease, black-rot, and Cedar-apple.

Table 1. A survey of existing plant leaf diseases detection

Authors	Algorithm	Descriptions	Orther
Sardogan, M. [3]	LVQ	Take the input image, process and classify it according to certain categories	Use RGB convolution for dataset and reLU, max pooling for output
Gadade, H. D. [4]	LVQ	Fully automatic disease analysis, detection and measurement	Using a classification approach to detect diseases, classifying diseases use color-based thresholding and calculate area percentage to measure severity
de Luna, R. G. [5]	Auto-encoder	Learn efficient data encodings in an unsupervised way	Use Alexnet and RCNN to speed up disease detection
Li, X. [6]	OTSU	Analyze foreground and background to locate disease	Use SVM, ResNet, VGG to compare and improve results after each execution
Jiang, D. [7]	k-Means	Split the data into multiple clusters, each cluster corresponds to a disease	Using Resnet-50 model to improve analysis and training
Indumathi, R. [8]	Random Forest	Randomize disease data to build Decision Tree and synthesize results	Image data is preprocessed and uses the K-Medoid system to find diseased areas
Chakraborty, S. [9]	Otsu thresholding	Process and separate the diseased part, then compare it with the original image in the dataset	Use histogram to process and balance images to increase accuracy
Kumar, S. [10]	K-means	Separating diseased objects on leaves, dividing into clusters and conducting analysis	Use SVM, GLCM to classify different diseases in dataset
Mekha*, P. [11]	Random forest	Accurate identification of segmentation information classification and statistics based on huge data sets	Compare the accuracy of algorithms to choose the right algorithm
Reddy, T. V. [12]	Cascade Inception	Disease identification through comparison with the previously trained dataset	Optimize the training process by using AlexNet and GoogLeNet

3 Proposed Method

Existing plant leaf disease detection systems fail to obtain both fast processing time and high accuracy for a commercial product. Therefore, in this research, we aim to apply Yolo-v5 architecture [13] to develop a robust plant leaf disease detection and classification for three types of disease: apple-scab disease, black-rot, and Cedar-apple. In this research, first, we study the benefit of auto-encoder [15]. Second, we introduce a robust feature-based auto-encoder method to encode and extract robust information/texture for training and detecting plant leaf disease. Figure 1 shows the proposed model by using Yolo-v5-architecture and Auto-encoder architecture. Our proposed robust feature T is calculated as follows:

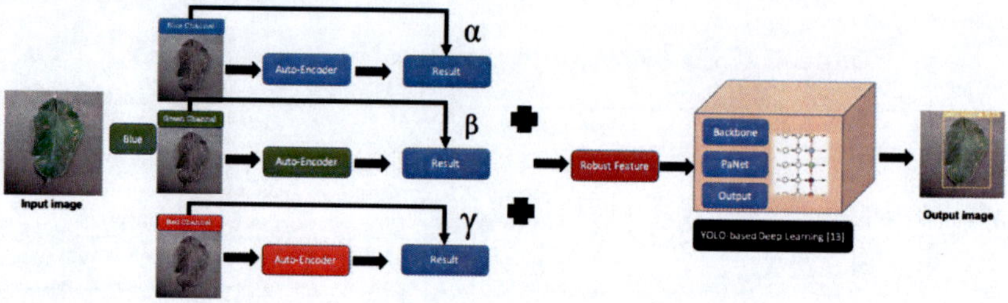

Fig. 1. The proposed system using Yolo-v5 [13] architecture to detect and classify leaf disease

$$
\begin{aligned}
&AE^{c}_{(n)}(I) = \frac{1}{1+e^{-\left(W_c^{(n)} * AE^{c}_{n-1}(I) + b_c^{(n)}\right)}} \\
&\mathrm{O}_{blue}(I_{blue}) = \alpha_{blue} AE^{c}_{(n)}(I) + (1-\alpha_{blue}) I_{blue} \\
&\mathrm{O}_{green}(I_{green}) = \alpha_{green} AE^{green}_{(n)}(I) + (1-\alpha_{green}) I_{green} \\
&\mathrm{O}_{red}(I_{green}) = \alpha_{red} AE^{red}_{(n)}(I) + (1-\alpha_{red}) I_{red} \\
&\mathrm{T} = \beta_{blue} \times \mathrm{O}_{blue}(I_{blue}) + \beta_{green} \times \mathrm{O}_{blue}(I_{blue}) \\
&+ (1-\beta_{blue}-\beta_{green}) \times \mathrm{O}_{red}(I_{red})
\end{aligned} \tag{1}
$$

where $\mathrm{O}_{blue}(I_{blue})$, $\mathrm{O}_{green}(I_{green})$, and $\mathrm{O}_{red}(I_{red})$ are the result after applying auto-encode $AE^{c}_{(n)}(I)$ on the image channel c.

4 Experimental Results

We use a total of 5740 images of PlantVillage dataset [14]. After the process of assigning labels to 3 diseases of apple trees: black-rot, apple-scab, cedar-apple. Next, we split the dataset into three parts: training Set (4,018 images), validation Set (1148 images), and testing Set (574 images). To make the training process go smoothly and save time, we took advantage of Google Colab's GPU. With equivalent epochs of 99, the batch of 64, the model we used is Yolo-v5 and finally got the result as Fig. 2. The dataset is organized in a balance way to help the proposed system obtain the best result in training.

First, assign labels to the images so that the diseased part of the leaves is shown specifically, the images are taken from many different light environments to increase accuracy when applied in practice. Then conduct training with Yolo-v5 model, will conduct training with the number of epochs pre-installed, with 5740 images, 1 epoch takes 1 min 38 s for ColabPro. And about 2 min 45 s for Colab free version. It also depends on the user's internet speed. After obtaining the training results, we proceed to use 10% of the dataset used for testing (574 images was not trained), the results obtained are about 81.28% in term of the detection rate, and 91.93% in term of the classification rate. While the original Yolo [13] achieved the detection rate of 78.85% and classification rate of 89.95%.

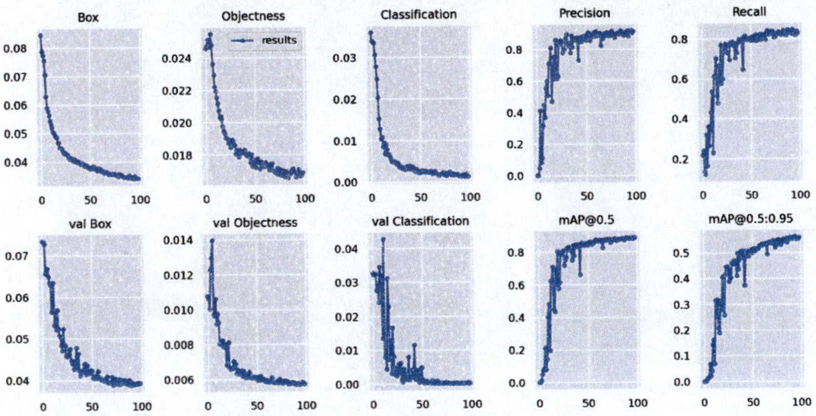

Fig. 2. Result after training on Yolo-v5

After the training process, the results obtained as shown in Fig. 3 include three diseases: apple scale disease, black rot disease, and cedar disease. The locations with translation are delimited by the Ground Truth bounding box. The Ground Truth bounding box is the contour that we assign labels to the object using the Roboflow website. Roboflow provides a lot of utilities that are especially suitable for data set management. To be able to conduct accuracy checks, we use IoU (Intersection over Union) - an evaluation metric used to measure the accuracy of object detection on a particular data set). If the IoU ratio between the Truth bounding box and the Prediction bounding box is greater than or equal to 0.5, the object is correctly recognized (True positive: TN). Conversely, if the IoU ratio is less than 0.5, the object is falsely identified (False positive: FP). And if the object is not recognized then it is False negative: FN.

(a) Apple-scab disease

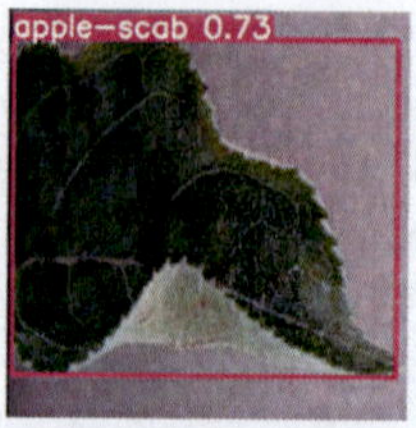

(b) Apple-scab disease

(c) Apple-scab disease

(d) Black-rot disease

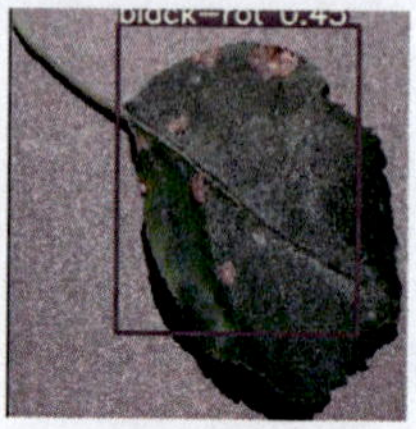

(e) Black-rot disease

(f) Black-rot disease

(g) Cedar-Apple disease

(h) Cedar-Apple disease

(i) Cedar-Apple disease

Fig. 3. Training results of 3 types of leaf diseases

Figure 4 is a graph showing Precision, Recall, and F1-score. Precision is reaching 0.831. The higher the score, the more positive the model predicts, the more positive. $Precision = \frac{TP}{TP+FP}$. Recall reached 0.98. The higher the Recall, the less the number of missed positive points. Recall = 1, i.e. all points labeled as Positive are recognized by the model. $Recall = \frac{TP}{TP+FN}$ However, to evaluate model quality, it is not possible to rely solely on Precision or Recall, instead one uses the F1 index as the harmonic mean. This index is calculated by the formula $\frac{2}{F1} = \frac{1}{precision} + \frac{1}{recall}$.

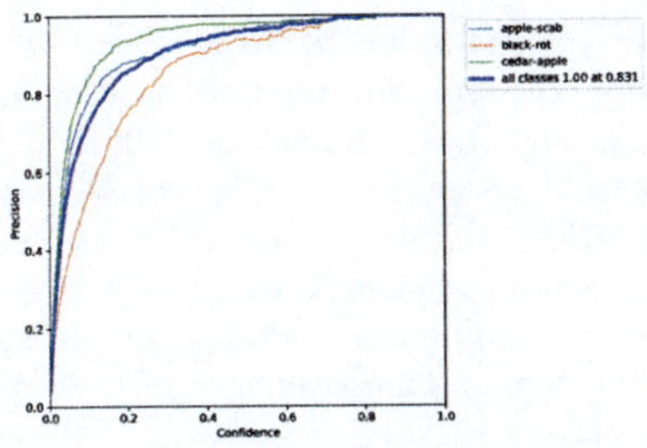

(a) Precision curve

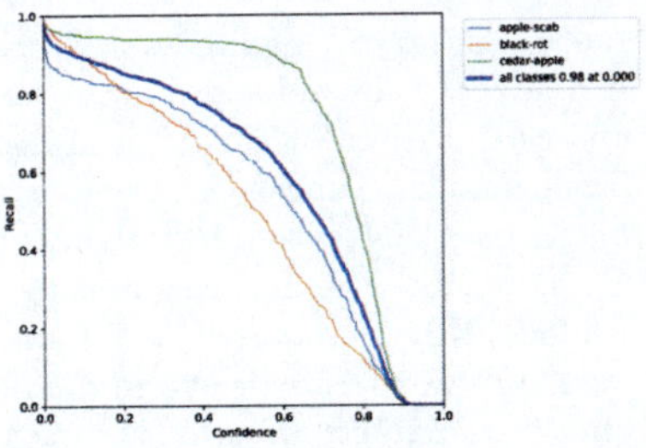

(b) Recall

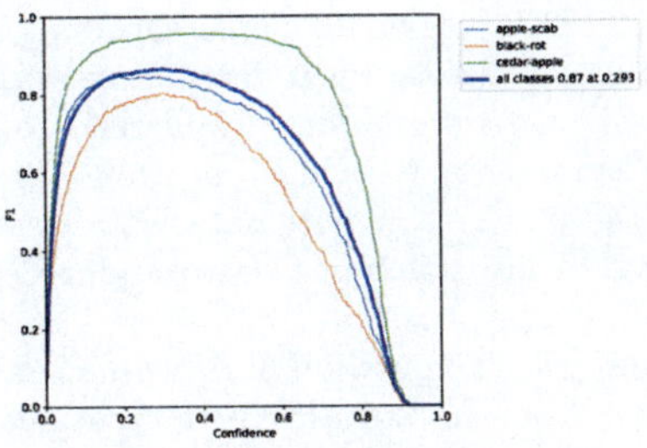

(c) F1-score

Fig. 4. Precision-Recall-F1

5 Conclusions

At the present time, the agriculture sector is still one of the important days, recognizing and being able to detect diseases in crops early can bring promotion to the agricultural industry.

In this case, we propose a leaf disease detection automatic using the Deep Learning technique using the Yolo-v5 algorithm [13] and the proposed stable feature. Up to the present time, when we compared to the 5740 tested images we have achieved 81,28% with detection rate and 91,93% with classification rate. Early detection of leaf diseases will be beneficial for the farmer helping the farmer to know the situation and solve the problems on the plants. However, the current method still has a limitation about the processing time of the encoding stage by using the auto-encoder. It is necessary to increase the performance of this prepossessing state by implementing GPU in the future. Our future plan is to increase the detection rate as well as the classification rate to a higher level to be able to improve and accelerate the detection of leaf diseases. In addition, we also plan to integrate our system into E-commerce to make our solution to farmers in near future.

References

1. World Trade Organization Homepage. https://www.wto.org/english/tratop_e/sps_e/spsagr_e.htm. Accessed 29 Oct 2021

2. Liu, P., Casey, S., Cadilhon, J.-J., Hoejskov, P.S., Morgan, N., Agriculture Group: A practical manual for producers and exporters from Asia. Regulations, standards and certification for agricultural exports, 1st edn. RAP Publication (2007)
3. Sardogan, M., Tuncer, A., Ozen, Y.: Plant leaf disease detection and classification based on CNN with LVQ algorithm. In: 2018 3rd International Conference on Computer Science and Engineering (UBMK), pp. 382–385 (2018)
4. Gadade, H.D., Kirange, D.K.: Tomato leaf disease diagnosis and severity measurement. In: 2020 Fourth World Conference on Smart Trends in Systems, Security and Sustainability (WorldS4), pp. 318–323 (2020)
5. de Luna, R.G., Dadios, E.P., Bandala, A.A.: Automated image capturing system for deep learning-based tomato plant leaf disease detection and recognition. In: TENCON 2018 - 2018 IEEE Region 10 Conference, pp. 1414–1419 (2018)
6. Li, X., Rai, L.: Apple leaf disease identification and classification using ResNet models. In: 2020 IEEE 3rd International Conference on Electronic Information and Communication Technology (ICEICT), pp. 738–742 (2020)
7. Jiang, D., Li, F., Yang, Y., Yu, S.: A tomato leaf diseases classification method based on deep learning. In: 2020 Chinese Control and Decision Conference (CCDC), pp. 1446–1450 (2020)
8. Indumathi, R., Saagari, N., Thejuswini, V., Swarnareka, R.: Leaf disease detection and fertilizer suggestion. In: 2019 IEEE International Conference on System, Computation, Automation and Networking (ICSCAN), pp. 1–7 (2019)
9. Chakraborty, S., Paul, S., Rahat-uz-Zaman, M.: Prediction of apple leaf diseases using multiclass support vector machine. In: 2021 2nd International Conference on Robotics, Electrical and Signal Processing Techniques (ICREST), pp. 147–151 (2021)
10. Kumar, S., Prasad, K., Srilekha, A., Suman, T., Rao, B.P., Vamshi Krishna, J.N.: Leaf disease detection and classification based on machine learning. In: 2020 International Conference on Smart Technologies in Computing, Electrical and Electronics (ICSTCEE), pp. 361–365 (2020)
11. Mekha, P., Teeyasuksaet, N.: Image classification of rice leaf diseases using random forest algorithm. In: 2021 Joint International Conference on Digital Arts, Media and Technology with ECTI Northern Section Conference on Electrical, Electronics, Computer and Telecommunication Engineering, pp. 165–169 (2021)
12. Vijaykanth Reddy, T., Sashi Rekha, K.: Deep Leaf Disease Prediction Framework (DLDPF) with transfer learning for automatic leaf disease detection. In: 2021 5th International Conference on Computing Methodologies and Communication (ICCMC), pp. 1408–1415 (2021)
13. YOLOv5. https://github.com/ultralytics/yolov5. Accessed Nov 2021
14. PlantVillage Dataset. https://github.com/spMohanty/PlantVillage-Dataset. Accessed 29 Oct 2021
15. Bengio, Y., Lamblin, P., Popovici, D., Larochelle, H.: Greedy layerwise training of deep networks. In: Proceedings of Advances in Neural Information Processing Systems, pp. 153–160 (2007)
16. Saba, D., Sahli, Y., Maouedj, R., Hadidi, A., Medjahed, M.B.: Towards artificial intelligence: concepts, applications, and innovations. In: Hassanien, A.-E., Taha, M.H.N., Khalifa, N.E.M. (eds.) Enabling AI Applications in Data Science. SCI, vol. 911, pp. 103–146. Springer, Cham (2021). https://doi.org/10.1007/978-3-030-52067-0_6

Reinforcement Learning for Developing an Intelligent Warehouse Environment

Van Luan Tran[1], Manh-Kha Kieu[2,3], Xuan-Hung Nguyen[1], Vu-Anh-Tram Nguyen[2], Tran-Thuy-Duong Ninh[2], Duc-Canh Nguyen[1], Narayan C. Debnath[4], Ngoc-Bich Le[5,6(✉)], and Ngoc-Huan Le[1(✉)]

[1] School of Engineering, Eastern International University, Thu Dau Mot, Binh Duong, Vietnam
{luan.tran,hung.nguyenxuan,canh.nguyen,huan.le}@eiu.edu.vn
[2] Becamex Business School, Eastern International University, Thu Dau Mot, Binh Duong, Vietnam
{kha.kieu,tram.nguyen,duong.ninh}@eiu.edu.vn
[3] School of Business and Management, RMIT University Vietnam, Ho Chi Minh City, Vietnam
[4] School of Computing and Information Technology, Eastern International University, Thu Dau Mot, Binh Duong, Vietnam
narayan.debnath@eiu.edu.vn
[5] School of Biomedical Engineering, International University, Ho Chi Minh City, Vietnam
[6] Vietnam National University Ho Chi Minh City, Ho Chi Minh City, Vietnam
lnbich@hcmiu.edu.vn

Abstract. Nowadays, warehouse optimization is one of the core components of logistics. With the development of artificial intelligence (AI) technology and the advancement of automation technology, building a smart warehouse is an important task. This paper presents the machine learning techniques and technologies for developing an intelligent warehouse. A reinforcement learning method is proposed to train a basic warehouse environment for an efficient storage policy. The experimental results help comprehend the building of the model of a neural network of the reinforcement learning algorithm and the characteristics of this technique. This study further helps to understand the basic concepts of machine learning techniques to develop an algorithm for smart warehouses.

Keywords: Smart warehouse · Reinforcement learning · Machine learning technique · Intelligent management systems · AI applications

1 Introduction

Recently, the fourth industrial revolution with the cornerstone of a combination of robots, artificial intelligence (AI), fast networking equipment, and big data has created many achievements in research for production and human life [1, 17, 24, 26]. Research and development of intelligence applications to reduce workforce and production optimization is an important task. Nowadays, human resource costs are constantly increasing, and businesses face tremendous pressure from fierce market competition [11]. Therefore, building a smart warehouse and intelligent management systems is becoming necessary

© The Author(s), under exclusive license to Springer Nature Switzerland AG 2022
A. E. Hassanien et al. (Eds.): AMLTA 2022, LNDECT 113, pp. 11–20, 2022.
https://doi.org/10.1007/978-3-031-03918-8_2

to optimize production and storage. Warehouses are typically used to keep business goods in storage. Large corporations can either construct their own warehouses or rent warehousing services from warehousing service providers. Warehouse service management determines e-commerce firms' success or failure in the global market. The warehouse is an essential component of the logistics business since its operational efficiency has a significant impact on the overall performance of the logistics industry. Inbound management, inventory management, and distribution management are the three major activities of a warehouse [10].

Warehouses play a critical role in the supply chain of food and agricultural goods, especially in the Vietnamese market. The warehouse is an essential component of Vietnamese logistics, including transportation, forwarding, warehousing, and other value-added services [3]. It is critical to increase the logistical competitiveness [9]. Investment in information technology (IT) is critical to improve Vietnam's logistics and warehousing competitiveness [7]. IT will add momentum to the transition from traditional warehouses to smart warehouses to optimize warehouse performance while reducing logistics and operating energy costs. Barcode, RFID, AR, AGV, IoT, WMS, and warehouse communication are technologies that can be combined with AI technology to make the warehouses smarter. AI might help warehouse operations change by automating activities, integrating information, and analyzing data in order to improve warehouse efficiency. As a result, AI applications in warehouses were promoted in order to enhance warehouse process improvement and in turn, improve Vietnam's logistics industry's competitive advantage. However, AI is a very complex field. Accordingly, each specific requirement needs to be addressed by particular needs. According to recent surveys, up to 85% of AI efforts fail to meet their expectations [25]. The testbed facility will be outfitted with cutting-edge machines and technologies that can mimic various AI investing situations, thus providing investors with the confidence and data they may need to make critical choices. Manufacturing companies may use testbeds to test new technologies in a live production setting before implementing them in the real world [21]. Testbeds also demonstrate their value in the academic world. In [8, 15, 19], the notion of testbed has aided the acquisition of practical and theoretical knowledge. In most universities, students are expected to deal with a significant amount of theory. On the other hand, testbeds provide options for students to collaborate and solve challenges while developing and building the testbed [15].

An attempt is made to build a smart warehouse at the Eastern International University (EIU) in Vietnam as simulated in Fig. 1. The EIU smart warehouse is a constantly developing environment where research and teaching may be smoothly linked, allowing bachelor and master students to gain hands-on experience while offering a realistic testbed for scientists. The main goal of testbeds is to bring together the goals of many stakeholders to examine how an innovative technology works in practice and how it may help businesses gain a competitive advantage. Therefore, the EIU Smart Warehouse provides various possibilities for businesses to test innovative solutions while also learning more about the academic environment. Both students and professors can get experience working on real-world projects. Consequently, it has the potential to facilitate more efficient technology transfer between the universities and industries.

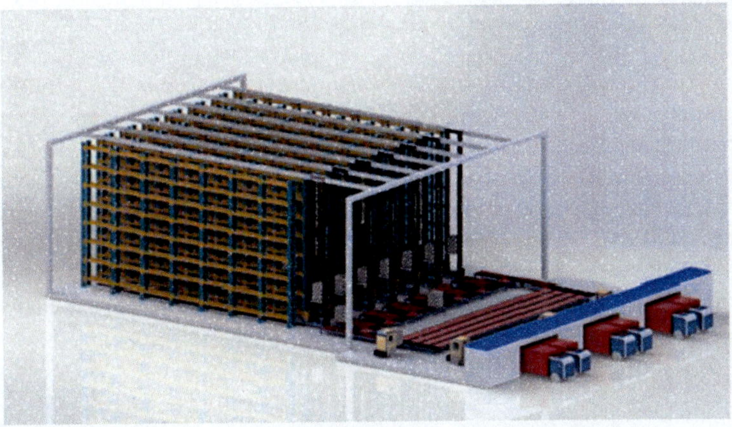

Fig. 1. Overview of the EIU smart warehouse with simulation.

This important project will assist enhance the quality of education in three EIU academic schools: Becamex Business School (BBS), School of Engineering (SOE), and School of Computing and Information Technology (CIT) in terms of effective teaching and learning. The project allows students from three schools to put theory into practice and make real-world selections. Students and researchers from BBS, SOE, and CIT can use the system to study a variety of topics including Motion control, PLC system, Industrial communication networks, Optimizing the sorting process, Intelligent systems, Warehouse Inventory management, Warehouse performance measurement, Applied computer vision, AI for the sorting process, Energy optimization, and Transaction time optimization. In terms of university business collaboration, this project acts as a testbed for firms to try out new technologies or solutions before putting them into practice. This is considered as an approach to promote the manufacturing process with significant cost savings and avoiding risks. Moreover, by maximizing the potential and value of local study, the EIU testbed benefits the local economy by attracting more investment and improving public service efficiency.

The EIU Smart Warehouse project is an application testing facility that uses a 1:10 miniature model (prototype) of a physical model with real-world features. The following are a list of hardware and software solutions:

- Hardware solution for smart warehouse.
  - Storage racks: 7 units can store up to 1372 pallets,
  - Automatic guided vehicle (AGV): 7 units,
  - Circulation conveyor system,
  - Sensor system (RFID) and technology solutions in smart warehouse management,
  - Controller (PLC) and control algorithm.
- Software solutions for smart warehouse.
  - Solutions to connect the physical controller system and management software,

– Solutions to optimize operating efficiency.

To develop an algorithm for a smart warehouse, an overview of machine learning techniques is presented with basic knowledge about four types of techniques: unsupervised learning, supervised learning, reinforcement learning, and imitation learning. This paper proposes the use of a reinforcement learning algorithm to perform an efficient storage policy and an optimization technique in a warehouse environment. An attempt is made to build a virtual warehouse environment and a robot with reinforcement learning for the optimization techniques. This experiment simulates and tests the robot using specified policies for reward and penalty with actions in a basic warehouse environment.

2 Machine Learning Techniques

Machine learning is a core part of artificial intelligence with many significant applications. Machine learning is diverse and complex, but in general, they are divided into four standard types, as shown in Fig. 2.

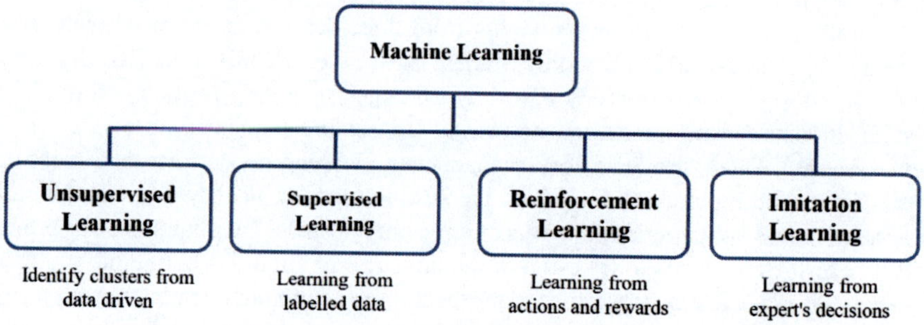

Fig. 2. Overview of machine learning technique using four types.

Unsupervised Learning is an algorithm that allows machines to learn on their own data to find similarities of features to identify the clusters or groups. This means that one only has the input dataset and do not label data. Unsupervised Learning is usually applied in clustering data as customer segmentation, targeted marketing, and recommended systems or dimensionality reduction of structure discovery, feature elicitation, meaningful compression, and big data visualization [5].

Supervised Learning is recently the most popular machine learning algorithms, especially computer vision and deep learning. This algorithm predicts new data output based on previously known training between the input data and labeled data (outcome). Supervised learning is usually used for image classification (object detection, semantic segmentation, instance segmentation), regression for weather forecasting, market forecasting, stock predictions, and estimating the object pose [24].

Reinforcement Learning is a machine learning technique that enables a machine to learn in a sequence of decisions, self-trained on reward and penalty for the actions it performs [4]. Machines learn in an interactive environment receiving some feedback from their actions and experiences. In reinforcement learning, the target is to explore a suitable action model to maximize the total cumulative reward. The reinforcement learning algorithm is popular in real-time training robot navigation, learning tasks, game AI, skill acquisition, inventory management in supply chains [23], and real-time decisions of the robot [13, 18, 22].

Imitation Learning is closely related to reinforcement learning techniques for learning from demonstrations of expert behavior [6]. Imitation learning techniques learn with decision policies from an expert while reinforcement learning is a policy that maximizes long-term reward with experiments from actions. Imitation learning has benefited from recent advancements in core machine learning techniques and the advancements of deep learning. This method depends on expert knowledge and the accuracy is also a subsidiary of expert demonstrations along with the expert policies. Imitation learning technique is usually applied in autonomous vehicles, autonomous driving systems [1], autonomous navigation [17], and training real-time robot manipulation [14].

3 Results and Discussion

The use of a reinforcement learning algorithm to train an efficient storage policy in a warehouse environment is proposed and described in this section. The task of reinforcement learning algorithm is to provide a diversity of methods to solve decision problems [20]. Reinforcement learning is proposed to solve inventory management problems [23] and supply chain optimization [2]. In [16], Kamoshida et al. proposed an automated guided vehicle route planning policy in a warehouse environment using deep reinforcement learning. They improved the efficiency of the picking activities limited by the five order data and one warehouse flatform.

This section establishes a simple warehouse environment to develop reinforcement learning to train a robot as simulated in Fig. 3. Reinforcement learning is an algorithm that builds robots self-trained on reward and punishment mechanisms.

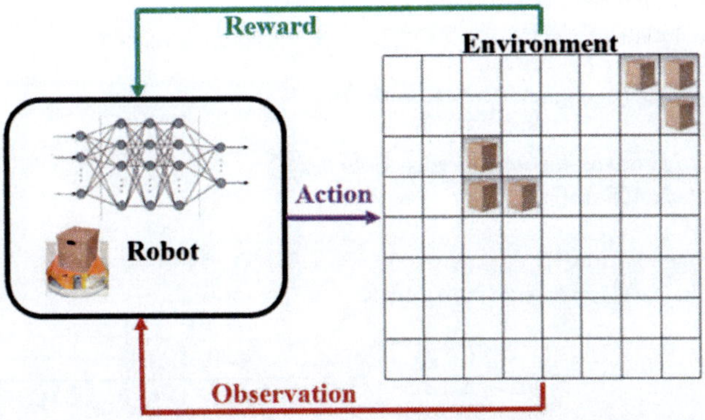

Fig. 3. Overview of the basic of reinforcement learning where the robot is learning in a simple warehouse environment.

The robot will be trained to gain maximum rewards and minimum punishment through observations in a basic warehouse environment. The simulated mobile robot moves can be described as follows. In the beginning, the robot is set to its initial state, and the mobile robot has the first observation information to initiate the first location in the warehouse coordinates. Robot is trained with four moving strategies having actions as Up, Down, Left, and Right. A neural network model is built with an input layer, hidden layer, and output layer. The input layer has the shape of GridColumnNum $*$ GridRowNum, that is the number of the storing location in the basic warehouse. The output layer is the number of actions. A Parametric Rectified Linear Unit (PReLU) activation function is applied for these neural networks. PReLU is a variable of ReLU, which is adaptively learning the parameters of the rectifiers. They help improve to solve model fitting with near-zero additional computational cost and less risk of overfitting [12].

A simple warehouse environment is created for the simulation as shown in Fig. 3. The reinforcement learning algorithm is trained as presented in Algorithm 1. In this training, the robot has tasks for finding the optimal paths and transporting the goods to the target locations in the warehouse. The target locations are the vacant position, which is the first priority for filling the remote location. For this algorithm, it is vital to build a policy of rewards and punishments for actions. The proposed method was trained with two policies. The first policy is the robot moving randomly to find the target location. If the robot moves to the occupied location or the visited location, it will get punishments. Otherwise, if the robot moves in the target locations, it will get rewards. The second policy is the robot moving randomly to find the target location, like the first policy. In addition, a reward policy is set up for each action. If the robot moves in the direction of the target, it will get a reward. Otherwise, it will get a punishment. In Fig. 4, the training for the reinforcement learning was done with 200 episodes for the first policies shown in Fig. 4 (a) and the second policy is shown in Fig. 4 (b). With the second policy, the robot learns more effectively in the basic warehouse environment.

Algorithm 1 Reinforcement learning algorithm in our simple warehouse.

```
int_papameters()
observation = int_state()
while True do
  determine_destination()
  action()
  determine_rewards_for_action_with_policies()
  experience.get_data()
  model.fit()
  update_observation()
  if episode > self.episode then break
  end if
end while
```

In this simulation, a Tkinter software was used, which is a graphical user interface (GUI) tool to build the simulation software. Tkinter software is the only framework that has been built into the Python library. The training for the reinforcement learning was done with 500 episodes based on this simulated environment, as shown in Fig. 5. In testing with the simulated environment, the robot moved to the targets following the basic warehouse policy. This solution with the policy can reduce inefficient exploration with simple techniques and better search accuracy of the path for the robot. The classic reinforcement learning method can execute relatively optimized policies for a robot but still have limitations in outperforming improvement. The reinforcement learning method using the policies of experts helps robots' study more effectively.

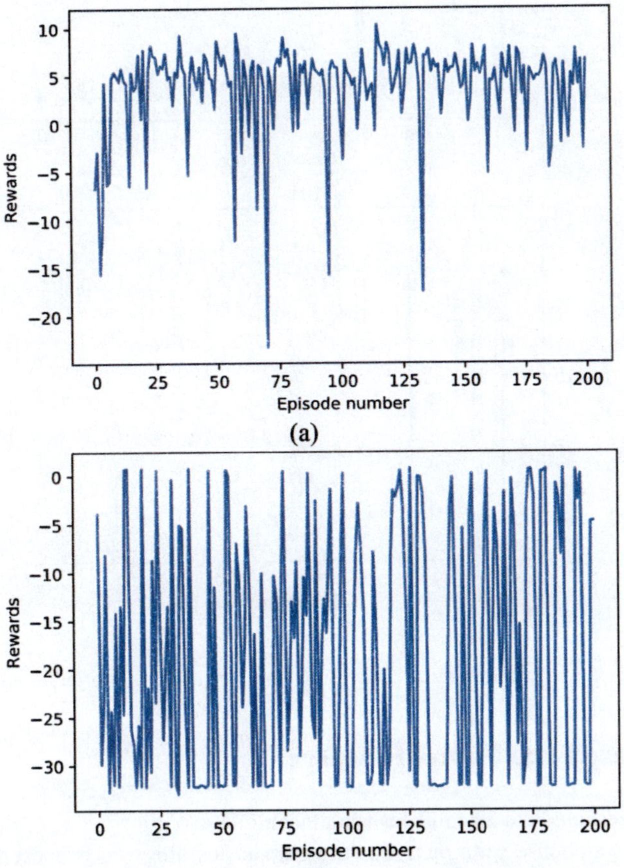

Fig. 4. Rewards per episode of the training with 200 episodes: (a) the first policy and (b) the second policy.

The process of discovery and policy are key in this classic reinforcement learning strategy. Our method has the drawback of making it difficult to handle large issues efficiently when many states and actions have yet to be practiced. Between the simulation and the real world, the noise environment is a challenge. In this paper, we built our model in a single-agent environment. So it's critical to create a deep reinforcement learning method for multi-agent complicated warehouse scenarios.

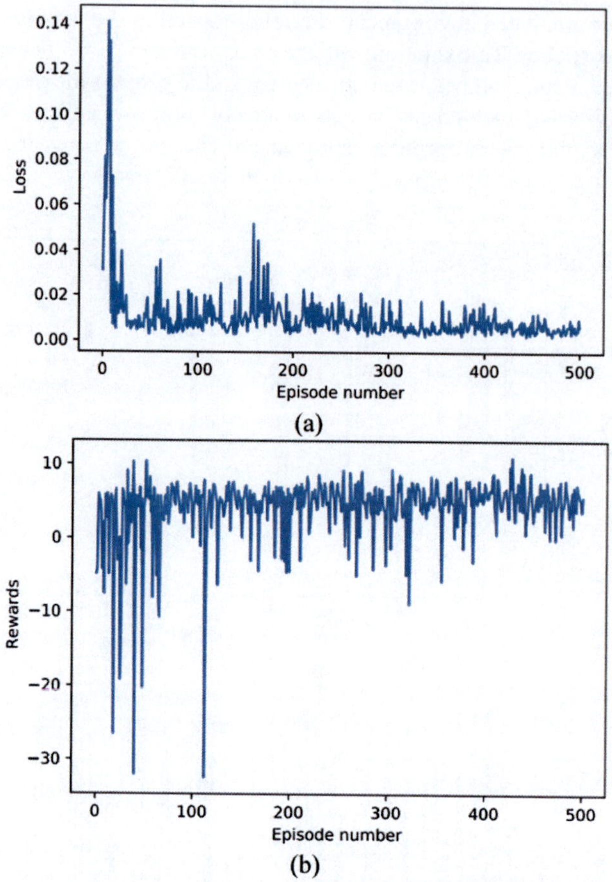

Fig. 5. 500 episodes results: (a) Loss per episode and (b) Rewards per episode

4 Conclusion and Future Research

This paper presented the machine learning technique to outline the algorithm for developing smart warehouse solutions. A reinforcement learning was proposed and trained in a simple warehouse environment with specified policies. A simulation environment was built to test the policies for rewards and penalties for the actions in training the reinforcement learning algorithm. In future research, the plan is to further study and develop a deep reinforcement learning algorithm for multi-agent complex warehouse environments to solve more challenging problems based on the experimental results.

Acknowledgments. This research is financially supported by Eastern International University, Binh Duong Province, Vietnam.

References

1. Ahmedov, H.B., Yi, D., Sui, J.: Brain-inspired deep imitation learning for autonomous driving systems. CoRR arXiv:2107.14654 (2021)
2. Alves, J.C., Mateus, G.R.: Deep reinforcement learning and optimization approach for multi-echelon supply chain with uncertain demands. In: Lalla-Ruiz, E., Mes, M., Voß, S. (eds.) ICCL 2020. LNCS, vol. 12433, pp. 584–599. Springer, Cham (2020). https://doi.org/10.1007/978-3-030-59747-4_38
3. Blancas, L.C., Isbell, J., Isbell, M., Tan, H.J., Tao, W.: Efficient logistics: a key to vietnams competitiveness. The World Bank Group (2014). https://EconPapers.repec.org/RePEc:wbk:wbpubs:16320
4. Bom, L., Henken, R., Wiering, M.A.: Reinforcement learning to train Ms. Pac-Man using higher-order action-relative inputs. In: Proceedings of the 2013 IEEE Symposium on Adaptive Dynamic Programming and Reinforcement Learning, ADPRL 2013, IEEE Symposium Series on Computational Intelligence (SSCI), 16–19 April 2013, Singapore, pp. 156–163. IEEE (2013). https://doi.org/10.1109/ADPRL.2013.6615002
5. Chang, J.R., Shrivastava, A., Koppula, H.S., Zhang, X., Tuzel, O.: Style equalization: unsupervised learning of controllable generative sequence models. CoRR arXiv:2110.02891 (2021)
6. Ciosek, K.: Imitation learning by reinforcement learning. CoRR arXiv:2108.04763 (2021)
7. Dang, V.L., Yeo, G.T.: Weighing the key factors to improve Vietnam's logistics system. Asian J. Shipp. Logist. **34**(4), 308–316 (2018). https://doi.org/10.1016/j.ajsl.2018.12.004
8. Falkenberg, R., et al.: PhyNetLab: an IoT-based warehouse testbed. In: 2017 Federated Conference on Computer Science and Information Systems (FedCSIS), pp. 1051–1055 (2017). https://doi.org/10.15439/2017F267
9. Gani, A.: The logistics performance effect in international trade. Asian J. Shipp. Logist. **33**(4), 279–288 (2017). https://doi.org/10.1016/j.ajsl.2017.12.012. https://www.sciencedirect.com/science/article/pii/S2092521217300688
10. Gijsbrechts, J., Boute, R., Zhang, D., Van Mieghem, J.: Can deep reinforcement learning improve inventory management performance on dual sourcing, lost sales and multi-echelon problems. SSRN Electron. J. (2019). https://doi.org/10.2139/ssrn.3302881
11. Hao, H., Jia, X., He, Q., Fu, S., Liu, K.: Deep reinforcement learning based AGVs real-time scheduling with mixed rule for flexible shop floor in industry 4.0. Comput. Ind. Eng. **149**, 106749 (2020). https://doi.org/10.1016/j.cie.2020.106749
12. He, K., Zhang, X., Ren, S., Sun, J.: Delving deep into rectifiers: Surpassing human level performance on ImageNet classification. CoRR arXiv:1502.01852 (2015)
13. Hilprecht, B., Binnig, C., Röhm, U.: Learning a partitioning advisor with deep reinforcement learning. CoRR arXiv:1904.01279 (2019)
14. Johns, E.: Coarse-to-fine imitation learning: robot manipulation from a single demonstration. CoRR arXiv:2105.06411 (2021)
15. Kaczmarczyk, V., Batn, O., Brad, Z., Arm, J.: An industry 4.0 testbed (self-acting barman): principles and design. In: IFAC-15th Conference on Programmable Devices and Embedded Systems PDeS 2018, vol. 51, no. 6, pp. 263–70 (2018). https://doi.org/10.1016/j.ifacol.2018.07.164
16. Kamoshida, R., Kazama, Y.: Acquisition of automated guided vehicle route planning policy using deep reinforcement learning. In: 2017 6th IEEE International Conference on Advanced Logistics and Transport (ICALT), pp. 1–6 (2017). https://doi.org/10.1109/ICAdLT.2017.8547000
17. Karnan, H., Warnell, G., Xiao, X., Stone, P.: VOILA: visual-observation-only imitation learning for autonomous navigation. CoRR arXiv:2105.09371 (2021)

18. Nasiriany, S., Liu, H., Zhu, Y.: Augmenting reinforcement learning with behavior primitives for diverse manipulation tasks. CoRR arXiv:2110.03655 (2021)
19. Ridolfi, M., Macoir, N., Gerwen, J.V.V., Rossey, J., Hoebeke, J., de Poorter, E.: Testbed for warehouse automation experiments using mobile AGVs and drones. In: IEEE INFOCOM 2019 - IEEE Conference on Computer Communications Workshops (INFOCOM WKSHPS), pp. 919–920 (2019). https://doi.org/10.1109/INFCOMW.2019.8845218
20. Rimélé, A., Grangier, P., Gamache, M., Gendreau, M., Rousseau, L.: E-commerce warehousing: learning a storage policy. CoRR arXiv:2101.08828 (2021)
21. Salunkhe, O., Gopalakrishnan, M., Skoogh, A., Fasth-Berglund, A.: Cyber-physical production testbed: literature review and concept development. Procedia Manuf. **25**, 2–9 (2018). https://doi.org/10.1016/j.promfg.2018.06.050
22. Sui, Z., Gosavi, A., Lin, L.: A reinforcement learning approach for inventory replenishment in vendor-managed inventory systems with consignment inventory. Eng. Manag. J. EMJ **22**, 44–53 (2010). https://doi.org/10.1080/10429247.2010.11431878
23. Sultana, N.N., Meisheri, H., Baniwal, V., Nath, S., Ravindran, B., Khadilkar, H.: Reinforcement learning for multi-product multi-node inventory management in supply chains. CoRR arXiv:2006.04037 (2020)
24. Tran, L.V., Lin, H.Y.: BiLuNetICP: a deep neural network for object semantic segmentation and 6d pose recognition. IEEE Sens. J. **21**(10), 11748–11757 (2021). https://doi.org/10.1109/JSEN.2020.3035632
25. Zhang, D., Pee, L.G., Cui, L.: Artificial intelligence in e-commerce fulfillment: a case study of resource orchestration at Alibabas smart warehouse. Int. J. Inf. Manag. **57**, 102304 (2021). https://doi.org/10.1016/j.ijinfomgt.2020.102304
26. Zhang, D., Zheng, Y., Li, Q., Wei, L., Zhang, D., Zhang, Z.: Explainable hierarchical imitation learning for robotic drink pouring. CoRR arXiv:2105.07348 (2021)

A Low-Cost Multi-sensor Deep Learning System for Pavement Distress Detection and Severity Classification

Mohamed A. Hedeya[1], Eslam Samir[1], Emad El-Sayed[1], Ahmed A. El-Sharkawy[1], Mohamed F. Abdel-Kader[1], Adel Moussa[1,2], and Rehab F. Abdel-Kader[1](✉)

[1] Faculty of Engineering, Port Said University, Port Said, Egypt
{mohamed.hedeya,eslam.samir,emad.elsayed,a.elsharkawy,mdfarouk, rehabfarouk}@eng.psu.edu.eg

[2] Schulich School of Engineering, University of Calgary, Calgary, Canada
amelsaye@ucalgary.ca

Abstract. Recent development in the transportation industry enforced the focus on timely pavement inspection and maintenance to preserve a sustainable transportation network. Manual pavement-distress detection systems are time-consuming, costly, and heavily dependent on the subjectivity and experience of the designated inspector. To overcome these challenges, computer vision algorithms incorporating machine learning models have been proposed as an alternative to conventional surveying techniques. This necessitates the need for accurate automated surveying systems capable of rapid data acquisition and efficient algorithms for pavement-distress detection and severity quantification. This paper presents a low-cost multi-sensor solution for the automatic detection and severity classification of pavement distress. The proposed system utilizes a sensor fusion strategy that combines RGB image data and depth information. A convolutional neural network (CNN) is trained to detect and classify 13 of the most common types of pavement distresses. The extracted distress regions of interest (ROI) are projected onto the 3D point cloud data where class-specific point-cloud processing is used to quantify and classify the severity of each detected distress. This optimizes the computational complexity of the proposed system and significantly reduces the processing time as only extracted regions will be subject to subsequent processing. Experimental results demonstrate the effectiveness and accuracy of the proposed system in detecting 13 of the most frequent types of pavement distress classes and accurate severity classification for pothole distresses.

Keywords: Distress identification and quantification · Convolutional neural networks · Pavement imaging · Pavement distress · Point cloud · Potholes

1 Introduction

Road networks are the largest strategic infrastructure that constitutes the backbone for social and economic developments. Various factors such as weather conditions and traffic

© The Author(s), under exclusive license to Springer Nature Switzerland AG 2022
A. E. Hassanien et al. (Eds.): AMLTA 2022, LNDECT 113, pp. 21–33, 2022.
https://doi.org/10.1007/978-3-031-03918-8_3

loads can adversely affect the pavement structure and result in different types of distresses [1–4]. Regular and accurate road-condition monitoring is essential to optimize resources allocation and prioritization for pavement preservation planning. Identifying the type, severity, and extent of the pavement distresses is an integral part of any road-condition monitoring system [4, 5]. Guidelines are provided in the pavement-condition index (PCI) handbooks for establishing the extent and severity of each pavement distress type [6], as distress severity classification differs from one distress type to the other. Some distress types require certain measures or calculations to be performed. On the other hand, the other distress classes require subjective color or visual inspections.

Conventional road-condition surveying utilizes manual inspection methods, which are time-consuming, prone to individual subjective evaluations, and have major safety concerns. Therefore, the development of automated pavement assessment systems is imperative to facilitate large-scale pavement monitoring projects. The design of automated pavement monitoring systems entails two major components. First, the selection and compilation of the appropriate on-vehicle road-condition monitoring sensors to acquire the required pavement data. Second, the utilization of efficient data processing and distress detection algorithms to identify and quantify the pavement distresses [1, 3–7].

Different types of detection sensors can be used in pavement-distress detection and severity classification such as cameras, 3-D laser scanners, sonars, accelerometers, and other vibration-based methods…etc. [1, 7]. The appropriate sensing devices to be incorporated in the detection system are closely linked to the type of distress acquired. The first thing that stands out is that with 3-D sensors, most of the distress types can be detected [8]. However, 3D clouds are sparse and have highly variable point density, due to factors such as non-uniform sampling of the 3D space, the effective range of the sensors, occlusion, and the relative pose. In addition, they require computationally intensive data processing and expensive data acquisition equipment.

In this paper, a low-cost automated deep learning pavement-distress detection and severity classification system is proposed. The proposed system uses multiple sensor modalities to combine the benefits of efficient pavement-distress detection and classification through 2D RGB images as well as accurate severity classification using the 3D point cloud data from depth sensors. The RGB-Camera/Depth sensor fusion is implemented through the projection of the detected distress region from the 2D RGB image onto the point cloud generated by the depth camera and hence use the point cloud to quantify and classify the severity of the detected distress. This optimizes the computational complexity and processing time. The effectiveness of the proposed system is demonstrated thru the detection and classification of 13 of the most common pavement distress types and the accurate severity classification of potholes.

The remainder of this paper is organized as follows: a survey of the related work presented in the literature is presented in Sect. 2. The proposed methodology is described in Sect. 3. The case study on pothole-severity classification is presented in Sect. 4. Experimental results and discussion of the performance of the proposed methodology are provided in Sect. 5. Finally, the paper is concluded in Sect. 6.

2 Related Work

The pavement distress detection problem has attracted the attention of many public agencies and researchers to develop effective methodologies to automate the process of distress detection, classification, and severity quantification aimed at specific types of distresses such as cracks, rutting, and potholes [1, 2, 5].

Automated pavement distress detection using 2D and 3D image acquisition systems remains an interesting research topic in the computer vision community. Image-based techniques are reasonably priced, provide an easy-to-use output, and the processing techniques are well-established. Radopoulou and Brilakis [3] proposed a pavement defects detection approach using cameras based on semantic texton forests algorithm. Jahanshahi et al. [4] used an RGB-D camera (Kinect sensor) to detect pavement defects using RANSAC and thresholding technique. Christodoulou et al. [9] presented a pavement patch detection approach using smartphone video images and vibration signals based on deep learning and support vector machine algorithms. However, vision-based methods have serious limitations such as vulnerability due to variations in surface textures, non-uniformity of the distress, and lack of appropriate illumination. Various image-based algorithms were proposed to overcome these challenges such as histogram equalization, image transformations, edge detection techniques, Markov methods, fuzzy set methods, k-means methods, Bayesian networks, and neural networks.

The detection accuracy of 3D point clouds acquired by LIDAR sensing devices outperforms image-based data as LIDAR sensing devices provide reliable depth information that can be used to accurately localize and detect most of the types of surface distresses. Using this kind of 3D data has better accuracy and ability to detect and quantify distress than the normal 2D images. Similar to the case of the digital 2D images, many traditional data processing, and segmentation techniques were used in processing 3D point cloud data. However, LiDAR point clouds are sparse and have highly variable point density, due to factors such as non-uniform sampling of the 3D space, the effective range of the sensors, occlusion, and the relative pose. In addition, they require computationally intensive data processing and expensive data acquisition equipment.

Potholes are bowl-shaped structural failures in the pavement surface caused by the contraction and expansion of rainwater that pervades into the ground. This is one of the most widespread road defects that can substantially affect driving comfort and pose a threat to vehicle and road safety. Various sensors have been utilized to detect potholes such as cameras, laser scanners, thermal cameras [10, 11]. Gupta et al. [12] proposed a pothole detection approach using thermal images. Dhiman and Klette [13] presented a deep learning approach to detect potholes using stereo-vision sensors. Wang et al. [14] employed wavelet energy field of the pavement images to detect and segment asphalt pavement potholes. Potholes have a significant vertical drop of the surface, which enables them to be recognized based on height differences measured by depth sensors. Ravi et al. [8] used high-grade LIDAR units to map and quantify potholes. Ouma [15] proposed an unsupervised classification approach based on spatial fuzzy c-means segmentation to detect potholes in RGB-D frames.

3 Proposed Methodology

In this paper, a fully automated system for pavement-distress detection and severity classification is developed. The general architecture for the proposed system is presented in Fig. 1.

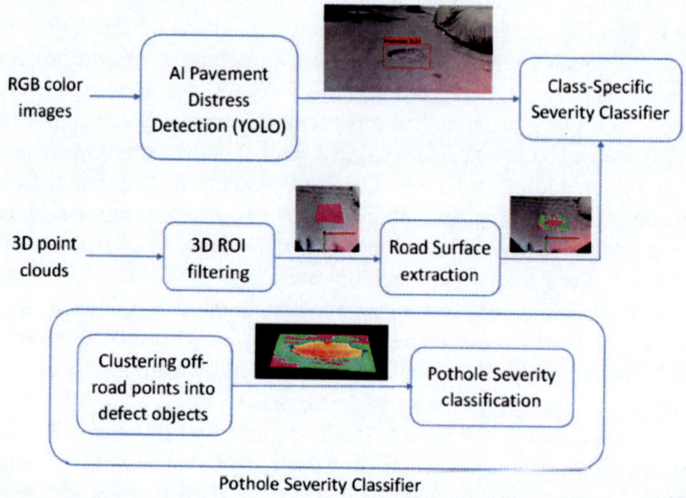

Fig. 1. Proposed System Architecture. The 'Magenta' color represents the 3D ROI points. The 'Green' colored points represent the segmented road surface. The point colors of the detected pothole clusters vary gradually from yellow (near-surface points) to red (deepest points). The cluster points are surrounded by the constructed concave hull (in blue), which is used to calculate the area of the pothole opening (at road surface)

3.1 Overall System Architecture

The proposed system utilizes a D455 Intel RealSense depth camera that generates both depth and RGB color images [16]. The camera incorporates an RGB color sensor and a stereoscopic depth module. The stereo vision module consists of a left imager and a right imager that capture the scene and send the captured data to the vision processor to calculate depth values for each pixel. The correspondence between the depth and RGB images is achieved by using a global shutter in the RGB sensor and matching its Field of View (FoV) with the depth field of view. The ideal range of this camera is between 0.6 m and 6 m. Figure 2 shows the custom mounting bracket that was used for the D455. According to the FoV of the camera, the bracket was located at the front of the surveying vehicle.

Fig. 2. Mounting bracket used for D455. The GoPro Camera on the same frame was not used in this study.

It should be mentioned that the proposed system architecture is not limited to potholes as it can detect up to 13 classes of pavement distresses. The severity classification stage can be modified and extended to pavement distresses other than potholes such as rutting and longitudinal cracks for which the severity classification is based on predefined measurements/thresholds. Other pavement distress types in which the severity classification is subjective/descriptive may require other classification methods to be incorporated. A brief description of the basic stages of the proposed system is presented in the following subsections:

3.2 Deep Learning Distress Detection

The first stage in the proposed system is utilizing the YOLOv3 convolutional neural network (CNN) on the custom-generated dataset. YOLOv3 is an object detector proposed by Farhadi and Redmon (2018) that uses Darknet-53 as the backbone CNN for feature extraction to perform real-time object detection [17]. Darknet-53 is more powerful than Darknet-19 and yet is more efficient than ResNet-101 or ResNet-152 [18]. YOLOv3 uses multiscale prediction, which means object detection is performed using multiple-scale feature maps as presented in Fig. 3. The YOLO detector is trained to score the image regions based on similarities to 13 of the most common pavement-distress classes such as rutting, cracks, patching, potholes, bleeding, corrugation, raveling & weathering, and bumps. Fusion of features from three-scale predictive outputs is used along with a predefined probability threshold to filter out most irrelevant image regions with low scores for improved detection accuracy.

The RGB image frame acquired by the D455 Intel RealSense depth camera is fed into the YOLO detector. The detection stage results in bounding boxes surrounding Regions of Interest (ROI's) in the image and identifying each pavement distress class.

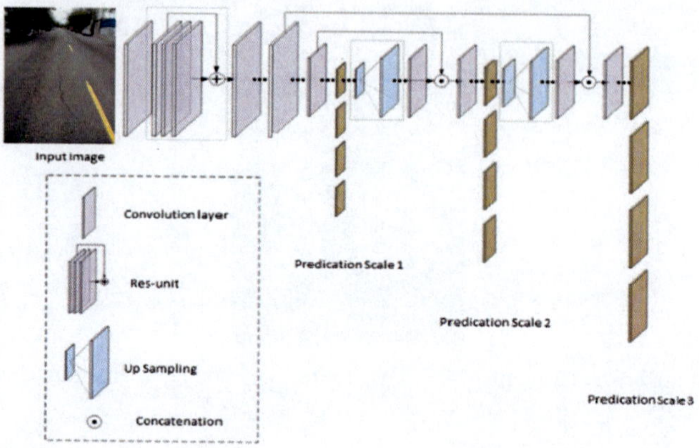

Fig. 3. The-YOLOv3-architecture for pavement-distress detection [17]

3.3 Dataset and Training Information

In the pavement-distress detection stage, a custom-collected dataset was utilized using RGB images from different sources and cameras. In addition to the images that were collected from Egyptian roads, we utilized images from two publicly-available benchmark datasets: The Pavement Image Dataset (PID) [19] and the RDD2020 dataset [20]. The PID dataset consists of 7,237 Google street-view images, while the RD2020 dataset includes 26,620 images collected from India, Japan, and Czech. Our experiments targeted 13 of the most common pavement-distress classes, including longitudinal cracks, alligator cracks, block cracks, bleeding, corrugation, patching, potholes, raveling, rutting, sealed cracks, bumps/sags, and transverse cracks. The annotations of the PID dataset had to be revised by renumbering the class IDs to follow the order of our 13 classes. On the other hand, an open-source annotation tool was used to manually annotate the RD2020 dataset images as well as the Egyptian road images based on the 13 classes. Samples of the various pavement-distress classes are presented in Fig. 4.

The following hyperparameters were utilized in model training: batch-size of 64, optimizer weight decay value of 0.0005, an initial learning rate of 0.001, and momentum of 0.9.

Class-Specific Severity Classifier

The pavement distress severity classification strategy depends on the distress type itself. For some pavement-distress classes, the severity classification is qualitative and subject to the individual visual assessment of the inspector.

For example, the Florida Department of Transportation (FDOT) classifies raveling into low, moderate, and severe levels based on the wearing of the aggregate and binder, the roughness of the surface texture, and the loss of aggregate [21]. Other distress classes such as cracks and potholes require quantitative assessment such as measurements for length, depth, area, and volume. In this paper, the proposed severity-classification framework was applied to the pothole severity-classification problem. However, it can be extended to other distress classes that require quantitative assessments.

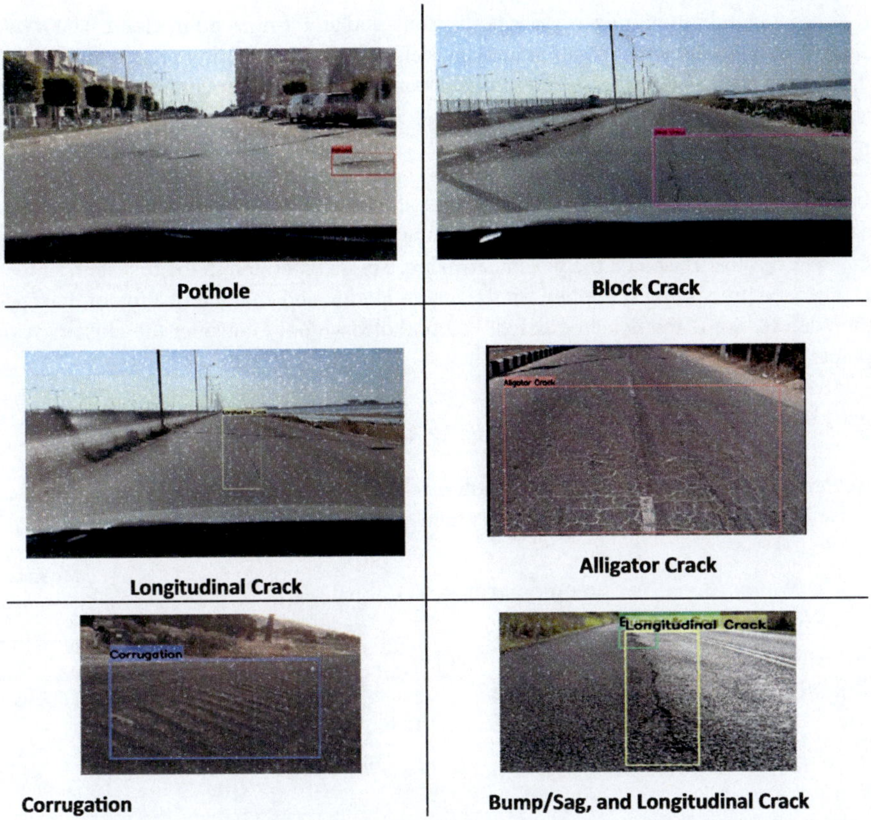

Fig. 4. Samples of pavement distress classes in the training dataset

3.4 Projection onto the Depth 3D Point Cloud and ROI Filtering

In this stage, we integrate information acquired from each of the RGB and depth images. The ROI estimates extracted from the RGB image using the YOLO distress-detection stage provide two great advantages. First, the 2D ROI bounding boxes can be projected onto the 3D point cloud depth image to perform 3D ROI filtering. This results in a significant reduction in the computational complexity and processing time as only filtered regions will be subject to subsequent processing. Vertices of the 2D bounding boxes are projected onto the 3D point cloud to determine the minimum and maximum points that we should use for ROI filtering. Second, the severity and density classifications of the different pavement-distress classes require different types of measurements. Therefore, knowledge of the class of the detected distress will allow the application of class-specific point-cloud processing.

Road Surface Extraction

The road surface points are automatically extracted from the 3D ROI by applying Random Sample Consensus (RANSAC) plane segmentation [22]. This step separates the

ROI point cloud into a *road-surface* point cloud and an *off-road* point cloud. The *road-surface* point cloud contains all points that belong to the best-fitting plane representing the road surface. The *off-road* point cloud comprises all the remaining points that lie above or beneath the road surface according to a pre-defined threshold.

Clustering of the Off-Road 3D Points
In this step, we apply Euclidean clustering to the off-road 3D points. The points within each detected cluster are then grouped and labeled according to their depth below/above the road surface. Based on the pavement-distress class identified in the first step, further class-based processing is applied for the quantification and characterization of that specific defect. So, if the detected defect is a pothole, we only consider the clusters with points below the road surface.

4 Case Study: Pothole Severity Classification

According to [6], the severity levels of potholes are defined based on the average diameter as well as the maximum depth of the pothole as presented in Table 1.

Table 1. Pothole's severity levels.

Maximum depth of pothole	Average diameter (mm) (in.)		
	100 to 200 mm (4 to 8 in.)	200 to 450 mm (8 to 18 in.)	450 to 750 mm (18 to 30 in.)
13 to ≤25 mm (1/2 to 1 in.)	L	L	M
>25 and ≤50 mm (1 to 2 in.)	L	M	H
>50 mm (2 in.)	M	M	H

Furthermore, the density of the pothole depends on the number of holes. As mentioned in [6], if the average diameter of the pothole is more than 750 mm, the area of the pothole is divided by 0.5 m^2 (5.5 ft^2) to find the equivalent number of holes. The following steps/calculations were performed to estimate the severity and calculate the density of the detected pothole (see Fig. 1):

1. After the clustering of *off-road* pixels, the depth of each point in the cluster is calculated as the vertical Euclidean distance to the detected road surface plane. The maximum Euclidian distance is used as the **maximum depth** of the pothole.
2. The cluster points are grouped based on their depth from the road surface.
3. The cluster points are projected onto the road surface plane. The projected point cloud is used to determine the concave hull that surrounds the pothole opening. The concave hull points are the blue-colored points surrounding the pothole cluster in Fig. 5. The concave hull is then used to calculate the **area of the pothole** at its opening (i.e., at its intersection with the road surface).

4. Principal component analysis (PCA) is applied to the cluster points to determine the dominant orientation of the pothole, and hence to determine the minimum bounding box of the cluster of pothole points. The length and width of the minimum bounding box were used to estimate the average diameter of the pothole. Figure 5 shows examples of the detected potholes. The endpoints that were used to estimate the two diameters were represented by 2 big blue points (first-diameter), and 2 big red points (second-diameter).
5. From the area of the pothole opening and the depth of each of the pothole cluster points the volume of the pothole is calculated. Volumetric information is very valuable for selecting the ideal repair approach as well as estimating the road repair cost.

5 Experimental Results

5.1 Results for the Distress Detection

Preliminary distress-detection results demonstrate that the pre-trained model yields accurate results in the detection and the classification of the 13 tested distress classes. However, the model requires more training as some false negatives were experienced with some of the distress classes that were severely underrepresented in the training dataset. Samples of the pavement-distress detection results are presented in Fig. 6. Samples of some of the detected potholes with various shapes and sizes are presented in Fig. 7.

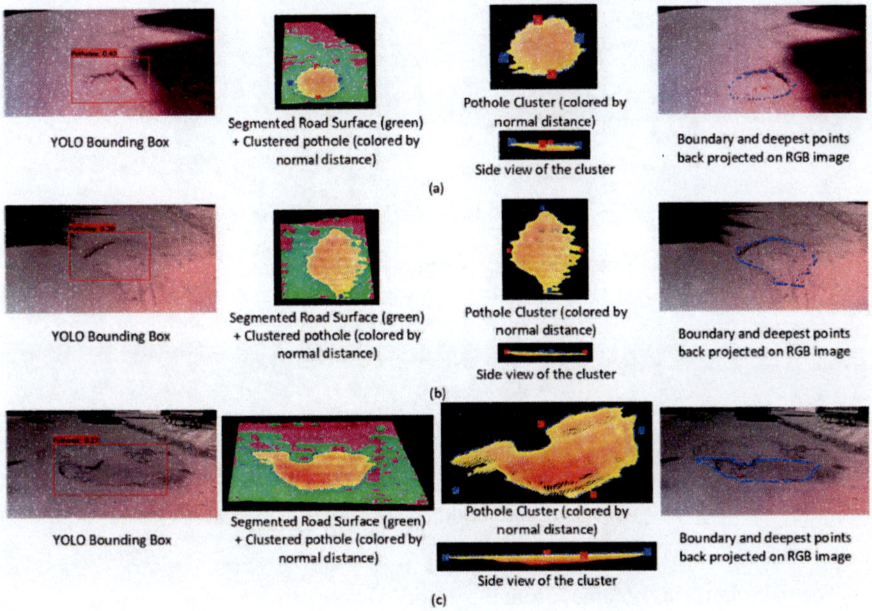

Fig. 5. Examples of severity classification workflow for detected potholes.

5.2 Results for Pothole Severity Classification

The performance of the proposed system was evaluated using several potholes with varying diameters, depths, and shapes as presented in Fig. 5. The depths of the tested potholes ranged from 4 cm to over 10 cm. Visual inspection of pothole-detection results validates the effectiveness of the proposed system for pothole detection and severity classification.

In the last column of Fig. 5, the pothole boundary points (the concave hull points) are back-projected onto the RGB image, for better visualization of the detected boundary. It's clear that the proposed system was able to efficiently detect the boundaries of various pothole shapes, and this was well reflected in the accuracy of the calculations that were used for estimating the pothole area and volume measurements.

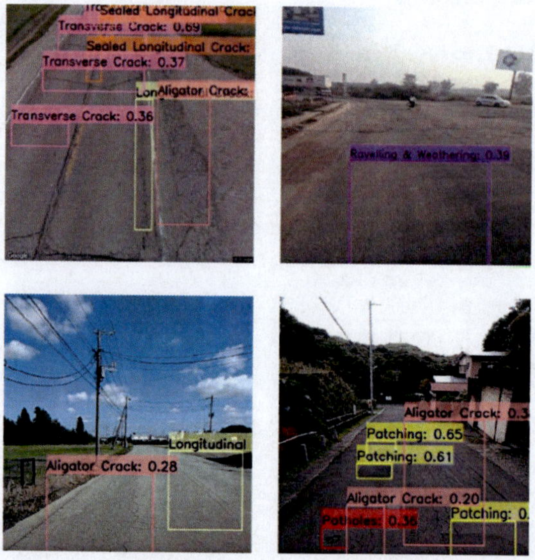

Fig. 6. Pavement distress detection results

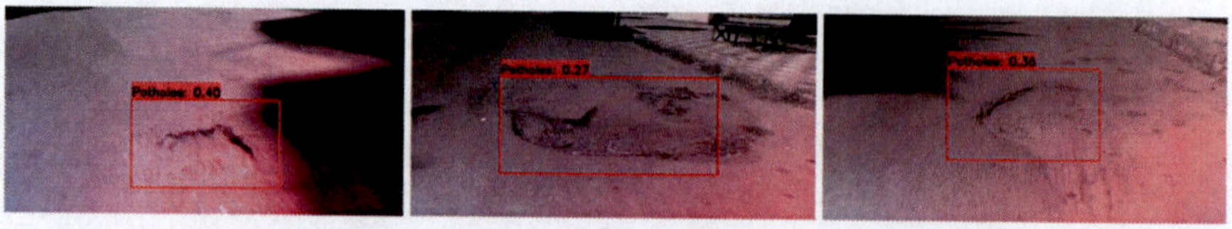

Fig. 7. YOLO detections of different pothole shapes and sizes.

The measurements obtained from the proposed system were compared to the results attained from a high-quality Leica P40 Terrestrial Laser Scanner and Cyclone Software. Table 2 shows a comparison between the volume and area estimates of a pothole as estimated by the proposed method using the low-cost RealSense D455 camera and the measurements obtained by the Leica P40 LIDAR and reconstructed by triangulation. Experimental results for the two systems are very close which validates the high accuracy of the proposed system and that it is comparable to results obtained with professional-quality systems.

Table 2. Volume and area estimations using the proposed system vs. using the Leica P40 Terrestrial Laser Scanner and Cyclone Software.

	Intel RealSense D455 depth camera and the proposed algorithm	**Leica P40 laser-scanner with tri-angulation applied**
	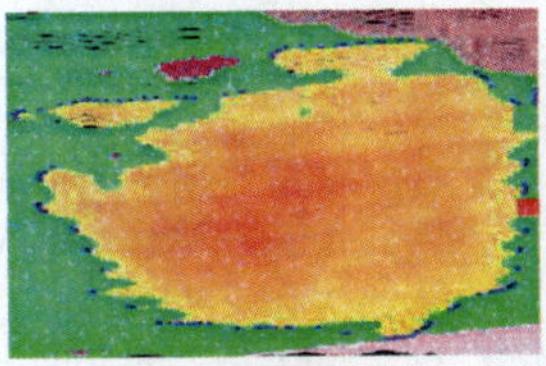	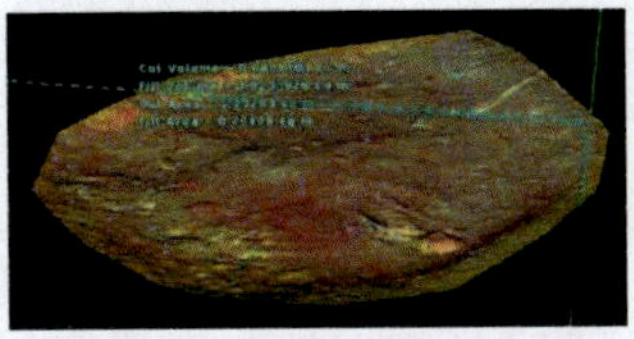
Volume	0.0237842 m^3	0.0233976 m^3
Area	0.748963 m^2	0.76498 m^2

6 Conclusion

This paper proposes an automated system for pavement-distress detection and severity classification. The proposed system incorporates an Intel RealSense D455 depth camera that generates both RGB and depth images. Initially, the YOLO v3 detector is utilized to detect and classify 13 classes of the most common types of pavement distresses from the RGB image.

The detection stage identifies the ROI areas in the color image that correspond to the detected distresses. Subsequently, the 2D ROI areas are projected onto the 3D-point cloud of the same frame allowing for 3D ROI filtering. The ROI extraction reduces the computational complexity as well as the processing time. Furthermore, knowing the distress type allows distress-specific severity classification methods as the measurements/analysis required for severity classification differ from one distress type to the other. In this work, we implemented pothole severity classification based on the maximum depth and average diameter measurements. The RANSAC plane segmentation algorithm is utilized to detect the road surface, and then Euclidean clustering is applied to off-road points to cluster them into 3D pothole clusters. The points in each cluster are then regrouped according to their depth with respect to the pavement surface. Visual results and estimated pothole measurements indicate that the proposed system was able to efficiently detect, quantify, and determine the severity of the detected potholes despite the use of a low-grade data acquisition solution. This paper presents a proof of concept of the capability of the proposed architecture that is not limited to classifying the severity of potholes only but can be extended for severity classification of other distress types in which the severity classification is based on predefined measurements/thresholds.

Acknowledgment. This research was funded by the Science and Technology Development Fund (STDF) grants for Artificial Intelligence, Research Project number: **42554.**

References

1. Ragnoli, A., De Blasiis, M., Di Benedetto, A.: Pavement distress detection methods: a review. Infrastructures **3**(4), 58 (2018)
2. Coenen, T.B., Golroo, A.: A review on automated pavement distress detection methods. Cogent Eng. **4**(1), 1374822 (2017)
3. Radopoulou, S.C., Brilakis, I.: Automated detection of multiple pavement defects. J. Comput. Civ. Eng. **31**(2), 04016057 (2017)
4. Jahanshahi, M.R., Jazizadeh, F., Masri, S.F., Becerik-Gerber, B.: Unsupervised approach for autonomous pavement-defect detection and quantification using an inexpensive depth sensor. J. Comput. Civ. Eng. **27**(6), 743–754 (2013)
5. Sjögren, L.: State of the art in monitoring road condition and road/vehicle interaction (2015)
6. Standard Practice for Roads and Parking Lots Pavement Condition Index Surveys. ASTM International, West Conshohocken, PA (2020). www.astm.org
7. Miller, J.S., Bellinger, W.Y.: Distress identification manual for the long-term pavement performance program (No. FHWA-HRT-13-092). Federal Highway Administration, Office of Infrastructure Research and Development, United States (2014)
8. Ravi, R., Habib, A., Bullock, D.: Pothole mapping and patching quantity estimates using LiDAR-based mobile mapping systems. Transp. Res. Record **2674**(9), 124–134 (2020). ASTM D6433-20
9. Christodoulou, S.E., Kyriakou, C., Hadjidemetriou, G.: Pavement patch defects detection and classification using smartphones, vibration signals, and video images. In: Mobility Patterns, Big Data and Transport Analytics, pp. 365–380. Elsevier (2019)
10. Kim, T., Ryu, S.K.: Review and analysis of pothole detection methods. J. Emerg. Trends Comput. Inf. Sci. **5**(8), 603–608 (2014)
11. Arjapure, S., Kalbande, D.R.: Review on analysis techniques for road pothole detection. In: Pant, M., Sharma, T.K., Verma, O.P., Singla, R., Sikander, A. (eds.) Soft Computing: Theories and Applications. AISC, vol. 1053, pp. 1189–1197. Springer, Singapore (2020). https://doi.org/10.1007/978-981-15-0751-9_109
12. Gupta, S., Sharma, P., Sharma, D., Gupta, V., Sambyal, N.: Detection and localization of potholes in thermal images using deep neural networks. Multimedia Tools Appl. **79**(35–36), 26265–26284 (2020). https://doi.org/10.1007/s11042-020-09293-8
13. Dhiman, A., Klette, R.: Pothole detection using computer vision and learning. IEEE Trans. Intell. Transp. Syst. **21**(8), 3536–3550 (2019)
14. Wang, P., Hu, Y., Dai, Y., Tian, M.: Asphalt pavement pothole detection and segmentation based on wavelet energy field. Math. Prob. Eng. **2017**, 1–13 (2017)
15. Ouma, Y.O.: On the use of low-cost RGB-D sensors for autonomous pothole detection with spatial fuzzy c-means segmentation. In: Geographic Information Systems in Geospatial Intelligence. IntechOpen (2019)
16. IntelSense: Intel® RealSense™ Depth Camera D455. IntelSense. https://www.intelrealsense.com/depth-camera-d455/. Accessed 22 Nov 2021

17. Redmon, J., Farhadi, A.: YOLOv3: an incremental improvement. arXiv preprint arXiv:1804.02767 (2018)
18. He, K., Zhang, X., Ren, S., Sun, J.: Deep residual learning for image recognition. In: Proceedings of the IEEE Conference on Computer Vision and Pattern Recognition, pp. 770–778 (2016)
19. Majidifard, H., Jin, P., Adu-Gyamfi, Y., Buttlar, W.G.: Pavement image datasets: a new benchmark dataset to classify and densify pavement distresses. Transp. Res. Rec. **2674**(2), 328–339 (2020)
20. Arya, D., et al.: RDD2020: an image dataset for smartphone-based road damage detection and classification. Mendeley Data, V1 (2021). https://doi.org/10.17632/5ty2wb6gvg.
21. Hsieh, Y.A., Tsai, Y.: Automated asphalt pavement raveling detection and classification using convolutional neural network and macrotexture analysis. Transp. Res. Rec. **2675**(9), 984–994 (2021). https://doi.org/10.1177/03611981211005450
22. Raguram, R., Frahm, J.-M., Pollefeys, M.: A comparative analysis of RANSAC techniques leading to adaptive real-time random sample consensus. In: Forsyth, D., Torr, P., Zisserman, A. (eds.) ECCV 2008. LNCS, vol. 5303, pp. 500–513. Springer, Heidelberg (2008). https://doi.org/10.1007/978-3-540-88688-4_37

Check for updates

An Intrusion Detection Model Based on Deep Learning and Multi-layer Perceptron in the Internet of Things (IoT) Network

Sally M. Elghamrawy[1(✉)], Mohamed O. Lotfy[2], and Yasser H. Elawady[1]

[1] Computer Engineering Department, Misr Higher Institute for Engineering and Technology, Mansoura, Egypt
Sally_elghamrawy@ieee.org, y.alawady@engmet.edu.eg
[2] Computer Engineering Department, British University in Egypt, Cairo, Egypt

Abstract. Recent years have seen a surge in the usage of deep learning as a critical and adaptable tool for the Intrusion Detection Systems (IDSs) and in the Internet of Things (IoT). Detecting intrusions using deep learning is compared to other techniques, such as machine learning, which have been used in the past, in this research. There is a lot of expertise, advice, and ongoing maintenance required for most solutions, including machine learning. The fundamental advantage of deep learning over other techniques is that it eliminates the majority of the feature extraction process while maintaining high accuracy, efficiency, and system reliability. In this paper, an intrusion detection model is proposed based on deep learning technique. A Multi-Layer Perceptron (MLP) Neural Network model is implemented. The KDDCUP 99' dataset is used in this research to test two deep learning architectures against each other and with previous works. Four hidden layers of ReLu activation, a softmax activation output, Adam optimizer, and early stopping validation loss monitoring are used in both topologies. Because they are implemented as multi-classification neural networks, they also use categorical cross entropy for loss function calculation. Form (10, 50, 10, 1) were the hidden layers in Model 1, while form (20, 20, 20, 1) were the hidden layers in Model 2. Model 1 architecture achieves a maximum accuracy of 99.88%, while Model 2 architecture achieves a maximum accuracy of 99.785%. The system's efficiency and accuracy were tested with regard to the size of the samples as a factor.

Keywords: Intrusion detection · Internet of Things · Deep learning · Neural network · Multi-layer perceptron · Anomaly detection · Data encoding

1 Introduction

When it comes to the Internet of Things (IoT), there are many different sorts of structures that all boil down to the same thing: "Multiple devices connected in a topology/hierarchy that best fits a given application." An attack on the IoT can take numerous forms, depending on its objective, due to the IoT's wide variety. Such attacks/intrusions are difficult to detect and involve a lot of time and effort [1, 2].

© The Author(s), under exclusive license to Springer Nature Switzerland AG 2022
A. E. Hassanien et al. (Eds.): AMLTA 2022, LNDECT 113, pp. 34–46, 2022.
https://doi.org/10.1007/978-3-031-03918-8_4

There are too many variables to consider to rely on a predetermined set of rules (e.g. if X is above a certain value then an attack has occurred). Attacks are dynamic and always changing. Rather than spending money on research and development, machine learning provides an opportunity to transfer a substantial portion of the upkeep process to the algorithm. Machine learning, on the other hand, is a profusion of approaches and algorithms because it is a difficult field with many variables [3].

The system's intent for operation and the transaction's purpose are both defined in a standard transaction specification. And as the Internet of Things' goal of modular and flexible communication emerges [3]. The boundaries of "regular transaction" extend to accommodate a huge and often difficult to define outline. The line between what is normal and what is aberrant will no longer be as clear-cut as it was [4]. Data flow in real-world applications can sometimes be inaccurately depicted in an experimental context. Noise is likely to be present in data gathered from real-world applications. As a result, data that appears to be normal but is usually irregular may be difficult to tell apart because of the noise. And if the abnormalities in a data stream are hostile in nature, the adversary entering the data will often try to adapt it. The enemy, on the other hand, will utilise the system's concept of normal as a cover for their data wherever possible (Chandola et al. 2009). Furthermore, the availability of data that precisely reflects the findings of specialists who have analysed the data and found it to be normal or malicious is often an issue [5, 6].

It is the goal of this research to gain and use knowledge in areas of communication technology and information science that are vital and developing. It intends to look at the latest developments in IoT, understand and analyze intrusion detection as an engineering challenge, and create an Intrusion Detection System that uses deep learning to solve it. Big data, programming, communication topologies, deep learning, and the current state of the field must be studied thoroughly.

2 Related Work

A variety of intrusions have evolved to become more difficult to detect. An intrusion detection system in an IoT system may not be able to keep up with the constant flow of data, and even if it can, the network may be too large and complicated to allow for continuous human supervision [4, 5]. An intrusion detection system must be in place that constantly monitors the network for signals of intrusion. Expanding on conditional programming is an early approach [7]. The goal is to develop an algorithm that can identify the boundaries set by a team of experts who will be responsible for routinely updating and maintaining it. Artificial Intelligence is defined as algorithms that use the expert's intelligence (in the form of conditions on which decisions are based) to make decisions [6]. A multi-agent system will then be installed on the various layers of the IoT to monitor and check for any anomalies. In terms of processing and network consumption, this is expensive and resource-intensive [8].

In [19], Farahnakian and Heikkonen used a deep auto-encoder based IDS where a deep learning neural network is used as both an encoder and a classifier. Auto encoders were used in the first four layers of the neural network. Their model managed to output a 95.53% accuracy for binary classification, and a 94.71% accuracy for multi-class classification.

In [22], Karatas et al. proposed IDS based on deep learning, showing that deep learning requires more resources in terms of computation. And in terms of data, deep learning can only function with high levels of accuracy in applications where data is overflowing. Deep learning's highest cost on computation is in the training stage, which is not a recurring cost.

Many research papers used multi-agents' systems, fuzzy systems and game theory for the intrusion detection system as showing in the following subsections.

2.1 Multi Agent Systems for IDS

Multi-agent models are decoupled from the monitored system and serve as a sensing layer for an Internet of Things (IoT) system. Host, detection, and network agents are deployed to various network nodes in these models [9]. The host keeps an eye on the detection agent's output and filters out the irrelevant data before sending it to the network agent via the terminal it is connected to. When a network layer fails, the network agents are there to keep the system running smoothly throughout the network [10]. A high-level network agent then transmits the detection information to the console for further processing after the network layer communicates through the hierarchy of network layers any data received from the host agent [11]. Sniffer agents were employed to detect intrusions on KDDCUP99 by Dhanalakshmi et al. 2017 in a slightly different topology [12]. They analyze TCP traffic and identify intrusions at all the terminals in the system. Next, the Host Filter Agent sorts packets based on communication protocol (TCP or UDP, for example) and performs the filtering action. Rule Based Agents are fed to Dhanalakshmi, which uses a combination of low-level and high-level networking agents to decide whether or not the final high-level packets meet the set of rules and are normal or an intrusion. When dealing with a large number of assailants, the system begins to suffer greatly in accuracy. When the number of attackers hits 30, the system loses around 50% of its effective detection accuracy, Going from 95% to 50% [12].

2.2 Fuzzy Systems for IDS

Clustering is the technique of grouping things by determining the clusters in which some data in a set is located. In the process of cluster analysis, data that is as comparable as possible is grouped together into clusters. Aside from that, it is also important to ensure that the definition generates as wide a gulf as possible between the meanings of distinct clusters. Clustering techniques can be studied and validated with the help of measures. They are based on the distance (or difference) between two points (or connectivity) [5, 9]. Two main clustering techniques exist: Fuzzy and Hard. A single cluster can be produced using the hard clustering or non-fuzzy clustering approach of data clustering. All data is grouped in a binary form using Anomaly detection, in which outliers are determined by their distance from the "typical" cluster, making this a straightforward application of the technique. The data might be either x or y, but not both x and y at the same time, in hard clustering to put it another way, hard clustering cannot be applied to data that has the potential for many output kinds. Clustering using fuzzy logic allows data to be divided into many groups. This means that instead of a binary approach, a record may be 0.5 x and 0.5 y. A cluster's membership is no longer a binary "yes" or "no," but rather a measure of how closely you fit in. This means that a data point may be part of the same cluster as other data points, but only to a lower extent. Instead, clusters are transformed into gradations with varying degrees of membership intensity for individual datapoints. Fuzzy-C-means clustering is the most commonly used method for fuzzy clustering. Once the appropriate number of clusters has been chosen, the algorithm generates random "membership degree" coefficients for each data point. Iterations of the algorithm begin to condense in an effort to minimize error and locate cluster centres (mean points in a cluster). Pre-processing is required for this type of database knowledge discovery to reach the stage where it can read the features of the KDD-Cup 99' dataset [12, 13]. Chandrasekhar and Raghuveer [9] created a pre-processing layer that standardized and binarized the dataset's features. A clustering technique (Fuzzy C-means) was employed, intending to partition the data into five clusters (one for each general attack type and one for the non-malicious record type) while changing the algorithm's weighting component to change how much it changes coefficients in each iteration. The top model was able to classify 5 clusters with an accuracy of 91.89% after numerous runs.

2.3 Game Theory Models for IDS

In a game theory model, attackers may be analysed mathematically and the competitive attitude of forecasting their actions based on the present network state is incorporated. The system could be better prepared for a specific attack if the odds were skewed in its favour. A Nash Equilibrium, where neither the system nor the attacker can change their plan again, ensures that this strategy remains predictable and thus easier to deal with. Because it relies on so many assumptions, a game theory model generally fails to deter attackers expecting to deal with it. However, the implementation of this system is costly and needs a significant number of assumptions. A Generative Adversarial Network (GAN) is an example of a game theory integration [14]. When two machine learning models are pitted against each other, one generates data and the other predicts how likely it is that the generative model will insert that data into the flow of real data. A Nash equilibrium is achieved when the Discriminator reaches its full capacity. As a result, it also shows that the Generator is now capable of producing data that cannot be distinguished from the genuine data flow. Using deep learning, a malicious generator may overwhelm the network with seemingly real data that doesn't deviate from the typical data flow, making it impossible for the discriminator to distinguish between legitimate and malicious data packets [15, 16].

Using a GAN model, Lee and Park, [17] created an optimized dataset. A sort of neural network called an Autoencoder was developed to enhance the dataset's features and reduce its dimensions using encoding and decoding. Prior to processing and training, this encoder aims to extract and optimize pertinent data. Their proposed GAN model, which was designed to address the issue of rare class imbalance, achieved a final performance of 98.17% using 40 hidden layer neurons.

It was a need for reducing the number of features in huge datasets that necessitated the introduction of deep learning into the intrusion detection systems area. A Deep learning technique was used to reduce features, increase computation efficiency, and improve data reliability [18]. Separate components of the proposed system are possible. There was a pre-processing stage that transformed the numbers into more understandable ones that could be fed directly into a neural network. This was done by scaling numerical data into a fractional system with a maximum of 1 and minimum of 0. Deep Belief Network (DBN), the Neural Network component, was then added to the system. Its anomaly detection component was a classifying SVM (support vector machine). The 41 features were reduced to five by the DBN. For binary classification, the final system had an accuracy rate of 92.84%. The detection of multiclass attacks was not addressed in this research. Deep learning has only recently begun to be used at the core of intrusion detection systems in present research [19]. Deep Learning networks have replaced feature reduction as the classifying component for finding anomalies. To do this, three types of Deep Learning models were deployed.

3 Architecture of the Proposed Intrusion Detection System

One of the primary objectives of the intrusion detection system is to improve the accuracy of the deep learning algorithm. An intrusion detection model is proposed based on deep learning technique and a Multi-Layer Perceptron (MLP) Neural Network model is implemented.

Layers are needed to optimize the workflow in this system. It is the pre-processing layer in the first place. Aims to reduce errors, optimize data, and remove anomalies in database structure are all part of this layer. Then comes feature engineering, which is the second layer. By assigning more weight to input that is directly relevant to the algorithm's decision-making process, this layer optimizes the algorithm. As the third layer, it includes the deep learning algorithm and the processes used to feed it and extract its output for evaluation or training purposes. Evaluation data and accuracy are generated by this layer as a last step. The architecture of the proposed intrusion detection system is shown in Fig. 1.

Two deep learning architectures are proposed, namely Model 1 and Model 2, to process the KDDCUP 99' dataset and measures their performance against each other and different works. Both architectures implement four hidden layers of ReLu activation, a softmax activation output layer, Adam optimizer, and an early stopping validation loss monitoring for supervised training. They also utilize categorical cross entropy for loss function calculation as they are implemented as multi-classification neural networks. Model 1 had hidden layers in the form (10, 50, 10, 1) and Model 2 had hidden layers in the form (20, 20, 20, 1). The Neural Network architectures had a max accuracy of 99.88% for Model 1 and a max accuracy of 99.785% for Model 2. The sample sizes were considered as a factor in the testing of the system's efficiency and accuracy.

3.1 Pre-processing and Feature Engineering

The final trained algorithm should be less affected by factors that could have a negative impact on the system's ability to improve on earlier efforts. According to the prior statement, each input in a deep learning framework, whether it be during training or otherwise, has a significant impact on how the algorithm performs. Furthermore, bias is created when a database is exposed to the same input more than once. Redundancies, on the other hand, are detrimental to the algorithm. Rescaling is another way to improve

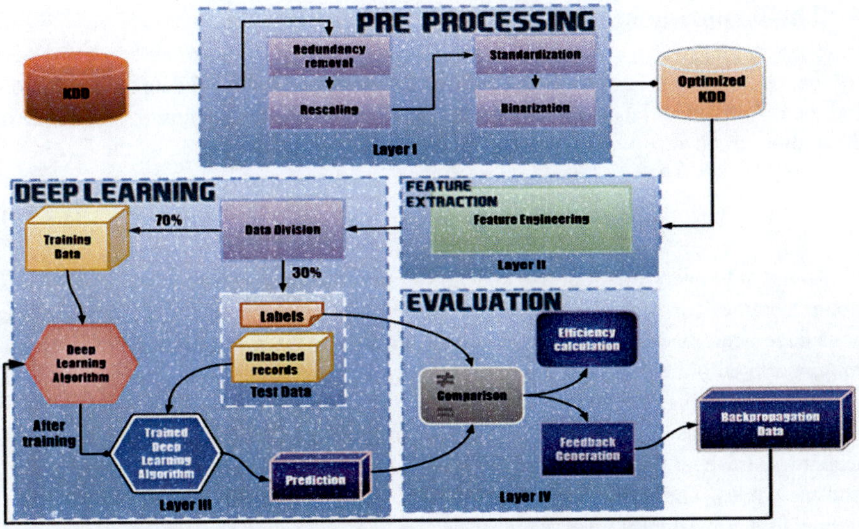

Fig. 1. The architecture of the proposed intrusion detection system

overall accuracy [19, 20]. A deep learning system will assign various weights to features measured on different scales, depending on the scale size. Because of this, rescaling is a crucial strategy for preventing a bias toward greater or smaller changes in comparison to other characteristics. Other preprocessing approaches, like as standardization and binarization, are also available for experimentation [21].

3.2 Deep Learning Layer

Deep learning layer includes a database division step that separates datasets for use in training and those for testing (initially into 70% and 30% respectively) [22, 23]. The training data contains records with the correct intrusion status indicated on them. In order to send the labels to the evaluation layer, the test data is divided into labels and unlabeled records, respectively. The deep learning algorithm is put through its paces in the training phase. Afterwards, the algorithm is tested with new data and a forecast is created and saved for evaluation.

3.3 Evaluation Layer

An algorithm's next run can be improved through the feedback cycle created by the Evaluation Layer. The algorithm's predictions are then compared against the labels it obtained from the database's division stage. That information is fed into a feedback generating function that generates distinct biassing and weight coefficient changes for deep learning algorithms based on the comparison [24, 25]. The recommended metrics for system, efficiency, and accuracy will be used in the evaluation layer.

4 The Experimental Results

After each model has been trained, the test data must be used to validate the predictions and the original data labels of the test data must be compiled and compared. Based on the number of correct predictions, the initial validation is determined by the Eq. 1:

$$Total\,Accuracy(\%) = \frac{Total\,correct\,predictions}{Total\,record\,count} \times 100 \quad (1)$$

As a result, every prediction is checked against the equivalent test record label and stored as either correct or incorrect. A counter keeps track of how many right suggestions have been made. As previously stated, the accurate prediction number is used in conjunction with the matrix at the following layer [26].

As shown in Fig. 2. The total accuracy results given are above 99% accuracy. Using a large 100% dataset, the first (10, 50, 10, 1) architecture achieved the highest possible accuracy. However, there was just a 0.44% difference in accuracy between the two extreme values. The results show that at lower sample sizes, the second (20, 20, 20, 1) architecture consistently gives a higher accuracy even though the computation did not require as much time (as seen in Fig. 3). As for the bigger dataset, the non-linear first architecture outperformed the linear second architecture by 3% points. In other words, the system was able to precisely detect an incursion, but it was also able to correctly classify it among the 22 different types of attacks.

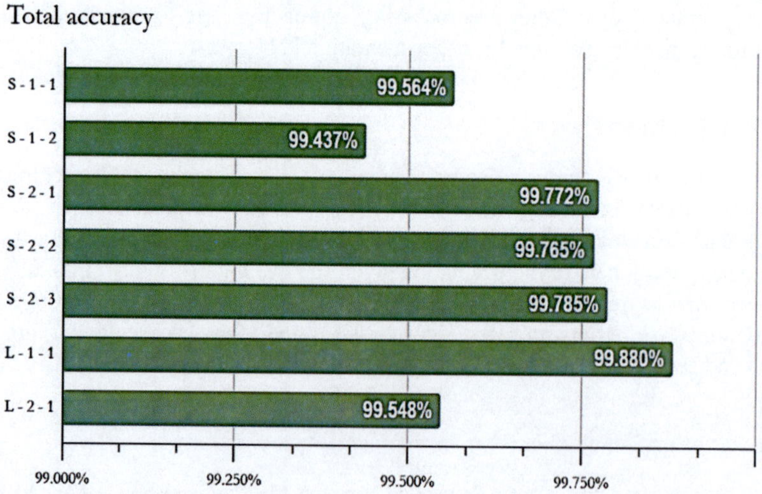

Fig. 2. Resulting accuracies S = small sub-dataset (10%) t, L = large dataset (100%), first index 1 = first architecture 2 = second, second index is the chronological run index

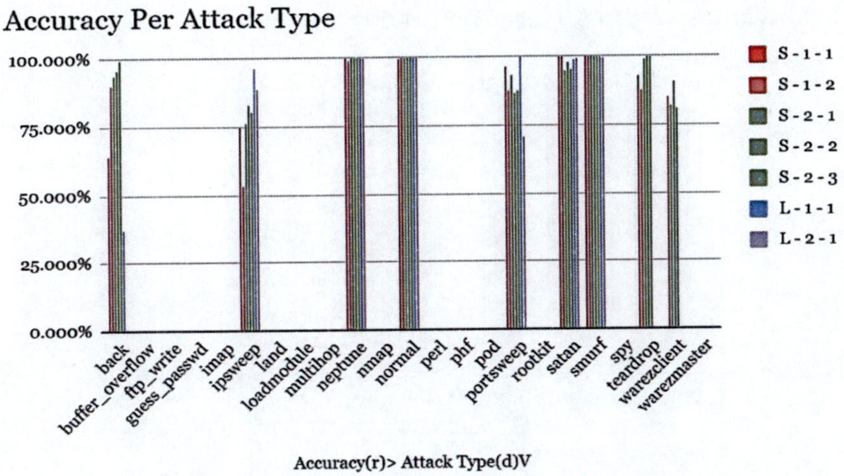

Fig. 3. Accuracy of systems in detecting certain attack types.

The individual accuracy graph in Fig. 4 shows the prediction accuracy of the Intrusion detection systems. A large number of attacks appear to be completely ineffective. In the table below, we display the number of times each of these attacks occurred in the dataset. There was no detection of any of the attacks presented below because of their low number in relation to the whole dataset, which prevented the neural networks from being properly trained to identify them. The cost function is not affected by their low numbers. This is because the neural network is continually aiming for the best possible accuracy in the shortest time possible. When there isn't much data to work with, it's difficult to combat this problem. It's also important to note that these attacks aren't affecting the neural network's ability to achieve a 99.8% accuracy level, thus they don't need to be trained. This is because the neural network doesn't need to learn to differentiate them in order to achieve this level of accuracy. Due to the obvious limited number of occurrences and the shuffled training and test data, it's possible that the algorithm wasn't tested on them, or that it was tested but never trained. After conducting the 25–75% test-train randomized splitting procedure, the data may have been entirely on one side of the split.

The rest of the individual accuracies are shown in Fig. 4. The most consistently detected with high accuracy across all architectures are the Normal (non-malicious records), Neptune attack, and smurf attack. They are all above 99.5% in all instances of the experiment. This is mostly due to the abundance of data in the records classified as these aforementioned types. Some of the rarer attacks are shown to be detected here. However, some neural networks have not detected certain types of them completely still. The smaller sample size neural networks managed to be the best at accurately detecting the rare types of attacks. This is largely due to the smaller sample ensuring that the smallest attacks are not cut out of the 10% sample as they are already rare and 10% of their numbers would further spread the gap between these attacks and the more common ones. Moreover, the 100% dataset Neural Networks performed poorly on all the rare attacks, reaching only 37.44% accuracy in detecting the "back" attack type in the first

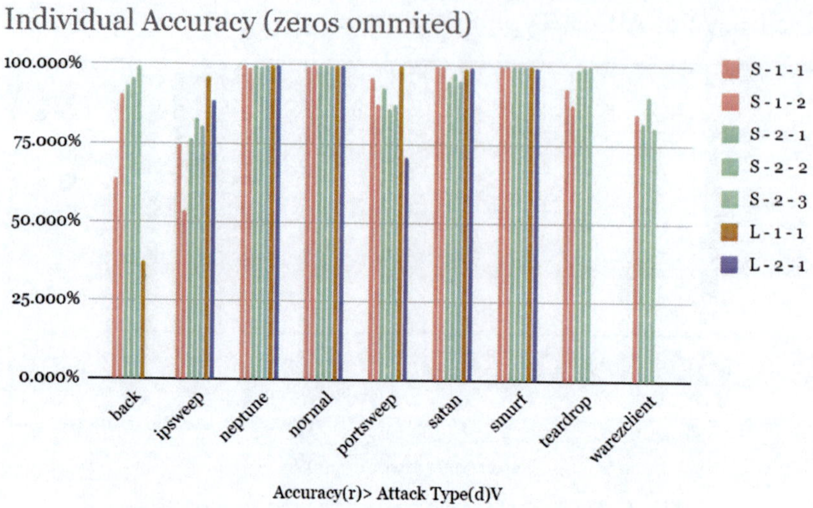

Fig. 4. Individual accuracies of detected classes (zero acc omitted)

architecture and 0% in the second. That, and a total accuracy of 0% in both "Teardrop" and "Warezclient" attack types in both architectures.

The computation time graph shows how the 100% sample size run of the first architecture takes almost 1000 s to compute the final neural network while reaching an accuracy of 99.88% in a real world application this is a realistic amount of time (16 min and ~40 s) to compute an intrusion detection system but the scaling of time relative to how little the actual difference is in terms of overall accuracy needs to be addressed. The first run of the second architecture using a small 10% subset of the dataset reached 99.772% and was computed in 2 min and 5 s, meaning that in these specific cases it took an extra 14 min and 35 s or ~800% stretch to only gain ~0.11% increase in accuracy. Even when compared to the least performing run which yielded 99.437%, the difference is still minute even though the higher accuracy was computed in 6× the time. It should be noted that even though the small sample Neural Networks performed lower in terms of overall accuracy, they learned to identify more rare attacks with higher accuracy due to the increase in relative ratio of their number vs. other attacks. And thus it is to be considered and applied ad hoc to the scenario, as for most cases the computation time is irrelevant with respect to reaching the highest accuracy possible. And in some cases the computation time or "the most efficient architecture for implementation" is more important than a 0.4% accuracy increase. It all depends on the application. Were this a medical application for example even a 0.01% accuracy increase is worth hours of computation time as it is a sensitive field (Fig. 5).

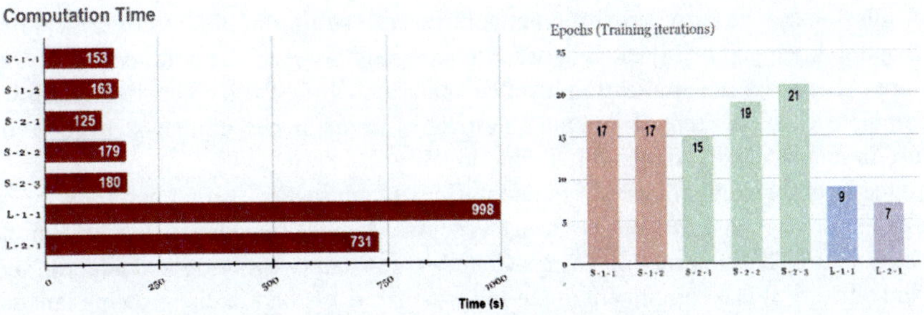

Fig. 5. The Computation time/s and number of Epochs of each dataset

5 Comparison Between Proposed Models and the Others

As shown in Table 1 the proposed models are out-performed the surveyed literature while maintaining a multiclass model. Even in the least accuracy runs with both the models consistently provide 99.4+% accuracy. The models when compared with each other show that a model trained on the 10% subset is more accurate at dealing with rare attacks while not losing much overall accuracy. The first model performed better in overall accuracy when trained with the 100% sample size. The second model consistently performed better than the first when trained with the 10% subset.

Table 1. Comparison between the proposed models and the others.

Model	Method	Accuracy	Detection type
Proposed model 1	**MLP**	**99.4%–99.88%**	**Multi-class**
Proposed model 2	**MLP**	**99.54%–99.785%**	**Multi-class**
Lee and Park, 2019 [17]	GAN	98.17%	Multi-class
Salama et al., 2011 [18]	DBN-SVM	92.84%	Binary classification
Farahnakian and Heikkonen 2018 [19]	DAE	94.71%	Multi-class
Dhanalakshmi et al. 2017 [12]	Multi Agent	95%–50%	Binary classification
Chandola, V., Banerjee, A., & Kumar, V. (2009) [6]	Fuzzy-C	91.89%	5 cluster identification

6 Conclusion

The goal of this study was to propose and design an IoT intrusion detection system using deep learning technique. In addition, the results of the proposed models are compared with other regularly used solutions. The Internet of Things (IoT) itself contains numerous vulnerabilities that can be exploited without effective intrusion detection. The transmission can be hijacked at the network layer, resulting in the loss of critical data. Perception layer can be hijacked in order to alter readings, which could result in system failure. Attacks on the application layer can cause anything from a network overflow to a complete system failure. There are a number of intrusion detection systems currently in use in industries that demand IoT functionality.

Due to the dynamic nature of the definitions of anomalies and normal transactions, the detection of such intrusions is a tough process. Many factors lead to this dynamic and diverse nature. The Internet of Things (IoT) is a constantly evolving and adapting field in and of itself. It is a component of the IoT's semantic vision that the growing language of communication between devices is to be produced in a flexible and modular manner that allows the inclusion of as many technologies as possible. As a result, it becomes increasingly difficult to tailor an intrusion detection system to the needs of a particular industry. As a result, the distinction between normal and outlier data becomes increasingly ambiguous. Furthermore, the adversarial nature of intrusion causes the intruders themselves to try to change their intrusion methods to meet the system's notion of normal so that they can pass undetected by the filters of an IDS. Real data, on the other hand, is far more unpredictable than any experimental simulation environment that may be used to train these algorithms. It is possible for real-world applications to encounter transactions that appear to be aberrant due to some noise, resulting in confusion or a false alarm in the system. IoT intrusion detection could benefit from the use of deep learning, which has the potential to be critical. Deep learning has a higher overall accuracy and cheaper deployment cost than multi-agent systems or game theory systems. It also relies less on expert maintenance. In the feature extraction phase, deep learning has an advantage over other machine learning algorithms because it requires little to no expert assistance in extracting patterns from databases. Deep learning has the drawback of requiring a vast amount of data to be taught. However, the field of IoT is overflowing with data. As a proof of concept, a database called KDDCUP99 was used in this experiment. If the database isn't properly processed, it's full of errors and omissions that can contaminate the algorithm and reduce its precision. A pre-processing/feature engineering layer, a deep learning layer, and an evaluation layer are all included in the proposed design. It is the goal of the pre-processing layer to eliminate any database errors that may have an adverse effect on the system. Secondly, two encoding techniques are implemented. The data can be encoded in two different ways. In order for the neural network to interpret changes in a feature's scale, z-score encoding is employed for the numerical data that needs to be scaled in similar ranges. Encoding text fields into binary representations of the various possible inputs to the current feature is called Text to Dummy encoding. Training and test data are separated in the deep learning layer and then the algorithm is trained using these test labels. As a result of this backpropagation cycle, the algorithm can learn from its mistakes and improve over time. The evaluation layer calculates the accuracy and efficiency of the system. This implementation is applied on two architectures of bottlenecked Neural Networks, an expanding and contracting size architecture (10, 50, 10, 1) and a linear architecture (20, 20, 20, 1) the two architectures are run through a small subset of the KDDCup 99' data set and the entire dataset as a form of comparative performance case study over sample size. The overall accuracy of the best performing model was 99.88%. The model that achieved this accuracy was the first architecture (10, 50, 10, 1) using the full KDDCup 99' dataset as its sample and took 16 min and 20 s to compute. However, the second architecture reached an overall accuracy of 99.772% in just 2 min and 5 s using a 10% subset as the sample. Additionally, that run also detected rarer attacks with a significantly higher accuracy than the best overall accuracy model. Thus, Future considerations include possible integrations of optimization algorithms or layers, integration of different deep learning topologies, increasing performance indices, expanding on training data and trying different datasets.

References

1. Aldweesh, A., Derhab, A., Emam, A.Z.: Deep learning approaches for anomaly-based intrusion detection systems: a survey, taxonomy, and open issues. Knowl.-Based Syst. **189**, 105124 (2020)
2. Allen, D.M.: The relationship between variable selection and data augmentation and a method for prediction. Technometrics **16**, 125–127 (1974)
3. Anthony, O., Odeyabinya, J., Emmanuel, S.: Intrusion detection in Internet of Things (IoT). Int. J. Adv. Res. Comput. Sci. **9**(1), 504–509 (2018)
4. Atzori, L., Iera, A., Morabito, G.: The Internet of Things: a survey. Comput. Netw. **54**(15), 2787–2805 (2010)
5. Bezdek, J.C.: Pattern Recognition with Fuzzy Objective Function Algorithms. ISBN 0-306-40671-3 (1981)
6. Chandola, V., Banerjee, A., Kumar, V.: Anomaly detection: a survey. ACM Comput. Surv. (CSUR) **41**(3), 1–58 (2009)
7. Wu, M., Jermaine, C.: Outlier detection by sampling with accuracy guarantees. In: Proceedings of the 12th ACM SIGKDD International Conference on Knowledge Discovery and Data Mining, pp. 767–772, August 2006
8. Deng, L., Li, D., Yao, X., Cox, D., Wang, H.: Mobile network intrusion detection for IoT system based on transfer learning algorithm. Clust. Comput. **22**(4), 9889–9904 (2018). https://doi.org/10.1007/s10586-018-1847-2
9. El-Ghamrawy, S.M., Eldesouky, A.I.: An agent decision support module based on granular rough model. Int. J. Inf. Technol. Decis. Mak. **11**(04), 793–820 (2012)
10. Chandrashekhar, A.M., Raghuveer, K.: Performance evaluation of data clustering techniques using KDD Cup-99 Intrusion detection data set. Int. J. Inf. Netw. Secur. **1**(4), 294 (2012)
11. El-Ghamrawy, S.M., El-Desouky, A.I., Sherief, M.: Dynamic ontology mapping for communication in distributed multi-agent intelligent system. In: 2009 International Conference on Networking and Media Convergence, pp. 103–108. IEEE, March 2009
12. Dhanalakshmi, K.S., Kannapiran, B.: Analysis of KDD CUP dataset using multi-agent methodology with effective fuzzy based intrusion detection system. J. Appl. Secur. Res. **12**(3), 424–439 (2017)
13. Tavallaee, M., Bagheri, E., Lu, W., Ghorbani, A.A.: A detailed analysis of the KDD CUP 99 data set. In: 2009 IEEE Symposium on Computational Intelligence for Security and Defense Applications, pp. 1–6. IEEE, July 2009
14. Garcia-Teodoro, P., Diaz-Verdejo, J., Maciá-Fernández, G., Vázquez, E.: Anomaly-based network intrusion detection: techniques, systems and challenges. Comput. Secur. **28**(1–2), 18–28 (2009)
15. Goodfellow, Bengio, Courville: This table-filling strategy is sometimes called dynamic programming, p. 214 (2016)
16. Goodfellow, I., et al.: Generative adversarial nets. In: Advances in Neural Information Processing Systems, pp. 2672–2680 (2014)

17. Lee, J., Park, K.: AE-CGAN model based high performance network intrusion detection system. Appl. Sci. **9**(20), 4221 (2019)
18. Salama, M.A., Eid, H.F., Ramadan, R.A., Darwish, A., Hassanien, A.E.: Hybrid intelligent intrusion detection scheme. In: Gaspar-Cunha, A., Takahashi, R., Schaefer, G., Costa, L. (eds.) Soft Computing in Industrial Applications, pp. 293–303. Springer, Heidelberg (2011). https://doi.org/10.1007/978-3-642-20505-7_26
19. Farahnakian, F., Heikkonen, J.: A deep auto-encoder based approach for intrusion detection system. In: 2018 20th International Conference on Advanced Communication Technology (ICACT), pp. 178–183. IEEE, February 2018
20. Soumyalatha, S.G.H.: Study of IoT: understanding IoT architecture, applications, issues and challenges. In: 1st International Conference on Innovations in Computing & Net-working (ICICN16), CSE, RRCE, May 2016. International Journal of Advanced Networking & Applications
21. Liu, H., Lang, B.: Machine learning and deep learning methods for intrusion detection systems: a survey. Appl. Sci. **9**(20), 4396 (2019)
22. Karatas, G., Demir, O., Sahingoz, O.K.: Deep learning in intrusion detection systems. In: 2018 International Congress on Big Data, Deep Learning and Fighting Cyber Terrorism (IBIGDELFT), pp. 113–116. IEEE, December 2018
23. Lin, P., Ye, K., Xu, C.-Z.: Dynamic network anomaly detection system by using deep learning techniques. In: Da Silva, D., Wang, Q., Zhang, L.-J. (eds.) CLOUD 2019. LNCS, vol. 11513, pp. 161–176. Springer, Cham (2019). https://doi.org/10.1007/978-3-030-23502-4_12
24. Sethi, P., Sarangi, S.R.: Internet of Things: architectures, protocols, and applications. J. Electr. Comput. Eng. **2017**, 1–25 (2017)
25. Xiao, L., Wan, X., Xiaozhen, L., Zhang, Y., Di, W.: IoT security techniques based on machine learning: how do IoT devices use AI to enhance security? IEEE Signal Process. Mag. **35**(5), 41–49 (2018)
26. Engy, E.L., Ali, E.L., Sally, E.G.: An optimized artificial neural network approach based on sperm whale optimization algorithm for predicting fertility quality. Stud. Inform. Control **27**(3), 349–358 (2018)

Transfer Learning and Recurrent Neural Networks for Automatic Arabic Sign Language Recognition

Elsayed Mahmoud(✉), Khaled Wassif, and Hanaa Bayomi

Faculty of Computers and Artificial Intelligence, Cairo University, Cairo, Egypt
sayedscu@gmail.com, {kwassif,h.mobarz}@fci-cu.edu.eg

Abstract. Arabic Sign Language (ArSL) is the most utilized for hearing and speech impairments in Arab countries. The recognition system of ArSL could be an innovation to empower communication between the deaf and others. Recent advances in gesture recognition using deep learning and computer vision-based techniques have proved promising. Due to a lack of ArSL datasets, the ArSL dataset was created. The dataset was then expanded using augmentation methods. This paper aims to create an architecture based on both Transfer Learning (TL) models and Recurrent Neural network (RNN) models for recognizing ArSL. The extraction of spatial and temporal data was accomplished by combining TL and RNN models. Furthermore, the hybrid models outperformed current architectures when tested on both the original and augmented datasets. More overall, the highest recognition accuracy of 93.4% was attained.

Keywords: Arabic sign language · Hand gesture · Video analysis · Transfer learning · Recurrent Neural Network

1 Introduction

The deaf community uses sign language (SL) as a native language; they use their hands, heads, and bodies to communicate. Furthermore, significant advances in deep learning would make Sign Language (SL) recognition simple and provide users with immediate definitions of their gestures.

There are many sign languages [1] such as American Sign Language (ASL), Indian Sign Language (ISL), British Sign Language (BSL), and Arabic Sign Language (ArSL) that is the primary language of the deaf in Arabic regions. Since there isn't an ArSL dataset available, the dataset for video-based Arabic sign language must be constructed.

Face shape, body position, hand movement, skin color, background, brightness, and camera angle are obstacles to developing an ArSL recognition system. Moreover, The most complex issue was extracting spatial and temporal information from videos. Therefore, there are several studies for understanding sign languages by various methods such as hand detection, face detection, and gesture recognition [2].

Transfer Learning (TL) and Recurrent Neural Networks (RNN) were Deep Learning's most innovative computer vision methods especially for video categorization [3]

© The Author(s), under exclusive license to Springer Nature Switzerland AG 2022
A. E. Hassanien et al. (Eds.): AMLTA 2022, LNDECT 113, pp. 47–59, 2022.
https://doi.org/10.1007/978-3-031-03918-8_5

(concentrated on labeling video clips automatically based on their semantic contents). In addition, it is used for many purposes such as human actions [4], intelligent control systems, human-computer interaction, and health care.

The following are the paper's key contributions:

1. The authors realized that generating a labeled Arabic sign language dataset is a big contributor that can serve as an effective kernel for related studies due to the scarcity of video-based Arabic Sign language datasets. This dataset would also be made available to other researchers.
2. Propose a unique architecture that combines (TL) and (RNN) models to extract video features and achieve optimal Arabic Sign Language identification.
3. Improve the outcomes by utilizing an augmented dataset that aids neural networks in learning and generalization.
4. Combine deep learning models with and without augmented datasets to demonstrate recognition accuracy.

The following is a summary of the paper's structure. Section 2 discusses several relevant techniques to sign language recognition architectures as well as the outcomes of their suggested methodologies; Sect. 3 goes through the dataset utilized. Section 4 describes the architectural models. The experiments and results are highlighted in Sect. 5. Section 6 concludes with a conclusion and a proposal for future improvements.

2 Related Work

Many studies have been published on utilizing deep learning to recognize sign languages, in which various neural networks were used to extract sign language features and good results were obtained.

Using a simple camera for image-based recognition, Tolentino et al. [5] employed CNN and skin-color techniques to categorize images to ASL alphabet, number, and static word recognition, achieving an average accuracy of 93.67%. Jiang et al. [6] used eight layers of CNN to identify Chinese sign languages in 1320 images, with a 90.91% success rate. Furthermore, Ameen and Vadera [7] used CNN to detect fingerspelling for letters of American Sign Language with a 90% accuracy rate, excluding for letters J and Z, utilizing over 60,000 images. Cayamcela and Lim [8] fine-tuned TL (GoogleNet and Alexnet) to train on 78,000 images in the RGB color of American Sign Language (ASL) and obtained 95% and 99% accuracy, respectively. Kamruzzaman [9] utilized CNN to accurately categorize 31 letters from the Arabic Alphabet images collection, achieving a 90% accuracy rate.

Using a Microsoft Kinect V2 camera, Beena et al. [10] employed CNN to categorize 33000 Kinect images of 24 alphabets and 0–9 digits in American Sign Language with a 94%. Saleh Aly et al. [11] used a Soft Kinect camera to acquire a photos dataset and used an unsupervised deep learning algorithm called PCANet to distinguish 28 alphabetic Arabic with a 99.5% recognition accuracy.

Furthermore, when it comes to video-based recognition, Shin et al. [12] trained CNN on Korean sign language videos that were transformed to 9 frames each video and categorized the videos into 10 words with an accuracy of around 84.5%. Rao et al. [13]

used CNN to train 200 Indian sign language words with five different signers at five different angles, resulting in a recognition rate of 92.88%. ElBadawy et al. [14] trained the 3DCNN model to recognize 25 gestures from the ArSL videos dataset, with a success rate of 90%.

In this phase, Transfer Learning is used, for the digits sign language dataset, Ozcan and Basturk [15] used a fine-tuned VGG19 model and an AlexNet model to RGB and RGB-D static gesture photos and achieved an accuracy of 94.8% and 94.2%, respectively. On 483 Turkish Sign Language (TSL) videos, Aktas et al. [16] employed the ResNet to detect and obtain 78.49% frame-level accuracy. Ji et al. [17] used VGG16 to develop two sign actions across a 60-min movie and achieved a 90% accuracy rate.

It has recently become feasible to automatically identify video-based datasets by combining CNN and RNN models. Vo et al. [18] employed the VGG16 model, then put the spatial features into long short term memory (LSTM) classifiers to detect 12 Vietnamese sign language words in video sequences, with a 95.83%. Elboushaki et al. [19] presented 3D-ResNets, which learned the spatiotemporal features of RGB and depth sequences and combined them with a ConvLSTM to capture the temporal features between them, achieving an accuracy of 87.66% without background subtraction and 88.33% with background subtraction that relied on the video's moving components. Liao [20] obtained 89% recognition accuracy using a deep 3DResNet and Bi-directional LSTM networks (B3D ResNet) to distinguish 500 vocabularies from videos of Chinese Sign Language.

3 Arabic Sign Language Dataset

Owing to the unavailability of an Arabic Sign Language benchmark dataset, the dataset was constructed to fit the recognition system using deep learning techniques, and it includes 20 signs in Arabic Sign Language, each of which was performed 100 times by various signers at varied speeds in identical settings in all videos for consistency. Figure 1 depicts some of the dataset's samples. The dataset is available to research purposes.

Fig. 1. Samples of ArSL dataset

The ArSL dataset was captured by three signers using a smartphone with a modest video camera with a display size of 840 * 540 and a resolution of 25 megapixels. In addition, videos were a total of 2000 videos with around 1.6 h across all signs. Each video was recorded between two and five seconds. Table 1 delves into the specifics of ArSL. These words were retrieved from the Arabic sign language's unified dictionary, which contains all of the Arabic language's vocabulary and its signs.

Table 1. Description of Arabic sign language dataset

Details/datasets	ArSL original dataset	ArSL augmentation dataset
Number of videos	2000 videos	4000 videos
Number of signers	3 signers	3 signers
Number of words	20 words	20 words
Number of videos per word	100 videos per word	200 videos per word
Number of seconds per video	2–5 s	2–8 s

Furthermore, as shown in Fig. 2, the original dataset was enriched by employing a combination of augmentation approaches that would double the number of videos, including adding noises, speeding up, slowing down, rotating, and translating.

Fig. 2. Augmented samples of ArSL signs

4 Methodology

Transfer Learning [21] is a prominent and promising topic in which the architecture of models and weights from a big annotation dataset such as ImageNet is transferred to a new project and attain high accuracy. Furthermore, it minimizes training time from days to hours. It also works well with tiny datasets. Despite the fact that ArSL video recognition differs from ImageNet raw image classification, it was able to extract spatial features.

On the other hand, the meaning of a sign is linked to the sequence of frames in a video, the process of understanding this sign is known as sign language recognition. Furthermore, with the spatial features of frames, we employed RNN to deal with the video to extract the temporal features of videos that were taken into account to improve the real-world video identification outcomes.

In the proposed network architecture, spatial and temporal features were extracted using Transfer Learning and Recurrent Neural Networks. The following are the three phases that were performed to create the hybrid technique TL and RNN.

4.1 Prepare the Dataset

An Arabic Sign Language dataset was created using a simple camera. To eliminate useless frames; first, 20% of the start and end of video frames were eliminated while converting each video into frames. Then the core frames were chosen on 10 equal time durations, which was 6 or 7 frames of each video, as illustrated in Fig. 3 as follows from the top left to the right bottom.

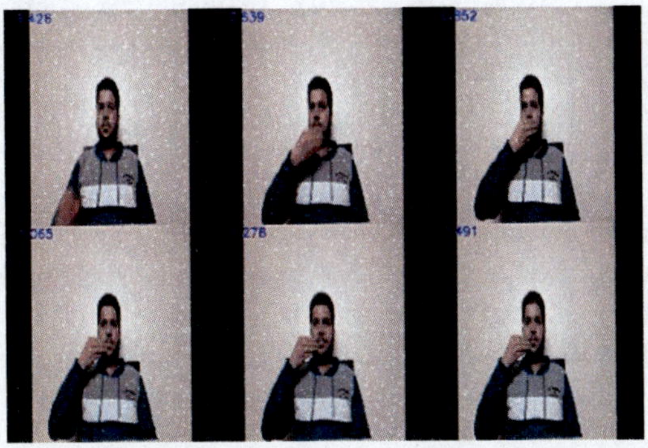

Fig. 3. The selected frames of the video

4.2 Extract the Spatial Features

Transfer Learning was used to apply what had been learned in earlier tasks to the current work. The features and weights were extracted from previously classified images and used to train on ArSL video recognition. As a consequence, VGG [22], ResNet

[23], MobileNet [24], DenseNet [25], NasNet [26], and EfficientNet [27] were identified as suitable models for extracting spatial information from video frames for ArSL identification. In a summary, the following TL models were illustrated:

The VGG16 architecture [22] won the 2014 ImageNet competition and received a top-five test accuracy of 92.7% in ImageNet. It consists of 16 layers (The additional layers boost the system's performance), 13 convolution layers with (3 × 3) kernel size, and 3 fully connected layers. VGG19 has raised the weight layer depth to 19. The VGG16 and VGG19 were used to reinitialize the weights and extract features from frames. However, because of the massive size of these models, they were extremely slow to train and used a lot of memory.

In the ILSVRC 2015 classification competition, the ResNet [23] came in first with a top-5 error rate of 3.57%. Because of the accuracy of the network saturates and rapidly declines once it reaches a certain limit. The training procedure will fail. ResNet came to aid the training process by incorporating a previous layer into the next layer, allowing residuals to be learned (the difference between the last layer and the current one). In addition, it prevents the vanishing gradient problem to occur. RESNET152V2 features 152 layers, making it eight times as deep as VGG19 but with fewer parameters. It also does batch normalization before applying each weight layer.

MobileNet [24] was released by Google in 2017 to increase accuracy while taking into account the restricted resources available for on-device or embedded apps. To control latency and accuracy, MobileNetV1 created a lightweight deep neural network utilizing depth-wise separable convolutions. MobileNetV2 uses inverted bottleneck blocks, and residual connections, and was employed as feature extraction. On a Google Pixel phone, MobileNetV2 models utilize 2× fewer operations, require 30% fewer parameters, and are 30–40% quicker than MobileNetV1 models, they attained greater accuracy. DenseNet [25] was introduced by Cornwell University, Tsinghua University, and Facebook AI Research (FAIR), in which each layer links to every other layer in a feed-forward method (concatenating outputs from the previous layers). The vanishing-gradient problem was solved with DenseNet, which also prompted the reuse feature and significantly decreased the number of parameters.

In 2017, the ImageNet competition was won by NASNet [26], Its goal is to figure out a network architecture that will give the greatest results on a specific task.CIFAR-10 designs have been deployed to ImageNet. CIFAR-10 revealed convolutional cells that learn well to ImageNet challenges. It reaches top 1 accuracy (82.7%) With lower floating point operations and parameters.

Google Research's Brain team's Mingxing and Quoc developed a family of EfficientNets [27] that range from B0 to B7. The model will expand when the network's breadth (the number of filters inside each layer), depth (the number of layers that data flows through), or resolution (which contains more and more detailed features) rises, resulting in a decrease in accuracy. With 66M parameters, 813 layers, and 37B FLOPS, EfficientNetB7 was able to strike a balance between the three last factors, achieving top-5 accuracy of 97.1% on the ImageNet dataset.

TL Models reinitialized the pre-trained weights of all layers during training to guarantee accurate training of the ArSL dataset and to extract more spatial features. It is, however, far more expensive in terms of both time and resources. Moreover, randomly deleting layers from the network was not an option. Table 2 presents the size of utilized models, the number of parameters, and the size of used frames.

Table 2. TL models properties

Models/properties	Size	Param.s	Input size
VGG16	528 MB	138M	224 * 224
VGG19	549 MB	143M	224 * 224
ResNet152V2	232 MB	60M	224 * 224
MobileNetV2	14 MB	3M	224 * 224
DenseNet201	80 MB	20M	224 * 224
NASNetMobile	23 MB	5M	224 * 224
EfficientNetB7	256 MB	66M	600 * 600

4.3 Extract the Temporal Features

The spatial characteristics from the previous phase were input into RNN models (memory-based neural networks) in the third step. RNN models used the memory of past frames to offer real-time predictions for ongoing frames. As a result, for the architecture of ArSL recognition, LSTM [28], BiLSTM [29], GRU [30], and BiGRU [31] were employed to extract temporal features with the optimum benefits of classification accuracy, reduced training time, and testing time, as shown in Fig. 4.

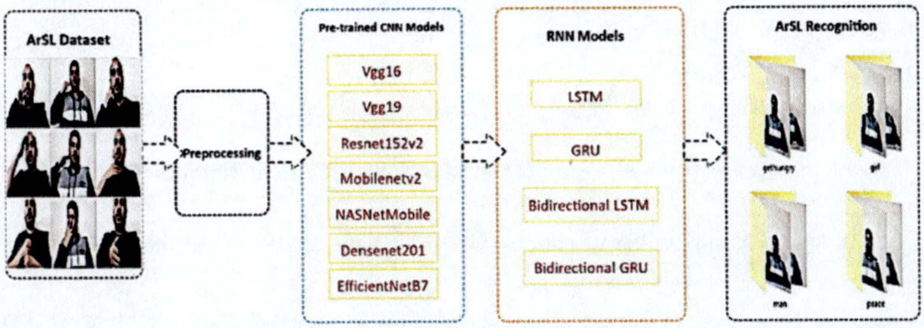

Fig. 4. The architecture of ArSL recognition.

Throughout sequence processing, LSTM [28] has three gates: The forged gate lets the network learn which state variables to remember/forget to decrease vanishing gradients, the input gate allows the network to turn off/on certain input values, and the output gate defines which values are permitted as an output from the cell during sequence processing.

To increase accuracy, BiLSTM [29] was used to capture spatial information and bidirectional temporal connections from the first and final frames. When numerous layers of BiLSTM are layered, the vanishing gradient issue develops. As a result, with some deep neural networks, it fails.

GRU [30] includes two gates to improve accuracy: the update gate recognizes what information to delete and what new information to add. Furthermore, the reset gate determines how much past data to erase.

Likewise, BiGRU [31] learned representations from previous and future time steps, allowing for a greater comprehension of the video while also reducing complexity.

4.4 Video Augmentation

Mixing certain augmented methods [32] to the original videos, such as add noises, shift, speed up, speed down, and rotate, was used to increase the amount of the original dataset, as well as to reduce the model's dependency on claimed features, avoid overfitting, and provide robust results. The preceding three stages (Prepare the dataset, Extract the spatial features, and Extract the temporal features) were carried out on the augmented dataset, which took the place of the original dataset.

5 Experimental and Results

5.1 Experiment Settings

The dataset was separated into three categories: training, validating, and testing, with percentages of 70%, 15%, and 15%, respectively. Furthermore, numerous attempts were made, such as picking all frames, selecting a number of fixed or dynamic frames randomly or consecutively, However, the result was a modest accuracy, and finally picking 6 or 7 frames of videos with a 10 equal time distribution to focus only on core frames of video without the unnecessary frames. This effort yielded the best results.

Tensorflow[1] (version 2.2.0) and Keras[2] (version 2.4.3) framework libraries were used to develop the architecture in Python (version 3.6). It was also trained on premium Colab GPUs using Keras.

Table 3. Description of the experiment for all techniques.

Number of classes	Number of epochs	Test size	Number of batches
20 classes	100–300 epochs	0. 3 of dataset	64 batches

As shown in Table 3, several efforts were made to improve the efficiency and performance of a combination of TL and RNN models. First, Early Stopping [33] was used to halt training between 100 and 300 epochs in order to improve performance without useless epochs.

1 https://www.tensorflow.org/.
2 https://keras.io/.

In addition, the ADAM optimizer [34] was utilized to do different computations and gradient re-scaling in order to alter the model's weights and get the lowest loss function possible. SGD and RMS [35] produced fewer outcomes. The training with ADAM appears to be the most effective.

While fine-tuning models saves time and data, it sometimes leads to fewer results, and the new architectures and weights make extracting features complex and unreliable.

Furthermore, Dropout [36] was employed with a 50% threshold to minimize overfitting and enhance neural network performance.

The Arabic Sign Language recognition hybrid models were evaluated for the validation of 15% of the original dataset and the augmented dataset, as well as predict labels of new videos. Finally, for each experiment, the recognition accuracy was measured using Eq. (1).

$$Accuracy = \frac{True\ Postives + True\ Negatives}{All\ Samples} \tag{1}$$

5.2 Models Results

The architecture, features, and weights of TL models were pre-trained for 2–3 weeks on ImageNet/CIFAR across many GPUs so that they could reuse. Moreover, RNN models were used to label the sign and were more efficient with the videos. Table 4 and Fig. 5 demonstrate that VGG16 + GRU, VGG16 + LSTM, and VGG19 + BiLSTM showed better results using the original dataset, with accuracy of 89.8%, 87.7%, and 87.2%, respectively.

Table 4. Comparison models with the original dataset.

Models	LSTM	BiLSTM	GRU	BiGRU
VGG16	**87.7**	86.8	**89.8**	86.5
VGG19	84.8	**87.2**	86.8	85.9
ResNet152V2	*60.2*	*59.6*	*59.9*	60.9
MobileNetV2	84.3	83.3	86.4	84.3
DenseNet201	80.7	80.8	81.1	81.3
NasNetMobile	69.4	70	73	70.7
EfficientNetB7	77	75.1	70.6	81.8

The EfficientNetB7 + GRU, EfficientNetB7 + BiGRU, and EfficientNetB7 + LSTM hybrid models also rated highest using the augmented dataset, with accuracy of 93.4%, 92.8%, and 92.7%, respectively, as shown in Table 5 and Fig. 6.

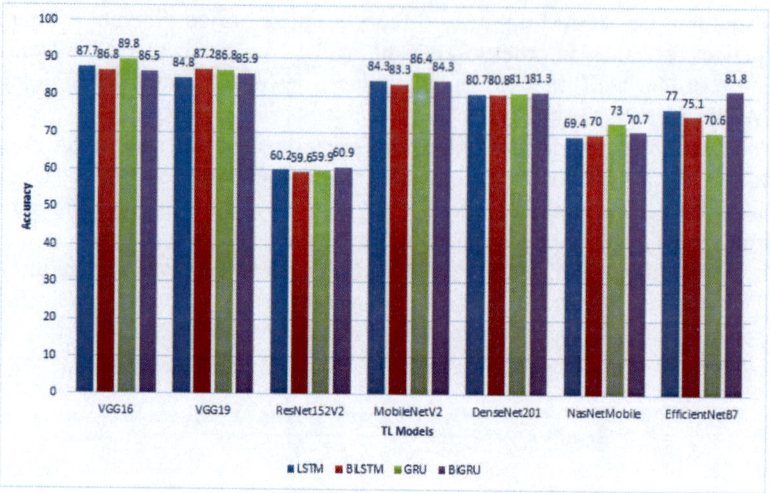

Fig. 5. Comparison of accuracy with original dataset

Table 5. Comparison models to the augmented dataset.

Models	LSTM	BLSTM	GRU	BGRU
VGG16	90	85.1	91.5	90.1
VGG19	88.5	90.7	84.9	86.9
ResNet152V2	*73.3*	*72.2*	76	*74.6*
MobileNetV2	91.5	90.6	90.4	92
DenseNet201	87.7	86.6	89.8	83.6
NasNetMobile	77	75.5	77.5	76.1
EfficientNetB7	**92.7**	91.4	**93.4**	**92.8**

The hybrid models were influenced by a variety of aspects during training, such as the size and efficiency of the TL models, GRU was a bit faster to train than LSTM since it had fewer operations, and BiGRU and BiLSTM moved in both directions, making them slower than GRU and LSTM.

The ArSL recognition challenges were effectively addressed by using the least amount of core frames from the original videos, utilizing suitable settings, employing TL and RNN models, and using augmented methods, all of which improved ArSL recognition accuracy and outperformed other traditional architectures.

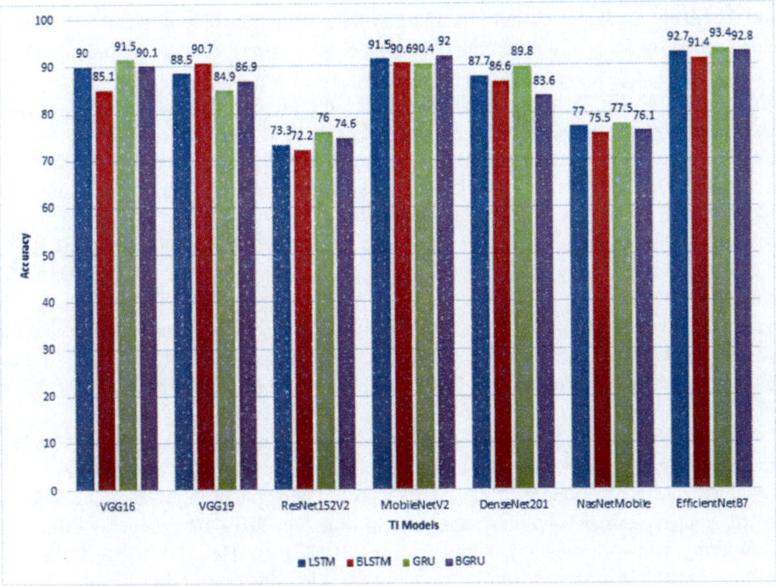

Fig. 6. Comparison of accuracy with the augmented dataset.

6 Conclusion and Future Works

This study employed a combination of the state-of-the-art deep learning algorithms to detect Arabic sign language automatically, allowing deaf people to communicate more easily with the outside world.

Transfer Learning was used to extract spatial features, as it is a technique that allows reuse weights and architecture of models pre-trained on a large dataset. To extract temporal characteristics from video-based datasets, RNN models were utilized.

VGG, ResNet, MobileNet, DenseNet, NasNet, and EfficientNet were combined with LSTM, BiLSTM, GRU, and BiGRU to increase ArSL recognition robustness using both the original and augmented datasets. Furthermore, the highest accuracy was 93.4%.

ArSL development is needed in the future to accommodate more ArSL words with complicated backgrounds and more signers.

Acknowledgments. The authors would like to express their gratitude to the signers who assisted in the creation of the dataset.

References

1. Bragg, D., et al.: Sign language recognition, generation, and translation: an interdisciplinary perspective. In: The 21st International ACM SIGACCESS Conference on Computers and Accessibility, pp. 16–31 (2019)

2. Xu, S., Liang, L., Ji, C.: Gesture recognition for human–machine interaction in table tennis video based on deep semantic understanding. Signal Process. Image Commun. **81**, 115688 (2020)
3. Wu, Z., Yao, T., Fu, Y., Jiang, Y.-G.: Deep learning for video classification and captioning. In: Frontiers of Multimedia Research, pp. 3–29 (2017)
4. Das, S., Chaudhary, A., Bremond, F., Thonnat, M.: Where to focus on for human action recognition? In: 2019 IEEE Winter Conference on Applications of Computer Vision (WACV), pp. 71–80. IEEE (2019)
5. Tolentino, L.K.S., Juan, R.S., Thio-ac, A.C., Pamahoy, M.A.B., Forteza, J.R.R., Garcia, X.J.O.: Static sign language recognition using deep learning. Int. J. Mach. Learn. Comput. **9**(6), 821–827 (2019)
6. Jiang, X., Lu, M., Wang, S.-H.: An eight-layer convolutional neural network with stochastic pooling, batch normalization and dropout for fingerspelling recognition of Chinese sign language. Multimedia Tools Appl. **79**(21), 15697–15715 (2020). https://doi.org/10.1007/s11042-019-08345-y
7. Ameen, S., Vadera, S.: A convolutional neural network to classify American sign language fingerspelling from depth and colour images. Expert Syst. **34**(3), e12197 (2017)
8. Cayamcela, M.E.M., Lim, W.: Fine-tuning a pre-trained convolutional neural network model to translate American sign language in real-time. In: 2019 International Conference on Computing, Networking and Communications (ICNC), pp. 100–104. IEEE (2019)
9. Kamruzzaman, M.: Arabic sign language recognition and generating Arabic speech using convolutional neural network. Wirel. Commun. Mob. Comput. **2020** (2020)
10. Beena, M., Namboodiri, M.A., Dean, P.: Automatic sign language finger spelling using convolution neural network: analysis. Int. J. Pure Appl. Math. **117**(20), 9–15 (2017)
11. Aly, S., Osman, B., Aly, W., Saber, M.: Arabic sign language fingerspelling recognition from depth and intensity images. In: 2016 12th International Computer Engineering Conference (ICENCO), pp. 99–104. IEEE (2016)
12. Shin, H., Kim, W.J., Jang, K.-A.: Korean sign language recognition based on image and convolution neural network. In: Proceedings of the 2nd International Conference on Image and Graphics Processing, pp. 52–55 (2019)
13. Rao, G.A., Syamala, K., Kishore, P., Sastry, A.: Deep convolutional neural networks for sign language recognition. In: 2018 Conference on Signal Processing and Communication Engineering Systems (SPACES), pp. 194–197. IEEE (2018)
14. ElBadawy, M., Elons, A., Shedeed, H.A., Tolba, M.: Arabic sign language recognition with 3D convolutional neural networks. In: 2017 Eighth International Conference on Intelligent Computing and Information Systems (ICICIS), pp. 66–71. IEEE (2017)
15. Ozcan, T., Basturk, A.: Transfer learning-based convolutional neural networks with heuristic optimization for hand gesture recognition. Neural Comput. Appl. **31**(12), 8955–8970 (2019). https://doi.org/10.1007/s00521-019-04427-y
16. Aktas, M., Gokberk, B., Akarun, L.: "Recognizing non-manual signs" in Turkish sign language. In: 2019 Ninth International Conference on Image Processing Theory, Tools and Applications (IPTA), pp. 1–6. IEEE (2019)
17. Ji, Y., Kim, S., Kim, Y.-J., Lee, K.-B.: Human-like sign-language learning method using deep learning. ETRI J. **40**(4), 435–445 (2018)
18. Vo, A.H., Pham, V.-H., Nguyen, B.T.: Deep learning for Vietnamese sign language recognition in video sequence. Int. J. Mach. Learn. Comput. **9**(4), 440–445 (2019)
19. Elboushaki, A., Hannane, R., Afdel, K., Koutti, L.: MultiD-CNN: a multi-dimensional feature learning approach based on deep convolutional networks for gesture recognition in RGB-D image sequences. Expert Syst. Appl. **139**, 112829 (2020)
20. Liao, Y., Xiong, P., Min, W., Min, W., Lu, J.: Dynamic sign language recognition based on video sequence with BLSTM-3D residual networks. IEEE Access **7**, 38044–38054 (2019)

21. Zhuang, F., et al.: A comprehensive survey on transfer learning. Proc. IEEE **109**(1), 43–76 (2020)
22. Simonyan, K., Zisserman, A.: Very deep convolutional networks for large-scale image recognition. arXiv preprint arXiv:1409.1556 (2014)
23. He, K., Zhang, X., Ren, S., Sun, J.: Deep residual learning for image recognition. In: Proceedings of the IEEE Conference on Computer Vision and Pattern Recognition, pp. 770–778 (2016)
24. Howard, A.G., et al.: MobileNets: efficient convolutional neural networks for mobile vision applications. arXiv preprint arXiv:1704.04861 (2017)
25. Huang, G., Liu, Z., Van Der Maaten, L., Weinberger, K.Q.: Densely connected convolutional networks. In: Proceedings of the IEEE Conference on Computer Vision and Pattern Recognition, pp. 4700–4708 (2017)
26. Zoph, B., Vasudevan, V., Shlens, J., Le, Q.V.: Learning transferable architectures for scalable image recognition. In: Proceedings of the IEEE Conference on Computer Vision and Pattern Recognition, pp. 8697–8710 (2018)
27. Tan, M., Le, Q.: EfficientNet: rethinking model scaling for convolutional neural networks. In: International Conference on Machine Learning, pp. 6105–6114. PMLR (2019)
28. Hochreiter, S., Schmidhuber, J.: Long short-term memory. Neural Comput. **9**(8), 1735–1780 (1997)
29. Cui, Z., Ke, R., Pu, Z., Wang, Y.: Stacked bidirectional and unidirectional LSTM recurrent neural network for forecasting network-wide traffic state with missing values. Transp. Res. Part C Emerg. Technol. **118**, 102674 (2020)
30. Chung, J., Gulcehre, C., Cho, K., Bengio, Y.: Empirical evaluation of gated recurrent neural networks on sequence modeling. arXiv preprint arXiv:1412.3555 (2014)
31. Lynn, H.M., Pan, S.B., Kim, P.: A deep bidirectional GRU network model for biometric electrocardiogram classification based on recurrent neural networks. IEEE Access **7**, 145395–145405 (2019)
32. Wen, Q., et al.: Time series data augmentation for deep learning: a survey. arXiv preprint arXiv:2002.12478 (2020)
33. Liang, H., et al.: DARTS+: improved differentiable architecture search with early stopping. arXiv preprint arXiv:1909.06035 (2019)
34. Kingma, D.P., Ba, J.: Adam: a method for stochastic optimization. arXiv preprint arXiv:1412.6980 (2014)
35. Postalcıoğlu, S.: Performance analysis of different optimizers for deep learning-based image recognition. Int. J. Pattern Recognit. Artif. Intell. **34**(02), 2051003 (2020)
36. Srivastava, N., Hinton, G., Krizhevsky, A., Sutskever, I., Salakhutdinov, R.: Dropout: a simple way to prevent neural networks from overfitting. J. Mach. Learn. Res. **15**(1), 1929–1958 (2014)

Robust Face Mask Detection Using Local Binary Pattern and Deep Learning

Loc Duc Quan[1], Duy Huu Nguyen[1], Thang Minh Tran[1], Narayan C. Debnath[2], and Vinh Dinh Nguyen[1](✉)

[1] FPT University, Can Tho 94000, Vietnam
{locqdce140037,duynhce140596,thangtmce140085,vinhnd18}@fpt.edu.vn
[2] School of Computing and Information Technology, Eastern International University, Thu Dau Mot, Vietnam
narayan.debnath@eiu.edu.vn

Abstract. The ongoing COVID-19 has caused a great amount of serious troubles for people around the world. Even though vaccination has been proven to be safe and highly effective against the COVID-19, it is far away to prevent thoroughly the spread of the disease and truly halt the pandemic. Therefore, we need to apply additional methods aside from vaccine injection, such as keeping the distance between people and always using the face masks during the ordinary conversations, in efforts to further reduce the COVID-19 contagion rate. To implement such methods, this research aims to investigate an efficient approach to detect and warn people that they should wear mask whenever they go to public places. Our proposed system studies the benefits of Local Binary Pattern (LBP) and deep learning model to provide accurate face mask detection and classification system. After comprehensive testing, we found that our system provided the detection rate up to 90% with the Kaggle, Face-Mask-Net, and our own datasets.

Keywords: Face mask detection · Local binary pattern · Deep learning

1 Introduction

Recently, the continuous COVID-19 outbreaks have become culprits to many serious issues on a global scale. According to WHO's global report, the cumulative number of infected cases has reached approximately 260 million people and there is already more than 5 million fatalities [1]. Vaccination is one of the most effective solution to safely prevent the spread of the COVID-19. However, the current most effective vaccines, namely Pfizer-BioNTech and Moderna vaccines, merely have about 95% efficacy against the SARS-CoV-2 [2], and their effectiveness does decline drastically after a period of time [3]. Hence, implementing other methods besides vaccination, for instance, maintaining social distancing and wearing face masks, is crucial to further help containing the spread of the

© The Author(s), under exclusive license to Springer Nature Switzerland AG 2022
A. E. Hassanien et al. (Eds.): AMLTA 2022, LNDECT 113, pp. 60–67, 2022.
https://doi.org/10.1007/978-3-031-03918-8_6

highly contagious disease. There is a myriad of studies showing that face masks are essential for protecting people from being infected with the COVID-19. N95 and conventional surgical masks can reach the effectiveness of 91% and 68% in blocking the virus transmission [4]. Consequently, maintaining social distancing and wearing face masks were made compulsory by many governments worldwide. A Face Mask Detection System using combination of Local Binary Pattern (LBP) and Deep Learning will provide a competent solution for the aforementioned governments to recognize and warn their citizens who do not wear face masks in public places. Notwithstanding, there is yet to be a real time system that integrates LBP into deep learning model and is capable of detecting higher resolution images with high accuracy. Therefore, to support creating a precise face mask detection system satisfying such requirements, this research studies an efficient method to detect faces that wear face masks, as well as classifying faces into two more groups namely not wearing masks and wearing masks incorrectly, by taking full advantages of the local binary pattern and deep learning model. The system were extensively tested on various datasets. The obtained results are greatly stable under many conditions. Detection rate reaches 90% with datasets from Kaggle, Face-Mask-Net, and datasets prepared by our research team. Our proposed paper has the following contributions: (1) We are the first one that applies the local patterns and deep learning techniques for detecting face mask. (2) The proposed system increases the accuracy of existing face mask detection system.

2 Related Works

Latterly, there are numerous researchers having tried to develop various methods for detecting face masks using machine learning-based techniques as mentioned in Table 1. A general consensus is to use deep learning models, a combination of them or with additional approaches, e.g. GLCM method and Haar Cascade classifier for face mask detection and classification. However, their image size, processing time, accuracy and high dependency on input data are not adequately good according to the current requirements in the industry. LBP were later implemented in more recent journals to revise the original methods and fix some of the problems. Nonetheless, there is still no proposed approach that incorporates LBP into deep learning model for face mask recognition.

Local Binary [5], that encode and extract stable local feature by evaluating the relationship of every pixel in the local region, have been widely used in various successful applications. Its simplicity to implement and efficiency in improving detection accuracy makes it a viable choice to incorporate into deep learning-based systems. Thus, it is further investigated to increase the precision of the proposed face mask detection system.

3 Proposed Method

First, we compute the local binary pattern [5] as follows:

Table 1. Survey on existing face mask detection systems

Survey Authors	Algorithms	Descriptions	Others
[10]	CNN	Using deep neural network to detect people wearing masks and emitting alerts when someone is not wearing a mask Curvature and GLCM	Collect more than 22000 images at 224 × 224 pixels, achieving accuracy up to 97%
[11]	CNN	Detecting face mask real time by deep learning and notify when there are violations	Dataset consists of 25000 images at 224 × 224 pixels and achieves up to 96% accuracy
[12]	MTCNN and LeNet	Combining deep Neural Network, cascade, and LeNet algorithms to detect face mask	Dividing dataset into 2 cases with mask and without mask. As a result, up to 98,21% accuracy is achieved
[13]	CNN	Using deep CNN to detect face mask	A trained system achieves up to 98,7% accuracy
[14]	MTCNN, FaceNet and SVM	Using SVM and deep CNN for detecting face mask.	With 8 scenarios, train accuracy is from 99,05% to 100% and test accuracy is from 47,43% to 98,5%
[15]	MobileNet and Global Pooling Block	Using Mobile-Net and Block-Pooling for face mask detection	With 2 datasets prepared by the research team, the system has an accuracy of 99–100%
[16]	Combine the benefits of cascade classifier and deep learning-based YOLO method	Combine the benefits of cascade classifier and deep learning-based YOLO method	Selecting 7000 images from dataset MAFA, the system can detect in real-time with 30 fps and achieve up to 90,1%
[17]	GLCM, SVM	Release a new dataset for face mask detection by using Surface Curvature and GLCM	Shorten the detection time and achieve an accuracy of 87.5%
[18]	Investigate the deep CNN-based method	Using deep CNN for detecting face mask	Collecting 20000 images including face mask and thumb. The system is trained by CNN and has an accuracy of up to 97.67%
[19]	SSD, CNN	Identifying people wearing masks to allow them to enter multiple places like malls, shops, hospitals	The identification device is developed with the mobile application. It will send a notification to the owner if someone is not wearing a mask

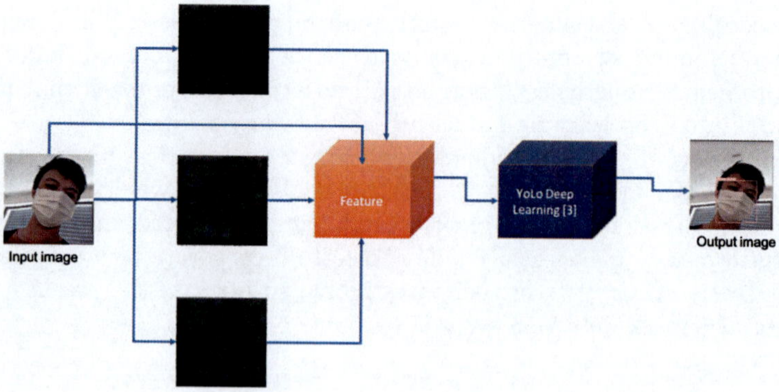

Fig. 1. The workflow that shows how our system work for recognizing face mask.

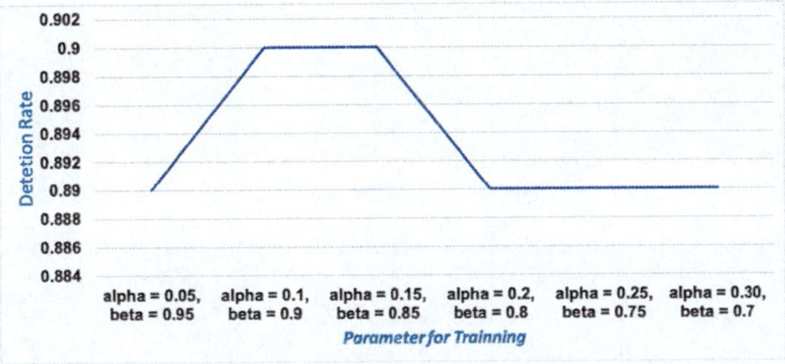

Fig. 2. Detection rates of our system by using various parameter for training

$$
\begin{aligned}
&F_{LBP}^{N,R}\left(I_c\left(x_c, y_c\right)\right) = \sum_{i=0}^{N-1} \varphi\left(I_m\left(x_m, y_m\right), I_c\left(x_c, y_c\right)\right) \times 2^i \\
&\varphi\left(I_m\left(x_m, y_m\right), I_c\left(x_c, y_c\right)\right) = \begin{cases} 1, I_m\left(x_m, y_m\right) > I_c\left(x_c, y_c\right) \\ 0, I_m\left(x_m, y_m\right) \leq I_c\left(x_c, y_c\right) \end{cases}
\end{aligned} \tag{1}
$$

where $I_c\left(x_c, y_c\right)$ is the center pixel intensity value, and $I_m\left(x_m, y_m\right)$ is the pixel value of neighboring pixels.

Second, we used the idea of Nguyen et al. [6] to create the training input features $T_{feature}^{B}\left[I\left(i,j\right)\right]$, $T_{feature}^{G}\left[I\left(i,j\right)\right]$, and $T_{feature}^{R}\left[I\left(i,j\right)\right]$ as follows:

$$
\begin{aligned}
&T_{feature}^{B}\left[I\left(i,j\right)\right] = \alpha_B \times F_{LBP,B}^{N,R}\left(I\left(i,j\right)\right) + \beta_B \times \left(I\left(x_c, y_c\right)\right) \\
&T_{feature}^{G}\left[I\left(i,j\right)\right] = \alpha_G \times F_{LBP,B}^{N,R}\left(I\left(i,j\right)\right) + \beta_G \times \left(I\left(x_c, y_c\right)\right) \\
&T_{feature}^{R}\left[I\left(i,j\right)\right] = \alpha_R \times F_{LBP,B}^{N,R}\left(I\left(i,j\right)\right) + \beta_R \times \left(I\left(x_c, y_c\right)\right) \\
&\alpha_B + \beta_B = \alpha_G + \beta_G = \alpha_R + \beta_R = 1
\end{aligned} \tag{2}
$$

where $F_{LBP,B}^{N,R}\left(I\left(i,j\right)\right)$ is the LBP's result at pixel $I\left(x,y\right)$. $I(i,j)$ is the magnitude value of the pixel i, j. $T_{feature}^{B}\left[I\left(i,j\right)\right]$, $T_{feature}^{G}\left[I\left(i,j\right)\right]$, and $T_{feature}^{R}\left[I\left(i,j\right)\right]$ are encoded results of the local region on three channel blue, green, and red, respectively.

Third, to create good face mask data for training, we used dlib and OpenCV libraries to generate facial landmark to put the virtual face mask between chin and nose as discussed in [20].

Finally, our proposed feature input and training data were inputted to the YOLO-based deep learning model [7] to detect and classify face masks from the input image.

Figure 1 shows the system architecture design of the our method.

4 Experimental Results

We used the deep learning-based YOLO algorithm and integrated the LBP features to verify the precision of the proposed system. We use two datasets to

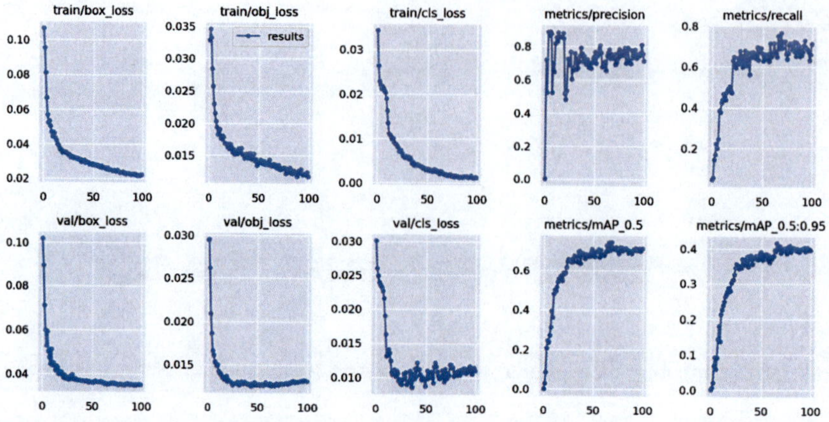

Fig. 3. The train and loss visualization results of our system.

verify the accuracy of our new algorithm, including Kaggle dataset (853 images) [21] and MaskFace-Net dataset (137,016 images) [9]. Our system were trained and tested using a virtual machine with 2.30 GHz CPU (2 cores), Tesla P100 16 GB GPU on Google Colab.

To evaluate the results of our proposed algorithm, we used the standard rate of detection based on intersection over union (IoU) as follows:

$$\mathrm{P}_{rate} = \frac{\sum\limits_{M}\sum\limits_{I} \Pi\left(DT_i, GT_i\right) \geq \phi}{\sum\limits_{M}\sum\limits_{I} GT_i} \tag{3}$$

where $\Pi\left(DT_i, GT_i\right)$ is used to measure the IoU between the detected result from our program DT_i and labeled ground truth data GT_i.

Figure 3 illustrates visualization accuracy of our proposed system in terms of training and loss functions. To verify which parameters are best for our method, we setup a test with the Kaggle dataset, as shown in Fig. 2. The over-fitting has been void by using the regularization techniques. Experimental results state that our system obtained the most accuracy when alpha parameter equals 0.1 and beta parameter equals 0.9. Figures 4, 5, and 6 show the face mask detection results with Kaggle dataset, MaskFace-Net dataset, and our own dataset, respectively, on our system. The results stated that our proposed method got adequately high accuracy (90%) under various environments due to the benefit of local binary pattern as discussed in [6]. These experimental results proved that our proposed system got the stable results under various lighting environment because the local binary patterns might produce robust texture information. However, the proposed method has limitations to estimate the predefined

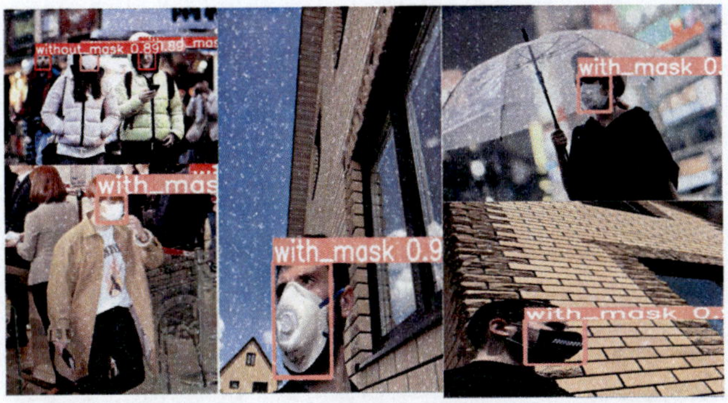

Fig. 4. The face mask detection results of our system with the Kaggle dataset.

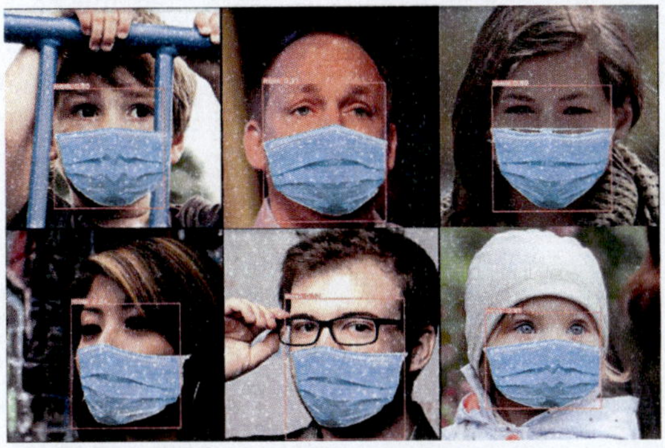

Fig. 5. The face mask detection results of our system with MaskFace-Net dataset.

Fig. 6. The face mask detection results of our system with our dataset.

parameters, they can only estimated by means of manually conducting experiments. Therefore, we aim to find a better solution to compute these parameters by considering the relationship between the local region features in future. In addition, the accuracy of the system is still not good under occlusion case or rotated mask due to the limitation of the base-network YOLOv5 as show in Fig. 7. We plan to evaluate more baseline deep neural network in future.

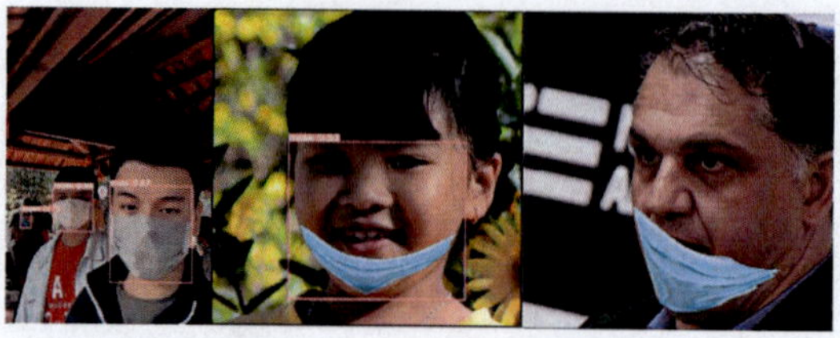

Fig. 7. The face mask detection results of our system for false detection cases.

5 Conclusion

The relentless COVID-19 surge has caused severe problems for every people around the globe. Thus, it is vital to find alternative approaches to hinder the spread of the pandemic. Therefore, we propose a novel method to alarm and remind people that the should wear face mask whenever they go outside for business. Although there is a great number of proposed methods to detect face masks in the previous years, there is yet to be one that investigates and further combines deep learning and LBP to increase the accuracy of existing face mask detection systems. The test results show that our system got a stable performance under various environments and conditions. Our proposed system improves the accuracy of the YOLO-v5 version for detecting face mask. In the future, the proposed face mask detection is going to be deployed on limited computational devices for commercial products, such as FPGAs.

References

1. WHO Coronavirus/COVID-19 dashboard. https://covid19.who.int/. Accessed 23 Nov2021
2. Kathy, K.: Comparing the COVID-19 vaccines: how are they different? https://www.yalemedicine.org/news/covid-19-vaccine-comparison. Accessed 23 Nov 2021
3. Burger, L.: British study shows COVID-19 vaccine efficacy wanes under Delta. https://www.reuters.com/business/healthcare-pharmaceuticals/british-study-shows-covid-19-vaccine-efficacy-wanes-under-delta-2021-08-18/. Accessed 23 Nov 2021
4. Feng, S., Shen, C., Xia, N., Song, W., Fan, M., Cowling, B.J.: Rational use of face masks in the COVID-19 pandemic. Lancet Respir. Med. **8**(5), 434–436 (2020)
5. Ojala, T., Pietikäinen, M., Harwood, D.: A comparative study of texture measures with classification based on feature distributions. Pattern Recogn. **19**(3), 51–59 (2016)
6. Nguyen, V.D., Tran, D.T., Byun, J.Y., Jeon, J.W.: Real-time vehicle detection using an effective region proposal-based depth and 3-channel pattern. IEEE Trans. Intell. Transp. Syst. **20**(10), 3634–3646 (2019)
7. Bochkovskiy, A., Wang, C.-Y., Liao, H.-Y.M.: YOLOv4: optimal speed and accuracy of object detection. arXiv:2004.10934 [cs.CV] (2020)
8. Google Colab homepage (2021). https://colab.research.google.com/

9. Adnane, C., Karim, H., Halim, B., Mahmoud, M.: MaskedFace-Net - a dataset of correctly/incorrectly masked face images in the context of COVID-19. Smart Health **19**, 100144 (2021). https://doi.org/10.1016/j.smhl.2020.100144
10. Militante, S.V., Dionisio, N.V.: Deep learning implementation of facemask and physical distancing detection with alarm systems. In: 2020 3rd International Conference on Vocational Education and Electrical Engineering (ICVEE), pp. 1–5 (2020). https://doi.org/10.1109/ICVEE50212.2020.9243183
11. Militante, S.V., Dionisio, N.V.: Real-time facemask recognition with alarm system using deep learning. In: 2020 11th IEEE Control and System Graduate Research Colloquium (ICSGRC), pp. 106–110 (2020). https://doi.org/10.1109/ICSGRC49013.2020.9232610
12. Rusli, M.H., Sjarif, N.N.A., Yuhaniz, S.S., Kok, S., Kadir, M.S.: Evaluating the masked and unmasked face with LeNet algorithm. In: 2021 IEEE 17th International Colloquium on Signal Processing and Its Applications (CSPA), pp. 171–176 (2021). https://doi.org/10.1109/CSPA52141.2021.9377283
13. Suresh, K., Palangappa, M., Bhuvan, S.: Face mask detection by using optimistic convolutional neural network. In: 2021 6th International Conference on Inventive Computation Technologies (ICICT), pp. 1084–1089 (2021). https://doi.org/10.1109/ICICT50816.2021.9358653
14. Ejaz, M.S., Islam, M.R.: Masked face recognition using convolutional neural network. In: 2019 International Conference on Sustainable Technologies for Industry 4.0 (STI), pp. 1–6 (2019). https://doi.org/10.1109/STI47673.2019.9068044
15. Venkateswarlu, I.B., Kakarla, J., Prakash, S.: Face mask detection using MobileNet and Global Pooling Block. In: 2020 IEEE 4th Conference on Information and Communication Technology (CICT), pp. 1–5 (2020). https://doi.org/10.1109/CICT51604.2020.9312083
16. Vinh, T.Q., Anh, N.T.N.: Real-time face mask detector using YOLOv3 algorithm and Haar Cascade classifier. In: 2020 International Conference on Advanced Computing and Applications (ACOMP), pp. 146–149 (2020). https://doi.org/10.1109/ACOMP50827.2020.00029
17. Lionnie, R., Apriono, C., Gunawan, D.: Face mask recognition with realistic fabric face mask data set: a combination using surface curvature and GLCM. In: 2021 IEEE International IOT, Electronics and Mechatronics Conference (IEMTRONICS), pp. 1–6 (2021). https://doi.org/10.1109/IEMTRONICS52119.2021.9422532
18. Amin, M.I., Adeel, H.M., Touseef, R., Awais, Q.: Person identification with masked face and thumb images under pandemic of COVID-19. In: 2021 7th International Conference on Control, Instrumentation and Automation (ICCIA), pp. 1–4 (2021). https://doi.org/10.1109/ICCIA52082.2021.9403577
19. Baluprithviraj, K.N., Bharathi, K.R., Chendhuran, S., Lokeshwaran, P.: Artificial intelligence based smart door with face mask detection. In: 2021 International Conference on Artificial Intelligence and Smart Systems (ICAIS), pp. 543–548 (2021). https://doi.org/10.1109/ICAIS50930.2021.9395807
20. Facial landmarks with dlib, OpenCV, and Python. https://www.pyimagesearch.com/2017/04/03/facial-landmarks-dlib-opencv-python/. Accessed 23 Nov 2021
21. Face mask detection dataset. https://www.kaggle.com/andrewmvd/face-mask-detection. Accessed 23 Nov 2021

Steganography Adaptation Model for Data Security Enhancement in Ad-Hoc Cloud Based V-BOINC Through Deep Learning

Ahmed A. Mawgoud(✉), Mohamed Hamed N. Taha, and Amira Kotb

Information Technology Department, Faculty of Computers and Artificial Intelligence, Cairo University, Giza, Egypt
aabdelmawgoud@pg.cu.edu.eg, mnasrtaha@cu.edu.eg, a.kotb@fci-cu.edu.eg

Abstract. Cloud computing's automation, scalability, and availability were vital features in the early days of digital transformation. Meanwhile, substantial concerns were expressed about cloud security and privacy. Due to the COVID-19 outbreak, several businesses have had serious issues speeding up their cloud migration efforts. This work intends to improve steganography in ad-hoc cloud systems using deep learning. This study is implemented in two phases.

Phase 1: The 'Ad-hoc Cloud System' concept and deployment method were created using V-BOINC, a tool that allows developers to bypass application-level security checks, the implemented ad-hoc cloud system was compared Amazon AC2 and showed high evaluation rate in some matrices.

Phase 2: We evaluate the data transmission security in ad-hoc cloud systems using a modified steganography with deep learning usage to replace or enhance an image-hiding system. In this study, the proposed model inputs data/images into the ad-hoc cloud system to guarantee high rate of data/image concealing. Statistically, a systematic steganography model hides lower message detection rates, the proposed deep steganography approach outperformed several attacks in the ad-hoc cloud environment.

Keywords: Cloud computing · Steganography · Encryption · Cloud cecurity · Deep learning

1 Introduction

During the COVID-19 period, cloud computing became one of the fastest developing technologies, providing automated and orchestrated solutions for both individuals and corporations [1]. As a result of this expansion, cybersecurity threats have recently escalated sharply. Cloud security is always a hot topic in edge computing. Many research subjects propose ways to improve cloud security. However, there is still plenty to be said about significant cloud security issues. The capacity to use current resources to create cloud services with less stable hosts than 'Grid Computing'. Instead, the 'Ad-hoc Cloud Model' notion may be comparable to 'Volunteer Computing', the ad-hoc cloud systems paradigm itself incorporating extra keys [2].

© The Author(s), under exclusive license to Springer Nature Switzerland AG 2022
A. E. Hassanien et al. (Eds.): AMLTA 2022, LNDECT 113, pp. 68–77, 2022.
https://doi.org/10.1007/978-3-031-03918-8_7

1.1 Ad-Hoc Cloud Computing

The concept of ad-hoc cloud can be defined as a co-operation of set of network nodes for utilizing its resources as a cloud structure for achieving a certain target. The volunteer cloud network consists of the integration of public resources along with distributed structural, various structures were proposed for providing the features of public cloud systems through utilizing existing resources in an ad-hoc model to initiate a small scale cloud system through volunteer resources [3].

1.2 Deep Steganography

The name 'Steganography' is the process of hiding texts or data using images; it was first used in the 15th century, when messages were physically hidden. Steganography is now a type of encryption. Steganography is problematic because it can alter the carrier's appearance and data. Initially, the information volume was hidden. Images have long been used to hide messages in text. The hidden data rate is measured in *bpp* (bits-per-pixel). This technique typically limits data to 0.4 *bpp* or less. Besides the carrier alteration, the *bpp* grows with message length, and the extent of adjustments depends on the original picture's resolution [4].

1.3 Contribution

This paper introduces a framework consists of two parts for the aim to:

- Encrypt the data transferred in the system by the integration of steganography with deep learning, the main focus is developing an ad-hoc cloud framework through absorbing services over irregular & unstable network. For both availability and efficiency, this study suggests the technology should be deployed on LAN networks. WAN networks can be used to assess security. Systems requiring high security may not be efficient for ad-hoc cloud architecture.
- Apply an approach for hiding data\image with N $\times$ N $\times$ RGB pixels throughout a cover image while retaining the cover artwork's original form "color channel = 8 bits". Aside from those studies, we have reduced the constraints by which the hidden image has been supplied without losing the image quality. To loss both reliability and carrier security is the key challenge here.

1.4 Paper Organization

The paper is organized as follows: Sect. 2 highlights the main challenges for considering steganography in ad-hoc cloud model, Sect. 3 illustrates the implementation model of ad-hoc cloud system and the training network approach for steganography, Sect. 4 defines the experiment implementation steps and performance measurement, and Sect. 5 discusses and analyses the output of both strengths and weaknesses. Section 6, it summarizes the overall paper idea and process.

2 Literature Review

This section will examine the idea of turning a virtualized infrastructure into an ad hoc cloud. The term ad-hoc cloud computing was recently used to describe the use of non-exclusive and intermittently accessible hosts and devices to construct distributed architecture. Ad-hoc cloud systems and volunteer computing have two main existing initiatives that will be discussed. In [5] the authors introduced the first proposed model of the ad-hoc cloud systems; to better utilization of the existing resources within a small scale network, reduce net energy usage, and allow organizations to run their own ad-hoc cloud systems. Their work shows one way to overcome the primary ad-hoc cloud research and project hurdles. As it can be shown in the Fig. 1 below.

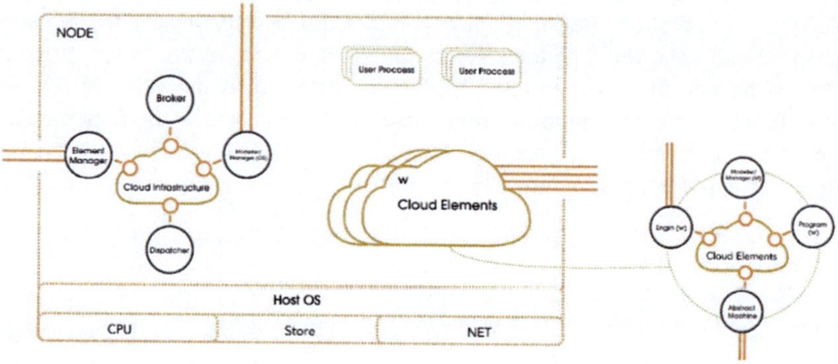

Fig. 1. An illustration of the planned node components between the cloud infrastructure and cloud elements.

Each cloud component described above has a modeler/manager module on the VM. This tracks resource usage and execution on the host. The host-side counterpart evaluates the cloud element's impact. Tasks are assigned to nodes utilizing 'Broker' and 'Dispatcher' architectures. System design module for ad hoc computing. Among other things, the proposed systems differ in scheduling algorithms, QoS guarantees, and how they incorporate reliability into an unstable infrastructure. The proposed method clarifies these contrasts.

In [6] the authors have created Legion software to create web portals for many purposes, including uploading processes to V-BOINC. By creating a cloud interface that talks with a Legion cloud service using SOAP. In order to connect to V-BOINC activities, legion builds and maintains duplicate data on the BOINC database. Because 'Legion' has so many activities, it is similar to the reason 'WS-PGRADE/gUSE' did not use to submit work to V-BOINC, they were either inadequate for our needs or did not deliver the basic capabilities we pursue. So we created a separate 'Job Submission System' for BOINC. Long-term data interchange between organizations and public cloud services increases the risk of unintended privacy breaches. Cloud platform security issues, information leakage, viruses, or other illicit behaviors are usually considered regular users by cybercriminals. Cloud services can monitor IT systems but are difficult to secure.

Concerns about privacy haven't slowed the spread of cloud computing or data centers [7]. Businesses must examine system security strategies to avoid unauthorized data transmission, service outages, and bad publicity. Other than cloud services, public APIs provide new security threats. Cyberattacks target cloud infrastructures. Cybercriminals frequently employ cloud-based penetration testing software to attack suspects' systems. The goal is to get personal data into networks. To avoid confusion with steganography, auto encoder networks are employed to compress images. During training, the network should learn to compress data in secret images to the cover image's lowest regions. Despite the recent positive contributions of deep neural networks in steganalysis, there are numerous initiatives to incorporate neural networks into concealing method [8].

3 Proposed Solution

These modules will be used to construct an ad hoc cloud infrastructure, then a detailed prototype design. Basic ad-hoc cloud architecture can be used. The high-level components are presented in Fig. 2 below. ad-hoc cloud is suspected of being hacked Implementing ad-hoc computing systems should follow these concepts.

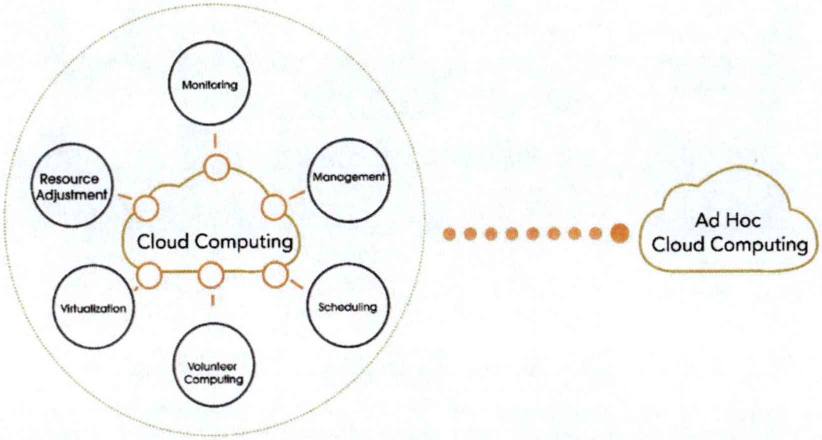

Fig. 2. Ad-hoc Cloud System model structure's six primary distinguishing characteristics

- **Ad-hoc Cloud Model:** The ad-hoc cloud paradigm borrows substantially from both public and private cloud system platforms.
- **Volunteer computing:** You can't monitor and govern a large number of distributed and unexpected resources with this cloud solution.
- **Virtualization:** To secure the resources & procedures for the host along with the 'Ad-hoc Cloud Job', volunteer resources must be protected.

- **Scheduling:** Due to the instability of ad-hoc computing systems, additional scheduling algorithms that integrate resource demand, availability, specification, and dependability must be devised.
- **Monitoring:** Volunteer computer infrastructures are required to supply data for these decision schedules.
- **Management:** Administrators can use infrastructure management to cloudlet, troubleshoot single hosts, or manage several ad-hoc hosts.
- **Resource Adjustment**: The viability depends on limiting cloud process intrusion. However, basic virtualization technologies and open-source tools can provide this functionality.

V-BOINC was used to build the ad-hoc cloud system. So we have BOINC's V-features and an early client-server architecture. These new features make V-BOINC an ad hoc cloud. These are BOINC features: 'VM Service' and 'Job Service'. 'Job Service' can get 'Cloud Jobs' from ad-hoc hosts and register with BOINC; because the V-BOINC client implements so many functions, an ad-hoc client's structure is more complex. Unlike the V-BOINC server, which serves volunteers via a V-BOINC node. Figure 3 give a detailed illustration of the (Ad-hoc Cloud Client) framework.

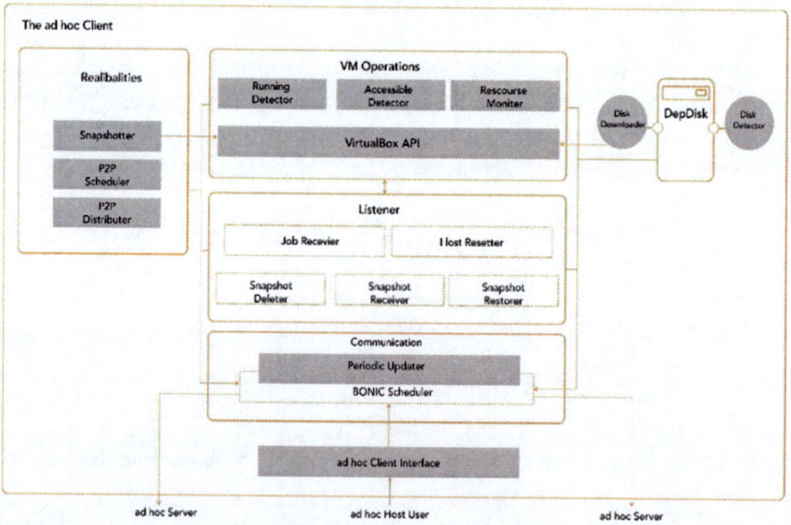

Fig. 3. VM Operations, BOINC Scheduler, DepDisk, and Reliabilities are the four primary components of the (Ad-hoc Cloud Client).

The ad-hoc client modules are the user interface, connectivity, listener, and reliability. Creating a user in the ad-hoc cloud is controlled through the BOINC Manager GUI. Meanwhile, the Listener component awaits any ad-hoc server instructions. The (Virtual Machines Services Component) manages all aspects of the VM-VirtualBox relationship. After a notification is issued, a 'job' schedule will be allocated to the host with high reliability using the 'VM Service'. The scheduler works based on: a) Previous overall cloud jobs executed. Previous 'cloud jobs' finished. b) The host failure rate (hardware or software), c) The guest VM error rate (configuration, installation, processing and termination). The determination of each host's dependability is given in Eq. 1 as illustrated below:

$$\text{Host Reliability} = \begin{cases} 0, & \text{if } NF = CA \\ 100, & \text{if } NF = 0 \\ \left(\frac{CC}{CA}\right) * 100 & \text{Otherwise} \end{cases} \quad (1)$$

$NF =$ Overall Ad-hoc Failure Rate (client-guest). $CA =$ Total planned jobs for ad-hoc host. $CC =$ Total completed jobs via 'Ad-hoc Host'. The VMs API uses the (Snapshot) function to create auto-checkpoints. Placing the snapshot files in the auto-assigned VM image folder. About 20 models of this network, with hidden layers. Our analysis found that five convolution layers utilizing 35 filters worked best. Finally, the picture transmitter uses the 'Reveal Network' as a decoder. It doesn't get the cover or the secret photos. The decoder removes the 'Cover Image' to reveal the secret data. Figure 4 below depicts the three networks (Prep, Hiding, and Reveal) during training, where *c* represents the cover picture and s represents the secret image.

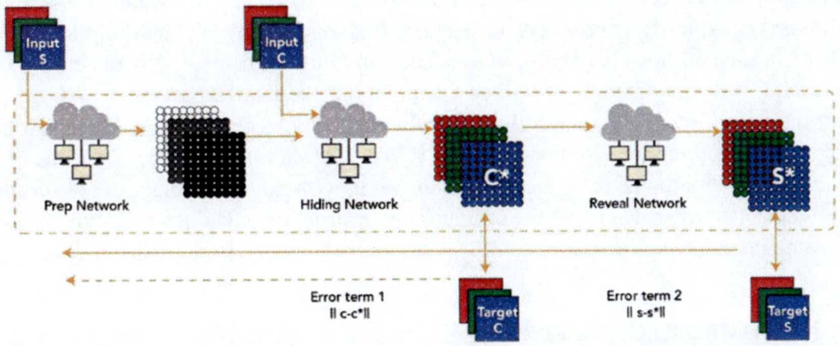

Fig. 4. The three networks (Prep Network, Hiding Network and Reveal Network) were trained as one network, Error Term $\|c - c^*\|$ affected both (Prep Network) and (Hiding Network) while Error Term $\|s - s^*\|$

This is done to prevent the hidden picture from being encoded using LSBs; to prevent the secret image restoration from being contained only in the LSB, the noise was reverse engineered periodically. Also, a practical investigation of network functions in the Fig. 5 below that represents three trained components through one network. Nevertheless, it is less difficult to divide them into three main points to describe them easily.

- **Left Part:** The training of the secret image using a sender.
- **Center Part:** Using the cover image for data hiding.
- **Right Part:** Represent network usage of hidden image reveal over a receiver.

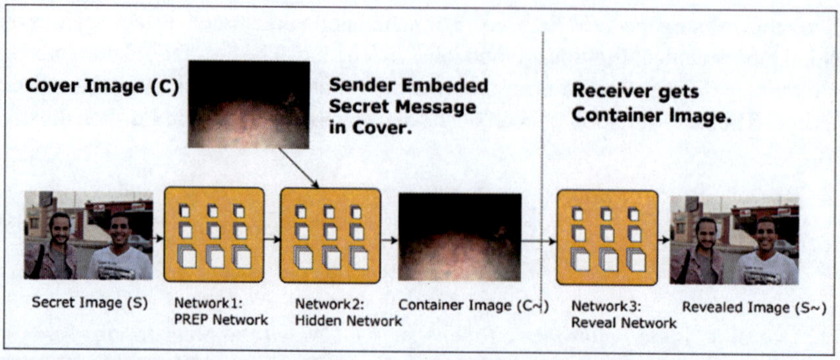

Fig. 5. The proposed system in divided in three main parts. *Part 1*: Preparing the secret image, *Part 2*: Concealing the image through the cover image and *Part 3*: utilizing the reveal network for the hidden image exposure.

4 Experiment

The implementation phase used eight ASUS ROG Strix GL702VS with AMD Radeon RX 580 4 GB processor, 32 GB DDR4 RAM, and 512 GB SanDisk SSD hard disc running Windows 10, the applications are BOINC 7.16.20 and VMware 15.2, and the optimum checkpoint rate results of 15 per hour for a minimum of 2.52 GB of transmitted data from each ad-hoc client receiver was chosen. In the worst case, the transmitter can send 8.2 GB data/hour to eight ad-hoc customers. As indicated previously, ADaM [9] built the 3 networks. The image evaluation might be replaced or restored. Thus, it can reduce the pixel variance total square loss. Used to build the networks. This was done to compare the outcome for cover image encoding with no secret image '= 0' through the same network, as it delivers the optimum reconstruction cover image failure through the same network.

5 Discussion and Analysis

The suggested model of ad-hoc cloud perception was analyzed for security and reliability using 9 nodes and 'Nagios Network Analysis Tool' for 14 days on 5 hosts. The optimum hour was determined by parsing the monitoring data and calculating each host's performance. If the ad-hoc server receives a notification, the divided check-points provided by the 'Ad-hoc Cloud Hosts' must be implemented. Overheads associated with regular interval checks, as well as the possibility of traffic handling, are generated via a resource network, as determined by the designed algorithm P2P.

The VM recovery performance expenses should be considered while limiting the (Cloud Job) time. The (Ad-hoc Scheduler) selects the nearest optimum (Ad-hoc Host) for recovery of the (Ad-hoc Guest), the (Ad-hoc Client) decompression process for the checkpoint, and finally the restoration.

Figure 6 below shows the time it takes to calculate the (Cloud Job's) affection; the (Ad-hoc Client) detects a non-functioning (Ad-hoc Guest) and recovers it on another (Ad-hoc Host) in around 60 s. When testing various types of attacks on the transferred images\data using deep steganography, the form of attack is more resistant to watermarks when unauthorized people may access them. These assaults are dangerous because they can alter the image's data and features, resulting in a distorted image. During the watermarked image transmission phase, various multiplicative sounds begin to accumulate. Several-watermarked photos were tested for durability using repeated noise attacks of varying densities (Watermarks) (NCC-BER). In this example (Variance = 0.005; Mean = 0; Noise Density = 0.05), the output is presented in Table 1 below, and it proves the proposed approach is strong enough to handle changes like watermarks imported from a random image.

The implementation of steganography along with deep learning shows great resilience to noise attacks and robustness to JPEG, Gaussian and Poisson attacks. Lower results can reveal certain performance limitations. As shown in Fig. 7 below, the proposed approach's output was compared to [4, 10, 11], and [12].

Table 1. Different attacks performance evaluation for watermarks extractions for the proposed deep steganography approach.

Noise Attack	**Poisson Noise**	**JPEG**	**Gaussian Noise**	**Sharp**
		Grey Scaling Results		
Watermarks Extraction				
Bit Error Rate	14.3578	7.9851	8.56932	0
Normalized Cross Correlation	0.9056	0.9163	0.9177	1
		Grey Scaling Results		
Watermarks Extraction				
Bit Error Rate	8.6531	2.0569	6.3248	0
Normalized Cross Correlation	0.9217	0.9816	0.9025	1

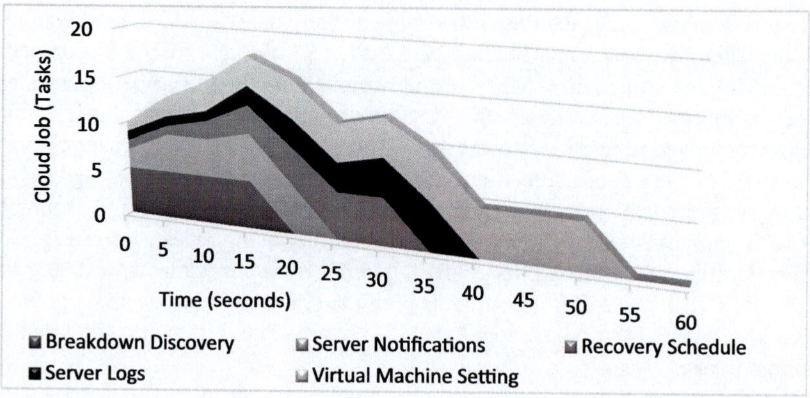

Fig. 6. The process of the (Ad-hoc Cloud) recovery process overheads through time in seconds

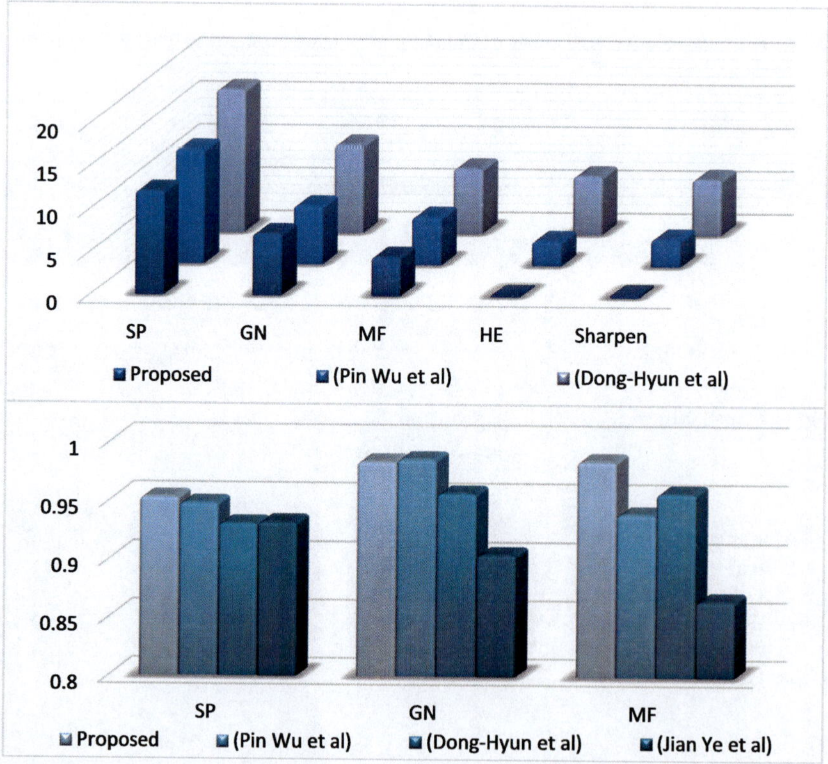

Fig. 7. The evaluation of 'Lena' image, a) A comparison of the (Bit Error Rate) for gray scaling photos with [10, 11], b) A comparison of the (Normalized Cross Correlation) for color scaling photos [10, 11, 12]

6 Conclusion

This study examined an approach for securing data in ad-hoc cloud systems using improved steganography with deep learning. The paradigm of ad-hoc cloud computing platform and its deployment approach were developed in this study. The end-hardware user's is used to launch cloud services on demand. This utility lets developers to circumvent application-level security checkpoints through V-BOINC VMs. Ad hoc cloud can improve network performance and use while cutting expenses. Those unable or unwilling for cloud systems migration can explore the possibilities of ad-hoc cloud solutions. This study expands on steganography and the use of relevant both data & image in another images. Attempts to use machine learning to augment or replace an image-hiding scheme have failed before. An algorithm that seamlessly integrates a color image into another was developed.

References

1. Tuli, S., Tuli, S., Tuli, R., Gill, S.S.: Predicting the growth and trend of COVID-19 pandemic using machine learning and cloud computing. Internet of Things **11**, 100222 (2020)
2. McGilvary, G.A., Barker, A., Atkinson, M.: Ad hoc cloud computing. In: 2015 IEEE 8th International Conference on Cloud Computing, pp. 1063–1068, June 2015. IEEE (2015)
3. Mawgoud, A.A.: A survey on ad-hoc cloud computing challenges. In: 2020 International Conference on Innovative Trends in Communication and Computer Engineering (ITCE), February 2020, pp. 14–19. IEEE (2020)
4. Zhang, C., Benz, P., Karjauv, A., Sun, G., Kweon, I.S.: UDH: universal deep hiding for steganography, watermarking, and light field messaging. Adv. Neural. Inf. Process. Syst. **33**, 10223–10234 (2020)
5. Kirby, G., Dearle, A., Macdonald, A., Fernandes, A.: An approach to ad hoc cloud computing. arXiv preprint arXiv:1002.4738 (2010)
6. Ríos, G.: Legion: an extensible lightweight framework for easy BOINC task submission, monitoring and result retrieval using web services. In: Proceedings of the Latin American Conference on High Performance Computing (2011)
7. Carlin, S., Curran, K.: Cloud computing security. In: Pervasive and Ubiquitous Technology Innovations for Ambient Intelligence Environments, pp. 12–17. IGI Global (2013)
8. Mawgoud, A.A., Albusuny, A., Abu-Talleb, A., Tawfik, B.S.: Localization of facial images manipulation in digital forensics via convolutional neural networks. In: Hassanien, A.E., Darwish, A., Abd El-Kader, S.M., Alboaneen, D.A. (eds.) Enabling Machine Learning Applications in Data Science. AIS, pp. 313–325. Springer, Singapore (2021). https://doi.org/10.1007/978-981-33-6129-4_22
9. Chilimbi, T., et al.: Project Adam: building an efficient and scalable deep learning training system. In: 11th USENIX Symposium on Operating Systems Design and Implementation, OSDI 2014 (2014)
10. Wu, P., Yang, Y., Li, X.: StegNet: mega image steganography capacity with deep convolutional network. Future Internet **10**(6), 54 (2018)
11. Kim, D.H., Lee, H.Y.: Deep learning-based steganalysis against spatial domain steganography. In: 2017 European Conference on Electrical Engineering and Computer Science (EECS), November 2017, pp. 1–4. IEEE (2017)
12. Ye, J., Ni, J., Yi, Y.: Deep learning hierarchical representations for image steganalysis. IEEE Trans. Inf. Forensics Secur. **12**(11), 2545–2557 (2017)

Performance of Different Deep Learning Models for COVID-19 Detection

Sara Hisham Ahmed(✉), Aya Hossam, and Basem M. ElHalawany

Electrical Engineering Department, Faculty of Engineering at Shoubra, Benha University, Cairo, Egypt
{sara.hesham21,Aya.ahmed,basem.mamdoh}@feng.bu.edu.eg

Abstract. Coronavirus disease (COVID-19) is one of the deadliest respiratory illnesses spread since the end of this century. Early coronavirus classification is critical for preventing the disease's fast spread and preserving the life of patients. Researchers focus on investigating the characteristics of the virus causing it and developing appropriate countermeasures. An extraordinary surge of pathogens has occurred, and significant efforts are made to combat the epidemic. Deep learning approaches have gained a lot of interest for medical diagnosis including the diagnosis and detection of COVID-19. Most of the intelligent radiology are utilizing Chest X-Rays (CXR) images and Computed Tomography (CT) images for detecting COVID-19. This paper provides an overview on deep learning approaches for COVID-19 classification employing several datasets, as well as data analytics on its global propagation.

Keywords: COVID-19 · Deep learning · CT images · Chest X-rays images · Image classification

1 Introduction

Since December 2019, a new virus from the corona family, also known as COVID-19, has spread from China to almost every nation across the world. The virus induces a severe, perhaps deadly respiratory illness. On March 12, 2020, the World Health Organization (WHO) declared it a global pandemic [1], which leads to a world-wide race in research and medical institutions to find a cure or vaccine, in addition to medical diagnosis and detection techniques. Computer-aided diagnosis (CAD) techniques have been used for several years for helping in detection of several illness, which make it an excellent choice in COVID-19 pandemic. Machine and Deep learning algorithms have witnessed a great attention for many applications including CAD, image processing, communication, and many other fields [1–3]. In CAD, can these algorithms can be used to estimate and forecast prevalence. By utilizing learning-based approaches from COVID-19 centric modeling, classification, and estimation, Deep learning has proven to be highly effective for aiding epidemic prediction, coronavirus detection, as well as information science and surveillance. According to WHO data, the global cumulative number of Coronavirus infections as of December 20, 2021, was 271,963,258 cases and

© The Author(s), under exclusive license to Springer Nature Switzerland AG 2022
A. E. Hassanien et al. (Eds.): AMLTA 2022, LNDECT 113, pp. 78–88, 2022.
https://doi.org/10.1007/978-3-031-03918-8_8

5,331,019 fatalities [4]. The American Continent has the greatest COVID-19 infection rate with 99,115,289 cases and 2,378,490 fatalities, while the African Continent has the lowest COVID-19 infection rate with 6,716,124 cases and 154,185 deaths. The European Continent, on the other hand, had 93,603,988 illness cases and 1,619,560 fatalities. In Asia, there were 44,799,253 infections and 716,076 fatalities as shown in Fig. 1.

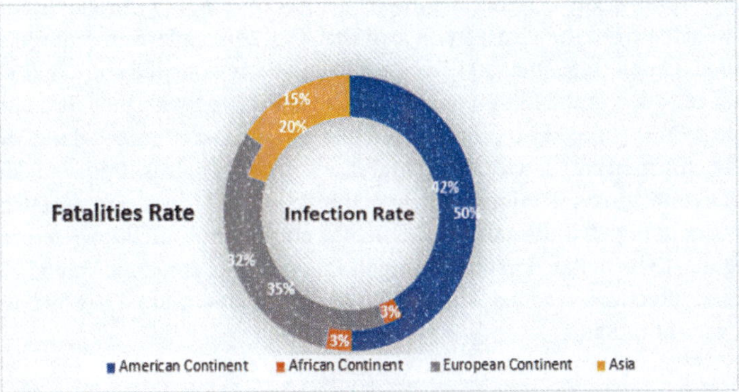

Fig. 1. COVID-19 statistics on infection and fatalities rates through continents

According to WHO, many countries around the world have reported confirmed cases. The United States of America recorded 49,833,416 infection cases with 792,371 death cases. Brazil declared 22,201,221 illness cases with 617,271 death cases. Russia stated 10,159,389 infection cases with 295,104 death cases. Germany announced 6,721,375 illness cases with 107,639 death cases. Japan denoted 1,729,401cases with 18,378 death cases. China avowed 129,430 infection cases with 5,698 death cases shown in Fig. 2.

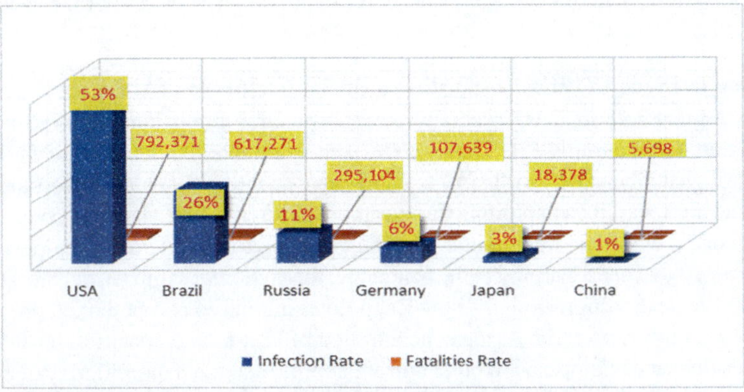

Fig. 2. COVID-19 statistics on infection and fatalities rates through nations

The Middle East was also affected by COVID-19, according to WHO statistics and research on the Coronavirus pandemic. Every government that restricted access to its ports and borders, as happened all around the world, affected the economy. Following the prevalence of the Coronavirus in Egypt, the country reported a cumulative total of infection cases of 370,819 confirmed instances of infection, with a total of 21,155 fatalities. In addition, Saudi Arabia reported 550,369 confirmed COVID-19 injuries and 8,856 deaths. While Tunisia also verified 719,903 infection cases and 25,443 deaths, Kuwait reported 413,790 definite cases and 2,466 fatalities. Libya denoted 379,328 injuries and 5,569 deaths. Bahrain avowed a total of 278,149 coronavirus infections and 1,394 deaths from the virus. Syria said that 49,423 people were injured and 2,823 people killed. Oman stated that 304,741 people had been wounded and 4,113 people had died. Qatar revealed that 245,690 people had been infected with the Coronavirus, with 613 people dying. Turkey also promulgated that the number of infected individuals was 9,118,424 confirmed cases, with 79,863 fatalities from this virus. Iraq also declared the number of Coronavirus infections, which totaled 2,088,833 cases, and the fatalities were 24,007 shown in Fig. 3. India announced that the coronavirus infection rate is constantly increasing due to its mutated nature during the stages of its epidemic spread, as the total of confirmed infections reached 34,726,049, and the deaths were 476,869 as a result of the Delta variant mutator.

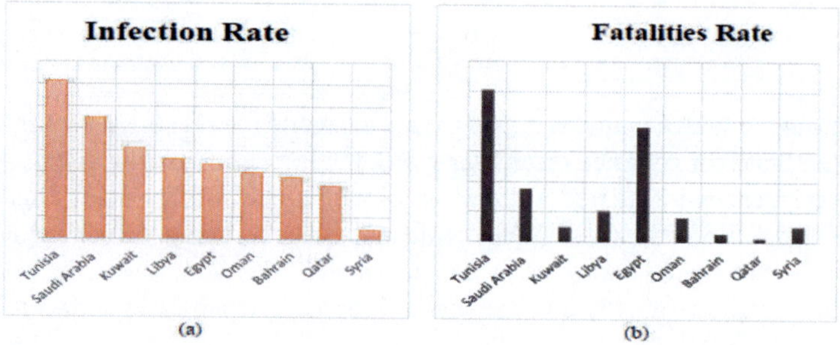

(a) (b)

Fig. 3. COVID-19 statistics through Middle East nations. (a) Infection Rates, (b) Fatalities Rate

Away from the COVID-19 statistics, artificial intelligence (AI) based systems are playing a great role in CAD specially those depending on artificial neural networks (ANN) and deep-learning [5]. These ANNs aim to imitate the behavior of the human brain, though their capacity to "learn" from massive quantities of data. On the other hand, deep learning is based on exploiting three or more-layered ANN. The main contributions of this work is classified into several parts that provide a brief literature review, a quick note on the role of deep learning in the pandemic, the epidemic's impact, debate, and conclusion. The next section explains how Deep Learning-based computerized pneumonia diagnosis is utilized to estimate the concentration of lesions and opacities in COVID-19 instances detected. Compared to other current approaches, such algorithms can evaluate chest x-rays and CT scan images findings in a short time.

2 Deep Learning (DL)

Deep learning enables computational models made up of several processing layers to learn data representations with varying degrees of abstraction [6]. ANN are the main building blocks to create DL models. These networks rely on the connectivity of artificial neurons. ANN created a large number of basic processing units that are linked with one another. The artificial neurons accept inputs from other components or other artificial neurons. Then the result is translated by a transfer function into the output after the inputs are weighted and combined [7]. One interesting characteristic of DL is its capacity to learn the proper representations, which allows the system to adopt the data from a deep level. Multiple layers are utilized to develop relevant models. In the fight against coronaviruses, creating novel and effective diagnosis and treatment approaches is critical, as it determines the success of treating COVID-19 illness.

2.1 The DL-Algorithms Steps in COVID-19 Diagnosis

Deep Learning is a significant tool in the fight against COVID-19. Deep learning relies on data gathered from many sources to understand how to generate new data in their thoughts. The models may match covid-19 data with pneumonia data and acquire a coronavirus detection result. The most crucial aspect of the deep learning model is data processing. The distinction between several data classes, such as Covid-19, pneumonia, and normal instances, improves the deep learning assessment model and distinguishes key features of each other. As shown in Fig. 4, the Deep learning model goes through several phases to detect Coronavirus quickly, including:

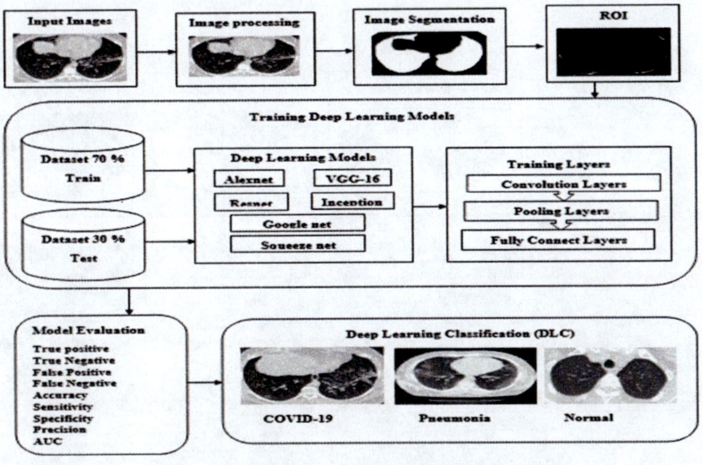

Fig. 4. The sequence of deep learning algorithms

- First: COVID-19 detection methods include Computed Tomography (CT) scan and chest x-ray images. These images of COVID-19 patients were gathered from various sources. Many dataset providers, including Kaggle, GitHub, Chex Pert, and others.

- Second: image datasets perform processing steps after collecting them as data cleaning. Data cleaning must be applied to remove obscure images and eliminate the causes of dataset exceptions. Additionally, all dataset images must be cropped and resized.
- Third: an image segmentation neural network help to extract features from the region of interest parts of an image. The decision-making mechanism may integrate these attributes to classify the region of COVID-19 infection. Image segmentation generates a sequence of segments that represent the whole image or a set of contours from the image as shown in Fig. 5. Each pixel is comparable in some way to the calculated attribute. Adjacent areas have drastically varying colors concerning the same feature.
- Fourth: model training is the process of providing data to a deep learning system to assist it in learning optimal values for all involved attributes. It consists of the sample output data besides the corresponding sets of input data that influence the result. The training model is used to process the input data and compare the processed output to the sample output. This iterative process is known as "model fitting."
- Fifth: the model is evaluated based on its ability to achieve the deep learning algorithm's aims. The performance of the model is determined by Confusion Matrix values. **A confusion matrix.** is a table used to explain the performance of a classification model on a collection of test data with known actual values. As illustrated in Table 1, each row of the matrix represents cases in actual data, whereas each column represents instances in a forecasting class. This matrix depends on four values as True Positive (TP), True Negative (TN), False Positive (FP), False Negative (FN). On the other hand, The Performance is evaluated using measures such as the accuracy, specificity, sensitivity, the F-measure, root mean square error (RMSE), and the Kappa statistic, by applying a confusion matrix.

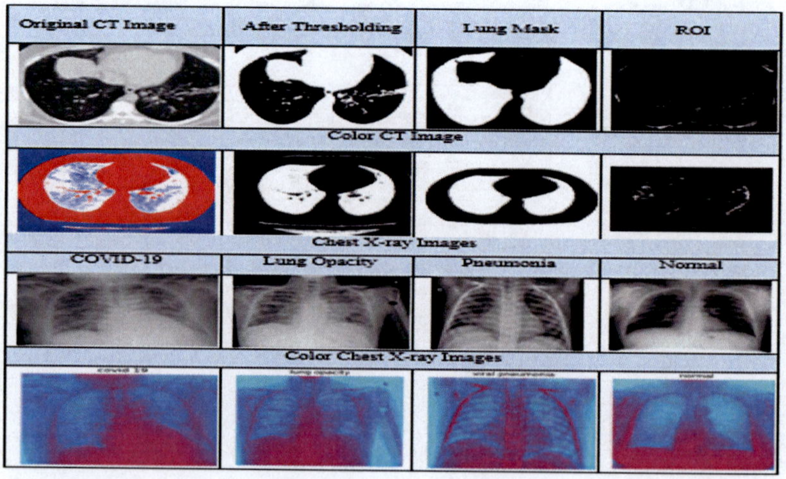

Fig. 5. Several types of data sets, including CT scans and chest X-ray images, and their colors represent the Region of Infection contours.

Table 1. Confusion matrix 2 × 2

	Predicted cases	
Actual cases	TP	FN
	FP	TN

Accuracy: is the ratio of correct predictions to the total number of Coronavirus cases evaluated. It can be determined by the following Eq. (1):

$$\text{Accuracy} = \frac{\text{TN} + \text{TP}}{\text{TP} + \text{FP} + \text{TN} + \text{FN}} \quad (1)$$

The Error rates: is defined as the performance statistic that informs of incorrect predictions without classifying between positive and negative forecasts. It can evaluated by Eq. (2):

$$Error\ rate = \frac{\text{FP} + \text{FN}}{\text{TP} + \text{FP} + \text{TN} + \text{FN}} \quad (2)$$

Sensitivity: also known as True Positive Rate (TPR) or Recall, is the proportion of patients who got a positive result on this diagnosis. It can estimated by Eq. (3):

$$Sensitivity = \frac{\text{TP}}{\text{TP} + \text{FN}} \quad (3)$$

Specificity: also known as True Negative Rate (TNR), is the proportion of patients who got a negative result on this diagnosis. It can be calculated by Eq. (4):

$$Specificity = \frac{\text{TN}}{\text{TN} + \text{FP}} \quad (4)$$

Precision: also known as Positive Predictive Value (PPV), is the positive results ratio in statistics and diagnostic tests that are a true positive. It can evaluated by Eq. (5):

$$Precision = \frac{\text{TP}}{\text{TP} + \text{FP}} \quad (5)$$

False Positive Rate (FPR). is the portion of negative cases identified improperly as positive instances to the total number of negative occurrences. It can evaluated by Eq. (6):

$$FPR = \frac{\text{FP}}{\text{FP} + \text{TN}} \quad (6)$$

In addition, a Receiver Operating Characteristic curve (ROC curve) is a graph that represents the diagnostic performance of a classifier system when its detection threshold is changed. It may assess by plotting the True Positive Rate (TPR) against the False Positive Rate (FPR) at various threshold levels. The classification threshold invariant describes as the Area Under the Curve (AUC), and this entire two-dimensional area is under the complete ROC curve. In the next part, various deep learning approaches are presented to detect SARS-COV-2 using a different dataset. Several methodologies are evaluated in each proposal.

2.2 DL-Models for COVID-19 Detection

Several research studies have been discussed below and summarized in Table 2 which used CT scans and Chest X-Ray (CXR) images to diagnose COVID-19 infection. DL approaches used many methodologies to discriminate these features between coronavirus images and others. The criteria for selecting these papers was based on the variety of used dataset type and the number of classes in each detection model. Deep Learning Novels used two types of datasets which are CT Scan Images and Chest X-ray Images. Some researchs have been depended on COVID-19 patient classification between two categories, while others worked with three classifications. Various novels presented the accuracy of different models with two or three classes of SARS-COV-2 images and others. The model's evaluation differs from each other. All the methodologies of each study were denoted with their best evaluations.

Wang et al. [8] proposed a fast COVID-19 diagnosis system employing CT scans. The dataset was separated into 70% training dataset and 30% test set. The authors used the AlexNet approach which achieved an accuracy of 90.90%, the sensitivity of 70.31%, and precision of 74.36% in 100 times.

Wu et al. [9] introduced a rapid SARS-COV-2 detection scheme. They collected 2000 CT images for the training set beside 230 CT images for the test set. They applied many CNN approaches. The best approaches achieved the same accuracy of 82.5%.

Almourish et al. [10] experimented with 746 CT images separated into 70% training set and 30% test set. They used five pre-trained CNN models to diagnose the novel by training the dataset with 32 batch sizes and 25 epochs. The best evaluation reported by ResNet-50 methodology which detected 92.8% COVID-19 positive infection.

Maghdid et al. [11] proposed a DL-based lung CT scans and CXR images identification system to identify patients with the corona-virus disease. The authors collected 339 CT images and 120 CXR images for the training set. Besides collecting 22 CT images and 50 CXR images for the test set using the CNN approach as AlexNet. The authors reported an assessment accuracy of 82% for CT scans and 98% for CXR images. They also found that 72% of CT scans and 100% of CXR images were sensitive to COVID-19 infection. They also claimed that the specificity of 100% of CT scans and 96% of CXR images was high.

Cortés et al. [12] collected 11,312 CXR images separated into 11,112 CXR radiographs for the training set and 200 CXR radiographs for the test set using the AlexNet approach. For training, the CNN model took 55 min and 42 s. This approach achieved an accuracy of 96.5%, a sensitivity of 98%, and a specificity of 91.7%.

Minaee et al. [13] collected 5K CXR images separated into 2000 CXR images for the training set and 3000 CXR images for the test set. They used four pre-trained CNN models to diagnose the coronavirus. The best evaluation was by the SqueezeNet methodology. This approach obtained a sensitivity of 98% and a specificity of 92.9%.

Naviwala et al. [14] employed CXR images separated into 2084 CXR images for the training set and 3100 CXR images for the test set. They used four pre-trained CNN models to diagnose the coronavirus, including VGG-16, AlexNet, ResNet-18, and Inception V3. The best evaluation was by the ResNet-18 methodology. This ap-proach achieved a sensitivity of 94% and a specificity of 99.3%, with an AUC of 98.8%.

Table 2. Deep learning approaches using CT and X-ray images

Reference	Data type	Number of classes	Train data size	Test data size	Methodology	The best methodology	The evaluation
Wang et al. [8]	CT	2	70% of the total data	30% of the total data	AlexNet	AlexNet	Accuracy (90.90%) Recall (70.31%) Precision (74.36%)
Wu et al. [9]	CT	2	2000	230	ResNet AlexNet VGG Squeeze Net Dense Net	ResNet VGG	Accuracy (82.5%), (82.5%)
Almourish et al. [10]	CT	2	70% of the total data	30% of the total data	ResNet-50 ResNet-101 Squeezenet1.0 VGG-11 Alex Net	ResNet-50	Accuracy (92.8%) Recall (92.5%) Precision (93.2%) F1-score (92.6%) Loss (0.28)
Maghdid et al. [11]	CT CXR	2	339 120	22 50	AlexNet	AlexNet	Accuracy (82%), (98%) Sensitivity (72%), (100%) Specificity (100%), (96%)
Cortés et al. [12]	CXR	2	11112	200	AlexNet	AlexNet	Accuracy (96.5%) Sensitivity (98%) Specificity (91.7%)
Minaee et al. [13]	CXR	2	2000	3000	ResNet18 ResNet50 Squeeze Net Densenet-121	Squeeze Net	Sensitivity (98%) Specificity (92.9%)
Naviwala et al. [14]	CXR	2	2084	3100	VGG-16 AlexNet ResNet18 Inception V3	ResNet18	Sensitivity (94%) Specificity (99.3%) AUC (98.8%)
Progga et al. [15]	CXR	3	90% of the total data	10% of the total data	VGG-19 Mobilenet-V2 VGG-16	VGG-16	Accuracy (98.75%) Recall (100%) Precision (98%) F1-score (99%)

(*continued*)

Table 2. (*continued*)

Reference	Data type	Number of classes	Train data size	Test data size	Methodology	The best methodology	The evaluation
Bhatia et al. [16]	CXR	3	2814	90	AlexNet DenseNet-121 Google Net ResNet-34 ResNet-18 Shuffle Net VGG-16 Squeezenet1.0 Squeezenet1.1	AlexNet Squeezenet1.1	Accuracy (97.78%), (98.89) Sensitivity (100%), (98.3%) Specificity (100%), (100%) Precision (100%), (96.55%)

Progga et al. [15] launched the SARS-COV-2 disclosure scheme. The CNN models studied with CXR images are separated into 90% training set and 10% test. The authors used three pre-trained CNN models to diagnose the coronavirus. The best evaluation was by the VGG-16 methodology. This approach reported an accuracy of 98.75%, a sensitivity of 100%, a specificity of 98%, with an F1-score of 99%.

Bhatia et al. [16] proposed DL approaches-based Chest X-Rays (CXR) identification system to identify patients with Coronavirus disease. The authors collected the CXR images from several sources. There were multiple classes as COVID-19, Viral Pneumonia, and Normal. For training, they used 2,814 images. For testing, there were 90 images. The authors utilized multiple pre-trained CNN models to diagnose the coronavirus. The best evaluation was by the AlexNet and Squeeze-net. These approaches resulted in an accuracy of 97.78% and 98.89%., respectively. These approaches also achieved a sensitivity of 100% and 98.3%.

3 Discussion

The quality of the training data limits the constructed DL models for diagnosis and prognosis using radiological imaging data. There are various public datasets accessible for researchers to train the DL models with respect to their objectives. However, these datasets are not large enough or of sufficient quality to train trustworthy models. Further, all studies employing publicly available datasets had a high or unknown risk of bias. However, if researchers throughout the globe submit their data for public evaluation, the amount and quality of these databases may be constantly increased. Because the quality of many COVID-19 datasets is dubious, it is likely more beneficial to the research community to develop a database with a systematic evaluation of contributed data. Its focus is on providing data of quality as a public database right away.

4 Conclusion

This systematic survey focuses on the deep learning literature employing CT and CXR imaging for COVID-19 diagnosis and prognosis, emphasizing the quality of the methodology used and the repeatability of the approaches. Deep learning algorithms have been active in forecasting potential concerns with COVID-19 growth. This study provided a recent summary and continuing advances in COVID-19 detection trends utilizing deep learning approaches with CT and X-ray images. This study also highlighted certain data prediction use cases. When combined with the deep learning approaches for data analysis, It an help to halt the epidemic. Present datasets such as CT scans and chest X-ray results discovered as part of this comprehensive research and the most prevalent concerns are addressed and emphasized. Finally, it would be advantageous for researchers to focus more on early illness prediction to prevent the spread of COVID-19.

References

1. Hossam, A., Magdy, A., Fawzy, A., Abd El-Kader, S.M.: An integrated IoT system to control the spread of COVID-19 in Egypt. In: Hassanien, A.E., Slowik, A., Snášel, V., El-Deeb, H., Tolba, F.M. (eds.) AISI 2020. AISC, vol. 1261, pp. 336–346. Springer, Cham (2021). https://doi.org/10.1007/978-3-030-58669-0_31
2. Hamed, M.I., ElHalawany, B.M., Fouda, M.M., Tag Eldien, A.S.: A new approach for server-based load balancing using software-defined networking. In: International Conference on Intelligent Computing and Information Systems (ICICIS), pp. 30–35 (2017)
3. Hashima, S., ElHalawany, B.M., Hatano, K., Kaishun, W., Mohamed, E.M.: Leveraging machine-learning for D2D communications in 5G/beyond 5G networks. Electronics **10**(2), 169 (2021)
4. WHO Coronavirus (COVID-19) Dashboard. (October 2021). Retrieved from World Health Organization: https://covid19.who.int/table
5. Fogel, D.B.: Review of computational intelligence: imitating life. Proc. IEEE **83**(11), 1588 (1995)
6. LeCun, Y., Bengio, Y., Hinton, G.: Deep learning. Nature **521**, 436–444 (2015)
7. Sharma, V., Rai, S.: Dev, A: A comprehensive study of artificial neural networks. Int. J. Adv. Res. Comput. Sci. Softw. Eng. **2**(10), 278–284 (2012)
8. Wang, T., Zhao, Y., Zhu, L., Liu, G., Ma, Z., Zheng, J.: Lung CT image aided detection COVID-19 based on Alexnet network. In: 2020 5th International Conference on Communication, Image and Signal Processing (CCISP), pp. 199–203. IEEE (2020)
9. Wu, X., Wang, Z., Hu, S.: Recognizing COVID-19 positive: through CT images. In: 2020 Chinese Automation Congress (CAC), pp. 4572–4577. IEEE (2020)
10. Almourish, M.H., Saif, A.A., Radman, B.M.N., Saeed, A.Y.A.: COVID-19 diagnosis based on CT images using pre-trained models. In: International Conference of Technology, Science and Administration (ICTSA), pp. 1–5. IEEE (2021)
11. Maghdid, H.S., Asaad, A.T., Ghafoor, K.Z., Sadiq, A.S., Mirjalili, S., Khan, M.K.: Diagnosing COVID-19 pneumonia from X-ray and CT images using deep learning and transfer learning algorithms. In: Multimodal Image Exploitation and Learning, vol. 11734, p. 117340E. International Society for Optics and Photonics (2021)
12. Cortes, E., Sanchez, S.: Deep learning transfer with alexnet for chest X-ray COVID-19 recognition. IEEE Lat. Am. Trans. **19**(6), 944–951 (2021)

13. Minaee, S., Kafieh, R., Sonka, M., Yazdani, S., Soufi, G.J.: Deep-COVID: predicting COVID-19 from chest X-ray images using deep transfer learning. Med. Image Anal. **65**, 101794 (2020)
14. Naviwala, M.H., Qureshi, R.: Performance analysis of deep learning frameworks for COVID 19 detection. In: International Conference on Digital Futures and Transformative Technologies (ICoDT2), pp. 1–6. IEEE (2021)
15. Ilma Progga, N., Hossain, M.S., Andersson, K.: A deep transfer learning approach to diagnose COVID-19 using X-ray images. In: International Women in Engineering (WIE) Conference on Electrical and Computer Engineering (WIECON-ECE), pp. pp.177–182. IEEE (2020)
16. Bhatia, N., Bhola, G.: Transfer learning for detection of COVID-19 infection using chest X-ray images. In: 2021 5th International Conference on Computing Methodologies and Communication (ICCMC), pp. 1602–1609. IEEE (2021)

Deep Learning-Based Apple Leaves Disease Identification Approach with Imbalanced Data

Hassan Amin[1(✉)], Ashraf Darwish[1], and Aboul Ella Hassanien[2]

[1] Faculty of Science, Helwan University, Cairo, Egypt
hassanmamin994@gmail.com
[2] Faculty of Computers and Artificial Intelligence, Cairo University, Giza, Egypt

Abstract. Plant diseases pose a significant threat to global food security. Rapid identification of infected plants can significantly impact the overall health of the plant crops and reduce the loss caused by infection spread. Deep learning technologies have been widely used to automate the process of plant disease detection from digital images and accurately identify infected plants promptly. This paper develops a hybrid model by utilizing deep neural networks and support vector machines to classify four classes of apple leaves, namely healthy, rust, scab, and multiple diseases, from digital images with an accuracy of 95.36%. The datasets used in this paper suffered from a class imbalance in its class representation; hence the random oversampling technique has been used to increase the number of samples in the minority class.

Keywords: Plant diseases · Convolutional neural networks · Imbalanced data · Deep learning · Support vector machines

1 Introduction

Agriculture leads the economy in many countries. Apples are fruits with a high demand for import and export to other countries [1]. However, apple crops are subjected to many diseases and infections, reducing crop yield and affecting many countries' economies and global food security. Two of the major diseases that can infect apple crops are Rust and Apple scab which can decrease the yield of the apple crops. Early detection of such diseases is crucial in limiting their spread across the crops and preventing further damage to production [2]. In many countries, the main method of identifying plant diseases is by manual inspection performed by experts. Experts will continuously monitor plant leaves to identify any symptoms, which is time-consuming and has high labor costs. Besides that, experts are not always available to cover all crops in the countries. Hence, automatic detection of disease symptoms will greatly benefit farmers and reduce labor costs and time. Recent advances in machine learning and deep learning fields enabled researchers to develop systems capable of automatically detecting apple disease accurately at an early

H. Amin, A. Darwish and A.E. Hassanien—Scientific Research Group in Egypt (SRGE) http://www.egyptscience.net.

© The Author(s), under exclusive license to Springer Nature Switzerland AG 2022
A. E. Hassanien et al. (Eds.): AMLTA 2022, LNDECT 113, pp. 89–98, 2022.
https://doi.org/10.1007/978-3-031-03918-8_9

stage, increasing the effectiveness and efficiency of the process and saving crops from further damages [3]. In [4], the authors proposed a system that detects crop diseases and alerts the farmers periodically with the findings to help them take rapid actions. Canny Edge Detection [5] was adapted to capture changes in the leaves' shapes and colors to identify the disease accurately. And in [6], the authors collected images of multiple plant species and performed feature extraction using machine learning. Afterward, different machine learning techniques were used to classify the extracted features, and the final model achieved a classification accuracy of 94%. Although machine learning techniques have been successfully employed in increasing the efficiency of the disease analysis in crops, they have some limitations when exposed to factors like variable lighting conditions of the crop images and the stage of the disease, which affects the accuracy of disease detection. Some of the advantages of deep learning techniques include their robustness to such variations. The deep learning model learns complex features from the images, making it less prone to such variations. Also, deep learning techniques eliminate the requirement of performing feature extraction [7], which is a complex process that requires expertise in the problem at hand.

Deep learning algorithms combine the steps of feature extraction and classification while extracting high-level features from the data incrementally. Another evident advantage of deep learning algorithms is their ability to work with large datasets [8]. While the increased number of samples in a dataset can improve the classification accuracy of a deep learning algorithm, the ratio of the number of samples per class can hugely impact its performance. A class with a very small number of samples relative to another class, referred to as the minority class, can be overseen by the deep learning algorithm and limit its ability to correctly distinguish between both classes, which is usually referred to as the data imbalance problem. This paper proposes a new model to classify apple leaf diseases based on convolutional neural networks (CNN) and support vector machines (SVM). Data imbalance in the utilized dataset was addressed using the random oversampling technique to increase the number of samples in the minority class.

The remainder of this paper is organized as follows: Sect. 2 introduces conceptual background concepts for the algorithms and the related methods that have been used in this paper. Section 3 presents the proposed classification model and explains the proposed approach in detail. Section 4 describes the model's experimental results. Finally, Sect. 5 concludes the paper, and future work is discussed.

2 Basics and Background

In this section, the data imbalance problem is defined, and the convolutional neural networks are explained alongside defining the transfer learning technique used to reduce training time and improve accuracy.

2.1 Data Imbalance

For a given dataset, when one class has considerably fewer samples than another class, it's referred to as having a class imbalance where the class with fewer samples is called the minority group. The class with the largest samples is called the majority group. Which introduces skewed results for the classification algorithm as a dataset with balanced classes produces better results. The large variation in the number of samples will force the classifier to prioritize learning features from the majority class and can, partially or fully, ignore the minority class [9]. Imbalanced datasets are commonly found in applications where the positive class exists infrequently. This skewed data representation can be attributed to the dataset being inherently imbalanced by naturally occurring data frequencies, such as in the medical diagnostics where most patients are inherently well or extrinsically imbalanced due to collection or storage procedures [10].

There are different methods of addressing the data imbalance problem, classified as data-level based, algorithmic-level based, and hybrid methods [11]. This research uses a data-level-based method, Random Over Sampling (ROS), to alleviate the data imbalance problem. Random oversampling (ROS) involves randomly selecting samples from the minority class, duplicating, and adding them to the training dataset. This operation is performed with replacement until an acceptable number of images are present in the minority class. This technique can be effective for algorithms affected by a skewed data distribution in the dataset.

2.2 Convolutional Neural Networks

CNN's are deep learning algorithms with multiple layers combined to extract and learn different features from images. CNN's are composed of two main bases: feature extraction and classification. The feature extraction base has three major types of layers: convolutional layers [12], activation layers [13], and pooling layers [14]. Being responsible for extracting various features from the images, feature extraction applies the convolutional operation to the image pixels, reducing it to a form that is easy to process, called feature maps, while keeping most of its original features. In contrast, the classification base is responsible for mapping the feature maps to a one-dimensional vector that corresponds to the correct classification of any given instance by using fully connected layers. Dropout layers [15] are one of the regularization techniques to improve generalization and introduce a probability by which some outputs of the layer are removed.

2.3 Transfer Learning

Transfer learning is a technique where a model trained for a specific task is used as a starting point to train another model [16]. Using transfer learning reduces the time required to train a new model, which can take days, if not weeks, to conclude. There are two primary approaches for transfer learning: (1) Develop a model. (2) Retrain a model. Developing a model approach starts by selecting a task similar to the target task but with abundant data and developing a model and training it on that task. After the model reaches an acceptable performance, it's used as a starting point to train another model for the required task.

While the retrain a model approach includes selecting an existing model that was previously trained on another task and adapting it to the new task at hand. The main aim of transfer learning is to utilize the knowledge gained by a model in a different domain and adapt it to the required domain.

3 The Proposed Approach

In this section, the proposed classification model is discussed, and the different phases of the proposed model are presented. Figure 1 shows a visual representation for the proposed framework, which is summed into 3 major phases: (1) data preprocessing phase, (2) training phase, (3) evaluation phase.

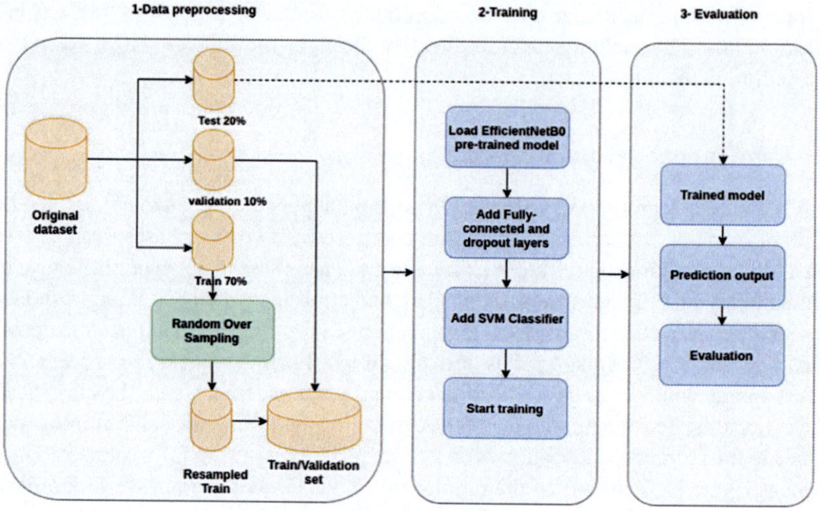

Fig. 1. The proposed model framework

3.1 Dataset Description

The dataset used in the experiment is a subset of the Plant Pathology 2020 Challenge dataset. The subset of the dataset is released on Kaggle, where it consists of 1821 high-quality, real-life images of multiple apple foliar diseases. The dataset contains four categories: healthy, rust, scab, and multiple, which includes leaves infected with more than one disease. Table (1) shows the number of images in each category.

Table 1. The number of images in each category.

Category	Number of images
Healthy	516
Rust	622
Scab	592
Multiple	91
Total	1821

3.2 Data Preprocessing Phase

The obtained images from the plant pathology 2020 challenge dataset were divided into 3 sets, namely training, validation, and testing, with 70%, 10%, and 20% of the dataset's original size, respectively. The training set is fed to the model to learn the complex features of the images. In contrast, the validation set is kept separate from the training set. It monitors the model's performance by feeding them to the model after each epoch and evaluating its performance. After the training has concluded, the test set is used to evaluate the model's overall performance on data that it didn't see before. To resolve the class imbalance problem in the dataset, ROS was used to increase the number of images in the minority class by randomly selecting instances of the minority class and duplicating them. Furthermore, to avoid overfitting, data augmentation techniques have been used on the training set to increase the variety in the images by performing a combination of horizontal flip, rotation, shearing, and zooming on the training data randomly. Finally, before the sets were used in the remaining phases, images were resized to 224 × 224 resolution, and pixel values were rescaled by dividing them by 256 to get values between 0 and 1.

3.3 Training Phase

In this phase, the baseline and the proposed models are constructed using the EfficientNetB0 pre-trained CNN. The top layers of the CNN were removed, and two fully connected layers were added with 1024 neurons each. The final classification layer was a softmax activated layer for the baseline model. The categorical cross-entropy loss function was used, Whereas the proposed model had a linear activated layer with L2 regularized kernel and used the categorical hinge loss function. After training finished for each model, the models were evaluated by testing them against the test subset to evaluate their performance.

3.4 Evaluation Phase

In this phase, four evaluation metrics are utilized to evaluate the performance of both the baseline and the proposed models, namely accuracy, precision, recall, and f1-score. Accuracy is the proportion of the correctly predicted instances out of all predictions usually calculated as a percentage by Eq. (1). Precision is the ratio between the number of correctly predicted instances of a specific category to the total number of predictions of that category and can be calculated as Eq. (2). The recall measures the correctly predicted instances of a specific category to all actual instances in that category which can be calculated as Eq. (3). F1-score is the weighted average of the precision and recall, calculated as Eq. (4).

$$Accuracy = \frac{number\ of\ correctly\ predicted\ instances}{number\ of\ all\ predictions} \tag{1}$$

$$Precision = \frac{number\ of\ correctly\ predicted\ instances\ of\ a\ category}{total\ number\ of\ predictions\ of\ the\ same\ category} \tag{2}$$

$$Recall = \frac{number\ of\ correctly\ predicted\ instances\ of\ a\ category}{total\ number\ of\ instances\ in\ the\ same\ category} \tag{3}$$

$$F1{-}score = \frac{2 \times precision \times recall}{precision + recall} \tag{4}$$

4 Experimental Results and Analysis

The Keras library was used to build the structure of the proposed approach. Keras is a high-level framework written in Python that enables researchers to build and develop deep learning models rapidly. This section discusses the results obtained by the proposed approach.

4.1 Data Imbalance Problem

Random oversampling (ROS) was used to address the data imbalance in the dataset by increasing the number of images in the minority category. Afterward, the generated images are used as the training set for the proposed model. The number of images before and after ROS for the training set can be viewed in Table (2).

Table 2. Number of images of the training set before and after resampling

Category	Before ROS	After ROS
Healthy	316	316
Rust	435	435
Scab	414	414
Multiple	63	200
Total	1273	1410

4.2 Data Augmentation

The ImageDataGenerator generator of the Keras framework was used to perform the augmentation by performing a combination of the horizontal flip, rotation, shearing, zooming techniques, then resizing and rescaling the images on the training dataset. Table (3) shows the corresponding values for each method used.

Table 3. Data augmentation values used for each augmentation techniques

Augmentation technique	Value
Zoom	20%
Shear	20%
Rotation	20°
Horizontal flip	True

4.3 Setup of the Experiment

A baseline model was set by training EfficientNetB0 on the original training dataset. All the training setup for the baseline and the proposed model were kept the same to minimize any variance in the training process. The training was initiated with a learning rate of 0.01 and was set to automatically decrease by 0.1 after every 4 epochs of no-improvement on the validation loss value. The training was developed to stop after 10 epochs of no-improvement on the validation loss value to avoid model's overfitting and was set to be trained for a maximum of 50 epochs. The categorical cross-entropy loss function was used to calculate the difference between the actual and predicted output in the baseline model.

In contrast, the categorical hinge loss function was used in the proposed model with L2 kernel regularize of value 0.01. The Adam optimizer was used to update the model's weights to reduce the loss value in both models. Table (4) summarizes the hyperparameters used for the proposed model.

Table 4. Hyperparameter values used for the proposed model

Hyperparameter	Value
L2	0.01
Dropout	0.3
Batch size	32

4.4 Evaluation of the Model

After training concluded, the model was evaluated by running it against the test set. The baseline model achieved a classification accuracy of 92.64%, while the proposed model outperformed the baseline by achieving a classification accuracy of 95.36% on the test set. Table (5) shows the accuracy and macro average values for precision, recall, and f1-score. At the same time, Table (6) and Table (7) show the precision, recall, and f1-score for each model regarding each category.

Table 5. Comparison between the baseline and the proposed model classification metrics

Category	Baseline	Proposed model
Accuracy (%)	92.64	95.36
Precision (macro avg) (%)	88	94
Recall (macro avg) (%)	85	84
F1-score (macro avg) (%)	87	87

Table 6. Precision, recall and f1-score for the baseline model

Category	Precision (%)	Recall (%)	F1-score (%)
Healthy	91	92	92
Multiple	73	58	65
Rust	96	98	97
Scab	93	93	93

Table 7. Precision, recall and f1-score for the proposed model

Category	Precision (%)	Recall (%)	F1-score (%)
Healthy	94	97	96
Multiple	89	42	57
Rust	95	99	97
Scab	97	98	98

From the results above, it's evident that the proposed model outperformed the baseline model in terms of accuracy and precision. The proposed approach achieved the highest performance concerning precision on all categories. The proposed model was also able to outperform the baseline model in terms of recall and f1-score on the healthy, rust and scab while it suffered on the multiple categories. While the proposed model had decreased value for the recall and f1-score on the multiple class than the baseline model, it improved precision results on the same class.

5 Conclusion and Future Work

This paper proposed a novel apple leaves classification model that can accurately identify diseased and healthy apple leaves from digital images using deep neural networks and support vector machines. Using a base model's accuracy, precision, recall, and f1-score metrics, the proposed approach was verified. The experimental results show that the proposed model can accurately classify healthy and diseased apple leaves. As future work, the proposed model's hyperparameter values can be optimized using one of the optimization algorithms instead of manually selecting such values.

References

1. Hossain, E., Hossain, M.F., Rahaman, M.A.: A color and texture based approach for the detection and classification of plant leaf disease using KNN classifier. In: 2019 International Conference on Electrical, Computer and Communication Engineering (ECCE), pp. 1–6. IEEE (2019)
2. Singh, V., Misra, A.K.: Detection of plant leaf diseases using image segmentation and soft computing techniques. Information processing in Agriculture **4**(1), 41–49 (2017)
3. Liakos, K.G., Busato, P., Moshou, D., Pearson, S., Bochtis, D.: Machine learning in agriculture: a review. Sensors **18**(8), 2674 (2018)
4. Badage, A.: Crop disease detection using machine learning: Indian agriculture. Int. Res. J. Eng. Technol. **5**, 866–869 (2018)
5. Canny, J.: A computational approach to edge detection. IEEE Trans. Pattern Anal. Mach. Intell. **PAMI-8**, 679–698 (1986)
6. Korkut, U.B., Göktürk, Ö.B., Yildiz, O.: Detection of plant diseases by machine learning. In: Proceedings of the 2018 26th Signal Processing and Communications Applications Conference (SIU), Izmir, Turkey, 2–5 May 2018, pp. 1–4 (2018)
7. LeCun, Y., Bengio, Y., Hinton, G.: Deep learning. Nature **521**(7553), 436–444 (2015)
8. Kumar, E.P., Sharma, E.P.: Artificial neural networks-a study. Int. J. Emerg. Eng. Res. Technol. **2**, 143–148 (2014)
9. Krawczyk, B.: Learning from imbalanced data: open challenges and future directions. Prog. Artif. Intell. **5**(4), 221–232 (2016)
10. Seiffert, C., Khoshgoftaar, T., Van Hulse, J., Napolitano, A.: Mining data with rare events: a case study, in: 19th IEEE International Conference on Tools with Artificial Intelligence, ICTAI 2007, pp. 132–141. IEEE (2007)
11. Johnson, J.M., Khoshgoftaar, T.M.: Survey on deep learning with class imbalance. J. Big Data **6**, 27 (2019)
12. Rawat, W., Wang, Z.: Deep convolutional neural networks for image classification: a comprehensive review. Neural Comput. **29**(9), 2352–2449 (2017)
13. Jang, J., Cho, H., Kim, J., Lee, J., Yang, S.: Deep neural networks with a set of node-wise varying activation functions. Neural Netw. **126**, 118–131 (2020)
14. Suárez-Paniagua, V., Segura-Bedmar, I.: Evaluation of pooling operations in convolutional architectures for drug-drug interaction extraction. BMC Bioinf. **19**(8), 39–47 (2018)

15. Srivastava, N., Hinton, G., Krizhevsky, A., Sutskever, I., Salakhutdinov, R.: Dropout: a simple way to prevent neural networks from overfitting. J. Mach. Learn. Res. **15**(1), 1929–1958 (2014)
16. Torrey, L., Shavlik, J.: Transfer learning. In: Handbook of Research on Machine Learning Applications and Trends: Algorithms, Methods, and Techniques, pp. 242–264. IGI global (2010)

Commodity Image Retrieval Based on Image and Text Data

Hongjie Zhang[1,2], Jian Xu[1,2], Huadong Sun[1,2], and Zhijie Zhao[1,2](✉)

[1] Heilongjiang Provincial Key Laboratory of Electronic Commerce and Information Processing, Harbin University of Commerce, Harbin, China
ZhaoZJ1963@163.com

[2] School of Computer and Information Engineering, Harbin University of Commerce, Harbin, China

Abstract. With the continuous popularization and development of the Internet, online shopping has gradually become people's main consumption mode. This paper mainly studies the image retrieval task of goods in e-commerce websites. The main purpose of applying image retrieval in shopping websites is to enable users to search for the expected goods more conveniently and accurately in the massive commodity information. Given the traditional image retrieval process, only the image or text of goods is used as the retrieval object. The query results obtained do not take advantage of the information relevance and complementarity between text and images, which loses the retrieval advantages of goods. To solve the problem, this paper first designs an end-to-end supervised learning algorithm to project heterogeneous data into a common metric space and apply traditional indexing schemes in this space to achieve efficient image retrieval. Secondly, a fusion method is proposed to give better data and higher weight according to the semantic capture quality of input features. Finally, an objective function is proposed, which can correctly embed the fusion features of image and text into their respective feature space, make the fusion features of the same kind of image and text closer to each other, and separate the dissimilar features. The experimental results show that the average accuracy of this method on the test set of commodity data is 70%, which is about 6% higher than the image content-based and text-based image retrieval methods, which proves the effectiveness.

Keywords: E-commerce · Commodity image retrieval · Deep learning

1 Introduction

One of the main goals of e-commerce websites is effective communication and interaction between online vendors and consumers to enhance consumer trust1. Unlike brick-and-mortar stores, e-commerce platforms only allow vendors to advertise and display their goods to consumers through web pages, so consumers' understanding of the degree of authenticity of goods often depends on the completeness of the commodity information delivered in the pages. In the face of massive e-commerce data, how to exclude irrelevant, redundant, or even false information and accurately and efficiently retrieve

© The Author(s), under exclusive license to Springer Nature Switzerland AG 2022
A. E. Hassanien et al. (Eds.): AMLTA 2022, LNDECT 113, pp. 99–111, 2022.
https://doi.org/10.1007/978-3-031-03918-8_10

satisfactory products when consumers are shopping online is not only an important factor to promote the development of e-commerce industry, but also one of the key areas of concern for merchants, consumers, and researchers, with great development prospects and application markets [23, 25]. Therefore, commodity image retrieval for e-commerce platforms has an important research value.

Because the traditional text-based commodity image retrieval method mainly completes the image retrieval indirectly from the perspective of text similarity2. However, relying only on text for commodity image retrieval may lead to repeated use of words in the text due to overlapping text descriptions of different commodities across categories. The commodity descriptions are similar, but the commodity images are completely different. According to the text similarity, other categories of goods will be returned to the user first, but they are very different visually. Similarly, content-based image retrieval (CBIR) is used to find similar images from large-scale datasets for query images. Generally, it uses representative features of query images with similar data and images to rank images for retrieval. However, only the underlying features of the commodity image are used as the retrieval conditions. Due to the lack of text information to describe the high-level concepts of the commodity, images with a similar visual appearance in the retrieval results, such as color, texture, or shape, will be returned preferentially because the distance between features is more matched. People are used to measuring whether the retrieval results meet the expectations at the semantic level. This is also one of the reasons for the inaccurate image retrieval results.

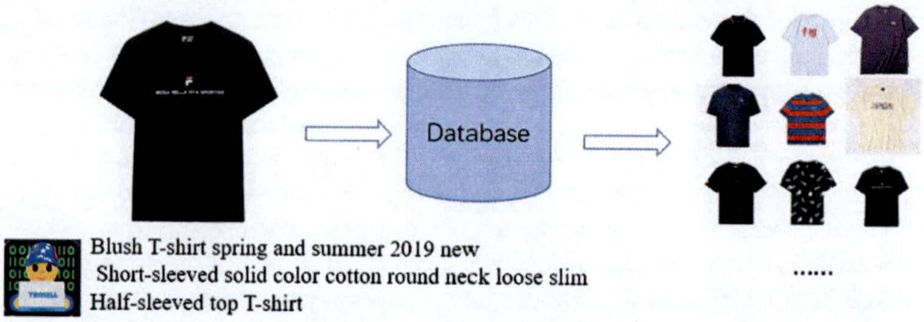

Fig. 1. Examples of querying product images using images and text

In [3], authors, through investigation and statistics on the commodity pages of e-commerce websites, found that the commodities in the e-commerce platform include the image and text description of the commodities simultaneously, mainly to increase the probability that users retrieve the commodities. Thus, the picture shows the commodity visually, and the text describes the high-level concept of the commodity. The information of the two modes depicts the visual and text feature space of the same commodity, respectively. At the same time, it also reflects the semantic relevance and information complementarity between the image and text description contained in the commodity. Therefore, this paper aims to represent the input query, that is, an image with the text description of the product as auxiliary information, and expect the system to return high-quality product images. Figure (1) shows examples of querying product images.

To clarify the products being sold on the website and attract consumers to buy them, general e-commerce websites require merchants to upload clear images. The commodity

Fig. 2. Amazon.com main image example

images on the website are divided into main images and attached drawings. Taking Amazon as an example, only the main images can be effectively displayed before entering the details page. Some relevant items can be added as attached drawings to show how the sold goods are used and cause purchase desire. So, the main picture is good or bad. It will directly affect the click rate of the item. The main picture of a commodity is often located in the center of the whole image so that consumers can intuitively obtain most of the information of the commodity. It stipulates: (1) The picture must have a pure white background; (2) There are words, signs, borders, watermarks, etc. in the main drawing; (3) Commodities must be at least 85% of the entire picture; and (4) The image must be clear, without mosaics or jagged edges, etc. Figure (2) shows amazon.com main image example.

2 Related Work

In this paper, we consider the technique of deep learning to accomplish commodity image retrieval. Commodity image retrieval involves many research fields and is an important visual problem. It includes input image, feature extraction, similarity measurement, and sorting [4]. Unlike traditional feature extraction methods, where a feature extractor is manually designed to convert the original pixel values into an appropriate underlying feature vector based on the designer's a priori knowledge, the feature representation has limited capability. In recent years, deep learning techniques have made great breakthroughs and achieved significant results in computer vision, natural language processing, and speech recognition [5].

To achieve fast retrieval of commodity images, authors [6] learns binary codes directly by adding a latent attribute layer with sigmoid output between two fully connected layers through a coarse-to-fine strategy for fast retrieval of clothing images. Authors in [7] collated a training set of 5000 labeled bag-like style images. They proposed a method to combine the features of different layers of convolutional neural networks for image content retrieval of bag-like styles achieved better results. Still, there are problems such as a small dataset for model training and insufficient migration ability of the model. Authors in [8] designed a network parameter learning method for similarity calculation of specific categories. According to the clothing features extracted by deep learning, a small network is trained to calculate similarity. The similarity measurement network learned in this "coarse-to-fine" way has better retrieval performance than deep features based on ImageNet training. However, this paper only carried out similarity learning based on offline CNN feature and did not explore the end-to-end retrieval model. In [9], authors use cross-entropy loss based on multi-label attribute data and triplet loss based on clothing same ID data, which are complementary in terms of retrieval effect improvement.

The combination of text and images has been extensively studied, especially in visual quizzing10, using joint input of image and text data after the literature [11] used LSTM for image and text fusion. Literature 12 proposes MCB, which projects text and images into a high-dimensional space and convolves them using the fast Fourier transform (FFT) to obtain fused features. A new decision-level fusion strategy is proposed in literature 13, where text is overlaid onto the original image associated with the text after text-CNN classification model extracts text features to achieve the fusion process of text and image [14]. High-quality joint representations are obtained by efficiently combining information from different patterns through vector connections [15, 16].

The heterogeneous relationship that exists between the underlying features of the two-model data. A gated multimodal learning model based on gating is proposed in the literature 18 to find a suitable intermediate representation based on the different data of the input. The simpler simple addition, maximum pooling, and self-encoder model ComposeAE based approaches are used [18] and [19]. Authors in [20] learn a shared latent space using a classification target with a shared classification that guarantees that images and text are mapped to the same space.

3 Method

In this paper, we analyze and study the methods of using both images and text to describe the same product in e-commerce websites, or users retrieve a certain commodity using multimodal data of images and text simultaneously. The essence is to learn a feature space that combines heterogeneous images and text into a uniform representation by an end-to-end depth model and measure the similarity between image and text pairs by Euclidean distance.

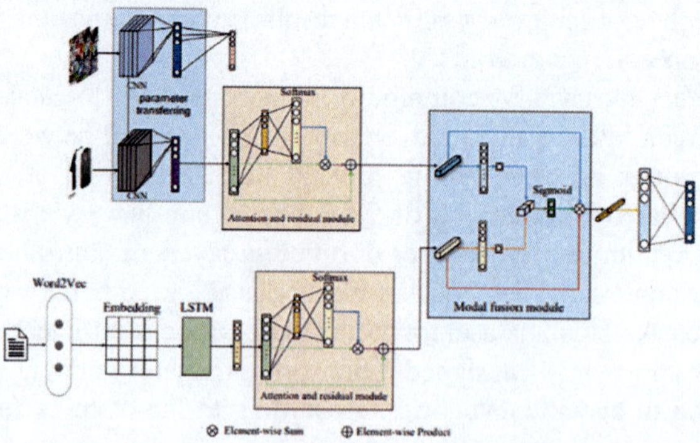

Fig. 3. Image and text-based image retrieval model for commodities

It can be seen intuitively from the Fig. (3) that it includes two branches, namely image and text network, and includes a feature fusion module. The whole model is trained end-to-end. To ensure that the heterogeneous data (images and text) map to a common space, this paper performs feature fusion and shares the last fully connected layer weights to force an effective fusion of the two data. The main purpose of the above model is to learn a nonlinear mapping function from the data space of images and text to the label space using image and text data that share a common label as input and label information as a supervised means. In which images use pre-trained VGG16 model, the text is mapped to text features using word2vec and LSTM, and a residual fully connected layer attention mechanism is designed to intersperse in two branches to avoid the problem of fixed weights in the fully connected layer. Suppose the image in the above figure is $x_i \in x, i = 1, 2, \ldots N$ and the text is t_i, the whole model can be represented as the mapping function f, then the fusion feature z_i of image and text can be represented as $f(x_i, t_i)$, and finally, the classification model of multimodal data is obtained. Classifying multi-model data is an intermediate step in the retrieval process. The purpose is to narrow the retrieval scope and improve the retrieval performance. The fundamental purpose is to get the features that integrate the two-mode information. The process of retrieval of commodity images based on image and text data is divided into three steps:

(1) Training classification neural networks for processing multi-mode data.
(2) The fused features of image and text are extracted from the penultimate fully connected layer in the trained network.
(3) Suppose the query sample is

$$f_{\text{quary}} = f(x, t) \tag{1}$$

calculate the similarity $d\left(f_{\text{quary}}, f(x_i, t_i)\right)$ between the sample features to be retrieved and all the sample features in the database, sort the obtained similarity and return several images of goods with higher similarity.

3.1 Image and Text Feature Fusion

In this paper, a weighted fusion method is proposed for the performance of different modal data in the task, which does not require artificial control of the size of the weights. The structure of the method is shown in Fig. (4). The model can automatically evaluate the importance of different modal branches to the final results during the training process, assign different weights to the features of different modalities, enhance the features that are useful to the results, and suppress the features that have little effect on the results, and finally obtain the final results by vector concatenation The final modal fusion result is obtained by vector concatenation. In addition, the simple vector connection can be seen as a special case of equal weighting values for different patterns.

$$\phi_{xt} = \text{Sigmoid}(W) \odot [\phi_x, \phi_t] \tag{2}$$

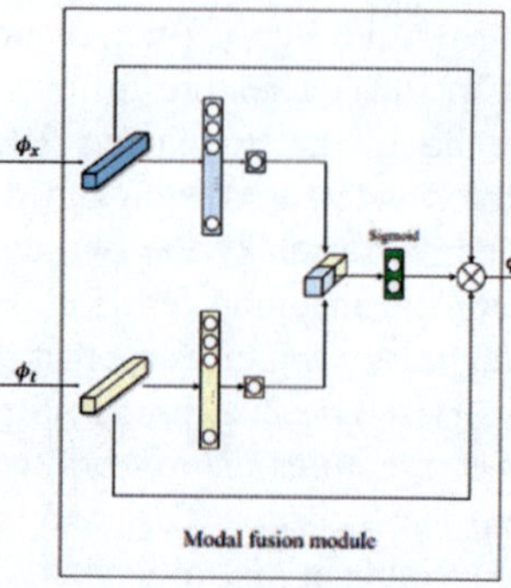

Fig. 4. Schematic diagram of image and text fusion method

$$W = f_{MLP}\left(\left[f_{MLP}(\phi_x), f_{MLP}(\phi_t)\right]\right) \tag{3}$$

where $[\cdot]$ is the Concatenate operation, $\phi_x, \phi_t, w_x, w_t \in R^d$ is the depth feature of commodity image and text respectively, $f_{MLP}(\cdot)$ is the fully connected layer with Relu function, and ϕ_{xt} is the fusion feature of image and text. $W \in R^{1\times 2}$ is the learnable parameter for balancing different data. To increase the nonlinear expressiveness of the data of each pattern, the features of both patterns are reduced in dimensionality and connected using vectors using full connectivity, the correlation between the patterns is obtained through an FC layer containing the number of neurons equal to the number of each pattern, and finally, the weights of the different branches between 0 and 1 are obtained by sigmoid.

3.2 Target Function

Since the overall goal of this paper is to achieve efficient retrieval by narrowing the range of similar samples based on classification networks, the classification process pays more attention to the supervision information of samples and is better at distinguishing features. The disadvantage is that it only considers the relationship between the distribution of samples in the feature space and the classification boundary and does not consider the relationship between sample points. Therefore, the retrieval effect is not ideal when the classification is successful. The retrieval process pays more attention to the relative distribution between sample points. When building the retrieval model, spatial constraints between samples are generally added to meet the retrieval requirements. More compact sample points of the same kind are required, and sample points of different categories have a certain interval in space. However, because of the weak supervisory information in the retrieval process, the association between images may be ignored when considering the spatial constraints of images, resulting in the model losing the ability to model the overall data.

Therefore, considering the advantages and disadvantages of the above classification and retrieval processes, this paper proposes an objective function for the online model that considers the advantages of both classification and retrieval, specifically: for each training batch, an objective function that incorporates the advantages of each of triplet loss and cross-entropy is designed so that the model, when completing the data retrieval tasks of both models, can both obtain discriminative joint features to distinguish multiple types of commodities, and also to learn a good metric for obtaining the universal association between commodities that do not appear in the training set when performing algorithm evaluation. Define the model to calculate the loss based on a batch during training, so assume that a batch size in the process is B, Z_i^+ is the target sample of Z_i and Z_j is the non-target sample of Z_i. N is the number of categories. The final objective function is expressed as the following.

$$L = \frac{1}{B}\left(\sum_{i,j=1,i\neq j}^{B} \log\left\{1+\exp\left\{\sigma\left(\mathrm{K}\left(\mathrm{Z}_i, \mathrm{Z}_\mathrm{i}^+\right) - \mathrm{K}\left(\mathrm{Z}_i - \mathrm{Z}_j^-\right)\right)\right\}\right\} - \sum_{i=1}^{B}\sum_{n=1}^{N} p(n)\log q(n)\right) \tag{4}$$

where σ is the sigmoid function, $p(\cdot)$, $q(\cdot) \in R^N$, $K(\cdot)$ denote the L_2 parametrization of the vector, and

$$\mathrm{Z}_i = f_{combine}(x_i, t_i)^{quary} \tag{5}$$

denotes the fused features of the final image and text.

4 Experiment

The experiments in this paper use Windows 10 operating system, the development language is python 3.8.1, and NVIDIA GeForce RTX 2080 Ti graphics card to accelerate the training process. The TensorFlow framework based on python is used for model building and parameter training. One of the advanced APIs (Keras) can quickly build various advanced models.

4.1 Evaluation Metrics

To evaluate the performance of this paper on image retrieval methods more fairly, accurately, and rigorously, the mean average precision (mAP)22 is used as the evaluation index of the algorithm in the testing phase, and the higher value of the result indicates better performance. The formula is as follows.

$$mAP = \frac{1}{N}\sum\nolimits_{i=1}^{N} AP_i \tag{6}$$

$$\mathrm{AP} = \frac{1}{T}\sum\nolimits_{r=1}^{R} P(r)\delta(r) \tag{7}$$

$$Precision = \frac{TP}{TP + FP} \tag{8}$$

where N is the total number of samples to be retrieved, T is the number of relevant samples in the retrieval set, R is the number of samples of interest, $\delta(r)$ is an indicator function that is 1 if the sample ranked r is relevant and 0 otherwise, TP is a true positive, and FP is a false positive. In this paper, to simulate the real retrieval environment, the samples to be retrieved are removed from the sample database during retrieval, so the ideal case of $R@1$ is 0.50, $R@5$ is 0.71, and $R@10$ is 0.7980.

4.2 Datasets

A Commodity Dataset (CD) has been collected and organized, which is now publicly available. The entire dataset is formed by laboratory members collected on Taobao and contains images and text captions that uniquely correspond to each image (Fig. 5).

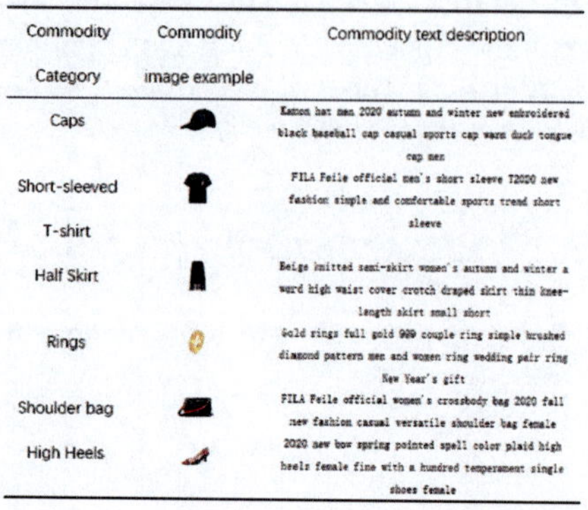

Commodity Category	Commodity image example	Commodity text description
Caps		Kanon hat men 2020 autumn and winter new embroidered black baseball cap casual sports cap warm duck tongue cap men
Short-sleeved T-shirt		FILA Feile official men's short sleeve T2020 new fashion simple and comfortable sports trend short sleeve
Half Skirt		Beige knitted semi-skirt women's autumn and winter a word high waist cover crotch draped skirt thin knee-length skirt small short
Rings		Gold rings full gold 999 couple ring simple brushed diamond pattern men and women ring wedding pair ring New Year's gift
Shoulder bag		FILA Feile official women's crossbody bag 2020 fall new fashion casual versatile shoulder bag female
High Heels		2020 new bow spring pointed spell color plaid high heels female fine with a hundred temperament single shoes female

Fig. 5. Selected examples of commodity datasets

CD has six parent categories and 38 subcategories, each containing 100 images. CD has four characteristics: (1) the products under the same subcategory are displayed from a uniform angle; (2) the background of the images is single, mainly in a solid color; (3) the products are mainly displayed in the center of the whole image; (4) the text title of each image mainly contains the category, brand, season, material, shape, year, etc. of the product. The CD has both image and text information, which researchers can use to conduct single-mode or multi-mode research work. The characteristics of the CD make it possible to eliminate the complicated pre-processing steps for the images in the library, which facilitates the research in this paper.

The division of the datasets and the percentage of each category.

Before the experiment, the commodity data set is randomly divided into training-set: verification-set: test-set = 0.72:0.2:0.08, and the number of each category in the test set is the same, as shown in Fig. (6).

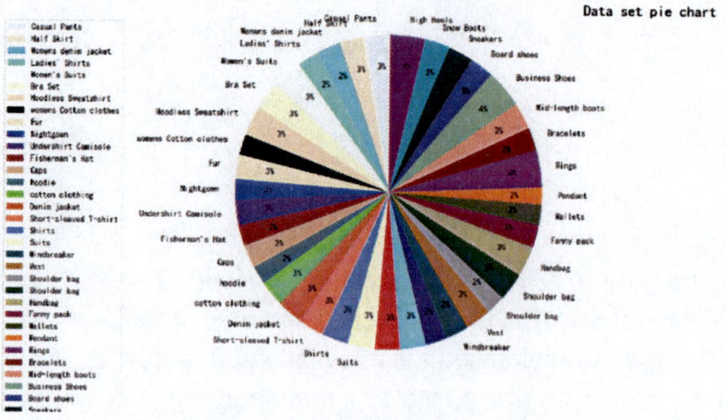

Fig. 6. Percentage of each category in the test set

4.3 Experimental Details

First, this paper uses a fully connected design that lacks a top layer for image branching, a VGG16 model pre-trained on ImageNet and kept frozen. A fully connected layer containing 512 neurons is set up to integrate abstract information of images in the network. This can speed up the overall model training and improve the training effect of the model. In addition, to solve the problem that the fully connected layer is insensitive to the location information of the input data, this paper designs a residual attention module consisting of 512 neurons in each of the two FC layers and a softmax activation function (the last FC layer of this module serves as the extraction layer of the final image features), to improve the performance of each neuron in the FC during the training process, enhancing the final results for neurons that are useful and suppress less useful neurons. Meanwhile, the robustness of the model is enhanced by randomly flipping, cropping, and panning the input images for each batch of training. The image size of the input model is $224 \times 224 \times 3$, and L2 regularization and Dropout are used to avoid the risk of overfitting in the experiment. It should be noted that the image input images are cropped in advance to exclude the influence of the background on the commodity images as much as possible.

Secondly, the Chinese text is deactivated and filtered, and then the exact word separation is performed using the jieba third-party Chinese word separation tool. Considering that using an external corpus will lead to the problem that the word2vec model lacks specialized vocabulary in the e-commerce domain and cannot find the corresponding word vectors, this paper only inputs the corpus of the commodity text dataset used in this paper into the Word2vec model for training, to obtain the targeted word vectors. At the same time, set the sequence length of each short text to [24] (the maximum length of each sentence does not exceed 21 words), the dimension of each word to 100, the maximum distance of the context, i.e., the window size, to 5, and use the skip gram model, with the minimum word frequency set to 3, and the rest are set by default. The trained word vector is used as the weight of the embedding layer to map each sentence to the corresponding value, and an LSTM unit and an FC layer containing 512 neurons are used as the final text feature.

Finally, the 512-dimensional feature vector generated from the last FC layer is used as the joint expression of image and text features. In the training process, Adam optimizer is used for algorithm optimization training. The method of reading data online, the generator, is adopted to increase memory utilization and avoid memory crashes caused by one-time reading. The reasonable batch size is 64, according to the experimental hardware. In this paper, Euclidean distance is used to measure the similarity between samples.

4.4 Experimental Results and Analysis

The following methods are used to compare the effectiveness of commodity image retrieval in this paper, and all methods use mAP as an evaluation criterion.

(1) As the most important object of comparison with the joint image and text-based retrieval model, the first is the use of images Image only: $\phi_{xt} = \phi_x$.

(2) Text only: $\phi_{xt} = \phi_t$.
(3) To use images and text as the final joint feature using vector concatenation $\phi_{xt} = f_{MLP}(\phi_x, \phi_t)$.
(4) Using cross-entropy as the loss function of the classification network, the other experimental settings are kept the same, where the combination of image and text features is done using vector concatenation and from the last FC layer as a joint feature: Cross-Loss.
(5) Only the gated connection method [3] has been used for the fusion of image and text features, i.e., the tanh activation function is used as a "gate" for image features and text features, and the summation method is used as the final joint feature.
(6) Using direct addition of images and text(Add): $\phi_{xt} = \phi_x + \phi_t$.
(7) Using late fusion i.e., separate classification models are used for image and text data.

Table 1. Comparison of mAP scores on the commodity dataset, with the highest scores in bold

Method	R@1	R@5	R@10
Image Only	0.4092	0.5894	0.6357
Text Only	0.4000	0.5608	0.6367
Cross-Loss	0.4360	0.6370	0.6769
Concatenate	0.4320	0.5740	0.6523
Gated Connection	0.4456	0.5780	0.6456
Add	0.4365	0.5701	0.6390
Late Fusion	0.4489	0.5978	0.6770
Ours	**0.4520**	**0.6470**	**0.7068**

The above table (Table 1) compares the retrieval performance of the eight methods on the commodity dataset. Firstly, it can be visualized that the mAP scores for joint retrieval of commodity images with both commodity images and text descriptions as input are higher than those using only images or text for all the number of samples of interest returned (R = 1, 5, 10), mainly because text descriptions play a good role as a complement to image information. Second, the comparison with connection, summation, or gating shows that the fusion method in this paper has certain superiority. Finally, the objective function proposed in this paper is helpful for the final commodity image retrieval accuracy, as shown by the fact that the objective function in this paper has a certain mAP score lead compared with the cross-entropy loss function with the mAP score reaching 70.68% at R = 10. And the feature-level fusion is more helpful to the final retrieval effect than the decision-level fusion.

To visualize the effectiveness of the method in this paper more intuitively, the test data of the commodity dataset was visualized in a two-dimensional plane by using the t-SNE [22] methods. As shown in Fig. (7), each data point in the high-dimensional space is

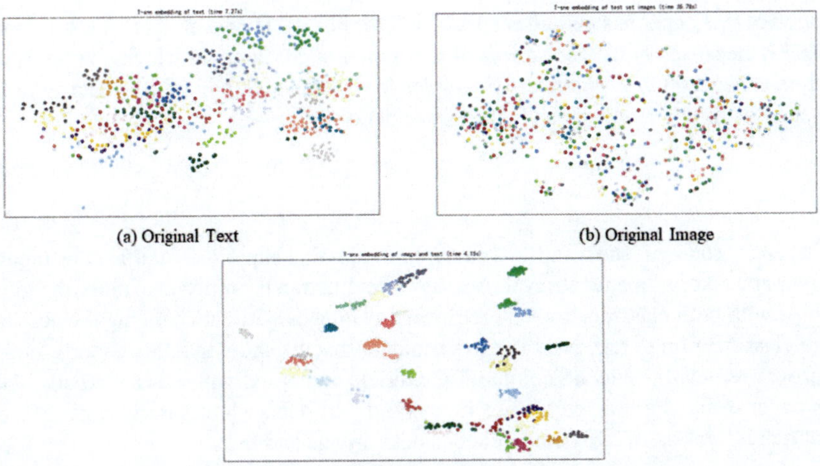

(a) Original Text (b) Original Image

(c) Image and Text

Fig. 7. t-SNE dimensionality reduction visualization analysis

visualized in the low-dimensional space. Similar samples are closer together in the graph and are represented as a cluster, with the same color markers representing samples under the same semantics. (a) represents the original text's word vectors mapped by word2vec. (b) represents the low-dimensional spatial visualization of the original pixel points for each class of images. (c) represents the joint feature visualization of images and text processed by the model. From the figure, we can intuitively see that the original space's text and image sample points are cluttered and very difficult to distinguish. After using the method modeled in this paper, only a small number of data clusters are mixed. Most of the sample points are perfectly contrasted and classified, which greatly improves the accuracy of the retrieval results and illustrates the method's effectiveness in this paper.

Fig. 8. Retrieval results from the commodity image dataset

Figure (8) shows the actual operation of the method in this paper, where the images and text descriptions of any selected men's trench coat and women's bustier skirt from the test set are used as the sample pairs to be queried. The retrieval results show that the method of this paper has certain accuracy and feasibility. However, there are still errors, probably because (1) the robustness of the model is not enough, which can be further will be enhanced (2) the number of samples in the self-built commodity dataset of this paper is insufficient, which can be further expanded.

5 Conclusion

This paper proposes a new end-to-end depth model that supports simultaneous input of images and text for image retrieval. An objective function is proposed to make the model trained with both classification and retrieval advantages without changing the accuracy from classification to retrieval. The experimental results show that this paper's method improves retrieval results on commodity datasets compared with other methods. Also, this paper designs a method to fuse image and text data, which provides an effective joint representation of two heterogeneous data, image and text.

Acknowledgment. This paper is supported by the Natural Science Foundation of Heilongjiang Province Project Funding (LH2021F036).

References

1. Jo, Y., Wi, J., Kim, M., Lee, J.Y.: Flexible fashion product retrieval using multimodality-based deep learning. Appl. Sci. **10**(5), 1569 (2020)
2. Hamiti, A., Hamiti, A.: A comparative study of text-based image retrieval and content-based image retrieval techniques. J. Capital Normal Univ. (Nat. Sci. Edn.) **33**(4), 4 (2012)
3. Arevalo, J., Solorio, T., Montes-y-Gómez, M., et al.: Gated multimodal units for information fusion (2017)
4. Sun, J., Yuan, F.: Content-based image retrieval technology. Comput. Syst. Appl. **20**(8), 5 (2011)
5. Hao, X., Zhang, G., Ma, S.: Deep learning. Int. J. Semant. Comput. **10**(03), 417–439 (2016)
6. Lin, K., Yang, H.F., Liu, K.H., et al.: Rapid clothing retrieval via deep learning of binary codes and hierarchical search. In: ACM on International Conference on Multimedia Retrieval. ACM (2015)
7. Luo, Z.: Combined with the characteristics of different layers of convolutional neural network, package commodity retrieval is carried out. Comput. Appl. Softw. **35**(1), 6 (2018)
8. Kiapour, M.H., Han, X., Lazebnik, S., et al.: Where to buy it: matching street clothing photos in online shops. In: IEEE International Conference on Computer Vision. IEEE (2015)
9. Huang, J., Feris, R.S., Chen, Q., et al.: Cross-domain image retrieval with a dual attribute-aware ranking network. In: IEEE International Conference on Computer Vision. IEEE (2015)
10. Qi, W., Teney, D., Wang, P., Shen, C., Dick, A., van den Hengel, A.: Visual Question answering: a survey of methods and datasets. Comput. Vis. Image Underst. **163**, 21–40 (2017). https://doi.org/10.1016/j.cviu.2017.05.001
11. Ren, M., Kiros, R., Zemel, R.: Exploring models and data for image question answering. Litoral Revista De La Poesía Y El Pensamiento, 2953–2961 (2015)
12. Fukui, A., Park, D.H., Yang, D., et al.: Multimodal compact bilinear pooling for visual question answering and visual grounding (2016)
13. Zahavy, T., Magnani, A., Krishnan, A., et al.: Is a picture worth a thousand words? A deep multimodal fusion architecture for product classification in e-commerce (2016)

14. Gallo, I., Calefati, A., Nawaz, S., et al.: Image and encoded text fusion for multimodal classification. In: 2018 Digital Image Computing: Techniques and Applications (DICTA) (2018)
15. Kiela, D., Bottou, L.: Learning image embeddings using convolutional neural networks for improved multi-modal semantics. In: Proceedings of the 2014 Conference on Empirical Methods in Natural Language Processing (EMNLP) (2014)
16. Guo, X., Wu, H., Cheng, Y., et al.: Dialog-based interactive image retrieval. arXiv preprint arXiv:1805.00145 (2018)
17. Misra, I., Gupta, A., Hebert, M.: From red wine to red tomato: composition with context. In: IEEE Conference on Computer Vision & Pattern Recognition. IEEE Computer Society, pp. 1160–1169 (2017)
18. Kiela, D., Grave, E., Joulin, A., et al.: Efficient large-scale multi-modal classification (2018)
19. Anwaar, M.U., Labintcev, E., Kleinsteuber, M.: Compositional learning of image-text query for image retrieval (2020)
20. Narayana, P., Pednekar, A., Krishnamoorthy, A., et al.: HUSE: hierarchical universal semantic embeddings (2019)
21. Wang, K., Yin, Q., Wei, W., et al.: A comprehensive survey on cross-modal retrieval (2016)
22. Wattenberg, M., Viégas, F., Johnson, I.: How to use t-SNE effectively. Distill **1**(10), e2 (2016)
23. Ting, S., Guohua, G.: Image retrieval method for deep neural network. Int. J. Sig. Process. Image Process. Pattern Recogn. **9**(7), 33–42 (2016). NADIA, ISSN 2005-4254 (Print); 2207-970X (Online). https://doi.org/10.14257/ijsip.2016.9.7.04
24. Bagri, N., Johari, P.K.: A comparative study on feature extraction using texture and shape for content-based image retrieval. Int. J. Adv. Sci. Technol. **80**, 41–52 (2015). NADIA, ISSN 2005-4238 (Print); 2207-6360 (Online). https://doi.org/10.14257/ijast.2015.80.04
25. Yang, D., Grice, S.: Research on the design of E-commerce recommendation system. Int. J. Smart Bus. Technol. **6**(1), 15–30 (2018). https://doi.org/10.21742/IJSBT.2018.6.1.02

Machine Learning Technologies

Artificial Intelligence Based Solutions to Smart Warehouse Development: A Conceptual Framework

Vu-Anh-Tram Nguyen[1], Ngoc-Bich Le[2,3], Manh-Kha Kieu[1,4], Xuan-Hung Nguyen[5], Duc-Canh Nguyen[5], Ngoc-Huan Le[5], Tran-Thuy-Duong Ninh[1(✉)], and Narayan C. Debnath[6(✉)]

[1] Becamex Business School, Eastern International University, Thu Dau Mot, Binh Duong, Vietnam
{tram.nguyen,kha.kieu,duong.ninh}@eiu.edu.vn
[2] School of Biomedical Engineering, International University, Ho Chi Minh City, Vietnam
lnbich@hcmiu.edu.vn
[3] Vietnam National University Ho Chi Minh City, Ho Chi Minh City, Vietnam
[4] School of Business and Management, RMIT University, Ho Chi Minh City, Vietnam
[5] Mechanical and Mechatronics Department, Eastern International University, Thu Dau Mot, Binh Duong, Vietnam
{hung.nguyenxuan,canh.nguyen,huan.le}@eiu.edu.vn
[6] School of Computing and Information Technology, Eastern International University, Thu Dau Mot, Binh Duong, Vietnam
narayan.debnath@eiu.edu.vn

Abstract. Responding to the vital role and robust development of smart warehouses, in this study, solutions for applying Artificial Intelligence (AI) to smart warehouse development, especially in Vietnam, were analyzed and proposed. The paper has investigated the factors affecting the application of AI for smart warehouse development in Vietnam through SWOT. Solution groups include (1) Investment decision support solutions that limit investment risks through solutions to test innovative approaches and algorithms, (2) Solutions for AI application in smart warehouse development, (3) Solutions to develop AI resources for smart warehouses. Therefore, by using SWOT analysis, this paper aims to generate value strategies AI-based solutions for the smart warehouse to improve the warehouse process efficiencies and strengthen the competitive advantage of the logistics industry in Vietnam.

Keywords: Artificial intelligence · Machine learning · Deep learning · AI application · Smart warehouse · Industry 4.0 · Digital transformation

1 Introduction

A warehouse is a place used to store goods for commercial purposes. Most warehouses in Vietnam are traditionally built and managed, leading to many limitations, such as

© The Author(s), under exclusive license to Springer Nature Switzerland AG 2022
A. E. Hassanien et al. (Eds.): AMLTA 2022, LNDECT 113, pp. 115–124, 2022.
https://doi.org/10.1007/978-3-031-03918-8_11

inefficient space management, inefficient operations, and inefficient material handling equipment [1].

With new technologies characterized by intelligent autonomous cyber-physical systems [2] and the explosion of e-commerce, warehouses worldwide have adapted to technological innovations and changes. According to the authors [3], a smart warehouse is designed to work at optimum efficiency and incorporates best practices and modern technology. The main point for smart warehouse applications is robust location, time-efficient communication planning, human activity detection, and multi-robot collaboration [4]. In [5], instead of designating fixed placements, the authors studied dynamic slotting for reducing delivery time. Several algorithms were employed to implement some features, including optimum order picking, optimal product positioning, and zone capacity picking [6].

In recent years, Artificial intelligence (AI) algorithms like Machine Learning or Deep Learning have been efficiently applied in any field of industry [7]. In manufacturing, more than half of Europe's large firms have implemented at least one AI use case in industrial processes [8]. SWOT analysis is applied to evaluate the benefits, drawbacks, opportunities, and risks of using AI to manage and operate warehouses in Vietnam. This paper provides a unique AI-based solution to address the difficulties of existing problems.

2 SWOT Analysis

2.1 Strengths

Continue with Growing Demand and Economy

E-commerce. The pandemic effect and the growing middle class have driven the spending behaviour changes and the acceleration of the e-commerce platform in Vietnam. Therefore, the need for additional warehouses in many provinces or cities of Vietnam will continue increasing for the manufacturers' expansion to ensure future supply despite the ongoing pandemic [9]. The rise of e-commerce has promoted the collaboration between manufacturing, industries, logistics and warehouses nationwide. This trend will foster the demand for warehouses to accommodate the new wave of the economy [10].

Manufacturing Bases Relocation. The global Covid pandemic situation is still complicated, leading to booming e-commerce and the influx of relocation out of the country like China. Diversifying production across multiple bases and inexpensive capital requirements for launching new manufacturing in Vietnam boost the well-located warehouses, logistics and industrial parks for investors in Vietnam [10]. The high value-added products and multinational manufacturers such as Pegatron, Foxconn should experience more expansion to Vietnam during the recent outbreak of the Covid-19 pandemic [11].

The growing industrial park in Vietnam is the first signal. There were 374 established industrial nationwide, while the remarkable increase of average occupancy reached up to 99%. As The Vietnam Logistics Business Association predicted, the scale of the industry in Vietnam will be $40–42 billion, with an annual growth rate of 14–16%. Therefore, in 2021 many provinces will try to attract investments for new industrial clusters.

Strong Commitment to Digital Transformation from the Government
The Vietnam government has paid significant attention to the role of technology applications through Decision No. 703, signed by the Prime Minister on June 7, 2019 [12]. Information technology (IT) plays the backbone in circulation and good distribution and connects the value chain parties. In recent years, consolidation of warehouses has been the trend with integrating all functions. Although the IT application level is still a challenge, IT enhancement is critical for the Vietnam government.

In Vietnam, low technology integration remains the main point of enhancing the competitiveness of the logistics industry [13]. The logistics industry has started to apply technology implications; many companies pay attention to using a technology solution to create advantages for the sector-specific industries. Warehouse management software is an essential application with nearly 70% of the surveyed companies [14].

Strong Investment in AI Primarily from Private Investment
In 2018, the AI industry grew more than 70% compared to 2017, equivalent to $200 billion [15]. Based on the AI Index report 2021 [16], the total investment in 2020 is $67,854 million. Ministry of Planning and Investment promote AI development, identifying as the key breakthrough technologies, to develop a national strategy for the 4th industrial revolution. In the strategy [17], AI is nominated as one of the top priority technologies for future development and investments.

High Warehouse Demand
Warehouse cost is expected to increase from 1.5% to 4% annually. Demand for factories and warehouses is also predicted to increase from 4 to 11% after the end of the Covid pandemic [18]. Due to the increased demand after the pandemic effect, the new wave of relocation, the rise of e-commerce, Vietnam's industrial and logistic real estate, namely developer of ready-built factories (RBF), warehousing and built to suit solutions, became destinations for investors [11].

The Supply of AI Human Resources is Thriving with Attractive Salaries
According to the AI Index 2018 report [15], the number of students in this field has increased continuously, four times from 2012 to 2018 in US and 16 times in seven years in Tsinghua University in China. IT Human Resources has 400,000 with 50,000 graduated students each year from more than 153 training schools in Vietnam. The average AI job posting monthly salary is $1,958 [19]. The ecosystem of AI is increasingly diverse with schools, institutes, communities such as VietAI, VinAi, QuynhonAI, etc.

2.2 Weaknesses

Infrastructure for AI Technology and Digital Transformation is Limited
The core foundation of AI is data and computing infrastructure. In Vietnam, computers with large computing capabilities and big data collection capabilities are concentrated in large universities, some research institutes and companies. In addition, the mechanism

for sharing data and computing infrastructure have not been built in a synchronous and systematic manner.

Weak Technology Driver to Change the Mindset
When implementing AI and digital transformation, it is possible to face the fear of change in the organization such as changing long-standing work habits, significant investment, pressure to Return on Investment (ROI). Leaders truly overcome the obstacles of thinking to adapt the wave of digital transformation and AI implementation.

Lack of Synchronous and Systematic Investment
The investment and development of a smart warehouse system and the application of AI to smart warehouse operation is still limited, not synchronized, and not systematic. Currently, the application of smart warehouses is only concentrated in large enterprises such as Amazon, Walmart, Alibaba, DHL, Vinamilk, etc.

2.3 Opportunities

Figure 1 below demonstrates the potential opportunities of applying AI to smart warehouses including (1) Cut down operating energy (2) Logistics cost reduction and (3) Advantages from a system automation perspective.

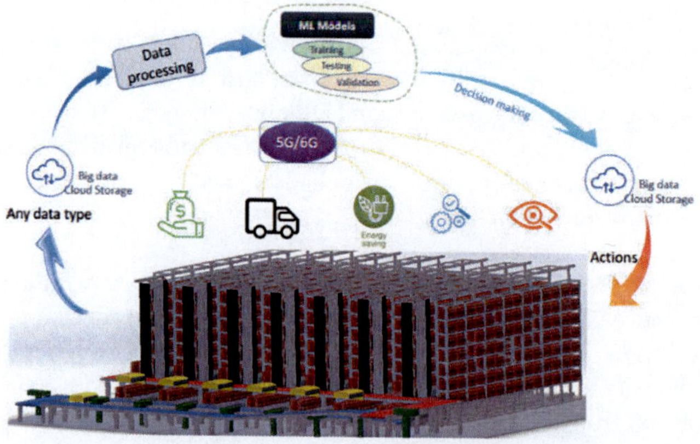

Fig. 1. Opportunities from applying AI to smart warehouse.

Cut Down Operating Energy
The reduction of warehouse operating energy consumption is based on cost reduction, profit increase, and legal compliance issues [20]. Reducing logistical costs and increasing efficiency are also beneficial. Consequently, the application of AI will save operating energy through optimal management and operation algorithms.

Logistic Cost Reduction
Warehouse management costs range from 10% to 41% of a business's total logistics

costs [21]. Smart warehouses, especially AI-powered smart warehouses, have significant potential to cut logistics costs. Applying AI models in finding the optimal sorting solution combined with considering more inputs and boundary conditions will significantly improve the goods arrangement.

Advantages from a System Automation Perspective
Automation in the smart warehouse brings many benefits to investors, such as reducing labour costs, picking up and delivery times, reducing errors when delivering goods, and improving overall management efficiency.

2.4 Threats

Risk of Cyber-Attacks and Security
Using data exchange and management characteristics over the network, hackers can efficiently perform AI system attacks by using AI technology. Therefore, the possibility of security problems exists due to the limited security of internet connection based communications. Furthermore, current encryption methods are no longer effective enough to guarantee a certain level of security.

Workforce Resource Limitations
According to a 2018 report of the Element AI Independent Research Laboratory in Montreal, Canada, only about 10,000 experts qualified to solve complex problems about AI in the world. In Vietnam, despite the attractive salary, the current AI human resources only meet 1/10 of the market.

3 Proposed Solutions and Current Approaches

The advantage of a SWOT is to generate the value strategies from the SWOT items. By combining four factors of SWOT analysis, four strategies could be inferred, including Attach strategy (SO), Improve strategy (WO), Defend strategy (ST) and Exit strategy (WT). This paper focuses on two active strategies described and analyzed in the previous section.

3.1 WO Strategy (Improve): Testbed as a Trial for Investment Decision

Integrating with supply chain partners will foster innovation by reducing the cost of investing in new technology, lowering the cost of knowledge sharing and training, and improving visibility [22]. The testbed centre is flexible enough to simulate various AI investment scenarios that allow investors to make accurate decisions.

Algorithm/System Evaluation
A testbed is a platform for evaluating the precision and repeatability of theories, simulated results, computational techniques, emerging technologies, and applications. The testbed provides a dynamic workplace free of the dangers and penalties associated with testing

in a live environment. A testbed may showcase new modules or real solutions, and it can comprise software, hardware devices, and network components [23].

Limit Investment Risk
Future warehouses are formed of modules integrated entities with the capacity for communication and the energy-aware operation linked to the handling and storage of traditional materials. However, the widespread application of logistics, embedded systems, and wireless communications leads to a new complexity in creating and assessing such systems. While theory and simulation techniques for each field are provided, numerous problems are frequently identified upon practical system implementation [24].

Evaluation of Investment Efficiency
Due to the limitation of operating space and minimum management costs, the warehouse system is always promoted to innovate towards digitization. Integrating complex solutions (recognition techniques, augmented reality technology, self-driving vehicles, etc.) into corporate practices requires testing and validation. This is challenging work in large, continuous, and difficult-to-control operating environments. In [25], the authors designed a warehouse automation experiment that focuses on the small-scale evaluation of mobile robotic units in a warehouse setting. The authors used Cyber-Physical Production System testbeds in the manufacturing sector to test, validate, and verify integration with real-world platforms [26].

3.2 WO Strategy (Improve): AI-Powered Solutions

AI and IoT Blended
IoT, which could collect a considerable amount of data from many different sources, is the perfect support for AI to create smart factories and decision-making with or without human intervention. In this hybrid approach, IoT derives data from devices interacting with each other using the internet, while AI makes the device smarter from its own data sets and experiences.

Operating Energy Optimization
The authors utilized an AI-based method in [27] to simulate operational CO_2 refrigerant-based industrial cooling systems for integration into a global energy management system. When using AI to optimize operational processes, energy consumption is expected at the allowable level. In [28], a deep reinforcement learning (DRL) algorithm is used for the real-time scheduling of AGVs for the flexible factory floor. From that result, the authors have extended the multi-objective optimization studies of real-time scheduling of AGVs.

AGVs/Forklifts Scheduling Optimization
AI algorithms can ultimately support the AGVs/Forklifts scheduling optimization. In [29], in the context of warehouse management, the authors built a DRL approach to allow a vehicle to choose and move to the nearest task by integrating the information of mapping and localization. The DRL approach was also used for planning routes for the

AGV picking system [30]. A new design methodology for DRL system was proposed and validated for autonomous navigation [31, 32].

Warehousing Management Optimization
Authors pay attention to the successful application of AI in the processes of goods storing, order picking, and order packing at Alibaba's Smart Warehouse. Several AI algorithms to analyze and use data quickly for real-time optimization and decision-making at Alibaba [33].

In the Goods Storing Process. Automatization tridimensional storehouses (ATS) is applied. At ATS, information about the total weight, three-dimensional dimensions of the goods, and the identity of the pallet are immediately identified and updated as inventory data. The AI algorithm then calculates the most efficient location to store pallets based on historical data.

In the Order Picking Process. "Order to man" (O2M) AGVs, "goods to man" (G2M) AGVs, and forklift AGVs are used for the order picking process. The warehouse management system (WMS) identifies the appropriate packaging box based on the order request, warehousing data and the packing algorithm and the AGVs will transport the necessary goods to workstations.

In the Order Packing Process. Once the order is confirmed, the order box will be packed under the instructions from the 3D packing algorithm. The AI algorithm ensured that the items are packed securely.

Steps (from Basic to Advanced) to Deploy AI Projects to Support the Operations of a Real Smart Warehouse [13]

Step 1: Deploy Easy-to-Succeed AI Applications into a Live Warehouse Environment. This is the essential step for real-time data processing from the warehouse.

Step 2: Put Down a Solid Foundation in Data Governance. The enterprise should design a data governance framework and analysis to provide a solid basis for scalability and promote future deployments.

Step 3: Scale AI Solutions Across the Smart Warehouse. AI applications at various levels in the warehouse may be installed and distributed across many sectors.

3.3 SO Strategy (Attack): AI Resource Development

Build an AI Network and Community for Smart Warehouses Systematically and Synchronously
Data shows the massive importance of AI applications. Consequently, Vietnam has been establishing AI networks and communities for many fields to promote the strength of community in data development. The most recent one is the Vietnam-Australia Artificial Intelligence Cooperation Network, launched in August 2021. It demonstrates the

necessity of systematical and synchronous networks and communication, especially the smart warehouses.

AI Resource Training

AI Training for Automation and Mechatronics Engineers: The foundation of AI is digital transformation, data, computing infrastructure, and automation foundation. Integrating AI competencies for Automation and Mechatronics majors will be very convenient with automation and mechatronics based knowledge.

AI Training for Logistics Bachelors: Recently, digital transformation has been booming in the logistics industry, especially during and after the Covid 19 pandemic. Therefore, integrating AI competency for logistics students to serve the smart warehouse field attracts significant attention due to the considerable market demand.

4 Conclusions and Future Research

This paper proposes solutions for the AI application in smart warehouses. SWOT analysis has been applied to assess many aspects, including economic, technical, market needs, resources, investment trends, and environment. From the results of the SWOT analysis, the solution groups were proposed, including (1) Investment decision support by limiting risk through test innovative approaches and algorithms, (2) Solutions for AI application in smart warehouse development, and (3) Solution to develop AI resources for smart warehouses. The potential future research would address the effectiveness of the proposed conceptual framework to real-life applications of smart warehouses through advanced AI resources and technology. Moreover, the current study has tremendous potential to develop new algorithms in AI, machine learning, and deep learning towards smart warehouses in the future.

Acknowledgement. This research is financially supported by Eastern International University, Binh Duong Province, Vietnam.

References

1. Kamali, A.: Smart warehouse vs traditional warehouse–review. CiiT Int. J. Autom. Auton. Syst. **11**(1), 9–16 (2019)
2. Tekinerdogan, B.: Engineering connected intelligence: a socio-technical perspective. Wageningen University, Wageningen, The Netherlands (2017)
3. Jabbar, S., Khan, M., Silva, B.N., Han, K.: A REST-based industrial web of things' framework for smart warehousing. J. Supercomput. **74**(9), 4419–4433 (2018)
4. Liu, X., Cao, J., Yang, Y., Jiang, S.: CPS-based smart warehouse for industry 4.0: a survey of the underlying technologies. Computers **7**(1), 13 (2018)
5. Papcun, P., et al.: Augmented reality for humans-robots interaction in dynamic slotting "chaotic storage" smart warehouses. In: Ameri, F., Stecke, K.E., von Cieminski, G., Kiritsis, D. (eds.) APMS 2019. IAICT, vol. 566, pp. 633–641. Springer, Cham (2019). https://doi.org/10.1007/978-3-030-30000-5_77

6. Cogo, E., Žunić, E., Beširević, A., Delalić, S., Hodžić, K.: Position based visualization of real world warehouse data in a smart warehouse management system. In: 2020 19th International Symposium INFOTEH-JAHORINA (INFOTEH), pp. 1–6 (2020)
7. Copeland, B.J., Proudfoot, D.: Artificial intelligence. In: Philosophy of Psychology and Cognitive Science, pp. 429–482. Elsevier (2007)
8. Capgemini Research Institute. www.capgemini.com
9. Ngoc, B.: Warehousing continues to rise unabated in local market Warehousing continues to rise unabated in local market. Vietnam Investment Rev. (2021)
10. Savills. https://industrial.savills.com.vn/2021/07/industrial-and-logistics-real-estate
11. Savills. https://www.savills.com/research_articles/255800/187576-1
12. Prime Minister: The Decision number 703/QD-Ttg: Building a competitive transport market in the direction of developing multimodal transport, connecting between different forms of transport, focusing on technology application. Information to minimize transportation costs (2019)
13. Ministry of Industry and Trade: Vietnam logistics report 2019: Logistics enhances the value of agricultural products (2019)
14. Ministry of Industry and Trade: Vietnam logistics report 2020: Reduced logistics costs (2020)
15. Shoham, Y., et al.: The AI Index 2018 Annual Report. AI Index Steering Committee, Human-Centered AI Initiative, Stanford University, Stanford, CA (2018)
16. Zhang, D., et al: The AI Index 2021 Annual Report. AI Index Steering Committee, Human-Centered AI Institute, Stanford University, Stanford, CA (2021)
17. National Strategy for the 4th Industrial Revolution. www.most.gov.vn
18. Phu My 3 SIP. http://www.phumy3sip.com/media-center/general-new/trends-in-vietnams-logistics-and-warehousing-market-refrigerated-warehouse
19. VietnamWorks InTECH. https://intech.vietnamworks.com/article/vietnamworks-cong-bo-bao-cao-thi-truong-nhan-luc-nganh-cong-nghe-thong-tin-thap-nien-2010-va-nam-2020-ra-mat-thuong-hieu-vietnamworks-intech
20. Official Journal of the European Union. https://eur-lex.europa.eu/legal-content/EN/TXT/HTML/?uri=CELEX:32009L0125&from=EN
21. Smartlog. https://gosmartlog.com/wp-content/uploads/2019/12/Bao-cao-logistics-viet-nam-2019.pdf
22. Siagian, H., Tarigan, Z.J.H., Jie, F.: Supply chain integration enables resilience, flexibility, and innovation to improve business performance in COVID-19 era. Sustainability **13**(9), 4669 (2021)
23. Boynton, P.: Measurement challenges and opportunities for developing smart grid testbeds. In: 10th Carnegie Mellon Conference on the Electricity Industry (2015)
24. Falkenberg, R., et al.: PhyNetLab: an IoT-based warehouse testbed. In: the Federated Conference on Computer Science and Information Systems, pp. 1051–1055 (2017)
25. Ridolfi, M., Macoir, N., Gerwen, J. V.-V., Rossey, J., Hoebeke, J., Poorter, E.D.: Testbed for warehouse automation experiments using mobile AGVs and drones. In: IEEE Conference on Computer Communications Workshops (INFOCOM WKSHPS), IEEE INFOCOM 2019 (2019)
26. Monostori, L.: Cyber-physical production systems: roots expectations and R&D challenges. Procedia CIRP **17**, 9–13 (2014)
27. Hu, H., Jia, X., He, Q., Fu, S., Liu, K.: Deep reinforcement learning based AGVs real-time scheduling with mixed rule for flexible shop floor in industry 4.0. Comput. Ind. Eng. **149**, 106749 (2020)
28. Salunkhea, O., Gopalakrishnana, M., Skoogha, A., Fasth-Berglund, Å.: Cyber-physical production testbed: literature review and concept development. In: 8th Swedish Production Symposium, SPS 2018, pp. 16–18 (2018)

29. Opalic, S.M., et al.: ANN modelling of CO_2 refrigerant cooling system COP in a smart warehouse. J. Cleaner Prod. **260**, 120887 (2020)
30. Li, M.P., Ganguly, A., Sankaran, P., Kwasinski, A., Kuhl, M.E., Ptucha, R.: Simulation analysis of a deep reinforcement learning approach for task selection by autonomous material handling vehicles. In: the 2018 Winter Simulation Conference, pp. 1073–1083 (2018)
31. Kamoshida, R., Kazama, Y.: Acquisition of automated guided vehicle route planning policy using deep reinforcement learning. In: 6th IEEE International Conference on Advanced Logistics and Transport (ICALT) (2017)
32. Hillebranda, M., Lakhania, M., Dumitrescu, R.: A design methodology for deep reinforcement learning in autonomous system. Procedia Manuf. **52**, 266–271 (2020)
33. Andersen, P.A., Goodwin, M., Granmo, O.C.: Towards safe reinforcement-learning in industrial grid-warehousing. Inf. Sci. **537**, 467–484 (2020)
34. Dan Zhang, L.G., Pee, L.C.: Artificial intelligence in e-commerce fulfillment: a case study of resource orchestration at Alibaba's Smart Warehouse. Int. J. Inf. Manage. **57**, 102304 (2021)

Long-Short Term Memory Model with Univariate Input for Forecasting Individual Household Electricity Consumption

Kuo-Chi Chang[1,3,6](✉), Elias Turatsinze[2](✉), Jishi Zheng[2], Fu-Hsiang Chang[4], Hsiao-Chuan Wang[5], and Governor David Kwabena Amesimenu[3]

[1] Department of Applied Intelligent Mechanical and Electrical Engineering, Yu Da University of Science and Technology, Miaoli County 361, Taiwan (R.O.C.)
albertchangxuite@gmail.com

[2] Fujian Province Key Laboratory of Automotive Electronics and Electric Drive, College of Transportation, Fujian University of Technology, Fuzhou 350118, China
tsinzelias10@yahoo.com

[3] Fujian Provincial Key Laboratory of Big Data Mining and Application, Fujian University of Science and Technology, Fuzhou, China

[4] Department of Tourism, Shih-Hsin University, Taipei City 116, Taiwan (R.O.C.)

[5] Institute of Environmental Engineering, National Taiwan University, Taipei 10617, Taiwan (R.O.C.)

[6] Department of Business Administration, North Borneo University College, 88400 Kota Kinabalu, Sabah, Malaysia

Abstract. Power load forecasting is becoming the key role in nowadays power distribution networks to understand the behavior of the electrical power systems where end users can predict the future trends of electricity usage and hence manage their residential appliances to reduce high electrical bills. Some researchers used traditional methods that cannot handle the problem of large-scale nonlinear time series data. The main objective of this paper is to conduct various optimized deep learning models for power load forecasting and presents the best performing model with high accuracy and the lowest possible Mean-Square Error by fine-tuning the parameters to achieve the best possible configurations of the model. Using hybrid Recurrent Neural Network techniques, optimal results have been achieved. The research was conducted on Long-short term memory, Long-short term memory encoder-decoder, Convolutional Neural Network Long-short term memory, Gated Recurrent Units, Convolutional Long-short term memory, and Bidirectional Long-short term memory. The training process was conducted using Google Colaboratory and the results showed that Convolutional Long-short term memory has the lowest Root-Mean Square Error among the 6 Long-short term memory networks time series forecasting models. The prediction plot of prediction and real values against epochs are almost inline.

Keywords: Power consumption · Long-short term memory · Forecasting · Deep learning · Electrical load

© The Author(s), under exclusive license to Springer Nature Switzerland AG 2022
A. E. Hassanien et al. (Eds.): AMLTA 2022, LNDECT 113, pp. 125–136, 2022.
https://doi.org/10.1007/978-3-031-03918-8_12

1 Introduction

The application of artificial intelligence in smart grids has gained extensive attention especially in monitoring and managing the electrical loads for sustainable stability of the power system [1]. Electrical energy cannot be stored in large quantities and the power system is exponentially expanding on the distribution networks due to the increased number of users [2]. Load forecasting helps utility companies and users to decide on electricity usage by reducing the economic consequences. Excessive power consumption is increasingly affecting the distribution networks due to the growth of the human population in the world with the advancement in higher technology and high electricity-dependent machinery [3]. Different researchers are putting more effort into how to manage and forecast future power consumption through the use of deep learning models. LSTM, encoder-decoder LSTM, CNN-LSTM encoder-decoder, ConvLSTM encoder-decoder, GRU, and Bi-LSTM are detailed in this work. Deep learning models try to better predict the future trends of energy use than the machine learning models and other traditional methods. Good prediction performance can be achieved by proposing the incorporation of GA to determine the number of LSTM units and window size [4, 5]. Electricity demand is increasing proportionally to the population growth and the evolution of the advanced technology where we are entering in the era of the internet of things that need to connect every single thing into the network and of course this is causing the huge use of electricity everywhere [2]. We need to make a future prediction to keep the balance on both sides [6]. The distributed generation being incorporated into the current power system can cause instability and unpredicted failures or blackout. To avoid this scenario, the distributed power generated need to be forecasted [7]. The rest of this paper is organized as follows: related works, deep learning models for load forecasting, results, and discussion, the conclusion with future recommended works, list of abbreviations and references.

2 Related Works

Electrical power consumption has increased due to the exponential growth of the population worldwide and this scenario affects the overall power system stability [8]. Due to this challenge, [9] AI-enhanced energy saving by forecasting the future trend of electricity can help in building a smart grid by sustaining energy usage. Load forecasting is the key aspect of the residential energy management system and [10] used multiple linear regression for predicting the future usage of electrical power. We need to forecast the power load to ensure the safety and stability of the power system production scheduling process and [11] used LSTM recurrent neural network for load forecasting. LSTM is one of the deep learning techniques that is mostly used in electrical power consumption and has low RMSE compared to SVM [12]. Modern technical equipment for measuring energy consumption is ubiquitous, which helps to reduce energy consumption, develop engineering and statistical analysis methods to effectively plan, predict, and monitor the increasing load on the grid [13]. Due to the infinite backtracking window, older RNN variants may be limited in the long-term dependence of the learning data, so new variants of LSTM were studied [14]. Those variants of LSTM have achieved a good performance compared to the older RNN such as ConvLSTM in [15], Multi-Step Short-Term Memory in [9], ensemble LSTM neural networks in [16], CNN with BiLSTM in [17], and CNN-GRU in [18].

The evolution and expansion of smart meter infrastructure in power systems have shown the introduction and deployment of short-term energy forecasting for improving the efficiency of household power usage [19–21]. Strong nonlinear modeling capabilities of computational intelligence methods simplified price prediction [22, 23]. The distributed generation into power system made prominently complex and uncertain where power load forecasting need to be considered for effective power system operations [19]. In [24] proposed LSTM and FFNN for electrical load forecasting to keep the balance between electricity demand and power generation. Traditional methods like Auto-Regressive Integrated Moving Average (ARIMA) in [25] are also applied for forecasting and depending on the scenarios, this ARIMA model can have good performance once optimized. When it comes to the wholesale electricity market, prediction is highly recommended and it is shown in [26] where ANN, SVM regression, and ARIMA models are compared to find the best method for industrial power consumption prediction and results showed that the SVM regression based on RBF has low Maximal Absolute Percentage Error (MaxAPE) of 21%.

Reliability, stability, self-healing, and flexibility are the main characters of smart grid and accurate prediction is an important aspect in the process of achieving a complete smart grid by using the deep learning models like LSTM as in [27] to help national grids plan for future expansion of the power system and [28] used LSTM to predict the peak electrical load. Recurrent neural networks specifically LSTMs are commonly used in modeling problems of time series and Convolutional neural networks are mostly used for feature extraction and compression of the model for data with spatial structure [29]. The combination of LSTM with CNN can improve the accuracy and performance of the forecasting model [30]. In some scenarios, the Linear regression model performs well over LSTM as the results proved in [31]. The multi-layer GRU network simplifies the structure and computation time over LSTM but with relatively the same performance [32].

3 Deep Learning Models for Load Forecasting

3.1 LSTM and LSTM-ED Neural Networks

LSTM is a special type of RNN that has the function of remembering the characteristics of the data sequence [33]. The traditional LSTM directly outputs a vector sequence while LSTM-ED has two sub-models where the first one reads and encodes the input sequence and then the second one will read that encoded input sequence for each step to make a prediction. The key role of encoder-decoder networks is that they make the model more sensitive to mutations by incorporating some external factors. The cell captures and stores the data flow. Besides, the cell links a module of past information to the module of the present one and again to the next cell. The data in each cell can be processed, filtered, or added for the next cell. The gate is based on the sigmoidal neural network layer so that the unit can optionally make the data pass or stop. Each layer generates 0 and 1 to represent the amount of each segment of the data that must pass through each cell. More specifically, the estimated zero value means "don't allow to pass through"; However, the estimation of 1 means "allow to pass through". RNN processes the input data sequence by using the internal state of the input data, which becomes a disappearing gradient problem, and it has a major negative impact on the accuracy of the model. The improved version of RNN is LSTM, which overcomes the problem of gradient disappearance by using LSTM [34].

Where x_t is the input at the current timestamp, h_t are the outputs of the hidden layer, Ø is the sigmoid function, C_t is the cell state. W_i, W_f, $\hat{w}_\partial$ and W_c are the weights for the input, output, forget gate, and memory cells, while B_i, B_∂, B_f and B_C are the biases, ∂_t is the output gate, i_t is the input gate. LSTM involves 3 types of gates that control the state of each cell and are explained here below:

1. **Forget gate:** It tells the cell state which information needs to be outputted or ignored by multiplying to a position in the matrix with 0 or 1.
2. **Memory gate:** It selects and tries to remember past information seen so far in the network and forget irrelevant information through a sigmoid layer followed by a tanh (The Hyperbolic Tangent Function) layer.
3. **Output gate:** Determine what will be the output of each cell and the resulting value will depend on the cellular status and filtration and newly added information.

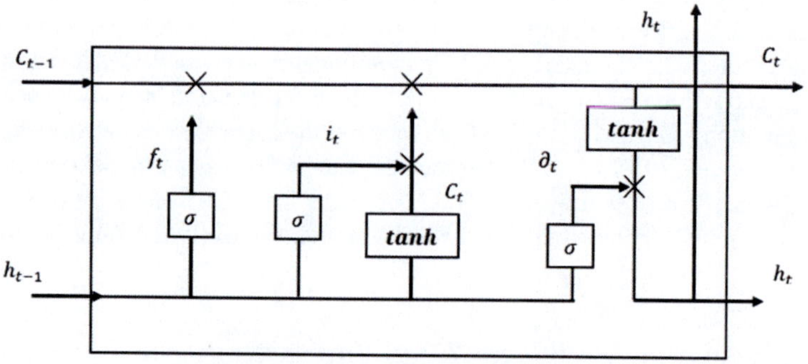

Fig. 1. Simplified architecture of LSTM

LSTM-ED is composed of an encoder and decoder where the input sequence needs to be encoded and decoded [35] (Fig. 1).

3.2 CNN-LSTM Neural Networks

A convolutional neural network can be used as an encoder to the LSTM model where this CNN automatically learns salient features and then LSTM decodes these features. Initial, The CNN model extracts features from the input data and then feeds it to the LSTM encoder to yield a Coding sequence. The encoded sequence is decoded by another subsequent LSTM decoder to advance it to the final dense layer for power consumption prediction [36]. CNN-LSTM refers to the addition of CNN layers before LSTM layers followed by a dense layer at the output as shown in the figure below (Fig. 2):

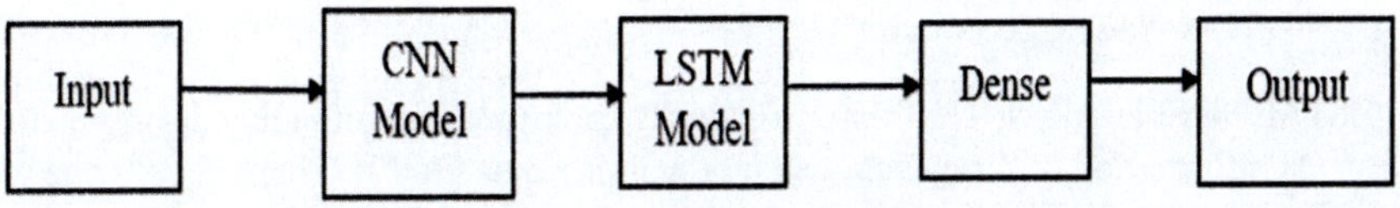

Fig. 2. CNN-LSTM architecture

CNN model is used in the preprocessing step by extracting important information and by mostly reorganizing the univariate input data to multi-dimensional batches [9]. In the second step, these reorganized batches become the input data to the LSTM units for prediction.

3.3 GRU Neural Networks

Gated recurrent units (GRU) is a new version of RNN almost similar to LSTM except that it only has two gates (reset gate and update gate) and the main difference with LSTM is that GRU has reduced the number of tensor operations that make GRU have high speed during the training process and less memory consumption [37]. GRU doesn't have an output gate and the update gate performs the works of input and forget gate of LSTM [38, 39].

Its architecture and governing formulas are here below (Fig. 3):

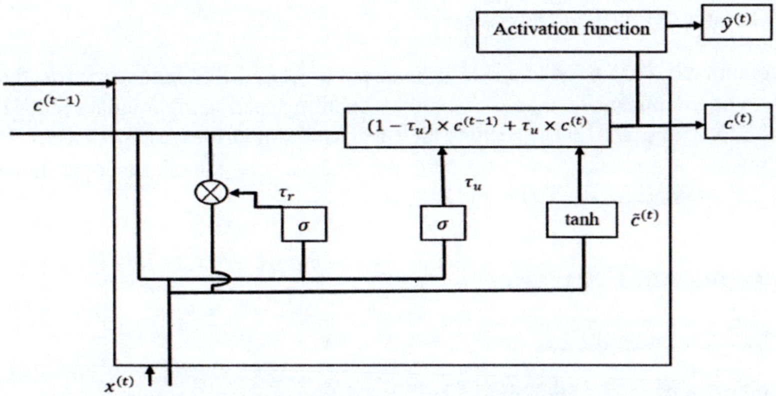

Fig. 3. Architecture of Gated Recurrent Unit (GRU)

$C(t)$ is the candidate activation and *bu*, *br*, and *bc*are the bias vectors. τ_u and τ_r are the update and reset gate respectively. When τ_r (ranging from -1 to 1) becomes very small it implies that it ignores more previous information and τ_u (ranging from 0 to 1) tending to be zero means that it retains more previous information.

3.4 BiLSTM Neural Networks

In LSTM, there is only LSTM that propagates backward, which makes it only get the previous information in series data when processing data [1, 40]. There is only one way in LSTM while in BiLSTM are two ways; the first one forwards the data while the second reverses the data so that the network can get the context information.

The reverse LSTM layer performs the same operation as that of the forward operation except that it is in the opposite direction for obtaining subsequent time data (Fig. 4).

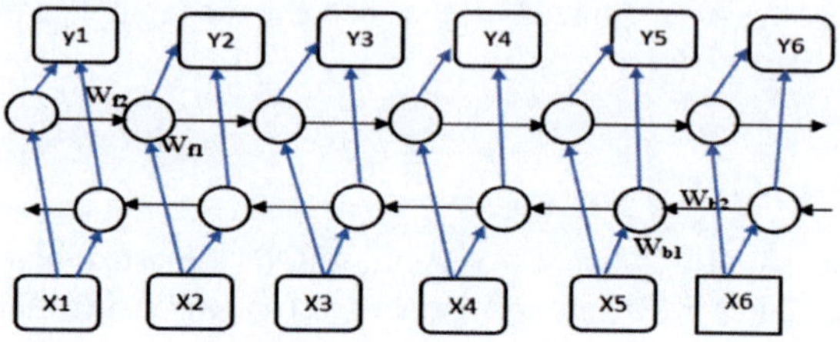

Fig. 4. Bi-LSTM architecture

h_f is the forward LSTM network output, h_b is the reverse LSTM network output, and y_i is the hidden layer final output.

3.5 ConvLSTM Neural Networks

Convolutional LSTM or convLSTM is a variant of CNN-LSTM that performs the convolution operations inside the LSTM cell by handling spatiotemporal data [20, 41–44]. ConvLSTM and CNN-LSTM seem to have the same similarities except that convLSTM has convolution embedded in its structure that make it uses convolution operation while CNN-LSTM concatenates CNN and LSTM.

4 Results and Discussion

4.1 Dataset Description

The objective of this research is to evaluate different deep learning models for load forecasting with general information about the usage of electricity in a given household.

The dataset contains 2075259 measurements. The data were collected from France in a house located in Sceaux (7 km from Paris) for 4 years (between December 2006 and November 2010) with one minute of sampling. The training process was conducted using the Google Colaboratory platform. The time series dataset is shown in Fig. 5.

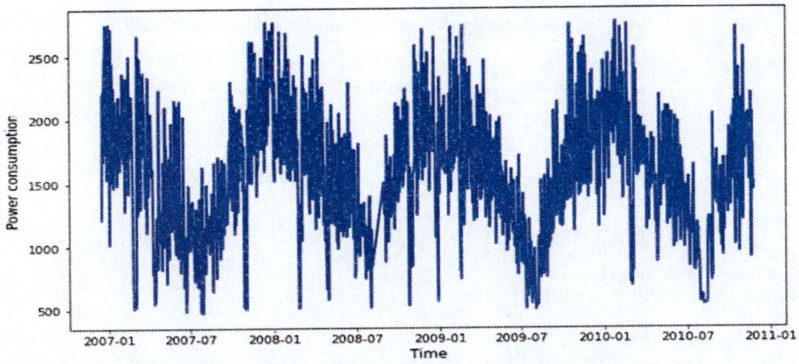

Fig. 5. Time series data for power consumption

4.2 Evaluation Metrics

Three types of evaluation metrics are mostly used for forecasting problems. The Root Mean Square Error (RMSE), Mean Square Error (MSE), and Mean Absolute Error (MAE) metrics. RMSE refers to the difference between the predictive value and the actual value. MSE refers to the average difference between the actual value of the model and the output value. Finally, MAE refers to the average absolute difference between prediction data and real data. In this paper, we have only used RMSE for comparing the models.

$$RMSE = \sqrt{\frac{1}{n}\sum_{1}^{n}(y - \hat{y})^2} \tag{1}$$

The comparison of LSTM, LSTM encoder-decoder, CNN-LSTM encoder-decoder, Gated Recurrent Units, Bi-directional LSTM, and Convolutional LSTM (convLSTM) are summarized in Table 1 according to the order of performance, and RMSE is adopted to be in the same unit (Kilowatt) as the active power and model with the lowest RMSE is the first out of the 6 models based on the experiment done in this paper.

Table 1. Evaluation metrics for different models

Model name	ConvLSTM	CNN-LSTM	LSTM-ED	BiLSTM	LSTM	GRU
RMSE	361.4	368.1	370.8	395.9	396.3	400.6

The experimental results showed that ConvLSTM is the first one out of the six models. The results can vary due to evaluation procedures and the stochastic nature of the model. This requires running the model several times to find the average RMSE. Variants of the LSTM network can accurately predict power consumption but encoder-decoder ConvLSTM seems to be the best one as is shown in Table 1. In Table 2, there is a comparison of forecasting techniques in terms of RMSE and it's clear that there is a big

Table 2. Comparing different techniques with our method

Model name	RMSE
ARIMA [42]	465.9 (Not scaled data)
RNN [20]	0.591
LSTM [20]	0.506
CNN-GRU [18]	0.31
Ours	0.29

difference between ConvLSTM and ARIMA model for the same dataset with RMSE of 361.4, and 465.9 respectively.

RNN has a big RMSE compared to its updated versions like LSTM or CNN-GRU. CNN-GRU is a hybrid of CNN and GRU (simplified LSTM) that has a high training speed. This paper was mainly focusing on the study of 6 LSTM time series models for forecasting individual household electricity consumption. As the experimental results showed, the RMSE of these models is in the range of 361.4 and 400.6 that is below the RMSE of ARIMA.

4.3 Prediction Results of ConvLSTM

Now that we have made comparison of the forecasting models and found that ConvLSTM outperformed all the models.

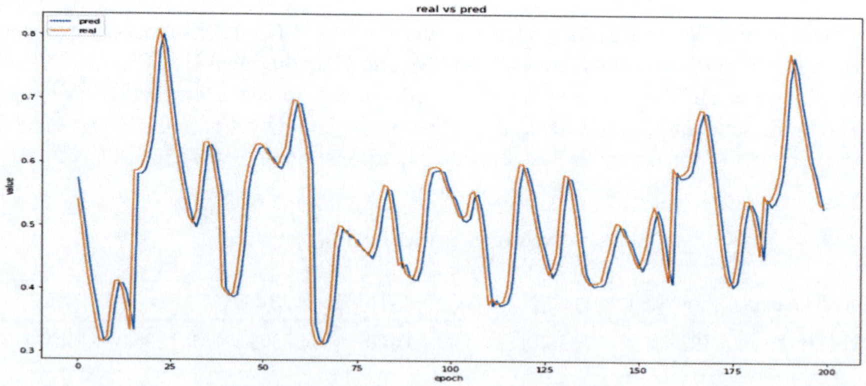

Fig. 6. Predicted values versus real values

The model R-square for training process is around 93.5% while for validation is 82%. The R-square cannot reach 100% because of different factors. Power consumption dataset used in this paper contains some missing data that is the major problem for not achieving the highest possible value of R-square. By plotting the prediction and real values against epochs as shown in Fig. 6, the results are acceptable because the actual results and the predicted results are almost inline.

4.4 Discussion of the Forecasting Models

LSTM is a type of recurrent neural network that has the function of remembering previous information. It has 3 gates (input, output, and forget gate). It is mostly used for large datasets but seems to not expose full memory and hidden layers and includes a memory cell that keeps information for a long time. GRU is almost similar to LSTM except that it has two gates (reset and update gate) that make it less complex and high training speed. It is mostly used for long sequence training samples and GRU can expose full memory and hidden layers. BiLSTM consists of two LSTMs that learn long-term dependencies that can be useful to learn from complete time series at each time step. It increases the amount of available data into the network and hence improves the performance of the algorithm. LSTM-ED is an RNN designed for sequence-to-sequence problems. It has two sub-models (encoder and decoder) where the first LSTM encoder reads the input sequence and encodes it and then the second LSTM decoder reads the encodes input sequence and outputs it to make one-step forecasting. CNN-LSTM is similar to the LSTM encoder-decoder except that the encoder here is the convolutional neural network (CNN) instead of LSTM where this CNN has to read the input sequence and capable of learning the salient features and then the LSTM decoder interpret output from CNN to make a prediction. ConvLSTM is the extension of the LSTM Encoder-decoder convolutional neural network that is used for spatial-temporal data. The only difference is that ConvLSTM has a simplified structure because CNN is embedded in the same architecture that makes it directly use convolutions for reading the input into the LSTM units and hence higher speed compared to CNN-LSTM. There are many factors that cannot allow the forecasting models to predict the exact values compared to the corresponding real values. One of these challenges is the missing information in the dataset where we have assumed the missing values to be the same as the corresponding previous datetime.

5 Conclusion and Future Work

This research proposed a hybrid of convolutional long-short term memory for dealing with univariate data for forecasting individual household power consumption. The research showed that this hybrid outperforms most of the current approaches such as ARIMA, support vector machine, traditional LSTM, CNN-LSTM, GRU, Bi-LSTM, LSTM-ED, and RNN. Modeling electrical power consumption in smart grids, residential buildings, and other power sectors is challenging work due to different variables that cannot easily be modeled like weather variables and the accompanying activities. Forecasting electrical power consumption primarily focuses on the total daily active power consumed for predicting the next couple of days or weeks. In this paper, six different variants of the LSTM network that are among the state-of-the-art forecasting algorithms nowadays have been detailed. The ConvLSTM encoder-decoder showed the lowest RMSE among them and concluded that it can be used for accurately forecasting the future trend of electricity. Further researches should be conducted for optimizing the ConvLSTM parameters using optimization algorithms such as GA, particle swarm optimization, and artificial bee colony algorithm and use data generated by IOT systems to improve the accuracy.

References

1. Du, J., Cheng, Y., Zhou, Q., Zhang, J., Zhang, X., Li, G.: Power load forecasting using BiLSTM-attention. IOP Conf. Ser. Earth Environ. Sci. **440**(3), 032115 (2020). https://doi.org/10.1088/1755-1315/440/3/032115
2. Turatsinze, E., et al.: Study of advanced power load management based on the low-cost internet of things and synchronous photovoltaic systems. In: Hassanien, A.E., Slowik, A., Snášel, V., El-Deeb, H., Tolba, F.M. (eds.) AISI 2020. AISC, vol. 1261, pp. 548–557. Springer, Cham (2021). https://doi.org/10.1007/978-3-030-58669-0_49
3. Ullah, F.U.M., Ullah, A., Haq, I.U., Rho, S., Baik, S.W.: Short-term prediction of residential power energy consumption via CNN and multi-layer Bi-directional LSTM networks. IEEE Access **8**, 123369–123380 (2020). https://doi.org/10.1109/ACCESS.2019.2963045
4. Chung, H., Shin, K.S.: Genetic algorithm-optimized long short-term memory network for stock market prediction. Sustainability **10**(10), 3765 (2018). https://doi.org/10.3390/su10103765
5. Lv, L., Kong, W., Qi, J., Zhang, J.: An improved long short-term memory neural network for stock forecast. MATEC Web Conf. **232**, 01024 (2018). https://doi.org/10.1051/matecconf/201823201024
6. Rahman, S., Alam, G.R.: Deep learning based ensemble method for household energy demand forecasting of smart home, pp. 18–20 (2019)
7. Pan, C., Tan, J.: Very short-term solar generation forecasting based on LSTM with temporal attention mechanism, pp. 267–271 (2019)
8. Kim, T.Y., Cho, S.B.: Predicting residential energy consumption using CNN-LSTM neural networks. Energy **182**, 72–81 (2019). https://doi.org/10.1016/j.energy.2019.05.230
9. Yan, K., Wang, X., Du, Y., Jin, N., Huang, H., Zhou, H.: Multi-step short-term power consumption forecasting with a hybrid deep learning strategy. Energies **11**(11), 1–15 (2018). https://doi.org/10.3390/en11113089
10. Peña-Guzmán, C., Rey, J.: Forecasting residential electric power consumption for Bogotá Colombia using regression models. Energy Rep. **6**, 561–566 (2020). https://doi.org/10.1016/j.egyr.2019.09.026
11. Elsworth, S., Güttel, S.: Time series forecasting using LSTM networks: a symbolic approach. arXiv, pp. 1–12 (2020)
12. Islam, M.R., Al Mamun, A., Sohel, M., Hossain, M.L., Uddin, M.M.: LSTM-based electrical load forecasting for Chattogram city of Bangladesh. In: 2020 International Conference on Emerging Smart Computing and Informatics (ESCI), ESCI 2020, pp. 188–192 (2020). https://doi.org/10.1109/ESCI48226.2020.9167536
13. Kalimoldayev, M., Drozdenko, A., Koplyk, I., Marinich, T., Abdildayeva, A., Zhukabayeva, T.: Analysis of modern approaches for the prediction of electric energy consumption. Open Eng. **10**(1), 350–361 (2020). https://doi.org/10.1515/eng-2020-0028
14. Lim, B., Zohren, S.: Time series forecasting with deep learning: a survey. arXiv (2020). https://doi.org/10.1098/rsta.2020.0209
15. Dairi, A., Harrou, F., Sun, Y., Khadraoui, S.: Short-term forecasting of photovoltaic solar power production using variational auto-encoder driven deep learning approach. Appl. Sci. **10**(23), 1–20 (2020). https://doi.org/10.3390/app10238400

16. Yan, K., Li, W., Ji, Z., Qi, M., Du, Y.: A hybrid LSTM neural network for energy consumption forecasting of individual households. IEEE Access **7**, 157633–157642 (2019). https://doi.org/10.1109/ACCESS.2019.2949065
17. Cnn, U., Le, T., Vo, M.T., Vo, B., Hwang, E., Rho, S.: Applied sciences improving electric energy consumption prediction
18. Sajjad, M., et al.: A novel CNN-GRU-based hybrid approach for short-term residential load forecasting. IEEE Access **8**, 143759–143768 (2020). https://doi.org/10.1109/ACCESS.2020.3009537
19. Ouyang, T., He, Y., Li, H., Sun, Z., Baek, S.: A deep learning framework for short-term power load forecasting. arXiv, pp. 1–8 (2017)
20. Chang, K.C., Chu, K.C., Wang, H.C., Lin, Y.C., Pan, J.S.: Agent-based middleware framework using distributed CPS for improving resource utilization in smart city. Future Gener. Comput. Syst. **108**, 445–453 (2020). https://doi.org/10.1016/j.future.2020.03.006
21. Chang, K.C., Chu, K.C., Wang, H.C., Lin, Y.C., Pan, J.S.: Energy saving technology of 5G base station based on internet of things collaborative control. IEEE Access **8**(1), 32935–32946 (2020). https://doi.org/10.1109/ACCESS.2020.2973648
22. Jiang, L., Hu, G.: Day-ahead price forecasting for electricity market using long-short term memory recurrent neural network. In: 2018 15th International Conference on Control, Automation, Robotics and Vision (ICARCV), pp. 949–954 (2018). https://doi.org/10.1109/ICARCV.2018.8581235
23. Chu, K.C., Horng, D.J., Chang, K.C.: Numerical optimization of the energy consumption for wireless sensor networks based on an improved Ant Colony Algorithm. IEEE Access **7**(1), 105562–105571 (2019). https://doi.org/10.1109/ACCESS.2019.2930408
24. Chandramitasari, W., Kurniawan, B., Fujimura, S.: Building deep neural network model for short term electricity consumption forecasting. In: 2018 International Symposium on Advanced Intelligent Informatics (SAIN), pp. 43–48 (2019). https://doi.org/10.1109/SAIN.2018.8673340
25. Taylor, J.W., McSharry, P.E.: Short-term load forecasting methods: an evaluation based on European data. IEEE Trans. Power Syst. **22**(4), 2213–2219 (2007). https://doi.org/10.1109/TPWRS.2007.907583
26. Babich, L., Svalov, D., Smirnov, A., Babich, M.: Industrial power consumption forecasting methods comparison. In: 2019 Ural Symposium on Biomedical Engineering, Radioelectronics and Information Technology (USBEREIT), vol. 2, pp. 307–309 (2019). https://doi.org/10.1109/USBEREIT.2019.8736640
27. Jiang, Q., Li, M.: Electricity power load forecast via long short-term memory recurrent neural networks, pp. 5–8 (2018). https://doi.org/10.1109/ICNISC.2018.00060
28. Sriwijaya, U., et al.: Peak load forecasting based on long short term memory, pp. 2019–2022 (2019)
29. Liu, P., Mo, R., Yang, J., Zhang, Y., Fu, X., Lan, P.: Forecasting using deep hybrid neural networks, pp. 159–164 (2019)
30. Farsi, B., Amayri, M., Bouguila, N., Eicker, U.: On short-term load forecasting using machine learning techniques and a novel parallel deep LSTM-CNN approach. IEEE Access **9**, 31191–31212 (2021). https://doi.org/10.1109/ACCESS.2021.3060290
31. Häring, T., Ahmadiahangar, R., Rosin, A., Korõtko, T.: Accuracy analysis of selected time series and machine learning methods for smart cities based on Estonian electricity consumption forecast, pp. 425–428 (2020)
32. Wang, Q., Lin, R., Zhao, Y., Zou, H.: Electricity consumption forecast based on empirical mode decomposition and gated recurrent unit hybrid model, pp. 1670–1674 (2020)
33. Gong, G., An, X., Mahato, N.K., Sun, S., Chen, S., Wen, Y.: Research on short-term load prediction based on Seq2seq model. Energies **12**(16), 1–18 (2019). https://doi.org/10.3390/en12163199

34. Hochreiter, S., Schmidhuber, J.: Long short-term memory. Neural Comput. **9**(8), 1735–1780 (1997). https://doi.org/10.1162/neco.1997.9.8.1735
35. Sutskever, I., Vinyals, O., Le, Q.V.: Sequence to sequence learning with neural networks. Adv. Neural Inf. Process. Syst. **4**, 3104–3112 (2014)
36. Khan, Z.A., Hussain, T., Ullah, A., Rho, S., Lee, M., Baik, S.W.: Towards efficient electricity forecasting in residential and commercial buildings: A novel hybrid CNN with a LSTM-AE based framework. Sensors (Switzerland) **20**(5), 1–16 (2020). https://doi.org/10.3390/s20051399
37. Khan, Z.A., Ullah, A., Ullah, W., Rho, S., Lee, M., Baik, S.W.: Electrical energy prediction in residential buildings for short-term horizons using hybrid deep learning strategy. Appl. Sci. **10**(23), 1–12 (2020). https://doi.org/10.3390/app10238634
38. Ribeiro, A.M.N.C., Do Carmo, P.R.X., Rodrigues, I.R., Sadok, D., Lynn, T., Endo, P.T.: Short-term firm-level energy-consumption forecasting for energy-intensive manufacturing: a comparison of machine learning and deep learning models. Algorithms **13**(11), 1–19 (2020). https://doi.org/10.3390/a13110274
39. Wu, L., Kong, C., Hao, X., Chen, W.: A short-term load forecasting method based on GRU-CNN hybrid neural network model. Math. Probl. Eng. **2020**, 1428104 (2020). https://doi.org/10.1155/2020/1428104
40. Wang, Y., Huang, M., Zhao, L., Zhu, X.: Attention-based LSTM for aspect-level sentiment classification. In: Proceedings of the 2016 Conference on Empirical Methods in Natural Language Processing, pp. 606–615 (2016). https://doi.org/10.18653/v1/d16-1058
41. Shi, X., Chen, Z., Wang, H.: Convolutional LSTM network: a machine learning approach for precipitation now casting. arXiv: 1506.04214v2 [cs. CV], pp. 1–12, 19 September 2015
42. Chang, K.C., Chu, K.C., Wang, H.C., Lin, Y.C., Pan, J.S.: Energy saving technology of 5G base station based on internet of things collaborative control. IEEE Access **8**, 32935–32946 (2020)
43. Chu, K.C., Chang, K.C., Wang, H.C., Lin, Y.C., Hsu, T.L.: Field-programmable gate array-based hardware design of optical fiber transducer integrated platform. J. Nanoelectron. Optoelectron. **15**(5), 663–671 (2020)
44. Chu, K.C., Horng, D.J., Chang, K.C.: Numerical optimization of the energy consumption for wireless sensor networks based on an improved ant colony algorithm. IEEE Access **7**, 105562–105571 (2019)

DNA-Binding-Proteins Identification Based on Hybrid Features Extraction from Hidden Markov Model

Sara Saber[1](✉), Uswah Khairuddin[2], and Rubiyah Yusof[2]

[1] Computer Engineering Department, Faculty of Engineering, Arab Academy of Science and Technology, Giza, Egypt
sarasaber@aast.edu

[2] Centre for Artificial Intelligence and Robotics, Malaysia-Japan International Institute of Technology, Universiti Teknologi Malaysia, Kuala Lumpur, Malaysia
{Uswah.kl,rubiyah.kl}@utm.my

Abstract. DNA-binding proteins (DNA-BPs) identification is very important for genome annotation as they have many applied and research applications in biological, bio-physical, and bio-chemical effects of antibiotics and steroids on DNA. This paper presents a proposed approach for DNA-BPs identification. This approach is based on hybrid features extraction and Hidden Markov Model (HMM). The protein sequence was encoded into a digital sequence, then divided the digital sequence into sub-frames. From the HMM profile, four features' groups were tested and compared each alone or combined; Amino Acid Composition (AAC), Auto Covariance Transformation (ACT), Ross-Covariance Transformation (CCT) and Best Tree Encoded (BTE). Random Forest (RF) classifier was used for matching the features between the enrolled features in the features database and the features of the tested sequence. The proposed approach has been tested using different four DNA-BPs datasets (PDB-1075, PDB-186, PDNA-543, and PDNA-316). The results show that the hybrid features extraction by combining all features gives the highest performance than using each features group alone. The accuracy of the proposed approach has been compared with the other available published identification methods for four tested DNA-BPs datasets. The results show that the proposed DNA-BPs identification approach provides higher accuracy than other published methods.

Keywords: DNA-BPs · HMM · AAC · BTE · Random Forest

1 Introduction

DNA is the cell blueprint which contains the main information which is coded for all the organisms. Thousands of proteins known as DNA-BPs are assisting DNA in performing its functions. DNA-BPs provide a variety of functions, including directing protein production, regulating cell development, and storing DNA in the nucleus. The structural composition of DNA is influenced by DNA-BPs. They also regulate and govern several biological activities like DNA transcription, replication, recombination, repair, and modification.

© The Author(s), under exclusive license to Springer Nature Switzerland AG 2022
A. E. Hassanien et al. (Eds.): AMLTA 2022, LNDECT 113, pp. 137–147, 2022.
https://doi.org/10.1007/978-3-031-03918-8_13

For identifying DNA-BPs, several experimental technical procedures can be applied, however, they are time-consuming and costly. Therefore, there is a critical need to replace existing experimental methods with suitable and efficient computational approaches. Several computational and statistical methods for identifying DNA-BPs have recently been presented as summarized in Table 1, but most of these methods are unable to offer invaluable knowledge for DNA-BPs identification.

Table 1. Summary of DNA-BP identification methods

Reference	Identification method basics	Dataset	Accuracy (%)
Qian1 et al. [1]	Multiple kernel learning-based on centered kernel alignment	PDB1075	84.19
		PDB186	83.7
Saber et al. [2]	Transfer Learning Models + RF	PDB1075	96.34
		PDB186	81.18
		PDNA-543	93.05
		PDNA-316	93.38
Li et al. [3]	Local features and long-term dependencies with primary sequences	PDB14189	82.81
Zhou et al. [4]	Long short-term memory and ensemble learning	PDNA-224	78.36
		DBP-123	80.51
		HOLO-83	79.19
Hu et al. [5]	Weighted Convolutional Features	Jackknife	86.84
Zou et al. [6]	Fuzzy twin support vector machines on self-representation	PDB1075	83.35
		PDB186	86.6
		PDB2272	78.79
Szilagyi and Skolnick [7]	RF and Gaussian naive Bayes	PDB186	76.90
Liu et al. [8]	Auto-cross covariance with ensemble learning	Authors	75.16
Qu et al. [9]	Mixed feature representation methods	PDB1075	77.43
		PDB186	81.58
Xu et al. [10]	PSSM with support vector machine (SVM)	PDB1075	79.96
		PDB186	79.96
Zhu et al. [11]	Position-specific scoring matrices (PSSM) and co-occurrence matrix	Yeast	97.06

(continued)

Table 1. *(continued)*

Reference	Identification method basics	Dataset	Accuracy (%)
		Human	98.95
		H. Pylori	89.69
Waris et al. [12]	PSSM with RF	Authors	92.3
Chowdhury [13]	Structural & evolutionary features + SVM	Jack-knife	90.18
Xu [14]	RF	Jack-knife	85.57
Zhang and Liu[15]	Position specific frequency matrix & distance-bigram transformation	PDB1075	81.02
		PDB186	80.65
Zhang et al. [16]	Evolutionary, structural, and physicochemical features	PDB186	80.9
Ma et al. [17]	Hybrid features and RF	Mainsett	89.56
Shen et al. [18]	Multi-scale local average blocks	PDNA-543	91.80
		PDNA-41	92.06
		PDNA-316	90.23
		PDNA-52	77.6
Krishna et al. [19]	Evolutionary features	DNA-Prot	81.83
		DNA binder	61.42
Gao et al. [20]	Threading based method	PDB186	59.7
Zhang et al. [21]	Bootstrap multiple CNN	PDNA-543	90.77
		PDNA-316	91.04

Because DNA-BPs have numerous interconnected physiological activities, their identification is regarded as a major challenge in genome annotation. DNA-BPs identification process means distinguishing the DNA-BPs (positive sample) from the non-DNA-BPs (negative sample), distinguishing the single-stranded DNA-BPs from the double-stranded DNA-BPs, or distinguishing the DNA-BPs from the Ribonucleic acid-binding proteins (RNA-BPs). Identification of DNA-BPs from the non-DNA-BPs is considered in this paper. DNA-BPs are proteins having DNA binding domains and interact with the major groove of B-DNA. Non-DNA-BPs are some structural proteins inside the chromosomes.

The main contribution of this paper combining and comparing the AAC, ACT, CCT and BTE features with HMM for DNA-BPs identification. The rest of the paper is organized as; the second section gives the materials and methods, the third section presents the results and discussion and the last section gives the concluding remarks.

2 Materials and Methods

The block diagram for identifying each DNA-BP sequence from the DNA-BPs dataset is shown in Fig. 1. Every protein sequence was encoded into digital sequence, then divided into frames, after that the features were extracted from each sequence and stored in features database. During testing, the same procedures were carried out and the RF classifier was used for matching the features between the enrolled features in the features database and the features of the tested sequence.

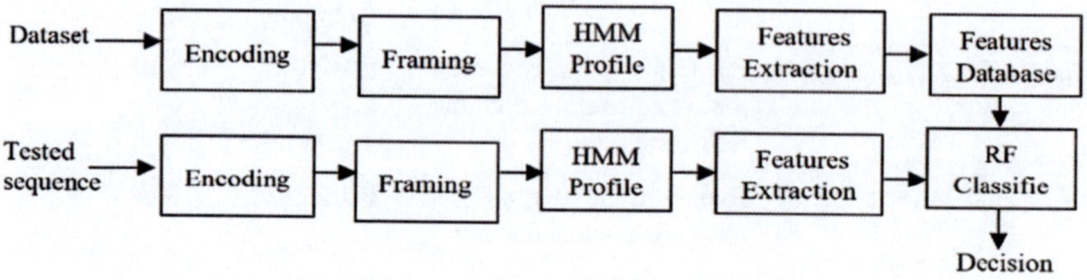

Fig. 1. Block diagram of the proposed DNA-BP identification approach

The steps of the proposed DNA-BP sequence will be presented with some details in the below subsections.

2.1 Datasets

The simulation carried out in this paper was implemented on the same DNA-BPs datasets used in [2] which are summarized in Table 2.

Table 2. Tested DNA-BPs datasets

Data set	PDB1075	PDB186	PDNA-543	PDNA-316
Source	Liu et al. [22]	Lou et al. [7]	Hu et al. [23]	Si et al. [24]
Total number of samples	1,075	186	144544	72718
Positive samples	525	93	9549	5609
Negative samples	550	93	134995	67109

2.2 Encoding

The DNA-BP sequence contains 20 amino acids represented using English capital letters. These letters were encoded into different binary numbers as shown in Table 3 to represents the different amino acids types such as:

MRGSAHVVIL →

01101000100100010000000010100110100101000101001010.

Table 3. DNA-BP sequence encoding

Amino acid	Letter	Binary number	Amino acid	Letter	Binary number
Alanine	A	00001	Leucine	L	01011
Arginine	R	00010	Lysine	K	01100
Asparagine	N	00011	Methionine	M	01101
Aspartic acid	D	00100	Phenylalanine	F	01110
Cysteine	C	00101	Proline	P	01111
Glutamine	Q	00110	Serine	S	10000
Glutamic acid	E	00111	Threonine	T	10001
Glycine	G	01000	Tryptophan	W	10010
Histidine	H	01001	Tyrosine	Y	10011
Isoleucine	I	01010	Valine	V	10100

2.3 Framing

The sequence lengths of the tested DNA-BPs datasets are different from 130 to 1350 letters, after encoding these numbers is multiplied by 5. The binary sequence is divided into the fixed size of frames to make it the stationary signal, the shorted sequence is completed with zero padding.

2.4 Hybrid Visual HMM Structure

HMMs are used for modeling the sequence of events in time. The model consists of a number of states, shown as circles in drawing as shown in Fig. 2.

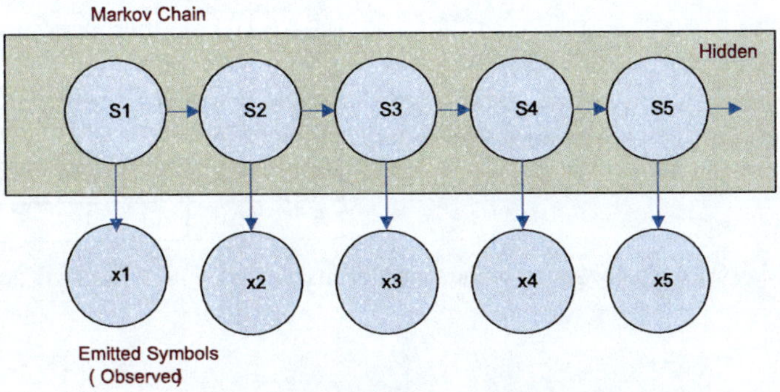

Fig. 2. HMM Graphic representation

HMM observations sequence is a piecewise stationary process, and it is a stochastic finite state automaton generating the observation string [25, 26]. Consider the sequence of observation vectors $O = \{O_1, \ldots\ldots, O_t, \ldots, O_T\}$. HMM consists of a number of N states $S = \{S_i\}$ and the observation string produced the result of calculating the vector O_t for each successive transition from the first state S_i to the last one S_j. Mathematically, HMM model can be represented as:

$$\lambda = [P_d.A.B] \tag{1}$$

where P_d is the probability distribution of the initial state. $A = \{A_{ij}\}$ is state transition probability between two state S_i and S_j, and $B = \{b_j(O_t)\}$ is the observation probability of emitting vector O_t at S_j.

The state statistical model is chosen to be *Gaussian Mixture Model* (GMM). It is considered hybrid model contains parametric and non-parametric density models. For parametric consideration, it has parameters and structure can be controlled the density behavior. For non-parametric consideration, it has many freedom degrees allowing the arbitrary density modeling. GMM density is weighted sum of Gaussian densities as:

$$P_{G,M}(x) = \sum W_m g(x, \mu_m, C_m) \tag{2}$$

where (m = 1…M) is the Gaussian component, M is the total number of Gaussian components. W_m represent the GMM weights. For K-dimensional densities, the K-dimensional Gaussian probability density function is:

$$g(x, \mu_m, C_m) = \frac{1 * e^{-1/2c_m^{-1}(x-\mu_m)(x-\mu_{m)}^T}}{(2\pi)^{k/2}|C_m|^{1/2}} \tag{3}$$

where μ_m is the mean vector, and C_m is the covariance matrix.

Figure 3 shows the used HMM graphic representations and transition matrix for the four classified classes.

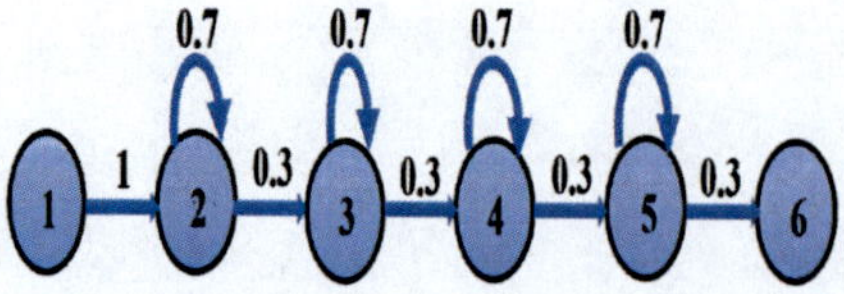

	1	2	3	4	5	6
1	0	1	0	0	0	0
2	0	0.7	0.3	0.1	0	0
3	0	0	0.7	0.3	0	0
4	0	0	0	0.7	0.3	0
5	0	0	0	0	0.7	0.3
6	0	0	0	0	0	0

Fig. 3. HMM representation of the transition matrix for the two classified classes.

2.5 Features Extraction

In this paper, four features groups were extracted; Amino Acid Composition (AAC), Auto Covariance Transformation (ACT), Ross-Covariance Transformation (CCT) and Best

Tree Encoded (BTE). AAC features can compute from the HMM profile, the following formula is used:

$$\overline{h_j} = \frac{1}{L}\sum_{i=1}^{L} h_{i,j} \quad (j = 1, 2, \cdots 20), \tag{4}$$

where L is the length of the protein sequence and $h_{i,j}$ represents the element at the ith row and j^{th} column of the HMM profile. In this way, 20 AAC features are obtained in total. ACT and CCT features are used to reflect the local sequence-order effect. These two techniques have been widely used to extract features from the HMM profile. Thus, in this work, ACT and CCT are also adopted to convert the HMM profile into two numerical vectors. BTE features are designed for normalizing the dynamic structure of best tree decomposition of the wavelet packets. After framing process, the frame is decomposed into wavelet packet tree through Wavelet Packet Decomposed (WPD) process, which consists of a number of levels each level contains many nodes. Then, Shannon entropy is calculated to obtain which best tree. After that, the images resulting from the entropy function that represents the best tree are normalized to grayscale images and divided into 9 parts. The Discrete Cosine Transform (DCT) is used to get the maximum two absolute values for each part of the image.

2.6 Classifier

The DNA-BP identification can is two classes (positive or negative) classification. One class is the positive DNA-BPs which means that, the DNA binding domains and interacts with the DNA. The second class is the negative non-DNA-BPs, which means that, the structural proteins may be found in the chromosomes. In this paper, the RF classifier presented by T. K. Ho [27] was used, where it gives good results in several approaches among the other classifiers [2, 13, 17]. RF consists of trees; each tree represents a subset of all possible attributes of the input features sequences. It constructs the decision ensemble in random trees based on the input features of the sequence, and the final decision is giving by voting about the results among the trees.

3 Results and Discussions

The proposed DNA-BP identification system is has been evaluated using four merits; accuracy, sensitivity, specificity which are percentages where the ideal classifier should give 100% of these three merits. The fourth merit is called Matthew's correlation coefficient (MCC) which has a range between -1 to $+1$; the ideal classifier should give $+$ 1 MCC. These four metrics were calculated using four numbers obtained by testing the proposed identification approach [28]. These numbers are True Positive (T_p), False Positive (F_p), True negative (T_n), and False negative (F_n). T_p number is the accumulative positive true results which mean that the system identifies the positive sample correctly (True). F_p number is the accumulative positive false results which mean that the system identifies the positive sample incorrect (False). T_n number is the accumulative negative true results which mean that the system identifies the negative sample correctly (True).

F_n number is the accumulative negative false results which mean that the system identifies the negative sample incorrect (False). The main four evaluation metrics can be calculated as:

$$\text{Accuracy} = \frac{T_P + T_N}{T_P + T_N + F_P + F_N} \times 100\% \tag{5}$$

$$\text{Sensitivity} = \frac{T_P}{T_P + F_N} \times 100\% \tag{6}$$

$$\text{Specificity} = \frac{T_N}{T_N + F_P} \times 100\% \tag{7}$$

$$\text{MCC} = \frac{(T_P T_N) - (F_P F_N)}{\sqrt{(T_P + F_P)(T_P + F_N)(T_N + F_P)(T_N + F_N)}} \tag{8}$$

The results of the proposed DNA-BPs identification approach for different features extraction methods are as shown in Table 4. The results show that, the hybrid features extraction by combining all features gives the highest performance than using each features group alone. This due to combining more features for DNA-BPs representations which increase the identification performance.

Table 4. Performance comparison between different features extraction method

Dataset	Features	Accuracy (%)	Sensitivity (%)	Specificity (%)	MCC
PDB1075	AAC	90.52	91.35	96.22	0.81
	ACT	93.61	93.87	95.83	0.89
	CCT	91.38	91.61	99.13	0.86
	BTE	96.74	95.97	98.93	0.92
	All features	**96.88**	**96.05**	**99.78**	**0.96**
PDB186	AAC	73.98	83.42	71.99	0.59
	ACT	79.51	88.22	75.33	0.67
	CCT	75.68	85.52	72.73	0.62
	BTE	80.60	91.49	76.30	0.71
	All features	**83.35**	**92.45**	**79.13**	**0.75**
PDNA-543	AAC	84.66	21.36	96.46	0.46
	ACT	88.73	34.92	98.78	0.53
	CCT	86.37	25.48	97.60	0.49
	BTE	91.83	49.35	97.44	0.64
	All features	**94.12**	**56.67**	**99.50**	**0.68**
PDNA-316	AAC	85.94	43.62	96.79	0.51

(continued)

Table 4. *(continued)*

Dataset	Features	Accuracy (%)	Sensitivity (%)	Specificity (%)	MCC
	ACT	89.59	53.08	98.68	0.64
	CCT	88.41	46.10	97.69	0.57
	BTE	92.66	65.58	99.68	0.69
	All features	**94.28**	**66.91**	**99.71**	**0.74**

Table 5 shows a comparison between the accuracy of the proposed DNA-BPs identification approach with the other available published identification methods for four tested DNA-BPs datasets. The results show that the proposed DNA-BPs identification approach provides higher accuracy than other published methods.

Table 5. Comparison of proposed approach with previous published methods

	PDB1075	PDB186	PDNA-543	PDNA-316
Saber et al. [2]	96.34	81.18	93.05	93.38
Qian1 et al. [1]	84.19	83.7	–	–
Zou et al. [6]	83.35	86.6	–	–
Zhang and Liu [15]	81.02	80.65	–	–
Qu et al. [9]	77.43	81.58	–	–
Xu et al. [10]	79.96	79.96	–	–
Szilagyi and Skolnick [7]	–	76.90	–	–
Zhang et al. [16]	–	80.9	–	–
Shen et al. [18]	–	–	91.80	90.23
Gao and Skolnick [20]	–	59.7	–	–
Zhang et al. [21]	–	–	–	91.04
Proposed approach	96.88	94.12	83.35	94.28

4 Conclusions

The paper presented an approach for DNA-BPs identification based on hybrid features extraction and HMM. The protein sequence was encoded into a digital sequence, then divided the digital sequence into sub-frames. From the HMM profile, four features groups were tested and compared; AAC, ACT, CCT and BTE. RF classifier was used for matching the features between the enrolled features in the features database and the features of the tested sequence. The proposed approach has been tested using different four DNA-BPs datasets (PDB-1075, PDB-186, PDNA-543, and PDNA-316). The results show that, the hybrid features extraction by combining all features gives the highest performance than using each features group alone. The accuracy of proposed approach has been compared with the other available published identification methods for four tested DNA-BPs datasets. The results show that the proposed DNA-BPs identification approach provides higher accuracy than other published methods.

References

1. Qian, Y., Jiang, L., Ding, Y., Tang, J., Guo, F.: A sequence based multiple Kernel model for identifying DNA binding proteins. BMC Bioinform. **22**, 291 (2021)
2. Saber, S., Khairuddin, U., Yusof, R., Madan, A.: DTLM-DBP: deep transfer learning models for DNA binding proteins identification. Comput. Mater. Continua **68**(3), 3563–3576 (2021)
3. Li, G., Du, X., Li, X., Zou, L., Zhang, G., Wu, Z.: Prediction of DNA binding proteins using local features and long-term dependencies with primary sequences based on deep learning. PeerJ **9**, e11262 (2021)
4. Zhou, J., Lu, Q., Xu, R., Gui, L., Wang, H.: EL_LSTM: prediction of DNA-binding residue from protein sequence by combining long short-term memory and ensemble learning. IEEE/ACM Trans. Comput. Biol. Bioinform. **17**(1), 124–135 (2020)
5. Hu, J., Rao, L., Zhu, Y.H., Zhang, G.J., Yu, D.J.: TargetDBP: enhancing the performance of identifying DNABinding proteins via weighted convolutional features. J. Chem. Inf. Model. **61**(1), 505–515 (2021)
6. Zou, Y., Ding, Y., Peng, L., Zou, Q.: FTWSVM SR: DNA binding proteins identification via fuzzy twin support vector machines on self representation. Interdisciplinary Sci. Comput. Life Sci. (2021)
7. Lin, C., et al.: Copy-move forgery detection using combined features and transitive matching. Multimedia Tools Appl. **78**(21), 30081–30096 (2018). https://doi.org/10.1007/s11042-018-6922-4
8. Park, C.-S., Choeh, J.Y.: Fast and robust copy-move forgery detection based on scale-space representation. Multimedia Tools Appl. **77**(13), 16795–16811 (2017). https://doi.org/10.1007/s11042-017-5248-y
9. Wang, X.-Y., Li, S., Liu, Y.-N., Niu, Y., Yang, H.-Y., Zhou, Z.: A new keypoint-based copy-move forgery detection for small smooth regions. Multimedia Tools Appl. **76**(22), 23353–23382 (2016). https://doi.org/10.1007/s11042-016-4140-5
10. Li, C., Ma, Q., Xiao, L., Zhang, A.: Image splicing detection based on Markov in QDCT domain. Neurocomputing **228**, 29–36 (2017)
11. Shen, X., Chen, H.: Splicing, image forgery detection using textural features based on the grey level co-occurrence matrices. IET Image Proc. **11**(1), 44–53 (2017)
12. Alahmadi, A.A., Hussain, M., Aboalsamh, H., Muhammad, G., Bebis, G.: Splicing image forgery detection based on DCT and LBP. In: Signal and Information Processing Conference. IEEE, Austin, TX, USA, pp. 253–256 (2013)
13. Jeronymo, D.C., Borges, Y.C.C., Coelho, L.S.: Image forgery detection by semi-automatic wavelet soft thresholding with error level analysis. Expert Syst. Appl. **85**, 348–356 (2017)
14. Muhammad, G., Hussain, M., Bebis, G.: Passive copy move image forgery detection using un-decimated dyadic wavelet transform. Digit. Investig. **9**, 49–57 (2012)
15. Isaac, M.M., Wilscy, M.: Image forgery detection based on wavelets and local phase quantization. Procedia Comput. Sci. **58**, 76–83 (2015)
16. Oommen, R.S., Jayamohan, M., Sruthy, S.: Using fractal, dimension and SVD for image forgery detection and localization. Procedia Technol. **24**, 1452–1459 (2016)

17. Al-Hammadi, M.H., Muhammad, G., Hussain, M., Bebis, G.: Curvelet transform and local texture based image forgery detection. In: Bebis, G., et al. (eds.) ISVC 2013. LNCS, vol. 8034, pp. 503–512. Springer, Heidelberg (2013).https://doi.org/10.1007/978-3-642-41939-3_49
18. Hayat, K., Qazi, T.: Forgery detection in digital images via discrete wavelet and discrete cosine transforms. Comput. Electr. Eng. **62**, 448–458 (2017)
19. Zhao, J., Guo, J.: Passive forensics for copy-move image forgery using a method based on DCT and SVD. Forensic Sci. **233**, 158–166 (2013)
20. Saleh, S.Q., Hussain, M., Muhammad, G., Bebis, G.: Evaluation of image forgery detection using multi-scale weber local descriptors. In: Advances in Visual Computing. ISVC 2013, Rethymnon, Crete, Greece. LNCS, vol. 8034, pp. 416–424. Springer, Heidelberg (2013). https://doi.org/10.1007/978-3-642-41939-3_40
21. Abdel-Basset, M., Manogaran, G., Fakhry, A.E., El-Henawy, I.: 2-Levels of clustering strategy to detect and locate copy-move forgery in digital images. Multimedia Tools Appl. **79**(7–8), 5419–5437 (2018). https://doi.org/10.1007/s11042-018-6266-0
22. Li, X., Sun, X., Liu, Q.: Image integrity authentication, scheme based on fixed point theory. IEEE Trans. Image Process. **24**(2), 632–645 (2015)
23. Zhu, Y., Shen, X., Chen, H.: Copy-move forgery detection based on scaled ORB. Multimedia Tools Appl. **75**(6), 3221–3233 (2015). https://doi.org/10.1007/s11042-014-2431-2
24. Rao, Y., Ni, J.: A deep learning approach to detection of splicing and copy-move forgeries in images. In: IEEE International Workshop on Information Forensics and Security (2016)
25. Kasban, H.: Fingerprints verification based on their spectrum. Neurocomputing **171**, 910–920 (2016)
26. Abozaid, A., Haggag, A., Kasban, H., Eltokhy, M.: Multimodal biometric scheme for human authentication technique based on voice and face recognition fusion. Multimedia Tools Appl. **78**(12), 16345–16361 (2018). https://doi.org/10.1007/s11042-018-7012-3
27. Hu, W.-C., Chen, W.-H., Huang, D.-Y., Yang, C.-Y.: Effective image forgery detection of tampered foreground or background image based on image watermarking and alpha mattes. Multimedia Tools Appl. **75**(6), 3495–3516 (2015). https://doi.org/10.1007/s11042-015-2449-0
28. Kasban, H., Nassar, S.: An efficient approach for forgery detection in digital images using Hilbert-Huang transform. Appl. Soft Comput. **97**, 106728 (2021)

Machine Learning Based Mobile Applications for Cardiovascular Diseases (CVDs)

Heba Y. M. Soliman[1(✉)], Mohamed Imam[2], and Heba M. Abdelatty[1]

[1] Port Said University, Port Said, Egypt
hebayms@eng.psu.edu.eg, heba.atty@gmail.com
[2] National Heart Institute, Cairo, Egypt
memam@msn.com

Abstract. Since Cardiovascular Diseases CVDs are the leading cause of mortality in the globe, health organizations aim to lower the number of people who die prematurely because of. m-Health is a popular type of digital health that depends mainly on mobile apps and has the potential to expand broadly in each hand. The mobile app development landscape has been changed by machine learning. In this paper, the importance, main characteristics, progress in publications and technical requirements of the most highly rated ML based CVDs mobile apps are presented, which could give clear guidance to developers in this area.

Keywords: CVDs · m-Health · ML

1 Introduction

Cardiovascular diseases CVDs are the largest cause of death worldwide, accounting for around 18 million deaths each year. More than 80% of the deaths previously mentioned occur in low- and middle-income nations due to inadequacies in preventive guiding legislation implementation. In high-income nations, however, ischemic heart disease has become the leading cause of mortality [1, 2]. The focus of all international efforts should be on prevention. Following the World Health Organization's WHO global action plan for non-communicable diseases [3, 4] the World Heart Federation WHF aims to reduce premature mortality attributable to CVD by 25% by 2025.

Cerebrovascular disease, hypertensive heart disease, ischemic heart disease, peripheral vascular disease, rheumatic heart disease, cardiomyopathies, arrhythmias, and other disorders are all included under the term CVD. CVDs are linked to a variety of risk factors, including behavioral ones like physical activity, poor nutrition, and smoking, as well as metabolic ones like diabetes, hypertension, dyslipidemia, and obesity [4].

Digital health (DH) is defined by the World Health Organization (WHO) as the use of information and communication technology to diagnose patients, track illnesses, teach healthcare workers, and monitor public health [5, 6]. There are several domains within digital health, including: (i) Mobile applications (m-Health) for patient screening. (ii) Telemedicine using wearable tools and sensors that might be used in conjunction with m-Health to track biological signals. (iii) Electronic Health Records (EHRs) which are

© The Author(s), under exclusive license to Springer Nature Switzerland AG 2022
A. E. Hassanien et al. (Eds.): AMLTA 2022, LNDECT 113, pp. 148–156, 2022.
https://doi.org/10.1007/978-3-031-03918-8_14

information databases that aid in the monitoring of healthcare practices and decision-making. (iv) Applications of Artificial Intelligence AI for autonomous medical data processing [7]. Digital health-based care, which is based on recent breakthroughs in information and communications technology, allows for considerable improvements after diagnosis.

Furthermore, with solutions such as mobile applications and monitoring devices for self-tracking and improving patient lifestyles, digital health might play an active part in primary prevention [8]. Mobile app developers rely on machine learning (ML) to construct strong algorithms that can analyze human behavior and aid users. Wearable device connectivity, cloud-based apps for better storage, and advanced cognitive interfaces are the primary elements predicted in CVDs mobile apps in the near future.

In this paper, ML based CVDs m-Health applications' encouraging factors, features, obstacles and progress in publications are presented in Sect. 2. In Sect. 3, a compact comparison between the highly rated CVDs mobile apps is clarified and tabulated. The technical requirements needed for more efficient future CVDs mobile apps are discussed in Sect. 4. And the paper's conclusion is clarified in Sect. 5.

2 ML Based m-Health for CVDs

Due to several factors, a larger deployment and use of digital health services is expected soon. These factors are: (i) increased life expectancy, which results in a larger cohort of chronic patients of older age, with associated reduced or impeded mobility; (ii) shortage of physicians and nurses to meet the future health needs of ageing population, despite an increase in health workers in the past decade; (iii) larger areas covered by wireless network connectivity, providing access to digital health resources and making their use more convenient, especially in rural and underpopulated areas [9].

The application of digital health technology in cardiovascular diseases (i.e., digital cardiology) dates to the late twentieth century, when mobile and internet connection was expanding. Initially, m-Health in cardiology focused on telephone-based call and text messaging strategies to promote medication adherence, lifestyle improvements, and smoking cessation. Smartphone-based technologies, machine learning, and low-cost, pocket-sized technologies, on the other hand, are being exploited at exponential rates during the technological revolution. While not all smartphone apps and digital health technologies have been proved to impact clinical change, the fast advancement of technology and big data is leading to more precise and creative methods for CVD prevention and treatment [10].

Accordingly, the market demand is being driven by increased public awareness of the benefits of healthcare management, the growing requirement for point-of-care diagnosis and treatment, and the growing focus on customized medicine. Rapid technological innovation in medical apps research creates both possibilities and concerns, such as the collecting of enormous volumes of data while maintaining the privacy and security of participants' data. However, the market's expansion will be limited by the market's strict rules. The global market for mobile medical apps was worth USD 3.65 billion in 2019 and is expected to grow to USD 17.61 billion by 2027. In 2019, the cardiology category had the biggest market share of 32.5% [11].

Despite all the positive characteristics listed above, there remain still obstacles to large-scale digital health adoption in cardiology. These impediments are either connected to the patient, the physician, or interoperability and technical difficulties.

Patient-related constraints arise from the fact that digital health innovations are currently mostly driven by technological considerations rather than patient wants and expectations. There is a dearth of patient engagement in design and co-creation. Other major challenges for physicians not to embrace this unique care delivery model include a lack of infrastructure, clarity in regulation and standards, incentives, expertise, and training among specialists in digital health instruments [12].

Accessibility of gathered data for visualization and analysis, as well as possible integration into the patient electronic health record for optimal use in the therapeutic process, remains an ongoing topic in terms of interoperability.

By managing calendars and notifying events, tracking typical actions, storing data, correcting spelling, responding to voice searches, providing lists of related outcomes, assisting digital marketing, assisting with intelligence gathering, and facilitating data accuracy and decision making, machine learning transforms the way people use mobile apps.

As a consequence, machine learning (ML) assists developers in the development of strong apps through data filtration, algorithmic training, model selection, parameter tuning, and prediction. Through character recognition, Natural Language Processing NLP, and predictive analysis, ML bridges the gap between knowing user behavior and

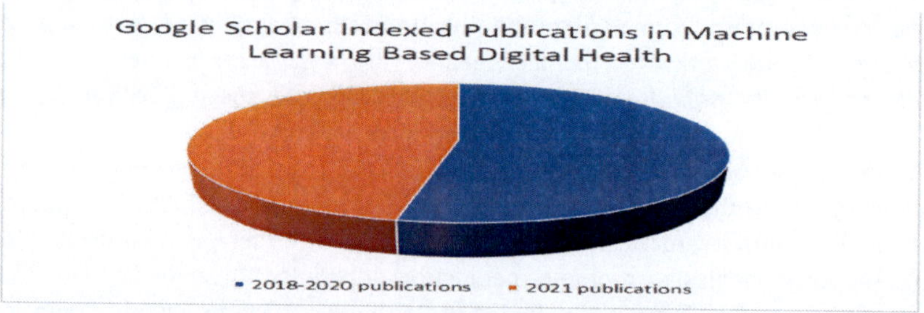

(a)

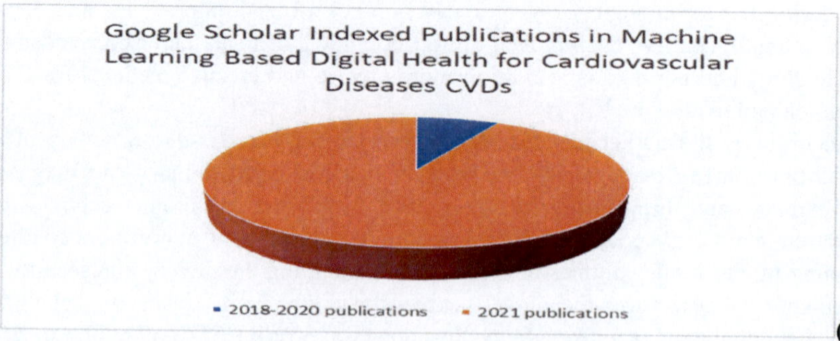

(b)

Fig. 1. Growth of number of Google Scholar indexed ML based Publications. (a) Digital Health. (b) Cardiovascular Disease CVDs.

using it to produce a personalized solution. As a result, it is expected that future apps will be more in line with the requirements and expectations of patients, physicians and service providers.

According to the Google scholar database, there has been a significant growth in the number of Google scholar indexed papers related to machine learning-based digital health. From 2021 till present, the number of publications (46800 publications) has been almost comparable to the total number of publications in 2018, 2019, and 2020 (52800 publications) [13].

The rise is greater for cardiovascular diseases, or CVDs. The number of publications in 2021 (16200 publications) is twelve times more than the total number of publications in 2018, 2019, and 2020 (1340 publications), Fig. 1. Accordingly, more work is encouraged in that field.

3 Characteristics of the Commercially Available CVDs Mobile Applications

In 2005, professor Brian Woodward directed the construction of one of the first mobile health systems. It was a one-of-a-kind device that allowed clinicians to watch patients' health and communicate data (such as blood oxygen saturation, blood pressure, glucose levels, and heartbeats) from a mobile phone to any hospital or clinic anywhere on the planet. The Apple App Store and Google play now contain thousands of healthcare applications [5].

The characteristics of CVDs mobile applications vary. Some of them rely on sensors and wearable devices, while others rely on data entry of the previously measured medical measures. Some can interpret and analyze the data entered, while others can merely save it for later use. The capability to share data between patients and service providers is also a key element.

Sensor-based apps are wirelessly connected to some cardiovascular health equipment in order to assess blood pressure, heart rate, heart rate recovery, heart rate variability, oxygen saturation, or to identify Atrial Fibrillation AF.

Sensor connectivity isn't required for all applications. Some merely serve as databases, sorting, storing, and analyzing data. Medical data is input numerically or uploaded in various file formats, then examined and processed to provide decisions or analysis graphs. For future analysis and processing, the data entered is stored on web servers or on the cloud.

The performance of a group of applications classified as of the best applications of the year 2020 in the United States of America USA has been analyzed according to the assessment of the Healthline platform, which evaluates medical services of various types in USA. The performance was evaluated by identifying the specifications available in these applications and determining their relative advantages and distinctive characteristics.

The applications chosen are Instant Heart Rate, Pulse Point Respond, Blood Pressure Monitor, Cardiio, Blood Pressure Companion, Kardia, Qardio, Fibricheck, Cardiac Diagnosis and Blood Pressure Tracker [5, 7].

Beside the main features listed above, the unique characteristics of each application that may distinguish it from the rest of the evaluated applications have been clarified in Table 1.

From the point of view of connectivity to external sensors, some applications rely on the embedded mobile camera as a sensor and direct intense light to measure the heart rate such as Instant Heart Rate, Cardio and Cardiac Diagnosis. Face detection technology is a choice for heart rate monitoring in the highly rated app Cardiio. Qardio app could be considered as a complete Internet of Things IoT platform. A set of external peripherals and sensors could be engaged for blood pressure measurement and Electrocardiography ECG.

Data processing and Analysis capability is an important measure to distinguish between the different applications. Kardia and Fibricheck give only an indication of normal or abnormal BP dependent on the measured value and they also classify the heart rate measured values into regular or irregular categories.

Cardiac Diagnosis classifies the measured values into three categories instead of two; normal, caution and risk. The decisions taken upon data processing and analysis could be shared with physicians and service providers either by exporting data in Excel or Pdf formats as done in Blood Pressure Tracker, Kardia, Blood Pressure Monitor and Pulse Point Respond.

In addition, some applications have distinctive characteristics that are not subject to the basic categories that we have previously exposed to. Instant Heart Rate gives the user a space to write notes about what is the user doing at time of the test. Some applications provide GPS and Geo-tracking services such as Pulse Point Respond and Qardio. Blood pressure Companion is one of few applications that deal with many patients' profiles at the same time. Instead of using a single measurement to take a decision, Qardio takes the decision according to the average value of three measurements.

Despite the diversity of these characteristics and their variations in various apps to give the user a variety of options, there are several more elements that must be considered in future applications to make them more acceptable for patients and health care professionals [8, 12, 14].

Table 1. Characteristics of the evaluated applications.

m-Health application	Sensor based	Data entry based	Data analysis and processing	Data storage	Data sharing	Other features
Instant Heart Rate	✓	✓		✓		– Phone's camera as a heart rate monitor – A note space to track what is the user doing at time of the test
Pulse Point Respond		✓			✓	– Connects patient with CPR Community – GPS based service

(continued)

Table 1. *(continued)*

m-Health application	Sensor based	Data entry based	Data analysis and processing	Data storage	Data sharing	Other features
Blood Pressure Monitor		√	√	√	√	– Export to email for the healthcare team
Cardii0	√			√		– Face detection technology to read heart rate – Finger sensor option (Photoplethysmography)
Blood Pressure Companion		√	√		√	– Advantage of dealing with many patients' profiles at the same time
Kardia	√		√	√	√	– Kardia band EKG device – Give an indication of normal or abnormal BP, an irregular heart beat AF – Sharing with healthcare providers
Qardio	√	√	√	√	√	– Triple measurement average – Slide show of photos for relaxation – Can be paired with an apple watch – Geo tracking
Fibricheck	√		√			– Gives a simple decision (regular- irregular) heart rhythm – Certified by FDA
Cardiac Diagnosis	√		√			– Directed intense light to measure heart rate – Output (risk level: normal- caution- risk)
Blood Pressure Tracker		√	√	√	√	– Long term tracking of BP – Export data in Excel or Pdf for sharing

4 Future Requirements

Based on the characteristics of the applications available to users, we conclude that there is no one integrated application that meets all the features and needs that a patient or a medical care provider would demand. Especially if there is a requirement to correlate what is acquired from these applications' data with patient data in a country's or region's integrated health system. There are apps that rely just on the patient's medical data being entered, which was measured using standard medical equipment that are not connected to the application, and there are applications that rely on a set of compatible wirelessly connected sensors to measure medical data [9].

How and how much data may be preserved, whether for analysis, making suitable judgments, or keeping it in preparation for providing the health system or health care providers, is also crucial in evaluating the success of these applications. Some applications store or download their own data in a central web database. Some applications instead access the cloud. Accordingly, there will be a need for medical apps that use Protected Health Information PHI, as in an electronic medical records service, to use secure authentication logins, good physical security of the server and encrypted hard drives. It is a good alternative for countries that have difficulties in providing sustainable internet access for their citizens to use apps with local databases for offline use.

From the analysis and processing point of view, several medical standards could be used to get predictions and decisions from the measured medical data, such as FRS-CHD Framingham Risk Score- Coronary Heart Disease, FRS- CVD Framingham Risk Score- cardiovascular disease and Atherosclerotic cardiovascular disease ASCVD risk [10]. Choosing the suitable medical standard is a very important factor to give clear predictions for the medical system.

In addition to the above-mentioned factors, there are a slew of others that must be considered while drafting a future m-health application. They include getting the necessary regulations, data security and privacy, technology platforms and third party software and services.

There are now tens of thousands of similar consumer apps accessible in app stores for public usage, with continued expansion expected. But the accreditation and regulations of these apps are not clear or standardized. Authorities like Food and Drugs Administration FDA should have a clear role in dealing with these apps [15]. Most applications have susceptible layers, including the operating system, the device, the network, and the servers that transport and store data. As the usage of mobile devices by patients grows, so do privacy issues. These risks range from inadvertent Protected Health Information PHI acquisition to electronic data transfer [16]. The rise of mobile phones and tablets has heightened these dangers. Consumers and healthcare have reaped significant benefits from new technology, but they also pose new concerns. The optimum platform for user experience, engagement, and resource investment (iOS, Android, or web-based) is a recurrent conundrum in app development. Thanks to newly developed technology, developers may now parse code into two independent platforms (for example, C# to iOS and Android) [11]. For a healthcare company looking to construct this sort of application, an API or other third-party software can give a much-needed shortcut. We would not have been able to construct such an application if we hadn't had the opportunity to link with this existing risk engine—not for a competition, and certainly not within the same deadline or budget [17].

5 Conclusion

Heart and circulatory problems are the leading cause of death worldwide, with around 18 million fatalities per year. By 2025, the World Heart Federation WHF hopes to reduce CVD-related premature mortality by 25%. Digital health with its several domains based on current developments in information and communications technology allows for significant improvements following diagnosis. Furthermore, digital health may play an active role in primary prevention with solutions like as mobile applications and monitoring devices for self-tracking and improving patient lives. The global market for mobile medical apps is continuously increasing to fulfill the demands. CVDs mobile applications come in a variety of shapes and sizes. Sensors are used by some, whereas data input is used by others. Some apps evaluate and analyze the data, while others can only keep it for later. A vital component is the capacity to transfer data between patients and service providers. According to the evaluation of the Healthline platform, the performance of a set of apps identified as the best applications of the year 2020 in the United States of America has been assessed and compared. Despite the range of these qualities and their modifications in various apps to provide the user with a variety of alternatives, there are a few more factors that should be considered in future applications to make them more acceptable for patients and health care providers. Getting the necessary regulations, data security and privacy, technology platforms and third party software and services are the main features that should be deeply considered in building the future m- Health applications for CVDs.

References

1. Narla, A.: Digital health for primary prevention of cardiovascular disease: promise to practice. Cardiovasc. Digit. Health J. **2**(59), 61 (2020)
2. Redfern, J.: A digital health intervention for cardiovascular disease management in primary care (CONNECT) randomized controlled trial. NPJ Digit. Med. **11**(1), 9 (2020)
3. Scott, C.: Best practices in digital health literacy. Int. J. Cardiol. **277**, 279 (2019). https://doi.org/10.1016/j.ijcard.2019.05.070
4. Patrick, D.: Technology Approaches to Digital Health Literacy, pp. 294–296. Elsevier, New York (2019). https://doi.org/10.1016/j.ijcard.2019.06.039
5. Healthline Homepage. https://www.healthline.com. Accessed 6 Aug 2021
6. Bussenius, H., Pedia, B.P.: Program: addressing pediatric blood pressure readings using a smartphone application. J. Nurse Pract. **11**(726), 729 (2015)
7. Maini, E., Venkateswarlu, B., Gupta, A.: Applying machine learning algorithms to develop a universal cardiovascular disease prediction system. In: Hemanth, J., Fernando, X., Lafata, P., Baig, Z. (eds.) ICICI 2018. LNDECT, vol. 26, pp. 627–632. Springer, Cham (2019). https://doi.org/10.1007/978-3-030-03146-6_69
8. Mohan, S.: Effective heart disease prediction using hybrid machine learning techniques. IEEE Access. **19**(81542), 81554 (2019). https://doi.org/10.1109/ACCESS.2019.2923707
9. Chayakrit, K.: Integration of novel monitoring devices with machine learning technology for scalable cardiovascular management. Nat. Rev. Cardiol. **18**, 75–91 (2020)

10. Asteggiano, R.: Survey on E-health knowledge and usage in general cardiology of the council of cardiology practice and the digital health committee. Eur. Heart J. **2**(342), 347 (2021)
11. Frederix, I.: ESC e-cardiology working group position paper: overcoming challenges in digital health implementation in cardiovascular medicine. Eur. J. Prev. Cardiol. **26**(1166), 1177 (2019)
12. Solberg, L.: Digital health in cardiology: time for action. Cardiology **145**, 106–109 (2020)
13. https://scholar.google.com/
14. Vervoort, D., Marvel, F.A., Isakadze, N., Kpodonu, J., Martin, S.S.: Digital cardiology: opportunities for disease prevention. Curr. Cardiovasc. Risk Rep. **14**(8), 1–7 (2020). https://doi.org/10.1007/s12170-020-00644-6
15. Santo, K., Redfern, J.: Digital health innovations to improve cardiovascular disease care. Curr. Atheroscler. Rep. **22**(12), 1 (2020). https://doi.org/10.1007/s11883-020-00889-x
16. Garg, N.: Comparison of different cardiovascular risk score calculators for cardiovascular risk prediction and guideline recommended statin uses. Indian Heart J. **69**, 458–463 (2017)
17. Weichelt, B.: Lessons learned from development of a mobile app for cardiovascular health awareness. Sustainability **13**, 1–13 (2021)

Regression Analysis for Remaining Useful Life Prediction of Aircraft Engines

Hala Mahmoud Sabry[1] and Yasser M. Abd El-Latif[2,3](✉)

[1] Post Graduate College of Computing and Information Technology, Arab Academy for Science, Technology and Maritime Transport, Cairo, Egypt
[2] College of Computing and Information Technology, Arab Academy for Science, Technology and Maritime Transport, Cairo, Egypt
Y.Abdellatif@sci.asu.edu.eg
[3] Faculty of Science, Ain Shams University, Cairo, Egypt

Abstract. Predicting Remaining Useful Lifetime (RUL) for deteriorating machines is helpful in avoiding unnecessary stoppages in production or usage of a product. This study predicts the (RUL) of the turbofan engine using deep learning techniques. The proposed model improved performance rate by combining networks with the genetic (GN), a one-dimensional convolutional neural network (CNN), long short-term memory (LSTM) with 3 dense layers and 2 dropout layers. Analysis for the experimental results demonstrates that the performance rate of the proposed model is high comparing to the different machine learning and deep learning techniques.

Keywords: Turbofan dataset · Aircraft engines · RUL estimation · Machine learning · Deep learning

1 Introduction

In the last few years, many researchers have focused on the predictive maintenance using Machine Learning (ML) and Deep Learning (DL) algorithms to predict temporal behavior and fault events from records to avoid breakdowns. However, there are many challenges because of the incipient of this progress, as well as the performance of applications rely on the suitable choice of the method. The turbofan engine is the aircraft heart and keeping it well maintained without faulty parts means decreasing the aircraft failures and unplanned stoppages by very high ratio. And as a top primacy, the safety of turbofan engines should be pursued as the imperfect components have increased the breakdowns of aircraft, and the turbofan engines malfunction have become critical for the safety of passengers. The aircraft's reliability will be ensured through successful maintenance.

One of the biggest challenges that are facing the airline is the maintenance for the aircrafts fleet. Since, the MRO (maintenance, repair, and overhaul) play a very important role in the airline's operation success. Due to the complexities in the MRO developing

© The Author(s), under exclusive license to Springer Nature Switzerland AG 2022
A. E. Hassanien et al. (Eds.): AMLTA 2022, LNDECT 113, pp. 157–168, 2022.
https://doi.org/10.1007/978-3-031-03918-8_15

new systems using new technologies is highly required to help the airlines in facilitating the required maintenance for their fleets.

Traditional maintenance systems are based on maintaining or replacing engine parts according to fixed schedule as per the parts lifetime that is by the visual inspection. This maintenance approach fails when require predicting the next occurrence failure or next faulty parts. The Scheduled maintenance is the traditional way of maintaining equipment and consists of a set of maintenance procedures defined at aircraft design time and that must be planned and performed periodically. However, whenever an unexpected failure occurs between two scheduled maintenance slots, the equipment becomes unavailable operationally until the necessary maintenance actions are performed. These unexpected failures can be a costly burden to the equipment owners because during the downtime they may not be able to provide the expected services to their clients. On the other hand, the goal of predictive maintenance is to prevent such unexpected equipment failures by continuously observing. It is not limited only to the unplanned maintenance or stoppage but it also it may lead to a disaster if this fault impacting the aircraft during the flight time.

The Remaining Useful Life prediction (RUL) is a method used to predict the performance of machines and obtain the time left before machines loses its operation ability [1].

The main methods that are used to the RUL prediction are the data-driven methods and the model-based methods [2]. The model-based method is a statistical method used to predict the future evolution of the degradation state as well as the RUL of the system, whereas the data-driven methods attempt to derive the degradation process of a machine from measured data using machine learning techniques [3]. These methods depend on the assumption that the statistical characteristics of data are relatively consistent unless a fault occurs. Based on historical data measured from machines the RUL predicted results are produced. Therefore, the data-driven methods accuracy prediction relies on not only the quantity but also the quality of historical data [2]. However, to collect qualified data in real cases is difficult. The goal of this paper is to predict the best performance rate with the usage of different regression techniques compared to the earlier chosen publications to the field of predictive maintenance.

The residuum of this work is organized as follows: Sect. 2 the related work on event prediction is presented. Whereas there are various methods for time series prediction that address the failures in predictive maintenance. In Sect. 3 the system content of turbofan engines is presented. Section 4 presents the proposed model. Section 5 presents the dataset description and the experimental results using multiple regressors for the prediction of best performance rate that could be obtained. Finally, the conclusion to demonstrate the finding and the future work are shown in Sect. 6.

2 Related Work

This section presents the related work through a comparison of recently published approaches regarding convenient datasets, and the usage of different machine learning and deep learning techniques in order to predict the best performance as well as the accuracy rate of the RUL. In a wide range of applications, the prediction in event logs is a well-studied problem [4]. Each application displays its own distinctive characteristics that have great effect on the design of the corresponding algorithm. Some application domains where failure prediction is important and prediction methods have been successfully applied are web servers [5], medical equipment [6, 7], and Hard drives [8, 9]. As there is a great development in the processing of big data, IOT, and storage, when the domain of expertise is not available the best approach is the Data-driven [10, 11]. When the model limitations are anonymous the RUL-based strategy shows better results in comparison to a conventional condition-level [12]. Also, in [13] conducted a study to predict machine failures using sensor data as well as to optimize predictive/corrective maintenance schedule for the next five days. The dataset used in this work is a historical data gathered from five datasets utilized by Fidan Boylu Uz in "Predictive Maintenance Modeling Guide" (Uz, 2016). Moreover, these datasets are combined in one large dataset. It consists of readings from 100 machines sensors. Therefore, the gradient boosting classifier gained the highest accuracy. Therefore, in [14] proposed a hybrid approach using a real-life aircraft Central Maintenance System (CMS) data to predict an aircraft component replacement in order to avoid unscheduled maintenance. The authors utilized the natural language processing (NLP) techniques, the Term Frequency - Inverse Document Frequency (TF-IDF) and Word2vec for the text vectorization and the pattern identification and the ensemble learning has been adapted to predict the rare engine's component failure as well as to avoid the data imbalanced issues that arises in the dataset when the distribution of the classes is not uniform. The proposed approach shows a better performance rate with approximately 10% rate than the Synthetic Minority Oversampling Techniques (SMOTE). Also, in [15] aimed to predict the RUL using the NASA turbofan dataset and compare the results with the root mean squared error (RMSE) and R-squared (R^2) measures using the machine learning techniques. The classification techniques are used for the prediction of the next n-days/cycle. The regression techniques are used to predict the resistance till the next failure. Therefore, the researchers proposed a comparative study on the pros and cons of the classification models versus the regression models to achieve for good precision results. The linear regression technique has been applied for the prediction of the RUL with the usage of the scikit-learn (SK-learn) library in Python. The model has been tested and trained in the ratio of 30:70 in sequence. While [16] with the usage of the NASA engines datasets the RUL has been predicted. The authors used the continual learning-based algorithm for fault prediction on transferring knowledge across decentral sub-datasets in order to predict large amount of data in a central data to facilitate the training. The architecture proposed is capable of effective learning on decentralized and smaller datasets without the need for centralized storage (cloud). The proposed in [17] identify a study to reduce the error between the actual and the predicted RUL to identify the operational factors that can affect the engine's component's reliability to evaluate whether if these can be utilized to reduce the number of failures. The limitations of this work are the difficultness to make a prior assessment of which operational factors and which distribution should be involved in the reliability analysis, using different techniques (KNN, Linear Regression, Decision Tree, SVM, Neural Network, Random Forest, and DNN) the results shows that the highest accuracy rate achieved with 84.2% by the key clustering. In [18] the researchers proposed a framework to predict when the component system will be at risk of failure due to the fault predictions that are not taking into consideration which led to unnecessary maintenance intervention. The dataset was gathered from a real date of 584 engines recorded between 2010 and 2015 of three airlines. Whereas the highest accuracy rate achieved is 93.29% using the Neural Network on the same dataset. Also, in [19] conducted a study to estimate the performance of the RUL of aircraft accurately and timely using the multilayer perceptron (MLP). The key problem is to mine the high dimension features with the internal relation hiding in the historical data.

3 Aircraft Engine System

The aircraft engine system has been built by NASA who developed the Commercial Modular Aero Propulsion System Simulation (C-MAPSS) which is a tool for simulating a realistic large commercial turbofan engine [1]. There are five main components in the system: Fan, Compressor (low-pressure compressor (LPC) and high-pressure compressor (HPC)), Combustion chamber, Turbine (high pressure turbine (HPT) and low-pressure turbine (LPT)) and the Nozzle. Each component's information is acquired through sensors.

This engine works through that the air divides into two main parts: the engine's essential first part is where the combustion will occur, whereas the second part is called "bypass air" which is the rest of air moved outside of the engine core through the duct. This bypass air cools the engine and makes it quieter. It's main duty to speed up and compress the air through the usage of airfoil-shaped spinning blades. To increase the pressure of the air vanes also called stators which are a non-moving airfoil shaped blades are utilized. When the air exits the compressor, it mixes up with the fuel and ignited. This air is like a spark, and it causes the fire in the combustor. Once the air makes its way to the combustor it flows through the turbine. The main final duct of the engine is the nozzle as it is the responsible part where the high-speed air shoots the back by transforming the internal energy of the working gas into impulsive force [20]. Figure 1 illustrates a diagram of the turbofan engine's components [21].

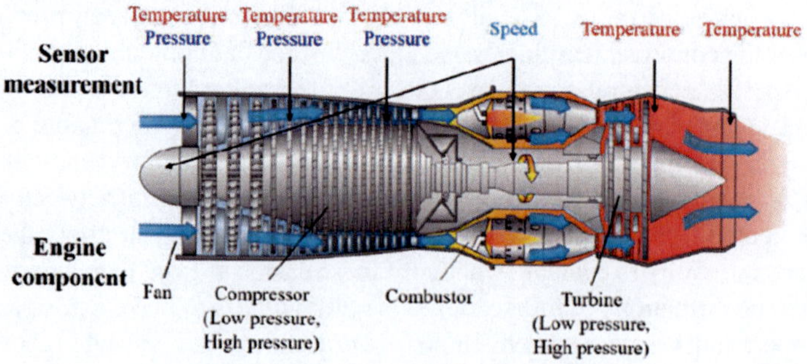

Fig. 1. Diagram of the turbofan engine.

4 Proposed Model for Predicting the RUL

This section illustrate that the proposed model is to develop a predictive maintenance system for jet engine that predicts the RUL of the next failure occurrence in the engine using the machine learning and deep learning algorithms and enhancing the system performance.

Figure 2 presents the proposed diagram which consists of five main components: Data collection, Pre-processing, Feature Extraction, and the algorithm stage where different regression techniques have been used for the prediction of the best performance rate that could be achieved.

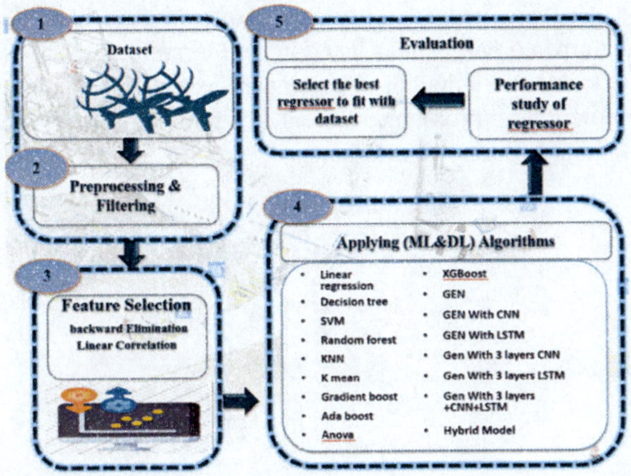

Fig. 2. The proposed diagram for predicting the RUL of aircraft engines.

The hybrid model is a merge of three algorithms: Genetic, CNN (1 layer) and LSTM (1 layer) with 3 dense layers and 2 dropout layers. The Genetic is a search-based algorithm utilized in machine learning to solve the optimization problems, and it is a subset of the evolutionary algorithms which often perform a well approximating solution to all kinds of problems. The advantage of using the Genetic algorithm is to predict the best hyper parameters such as: the number of nodes in each layer, number of kernels in CNN, number of cells in LSTM, and the type of activation used. Whereas the model consists of two parameters: (i) parameters, and (ii) hyper parameters. The model optimizer is the responsible one for the optimal value of the parameters (weights), and the hyper parameters are inserted manually in the system for that a metaheuristic algorithm to predict the optimal values. For that, the Genetic algorithm was essential to be used to predict the best rate for the hyper parameters. The dropout layers are used to avoid the overfitting in the data as to predict new data in the model. The CNN is a deep learning technique mostly applied in analyzing visual images and it consists of two aspects: (i) the feature extraction phase that has been used with one convolutional layer and ReLU activation function that enhanced the model performance, and (ii) the classification phase. For the pros the learning of accurate insights and patterns from the provided data depends on the cleansing and how well the data is structured. Finally, the LSTM algorithm is a deep learning algorithm that could process an entire sequence of data points as well as processing and classifying the data based on time series. As long as the dataset utilized in this work is a time series data, so the LSTM was the best technique to be used with three dense layers and two dropout layers in the hybrid model. The disadvantage of the LSTM is that it takes a lot of time in the real time prediction.

There are many accuracy-based evaluation used to evaluate regression models for RUL. Root mean squared error (RMSE) is used to evaluate the trained final model's RUL prognosis performance. Both RMSE and mean squared error (MSE) are given the same result when the estimated RUL value is less than or greater than the actual RUL. The learning goal is to obtain zero converged value, the smallest error for each engine's RUL prognosis, which implies that the prediction result is the same as the actual RUL. MSE tends to be overstated because it deals with square values. Therefore, the accuracy of the RMSE is compared with values listed in the literature [23].

The R-squared (R^2) coefficient of determination is a statistical measure of how well the regression predictions approximate the real data points. An R^2 of 1 indicates that the regression predictions perfectly fit the data.

$$\mathrm{MSE} = \frac{1}{n}\sum (y - y')^2 \tag{1}$$

$$\mathrm{MAE} = \frac{1}{n}\sum\nolimits_{i=1}^{n} |x_i - x| \tag{2}$$

$$R^2 = 1 - \frac{SS_{RES}}{SS_{TOT}} = 1 - \frac{\sum_i (y_i - \hat{y}_i)^2}{\sum_i (y_i - \overline{y})^2} \tag{3}$$

5 Experimental Results and Discussion

The NASA's C-MAPSS dataset is used to give characteristics for a turbofan engine's RUL. Therefore, C-MAPSS dataset with several features was used for the experiment, which simulated the turbofan engin's state. In this section, we introduce the C-MAPSS dataset that contains four datasets from turbo fan jet engines which includes Run-to-Failure simulated data. These datasets contain 24 multivariate time series signals. These multiple multivariate time series consist of noise when depicting real data and illustrating

Table 1. The number of features and rows in each dataset

File Name	Type	Features	Row	File Name	Type	Features	Row
Dataset 1	Test	26	13,096	Dataset 3	Test	26	16,596
	Train	26	20,631		Train	26	24,720
Dataset 2	Test	26	33,991	Dataset 4	Test	26	41,214
	Train	26	53,759		Train	26	61,249

Table 2. Turbofan engine dataset

#Column	Symbol	Types	Descriptions	Units
1	–	Integer	ID	–
2	–	Integer	Time in cycle	–
3	–	Double	Operational setting1	–
4	–	Double	Operational setting2	–
5	–	Double	Operational setting3	–
6	T2	Double	Total temperature at fan inlet	°R
7	T24	Double	Total temperature at LPC outlet	°R
8	T30	Double	Total temperature at HPC outlet	°R
9	T50	Double	Total temperature at LPT outlet	°R
10	P2	Double	Pressure at fan inlet	psia
11	P15	Double	Total pressure in bypass-duct	psia
12	P30	Double	Total pressure at HPC outlet	psia
13	NF	Double	Physical fan speed	Rpm
14	NC	Double	Physical core speed	Rpm
15	Epr	Double	Engine pressure ratio	---
16	ps30	Double	Static pressure at HPC outlet	psia
17	phi	Double	Ratio of fuel flow to Ps30	pps/psi
18	Nrf	Double	Corrected fan speed	Rpm
19	NRc	Double	Corrected core speed	Rpm
20	BPR	Double	Bypass ratio	–
21	far B	Double	Burner fuel-air ratio	–
22	htBleed	Double	Bleed enthalpy	–
23	Nf_dmd	Double	Demanded fan speed	Rpm
24	PCNFR_dmd	Double	Demanded corrected fan speed	Rpm
25	W31	Double	W31 HPT coolant bleed	lbm/s
26	W32	Double	W32 LPT coolant bleed	lbm/s

Table 3. MAE, R^2, RMSE values comparison using ML and DL algorithms

Algorithm	Complexity Time	Dataset 1			Dataset 2			Dataset 3			Dataset 4		
		MAE	R2	RMSE	MAE	R2	RMSE	MAE	R2	RMSE	MAE	R2	RMSE
Linear regression	O(K)	42	0.61	29	54	0.66	30.5	49	0.32	44.2	37	−0.2	38.7
Decision tree	O(d)	29.04	−0.12	27.9	33	−0.04	33.2	33	−0.3	27.1	21	−0.54	44.1
SVM	O(K*S^2 + S^3)	40.5	−0.31	46.5	47	0.2	30.1	53	−0.11	59.9	41	−0.44	33.2
Random Forest	O(t*K*S*log(S)	16.72	0.62	24.22	22	0.69	28.5	23	0.7	30.1	13.9	0.21	32.2
KNN	O(K*S')	17.8	−0.57	29.3	21.4	−0.32	33.2	24.4	−0.37	33.1	15.7	−0.65	42.2
K−Mean	O(S^2)	42.5	−0.36	78.4	47.9	−0.22	89.4	46	−0.21	72.5	41.2	−0.68	94.9
Gradient boost	O(q*S^3)	19.7	−0.64	26.3	26.3	0.23	32.4	35.3	0.3	29.6	16.8	−0.67	31.5
Ada boost	O(S*K^2)	–	–	27.2	–	–	32.9	–	–	38.1	–	–	31.9
A NOVA	–	–	–	32.9	–	–	39.9	–	–	34.5	–	–	39.3
XGBOOST	O(t*d*S*log(S))	13.88	0.69	19.8	17	0.71	26.2	12.2	0.53	27	11.3	0.47	30.1
Genetic (Gn)	O(T)	16.1	0.67	16.95	18.7	0.68	11.94	23.7	0.72	10.34	14.4	0.38	14.2
CNN	O(S)	18.85	0.43	15.9	16.77	0.56	20.1	13.14	0.51	20.3	13.41	0.36	20.13
LSTM	O(W)	16.88	0.68	13.7	12.51	0.7	15.6	10.29	−0.66	15.8	12.52	0.61	15.11
Gn + CNN	O(S + T)	14.32	0.52	10.28	15.8	0.55	10.86	10.27	0.45	10.27	13.3	0.38	11.11
Gn + LSTM	O(W + T)	14.28	0.68	10.13	15.1	0.7	10.41	10.14	0.65	10.14	14.1	0.43	10.51
Gn + CNN + LSTM	O((S*W)^T)	13.23	0.74	9.54	12.95	0.76	9.1	13.04	0.75	9.27	12.81	0.79	9.07

$W = 4*h + 4*h\hat{}2 + 3*h + h*k$, and $T = g*(p*q + p*q + p)$
where S is number of samples, K is number of features, h number of hidden units, g is number of generation, p number of population, q size of individual, t is number of trees and d is depth of tree
In our experiment model: $S = 20361$, $S' = 1000$, $k = 25$, $h = 128$, $g = 20$, $p = 10$, $q = 25$, $t = 100$, $d = 150$

the degradation until the engine fails as a time-series trajectory. Several engine's sensor information refers to a direct deterioration, but others consist of few performances degradation's information. Each dataset is divided into test and training subsets consist of 26 columns. The C-MAPSS dataset contains 21 sensors that measure turbofan engine information and status based on three operating settings [22]; the row in the dataset is a snapshot of data taken during a single operational cycle. Each engine gives a different time series. It operates normally at the start of each time series, and at various time points it evolves a fault. Table 1 shows the number of features and rows in each dataset, there are diversity in the number of rows in each dataset to predict the RUL. Table 2 shows the meaning of each column in the dataset.

At the beginning the engine's start is normal, and then after several cycles a fault is developed which cause failure to the engine. The dataset utilized consist of numerical files for the testing set 30%, training set 70%, and predicting the performance rate. The training set is where the error volume grows until the engine's system fails while the test set before the system's failure the time series ends sometimes [22].

Every run cycle, the sensors of the turbofan engine log all the records and the measures to the database. Each sensor has its own threshold that is based on the system design and all records should be within this threshold in the normal operation. The dataset records all these measurements for each operational flight. Once one of the sensors reads exceeds its threshold it raises a flag that indicate there exists a fault in this part of the turbofan engine that needs human intervention. The RUL prediction is based on these measurements and faults to build a prediction algorithm that can predict the failure before it occurs.

Accordingly, a comparative analysis of different regression algorithms on the NASA dataset has been performed. Some algorithms show good performance whereas some other algorithms perform poorly. Different algorithms are used to predict the best performance rate on the same dataset to reduce the error between the real and the predicted RUL values. In order to improve the performance rate a hybrid model is presented to predict the remaining life cycle (RUL) for the next failure occurrence in the engine. The proposed model managed to achieve the best performance in predicting the RUL comparing to the others algorithms. Table 3 shows different algorithms that have been applied to the experiment and tested the results by using the three accuracy-based evaluations MAE, R^2 and RMSE on the same four datasets to measure the performance. It is obvious that the proposed model is achieved the best results comparing to the other algorithms (Fig. 3).

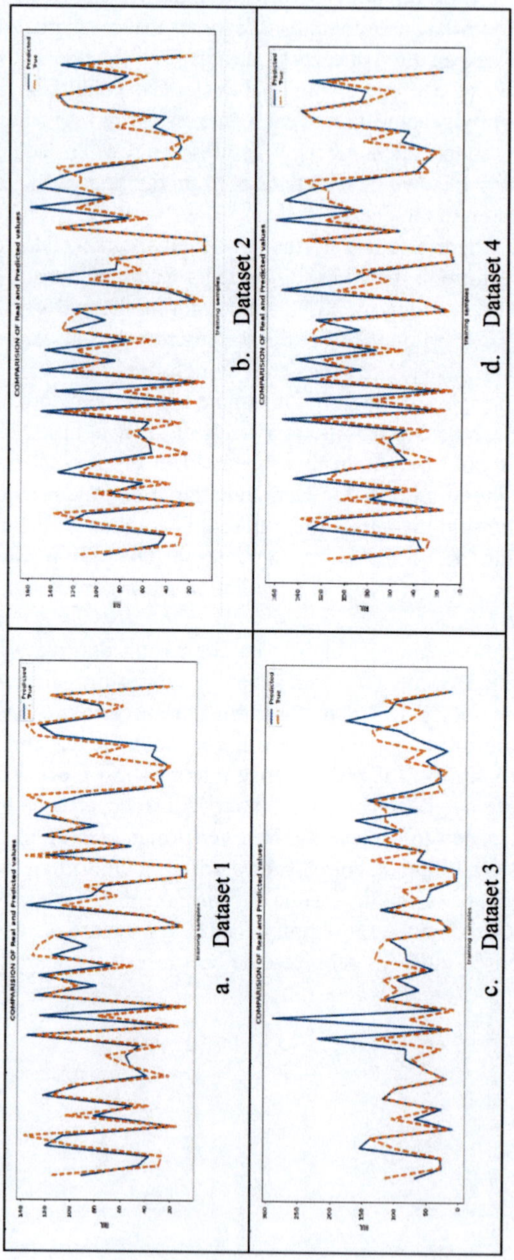

Fig. 3. Visualization of Actual vs. Predicted (RUL) values of the four datasets

6 Conclusion and Future Work

A hybrid model for Regression Analysis for the RUL prediction of aircraft engines has been presented to increase the safety of employees and reduce the loss of production time by predicting the RUL before the next failure occurrence. The hybrid model is a combined of Genetic, CNN (1 layer), LSTM (1 layer) with 3 dense layers and 2 dropout layers. Compared to the other researchers some of them only used the classification techniques for the prediction of the RUL in contrary with the others who utilized the regression techniques to predict the best performance rate. With the usage of various Machine Learning and Deep Learning techniques the results show that the hybrid model presented achieved the best performance rate compared to the other techniques. For future work we are aiming to use the classification techniques for predicting the (RUL) of aircraft engines.

References

1. Jay, L., Fangji, W., Wenyu, Z., Masoud, G., Linxia, L., David S.: Prognostics and health management design for rotary machinery systems— Reviews, methodology and applications. Mech. Syst. Signal Proc. **42**(1−2), 314−334 (2014)
2. Liu, J., Wang, W., Ma, F., Yang, Y.B., Yang, C.S.: A data-model-fusion prognostic framework for dynamic system state forecasting. Eng. Appl. Artif. Intell. **25**(4), 814–823 (2012)
3. Jieliu, L.W.W. Fared, G.: A multi-step predictor with a variable input pattern for system state forecasting. Mech. Syst. Sign. Proc. **23**, 1586–1599 (2009)
4. Felix, S., Maren, L., Mrioslaw, M.: A survey of online failure prediction methods. ACM Comput. Surv. (CSUR) **42**(3), 10 (2010)
5. Ke, Z., Jianwu, X., Martin, R.M., Guofei, J., Konstantinos, P., Hui, Z.: Automated it system failure prediction: A deep learning approach. In: IEEE International Conference on Big Data. IEEE, pp. 1291–1300 (2016)
6. Sipos, R., Fradkin, D., Moerchen, F., Wang, Z.: Log-based predictive maintenance. In: Proceedings of the 20th ACM SIGKDD International Conference on Knowledge Discovery and Data Mining. ACM, pp. 1867–1876 (2014)
7. Yuan, Y., Shiyu, Z., Crispian, S., Kamal, M., Yibin, Z.: Event log modeling and analysis for system failure prediction. IIE Trans. **43**(9), 647–660 (2011)
8. Junbo, S., Qiang, Z., Shiyu, Z., Xiaofeng, M., Mutasim, S.: Evaluation and comparison of mixed effects model based prognosis for hard failure. IEEE Trans. Reliab. **62**(2), 379–394 (2013)
9. Joseph, F.M., Gordon, F.H., Kenneth, K.D.: Machine learning methods for predicting failures in hard drives: a multiple-instance application. J. Mach. Learn. Res. **6**, 783–816 (2005)
10. Saxena, A., Kai, G., Simon, D., Eklund, N.: Damage propagation modeling for aircraft engine run-to-failure simulation. In: International Conference on Prognostics and Health Management. IEEE, pp. 1–9 (2008)
11. Valliappan, S., Bagavathi Sivakumar, P., Ananthanarayanan, V.: Efficient real-time decision making using streaming data analytics in IoT environment. In: Kamal, R., Henshaw, M., Nair, P.S. (eds.) International Conference on Advanced Computing Networking and Informatics. AISC, vol. 870, pp. 165–173. Springer, Singapore (2019). https://doi.org/10.1007/978-981-13-2673-8_19
12. Phuc, D., Eric, L., Alexandre, V., Beno, I.: Remaining useful life (RUL) based maintenance decision making for deteriorating system. In: IFAC Proceedings Volumes (IFACPapersOnline) (2018). https://doi.org/10.3182/20121122-2-ES-4026.00029.

13. Eman, O., Maher, M., Andrei, S.: Machine learning and optimization for predictive maintenance based on predicting failure in the next five days., In: In Proceedings of the 10th International Conference on Operations Research and Enterprise Systems (ICORES 2021), pp. 192–199 (2021)
14. Maren, D.D., Zakwan, S., Ian, K.J.: An integrated machine learning model for aircraft components rare failure prognostics with log-based dataset. ISA Trans. **113**, 127–139. https://doi.org/10.1016/j.isatra.2020.05.001.2020
15. Saranya, E., Sivakumar, P.: Data-driven prognostics for run-to-failure data employing machine learning models. In: Proceedings of the Fifth International Conference on Inventive Computation Technologies (ICICT-2020), pp. 528–533 (2020)
16. Maschler, B., Vietz, H., Jazdi, N., Weyrich, M.: Continual learning of fault prediction for turbofan engines using deep learning with elastic weight consolidation. In: IEEE International Conference on Emerging Technologies and Factory Automation (ETFA)(2020). https://doi.org/10.1109/etfa46521.2020.92119025th2020
17. Verhagen, W., Deboer, L.W.M.: Predictive maintenance for aircraft components using proportional hazard models. J. Ind. Inf. Int. **12**, 23–30 (2018). https://doi.org/10.1016/j.jii.04.004,2018
18. Marcia, B., Sanakra, R., Ivo p de Medeiros, C., Nascimento, H.P., Elsa, M.P.H: Forecasting fault events for predictive maintenance using data-driven techniques and ARMA modeling. Comput. Ind. Eng. **115**, 41–53. (2018). https://doi.org/10.1016/j.cie.2017.10.033
19. Caifeng, Z., Weirong, L., Bin, C., Dianzhu, G., Yijun, C.: A data-driven approach for remaining useful life prediction of aircraft engines. In: 21st International Conference on Intelligent Transportation Systems (ITSC) (2018)
20. https://www.boldmethod.com/learn-to-fly/aircraft-systems/how-does-a-jet-engine-turbofan-system-work-the-basics/. Accessed 2021
21. Changwoo, H., Changmin, L., Kwangsuk, L., Minseung, K., Dae, E.K., Kyeon, H.: Remaining useful life prognosis for turbofan engine using explainable deep neural networks with dimensionality reduction. Sensors **20**(22), 6626 (2020). https://doi.org/10.3390/s20226626
22. NASA Turbofan Jet Engine Data Set", Kaggle.com (2021). https://www.kaggle.com/behrad3d/nasa-cmaps. Accessed 26 Apr 2018
23. What are the differences between MSE and RMSE | i2tutorials, i2tutorials (2021). https://www.i2tutorials.com/differences-between-mse-and-rmse/#:~:text=MSE%20(Mean%20Squared%20Error)%20represents,is%20to%20actual%20data%20points.&text=RMSE%20(Root%20Mean%20Squared%20Error,the%20square%20root%20of%20MSE. Accessed 21 Sep 2021
24. Amgad, M., Shakirah, T., Sheraz, N., Rao, A., Izzatdin, A.: Data-driven deep learning-based attention mechanism for remaining useful life prediction: case study application to turbofan engine analysis. Electronics **10**(20), 2453, https://doi.org/10.3390/electronics10202453 Accessed 2021

Applying Machine Learning Technology to Perform Automatic Provisioning of the Optical Transport Network

Kamel H. Rahoma[1] and Ayman A. Ali[2](✉)

[1] Electrical Engineering Department, Faculty of Engineering, Minia University, Minya, Egypt
[2] Telecom Egypt, Minya, Egypt
aymanelmelm@gmail.com

Abstract. The Optical Transport Network (OTN) is considered valuable assets to any telecom operator. One of the most crucial difficulties to the owners of the OTN is how they can supervise all of these assets in the Optical Transport Network (OTN) smartly and efficiently. Controlling the operational tasks of the OTN intelligently will overcome the wrongs behavior of the human interventions while they are provisioning the optical network and will enhance the performance of the communication networks to the end customers. One of the most encouraging technologies which can assist in the administrating the assets of the OTN is the Artificial Inelegance (AI), that can be employed in the provisioning of the optical network to perform several of the everyday tasks in the network instead of using human interventions. In this paper, for the first time, we present the machine learning (ML) as a branch of the AI technology to handle and perform the routine tasks in the OTN of Egypt in smart and automated techniques. The expected outcomes from practicing ML in managing OTN show that the time of the fault localization will be reduced from average 40 min to about 10 min and consequently this will decrease main time of repair (MTTR) by about 30 min, the number of the customer's tickets will be lowered by about 25%, and the number of network faults will be decreased by about 75% as a result of performing the preventive maintenance tasks of the network in an automated technique, and the reply time to the clients is expected to be reduced from average 50 min to about 5 min only. These expected results prove that in the next future the artificial Inelegance with its branches will perform a significant function in managing and supervising the optical core network around the world and possibly all the communication networks will be managed by the same intelligent umbrella, this will make a significant role in optimizing the resources of the optical core networks by using intelligent and centralized platform to perform the needed tasks in the network without any human interventions, which will reduce the operational cost and will maximize the ROI from the OTN.

Keywords: Optical Transport Network (OTN) · Artificial Inelegance (AI) · Machine learning (ML) · Intelligent centralized platform · Human intervention · Everyday tasks

© The Author(s), under exclusive license to Springer Nature Switzerland AG 2022
A. E. Hassanien et al. (Eds.): AMLTA 2022, LNDECT 113, pp. 169–177, 2022.
https://doi.org/10.1007/978-3-031-03918-8_16

1 Introduction

With the coming 5G technology, the telecom operators will need a huge number of the high-capacity circuits to link their sites with each other's, The optical transport network (OTN) provides the solutions to the operators to have these circuits with the required high capacities which reach to 400 Gb/s for each circuit in the long-distance optical network. The OTN is extending. All over the country area, and it includes thousands of transmission vices from several venders and thousands of kilometers from optical fiber cables. The optical transport networks are playing the essential role in the communication infrastructure of the core network in Egypt, and any shortage in performance of the optical network provisioning will be reflected directly on the investments in the Communication industries in many business fields in Egypt. The OTN is controlled by multiple network management systems (NMS's) according to the type of the vendor of the transmission devices. The huge variations in the enormous number of the equipment types in the OTN along with the different types the transmission technologies, the different capacities, and the vendors make the administration and the supervision of this network very difficult to handle by using the traditional techniques. As a result of this huge amount of equipment and the need to perform a large number of tasks in the OTN to keep the network in healthy quality, there are multiple tasks maybe not performed and observed in the proper time in case of depending only on the human eyes, Consequently the number of the faults, the time to locate faults, the response time to the customers, and the number of customer's complains will increase proportionally according to the network capacities. All of these problems in managing the network increase the cost of the operations and will affect the performance and the quality of the communication networks to the end-users. The prior studies show that artificial intelligence (AI) has the encouraging expectation in supervising the different network applications in a smart and automatic method such administrations of the optical network [1]. Many papers illustrated how to employ artificial intelligence procedures to secure the failing status in the OTN by following double-exponential smoothing and support vector machine (DES-SVM) techniques [2]. Further investigations employed the graph-based correlation (GBC) heuristic in discovering the fault location in the OTN [3]. A long-short-term memory (LSTM) system with the deep learning procedure was proposed by several subjects as an answer to constantly evaluate the optical signal to noise ratio (OSNR) [4]. Less previous studies were made concerning the application of the AI methods in the administration of the configurations of the OTN to execute automatic provisioning of the optical transport network, an example of these cases examined how to manage the efficient routing and the spectrum appointment with the clasp time among the different OTN sites to discover the various fitting routes in the Dense Wave Division Multiplexing (DWDM) system of the OTN [5]. The article gives for the first time a smart representation to manage the necessary conventional responsibilities in the optical transport network. automatically, for clearness, just one task is intended to be done in this paper, and this is achieved by exercising the ways of the Artificial Neural Network (ANN) to perform the processes of monitoring the OSNR of the different optical links and locating the position of any optical faults in the OTN without any human interventions. The model is proposed to notify the administrators of the optical network about any signal degradation in the different optical links inside the network. The proposed model can be used to identify the cause of the optical faults by automatic enabling to the optical time-domain reflect meter (OTDR) which is built in the new generations of the transmission equipment to measure the attenuations of optical fiber cable in the detected section. The paper is arranged as follows: Sect. 2 presents the current challenges in the management of the optical transport network; Sect. 3 proposes smart models of the routine tasks in the optical transport network; Sect. 4 discusses the expected outcomes and finally, Sect. 5 presents the conclusion.

2 The Challenges in the Current Model of the Supervision of the OTN

Figure 1 shows a practical small model of the core optical transport network. The model consists of many DWDM and OTN devices with different transmission technologies, the capacity of every network area reaches to more than 8 Tb/s with 200 Gb/s for every wavelength and managed by several NMS's based on the vendor types. The administration of such heterogeneous and complexity OTN network is restricted by the human aspects to perform the required operational assignments, where the performance of the outputs of these tasks depends on the experience of the performed person and as a result of these outcomes the quality of the provided services to end-users will be fluctuated according to the personal acts in the operational tasks of the OTN, which is not accepted from the significant customers in the communication markets especially with the needs to implement the service level agreement (SLA) between the network operators and the important customers. The quality of the provided services depends essentially on performing preventive maintenance (PM) of the OTN. One of the important routine tasks of the PM in the OTN is the periodic monitoring of the OSNR values of the optical links in all the network, which keeps the high quality of the proved services from the optical network. The first difficulty in the popular design of OTN administration is whereby to observe thousands of the documents of the OSNR states which are correlated to a tremendous amount of the optical connections by the observing of personal eyes only. With this huge amount of the OSNR reports in the change management database (CMDB) of the NMS, it's very challenging to recognize which optical links require to be fixed as a proactive step [4]. An added complexity in the management of the OTN is how to manage the location of the faults and how simply identifies the root cause of this fault in the precise period [6]. In the conventional methods, the fault localization depends completely on the observations of the individual eyes to hundreds of the alarms which are listed in the CMDB of the multiple NMS's. The points which are employed in the fault localization state depend significantly on the background of the administered persons. These methods of the fault resolutions affect straight the operation expenses (OPEX) of the OTN by increasing the meantime to restore (MTTR) and providing the payments from the possibility of breaking the SLA with the valuable customers. One further difficulty in the actual model of the OTN management is the failure to achieve complete befits from several superior features which are possible in the innovative productions of the OTN equipment, one of these features is the availability to estimate the distance of fiber cable cut by using the build-in OTDR, where the obstacle in using this build-in OTDR is that it should be authorized manually by the responsible person to measure the distance of the cable cut, and this technique increases the time to recover any cable cut in the network or the time to identify the root cause of the services degradations in the optical network. This challenge can be formulated in other words as there are no any correlations between the raised alarms in the optical network and the internal OTDR of the OTN equipment which can be used to reduce the MTTR in the optical network.

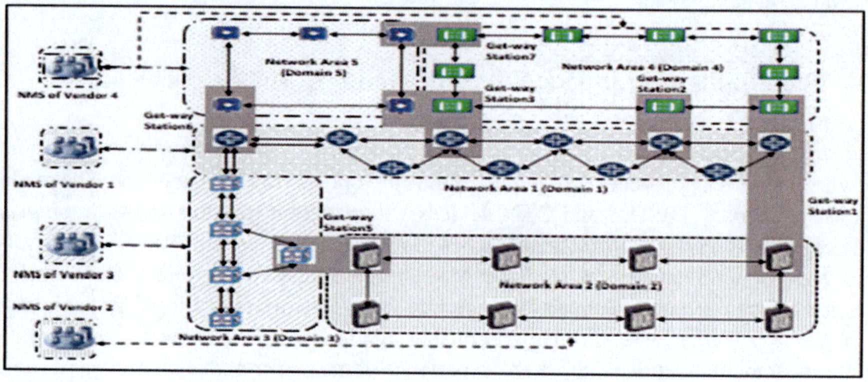

Fig. 1. Practical model of the optical transport network

3 Proposed Model for the Automatic Provision of the OTN

The proposed model is built by using the ML and ANN to perform the routine tasks in the OTN instead of the human observations such monitoring the ONSR values of all - optical links in the optical network. This will be done by three phases by and by using the dataset of the OSNR values of the different optical links which are stored in The CMDB of the NMS's, part of this dataset is used to train the proposed model by finding the correlation rules between the different variables of the alarm lists such primary and secondary alarms and the variables of the optical specifications for every optical link. The part of the data set which are used in the training phase is recorded in the CMDB of the different NMS'S and corresponding to the reals values in the different network elements (NE's) in the network [7, 8]. As shown in Fig. 2 the training Data from CMDB is classified into 2 parts, the first one is related to the types of the alarms and the 2'nd type of data set is related to the different input and output variables of the optical laser power of the optical links as shown in Table 1. There are direct associations among the output variables such as OSNR and bit error rates (BER) with the input variables as following equations [9]:

$$\theta_{att} = r \tag{1}$$

$$\theta_{rx} = 1 - r \tag{2}$$

$$OSNR_{TX} = \frac{P_T}{P_Q F_T} \quad (3)$$

$$OSNR_{RX} = \frac{P_R}{P_Q}\left(\frac{P_N}{P_Q} + \theta_{rx} F_R - 1\right) \quad (4)$$

$$\frac{OSNR_{RX}}{OSNR_{att}} = \frac{\theta_{att}}{\theta_{rx}} = \frac{r}{1-r} \quad (5)$$

Where: $OSNR_{TX}$ is the OSNR at the transmitter, $OSNR_{RX}$ is the OSNR at the receiver, OSNRatt is the OSNR the receiver and with attenuations, PT is the transmit power at the receiver, PQ is the Quantum noise power at 12.5 G, PN is the noise power, and FT is the noise figure at the transmitter side.

The plan includes 3 layers of the ANN the leading layer is the input variables and consists of n parameters and d ports, the next layer is defined for the connections among input variables and output variables with m powers and lastly, the final layer is the value of the output layer as illustrated in Fig. 5 [8].

$$OSNR = 10dB * log_{10}\frac{S}{N} \quad (6)$$

$$10log_{10}(BER) = 10.7 - 1.45(OSNR) \quad (7)$$

$$Y_{1k} = f(\sum_{n=1}^{n=d}\sum_{i=1}^{i=m} a_i x_n^k + w_0 \quad (8)$$

$$Y_{2k} = f(\sum_{n=1}^{n=d}\sum_{i=1}^{i=m} a_i x_n^k + w_1 \quad (9)$$

where: S denotes the linear optical power, N denotes the linear noise power, Y_{1k} (x) is the predicted output of the OSNR in the gate 1, Y_{2k} (x) is the coefficient value of the hidden layers, a_i is the coefficient value of the hidden layers, xk is the input of n variables for port number k in the network, w_0 is the fixed weight among the input variables and output OSNR, and w_1 is the fixed weight among the input variables and the output BER.

Phase 2 is the generalization stage, where part of data set which is not practiced in the training phase is employed to investigate the error ratio between the actual outputs and the predicted outputs from the system according to Eq. 10 of the root means square errors (RMSE).

$$RMSE = \sqrt{\frac{1}{N}\sum_{K^{\sim}=1}^{K^{\sim}=N}} \left(r_{\widetilde{K^{\sim}}} - r_{K^{\sim}}\right) \quad (10)$$

Phase 3 is the implementation stage, and for the simplicity is done by using 3layer artificial neural networks (ANN).

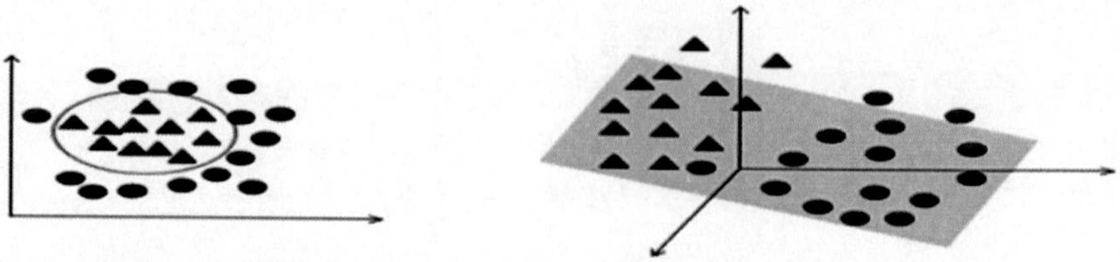

Fig. 2. Classification of the data set of the CMDB

Table 1. Variables of the optical laser power

Parameter	Transmitter port	Receiver port	Amplifier card
Optical power	Input	Input	Input
BER		Output	
OSNR	Output	Output	Output
Amplifier gain			
Fiber span attenuation		Input	
Line rate	Input	Input	

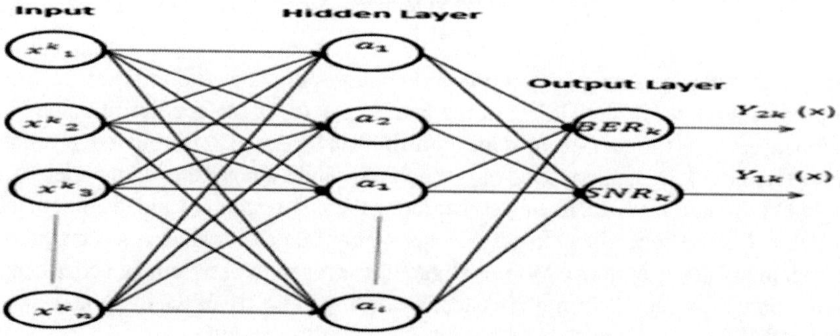

Fig. 3. Layers of the ANN proposed model

4 Results and Discussion

The ML model recognizes any variations in the OSNR parameters from its experience about the network parameters at the receive section and bit error rate at the transmit section, and it will estimate the values of the BER according to the practical values of the different variables from the CMDB of the NMS. In case of the estimated value of the BER is exceeded the threshold value which is determined by the administrator of the network the model will take more than one action to notify the administrator by the impacted section via SMS or E-mail. One of these actions also the ANN output will enable the built-in OTDR of the NE's in the sections which have had bad performance in the BER. After that, the output of the measurements of the OTDR will send automatically to the operation and maintenance (O&M) team to take the needed corrective actions as an example of Fig. 3. The automatic detections of any variations in the OSNR and BER will shortage the time of the discovering phase by long percentage rather than using human eyes for this job, and according to the statics from many vendors who work in this field in Egypt, using any type of the automation in performing operational duties in the OTN will reduce the reduce the time of detecting any faults from average 40 min to about 10 min and consequently this will decrease main time of repair (MTTR) by about 30 min, the number of the customer's tickets will be lowered by about 25%, and the number of network faults will be decreased by about 75% as a results of performing the preventive maintenance tasks of the network in a proactive and automated technique, and the reply time to the clients is expected to be reduced from average 50 min to about 5 min. to test the model of detecting any variations in the OSNR and BER by using ANN of the ML as proposed in our model, 500 records from practical OTN about the different variables in the system were used in the SPSS to find the model of the ML, we found The relations between the different parameters according to Eq. 11, Table 2, and Fig. 4. All of these outputs illustrate the correlations between the different input variables and the outputs and explain the linear relationships between the input variables and the output which can be used to trigger any other model to perform certain tasks automatically.

Table 2. Relations between input and output variables

Model		Unstandardized coefficients		Standardized coefficients	T	Sig.
		B	Std. error	Beta		.771
1	(Constant)	2.623E−9	.000		.291	.000
	BER	1000.000	.000	1.000	5090700.061	.000
2	(Constant)	1.856E−5	.000		5.153	.000
	BER	999.977	.004	1.000	227553.304	.000
	SNRt	−4.776E−7	.000	.000	−5.152	.000

$$BER^{\sim} = 0.357 + 999.997 * BER - (4.776E - 7)OSNRt \quad (11)$$

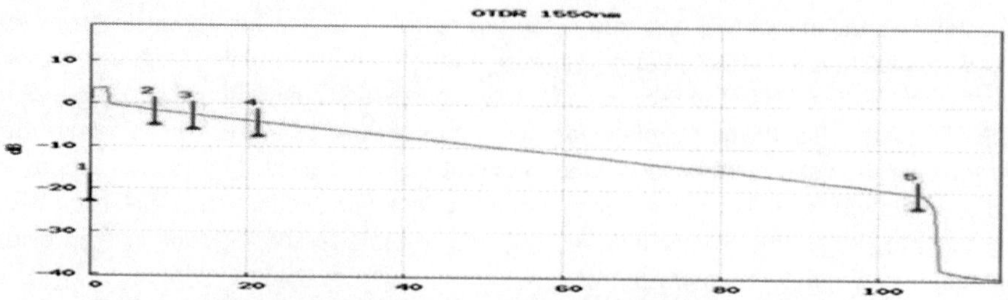

Fig. 4. Example of the OTDR output

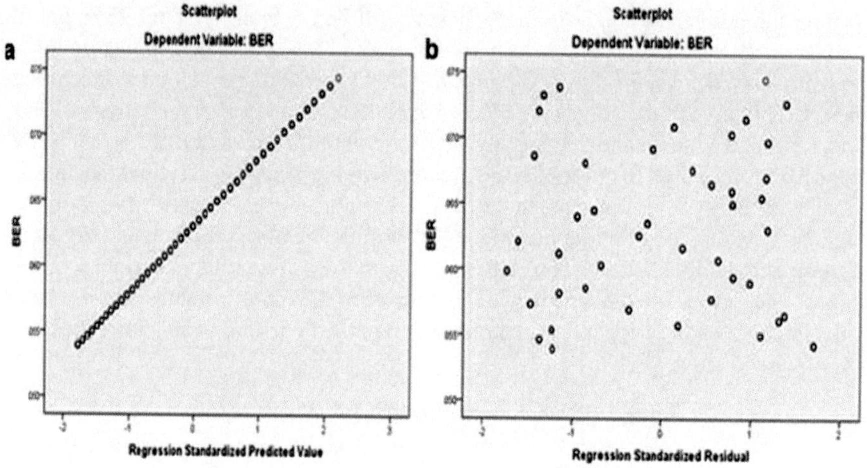

Fig. 5. Correlations between different variables

5 Conclusion and Future Work

The future work of this work is to build an intelligent unified platform to make all the operational tasks in the optical transport network by using deep learning and software-defined network (SDN) technologies with only one centralized controller. As discussed in this paper, one of the promising technologies which can help in administration, the operations and the maintenance of the OTN is the Artificial Inelegance (AI) that can be employed in the provisioning of the optical network to perform many of the daily tasks in the network rather than the human. In this paper, we offered the machine learning (ML) as a part of the AI technology to manage and execute the routine tasks in the OTN of Egypt in smart and automatic procedures. The expected results from functioning ML in maintaining the OTN show that the time of the fault localization will be reduced from average 40 min to about 10 min and consequently this will decrease main time of repair (MTTR) in the by about 30 min, the number of the customer's tickets will be lowered by about 25%, and the number of network faults will be decreased by about 75% as a results of performing the preventive maintenance tasks of the network in an automated technique, and the reply time to the clients is expected to be reduced from average 50 min to about 5 min only. These expected results demonstrate that in the next future the AI will perform an important function in managing and supervising the optical core network around the world and possibly it will lead the management and the operations of all the communication networks to be done under the same intelligent umbrella, this will make a significant role in optimizing the resources of the optical core networks by using intelligent centralized platform to perform the needed tasks in the network without any human intervention, which will reduce the operational cost and will maximize the ROI from the OTN.

References

1. Gao, J.: Machine learning applications for data center optimization (2014)
2. Wang, Z., et al.: Failure prediction using machine learning and time series in optical network. Opt. Express **25**(16), 18553–18565 (2017)
3. Panayiotou, T., Chatzis, S.P., Ellinas, G.: Leveraging statistical machine learning to address failure localization in optical networks. IEEE/OSA J. Opt. Commun. Netw. **10**(3), 162–173 (2018)
4. Wang, Z., Yang, A., Guo, P., He, P.: OSNR and nonlinear noise power estimation for optical fiber communication systems using LSTM based deep learning technique. Opt. Express **26**(16), 21346–21357 (2018)
5. Jia, W.B., Xu, Z.Q., Ding, Z., Wang, K.: An efficient routing and spectrum assignment algorithm using prediction for elastic optical networks. In: 2016 International Conference on Information System and Artificial Intelligence (ISAI), pp. 89–93. IEEE, June 2016
6. Ayoubi, S., et al.: Machine learning for cognitive network management. IEEE Commun. Mag. **56**(1), 158–165 (2018)
7. Tzelepis, D., et al.: Advanced fault location in MTDC networks utilising optically-multiplexed current measurements and machine learning approach. Int. J. Electr. Power Energy Syst. **97**, 319–333 (2018)
8. Yan, S., et al.: Field trial of machine-learning-assisted and SDN-based optical network management. In: Optical Fiber Communication Conference. Optical Society of America, p. M2E.1 (2019)
9. Pointurier, Y.: Design of low-margin optical networks. J. Opt. Commun. Netw. **9**(1), A9–A17 (2017)

Check for updates

Robo-Nurse Healthcare Complete System Using Artificial Intelligence

Khaled AbdelSalam, Samaa Hany(✉), Doha ElHady, Mariam Essam, Omnia Mahmoud, Mariam Mohammed, Asmaa Samir, and Ahmed Magdy

Electric Engineering Department, Faculty of Engineering, Suez Canal University, Ismailia, Egypt
khaled.abdelsalam@eng.suez.edu.eg, samaa20298@gmail.com

Abstract. Significant with COVID-19 pandemic breakout, and the high risk of acquiring this infection that is facing the Healthcare Workers (HCWs), a safe alternative was needed. As a result, robotics, artificial intelligence (AI) and internet of things (IoT) usage rose significantly to assist HCWs in their missions. This paper aims to represent a humanoid robot capable of performing HCWs' repetitive scheduled tasks such as monitoring vital signs, transferring medicine and food, or even connecting the doctor and patient remotely, is an ideal option for reducing direct contact between patients and HCWs, lowering the risk of infection for both parties. Humanoid robots can be employed in a variety of settings in hospitals, including cardiology, post-anesthesia care, and infection control. The creation of a humanoid robot that supports medical personnel by detecting the patient's body temperature and cardiac vital signs automatically and often and autonomously informs the HCWs of any irregularities is described in this study. It accomplishes this objective thanks to its integrated mobile vital signs unit, cloud database, image processing, and Artificial Intelligence (AI) capabilities, which enable it to recognize the patient and his situation, analyze the measured values, and alert the user to any potentially worrisome signals.

Keywords: Healthcare · Nursing robot · Infection control · Humanoid robot · Artificial intelligence · Internet of Things · Face recognition

1 Introduction

The nursing profession has seen significant technological and cultural changes in recent decades. With the increased usage of all forms of technical and communication developments, as well as the rapid proliferation of developing digital technologies such as artificial intelligence throughout all sectors of society, the nursing profession is nearing a critical period in its history. What effect will today's technological pressures have on the nursing profession, as well as nursing's future role and function? What distinguishes this occupation from other forms of caregiving? What value(s) do technology breakthroughs bring to existing nursing values while also revealing completely new ones? The profession as we know it could be dealing with a nagging emotional issue.

© The Author(s), under exclusive license to Springer Nature Switzerland AG 2022
A. E. Hassanien et al. (Eds.): AMLTA 2022, LNDECT 113, pp. 178–191, 2022.
https://doi.org/10.1007/978-3-031-03918-8_17

The HCWs' safety is always in jeopardy. Even though health workers make up less than 3% of the population in most nations and less than 2% in practically all low- and middle-income countries, they account for roughly 14% of COVID-19 cases reported to the World Health Organization (WHO). In some countries, the proportion can be as high as 35%. However, data availability and quality are limited, and it is not possible to establish whether health workers were infected in the workplace or in community settings. Thousands of health workers infected with COVID-19 have lost their lives worldwide. In addition to physical risks, the pandemic has placed extraordinary levels of psychological stress on health workers exposed to high-demand settings for long hours, living in constant fear of disease exposure while separated from family and facing social stigmatization. Before COVID-19 hit, medical professionals were already at higher risk of suicide in all parts of the world.

To save HCWs from the risk, there are some questions asked:

- How to decrease the direct contact between the HCWs and patients?
- How can the limited nursing stuff cover up the huge numbers of patients?
- How to release the pressure on the HCWs due to automatic vital signs units' shortage in some hospitals?
- How to decrease the vital signs units' numbers required in low-funded hospitals?

The rise of AI, IoT, and robotics has catapulted humanity into a new era of technological advancement and possibility. While AI is widely utilized in health care to assist with data analytics and clinical decision making, the potential for AI-driven digital health technology to influence the interaction between nurses and their patients should not be overlooked. As the popularity of this technology grows, so does the desire to apply it to the field of medicine.

1.1 Related Work

An early start for applying robotics in the medical field was in 1985, when Dr. Yik San Kwoh presented to the world PUMA 560, the first robotic arm to assist in surgery by successfully placing a needle for a human brain biopsy using Computed Tomography (CT) for guidance [14]. Further in 2000, the da Vinci Surgery System was the first robotic surgery system approved by the Food and Drug Administration (FDA) for general laparoscopic surgery [15]. However, the usage of this technology hasn't only been in the surgical field but also the nursing field. The National Chiao Tung University has developed a nursing robot named "ROLA" that can give the required care for the elderly daily life [16]. Besides, in 2013, Japan's Panasonic Company produced an intelligent robot to assist nursing workers, called HOSPI-R, a robot that can transport medicines under 20 Kg once a time [17]. In short words, nursing robots are being developed rapidly with the improvement of healthcare and robot technology, and it has brought humanity new chances, as the nursing robot can not only improve the nursing efficiency, but also enhance the self-nurse ability and adaptability of the patients and elderly. Therefore, nursing robots have become one of the hot research topics in recent years.

2 Research Method

Robot Nurse Assistant (RONA), a fully integrated and fully autonomous robot nurse, that serves as an interface for doctors to use over distances to communicate with patients. RONA was designed to assist the HCWs in some sections of the hospital as it contains a mobile vital signs unit that is connected to a cloud database through which autonomous regular patients' checkups can be implemented, and the HCWs can follow up the results of these checkups through his smart phone application with no need to a direct contact with the patients, besides being alerted for any irregular checkup results.

2.1 Software Implementation

Gathering the AI and the IoT technologies, RONA was designed to be completely autonomous, relying on computer vision for patients' face recognition, and using intelligence to detect critical cases and alarm the doctors, besides using the cloud to send and store data so that it is never lost. To apply this, RONA was built out of seven subsystems; each subsystem had a specific purpose. These were the subsystems:

2.1.1 Motion System

RONA is a self-operating robot that autonomously moves towards the patients' rooms based on a schedule prepared by the doctor. It recognizes and follows a line placed on the floor for each patient's room using a line follower technique through the hospital map stored in it. The track is made up of a black tape on a white background. RONA can also detect barriers using an ultrasonic sensor to determine whether an object is within a predetermined range of the robot.

2.1.2 Patient Identification System

AI played an essential role in RONA's patient identification system, as the system relied on image processing using deep convolutional neural networks (DCNN) for face detection and recognition. In face detection, a high-performance face detector named AInnoFace [13] was used, developed by AInnovation Technology Ltd, Beijing, China, through equipping the popular one-stage RetinaNet with some enhancements. In face identification, ArcFace pre-trained deep learning face identification model [12], was used (Fig. 1).

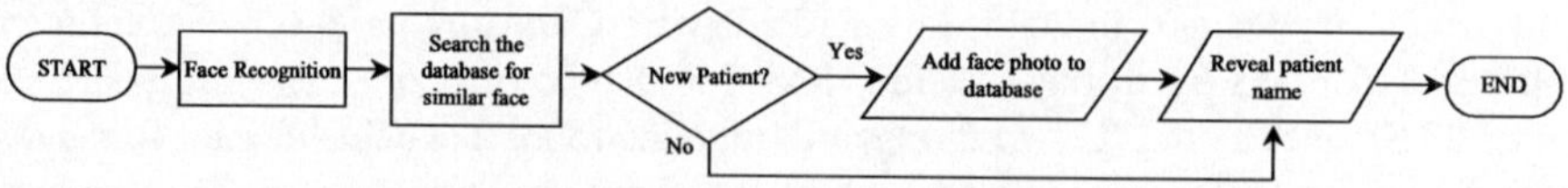

Fig. 1. Patient identification system operation flowchart

2.1.3 Medical System

RONA is an autonomous nurse assistant who follows up every patient's health in the hospital and checks vital signs by measuring them on a regular basis and alerting hospital staff by sending emails or mobile messages to their personal phones if any abnormal measurements was taken for the patient, and here is where the intelligence lays, RONA can detect the abnormal heart rate, temperature, Oxygen Percentage (SpO_2) measurements autonomously through comparing it to the reference values given by the medical staff.

2.1.4 Pharmacy Access System

To hold pharmaceuticals, RONA was given a three-degree-of-freedom arm. RONA goes to the pharmacy and distributes medicines to patients according to the mobile application schedule, eliminating the need for nurses to go there personally (Fig. 2).

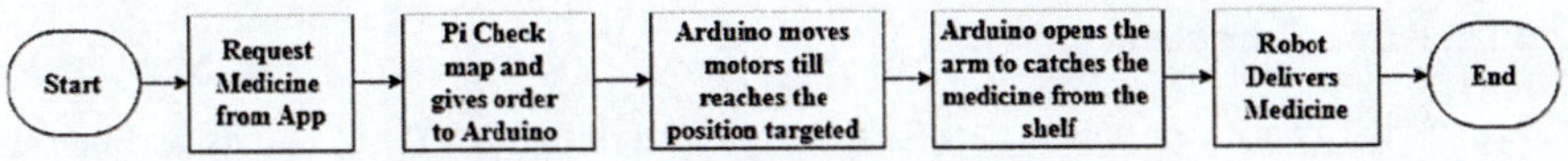

Fig. 2. Pharmacy access operation flowchart

2.1.5 Environment Safety System

This subsystem was included into RONA to ensure a safe environment for hospital patients. The system is based on two sensors that ensure smoke-free air, as well as the appropriate temperature and humidity.

2.1.6 User Interface (UI) System

The UI system is separated into two primary subsystems: the graphical user interface (GUI) and the sound system, both of which are used to connect with humans and provide them with necessary steps and directions.

2.1.7 IoT System

I. Mobile Application (App): it simplifies scheduling appointments and monitoring patients by updating their vital signs. The HCWs can access the patients' data received by the robot and receive emergency warnings if any result exceeds the safe line, and then follow the appropriate treatment plan.
II. Realtime Database: The Realtime Database (Firebase) is a cloud database that the robot uses to upload and save the data it collects (Fig. 3).

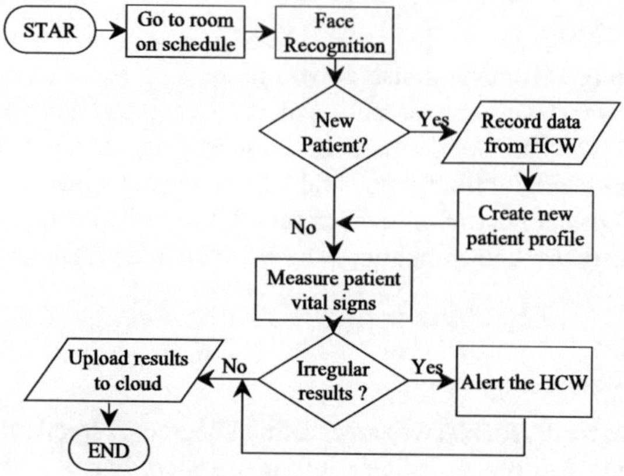

Fig. 3. System basic operation flowchart

2.2 Hardware Implementation

The main hardware components are Raspberry Pi 4B with 8 GB RAM and its peripherals along with an Arduino Mega 2560. The following list contains, in detail, all the hardware parts used in the robot:

i. Motors & Drivers:

   a. DC Motor
   b. VNH5019 Motor Driver

ii. Motion Guide Sensors:

   a. IR Sensor
   b. Ultrasonic Sensor

iii. Biomedical Sensors

   a. MLX90614 Non-Contact Infrared (IR) Temperature Sensor
   b. Maxim Integrated MAX30102 Pulse Oximeter & Heart-Rate Sensor
   c. ECG 3 Lead Sensor

iv. HC – 05 Bluetooth Module
v. DHT11 Sensor
vi. MQ2 Gas Sensor
vii. 9 g servo
viii. Raspberry Pi Camera v2

2.2.1 MLX90614 Sensor

A MLX90614 sensor was used to measure IR, which works on the Seebeck effect as shown in Fig. 4, which states that electromotive force is developed across two points of an electrically conducting material when there is a temperature difference between them. The sensor is made up of an array of thermocouples that form a thermopile through which the IR radiation is gathered and the temperature body is detected without contact.

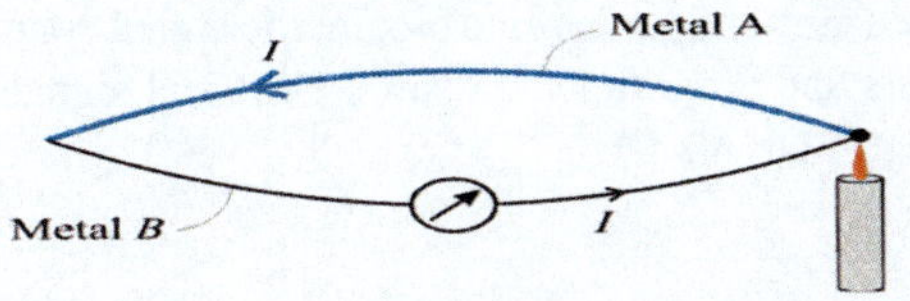

Fig. 4. The Seebeck effect

2.2.2 MAX30102 Sensor

A protein called Hemoglobin (Hb) is the principal carrier of oxygen across the veins and arteries in the human body. When Hemoglobin carries an oxygen particle, its color changes, and the light absorption changes as well. The oxygen percentage in the blood can be estimated using the following calculation based on the absorption rate of two light sources with different wavelengths (Fig. 5).

$$SpO_2 = \frac{HbO_2}{HbO_2 + Hb}$$

at SpO2: Oxygen %
HbO2: Oxyhemoglobin
Hb: Deoxyhemoglobin.

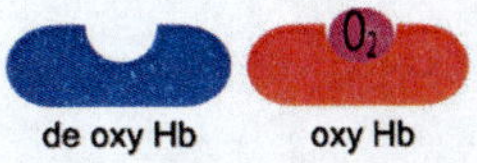

Fig. 5. Oxygen saturation calculation from Hb

The MAX30102 sensor, which consists of two Light Emitting Diodes (LEDs), one emitting IR and the other emitting red light, was used to calculate percentage of the Oxygenated Hemoglobin (HbO_2) and Hb, and hence SpO_2, using the ratio of the received IR and red rays.

The pulse rate of the heart can also be detected using the MAX30102 sensor, which is based on the same feature of hemoglobin, aside from the fact that the Oxyhemoglobin rate increases in the blood on the heartbeat and declines till the next beat. However, only one of the two LEDs is turned on when measuring the pulse rate, thus the rate of absorption is used, leaving us with the equation: Absorption Peaks/Minute = Pulse Per Minute (PPM) (Fig. 6).

2.2.3 ECG

An Electrocardiogram (ECG) is a test that checks the rhythm and electrical activity of the heart. The electrical signals produced by the heart each time it beats are collected through three electrodes attached to the skin in a triangular shape and sent to the AD8232

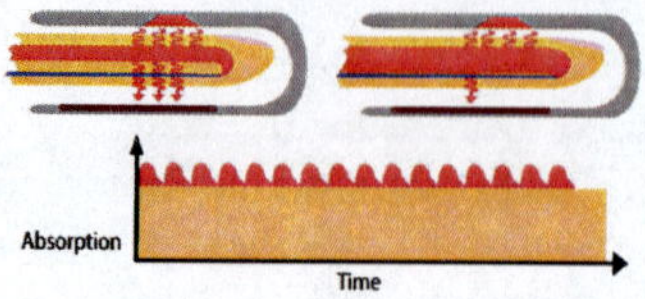

Fig. 6. Pulse rate working principle

module, an integrated front-end for cardiac bioelectric signal conditioning and heart rate monitoring, that filters and amplifies the heart's electrical signals to be plotted clearly and monitored through RONA's GUI.

2.2.4 DH11 Sensor

As water vapor is absorbed by the substrate, the ions are released, increasing the conductivity between the electrodes. The relative humidity affects the change in resistance between the two electrodes. The resistance between the electrodes reduces as the relative humidity rises, whereas the resistance rises as the relative humidity falls.

2.3 External Design Implementation

The external design was based on some principles that aimed to assure friendly user interaction and ability to accomplish the autonomous missions. These principles are:

1. Simplicity: robots in hospitals are designed for use by healthcare workers, physicians, and other professionals who lack engineering or debugging skills. As a result, designers must always provide simple construction, ease of handling, and quick maintenance for long-term usage of such equipment.
2. Humanization: external Body is a humanoid design that was chosen to make patient interaction more pleasant. RONA features a humanoid design to make it more friendly to patients; as a result, it exhibits human-like mobility and manipulation skills. A humanoid robot is anticipated to operate or help people in a human-centered setting without the need to adapt or adjust the surroundings.
3. Intelligence: AI, like much prior technology discourse, has become a catch-all term in nursing, referring to a wide range of gadgets or digital systems that mimic human cognitive processes such as problem-solving, decision-making, and training [18]. An excellent example of this is the facial recognition feature of the Raspberry Pi camera, in which RONA employs image processing technology to determine whether the patient has a database in the hospital or not, allowing RONA to determine whether he is a new patient or has old data.
4. Safety and Stability: This is one of the most important needs in healthcare technology, because the operator's safety is paramount while operating a robot in a medical setting. Even within the hospital, it should be safe for the operator, medical staff, health care professionals, and patients to be near a robot without posing a threat to anyone (Figs. 7 and 8).

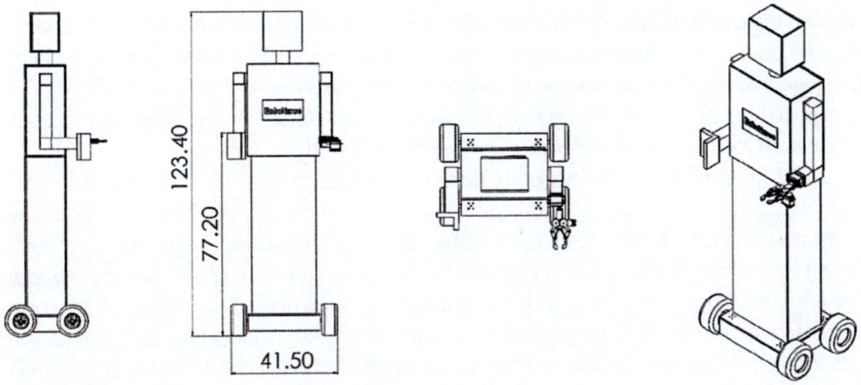

Fig. 7. Drawings

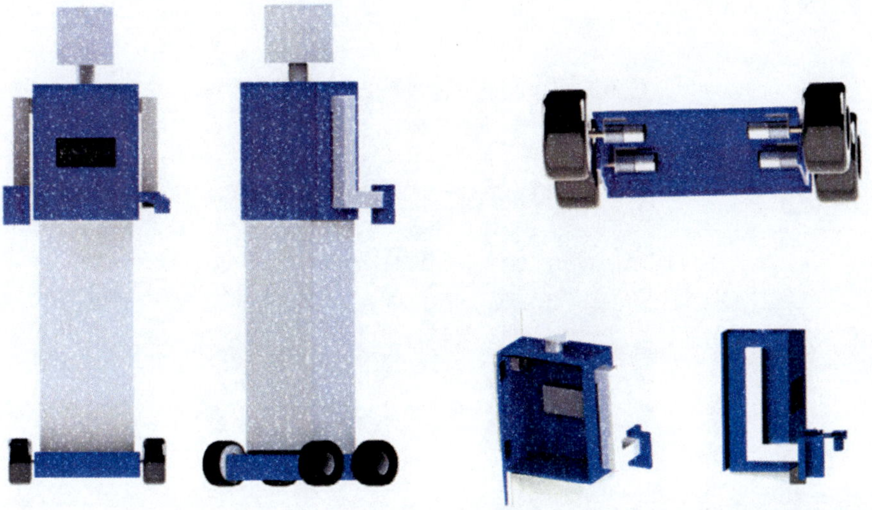

Fig. 8. CAD design

3 Results and Discussions

RONA, because of the methods mentioned above, is a fully automated robot that, using its autonomous features, can bring new safety levels to the medical staff in the hospitals through decreasing the direct contact required to follow up the patients' cases in some hospital sections like post-anesthesia and infection control. This will also relieve some of the repeated mission's headache of the nurses, as it moves autonomously based on a schedule to measure the vital signs of the patients and send it to the cloud that can be accessed on any registered account from the medical staff, besides the alarming system for emergency that assures the medical staff never miss a critical case. In short words, RONA is a robot that can allow the medical staff to keep a close eye on the patients without even leaving their offices. RONA uses ArcFace deep learning model to do face identification and AInnoFace model for face detection. With the help of TensorFlow Lite and Jetson Nano, it is easier to deploy such costly models on small devices. It's also possible to make use of the IoT system to use the cloud for such operation intensive models. RONA is flexible enough to cope with different scenarios.

In addition to that, RONA's body design resulted in a user-friendly interface with a light weight of 15 kg. The face recognition accuracy ranged around 99,38%. The power consumption ranged around 3 A/h at 100% operation mode where the robot keeps moving from room to room without stopping. While at average mode, a 12 A/h battery is enough for 6 h operating without the need to recharge. The medical measuring process for the patient takes around 3–5 min maximum. However, RONA has some limitations. First one is the inability to autonomously graph the electric cardio activity and only introduced a mobile ECG unit that requires some assistance from the nurse, and the second is that it doesn't contain a strong natural language processing ability which may introduce some interaction hardship. Below are some figures of RONA's result (Figs. 9, 10, 11, 12, 13, 14, 15, 16 and 17).

Fig. 9. Final external shape

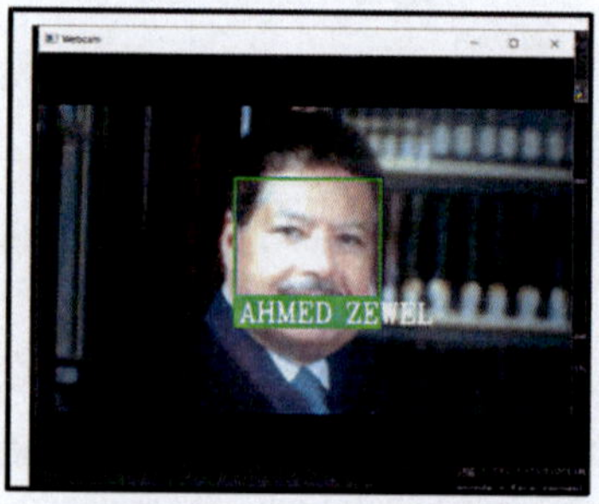

Fig. 10. Face recognition

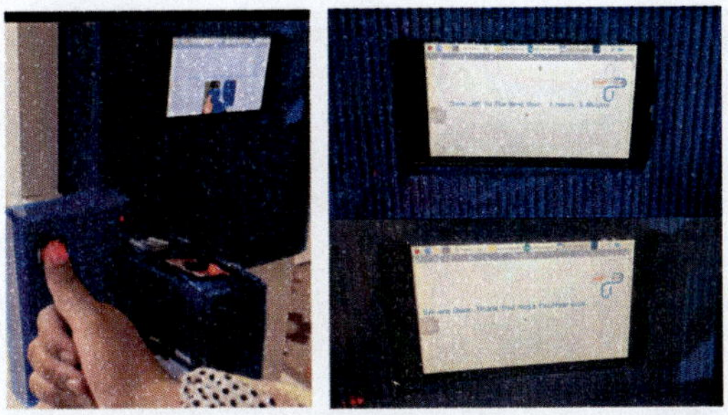

Fig. 11. Patient machine interaction

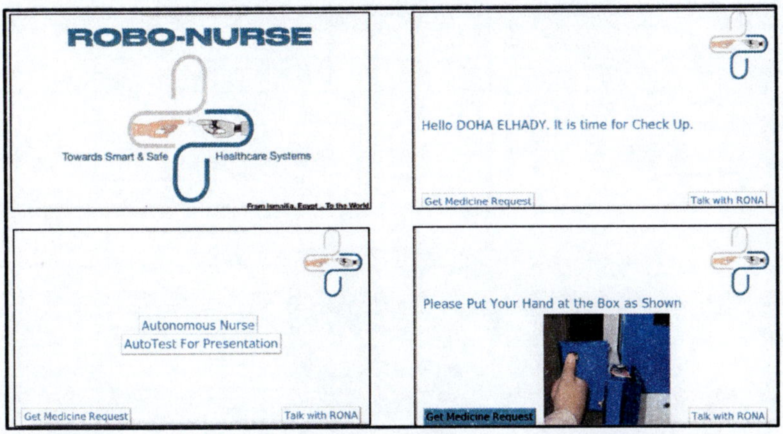

Fig. 12. GUI

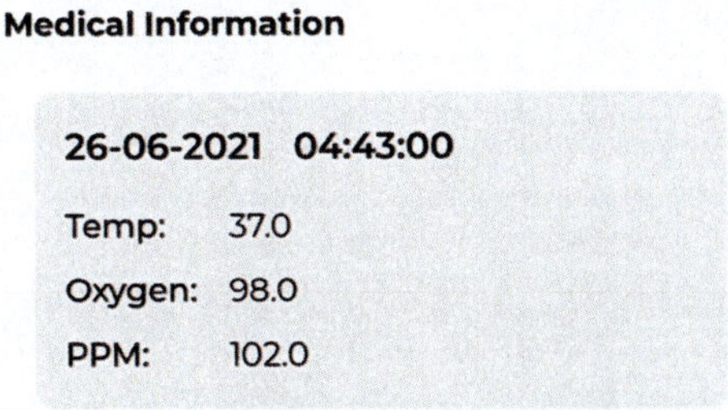

Fig. 13. Medical sensors readings on app

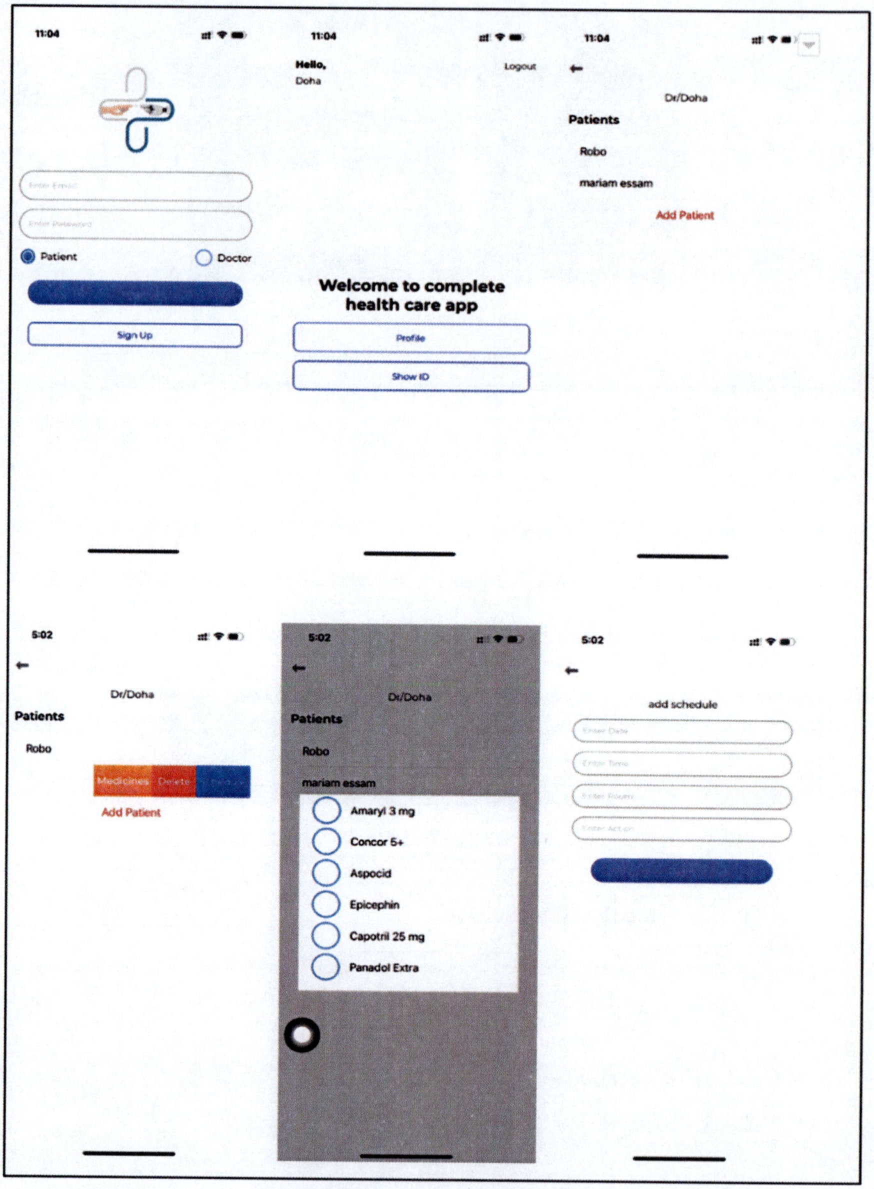

Fig. 14. Mobile application structure

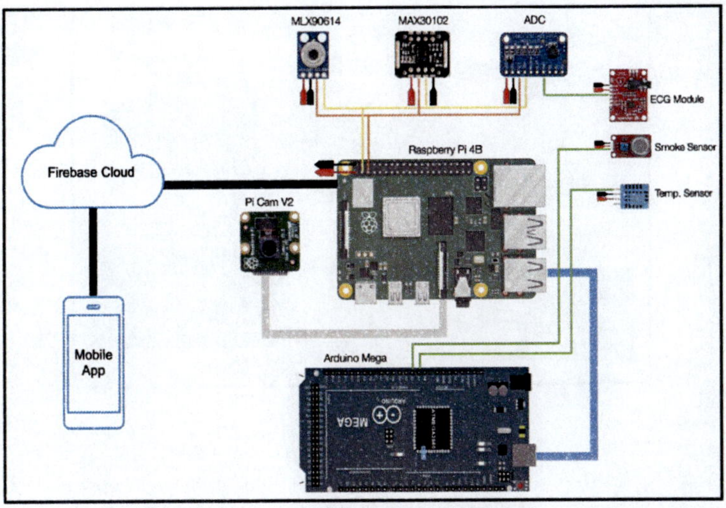

Fig. 15. Data acquisition system integration diagram (SID)

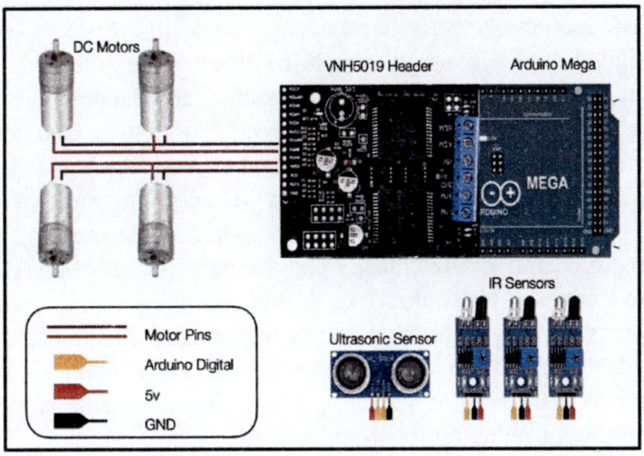

Fig. 16. Motion SID

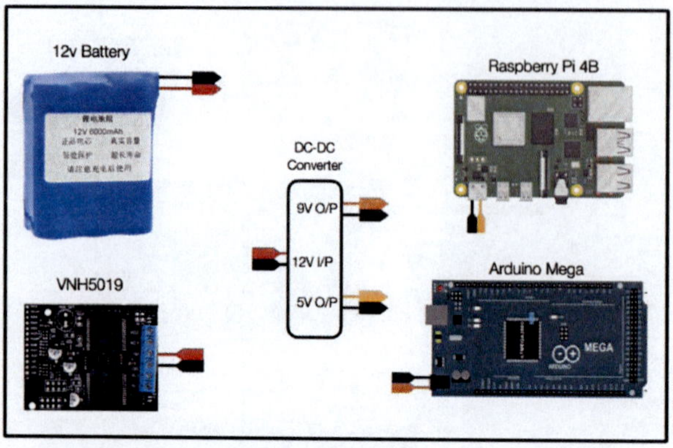

Fig. 17. Power SID

4 Conclusion

This paper aimed to present a model of a Robot Nurse, designed to assist the HCWs in some sections of the hospital. The robot design was selected to be a humanoid one, propelled by DC motors using electrical energy stored in a Lithium-ion battery. The robot was designed to obtain a light weight of 15 kg. The patient identification was based on a high accuracy image processing algorithm, and the motion was based on an efficient line following algorithm. The database was chosen to be Google Firebase to be easily connected to the App through the internet. There is no doubt that healthcare today is an exciting, rapidly developing, and changing industry. Therefore, it is necessary to continuously apply some developments and improvements to the project to keep it in line with the requirements of the industry and the needs of patients and HCWs. Some of these improvements so far could make RONA recognize different languages using natural language processing, determine the patient's gender, and increase Motors torque for more weight, so it could deliver more medications, heavy supplies…etc. RONA will be able to take over tasks such as delivering food, and transferring or moving patients, identify joy, sadness, anger and surprise and all other expressions to maintain the mental health of the patient, and interpret non-verbal language, such as the tilt of the head, a frown, a smile, and tone of voice.

Acknowledgment. The authors would like to acknowledge "SCU Racing Team" from the faculty of Engineering, Suez Canal university for providing the required facilities and environment to complete the research work. The authors would also like to thank Prof. Dr. Sahar Farouk and Prof. Dr. Mohamed Khalil for their guidance during developing the medical system and Eng. Mostafa Samir for his support during developing the IoT system.

References

1. Booth, R., Strudwick, G., McMurray, J., Chan, R., Cotton, K., Cooke, S.: The future of nursing informatics in a digitally-enabled world. In: Hussey, P., Kennedy, M.A. (eds.) Introduction

to Nursing Informatics. Health Informatics, vol. 16, pp. 395–417. Springer, Cham (2021). https://doi.org/10.1007/978-3-030-58740-6_16

2. Buchanan, C., Howitt, M., Wilson, R., Booth, R., Risling, T., Bamford, M.: Nursing in the Age of Artificial Intelligence: Protocol for a Scoping Review. JMIR Res. Protocols Med. Psychol. **9**(4), e17490 (2020)
3. Taylor, R.H., Menciassi, A., Fichtinger, G., Fiorini, P., Dario, P.: Medical robotics and computer-integrated surgery. In: Siciliano, B., Khatib, O. (eds.) Springer Handbook of Robotics. Springer Handbooks, pp. 1657–1684. Springer, Cham (2016). https://doi.org/10.1007/978-3-319-32552-1_63
4. Hameed Khan, Z., Siddique, A., Lee, Ch.W.: Robotics utilization for healthcare digitization in global COVID-19 management. MDPI Int. J. Environ. Res. Public Health **17**(11), 3819 (2020)
5. Jiang, J., Huang, Z.: Research progress and prospect of nursing robot. ResearchGate Recent Patents Mech. Eng. **11**(1), 41–57 (2018)
6. Liu, F., Xu, W., Huang, H., Ning, Y., Li, B.: Design and analysis of a high-payload manipulator based on a cable-driven serial-parallel mechanism. ASME J. Mech. Robot. **9**, (JMR-19-1033) (2019)
7. Fannin, R.: The rush to deploy robots in China amid the coronavirus outbreak. J. Power Electron. (2020)
8. Satale, K., Bhave, T., Chandak, C., Patil, S.A.: Nursing robot. Int. Res. J. Eng. Technol. (IRJET) **7**(12) (2020)
9. Abutaleb, A., Alsabhani, J., Alkinani, S., Alkaydi, S., Alghamdi, S., Bensenouci, A.: Design and implementation of a nurse robot. In: Proceedings of the International Conference on Industrial Engineering and Operations Management Dubai, UAE, 10–12 March 2020
10. Kyrarini, M., et al.: A survey of robots in healthcare. Technologies **9**, 8 (2021). https://doi.org/10.3390/technologies9010008
11. Eggleston, K., Lee, Y.S., Iizuka, T.: Robots and labor in the service sector: evidence from nursing homes. National Bureau of Economic Research, 1050 Massachusetts Avenue Cambridge, MA 02138, January 2021
12. Deng, J., Guo, J., Xue, N., Zafeiriou, S.: ArcFace: additive angular margin loss for deep face recognition. arXiv:1801.07698 v3 [cs.CV], 9 February 2019
13. Zhang, F., Fan, X., Ai, G., Song, J., Qin, Y., Wu, J.: Accurate face detection for high performance. arXiv:1905.01585 v3 [cs.CV], 24 May 2019
14. Kwoh, Y.S., Hou, J., Jonckheere, E.A., Hayati, S.: A robot with improved absolute positioning accuracy for CT guided stereotactic brain surgery. IEEE Trans. Biomed. Eng. **35**(2), 153–160 (1988)
15. Otero, J.R., Paparel, P., Atreya, D., Touijer, K., Guillonneau, B.: History, evolution and application of robotic surgery in urology. Urol. Robot. Surg. Arch. Esp. Urol. **60**(4), 335–341 (2007)
16. Wu, A.H.: Structural design and kinematics analysis of a comprehensive nursing service robot. MSc Dissertation, Tianjin University of Technology, Tianjin, China, December 2015
17. Jiang, J., Huang, Z., Huo, B., Zhang, Y., Song, S.: Research progress and prospect of nursing robot. Recent Patents Mech. Eng. **11**, 41–57 (2018)
18. Russell, S., Norvig, P.: Artificial Intelligence A Modern Approach Book. Prentice Hall Series in Artificial Intelligence. Prentice Hall, Upper Saddle River (2010)
19. Rayes, A., Salam, S.: Internet of Things from Hype to Reality, The Road to Digitization, 2nd edn. Springer, Cham (2022). https://doi.org/10.1007/978-3-030-90158-5
20. Robotic Nurse Assistant Market Size, Share & Trend Analysis Report By Product (Independence Support, Daily Care & Transportation and Pharma Automation), By End Use, By Region and Segment Forecasts, 2019–2026 (2019)

Resolving Context Inconsistency Approach Based on Random Forest Tree

Mohamed Hamed(✉), Hatem Abdelkader, and Amira Abdelatey

Information System Department, Faculty of Computers and Information, Menoufia University, Shebeen El-Kom 13829, Egypt
m.h.shaban128@gmail.com,
{hatem.abdelkader,amira.mohamed}@ci.menofia.edu.eg

Abstract. In modern and automated environments, billions of devices are communicating over the Internet in a robust and automated pattern. These devices used for sending information about surrounding objects and the environment, we call this information "Context". Contexts can help applications to be more context-aware (the ability of an application to perceive contexts and act automatically upon them). Sometimes context becomes inaccurate, noisy, or inconsistent due to some malfunction or transmission error. Several approaches have been suggested to solve inconsistent contexts by providing a context inconsistency resolution mechanism. This paper proposes a new approach to increase resolution accuracy by building a random forest tree model. The model is used to predicate which resolution method to use on context inconsistency based on previous knowledge. The results show that the proposed model is more effective in protecting correct contexts, removing bad contexts, and improving the functionality of context awareness.

Keywords: Internet of Things · Context awareness machine learning · Random forest

1 Introduction

The Internet of Things allows people and things to be Fully connected, ideally using any path/network and any service [1]. Rapid development of sensor networks and radio frequency identification (RFID) has led to a tremendous number of devices connected to the Internet sharing their data over time with each other. The IoT paradigm will surround users with smart and comfortable information about their environment. IoT allows every object to communicate and act upon environmental information (Context), which will offer the ability to understand and infer more environmental contexts. [2] IoT paradigm has three different definitions: Internet, things, and semantic oriented visions. The Internet and things visions deal with the sensors and the enabling technologies for connectivity.

The semantic vision takes great interest in representing, interconnecting, and organizing information generated by the IoT. In this context, middleware plays a significant role. Middleware is a software layer that hides a lot of functionality

© The Author(s), under exclusive license to Springer Nature Switzerland AG 2022
A. E. Hassanien et al. (Eds.): AMLTA 2022, LNDECT 113, pp. 192–199, 2022.
https://doi.org/10.1007/978-3-031-03918-8_18

needed by IoT. One of the main features of middleware is context awareness. Context-awareness is one of the most essential infrastructure requirements in IoT.

Context-awareness functionality [2–5] means the ability of an application to perceive contexts and act automatically upon it. But contexts are often defective and happen due to poorly functioning sensors or network transmission errors. Inaccurate and noisy context may cause applications to behave incorrectly and deviate from their original application specification.

Context inconsistency happens when two or more contexts describing the same environment are not giving the same information (e.g., two contexts coming from sensors deployed in a patient room, one may state that the patient is sleeping while the other says he is out). Thus, in recent years, context inconsistency detection has received attention [15,16]. Besides inconsistency detection, resolving defect contexts is also an important issue. Different approaches addressed context inconsistency resolution, which is the main goal of this paper.

The rest of the paper is organized as follows. Section 2 presents previous works related to inconsistency resolution. Section 3 offers our proposed approach. Section 4 shows experimental results and evaluations conducted. Section 5 concludes this paper with suggested future directions.

2 Related Work

Several research works addressed inconsistency in different fields such as medical [18,19] and industrial [2,14,15] fields. Different strategies have been studied in resolving consistency. Bu Yingyi et al. [6] has proposed a drop all strategy. The drop-all strategy adopts removing all contexts related to inconsistencies. Drop all strategy does not consider whether the contexts being removed are correct or not. It takes a cautious method in working with incorrect contexts. This leads to destroying more contexts than needed to be deleted.

P. Insuk. [7] attempt to follow user preferences or policies to solve inconsistencies. Chomicki et al. [8] proposed a strategy of the drop latest contexts. This strategy leads to the discovery and removal of the latest context. the drop latest assumes that this context is incorrect as it broke the data integrity and triggered an inconsistency. Therefore it is more likely to be inconsistent. Chang Xu et al. [9] follow a different strategy for inconsistency resolving. They take the approach of not considering a context incorrectly immediately; instead, it waits for more information to examine. Drop bad strategy compares context with past contexts and future contexts. This proves to be more efficient but at the expense of it leave context available until it can be marked as bad context. Chen et al. [10] combined low-level context inconsistencies with high-level system error recovery mechanism.

In [11,12], a strategy can resolve context inconsistencies by choosing the solution that minimizes the side effects. The side effect approach was highly dependent on user choice that may guide the resolution to the desired goal but not correct context inconsistency resolution. Xiang Li [13] utilize ontologies and formulas of time and distance reliability to form a hybrid method.

This paper proposes a new modeling approach, which builds a learning model to choose the best resolution strategy. The training of the model is based on previously solved inconsistencies features and their top-performing methods. So, the selection of strategy is the model's role, based on a historically suitable strategy. This model is intended to solve inconsistency in real-time and was evaluated against these real-time methods drop bad, drop latest, and drop all.

3 Proposed Approach

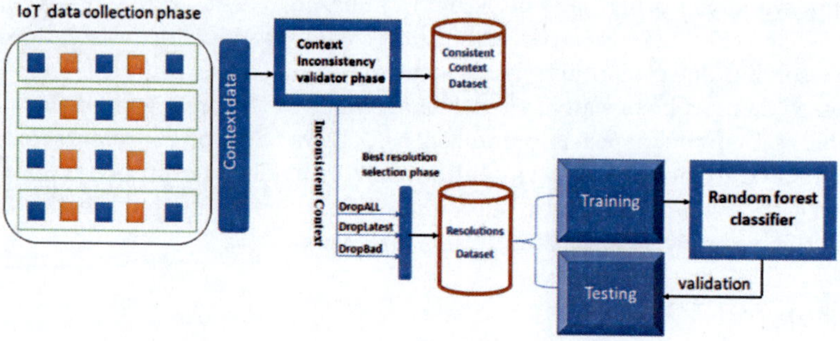

Fig. 1. Proposed model approach for addressing inconsistency

3.1 IoT Data Collection Phase

The proposed modeling approach starts with IoT devices, which are the data source. The proposed approach follows some steps, as shown in Fig. 1. The first step is providing data through IoT devices. To illustrate our proposed modeling approach, we have established a simulated environment of a warehouse, the environment record data coming from sensors deployed to each location in the warehouse. These data represent the location contexts of several loaders and items found in the warehouse. For each passing time, we record these contexts in the dataset. A controlled number of applications were randomly generated over time to move items by loaders to different locations and are executed by order of the time they came. The environment ran for about 100000-time increments and had 1000 applications performed by 10 loaders on 10 items.

Random errors of 40% are applied to the context dataset to simulate sensors malfunction. Each time passing, a detect context inconsistency is used by checking contexts coming in that time against several rules designed according to the environment.

3.2 Context Inconsistency Validator

The validator's purpose is to detect context inconsistencies by validating each coming context against a set of rules tailored to the environment. The consistent context data is preserved in a consistent context dataset and can be used by upcoming applications. On the other hand, contexts that break environment rules are supplied to the three resolution strategies and solved by each strategy.

3.3 Best Resolution Selection

In the proposed approach, three resolution strategies are applied to all inconsistencies detected in the environment. An evaluation is conducted for each resolution strategy result to reveal the most suitable one using the F-score [21] formula. F-score is the harmonic mean of the precision and recall and is defined by:

$$F_1 = \frac{\text{Precision} \times \text{Recall}}{\text{Precision} + \text{Recall}}$$

The best-applied resolution strategy for each inconsistency is chosen for the resolution dataset. The resolution dataset contains features of different inconsistencies detected and their suitable selected resolution strategy.

3.4 Random Forest Tree

Random Forest is a supervised algorithm that can be used for both classification and regression problems [20]. Random forest depends on decision trees. The main idea behind random forest is to create multiple decision trees from the training dataset and use average voting between these decision trees to classify new data. A random forest tree number of features are selected randomly to build each tree. The technique behind Random forest gives it more prediction accuracy than a decision tree.

The resolution dataset is the main component for modeling the learning model in the proposed modeling approach. The resolution dataset contains inconsistent data acquired from IoT devices. These inconsistent context data was resolved by each of the three resolution strategies and their top-performing method was selected as class attribute.

The collected dataset was used as a training dataset to build a model. The model predicates the class attribute of which resolution strategies to use and will be used for future coming inconsistency incidents based on their features. Our model was built using a random forest classification algorithm.

4 Experimental Results and Evaluations

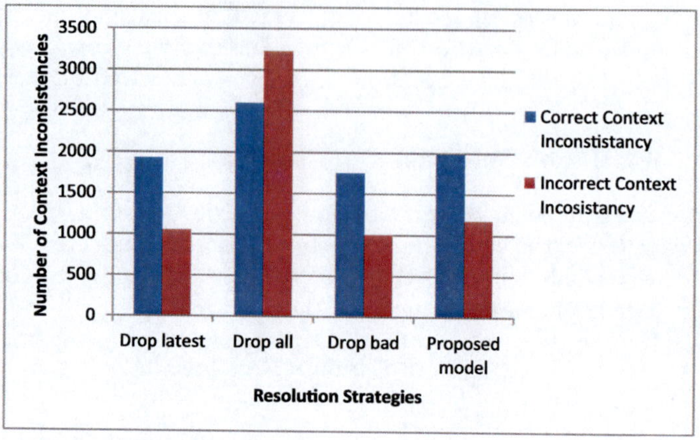

Fig. 2. Total number of correct and incorrect context inconsistencies resolved

The model was tested using the cross-validation [22] method divided into 10 folds and it showed a 94.54% accuracy result. Accuracy is defined as the total number of correct predictions. A confusion matrix was built to show the result of testing the model:

Table 1. Confusion matrix

		Predicated		
		DropLatest	**DropBad**	**DropAll**
Actual	**DropLatest**	728	45	13
	DropBad	63	627	5
	DropAll	11	10	1192

Figure 2 shows that although the drop all resolution strategy is incorrectly marking contexts participating in context inconsistencies yet. It tends to remove a lot more contexts wrongly marked as bad contexts. These incorrect contexts marked as bad may become handy while moving objects in the environment. That is because it removes all contexts participating in the inconsistency regardless of whether they are correct. On the other hand, our proposed model shows better results than drop latest and drop bad while more effectively maintaining correct contexts. Also, it does not require for all the incidents to wait for more information to decide which context is worse like in drop bad resolution strategy.

Figure 3 shows the number of undetected inconsistencies by each resolution strategy. We found that drop all again is more effective in detecting all inconsistencies. However, still, it suffers the most from removing correct contexts in contrast to our model, which still shows a balanced performance against all other methods.

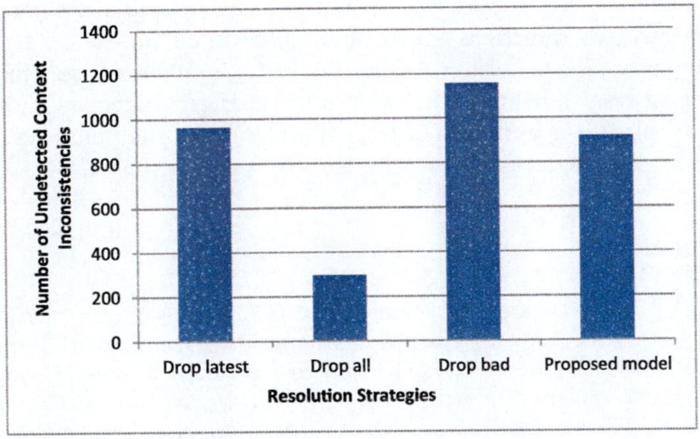

Fig. 3. Total number of un detected context inconsistencies

Figure 4 shows the total number of movements made by items and loaders in the environment. In a production use case, as the number of movements grows it will cost more money and time. Figure 3 shows that our model holds the same performance as drop latest and drop bad while being more cost-effective than drop all method.

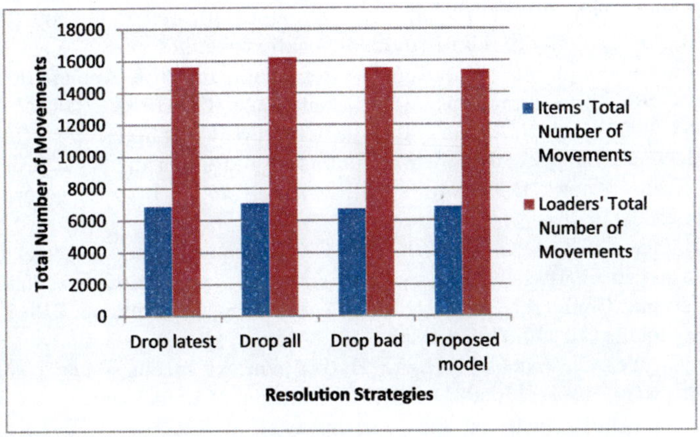

Fig. 4. Total numbers of items and loaders movements

5 Conclusion and Future Directions

This paper proposes a new modeling approach in resolving context inconsistencies. We have collected a resolution data set of best-performing resolution strategies. Then we trained our model using a random forest algorithm. The results show that our model is more effective in resolving inconsistencies with the least loss of correct contexts compared to other resolution techniques.

Future research directions are to investigate more the cold start problem related to our approach. Also, we need to investigate different machine learning techniques to resolve inconsistencies in automated environments based on the Internet of things. As systems start for the first time, they must use previous strategies until enough resolution data is ready to train the model.

References

1. Bassi, A., Horn, G.: Internet of things in 2020. In: Proceedings of the Joint European Commission/EPoSS Expert Workshop on RFID/Internet-of-Things (2008)
2. Atzori, L., Iera, A., Morabito, G.: The internet of things: a survey. Comput. Netw. **54**(15), 2787–2805 (2010). https://doi.org/10.1016/j.comnet.2010.05.010
3. Makris, P., Skoutas, D.N., Skianis, C.: A survey on context-aware mobile and wireless networking: on networking and computing environments' integration. IEEE Commun. Surv. Tutor. **15**(1), 362–386 (2013). https://doi.org/10.1109/SURV.2012.040912.00180
4. Dinh, L.T.N., Karmakar, G., Kamruzzaman, J.: A survey on context awareness in big data analytics for business applications. Knowl. Inf. Syst. **62**, 3387–3415 (2020). https://doi.org/10.1007/s10115-020-01462-3
5. Chelloug, S.A., El-Zawawy, M.A.: Middleware for internet of things: survey and challenges. Intell. Autom. Soft Comput. (2017). https://doi.org/10.1080/10798587.2017.1290328
6. Bu, Y., Gu, T., Tao, X., Li, J., Chen, S., Lu, J.: Managing quality of context in pervasive computing. In: 2006 Sixth International Conference on Quality Software (QSIC 2006), pp. 193–200 (2006). https://doi.org/10.1109/QSIC.2006.38
7. Park, I., Lee, D., Hyun, S.J.: A dynamic context-conflict management scheme for group-aware ubiquitous computing environments. In: 29th Annual International Computer Software and Applications Conference (COMPSAC 2005), vol. 2, pp. 359–364 (2005). https://doi.org/10.1109/COMPSAC.2005.21
8. Chomicki, J., Lobo, J., Naqvi, S.: Conflict resolution using logic programming. IEEE Trans. Knowl. Data Eng. **15**(1), 244–249 (2003). https://doi.org/10.1109/TKDE.2003.1161596
9. Xu, C., Cheung, S.C., Chan, W.K., Ye, C.: Heuristics-based strategies for resolving context inconsistencies in pervasive computing applications. In: 2008 the 28th International Conference on Distributed Computing Systems, pp. 713–721 (2008). https://doi.org/10.1109/ICDCS.2008.46
10. Chen, C., Ye, C., Jacobsen, H.-A.: Hybrid context inconsistency resolution for context-aware services. In: 2011 IEEE International Conference on Pervasive Computing and Communications (PerCom), pp. 10–19 (2011). https://doi.org/10.1109/PERCOM.2011.5767574

11. Xu, C., Cheung, S.C., Chan, W.K., Ye, C.: On impact-oriented automatic resolution of pervasive context inconsistency. In: The 6th Joint Meeting on European Software Engineering Conference and the ACM SIGSOFT Symposium on the Foundations of Software Engineering: Companion Papers (ESEC-FSE Companion 2007), pp. 569–572. Association for Computing Machinery, New York (2007). https://doi.org/10.1145/1295014.1295043
12. Xu, C., Ma, X., Cao, C., Lu, J.: Minimizing the side effect of context inconsistency resolution for ubiquitous computing. In: Puiatti, A., Gu, T. (eds.) MobiQuitous 2011. LNICST, vol. 104, pp. 285–297. Springer, Heidelberg (2012). https://doi.org/10.1007/978-3-642-30973-1_29
13. Li, X., Li, G.: A hybrid context inconsistency resolution method. In: 2015 7th International Conference on Intelligent Human-Machine Systems and Cybernetics, pp. 73–76 (2015). https://doi.org/10.1109/IHMSC.2015.121
14. Domingo, M.G., Forner, J.A.M.: Expanding the learning environment: combining physicality and virtuality - the internet of things for eLearning. In: 2010 10th IEEE International Conference on Advanced Learning Technologies, pp. 730–731 (2010). https://doi.org/10.1109/ICALT.2010.211
15. González, G.R., Organero, M.M., Kloos, C.D.: Early infrastructure of an internet of things in spaces for learning. In: Eighth IEEE International Conference on Advanced Learning Technologies, 2008, pp. 381–383 (2008). https://doi.org/10.1109/ICALT.2008.210
16. Zhang, D., Huang, H., Lai, C.F., et al.: Survey on context-awareness in ubiquitous media. Multim. Tools Appl. **67**, 179–211 (2013). https://doi.org/10.1007/s11042-011-0940-9
17. Raychoudhury, V., Cao, J., Kumar, M., Zhang, D.: Middleware for pervasive computing: a survey. Pervas. Mobile Comput. **9**(2), 177–200 (2013). https://doi.org/10.1016/j.pmcj.2012.08.006
18. Zhao, W., Wang, C., Nakahira, Y.: Medical application on internet of things. In: IET International Conference on Communication Technology and Application (ICCTA 2011), pp. 660–665 (2011). https://doi.org/10.1049/cp.2011.0751
19. Hu, F., Xie, D., Shen, S.: On the application of the internet of things in the field of medical and health care. In: 2013 IEEE International Conference on Green Computing and Communications and IEEE Internet of Things and IEEE Cyber, Physical and Social Computing, pp. 2053–2058 (2013). https://doi.org/10.1109/GreenCom-iThings-CPSCom.2013.384
20. Breiman, L.: Random forests. Mach. Learn. **45**, 5–32 (2001). https://doi.org/10.1023/A:1010933404324
21. Sammut, C., Webb, G.I.: Encyclopedia of Machine Learning, 1st edn. Springer Publishing Company, Incorporated (2011)
22. Refaeilzadeh, P., Tang, L., Liu, H.: Cross-validation. In: Liu, L., Özsu, M.T. (eds) Encyclopedia of Database Systems. Springer, Boston (2009). https://doi.org/10.1007/978-0-387-39940-9_565

Arduino Line Follower Using Fuzzy Logic Control

Kuo-Chi Chang[1,3,5], Shoaib Ahmed[2,3](✉), Zhang Cheng[2,3], Abubakar Ashraf[2,3], and Fu-Hsiang Chang[4]

[1] Department of Applied Intelligent Mechanical and Electrical Engineering, Yu Da University of Science and Technology, Miaoli 361, Taiwan, Republic of China

[2] School of Information Science and Engineering, Fujian University of Technology, Fuzhou, China
eshoa1996@outlook.com

[3] Fujian Provincial Key Laboratory of Big Data Mining and Applications, Fujian University of Technology, Fuzhou, China

[4] Department of Tourism, Shih-Hsin University, Taipei 116, Taiwan, Republic of China

[5] Department of Business Administration, North Borneo University College, 88400 Kota Kinabalu, Sabah, Malaysia

Abstract. The increasing demand of today's scenario for various sectors of life is its automation. The dream of affordable and productive labor made it day and night for the researchers of the time to get the dream to reality. However, due to the absence of compact processors to perform the calculations that were also required, the challenge was futile. Today, the microprocessor Arduino line technology is thousands of times more advanced than what occurred back in this technologically advanced world. In this project we are using four geared DC motors of 3 V and for line detection we are using two IR sensors. L298N motor driver IC is used to fulfill supply voltage for dc motor because Arduino UNO cannot provide the desire Voltage. In AT mega 328 microcontroller put on Arduino UNO, the Fuzzy logic control algorithm is implemented as the brain of line follower robot and it is programmed to follow a particular route. The goal of the project is to create a robotic machine which will follow a defined path. Path can be black or white and can be controlled using PWM (Pulse Width Modulation This technology will be used in few years in a fast way due to its automation purpose. This project will be useful for hospital patients and especially for blind people it can be beneficial for industrial state and reduced the overall cost of industry. Through line follower we can manage to work in the environment where human cannot work easily like in extreme heat.

Keywords: Automotive vehicle · Embedded system · Line follower · Logical control

1 Introduction

In the twenty first century every person need automation in various sectors of life. A great deal of development has been done to minimize the work load from human [1].

© The Author(s), under exclusive license to Springer Nature Switzerland AG 2022
A. E. Hassanien et al. (Eds.): AMLTA 2022, LNDECT 113, pp. 200–210, 2022.
https://doi.org/10.1007/978-3-031-03918-8_19

People often use vehicle to move from one place to another under the supervision of driver. In salt mines railway tracks are used to carry goods from one place to another which are costly and hard to maintain. In this project our main focus is to automate this sectors by making such a machine which will capable of following specific path Technology for operating buses and other public transit services has been proposed [2], and could end up as part of the highway navigation of autonomous vehicles.

First and foremost, no robot could have been built to completion without a stronghold on the microcontroller used. In recent years a great deal of time and effort have been spent on developing systems to enable an autonomous robot to follow a marked path using a vision system [3]. Not surprisingly, the majority of this research has been towards modifying, or designing from scratch, a full-sized road vehicle so that it can drive on ordinary roads without human supervision. Due to the large amount of space available in an ordinary road vehicle, high performance computers can be used to perform complex image processing and, typically, to maintain a mathematical model of the vehicle and the environment. Research into autonomous driving using smaller robots typically follows one of two approaches. In the first approach a mathematical model of the vehicle and its surroundings is generated, tested in simulation, and then applied to a robot built specifically for the purpose [4]. In the second approach a combination of a visual seroving system and a kinematic model is used, again the robot is typically designed around the solution technique due to the size of these robots, the processing resources available are quite limited so simpler models and techniques, such as visual servoing, are used to reduce the processing load [5].

This project is useful for hospital patients and for blind people and can be used as tourist guide in the museum. We are using four geared dc motors of 3 V each and the core of line follower machine will be microcontroller ATmega328p and all decision making is done through programming fed in microcontroller [6]. L298N motor driver Ic is used to fulfill supply voltage for four geared dc motor as they required 12 V and Arduino can only supply 5 V. So we add motor driver to drive the dc motor and overcome the deficiency. Line detection is done through IR sensor module So in this project Arduino Line Follower will be following black line on white surface and will trace the line as long as it is darker than its surrounding and will have the potential to take various degree of turns and it's simply the concept of Autonomous vehicle that's don't require any sort of supervision in general [7].

The following sections contain a brief overview of the research done in path following, including autonomous driving, and the research in visual servoing that can be applied to path following for autonomous robots.

2 Methodology

2.1 Lab Simulation

In the projects Proteus design lab is used for Simulink purposes. Proteus is simulating software that helps in attach many components with the ATmega328p. Like resistors, capacitors, LEDs, Motor driver IC, DC Motors and these are just fit. It has a complete library and everything that is ever need can be found here. We have used proteus version 8 for our project schematic diagram have shown Fig. 1.

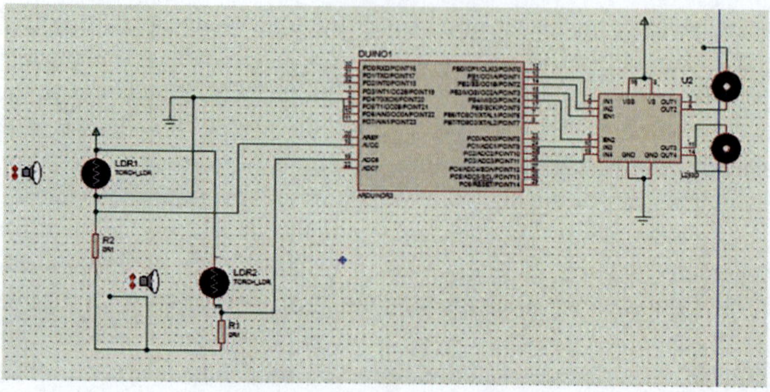

Fig. 1. Schematic diagram

The main features incorporated into the hardware are given below Table 1.

Table 1. Required components

No	Name	Quantity
1	The ATmega328p microcontroller	1
2	The voltage regulator and supporting components	1
3	Arduino UNO Board	1
4	Crystal oscillator (16 MHz)	1
5	The H-bridge motor control IC (L298N)	1
6	Motors, with coupled reduction gears	4
7	12 V LIPO battery	1
8	The LM393 comparator IC	1
9	Sensors	2

2.2 The ATmega328p Microcontroller

As it is a RISC (Reduced Instruction Set Computer) processor and is best suited for real-time operations, the ATmega328p microcontroller was used. It's also known as the line follower computer brain. It has 14 digital input/output pins, 6 analog inputs, a 16 MHz ceramic resonator, a USB link, a power port, an ICSP header, and a reset button (of which 6 can be used as PWM outputs) [8].

2.3 Voltage Regulator

An electrical regulator configured to automatically maintain a 32 is a voltage regulator level of constant voltage. It transforms +5 V to a positive voltage (7–29 V). Heat sink supplied in the center to release heat produced as a result of a fall through the IC. The input voltage is roughly 5 to 18 V, the ground voltage is 0 V and the controlled output is +5 V (Fig. 2).

Fig. 2. Voltage regulator

2.4 Circuit Diagram Explanation

We used two IR sensors for detecting the line in this Line Follower Robot circuit and a Comparator IC for comparing voltages [9]. The non-inverting mode configured comparator and the 10 K potentiometer are attached to its inverting terminal for reference voltage adjustment and the output of the IR receiver is directly connected to the non-inverting pins of both comparators. One Red LED is connected at output of in the sensor board when this led blinks then it means our sensor is working, then signal goes to microcontroller IC which is programmed and gives the output to the motor driver IC which rotates the motor s as per the programming of microcontroller IC (Fig. 3).

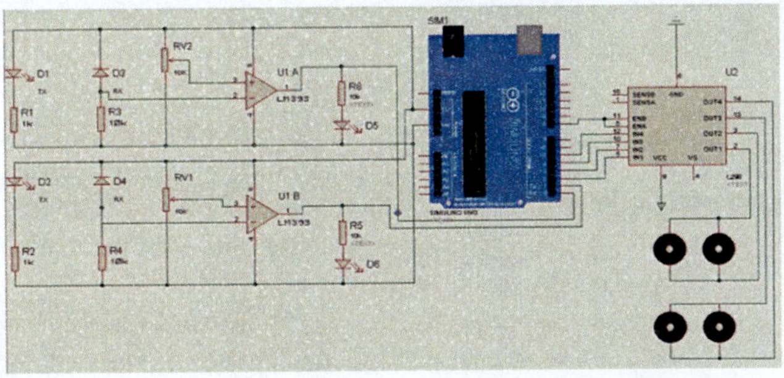

Fig. 3. Circuit diagram of Arduino line follower

Complete circuit diagram for Arduino line follower robot is shown in the above figure. It can be observe that output of comparators LM393 is directly connected to Arduino digital pin number 0 and 1. And the input pins 5, 7, 10 and 12 of the motor driver are connected. Digital Arduino pin numbers 2, 3, 4 and 5, respectively. And two motors are connected to the output pins of motor drivers 14 and 13, and two motors are connected to the output pins of pins 3 and 2. The connection of the dc motor is series.

2.5 Microcontroller-Motor Driver IC Interface

In this project we are using 4 dc motors of 3 V each and microcontroller can give maximum output of 5 V so L298N dual H-bridge module is used to control the DC motors [10]. It consists of four input switches and two enable PWM signals. It is connected to ATmega328 digital pin 2, 3, 4, and 5 for motor driver and enable pin 8 for motor PWM. The microcontroller can read signals from the logic of input switches.

2.6 Microcontroller-IR Sensor Module Interface

IR sensor module is connected to Arduino UNO digital pin number 0 and 1 respectively and need supply voltage of 5 V. Black line sensing is done through IR sensor and when IR tx hit the bright surface and light reflect back then robot will move. Sensing process is always done through digital signals and microcontroller can't read these signals so in this case we have used Comparator IC LM393IC.

2.7 Microcontroller-Variable Resistor Interface

Microcontroller is the heart of the whole line follower machine and it has only the decision-making ability and variable resistor is one of the main components that will be used in this project and the speed of dc motor is 125 rpm and some time we need to control speed of dc motor so we use variable resistor in this case variable resistor plays important role to control speed using PWM pins of the Arduino Uno [11]. Variable resistor have three pins Gnd, Vin and Vout respectively. Vin pin is connected to analog pin Ao and Vout is connected to Arduino digital pin 9 which is also known as PWM pin of Arduino we use PWM pin because we manually set our required speed we adjust speed according to our requirements.

2.8 Arduino IDE Interface with Microcontroller

Arduino IDE is the environment provided for microcontroller in this case we use ATmega328p and Microcontroller is hardware and we need a software platform to run code in this case Arduino IDE is used to compile code and feed this code in microcontroller to run line follower machine [12] we just need a USB cable and a pc with Arduino IDE in order to compile our required code "A "sketch" is considered a program or code written for Arduino. Arduino programs are written in C or C++. The Arduino IDE comes with a "Wiring" software library from the initial Wiring project, which makes it much simpler for certain popular input/output operations.

3 Summary of Methodology

In this section we will give overview that what we have actually done during the project firstly we have selected a 4wd robot chassis and 4 geared dc motor of 3 V each and ATmega328p microcontroller for decision making ability and to detect the line we have used 2 IR sensors Module that have the ability to sense the black line and comparator IC to convert analog signal to digital because sensing is always done through analog signal and microcontroller can understand only digital signal. Variable resistor is being used to control output voltage level using PWM pins of the microcontroller ATmega328p and to connect the whole circuit of the line follower machine we have used Male to Male and female to male jumper wires and mini breadboard to work as an extension for wiring to the whole circuit. L298n is a motor driver IC that is highly required to drive the dc motor because every motor is of 3 V each and microcontroller cannot fulfill 12 V because it can only give power of 5 V which is not enough l298n helps to drive the motor and we have done our wiring in series connection because in series we divide voltage and current remains the same and to power up the whole circuit we have used lipo battery of 12 V because it give us maximum output power which is not achievable in lithium ion battery in case of our project. We have used a power button which can support up to 3 A to easily power the circuit rather than connecting power jack every time to microcontroller. We have adjusted the IR sensors at the front end of the robot chassis which can sense the line from distance of 2.8 cm and give feedback to the microcontroller and microcontroller can give instruction to dc motors according to the feedback provided by IR sensors.

4 Physical Modeling

In this section we will discuss briefly about line follower Physical hardware structure and working principle of the machine and we will also discuss about its embedded quality of performing specific task at a particular time.

4.1 Block Diagram

Microcontroller (Atmega328), IC Comparator (LM393) [13–15], IR Sensors and IC Motor Driver modules (L298N) Block diagram clarifies that the Left and Right sensor sense the black line on white surface. LM393 Comparator is used to compare different voltage level. ATmega 328P is used as a backbone for the whole process motor driver IC (Fig. 4).

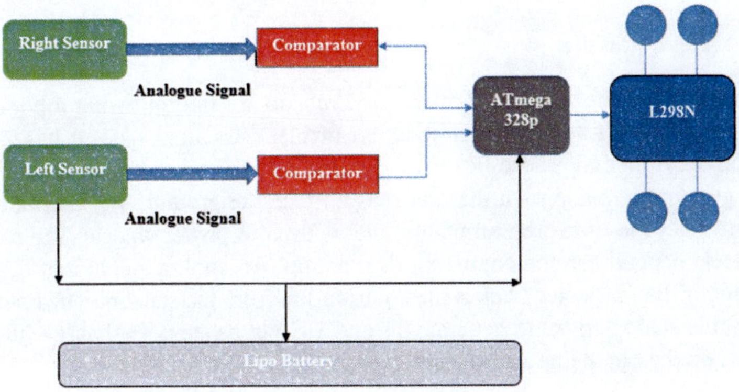

Fig. 4. Block diagram of Arduino line follower

4.2 Flow Chart

The flow chart initializes the process by sensing the black line on white surface with the help of IR sensor. In this Project we have used two IR sensor and according to the logic defined when we get voltage level of zeros from both the sensor the line follower machine it will send message to microcontroller and machine will move to the forward direction same as the case for left turn when left sensor voltage level is high it will move left side and vice versa for right side turn. If the voltage level of both right and left sensor is high, then it will stop the line follower machine (Fig. 5).

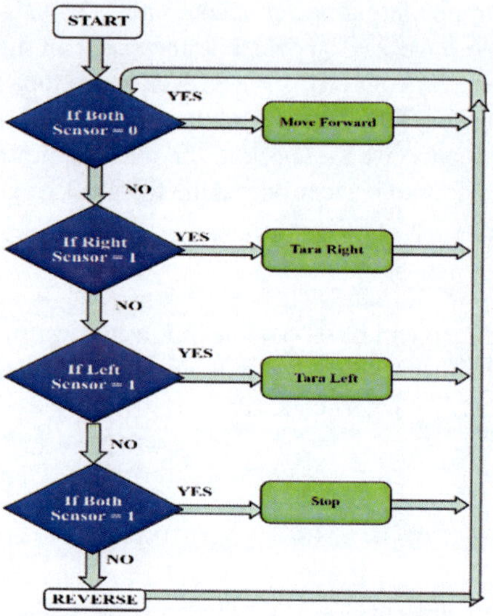

Fig. 5. Flow chart of Arduino line follower

4.3 Working Principle

In order to monitor the whole operation, the Embedded Line following robot uses four motors [15, 16], [17, 18]. For identification of black tracking tape, it has 2-infrared sensors on the front end. These IR sensors are used by the robot to detect the rows, and the configuration is made such that the sensors face the ground. We internally have a processor OTP (one time programmable) that is used to power wheels. The rotation of these wheels depends on the comparator's reaction. An analog signal that depends on the amount of light reflected back is the contribution from the sensors. The comparators are given this analog signal to generate 0 s and 1 s. Let us suppose that it reads 1 when a sensor is on the black line and it reads 0 when it is on the bright surface.

Straight Direction. When the left and right sensor responses are high, we should expect our robot to drive in a straight direction. That is, the left and right sensor will both be on the white surface according to our arrangement, and when the sensor sense the surface it will reflect the radiation back and line follower machine will move towards straight direction (Fig. 6).

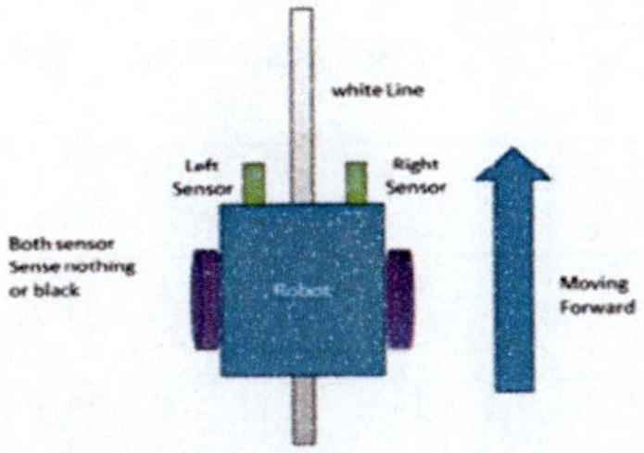

Fig. 6. Robot straight direction

Right Curve. The responses will change when a right curve is found on the line, i.e. the reaction of the right will become low as the sensor faces the black line and the response of the left sensor will be high. The control of the wheels is altered by this information changed i.e. the right wheel is held and the left wheel is made to move freely until the response from the left sensor becomes low. Then the same process repeats again (Fig. 7).

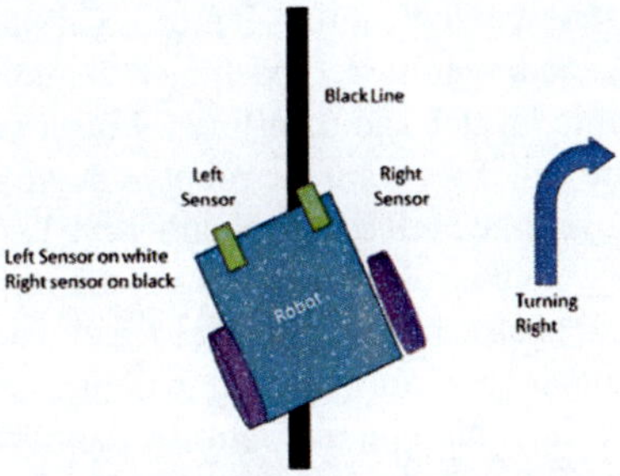

Fig. 7. Robot right direction

5 Result and Analysis

The aim of the line following the robot is to follow a line on its defined direction, which is obtained using IR sensors that detect the line and send the data to the comparator LM393 and then to the H bridge that controls the wheel's operation. The other tasks are managed by the microcontroller. In this project we have design the logic level according to our code we have set 0 for HIGH and 1 for LOW when logic level become zero it reflects back the rays and when logic level become 1 it absorbs the rays and as a result it transforms from high to low (Table 2).

Table 2. Logic design of Arduino line follower

Input		Output				Movement
Left sensor	Right sensor	Left Motor		Right Motor		Of Robot
LS	RS	LM1	LM2	RM1	RM2	
1	1	1	1	1	1	STOP
1	0	0	1	1	1	LEFT
0	1	1	1	0	1	RIGHT
0	0	0	1	0	1	FORWARD

Summary of Result

In this section we will give overview what we have achieved during the result by the line follower machine and it can be observe clearly that all the result depends upon the logic levels sense by the IR sensor and feedback provided to the microcontroller and then the instruction given to dc motors by Atmega328p microcontroller we can see if we have logic level 1 it means Off state and logic level 0 it means On state when both sensors have logic level 0 it means Arduino line follower machine will move in forward direction and when both sensor have logic level 1 and then line follower machine will stop because sensor will be on black line and if we want the robot to move in right direction the right sensor should have logic level 1 and left sensor should have logic level 0 same as the case for the left direction if we want the robot to move in this direction we need to have right sensor logic level 0 and left sensor need logic level 1 and we have used comparator IC to compare voltage level and then perform analog to digital conversion because sensing is always done in analog form which microcontroller cannot handle and ATmega328p only can understand digital value (Figs. 8, 9, 10, and 11).

Fig. 8. Robot in forward direction

Fig. 9. Robot right turn

Fig. 10. Line follower stop mode

Fig. 11. Robot left turn

6 Conclusion

Arduino Line Followers Robot has been successfully completed and tested with the incorporation of the functionality of each hardware node for its production using a microcontroller for automation purposes. The existence of each block has been reasoned out and carefully positioned, thereby leading to the unit's best function in industrial state we can used line follower machine to carry goods from one manufacturing plants to another without any sort of supervision because it is guided vehicle. Smarter versions of line followers can be used to deliver mails within office building and deliver medications in a hospital system. The line follower robots can also be improvised by using RFID tags. Voice control robot can be achieved using some vocal commands. By using the robot in real time applications, a health care system can be managing in an effectively way.

References

1. Springer, V.: Robotic process automation. Controlling **32**(1), 69–71 (2020)
2. Winston, C.: A new route to increasing economic growth: reducing highway congestion with autonomous vehicles. SSRN Electron. J. (2018)
3. Robot vision image processing and path planning. Int. J. Recent Trends Eng. Res. **4**(10), 55–58 (2018)
4. Heins, P., Jones, B., Taunton, D.: Design and validation of an unmanned surface vehicle simulation model. Appl. Math. Model. **48**, 749–774 (2017)
5. Frederick, E.: Watch a robot made of robots move around. Science (2019)
6. Roslidar, R., Mufti, A., Akbarsyah, H.: Perancangan robot light follower ATmega 328P. J. Rekay. Elektrik. **13**(2), 103 (2017)
7. Lager, M., Topp, E.: Remote supervision of an autonomous surface vehicle using virtual reality. IFAC-PapersOnLine **52**(8), 387–392 (2019)
8. Karpov, E., Kuznetsova, E.: Software and Hardware Implementation of the converter of control actions based on the microcontroller ATmega328p. Intellekt. Sist. Proizv. **16**(4), 95 (2019)
9. Design and implementation of line follower mobile robot. J. Crit. Rev. **7**(14) (2020)
10. Yousef, A.Y., Mostafa, M.: Dual DC motor speed control based on two independent digital PWM signals using PIC16F877A microcontroller. Indonesian J. Electric. Eng. Comput. Sci. **2**(3), 592 (2016)
11. Kozovsky, M., Blaha, P.: High speed operation tests of resolver using AURIX microcontroller interface. IFAC-PapersOnLine **51**(6), 384–389 (2018)
12. LabView interface with Arduino robotic ARM. Int. J. Sci. Res. **4**(11), 2423–2426 (2015)
13. Chang, K.C., Chu, K.C., Wang, H.C., Lin, Y.C., Pan, J.S.: Energy saving technology of 5G base station based on internet of things collaborative control. IEEE Access **8**, 32935–32946 (2020)
14. Chang, K.C., Chu, K.C., Wang, H.C., Lin, Y.C., Pan, J.S.: Agent-based middleware framework using distributed CPS for improving resource utilization in smart city. Futur. Gener. Comput. Syst. **108**, 445–453 (2020)
15. Chu, K.C., Chang, K.C., Wang, H.C., Lin, Y.C., Hsu, T.L.: Field-programmable gate array-based hardware design of optical fiber transducer integrated platform. J. Nanoelectron. Optoelectron. **15**(5), 663–671 (2020)
16. Chu, K.C., Horng, D.J., Chang, K.C.: Numerical optimization of the energy consumption for wireless sensor networks based on an improved ant colony algorithm. IEEE Access **7**, 105562–105571 (2019)

Evaluating Adaptive Facade Performance in Early Building Design Stage: An Integrated Daylighting Simulation and Machine Learning

Basma N. El-Mowafy(✉), Ashraf A. Elmokadem, and Ahmed A. Waseef

Architectural Engineering and Urban Planning Department, Faculty of Engineering, Port Said University, Port Said, Egypt
Basma_nashaat@eng.psu.edu.eg

Abstract. Globally, adaptive architectural systems are particularly considered as intelligent devices, that adapt the building to the outdoor environment or user needs. Their responsive reaction is taken as mechanical transformations based on environmental changes to achieve a comfortable indoor environment. They are controlled through computational methods, starting from interactive investigation devices, and automating optimization to achieve adequate suitable configurations. So, the paper studies the adaptive behavior of different kinetic facades that aiming to discover the suitable solutions that attain the best daylighting performance. It uses machine learning algorithms that helps architects to indicate the most suitable shading systems and exclude unfitting ones. The resulting systems are simulated by Diva for Rhino to compare their daylighting performance to figure which system is the most effective one in this case. This step is essential to validate the used algorithm and test the performance of the recommended models. The findings indicated that machine learning can be effectively used to select the suitable adaptive façade in the design stage.

Keywords: Adaptive façade · Daylighting · Machine learning · Simulation

1 Introduction

Global energy efficiency and sustainability in building design have become the main concern, buildings consume above one-third of global energy consumption and produce 40% of entire CO^2 emissions [1, 2] Consequently, the study of energy efficiency in buildings become the main motivation for several researchers, who have highlighted innovations through computer technology and machine learning [3]. The development of these technologies offers more instinctive tools that optimize and visualize building performance, to reduce the constant effort of architects to improve design productivity [4]. For instance, Daylighting is an important design component; it can reduce the cost of energy consumption that is used in heating and artificial lighting [5, 6]. As well, it has a huge effect on the physical and mental health of residents. Many studies approved that a suitable daylighting design can increase the productivity of users and enhance

© The Author(s), under exclusive license to Springer Nature Switzerland AG 2022
A. E. Hassanien et al. (Eds.): AMLTA 2022, LNDECT 113, pp. 211–223, 2022.
https://doi.org/10.1007/978-3-031-03918-8_20

their health [7]. However, the effective daylighting design must take into concern some matters like glare and unwanted heat gain [8].

Nowadays, the common design of office building façade contains large areas of glass. This design needs external shading devices to enhance indoor daylight distribution. One of the most promising trends in façade design is adaptive façade systems. The main benefits of an adaptive façade are achieving the occupants' comfort and decreasing energy consumption [9]. Different adaptive façade design aims to achieve an active and selective barrier that can manage mass transfer between the building and exterior, thermal insulation, natural ventilation, daylight, and harvesting solar energy [10].

But the simulation of these systems is a little complicated and needs a lot of time, especially to simulate all available forms. Therefore, designers need to do significant simulation efforts to explore alternative designs and predict the most suitable one that matches the designed building [11]. However, applying such simulation tasks in the early design stage is hard, as designers usually don't have enough time to test all available forms of adaptive facades [6].

Subsequently, designers can use statistical methods to decrease the number of available forms of adaptive facades. Nonetheless, these traditional methods have the main disadvantage: they can't capture non-linear patterns between potential influencing features and building performance [12]. Accordingly, machine-learning methods have been progressively used to solve non-linear matters as an alternative to traditional statistical methods in the field of engineering [13]. Machine-learning methods include several algorithms like artificial neural network (ANN), support vector machine (SVM), k-nearest neighborhood (KNN), and decision tree (DT) [14].

2 Related Works

Many researchers focused on the application of ML on predicting building energy performance like Olu-Ajayi et al. (2021), Ding et al. (2021), Seyedzadeh et al. (2020), and Fan et al. (2019) [1, 13, 15, 16]. The results generally permitted that the prediction of this technique has a better performance than traditional linear regression approaches. As an example, Wang (2021) indicates that current studies in ML-based building energy efficiency transform technical achievements into successful market applications [2].

A lot of studies compared the performance of different algorithms in predicting building performance or solving problems in the design stage. As Ye et al. 2021 introduced a framework that can recognize the critical influential building-oriented features. The study applied ML at the city-block-level by comparing the predictive performances of LR, KNN, SVM, DT, and RF models [14]. Similarly, Olu-Ajayihi (2021) compared the performance of nine machine learning techniques: DNN, ANN, SVR, DT, GB, LR, Stacking, RF, and KNN in predicting annual energy consumption [17].

Other researchers used the Ml algorithm as a classification tool. As an example, Zhou et al. (2021) projected a ML-based method to classify building structures. The study used twenty-nine features to characterize buildings and evaluated 12 popular ML algorithms [18]. On the other hand, many researchers focused on Ml as a tool in designing smart buildings. As Alanne et al. (2021) discussed the method of training building-integrated AI using machine learning applications. The research highlights the ability of the reinforcement learning (RL) technique and autonomous AI agents to control building and energy management processes [19].

Several studies have comprehensively combined simulation programs with machine learning. As an example, Feng et al. (2019) developed a quantitative technique that uses parametric design technology and machine learning algorithms to evaluate buildings' environmental performance in early decision stages [11].

Alternatively, many studies applied Ml in predicting and enhancing daylighting performance in buildings. Lin and Tsay (2021) developed a daylight predictive model that can predict daylight performance in the building design stage [6]. Also, Li and Lou, 2018 used Machine Learning techniques to predict solar radiation and daylight illuminance and classify the typical skies [20]. Also, Ayoub, (2020) introduced a review on using machine learning in predicting daylighting performance inside buildings. The study confirmed that MLAs can achieve fast and correct predictions [21].

Despite the significance of the above-mentioned efforts, limitations can be seen in two aspects. First, recognition of daylighting performance and glare control has not been explored. Compared to other building characteristics such as energy consumption. Second, the existing efforts didn't focus on adaptive façade systems as a new approach in facade design. Therefore, to address the above limitations, this research proposes a ML-based framework for selecting the most suitable faced systems, integrating with simulation to validate the resulting systems.

3 Building as a Machine and Machine Learning in Architecture

Le Corbusier described the house as a machine for living in' (Le Corbusier 1923) [19]. Then in the 1970s, computer and telecommunication technology had changed human life [22]. When Cedric Price invented "The Generator Project" in 1976, in collaboration with John and Julia Frazer. They designed a computer program that can procedure a layout of the site that can respond to changing requirements [23]. After few years, the "Autographic" company was instituted by John Frazer, and the first microcomputer-based design systems in the world was invented in 1979 [24].

The age of kinetic architecture started in the twenty-one century, as a lot of kinetic buildings are designed and built in recent years. For example, Milwaukee Art Museum by Santiago Calatrava in the United States has a kinetic structure shade and controls the temperature and light of the entrance [25, 26].

Machine learning (ML) targets recognizing relevant spatiotemporal patterns and information by statistical algorithms that use a dataset's variables to identify applicable spatiotemporal patterns [16]. Nevertheless, ML techniques have a key drawback, which is the running process requires impartial databases for inputting, and outputting to allow all possible varieties for training, validation, and testing [20]. On the other hand. The main stages of the machine learning process are described as follows [21]:

1. **Data collection:** It involves collecting data, this data can be historical, experimental, observational, or simulation-derived data, before creating the model.
2. **Data preparation:** It includes data normalization, scaling, and randomization.
3. **Data exploration and visualization:** this step uses statistical and visualization techniques to exam any relevant relationships between the input variables.

4. **Data pre-processing:** it splits the data into three parts. The largest part involves model training; the second part is the model testing part to evaluate its performance through the training; the last part is the validation part that offers additional model development.
5. **Prediction:** it is the target of machine learning.

The building's adaptability depends on its ability to learn. This learning can be achieved by AI training using machine learning algorithms and sharing knowledge between humans and a building [19]. As an example, many researchers have configured index parameters that can quantify the facade system design with architects' aesthetic preferences. These parameters conclude the advantages and disadvantages of the available forms, by comparing them with the pre-defined visual standards of the architects [4].

Another example of the application of ML in architecture is building recognition, which aims to utilize ML methods to categorize buildings that have similar characteristics (e.g., architecture styles, functions, roof shapes, and age). Generally, there are three approaches to address a building recognition problem: clustering, regression, and classification [18].

4 Adaptive Facade

An adaptive facade is defined as an intelligent facade that can adapt and respond to changes in environmental conditions. Adaptive facades can be applied to the functions of enhancing daylight quality and thermal heat performance in interior spaces [28]. This concept involves using smart materials, sensors, microprocessors, actuators, and management systems. Therefore, it depends on high-tech technologies and techniques, and applying computer programming, electrical and mechanical engineering.

The kinetic system depends on three main procedures: Input (sensors), process (logical unit), and output (actuators), also it needs a management system in most cases [29] as demonstrated in Fig. 1.

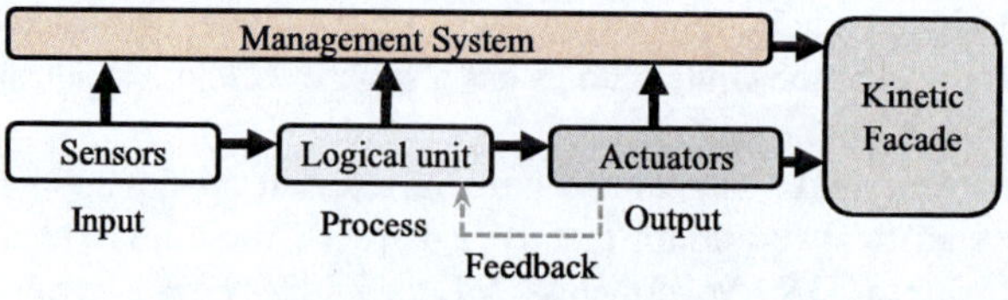

Fig. 1. Kinetic facade idea

5 Methodology

5.1 Data Collection: Available Forms of Kinetic Façade Systems

The research collected data about 18 kinetic facade models. The models are categorized into three main types as seen in Fig. 2:

1. Dynamic Glazing Skin: this type depends on a glass layer that has a changeable appearance.
2. Kinetic Shading Skin: it depends on mechanics that control shading devices in the outer skin of the building.
3. Middle Kinetic Skin: it depends on two glazing layers and a kinetic layer between them.

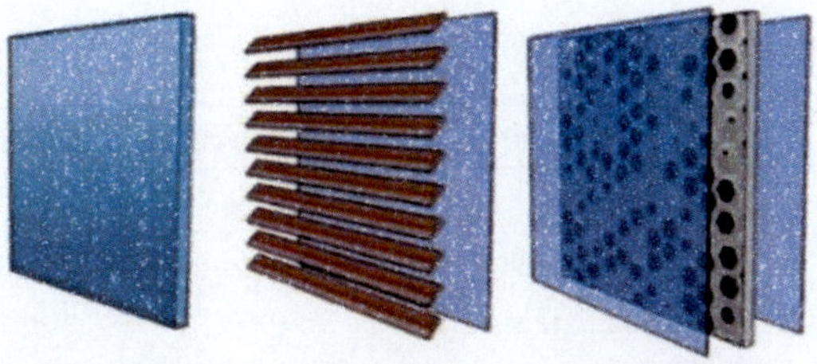

Fig. 2. Adaptive façade systems for daylighting control and solar heat gain control

5.2 Data Preparation: Applying System Possibility Scores

In the facade design stage, building design factors affect façade design variables. Therefore, affects how designers choose the most suitable kinetic systems. On the other hand, the main five design factors include several sub-variables. As a result, 22 variables are extracted to define every kinetic facade model.

These variables were identified for every model of the 18 kinetic facade models. The score of every model depends on the possibility of implementing the defined design variables with every system. Relations with 0 point means that this system can't be applied to the selected variables, and it will be excluded. As an example, if the Kinetic layer material is chosen as steel, it can't be applied in the case of dynamic glazing skin models. Therefore, these systems will be excluded despite their final score. Likewise, 1 point score means that this system is not recommended with this variable but it can be applied. When a model takes 2 points score, that means that applying this system is possible with the selected variable. Finally, the 3 points score means that this system is a recommended one to be applied with this design variable. After calculating this score of all variables of every façade model, they are compared by the defined variable scores in the case study as an application of Ml algorithms, as demonstrated in the following section.

5.3 Data Exploration and Case Study Setup

K-Nearest Neighbor (KNN) algorithm is used to predict the most efficient kinetic façade that suits the case study. The KNN algorithm is an advanced method to discover a solution to classification difficulties [30]. It is a non-parametric machine learning method that uses similarity to predict results based on the k nearest training samples [1]. The case study model is defined with its dimensions and properties. Consequently, the simulated façade width is 4 m. Also, the office depth and height are assumed to be 6 and 4 m respectively. On another hand, the simulated façade is a curtain wall with a kinetic shading system. Table 1 describes the Façade design factors and their aspects, which are configured at this stage. Some aspects are concluded based on the selected site, and the rest of them are assumed as they are the architect's decisions.

Table 1. Concluded and assumed façade design determinants

Determinants	Façade variables		Selected aspects	
Building design	Façade style		Modern	Assum.
	Building height		Medium height 5–12 stories	Conc.
	Used materials (kinetic layer)		PTFE	Assum.
Structure	Substructure spacing		2–4 m	Conc.
	Unit height		2–4 m	Assum.
	Substructure direction		Two-ways	Assum.
	Used materials		Light materials	Assum.
Orientation	Louvers direction		Vertical	Assum.
	Used materials	Glass	High UV value	Assum.
		Louvers	Low thermal conductivity	Assum.
Building properties	Façade orientation		South	Conc.
	Daylighting requirements		Office spaces: DF = 1–4%	Conc.
	Façade application		Daylighting & glare control + Solar heat control	Conc.
	Motion type		Translation + Rotation	Assum.
System properties	Energy consumption		Not specified (High + Medium + Low or no)	Assum.
	Control		Ext.	Assum.
	No. of Layers		3 or more	Assum.
	Layers arrange		Glass in	Assum.
	Layers space		less than 1 m	Assum.
	Mechanical parts		Not specified (High need + Medium need + Low need)	Assum.

(*continued*)

Table 1. (*continued*)

Determinants	Façade variables	Selected aspects	
Occupants needs	Layers space usage	No balcony, garden, or walkway	Conc.
	Privacy needs	Not specified (Fully + Not fully)	Conc.
Owner requirements	Installation cost	Reasonable	Assum.
	Lifetime cost	Reasonable	Assum.

5.4 Prediction Stage: Applying the KNN Algorithm as a Selective Filter

To filter the available forms of adaptive facades, the research employed the KNN algorithm, which is a simple ML algorithm and a good procedure for a problem with few datasets. Then, the previous factors are entered as inputs, KNN algorithm is used to define the similarity between the factors of the new case and the 18 available models Finally, it defines the most similar cases based on the summation of Euclidean distance for all design variables as demonstrated in Fig. 3.

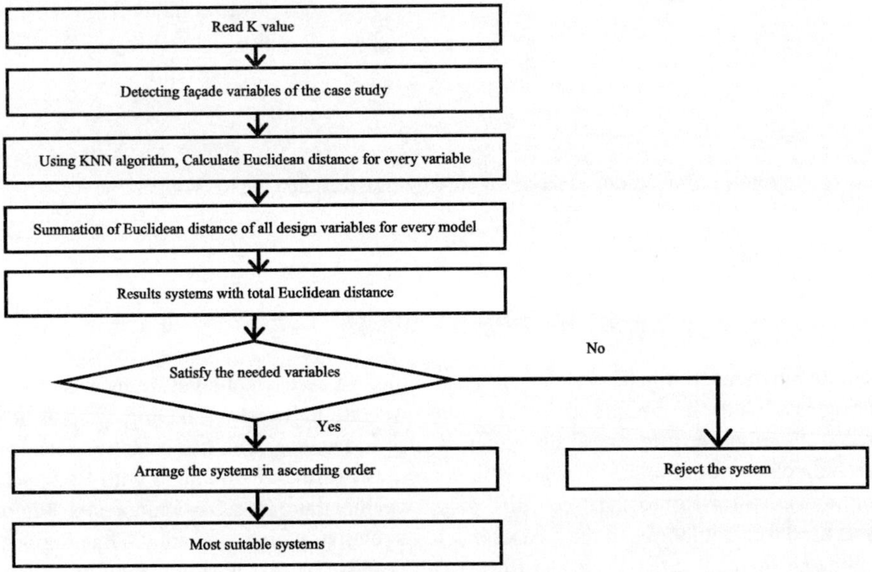

Fig. 3. KNN algorithm

The KNN classifier applies the Euclidean distance to differentiate the facades models. Further, it can be written as in Eq. 1 [30].

$$Dis(y, x) = \sqrt{\sum\nolimits_{i=1}^{n} (y_i - x_i)^2} \tag{1}$$

where,

y_i is the variable of designing every kinetic façade model.
x_i is the defined variables of designing the case study.
n is a number of variables.

Built on the summation of the Euclidean distance of all variables for each model, the models were sorted in ascending order. The final values of all models are illustrated in Fig. 4. But the models with red color are rejected as shown in the data preparation step (Sect. 3.2).

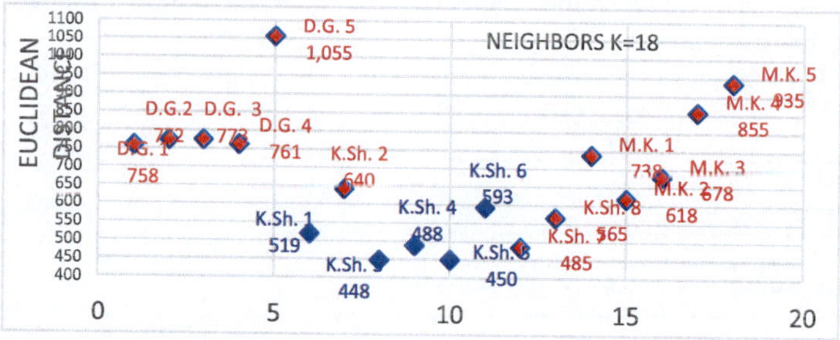

Fig. 4. The closest neighbors to the new case based on the KNN algorithm

6 Systems Modeling and Simulation

The recommended kinetic facades are re-modified to be horizontal systems to suit the design variables, as a requirement for south orientation. The five shading systems are reconfigured with different angle values as kinetic facades. The five adaptive facades are modeled with Grasshopper and Rhinoceros software and simulated with Radiance through Diva-for-Rhino that used Cairo, Egypt weather data. Diva for- Rhinoceros plugin was used for simulations of the 5 cases due to its relatively high efficiency. The plugin is validated by many existing types of research [31–34]. So, it is accurate enough for our research.

The simulated office model is an office room, and the following limitations are entered as inputs in the Diva plugin.

- Location: Egypt, Cairo
- Façade orientation: South (the positive y-axis is the north direction in Rhino)
- Furniture: It contains one door (1 * 2.2 m^2), six desks, and six monitor screens on them.
- Analysis surface: net of nodes distributed every 45 cm at height of 85 cm.
- Window: a curtain wall with an area (4 × 4 m).
- Kinetic layer: it takes certain positions with its rotation angle (30°, 60°, 90°)
- Used materials: they are set as demonstrated in Table 2.

Table 2. Used materials in Diva for Rhino

Item	Used material	Information
Wall	Generic interior wall 50	Reflectivity of 50%
Celling	Generic celling 70	Reflectivity of 80%
Floor	Generic floor 20	Reflectivity of 20%
Window	Double pane low E 65	vis = 0.65 - SHGC = 0.28 U-Value = 1.63 W/m^2K
Window frame	Sheet metal	80% Specularity - 90% reflectance
Kinetic layer	Simple Metal 0.8	50% Specularity - 60% reflectance
Door	Generic furniture 50	Reflectivity of 50%
Furniture	Generic furniture 50	Reflectivity of 50%

7 Results and Discussion

Next histograms are based on Daysim simulation report for the five simulated kinetic shading systems, which is an output from Diva-for-Rhino simulation. Table 3 presents the performance analysis of the shading skins.

The previous analysis reveals that the most effective system in daylighting quality is Kinetic shading 3 with accepted glare level. While kinetic shading 1 and Kinetic shading 6 haven't caused any glare inside the space, but has less effectible daylighting performance, as proved in Table 3.

Table 3. Comparison of measurement results for B.F. and five kinetic shading systems, based on Daysim simulation report

<table>
<tr><th>Parameter</th><th>Graph</th><th>Comment</th></tr>
<tr><td>Mean Daylight Factor (%)</td><td>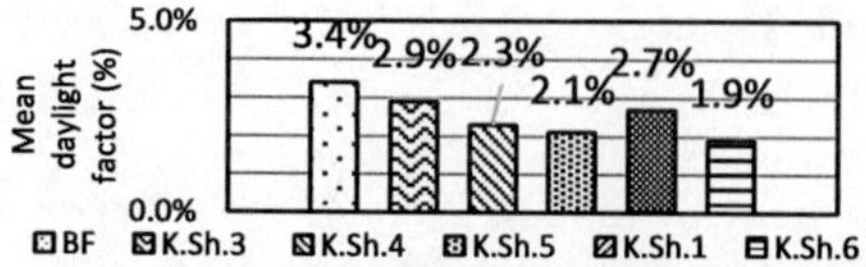
</td><td>The mean daylight factor reaches the highest level in case of K.Sh.3, followed by K.Sh.1 and K.Sh.4. Also, K.Sh.3 and K.Sh.1 have the best DF distribution.</td></tr>
<tr><td>Mean Illuminance at 21 Sept 9:00 am (Lux)</td><td>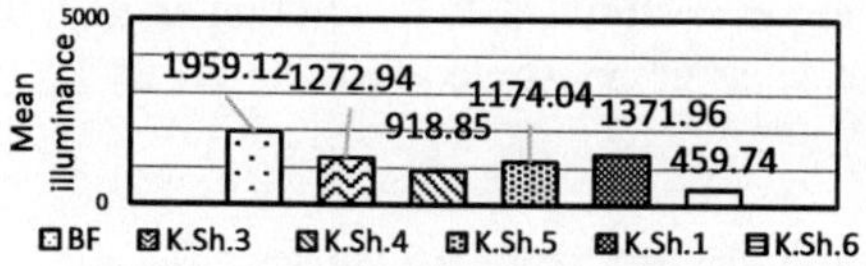
</td><td>All systems achieve the required level of mean illuminance except K.Sh.6. Mainly, K.Sh.1 and K.Sh.3 achieved the highest level of mean illuminance.</td></tr>
<tr><td>Mean daylight autonomy for active occupant behavior</td><td>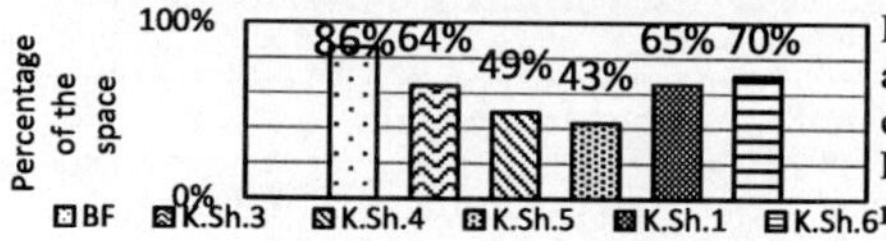
</td><td>K.Sh.6 system achieved 70% as a maximum mean value compared to 65% and 64% for K.Sh.1 and K.Sh.3 respectively.</td></tr>
<tr><td>Percentage of the space with a daylight autonomy larger than 50% for active occupant behavior</td><td>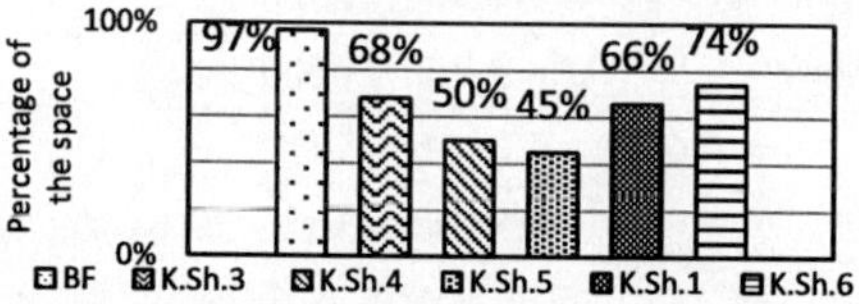
</td><td>K.Sh.6 system reaches 74%. Then, K.Sh.3 and K.Sh.1 have close values for percentage of the space with a daylight autonomy larger than 50%.</td></tr>
<tr><td>Continuous Daylight Autonomy (cDA) analysis</td><td>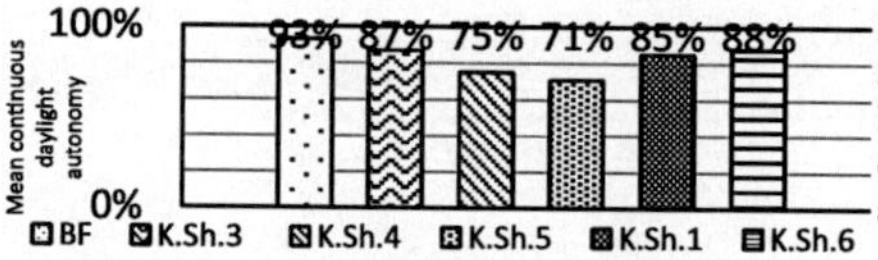
</td><td>K.Sh.3 system achieved almost the same performance level as K.Sh.6 and K.Sh.1 in continuous daylight autonomy analysis.</td></tr>
<tr><td>Daylight Factor (DF) analysis</td><td>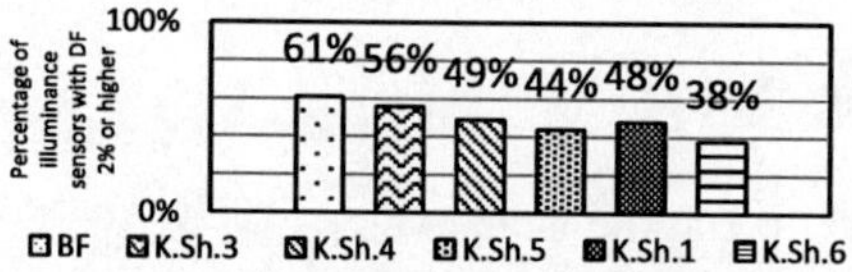
</td><td>In daylight factor analysis, K.Sh.3 system achieved the highest performance level by value 56% in the percentage of sensors with daylight factor of 2% or higher.</td></tr>
<tr><td>Useful Daylight Illuminance (UDI)</td><td>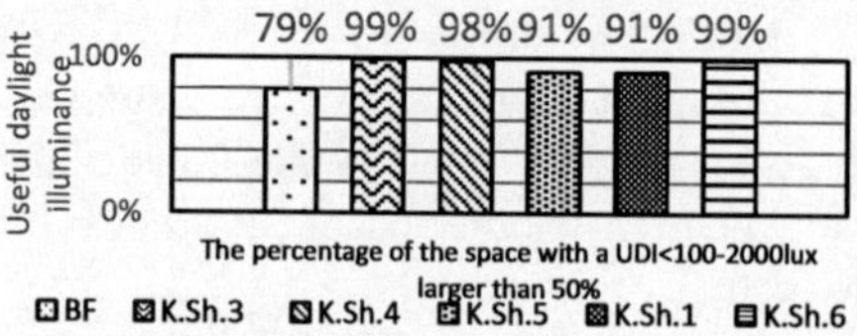
</td><td>Useful daylight illuminance value in case of K.Sh.3 equals to K.Sh.6 value. Also, other systems attained better performance than the base façade case.</td></tr>
<tr><td>Glare Percentage (% of occupied Hours)</td><td>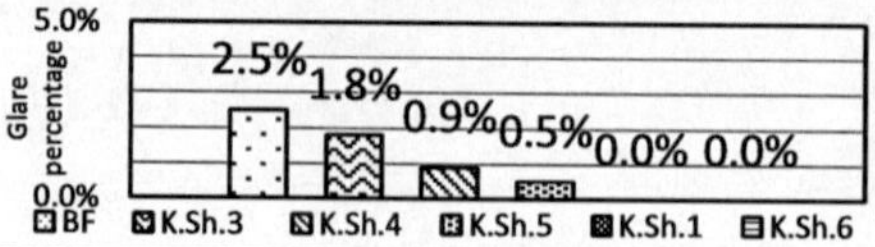
</td><td>The zero-glare level is achieved by K.Sh.1 and K.Sh. 6. But other systems reached lower values than the base façade case.</td></tr>
</table>

8 Conclusion

The façade design outlines the building character and performance, and affects the daylighting performance in the interior spaces. The form decision in the design stage guides the designer to the best daylighting performance and distribution. In particular, the idea of adaptive facades aims to simulate the dynamic nature of daylighting and the changed occupant's positions. As it has a real-time control with dynamic configuration in order to achieve better daylight performance and visual comfort, and reduce glare in the indoor spaces. In brief, adaptive faced investigate the best configuration of facade elements in a real-time reaction.

This research advanced a daylight predictive framework that uses ML algorithm to predict the suitable façade systems early in the design stage and the proposed method was validated by the simulation step. The key of the new method was obtaining Ml algorism to minimize the number of suitable systems by define the nearest neighbors to the new case. The use of the proposed algorithm rather than simulating all systems decrease the needed time of simulation, that is needed to validate all available kinetic systems. In addition to the method's utility in early design applications, the overall framework introduces an innovative way to combine advantages of both simulation process and machine learning algorithms.

The simulation results proved that KNN algorithm made the right prediction of the suitable system. Overall, the paper answers a significant question: how to select the most suitable kinetic shading system for a certain façade with the help of ML? The answer is introduced by a case study for an office façade in a certain location, its data was analyzed with KNN algorithm to specify the nearest neighbor. Then, the resulted systems were simulated to evaluate their daylighting performance.

References

1. Olu-Ajayi, R., Alaka, H., Sulaimon, I., Sunmola, F., Ajayi, S.: Building energy consumption prediction for residential buildings using deep learning and other machine learning techniques. J. Build. Eng. **45**, 103406 (2022). https://doi.org/10.1016/j.jobe.2021.103406
2. Wang, Z., Liu, J., Zhang, Y., Yuan, H., Zhang, R., Srinivasan, R.S.: Practical issues in implementing machine-learning models for building energy efficiency: moving beyond obstacles. Renew. Sustain. Energy Rev. **143**, 110929 (2021). https://doi.org/10.1016/j.rser.2021.110929
3. Pham, A.D., Ngo, N.T., Ha Truong, T.T., Huynh, N.T., Truong, N.S.: Predicting energy consumption in multiple buildings using machine learning for improving energy efficiency and sustainability. J. Clean. Prod. **260**, 121082 (2020). https://doi.org/10.1016/j.jclepro.2020.121082
4. Fan, Z., Liu, M., Tang, S.: A multi-objective optimization design method for gymnasium facade shading ratio integrating energy load and daylight comfort. Build. Environ. **207**, 108527 (2021). https://doi.org/10.1016/j.buildenv.2021.108527
5. Edwards, L., Torcellini, P.: A literature review of the effects of natural light on building occupants a literature review of the effects of natural light on building occupants. Contract **55** (2002)
6. Lin, C.H., Tsay, Y.S.: A metamodel based on intermediary features for daylight performance prediction of façade design. Build. Environ. **206**, 108371 (2021). https://doi.org/10.1016/j.buildenv.2021.108371

7. Knoop, M., et al.: Daylight: what makes the difference? Light. Res. Technol. **52**, 423–442 (2020). https://doi.org/10.1177/1477153519869758
8. Hosseini, S.M., Mohammadi, M., Guerra-Santin, O.: Interactive kinetic façade: improving visual comfort based on dynamic daylight and occupant's positions by 2D and 3D shape changes. Build. Environ. **165**, 106396 (2019). https://doi.org/10.1016/j.buildenv.2019.106396
9. Shen, L., Han, Y.: Optimizing the modular adaptive façade control strategy in open office space using integer programming and surrogate modelling. Energy Build. **254**, 111546 (2022). https://doi.org/10.1016/j.enbuild.2021.111546
10. Alkhatib, H., Lemarchand, P., Norton, B., O'Sullivan, D.T.J.: Deployment and control of adaptive building facades for energy generation, thermal insulation, ventilation and daylighting: a review. Appl. Therm. Eng. **185**, 116331 (2021). https://doi.org/10.1016/j.applthermaleng.2020.116331
11. Feng, K., Lu, W., Wang, Y.: Assessing environmental performance in early building design stage: an integrated parametric design and machine learning method. Sustain. Cities Soc. **50**, 101596 (2019). https://doi.org/10.1016/j.scs.2019.101596
12. Ding, C., Chen, P., Jiao, J.: Non-linear effects of the built environment on automobile-involved pedestrian crash frequency: a machine learning approach. Accid. Anal. Prev. **112**, 116–126 (2018). https://doi.org/10.1016/j.aap.2017.12.026
13. Fan, C., Xiao, F., Yan, C., Liu, C., Li, Z., Wang, J.: A novel methodology to explain and evaluate data-driven building energy performance models based on interpretable machine learning. Appl. Energy. **235**, 1551–1560 (2019). https://doi.org/10.1016/j.apenergy.2018.11.081
14. Ye, Z., Cheng, K., Hsu, S.C., Wei, H.H., Cheung, C.M.: Identifying critical building-oriented features in city-block-level building energy consumption: a data-driven machine learning approach. Appl. Energy. **301**, 117453 (2021). https://doi.org/10.1016/j.apenergy.2021.117453
15. Ding, Y., Fan, L., Liu, X.: Analysis of feature matrix in machine learning algorithms to predict energy consumption of public buildings. Energy Build. **249**, 111208 (2021). https://doi.org/10.1016/j.enbuild.2021.111208
16. Seyedzadeh, S., Pour Rahimian, F., Oliver, S., Rodriguez, S., Glesk, I.: Machine learning modelling for predicting non-domestic buildings energy performance: a model to support deep energy retrofit decision-making. Appl. Energy **279**, 115908 (2020). https://doi.org/10.1016/j.apenergy.2020.115908
17. Olu-Ajayi, R., Alaka, H., Sulaimon, I., Sunmola, F., Ajayi, S.: Machine learning for energy performance prediction at the design stage of buildings. Energy Sustain. Dev. **66**, 12–25 (2022). https://doi.org/10.1016/j.esd.2021.11.002
18. Zhou, P., Chang, Y.: Automated classification of building structures for urban built environment identification using machine learning. J. Build. Eng. **43**, 103008 (2021). https://doi.org/10.1016/j.jobe.2021.103008
19. Alanne, K., Sierla, S.: An overview of machine learning applications for smart buildings. Sustain. Cities Soc. **76**, 103445 (2022). https://doi.org/10.1016/j.scs.2021.103445
20. Li, D.H.W., Lou, S.: Review of solar irradiance and daylight illuminance modeling and sky classification. Renew. Energy. **126**, 445–453 (2018). https://doi.org/10.1016/j.renene.2018.03.063
21. Ayoub, M.: A review on machine learning algorithms to predict daylighting inside buildings. Sol. Energy. **202**, 249–275 (2020). https://doi.org/10.1016/j.solener.2020.03.104
22. Sherbini, K., Krawczyk, R.: Overview of intelligent architecture khaled sherbini 1 and robert krawczyk **2**, 137–152 (2004)
23. Elmokadem, A., Ekram, M., Waseef, A., Nashaat, B.: Kinetic architecture: concepts history and applications. Int. J. Sci. Res. **7**, 752 (2016). https://doi.org/10.21275/ART20181560
24. Trajkovsk, A.: How Adaptive Component Based Architecture Can Help with the Organizational Requirements of the Contemporary Society? (2014)

25. Shields, R.: The Advancement of the Application of Kinetics in Landscape Architecture (2019)
26. Zenter, Ö., Tayfun Yildirim, M.: Proposal of bio-inspired kinetic mechanisms utilization for providing functional flexibility in architectural design. Keywords Architect. Design Kinet. Architect. Biomimet. Architect. Funct. Flexib. **8**, 643–656 (2020)
27. Zhang, X.-D.: A Matrix Algebra Approach to Artificial Intelligence. Springer, Singapore (2020). https://doi.org/10.1007/978-981-15-2770-8
28. Tuma, F., Nassar, A.K., Kamel, M.K., Knowlton, L.M., Jawad, N.K.: Students and faculty perception of distance medical education outcomes in resource-constrained system during COVID-19 pandemic. A cross-sectional study. Ann. Med. Surg. **62**, 377–382 (2021). https://doi.org/10.1016/j.amsu.2021.01.073
29. Nashaat, B., Waseef, A.: Responsive kinetic façades: an effective solution for enhancing indoor environmental quality in buildings. In: The First Memaryat International Conference (MIC 2017) Architecture of the Future: Challenges and Visions. Saudi Arabia (2017)
30. Nayak, S.K., Panda, M., Palai, G.: Realization of optical ADDER circuit using photonic structure and KNN algorithm. Optik (Stuttg) **212**, 164675 (2020). https://doi.org/10.1016/j.ijleo.2020.164675
31. Marzouk, M., ElSharkawy, M., Mahmoud, A.: Optimizing daylight utilization of flat skylights in heritage buildings. J. Adv. Res. (2021). https://doi.org/10.1016/j.jare.2021.06.005
32. Sun, Y., Liu, X., Qu, W., Cao, G., Zou, N.: Analysis of daylight glare and optimal lighting design for comfortable office lighting. Optik (Stuttg) **206**, 1–5 (2020). https://doi.org/10.1016/j.ijleo.2020.164291
33. Keshavarzi, M., Caldas, L., Santos, L.: RadVR: a 6DOF virtual reality daylighting analysis tool. Autom. Constr. **125**, 103623 (2021). https://doi.org/10.1016/j.autcon.2021.103623
34. Le-Thanh, L., Le-Duc, T., Ngo-Minh, H., Nguyen, Q.H., Nguyen-Xuan, H.: Optimal design of an Origami-inspired kinetic façade by balancing composite motion optimization for improving daylight performance and energy efficiency. Energy **219**, 119557 (2021). https://doi.org/10.1016/j.energy.2020.119557

Check for updates

LTE Downlink Scheduling with Soft Policy Gradient Learning

Mona Nashaat(✉), Islam E. Shaalan, and Heba Nashaat

Electrical Engineering Department, Port Said University, Port Said 42523, Egypt
{monnashaat,is_shaalan,hebanashaat}@eng.psu.edu.eg

Abstract. Long Term Evolution (LTE) system is an essential technology proposed to achieve higher data rates while satisfying the user requirements and minimizing transmission latency. Evaluating LTE systems considers various factors, such as efficiency, fairness, system throughput, and delay. The diversity of these factors usually dictates the design of scheduling algorithms. Since satisfying all these factors in one system is impossible, many scheduling algorithms are proposed to handle resource allocation in the LTE system while ensuring the required Quality of Service (QoS). Moreover, since machine learning techniques have proven their success in handling time series prediction, the scheduling problem can be seen as an infinite-horizon Markov decision process to find the optimal scheduling strategy. Hence, in this context, this paper proposes an intelligent resources allocation algorithm that considers the QoS requirements while optimizing the scheduling policy. The model employs a deep soft policy gradient algorithm to analyze the state space and choose the correct scheduling action to learn the optimum scheduling policy. To evaluate the proposed technique, extensive experiments are performed to compare it to the state-of-the-art scheduling algorithms within a simulated environment. The experimental results show that the proposed model can decrease the network delay while improving the delay fairness and the overall network throughput.

Keywords: LTE · Resource allocation · Scheduling · Reinforcement learning

1 Introduction

With demands for internet services growing exponentially, several challenges have evolved regarding the design of wireless networks [1]. Users use wireless networks for various applications, including e-mail, file-sharing, voice-over IP, and multimedia applications. However, the limited network resources with the increasing demand for wireless services have exerted new challenges on the design of cellular networks. For example, the massive rise of multimedia applications has enforced rigorous Quality-of-Service (QoS) requirements to accommodate bandwidth-intensive content required by these applications [2]. Therefore, Long Term Evolution (LTE) was initially introduced to improve the performance of wireless networks. LTE systems depend on Orthogonal Frequency Division Multiple Access (OFDMA) to allow multiple access to network

© The Author(s), under exclusive license to Springer Nature Switzerland AG 2022
A. E. Hassanien et al. (Eds.): AMLTA 2022, LNDECT 113, pp. 224–236, 2022.
https://doi.org/10.1007/978-3-031-03918-8_21

resources and enhance bandwidth efficiency. In downlink transmission, OFDMA combines Time Division Multiple Access (TDMA) and Frequency Division Multiple Access (FDMA) to share the radio resources among users. It splits the network bandwidth into various narrow-band subcarriers and allocates them to users according to the system scheduling strategies [3].

Therefore, In downlink transmission, the time domain is split into radio frames, subframes, and slots. Radio frames have a length of 10 ms which are further divided into ten equally sized subframes. Consequently, each subframe consists of two adjacent slots of 0.5 ms. The smallest resource unit that can be assigned to a user is called a Resource Block (RB), which comprises a group of twelve contiguous subcarriers with 180 kHz in the frequency domain. In other words, at each 1 ms, the base station in LTE networks takes a scheduling decision to assign resources (in units of RBs) among users [4]. These decisions are made by a scheduling algorithm that tries to consider Qos requirements while managing the limited radio resources effectively. Therefore, many techniques are proposed to manage and control the available radio resources [5].

Furthermore, different scheduling techniques depend on optimization algorithms [6] to handle the tradeoff between efficiency and fairness among users. For example, Anand et al. [7] design the wireless system as a queuing network-based model and optimize the arrival rate to enhance the system performance. Alternatively, other techniques [8] are proposed to consider energy consumption in downlink transmission. The models aim to optimize energy consumption with different types of traffic.

Alternatively, predictive models are proposed for a wide range of applications in communication systems [9]. Since machine learning plays an essential role in integrating intelligence within an existing strategy, it provides promising solutions to many applications in wireless networks. For example, recent research applied machine learning techniques to network monitoring [10], transmission power prediction [11], and traffic type prediction [12]. As for scheduling in LTE, since traditional optimization techniques can get too complicated to use, machine learning can analyze various requirements and recommend scheduling decisions.

Consequently, Reinforcement Learning (RL) and Deep Learning (DL) are the most suitable techniques to optimize scheduling decisions in wireless networks. Hence, models can be built to accommodate different requirements without relaxing the system parameters or approximating the problem. Therefore, recent studies [ref] have proposed scheduling techniques for wireless communication that either utilize reinforcement learning deep neural networks. However, a closer look at these techniques shows some challenges. For example, RL techniques are not guaranteed to work with larger dimensions environments [13]. Therefore, most of these studies have adopted small environments with little traffic.

Hence, this paper presents Downlink Soft Policy Gradient (DSPG) scheduler, a learning algorithm to find the optimal scheduling policy in downlink transmission. The policy is selected to achieve the minimum delay and enhance the QoS requirement. Thus, we implement a deep reinforcement learning algorithm that applies a soft policy gradient learning to find the optimum policy. The model is designed to work with high-dimensional data and accommodate large density networks. An extensive set of experiments is conducted to evaluate the model with state-of-the-art techniques. The empirical evaluation shows that DSPG scheduler can lower the buffering delay to 54.57% compared to the baseline techniques.

The rest of this paper is organized as follows: Sect. 2 presents the research background in which we discuss downlink resource allocation in LTE. Section 3 summarizes the related work. The proposed scheduling algorithm is discussed in Sect. 4, while Sect. 5 describes the simulation implementation of the DSPG scheduler and compares it with other algorithms. Finally, Sect. 6 is devoted to the main conclusion and future work.

2 Downlink Resource Allocation in LTE

Downlink transmission is initially introduced to support higher transmission rates. The scheduling algorithm manages the network resources and delivers the receivers with their requested resources to maintain connections with an evolved node base station (eNodeB). However, as it also tries to consider the QoS requirements, scheduling is considered as challenging task for many reasons. For one reason, scheduling is not a part of network standards and, therefore, is left to the network operator to design and function. Consequently, there is no single scheduling algorithm that can handle all various types of inhomogeneous traffic. Furthermore, with higher demand for real-time multimedia applications, the scheduling problem is further characterized by rigid constraints such as packet delay and bandwidth. Therefore, the inconsistency between data rates and QoS requirements associated with different traffic types can be problematic [14].

The downlink transmission handles the communication from base stations (eNodeB) to user equipment (UE). As mentioned before, the time domain is split into frames of 10 ms duration each. A resource element is defined as the smallest time-frequency resource unit in the system. The smallest resource unit a scheduler can assign to a UE in a given transmission time interval (TTI) is called a scheduling block (SB) which consists of two consecutive RBs. In LTE networks, each attached UE informs the eNodeB with Channel Quality Indicator (CQI) via uplink transmission, which quantifies the UE estimation of its link quality. These estimations are repeatedly sent with intervals of tens of TTIs. Then, eNodeB starts the downlink scheduling process and allocates resources to UEs based on its scheduling and resource allocation function. An overall illustration of the downlink transmission process is depicted in Fig. 1. As the figure shows, eNodeB allocates the available RBs to U requestions UEs. In every TTI, eNodeB schedules SB to UEs based on QoS configuration and CQI parameters. The packet scheduler in eNodeB monitors the buffer status of all UEs and handles the QoS associated with the buffered data. While considering the buffer status, the packet scheduler also takes into account the arrival of new data (arrived bits) at every TTI. Then, it utilizes an RB allocation map based on the adopted scheduling policy to assign SB to suitable UEs. As a result, the eNodeB schedules some bits for transmission via downlink to UEs. These transmitted bits can be seen as scheduling actions (a_i) taken by the scheduler in eNodeB [4].

Various scheduling algorithms are proposed in the literature [15] based on different characteristics and allocation policies. For example, while some algorithms try to maximize the overall network throughput, others focus on enhancing fairness among users. One of the most commonly used scheduling algorithms is the maximum throughput (MT) algorithm. The algorithm allocates network resources based on the CQI information. It aims to maximize the throughput by assigning RBs to users with the best CQI.

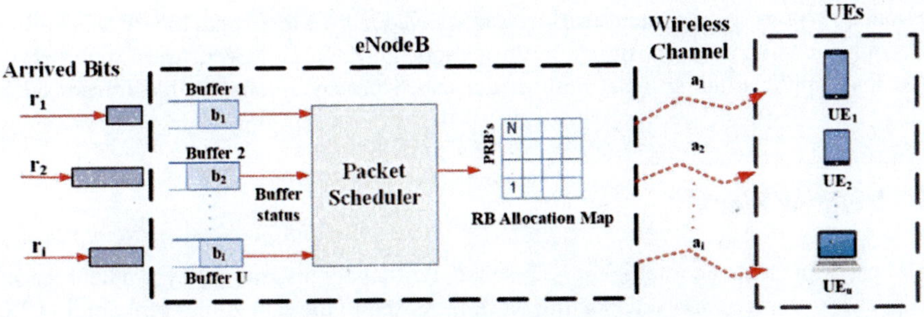

Fig. 1. Downlink transmission process.

Therefore, it allocates k resource blocks the ith UE number if it satisfies the following equation [16]:

$$i = argument\ \max_u[m_{u,k}(n)] \tag{1}$$

where the $m_{u,k}(n)$ is calculated based on the CQI feedback from the UE to eNodeB. According to Shanon law, the upper limit capacity of data rate can be calculated as follows [16]:

$$d_{u,k}(n) = log_2[1 + SIR_{u,k}(n)] \tag{2}$$

where $SIR_{u,k}$ is the signal to noise ratio of each UE on every RB. Therefore, MT tries to maximize the overall throughput in the cell by assigning every RB should to the UE, with the most remarkable capacity in its transmission channel [6]. The metric of the MT algorithm is formally defined as:

$$m_{u,k}^{MT}(n) = d_{u,k}(n) \tag{3}$$

Alternatively, the Proportional Fair (PF) scheduling algorithm tries to balance the overall throughput of the cell and throughput fairness between all UEs. It considers the previous throughput of each UE and schedules a UE when its current channel quality is high compared to its average CQI over time. Therefore, the metric of the PF algorithm can be defined as [16]:

$$m_{u,k}^{PF}(n) = \frac{d_{u,k}(n)}{TH_u(n-1)} \tag{4}$$

with:

$$TH_u(n) = \mu TH_u(n-1) + (1-\mu)th_u(n) \qquad where\ 0 \le \mu \le 1 \tag{5}$$

where $th_u(n)$ represents throughput achieved during the nth TTI for user uth UE, th_u indicates the previously achieved throughputs of the uth user during all previous TTIs. The throughputs are computed over a time window presented by parameter μ.

Overall, although the MT algorithm can enhance the cell capacity, it comes at the expense of fairness. Therefore, UEs with bad channel quality (i.e., located far from eNodeB) will likely experience service starvation. On the other hand, the PF scheduling algorithm can achieve considerable throughput fairness. However, users with delay-sensitive applications such as multimedia can experience considerable transmission delay.

3 Related Work

The paper presents an intelligent LTE scheduler that adopts soft policy gradient learning. Therefore, we surveyed existing work proposed to the scheduling problem in LTE downlink. In addition, the section summarizes previous solutions that apply predictive models and deep learning to downlink scheduling.

Many algorithms are proposed to allocate the radio resources in downlink transmission. For example, in [4], a resource allocation algorithm based on the Dragonfly metaheuristic technique is proposed. The algorithm considers the QoS requirements of the users and energy efficiency to select the best scheduling algorithm. Furthermore, another optimization technique is presented in [17], which utilizes a genetic algorithm for resource allocation by considering the objectives of minimizing the deployment cost and improving the quality of service performance.

Also, deep learning has been applied to a wide range of practical applications in LTE scheduling. For example, as for resource allocation, Q-learning has been used to minimize the energy consumption on the mobile terminal side under the application latency constraint. Previous research [18] considers the optimal offloading decision being jointly performed with the radio resource allocation. Also, another work [19] uses Q-learning to develop an algorithm for making dynamic handoff decisions. The algorithm [19] learns the optimal handoff policy from the user's experience and performs close to an optimal oracle algorithm. Moreover, other research [20] explores a joint task, spectrum, and transmit power allocation problem. In this research [20], the base stations (BSs) are outfitted with mobile-edge computing (MEC) servers to offer computational and communication services to users simultaneously.

On the other hand, in [21], a complete understanding of how machine learning is used to achieve dynamic resource allocation in vehicular networks is presented. Also, [22] provided a machine learning-based resource allocation approach that enhances video service quality for a larger number of users in a tremendous increase in Internet traffic. Also, a distributed risk-aware Machine learning solution for radio resource management is proposed in [23]. The study presents a hybrid orthogonal/non-orthogonal radio resource slicing, and the spectrum required for achieving the delay/reliability criterion is proactively regulated. Furthermore, in [24], a machine learning framework for resource allocation assisted by cloud computing is presented. The current scenario of the network is compared to prior scenarios to determine the most comparable one. The near-optimal solution in the most similar historical case is used to allocate radio resources for a given scenario.

Nevertheless, none of these approaches has applied policy gradient to find the optimal policy for the scheduling problem. Instead, existing work has either applied traditional heuristic-based optimization techniques [4, 25] or utilized Q-learning techniques to frame the scheduling process as a learning problem [21–23]. Thus, since policy gradient techniques have shown superior performance in large continuous action spaces, their capabilities are yet to be explored in the LTE environment to resolve the scheduling challenge, which is what this research precisely tries to explore.

4 DSPG Scheduler: The Proposed Scheduling Algorithm

In this section, the architecture of the proposed scheduler is discussed. This section is divided into two subsections. First, we formulate the problem statement for downlink scheduling. Then, in the second subsection, the design of the proposed solution is described.

4.1 Problem Statement

In the LTE downlink scheduling process, we consider an eNodeB that tries to allocate recourse to UEs every TTI as it checks the buffer status and the CQI reports of all UEs. Also, the proposed model considers the traffic type to be delay-sensitive multimedia applications. Thus, the scheduler aims to determine traffic priorities and decide which amount of bits to be scheduled first. Similarly, in reinforcement learning, an agent learns optimal policies by applying trial-and-error activities in a stochastic environment to maximize a rewarding objective [ref]. Hence, a scheduling model can be trained to learn an optimal policy π^* that maps the UEs states to scheduling actions. Scheduling actions involve scheduling $a_i \in A$ bits from the ith UE for transmission. Thus, we formulate the scheduling algorithm as a policy search problem in an infinite-horizon Markov decision process (MDP) to find the optimal policy.

The optimal policy $\pi^*: S \rightarrow A$ minimizes the average delay for all the UEs where S represents the state space, and A is a set of scheduling actions to transmit bits from a given UE. In LTE network, the state space S_i for the ith UE is represented by the number of bits in its buffer b_i. Therefore, we define the buffer size as a penalty rule as:

$$c_i^n(s_i^n, a_i^n) = [\max(b_i^n - a_i^n, 0) + r_i^n] - b_i^n] \tag{6}$$

where s_i^n and a_i^n represent the state space and the scheduling action (the number of bits scheduled for transmission) of the ith UE at the nth TTI. Also, r_i^n is the newly arrived bits at the nth TTI. Since decreasing the long-term average of buffer cost is equivalent to the UE's delay, we can define the total cost in a given time slot n as the total buffer cost of all the UEs as:

$$c^n(s^n, a^n) = \sum\nolimits_{i=1}^{U} c_i^n(s_i^n, a_i^n) \tag{7}$$

Hence, we can formally define the value function for each state as:

$$V^\pi = E_\pi\left[\sum\nolimits_{n=0}^{\infty} \gamma^n c^n(s^n, a^n)|s = s^0\right] \tag{8}$$

The value function represents the expected value obtained by the policy for each state s^n where E_π is the expectation under policy π, γ is a discount factor and c^n is the cost presented in (7). We adopt a deep soft policy gradient with a soft Bellman equation [26] to ensure stable learning. Therefore, we represent the policy by the parametric policy function, which chooses an action a in a state s. During training, the system parameters are updated by sampling the stochastic policy in the direction that minimizes the expected penalty. Hence, the scheduling algorithm is trained to find the optimal scheduling policy as:

$$\pi^* = \min_{\pi} V^{\pi} \tag{9}$$

where π^* represents the optimal policy and should satisfy the soft Bellman equation as in [26]. Implementing the soft policy gradient approach to find the policy ensures learning a policy even with large state and action spaces, which is a holding assumption in scheduling problems. Even though soft policy gradient has been mainly applied to continuous action spaces, previous research proves its effectiveness with discrete action spaces [26].

4.2 Model Design

DSPG scheduler is implemented using the Soft Policy Gradient algorithm [ref Soft Policy Gradient Method for Maximum Entropy Deep Reinforcement Learning]. The model applies a deep neural network with an off-policy deterministic policy. The gradient algorithm is proven to solve a wide range of challenging continuous tasks with large action spaces. The algorithm adopts an approximation of the expectation of actions in an Actor-Critic framework. An overview of the proposed framework is shown in Fig. 2. As shown in the figure, the framework consists of two networks. First, the policy structure network is known as the actor. The network maintains a parameterized actor function that stochastically maps states to all possible actions. The actor is updated with soft policy gradient to learn the maximum entropy policies as in [26].

Secondly, the critic network is trained to assess the selected policy by optimizing the soft loss as follows:

$$\mathcal{L}(\omega) = \mathrm{E}\left[\left(Q^{\theta}\left(\mathrm{s}^{\mathrm{n}}, \mathrm{a}^{\mathrm{n}}\right) - y^{n}\right)^{2}\right] \tag{10}$$

where ω and θ are the parameters of the actor and critic networks, respectively. Q^{θ} is the critic function which learns the action-value function and y^n denotes the target value obtained from the target networks.

Since the outer expectation of soft Bellman [26] relies on the state transition, the algorithm utilizes previous off-policy transitions to learn the soft Q-function using a replay buffer and target networks [27]. The replay buffer can be seen as a cache that is used to store transition history. Therefore, at each time slot, the actor and critic networks are learned by sampling previous transitions, and the parameters are optimized by minimizing the soft loss.

Furthermore, the model uses target networks to stabilize the learning process and assure convergence [ref], so values calculated by the network are considered to update network equations. Hence, copies of the networks (both actor and critic) are created and updated slowly, improving stability.

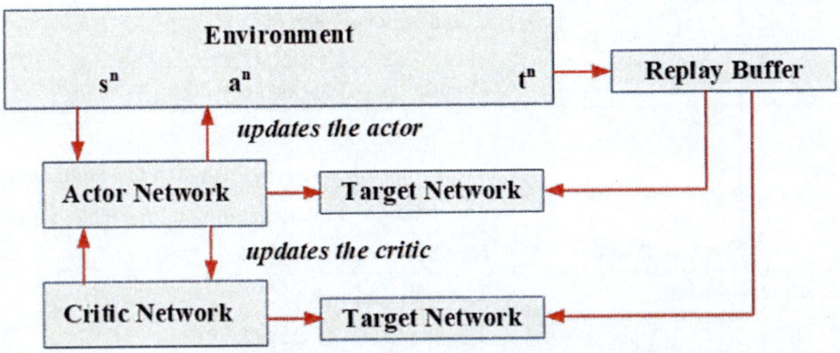

Fig. 2. The overall structure of DSPG scheduler.

To learn the model parameters, we use Adam [28] with learning rates $5 * 10^{-6}$ and $5 * 10^{-5}$ for the actor and critic networks, respectively. Both networks are built as fully connected feed-forward neural networks with four hidden layers and ReLU activation functions for hidden layers. For output layers, the sigmoid function is used for the mean and standard deviation.

5 Simulation Implementation and Results

This section presents the empirical results obtained from the simulation to compare the proposed model with two baseline scheduling techniques. The experimental evaluation aims to validate the following claim:

- Training a reinforcement learning agent to make scheduling decisions can yield a high-quality scheduling strategy compared to traditional scheduling techniques. The results obtained from the simulation show that DSPG scheduler can outperform the baseline techniques and lower the delay by up to 54.57%. As for throughput fairness, the proposed technique can enhance the fairness by 10.96% on average compared with the PF scheduling algorithm.

The experiments include two traditional scheduling algorithms: the MT algorithm and the PF scheduler. The simulation environment is developed in Python. The basic configurations of the simulation are summarized in Table 1. As the table shows, the simulation considers a hexagonal grid with 6 BSs with a central BS that performs the scheduling decisions to allocate resources among UEs under its coverage. In the experiments, we consider three different scenarios, 6 UEs/10 RBs (low), 20 UEs/50 RBs (medium), and 30 UEs/100 RBs (large-scale). Moreover, the experiments apply two types of data arrival rates. The first type assumes all UEs have the same arrival rate (50 Kbps), and the second type with UEs with variable arrival rates (ranging from 10 to 100 Kbps).

The proposed technique is implemented as described in the previous section. Both actor and critic networks are fully connected feed-forward neural networks with two hidden layers. The hidden layers apply ReLU as the activation function. The training data

Table 1. Simulation parameters.

Notation	Value	Description
U	{6, 20, 30}	The number of UEs
eNodeBs	7	The number of eNodeB (including the central one)
TTI	1 ms	The duration of the transmission time slot
RBs	{10, 50, 100}	Number of resource block
Cell Radius	1 km	The radius of the cell
Bandwidth	{1.4, 10, 20} MH	The bandwidth of the network
Arrival Rate	50 or 10:100 Kbp	Rate of newly arrived bits

comprises the samples collected by running the simulation for 400 epochs of 20,000-time slots, each with a minibatch of 32 samples. The results are also averaged over 50 epochs of 50,000-time slots each. The experimental evaluation considers four performance metrics: buffering delay, delay fairness, throughput, and throughput fairness. Fairness is computed as in [26] as:

$$J^{fairness} = \frac{\left(\sum_{u=1}^{U} x_u\right)^2}{U\sum_{u=1}^{U} {x_u}^2} \tag{11}$$

where U is the number of users, and (x) represents the parameter with which the fairness index is calculated. In our case, the fairness index is calculated for throughput and the buffering delay.

Figure 3 illustrates the performance metrics achieved by DSPG scheduler and the comparing algorithms under the simulation scenarios with fixed data arrival rates. The figures illustrate that the proposed model outperforms the traditional scheduling algorithms in all the metrics. As for the MT algorithm, the proposed algorithm reduces the delay by 47.29%, 45.65%, and 54.57% when compared in the low, medium, and large scale scenarios, respectively. The largest improvement occurs with the large-scale scenario (30 UEs/100 RBs), proving the proposed algorithm's scalability to accommodate higher numbers of UEs. Similarly, DSPG scheduler reduces the delay compared to the PF algorithm in all scenarios, with an average reduction of 8.31%.

Furthermore, the model shows superior performance in throughput than the MT and PF algorithms, with average improvements of 12.29% and 28.93%, respectively, across all scenarios. Likewise, the results show more significant improvements when we scale up the number of UEs and RBs. For example, while the model could improve the delay fairness in the low-scale scenario with 18.46% compared to the MT algorithm, it achieved 41.89% improvement compared to the same algorithm with the large-scale scenario.

Alternatively, the simulation results with variable data arrival rates are depicted in Fig. 4. As for delay fairness, the MT algorithm consistently achieved the worst results since it focuses on maximizing the throughput without considering the fairness. Thus, DSPG scheduler could enhance the delay fairness with 33.57%, 40.53%, and 44.44%

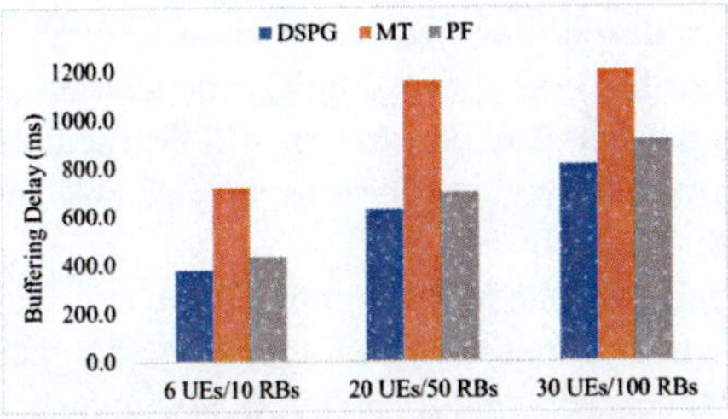

(a) Buffering Delay vs. Number of Users and Different RBs

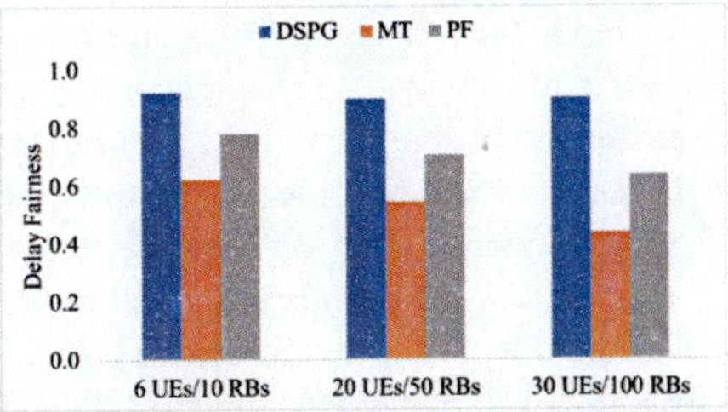

(b) Delay Fairness vs. Number of Users and Different RBs

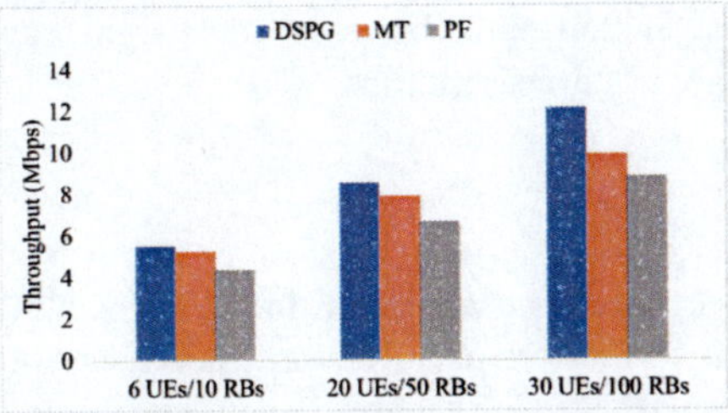

(c) Throughput vs. Number of Users and Different RBs

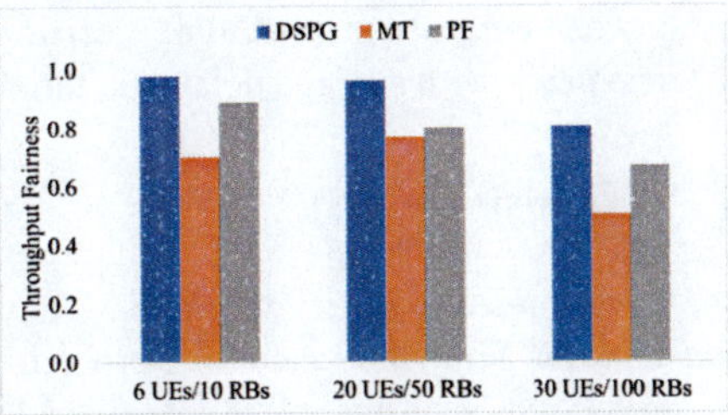

(d) Throughput Fairness vs. Number of Users and Different RBs

Fig. 3. Performance with fixed data arrival rates

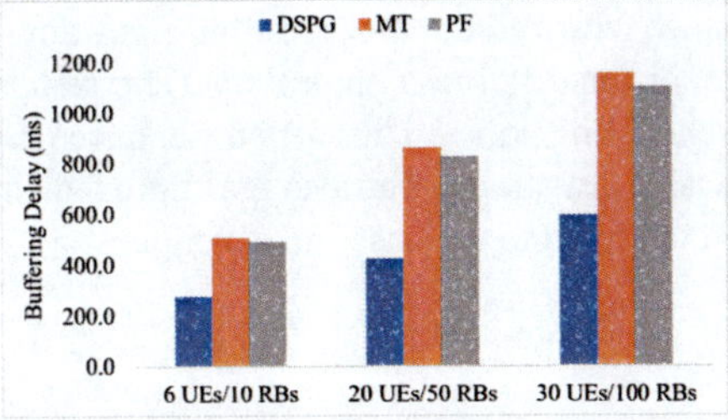

(a) Buffering Delay vs. Number of Users and Different RBs

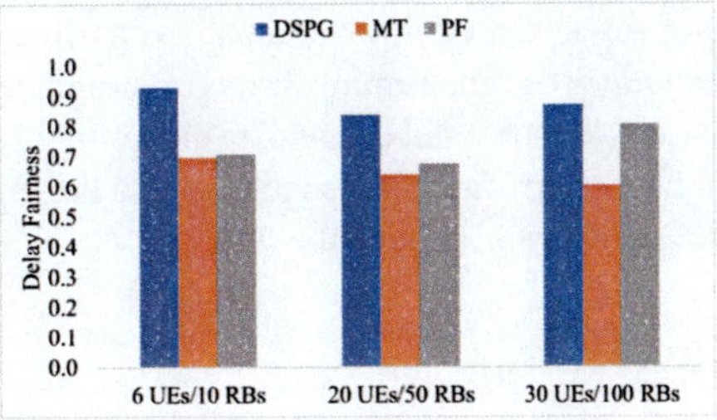

(b) Delay Fairness vs. Number of Users and Different RBs

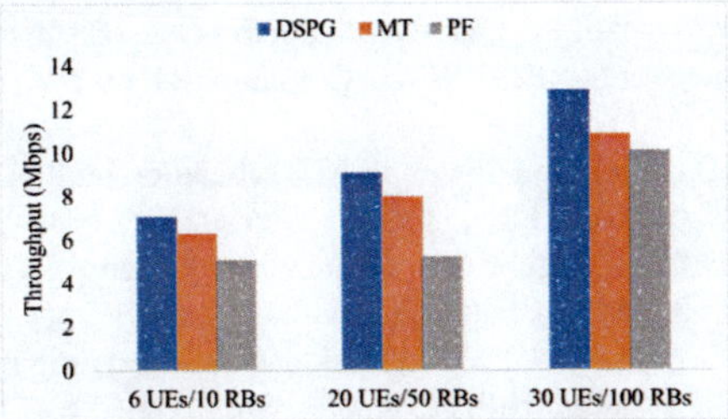

(c) Throughput vs. Number of Users and Different RBs

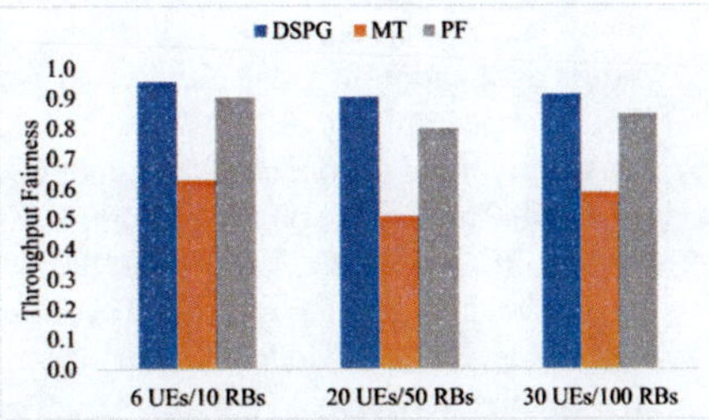

(d) Throughput Fairness vs. Number of Users and Different RBs

Fig. 4. Performance with variable data arrival rates

compared to the MT algorithm in all scenarios. In addition, Fig. 4d shows the throughput fairness achieved by the three scheduling algorithms. As shown, DSPG scheduler model surpasses the baseline techniques in throughput fairness. For example, in the medium-scale scenario, the proposed technique improves throughput fairness with 77.56% and 12.89% compared to the MT and PF algorithms, respectively.

Conversely, the proposed algorithm outperforms the MT algorithm in the low-scale scenario, with an improvement of 51.75% in throughput fairness. Moreover, with more RBs and UEs, the proposed model could enhance resource fairness by 55.54% compared to the same algorithm. Overall, the simulation results show the same trend for both fixed and variable data arrival rates. As the number of UEs and RBs scales up, the model performance shows more significant improvements. Also, the model considers the buffer status when evaluating scheduling decisions. Therefore, it shows the most significant enhancements with average delay and throughput fairness metrics.

6 Conclusions

The paper presents DSPG scheduler, a novel scheduling algorithm for LTE downlink transmission. First, the scheduling problem is formulated as a Reinforcement Learning problem in which an agent (the scheduler) learns the optimal scheduling policy. Second, the algorithm is trained to consider the buffer status when evaluating scheduling actions. The model adopts a deep soft policy gradient with the soft Bellman equation to enhance the model scalability to work with large-scale environments. Hence, the scheduling strategy is learned by sampling the action space to minimize the overall delay. To evaluate the proposed model, we compare its performance with traditional scheduling algorithms, including the maximum throughput and proportional fairness algorithms. The results show that the model can realize an optimum scheduling policy. Thus, it could enhance the network performance, especially with large-scale networks with an average improvement of 26.72%, and reduce the delay by 40.07% compared to the baseline techniques.

References

1. Kayali, M.O., Shmeiss, Z., Safa, H., El-Hajj, W.: Downlink scheduling in LTE: challenges, improvement, and analysis. In: 2017 13th International Wireless Communications and Mobile Computing Conference (IWCMC), pp. 323–328 (2017)
2. Zheng, K., Zhang, X., Zheng, Q., Xiang, W., Hanzo, L.: Quality-of-experience assessment and its application to video services in lte networks. IEEE Wirel. Commun. **22**(1), 70–78 (2015). https://doi.org/10.1109/MWC.2015.7054721
3. Van den Eynde, J., Blondia, C.: A minimal delay violation downlink LTE scheduler. In: 2021 IEEE 46th Conference on Local Computer Networks (LCN), pp. 387–390 (2021)
4. Nashaat, H., Refaat, O., Zaki, F.W., Shaalan, I.E.: Dragonfly-based joint delay/energy LTE downlink scheduling algorithm. IEEE Access **8**, 35392–35402 (2020)
5. Capozzi, F., Piro, G., Grieco, L.A., Boggia, G., Camarda, P.: Downlink packet scheduling in LTE cellular networks: key design issues and a survey. IEEE Commun. Surv. Tutor. **15**(2), 678–700 (2013). https://doi.org/10.1109/SURV.2012.060912.00100
6. Elhadad, M.I., El-Shafai, W., El-Rabaie, E.-S.M., Abd-Elnaby, M., Abd El-Samie, F.E.: Enhanced fair earliest due date first scheduling strategy for multimedia applications in LTE downlink framework. Int. J. Commun. Syst. **33**(6), e4190 (2020)

7. Anand, A., de Veciana, G.: Resource allocation and HARQ optimization for URLLC traffic in 5G wireless networks. IEEE J. Sel. Areas Commun. **36**(11), 2411–2421 (2018). https://doi.org/10.1109/JSAC.2018.2874122
8. Battistella Nadas, J.P., Onireti, O., Souza, R.D., Alves, H., Brante, G., Imran, M.A.: Performance analysis of hybrid ARQ for ultra-reliable low latency communications. IEEE Sens. J. **19**(9), 3521–3531 (2019). https://doi.org/10.1109/JSEN.2019.2891221
9. Mao, Q., Hu, F., Hao, Q.: Deep learning for intelligent wireless networks: a comprehensive survey. IEEE Commun. Surv. Tutor. **20**(4), 2595–2621 (2018). https://doi.org/10.1109/COMST.2018.2846401
10. Zoha, A., Imran, A., Abu-Dayya, A., Saeed, A.: A machine learning framework for detection of sleeping cells in LTE network (2014)
11. Falkenberg, R., Sliwa, B., Piatkowski, N., Wietfeld, C.: Machine learning based uplink transmission power prediction for LTE and upcoming 5G networks using passive downlink indicators. In: 2018 IEEE 88th Vehicular Technology Conference (VTC-Fall), pp. 1–7 (2018)
12. Mamman, M.: Downlink scheduling algorithm over LTE networks using genetic algorithm. In: Advances in Materials and Processing Technologies, pp. 1–12 (2020)
13. Nomeir, M.: Machine Learning Based Critical Resource Allocation in Mixed-Traffic Cellular Networks (2021)
14. Monikandan, B.S., Sivasubramanian, A., Babu, S.P.K., Prasanna Venkatesan, G.K.D., Arunachalaperumal, C.: Channel aware optimized proportional fair scheduler for LTE downlink. Peer-to-Peer Netw. Appl. **13**(6), 2135–2144 (2020)
15. Nwawelu, U., Ani, C., Ahaneku, M.: Comparative analysis of the performance of resource allocation algorithms in long term evolution networks. Niger. J. Technol. **36**(1), 163–171 (2017)
16. Liu, S., Zhang, C., Zhou, Y., Zhang, Y.: Delay-based weighted proportional fair algorithm for lte downlink packet scheduling. Wirel. Pers. Commun. **82**(3), 1955–1965 (2015)
17. Devarasetty, P., Reddy, S.: Genetic algorithm for quality of service based resource allocation in cloud computing. Evol. Intel. **14**(2), 381–387 (2019). https://doi.org/10.1007/s12065-019-00233-6
18. Li, H., Gao, H., Lv, T., Lu, Y.: Deep Q-learning based dynamic resource allocation for self-powered ultra-dense networks. In: 2018 IEEE International Conference on Communications Workshops (ICC Workshops), pp. 1–6 (2018). https://doi.org/10.1109/ICCW.2018.8403505
19. Tabrizi, H., Farhadi, G., Cioffi, J.: Dynamic handoff decision in heterogeneous wireless systems: Q-learning approach. In: 2012 IEEE International Conference on Communications (ICC), pp. 3217–3222 (2012). https://doi.org/10.1109/ICC.2012.6364194
20. Wang, S., Chen, M., Liu, X., Yin, C., Cui, S., Vincent Poor, H.: A machine learning approach for task and resource allocation in mobile-edge computing-based networks. IEEE Internet Things J. **8**(3), 1358–1372 (2021). https://doi.org/10.1109/JIOT.2020.3011286
21. Nurcahyani, I., Lee, J.W.: Role of machine learning in resource allocation strategy over vehicular networks: a survey. Sensors **21**(19), 6542 (2021). https://doi.org/10.3390/s21196542
22. Comşa, I.-S., et al.: A machine learning resource allocation solution to improve video quality in remote education. IEEE Trans. Broadcast. **67**(3), 664–684 (2021). https://doi.org/10.1109/TBC.2021.3068872
23. Azari, A., Ozger, M., Cavdar, C.: Risk-aware resource allocation for URLLC: challenges and strategies with machine learning. IEEE Commun. Mag. **57**(3), 42–48 (2019). https://doi.org/10.1109/MCOM.2019.1800610
24. Wang, J.-B., et al.: A machine learning framework for resource allocation assisted by cloud computing. IEEE Network **32**(2), 144–151 (2018)

25. Sabagh, M.R., Dianati, M., Imran, M.A., Tafazolli, R.: A heuristic energy efficient scheduling scheme for VoIP in 3GPP LTE networks. In: 2013 IEEE International Conference on Communications Workshops (ICC), pp. 413–418 (2013). https://doi.org/10.1109/ICCW.2013.6649269
26. Shi, W., Song, S., Wu, C.: Soft policy gradient method for maximum entropy deep reinforcement learning. arXiv preprint arXiv:1909.03198 (2019)
27. Haarnoja, T., Zhou, A., Abbeel, P., Levine, S.: Soft Actor-Critic: Off-Policy Maximum Entropy Deep Reinforcement Learning with a Stochastic Actor (2018)
28. Zhang, Z.: Improved adam optimizer for deep neural networks. In: 2018 IEEE/ACM 26th International Symposium on Quality of Service (IWQoS), pp. 1–2 (2018)

Predicting the Road Accidents Severity Using Artificial Neural Network

Saeed Al Mansoori(✉) and Khaled Shaalan

Faculty of Engineering and IT, The British University in Dubai, Dubai, United Arab Emirates
saeed.almansoori@mbrsc.ae

Abstract. Safety on roads is one of the most pressing and challenging concerns in most countries since road accidents lead to a great number of causalities and disabilities as well as serious loss of wealth resources. Millions of people lose their lives and others suffer from severe injuries that result in destructive disabilities which prevent them from practicing their normal life and actively contribute in their societies. Additionally, road accidents undermine infrastructure and imposes grave harm on materialistic possessions through destroying roads, means of transports and other urban amenities. Based on the recent statistics, the UK is one of the most suffering countries from the increasing numbers of road accidents with a high degree of fatality, disability and money-oriented losses. This research paper proposes a mechanism to predict the level of injuries severity (fatal or slight) caused by road accidents in the UK using Artificial Neural Network (ANN). It also aims to investigate and analyze the dataset attributes that have a significant correlation with road accidents in order to gain a profound insight of this problem to predict the road accidents severity and come up with conclusions and practical recommendations to alleviate road accidents repercussions and dangers. The dataset used in this study is taken from Kaggle, a free online open source website, which comprises all the details of the recorded accidents that have occurred in the UK, published by the UK Transports Department, in 2015. The experimental results revealed that the proposed ANN prediction model achieved a reasonable classification accuracy of 76%.

Keywords: Machine learning · Accident severity · Artificial neural network · Accuracy · Precision · Recall

1 Introduction

Road accidents are a significant cause of fatalities, disabilities as well as property and economic losses on daily basis all over the globe; therefore, it is an issue of a major concern for the entire international community. These accidents are entailed due to four main factors that are human, vehicle, road and environmental factors [1]. Examples of human related factors are age and gender while the environmental factors influencing road accidents include weather conditions, light conditions, road surface and topology, etc. As per the World Health Organization (WHO) report, approximately 1.3 million

© The Author(s), under exclusive license to Springer Nature Switzerland AG 2022
A. E. Hassanien et al. (Eds.): AMLTA 2022, LNDECT 113, pp. 237–249, 2022.
https://doi.org/10.1007/978-3-031-03918-8_22

people lose their life every year, in other words 3,287 deaths per day. In addition, from 20–50 million people suffer from serious injuries and disabilities. The report also confirms that road injuries are a leading cause for children's death (below 5 years old). From the economic point of view, road accidents cost the world approximately 518 billion dollars forming 2% of the annual Gross Domestic Product (GDP) of each individual country [1]. The rate of road accidents and the degree of their severity vary among countries. Unexpectedly, the UK has one of the most advanced road networks; however, it suffers from high increasing levels of accidents. According to the British police records, 1,770 people were killed in road accidents in the UK by the end of June 2018 which represents an increase of 4% when compared with the number of fatalities in 2013. Furthermore, the number of the people seriously injured due to road accidents were 26,610 in June 2018 with an increase of 3,607 victims when compared with the people got injured in 2013. The year 2017 witnessed a remarkable rise in the number of causalities in the road accidents in Great Britain as there were 129,983 accidents reported to the police through STATS19 data collection system, 24,210 of which were serious leading to 1,793 causalities.

In addition, the recent statistics shows that the number of cars accidents was the highest in UK forming 45% of road deaths since cars drivers form more than 80% of the road users in Britain. Pedestrians came in the second position as they were 25% of road accident victims since they could be killed by all means of transport while walking or crossing roads. The third position was represented by the motorcyclists as they were about 19%, while the pedal cyclists came in the 4th position with 6% of the total deaths in roads accidents [1].

Although it is important to assert the impossibility of totally eliminating the occurrence of road accidents; however, applying certain practices can reduce the intensity of accidents and the severity of injuries associated with them. To increase safety on roads and improve the performance of traffic systems, it is essential to analyze and identify the factors (i.e. attributes) that are more significant and correlated to road accidents and build a prediction model for road accidents severity. Machine learning has been introduced as an effective artificial intelligent solution to design prediction models by permitting the system to learn from the existing cases.

This research paper aims at analyzing and identifying the key factors that influence the traffic accidents in the UK as well as proposing a model to predict the degree of injuries' severity caused by road accidents based on 8,500 accident records that had occurred in Great British in 2015. The frequency based analysis, Information Gain and Gain Ratio measures are utilized to shortlist the key factors contributing to road accidents while the Artificial Neural Network (ANN) is used to build a model to predict the accidents severity. Accordingly, the proposed model has nine input parameters stated in Sect. 3 while the output parameter is the accident severity level which predicts whether the accident is slight or fatal.

The remaining part of this research is organized as follows. Section 2 reports the related works in this domain. Section 3 introduces the dataset used to carry out this study. Section 4 explains the proposed methodology to analyze the data and build a prediction model for road accidents severity by employing Artificial Neural Network

(ANN). Section 5 presents the results and findings of this research. The conclusion is stated in Sect. 6.

2 Literature Review

Several studies have been conducted on the prediction of the road accident severity domain using machine learning techniques such as neural network, Bayesian network, decision tree, support vector machine, association rules and clustering. This section reports some recent studies aiming at analyzing road accidents factors and predicting the associated degree of severity.

For instance, [2] developed a prediction model to estimate the injury severity caused by road accidents using an artificial neural network based on 5,973 records for the accidents that took place in Abu Dhabi between 2008 and 2013. They classified accidents severity into four main classes that are minor, moderate, severe and death. The authors used WEKA software to build two neural network classifiers based on the accidents dataset. The first classifier was trained and assessed using the entire dataset while the second classifier was trained with 90% of the data and tested with the remaining 10%. The simulation results demonstrated that the neural network classifiers were capable of predicting the accident severity with an acceptable accuracy. The overall performance of their prediction model gave 81.6% for training data and 74.6% for testing. Then, they applied a k-means algorithm to improve the accuracy of the prediction by splitting the accidents dataset into three clusters. Accordingly, a significant improvement has been noticed in the prediction accuracy and for the training data in particular.

Similarly, [3] proposed a novel approach to predict traffic accidents severity using a deep learning model based on a Recurrent Neural Network (RNN). The authors developed their model relying on 1,130 records of accidents that took place on the North-South Expressway, Malaysia throughout a period of six years between 2009 and 2015. A systematic grid search was used to test various network designs and structures to obtain an optimal network for forecasting the injury severity of road accidents. The chosen network structure consists of five layers that are the input layer, the Long-Short Memory layer, two dense layers and a Softmax layer. Afterwards, the dropout technique with probability, $P = 0.3$, was used to eliminate the model complexity for over-fitting prevention. In addition, a Stochastic Gradient Descent (SGD) technique was utilized to train the network within Tensorflow context. To identify the advantages and the drawbacks of their proposed RNN model, they compared it with two well-known models namely, Multilayer Perceptron (MLP) and Bayesian Logistic Regression (BLR) models. Accordingly, the findings of the comparative analyses depicted that the proposed RNN model gave better performance in terms of validation accuracy of 71.77% in comparison with 65.48% and 58.30% for MLP and BLR respectively. Thus, the RNN is considered a promising tool to solve such a critical problem.

Moreover, [4] aimed to build a model for predicting the severity caused by road accidents using decision trees and neural networks. Their research was conducted based on a real dataset of 417,670 accidents records occurred in the USA covering the period of 1995–2000. This set of data is collected from the National Automotive Sampling System. The outcomes of the simulation demonstrated that the decision tree performs much better than the neural network in predicting all classes of injury, namely zero injury, probable injury, mild injury, incapacitating injury and deadly injury. In addition, the study highlights the main factors impacting deadly injury that are the use of seat belt, the light condition of the road along with the drinking of alcohol. Their experiments also revealed that the decision tree based model performed better predictions for two injury classes that are the deadly injury and the zero injury compared to the other classes with an accuracy of 89.46% and 67.54% respectively. Finally, the authors pointed out that the models might have come up with better and more precise results if the attribute "actual speed at the accident occurrence" is used, as 67.68% of the data records lacks speed information.

Likewise, [5] compared two modeling algorithms for predicting the injury severity due to traffic accidents. The employed models are Bayesian Network and regression. In this study, the authors define the term "severity" according to three main indicators that are the number of mortalities, the number of injuries as well as the property damage. These severity indicators are investigated via the two models; thus, the factors influencing the severity are recognized. The results demonstrated that the Bayesian Network gave higher accuracy compared to the regression model in predicting the severity of traffic accidents. This study was conducted based on 2,246 accidents records occurred in Jilin province, China in 2010. The dataset contained a valuable variety of attributes including accident location, accident instance of time, vehicle condition, weather condition, road geometrics, road surface condition, etc.

Recently, [6] discussed a novel severity prediction model based on a convolutional neural network. Their algorithm focuses on combining the relationships among the accidents features by applying a new concept called the feature matrix to gray image (FM2GI). This concept aims at obtaining deeper feature correlations between data; hence it converts a one dimensional feature into a gray scale images format that contains combination relationships depending on the weights of the features. The proposed model is fed with the gray scale images as input variables, while the output represents the degree of injury severity whether it is fatal, serious or slight. The authors tested their model based on 21,436 accidents records took place in Leeds, UK over 8 years' time span from 2009 to 2016. To validate the efficacy of the suggested model, it was compared with three machine learning methods and six statistical models. The machine learning methods are; neural networks, 1D convolution and Long Short-Term Memory-Recurrent Neural Network while the statistical models are decision trees, support vector machines, Naïve Bayes', K-nearest neighbor, logistic regression and gradient boosting. The experimental results revealed that the proposed model provide better performance in comparison with the competitive models.

Finally, [7] applied the ordered probit regression model to predict the accidents severity encountered by vehicle occupants. The proposed model is efficient in assessing the influence of the elements involved in road accidents. The data utilized to conduct this study were supplied by the National Road Safety Authority which contains all the accidents happened in Portugal over a period of five years from 2007 to 2011. During the study time span, there were 6,752 accidents records, of which 5,021 caused by drivers, 1,183 related to passengers while pedestrians caused 548 of them. Since the study focuses on analyzing and predicting the injury severity caused by vehicle occupants, the other records were ignored which leads to using only 4,528 records to assess the performance of the proposed model. Out of this dataset, 41.63% denotes no injury cases, 56.85% represents slight injury, 1.10% indicates serious injury while fatal injury forms 0.42%. Due to the minimal rates of the two most severe injury cases, the authors combined the serious and fatal injury cases into a single severity level resulting in three levels of injury severity that are no injury, slight injury and serious or fatal injury. The outcomes of this study indicate that some kinds of traffic accidents, particularly rollovers, run-offs and colliding with fixed objects seem to be the principal factors for the most injury severity. In addition, travelers using light motor vehicles on a dual road of a dry surface are inclined to encounter more serious injuries than the travelers of heavy vehicles on a one-sided road of a wet surface. On the other hand, the study points out that the drivers' seat is the most secured place in vehicles leading to the least severe injury level in case of accidents occurrence. Furthermore, the analysis of the study pinpoints that females are more likely to have more fatal injuries than males. The research finds out that accidents in the urban areas lead to the lowest level of injury severity in Portugal. Also, the researchers prove the nullity of the general beliefs that the increase of the occupants' age increases the severity of injury in road accidents and imposing speed limits decreases severe injury levels. Moreover, the research affirms that males involved in road accidents suffer from less severe injuries than females although they tend to drive more violently and at a higher speed due to their technical skills in taking more defensive measures to avoid serious injuries.

3 Dataset

The road accidents data used for conducting this study were obtained from the UK Transportation Department. The dataset consists of 8,500 accidents records which took place on the rural and the urban UK roads in 2015. Each of these records has 31 different attributes gathered during the time of the accidents a long with one feature for the class label, accident severity. Table 1 illustrates the description of the selected attributes to be used in this study. These attributes are categorized into three main groups: (1) Road characteristics (Road type, Road class, Junction control, Road surface conditions, Urban or rural area and Speed limit), (2) Environmental characteristics (Light conditions and Weather conditions) and (3) Vehicle characteristics (No. of vehicles involved in the accident). To analyze the factors (i.e. attributes) influencing the severity of collisions, the dependent variable, accident severity, was classified into two main categories, fatal and slight accidents. Table 2 shows a brief description of both categories.

Table 1. Selected attributes description

Feature name	Feature description
No. of vehicles	Number of vehicles intricate in the road accidents
Speed limit	The road speed limitation where the accident took place
Road type	**1**: Single carriageway **2**: Dual carriageway **3**: One-way street **4**: Roundabout **5**: Slip road
Junction control	**1**: Not at junction or within 20 m **2**: T or staggered junction **3**: Mini-roundabout **4**: Crossroads **5**: Private drive or entrance
Light condition	**1**: Daylight **2**: Darkness – lights lit **3**: Darkness – lights unlit **4**: Darkness – no lighting
Road surface condition	**1**: Dry **2**: Wet or damp **3**: Snow **4**: Frost or ice **5**: Flood over 3 cm deep
Road class	**1**: A **2**: B **3**: C **4**: A(M) **5**: Motorway
Weather condition	**1**: Fine no high winds **2**: Fine and high winds **3**: Raining no high winds **4**: Raining and high winds **5**: Fog or mist
Urban or Rural area	**1**: Urban **2**: Rural

Table 2. Accident severity classes description

Accident severity class	Class description
Fatal	An accident that results in human death
Slight	An accident that results in minor injuries with no killed or seriously injured cases

4 The Proposed Methodology

This section discusses the proposed model based on Artificial Neural Network (ANN) to predict the level of injury severity caused by road accidents. Figure 1 illustrates the main phases of the proposed ANN model. Initially, the dataset is preprocessed to normalize the data, fill the missing values and to select the attributes of the highest impact on causing accidents. Afterwards, the prediction model relying on artificial neural network is developed using nine input variables and one output variable. The input variables used to feed the ANN are weather conditions, light conditions, road conditions, road speed limit, road class, road type, road junction control type, the number of vehicle(s) involved in the accident, in addition to urban or rural area while the resultant output from the ANN model is the level of accident severity (fatal or slight). Finally, the performance of the proposed ANN model is assessed using three evaluation measures that are accuracy, precision and recall [8].

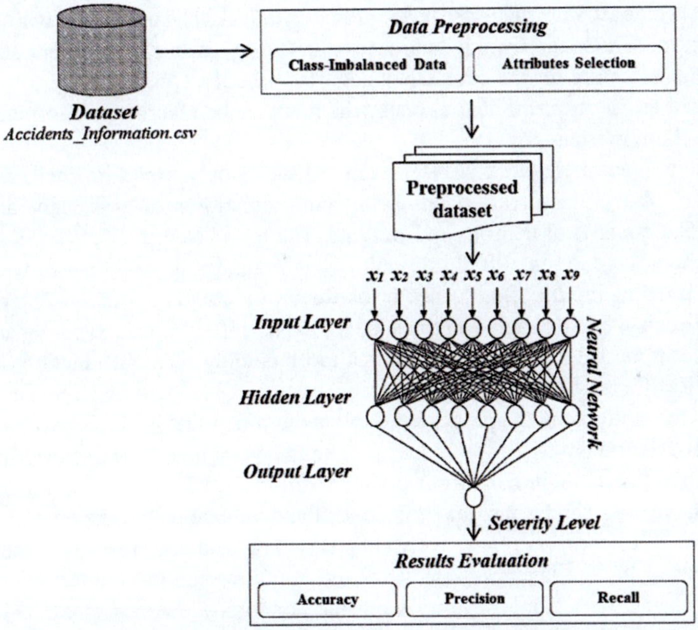

Fig. 1. The phases of the proposed methodology

Data pre-processing is a critical and essential step for handling and preparing data for the analysis phase before using the machine learning techniques. It consists of various steps such as cleaning, integration, normalization, reduction, transformation, feature extraction, selection [9], etc. In this study, Class-Imbalanced data algorithm as well as attribute (i.e. feature) selection task are applied on the dataset, and they are briefly explained in the below sub-sections.

It is a common problem in data classification induced by the skewed distribution (i.e. disproportionate ratio) of samples among classes. The minority samples are ignored in most of the classification techniques resulting in low accuracy prediction model. This problem can be handled through applying various approaches including sampling methods, cost-sensitive methods or kernel and active learning methods. In this study, sampling methods are utilized to solve this issue by resampling the imbalanced dataset. The resampling approach can be implemented either by under-sampling the classes that has the majority of samples, over-sampling the classes that has the minority of samples, or the mixture of both over and under sampling techniques.

The dataset used in this study is imbalanced where the majority of samples represent the cases of slight injury severity while the minority of samples belongs to fatal injury severity class. In such manner, the prediction model will be biased towards the major class, slight accident severity, and will predict every case as it belongs to it. To overcome this obstacle and to enhance the classification accuracy of data, a hybrid of over-sampling and under-sampling techniques is applied. This task is conducted using R tool resulting in 4,250 records for each class of accident severity (fatal and slight).

It is a process of selecting the significant factors influencing the accidents severity from the accidents information dataset which is effectively used in constructing the prediction model. Generally, the dataset used in this study has 31 different attributes a long with one feature for the class label, accident severity. Out of these attributes, nine are selected as the most influential ones which are ranked based on Information Gain and Gain Ratio measures.

Artificial Neural Network (ANN) is one of the widely used branches of machine learning in solving classification, clustering and prediction problems. Generally, it is a function that consists of neurons and weights. The term "Neuron" or "Node" as called in other studies, represents the basic unit of computation in ANN while weights are real values indicating the significance of particular inputs to the output. The nodes are linked to each other through weighted connections [10]. Various types of activation functions can be used based on the application including sigmoid function, threshold function, ReLU (rectified linear unit) function, hyperbolic tangent function, etc. Figure 2 illustrates the structure of the proposed ANN prediction model which is developed using MATLAB. It is composed of nine nodes in the input layer which are the selected attributes including road class, light conditions, road conditions, road type, speed limit, urban or rural area, weather conditions, junction control and vehicle count, denoted in Fig. 2 as $\boldsymbol{i_1, i_2, i_3, i_4, i_5, i_6, i_7, i_8, i_9}$ respectively [11]. Based on trial and error experimentations, the number of hidden layers is selected to be 50. However, the output layer has one node expressing the accident severity level (fatal or slight), denoted as y. The input and hidden layers are connected using a set of coefficients (i.e. weights) termed as $\boldsymbol{W_{ij}}$, where i indicates the input layer and j indicates the hidden layer. Likewise, the hidden and output layers are connected using weights denoted as $\boldsymbol{W_{jn}}$, where $\boldsymbol{n}$ indicates the output layer [12–14]. In this study, 80% of data were used for training while 20% were utilized for testing purposes.

In this study, three performance indicators were utilized to assess the effectiveness of the proposed ANN model in predicting the level of injury severity caused by road accidents. These measures are accuracy, precision and recall. The accuracy of the proposed model is defined as a ratio of the correctly predicted severity levels to the total number of observations. It is calculated as shown in Eq. 1. The precision is another validation

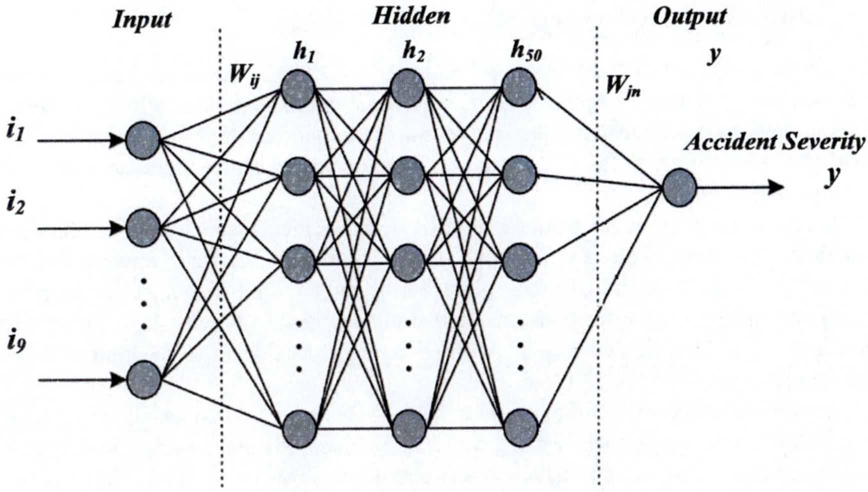

Fig. 2. Structure of the proposed ANN prediction model

parameter; it is identified as a criterion for measuring exactness. For instance, this measure can answer questions like, out of all road accidents severity predicted as fatal, how many of them are actually fatal? Mathematically, the precision is calculated as shown in Eq. 2, where true positive (TP) is the number of correctly predicted severity levels, also known as the sensitivity of a classification model, while False Positive (FP) and False Negative (FN) indicate the number of mistakenly predicted event and no-event values. On the other hand, the recall parameter is used to answer questions such as out of all road accidents that are truly fatal, how many accidents are predicted as so? Equation 3 is used to measure the recall parameter.

$$Accuracy = \frac{No.\ of\ correct\ predictions}{Total\ No.\ of\ predictions} \tag{1}$$

$$Precision = \frac{TP}{TP + FP} \tag{2}$$

$$Recall = \frac{TP}{TP + FN} \tag{3}$$

5 Results and Discussions

The results and findings of this research paper are discussed briefly in the following two sub-sections. Section 5.1 discusses the findings of analyzing the main attributes influencing the road accidents severity whereas Sect. 5.2 elaborates on the performance of the proposed ANN model in predicting the accident severity level.

5.1 Attributes vs Accident Severity

As shown in Fig. 3(a), the highest percentage of accidents occurred during a clear weather, out of which 27% of accidents were fatal while 21% were slight accidents. This is obvious since clear weather is the most common one in different parts of the world and particularly in the UK; therefore, more accidents can be recorded under this weather condition category.

On the other hand, it can be observed that slight accidents were often to occur than fatal accidents during rainy and foggy weathers. Figure 3(b) illustrates that most of the UK accidents occurred in 2015 took place during the daylight with approximately an equal percentage of 24% for both fatal and slight accidents while the least number of accidents occurred when unlit lights with percentages less than 5% for both accident severity levels.

Figure 3(c) demonstrates that most fatal and slight accidents happened on wet roads and less likely to happen when there was a flood since this phenomenon is very rare to happen and people do not drive in such conditions. This study shows that most of the fatal and slight accidents took place on the roads that have speed limit of 30 mph.

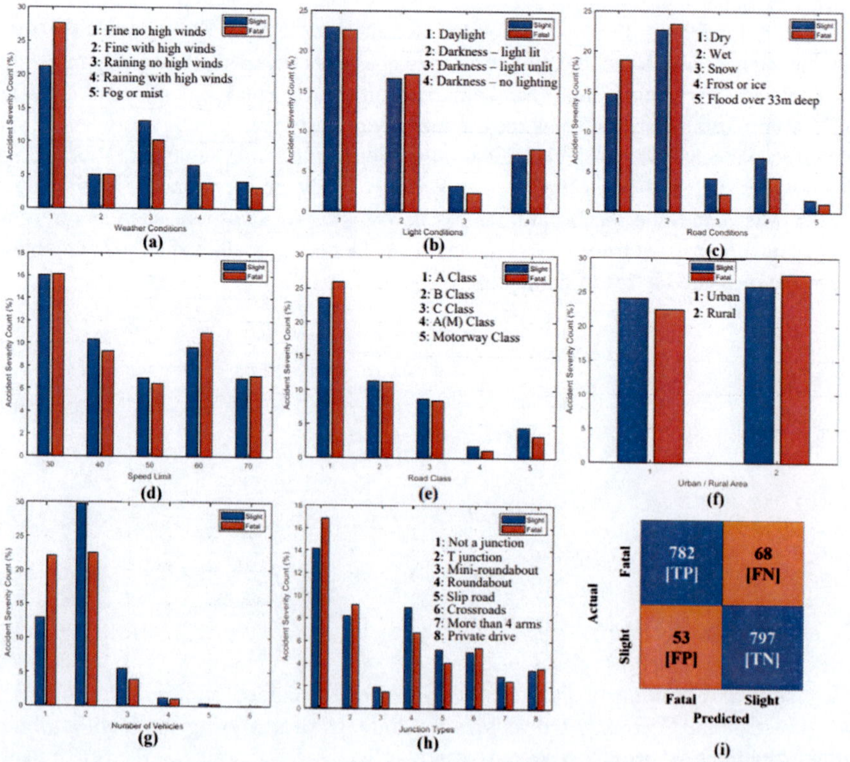

Fig. 3. (a) Weather Conditions, (b) Light Conditions, (c) Road Conditions, (d) Road Speed Limit, € Road Class, (f) Urban or Rural Area, (g) Number of Vehicles, (h) Road Junction Type, (i) Confusion Matrix

However, the roads with a speed limit of 50 mph experienced the least fatal and slight accidents as shown in Fig. 3(d).

According to the analysis results (Fig. 3(e)), it can be noticed that most of the accidents occurred on A Class roads, out of which fatal accidents form approximately 26% of the total accidents while slight accidents form roughly 24% of the accidents. However, B Class, C Class and Motorway Class witnessed different levels of fatal and slight accidents of the following approximate proportions: more than 10%, less than 10% and lower than 5% respectively.

Figure 3(f) illustrates that most of the fatal and slight accidents took place on rural area roads with a percentage of occurrence above 25% for each of the accident severity classes. However, the urban area roads experienced a lower percentage of slight and fatal accidents (below 25%).

Based on the analysis results shown in Fig. 3(g), it can be noticed that most of the fatal and slight injuries were caused as a result of collisions between two vehicles followed by accidents caused via one vehicle, three vehicles, four vehicles, five vehicles and six vehicles respectively. Figure 3(h) shows that most of the fatal and slight accidents do not occur at junctions while the accidents took place at mini-roundabout were very rare to happen with a percentage of less than 3% for each fatal and slight accident.

5.2 Accident Severity Prediction Results

The experimental results exposed that the proposed model is effective in predicting the injury severity level caused by road accidents with a promising accuracy of 92%. The confusion matrix of the proposed CNN model is shown in Fig. 3(i).

6 Conclusion

This research paper proposes an effective mechanism to analyze and monitor a critical global issue that causes losses in lives and assets on daily bases, that is road accidents. It also aims at predicting the injury severity level caused by road accidents using Artificial Neural Network (ANN) technique. The prediction model is constructed based on a dataset composed of 8,500 accident records with 31 attributes and one class label (accident severity), covering the accidents occurred in UK in 2015. Prior to training the ANN model, the raw data underwent several pre-processing steps that are cleaning, normalizing as well as applying sampling techniques, using R tool, to overcome a Class-Imbalanced data issue induced by the skewed distribution (i.e. disproportionate ratio) of injury severity samples; fatal and slight. Furthermore, Information Gain and Gain Ratio measures are utilized to rank the attributes according to their impacts on accidents severity; hence, nine attributes were selected as input variables to the ANN model. These attributes are weather conditions, light conditions, road conditions, road speed limit, road class, road type, road junction control type, the number of vehicle involved in the accident, as well as urban or rural area. Out of the dataset, 80% were used for training the ANN model while the rest of data was used for testing purposes. The ANN prediction model is developed using special toolboxes in MATLAB. The following findings were derived from the outcomes of the data analysis and the evaluation of the prediction model proposed in this paper:

- The highest percentage of fatal and slight accidents occurred during a clear weather while the lowest percentage of accidents happened in rainy and foggy weathers.
- Most accidents in the UK occurred during the daylight with an equal percentage of 24% for fatal and slight accidents while the unlit roads experienced the lowest percentage of accidents.
- The roads with 30 mph speed limit witnessed the highest percentage of the fatal and slight accidents, whereas the roads with 50 mph speed limit had the least accidents.
- Out of the total accidents, 26% of fatal accidents and 24% of slight accidents occurred on A Class roads in UK.
- Rural area roads witnessed fatal and slight accidents with above 25% for each of them, while the urban area roads experienced below 25% for fatal and slight accidents.
- The experimental results exposed that the proposed model is effective in predicting the injury severity caused by road accidents with a promising accuracy, precision and recall of 76%, 0.765 and 0.755 respectively.

References

1. Tian, R., Yang, Z., Zhang, M.: Method of road traffic accidents causes analysis based on data mining. Int. Conf. Comput. Intell. Softw. Eng. (2010)
2. Alkheder, S., Taamneh, M., Taamneh, S.: Severity prediction of traffic accident using an artificial neural network: traffic accident severity prediction using artificial neural network. J. Forecast. **36**(1), 100–108 (2016)
3. Sameen, M., Pradhan, B.: Severity prediction of traffic accidents with recurrent neural networks. J. Appl. Sci. **7**(6), 476–8480 (2017)
4. Chong, M., Abraham, A., Paprzycki, M.: Traffic accident data mining using machine learning paradigms. In: Proceedings of the Fourth International Conference on Intelligent Systems Design and Applications, pp. 415–420 (2004)
5. Zong, F., Xu, H., Zhang, H.: Prediction for traffic accident severity: comparing the Bayesian network and regression models. Math. Prob. Eng. 1–9 (2013)
6. Zheng, M., et al.: Traffic accident's severity prediction: a deep-learning approach-based CNN network. IEEE Access **7**, 39897–39910 (2019)
7. Garrido, R., Bastos, A., de Almeida, A., Elvas, J.P.: Prediction of road accident severity using the ordered probit model. Transp. Res. Procedia **3**, 214–223 (2014)
8. Abdel-Aty, M., Abdelwahab, H.: Analysis and prediction of traffic fatalities resulting from angle collisions including the effect of vehicles' configuration and compatibility. Acc. Anal. Prevent. **36**(3), 457–469 (2003)
9. Al-Ghamdi, A.S.: Using logistic regression to estimate the influence of accident factors on accident severity. Accid. Anal. Prev. **34**(6), 729–741 (2002)
10. Chang, L., Wang, H.: Analysis of traffic injury severity: an application of nonparametric classification tree techniques. Accid. Anal. Prev. **38**(5), 1019–1027 (2006)
11. Gaid, M.L., Salloum, S.A.: Homomorphic encryption. In: Hassanien, A.E., et al. (eds.) Proceedings of the International Conference on Artificial Intelligence and Computer Vision (AICV2021), pp. 634–642. Springer, Cham (2021). https://doi.org/10.1007/978-3-030-76346-6_56
12. Gaid, M.L., Yousuf, H., Salloum, S.A., Shaalan, K.: Implementing sequence to sequence neural networks using C#.Net. In: Hassanien, A.E., et al. (eds.) Proceedings of the International Conference on Artificial Intelligence and Computer Vision (AICV2021), pp. 742–753. Springer, Cham (2021). https://doi.org/10.1007/978-3-030-76346-6_66

13. Yousuf, H., Lahzi, M., Salloum, S.A., Shaalan, K.: A systematic review on sequence-to-sequence learning with neural network and its models. Int. J. Electric. Comp. Eng. **11**(3), 2315 (2021)
14. Yousuf, H., Al-Hamad, A.Q., Salloum, S.: An Overview on CryptDb and Word2vec Approaches

Predicting the Intention to Use Audi and Video Teaching Styles: An Empirical Study with PLS-SEM and Machine Learning Models

Khadija Alhumaid[1], Raghad Alfaisal[2], Noha Alnazzawi[3], Aseel Alfaisal[4], Naimah Nasser Alhumaidhi[5], Mohammad Alamarin[6], and Said A. Salloum[7](✉)

[1] Student Services, Rabdan Academy, Abu Dhabi, UAE
[2] Faculty of Art, Computing and Creative Industries, Universiti Pendidikan Sultan Idris, Perak, Malaysia
[3] Computer Science and Engineering Department, Yanbu University College, Royal Commission for Jubail and Yanbu, Yanbu Industrial City, Yanbu, Kingdom of Saudi Arabia
[4] Department of Languages and Translation, Arar Community College, Northern Border University, Arar, Kingdom of Saudi Arabia
[5] Educational Administration Department School of Education, Taibah University, Al Madinah Al Munawwarah, Medina, Kingdom of Saudi Arabia
[6] Industrial Engineering, The Hashemite University, Zarqa, Jordan
[7] School of Science, Engineering, and Environment, University of Salford, Salford, UK
S.A.S.Salloum@edu.salford.ac.uk

Abstract. Lately, there has been an increasing trend of using audio and video content on different online platforms. This trend of incorporation of audio and video content in online interaction was specifically beneficial for teachers and students who experienced better teaching and learning owing to the audio and visual data that offered them a richer interactive environment. The educational domain was positively affected by this practice. It is important to note that there is a dearth of empirical studies that use a conceptual model based on the factor of acceptance to offer insight into the acceptance of this video and audio content. Fortunately, this dearth in literature has been addressed by the current study that contributes in exploring the factors like speed and vividness, perceived concentration, perceived ease of use, perceived usefulness and sheds light on the impact of these factors on the acceptance of audio-video material. This study used the survey method for data collection and subsequently evaluated the research model. The Partial least squares-structural equation modelling (PLS-SEM) and machine learning models were used for this study. The study used a conceptual model to reveal the intentions of the students for the incorporation of audio-visual content on online platforms. The acceptance of technology by students was investigated concerning TAM constructs as well as the predictors like perceived concentration, speed, and vividness. Lastly, the study suggested theoretical and practical implications in the recommendations section. The recommendations were specifically associated with technology users, designers, and change managers.

Keywords: Audio and video content · Machine learning · Structural equation modelling · Technology Acceptance Model

© The Author(s), under exclusive license to Springer Nature Switzerland AG 2022
A. E. Hassanien et al. (Eds.): AMLTA 2022, LNDECT 113, pp. 250–264, 2022.
https://doi.org/10.1007/978-3-031-03918-8_23

1 Introduction

Effective information delivery is a skill used in teaching, which is characterized by online platforms [1–3]. The learning objectives cannot be accomplished if the instructor fails to deliver the material to the students. Hence, different audio and video material that suits the development of the digital era can be incorporated into the interactive learning media [4]. The information technology system will assist the transfer of audio and video material in a helpful manner [5–7]. As per [8, 9], the subject experts have broadly acclaimed the media variations and their significance in enabling interactive learning to inspire the concerns and thoughts of students. The considerable effects of the audio and video material in accomplishing the milestones have been emphasized by the earlier research studies that underscored the audio and video material within the educational domain [7, 10–14]. Likewise, the significance of audio and visual material in employing medial and economical objectives has been focused on by the studies in the non-educational domain [15–19].

Besides determining an ultimate objective of measuring the efficiency of concentration, vividness and speed, the present study aims to establish the impact of audio and visual media on a user's acceptance of e-learning platforms. Under the earlier works, the exposure of online platform users is positively affected by the users' perceived concentration. Hence, a promising environment is realized, where they can easily measure their insight with a newly taught material [20, 21]. In addition, to accept the audio and visual material, investigating the effects of vividness and speed is another goal of this study. The richness of the content and the quality of the visual display is thickly correlated with the vividness [22]. The fact is realized by this study that e-learning users should appear as an educator or learners to assess the online learning platform. Thus, the acceptance model – TAM and other external factors including vividness and speed, and perceived concentration are unified by this study. The adoption of the conceptual model successfully measuring the acceptance of audio and video material used in online learning platforms is the key difference of this study from earlier studies.

According to the existing literature, there is no sufficient empirical research, which supports the implementation of audio and video material used in online learning platforms in the educational domain in the UAE, and this phenomenon indicates the fact that the researchers failed to gain an insight with the factors influencing students' actual use. Regarding the methodology for evaluating the theoretical models, most of the technology acceptance studies have employed structural equation modeling (SEM). Therefore, this research entails two-fold objectives. Firstly, the Technology Acceptance Model (TAM) will be incorporated to evaluate the acceptance of audio and video material used in online learning platforms [23, 24]. Secondly, by employing the ML algorithms and PLS-SEM, the researchers will evaluate it by authenticating the theoretical model. This study is attributed to a conceptual framework, which is one of its key worth that underlines the thick bond between the construct of speed and vividness, perceived concentration, and the TAM constructs of perceived usefulness and perceived ease of use. Ultimately, the researchers will understand the ulterior motives of using audio and video material in an educational environment where enabling and increasing the productivity of instructors and students is the primary objective.

2 Theoretical Framework

2.1 Technology Acceptance Model (TAM)

The computer-based technology acceptance and adoption research heavily relies on TAM in diverse domains. According to planned behavior theories and reasoned action, [25] established TAM to rationally explain the organizational and technological factors and challenges that are likely to affect users' acceptance of technology. In particular, TAM originally aims to reveal the causal relations between the beliefs encouraging the technology acceptance and adoption, specifically, perceived ease of use, perceived usefulness, attitudes, and intentions to use a system [25]. Contrary to TRA, the organizations employing TAM have demonstrated theoretical justifications and a practical indication of TAM's viability to predict the direct association between key beliefs, i.e., perceived usefulness and ease of use and behavioral intentions to use a technology system. In the latest shape of TAM, the direct Belief-Intention relationship is indicative of the fact that individuals are likely to build positive intentions to utilize a specific system within organizations when they possess positive beliefs about the perceived ease of use and perceived usefulness of the system [25].

Besides its theoretical objectives, TAM fulfills practical values by developing parsimonious measurement scales for predicting system use in organizations [25]. According to [25], two factors are decisive determinants of a user's intention to use a technological system, namely, perceived usefulness (PU) and perceived ease of use (PE). First, PU refers to the degree to which a potential technology user believes that using a particular system of interest offers a better advantage or is a more effective substitute over other methods. Thus, a system is expected to enhance a user's job performance. Second, on the other hand, perceived ease of use designates the extent to which a potential user believes that using the target system would be free of great efforts or difficulty, and hence, the degree to which a user would be capable of harnessing the technology in question to achieve his or her job tasks smoothly, successfully and efficiently [25]. Furthermore, TAM hypothesizes a causal linkage between PU and PE, assuming that PU influences PE [25]. The casual paths that TAM hypothesizes have been empirically substantiated in a wide range of end-user computing technologies and user populations. Previous related studies have indicated that PU and PE are positively and significantly related to the intention to use learning materials that are designed using multimedia and sensorial rich resources, especially in virtual and augmented reality settings (e.g., [10, 26–28]). Therefore, the current study makes the following hypotheses:

H1: Perceived Usefulness (PU) positively affects the intention to use audio and video learning materials (AA).

H2: Perceived Ease of Use (PE) positively affects the intention to use audio and video learning materials (AA).

2.2 Flow Theory

2.2.1 Concentration

To observe the finest experiences of deep absorption in engaging self-rewarding activities, [29] formulated the flow theory, wherein an individual perceives different challenges in accordance with their proficiencies and abilities. Concerning a state of flow, people do possess clear proximal goals, where they get instant feedback to fine-tune their progress [30]. It can also be said that a holistic sensation experienced by people while doing an activity with total involvement is referred to as the flow [31]. As per the flow theory, when an experience completely occupies the individual, they tend to give their best output. Thus, they overlook the awareness of surrounding factors such as noises, fatigue, hunger, or other factors and become emotionally and mentally detached. The intrinsically motivated subjective experience phenomena basically originate from the flow theory. Moreover, being the major reasons to engage in an activity, these experiences are enjoyable and intrinsically rewarding. A subjective mental condition with the deep involvement of individuals in an engaging activity can be described as the flow, wherein an individual is simultaneously performing many tasks with controllable challenges, such as attempting various goals, constantly processing feedback besides adjusting the actions based on that feedback [32]. An intensive and focused concentration is also indicative of the state of flow, for example, what actions are being performed in a moment, integrating those actions besides illustrating awareness in a state of stability together with a sense of control over one's action, loss of reflective self-awareness, perceiving the experience as being intrinsically motivating as well as distortion of temporal experience [32].

With respect to virtual and e-learning, it is presumed by the flow theory that a user possesses perceived concentration, which persuades their intention to adopt virtual learning platforms. Hence, the constructs like "cognitive absorption" and "perceived concentration" can be measured through the flow theory, which is portrayed as digitized and virtual learning environments. Foregoing in view, the learning experience or an object is exclusively focused by an individual learner, where, he/she loses consciousness of all other things except the experience [10]. Empirical evidence confirming positive relationships between continued intention to use e-learning systems and flow experience has also been substantiated by the earlier studies (e.g., [33–35]). Hence, the existing study hypothesizes that;

H3: Perceived level of concentration (PC) positively affects the intention to use audio and video learning materials (AA).

2.3 Virtual Reality Attributes

2.3.1 Vividness

Concerning the telepresence view, one of the vital attributes of virtual reality is vividness. Virtual reality is referred to as a skill intervened in a technologically simulated situation. The presence and telepresence are believed to be the two dimensions of virtual reality. The perceived presence in a setting mediated by a communication technology medium is referred to as presence [36]. Whereas, the sensations of vividness and interactivity illustrate the virtual reality from the telepresence view [37]. As per [36], vividness refers to *"the representational richness of a mediated environment as defined by its formal features, that is, how an environment presents information to the senses" (p. 11).* Moreover, [36] asserts that vividness is *"the ability of technology to create a sensorially rich mediated environment" (p. 80).*

The quality of the procedural learning experience online is improved by a sense of vividness, as the representational richness of information is enhanced by the vividness, for instance, comparison between video materials and text materials [38, 39]. The visual aid is used to drive the vividness of video materials in such a way that a procedural learning experience provides to users through visual realistic and dynamic demonstrations is likely to strengthen the cognition process [11]. Until now, the positive link between the perceived usefulness of video learning materials and a sense of vividness has been corroborated by the earlier literature with the help of YouTube, for instance [11]. The vividness is associated with aesthetic appeal in online shopping. Hence, the actual products are visually displayed through imaginary objects in a way that establishes clear pictures in the minds of customers, thus it strongly influences them to buy [40]. The technology attributes, such as perceived usefulness, ease of use, and perceived enjoyment of augmented reality mobile applications [28] have positively been affected by the vividness. Likewise, the present study presumes that:

H4: Perceived level of vividness (SV) positively affects the intention to use audio and video learning materials (AA).

Based on our literature review search, a research model as shown in Fig. 1.

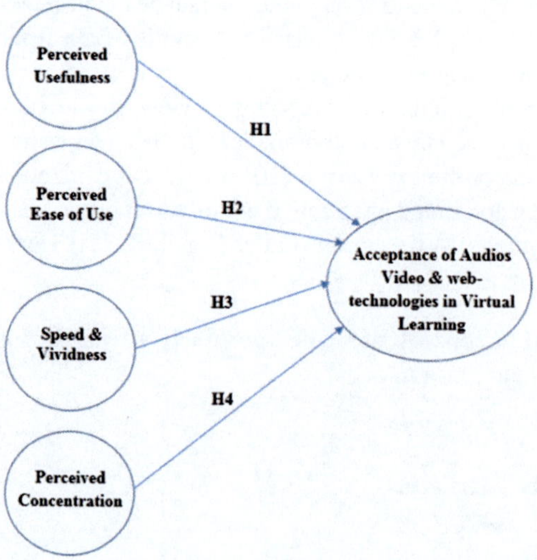

Fig. 1. Research model.

3 Research Methodology

3.1 Data Collection

The students of UAE universities were considered and the data was collected through online surveys from 20th October 2021 to 30th December 2021 during the fall semester of 2021/2022. Almost 500 questionnaires were distributed among random students, where 425 students completed this questionnaire and the percentage of the respondents remained 85%. These 425 responses of questionnaires were correct and authentic as per the report of the team. As far as the given sample size is concerned, the 425 valid responses were suitable according to [41], like the anticipated sampling size for a population of 1500 in 306 respondents. While considering negligible details, the sample size of 425 was comparatively higher. Hence, in this case, the sample size corresponded well with structural equation modelling and the evaluation [42], which was employed to corroborate the hypotheses.

3.2 Personal/Demographic Information

Correspondingly, the study sample contained 52% of males and 48% of females. The age of students was from 18 to 29 years for 65% of the respondents, while the remaining 35% were greater than 29 years. Most of the respondents were well-informed university graduates. In addition, only 10% of the respondents were having a doctoral degree, while 21% were master's degree holders and the 69% of them had a bachelor's degree. As recommended by [43], the "purposive sampling approach" was employed, when the respondents are ready to volunteer and they are approachable without any difficulty. The students from various age groups, diverse disciplines/levels, and many colleges constituted this study sample. Moreover, the IBM SPSS Statistics application was used to calculate the demographic data.

3.3 Study Instrument

A survey instrument has been introduced to validate the hypothesis. To measure the five constructs in the questionnaire, the research contained 14 items. To enhance the effectiveness of the paper, the questions from the prior studies were revised and altered by the authors.

3.4 Survey Structure

Questionnaires for the survey were given to students which constituted 3 parts of the study.

- The first segments collected personal information such as the demographics of the respondents.
- The second segment consisted of two items which constituted the main question of Acceptance of Audio and Videos in Virtual learning.

- There were 12 items in the third segment which show the Perceived Ease of Use, Perceived Usefulness, Perceived Concentration, and speed, and vividness.

The Likert Scale was used to determine the degree such as, strongly disagree (1), disagree (2), neutral (3), agree (4), and strongly agreed (5).

4 Findings and Discussion

4.1 Data Analysis

The partial least square-structural equation modeling (PLS-SEM) with the help of Smart-PLS V.3.2.7 software [44] is used in this paper for data analysis. A two-step assessment approach was used for studying the collected data is comprised of the structural and the measurement models [45]. The benefits of using PLS-SEM are multiple for this study because this paper aims to establish a current theory and the PLS-SEM model is the most appropriate choice [46], also by using the PLS-SEM we can deal with complex studies (investigative) [47]. The PLS-SEM model can analyze the model as a single unit rather than separating them into different groups [48]. Finally, the PLS-SEM model can provide accurate measurements because it provides concurrent analysis for the structural and measurement models [49].

4.2 Convergent Validity

The suggestions from the source [44] are that we need to take into consideration the construct reliability (including composite reliability (CR), Dijkstra-Hensley's (PA), and Cronbach's alpha (CA)) and validity (including convergent and discriminant validity) into account for the evaluation of the measurement model. While figuring out the construct reliability, in Table 1 it is proven that Cronbach's alpha (CA) has values ranging between 0.815 and 0.890 which are greater than the threshold value of 0.7 [45]. Furthermore, according to Table 1, the results show that the values of composite reliability (CR) range between 0.801 and 0.888, which is larger than the ideal value of 0.7 [46]. It is suggested to researchers to use the Dijkstra-Henseler's rho (pA) reliability coefficient for the evaluation and documentation of construct reliability [44]. Like CA and CR, the values 0.7 or greater than 0.7 should be seen in the reliability coefficient ρA during the investigation, while for greater stages of research, the value 0.80 or 0.90 are recommended to be used [45, 47, 48]. According to Table 1, the reliability coefficient ρA of each of the measurement construct is greater than 0.70. These results grant credibility to construct reliability and finally, it is considered that all constructs are free from defects and are accurate.

In order to measure the convergent validity, it is required that the average variance extracted (AVE) should be tested as well as factor loading should also be tested [44]. The outcomes are shown in Table 1 and the rate of 0.7 for all factor loadings was different.in Table 1 it is seen that AVE values that are generated range between 0.611 and 0.731 which is greater than the threshold value of 0.5. Considering the expected results, we can effectively attain the convergent validity of constructs.

4.3 Discriminant Validity

To measure the discriminant validity, two criteria require measurement which is the Fornell-Larker criterion, and the Heterotrait-Monotrait ratio (HTMT) [45]. The Fornell-Larker condition, affirmed the prerequisites as shown in the results of Table 2 as all the AVEs alongside with their square roots are larger than their relationship with other structures from the source [54].

Table 3 given below has the HRMT ratio showing the threshold value to be higher than the values of individual constructs [55]. The introduction of the HTMT ratio was due to such circumstances as with its help we can measure the discriminant validity. The results from this study showed that there are no inconsistencies with the validity and reliability of this measurement model, allowing the structural model to be assessed with supplementary use of the data.

Table 1. Convergent validity results which assures acceptable values (Factor loading, Cronbach's Alpha, composite reliability ≥ 0.70 & AVE > 0.5).

Constructs	Items	Factor loading	Cronbach's Alpha	CR	PA	AVE
Acceptance of audio and videos in virtual learning	AA1	0.837	0.880	0.853	0.854	0.625
	AA2	0.852				
Perceived ease of use	PE1	0.767	0.815	0.829	0.833	0.714
	PE2	0.758				
	PE3	0.720				
Perceived usefulness	PU1	0.729	0.828	0.801	0.824	0.611
	PU2	0.889				
	PU3	0.880				
Perceived concentration	PC1	0.824	0.890	0.888	0.831	0.719
	PC2	0.768				
	PC3	0.769				
Speed and vividness	SV1	0.728	0.838	0.824	0.859	0.731
	SV2	0.762				
	SV3	0.789				

Table 2. Fornell-Larcker scale.

	AA	PE	PU	PC	SV
AA	**0.902**				
PEOU	0.519	**0.881**			
PU	0.620	0.295	**0.879**		
PC	0.324	0.572	0.654	**0.832**	
SV	0.446	0.438	0.502	0.580	**0.815**

Table 3. Heterotrait-Monotrait Ratio (HTMT).

	AA	PE	PU	PC	SV
AA					
PE	0.527				
PU	0.138	0.428			
PC	0.182	0.523	0.553		
SV	0.475	0.424	0.422	0.523	

4.4 Hypotheses Testing Using PLS-SEM

The structural equation model with the Smart PLS having the maximum likelihood estimation was used to determine the theoretical constructs of the structural model [49, 50]. The hypotheses were studied which can be seen in Table 4 and Fig. 2 showing that this structure had very high predictive power [51]. The percentage of variance as seen in the table and figure shows that it is about 77.5% for Acceptance of Audio and videos in Virtual learning. The outcomes generated by the PLS-SEM technique show evidence, that the developed hypotheses expressed as the beta (β) values, t-values, and p-values in Table 5 are accurate. All of the authors have supported all the hypotheses. Based on data analysis hypotheses H1, H2, H3, and H4 were supported by the empirical data. Perceived Usefulness (PU), Perceived Ease of Use (PE), Speed and Vividness (SV), and Perceived Concentration (PC) has significant effects on Acceptance of Audio and Videos in Virtual Learning (AA) ($\beta = 0.592, P < 0.001$), ($\beta = 0.558, P < 0.001$), ($\beta = 0.722, P < 0.001$), and ($\beta = 0.305, P < 0.05$) respectively; hence H1, H2, H3 and H4 are supported.

Table 4. R^2 of the endogenous latent variables.

Constructs	R^2	Results
AA	0.775	High

Table 5. Hypotheses-testing of the research model (significant at $p^{**} \leq 0.01$, $p^{*} < 0.05$).

H	Relationship	Path	*t*-value	*p*-value	Direction	Decision
H1	PU → AA	0.592	14.228	0.000	Positive	Supported**
H2	PE → AA	0.558	10.189	0.000	Positive	Supported**
H3	SV → AA	0.722	9.210	0.001	Positive	Supported**
H4	PC → AA	0.305	3.567	0.023	Positive	Supported*

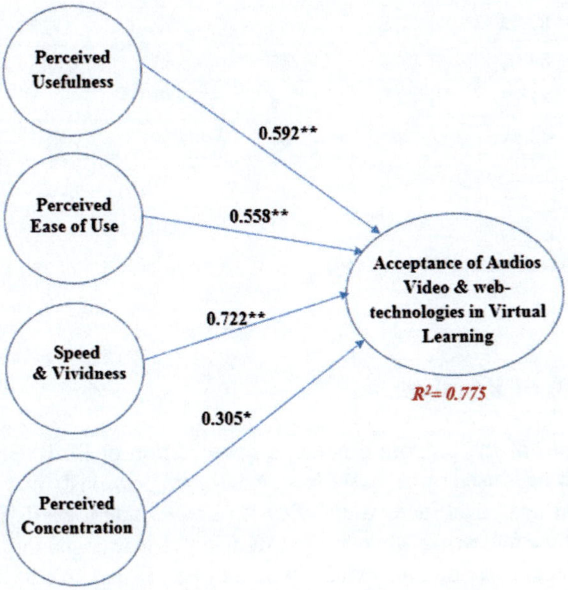

Fig. 2. Path coefficient of the model (significant at $p^{**} \leq 0.01$, $p^{*} < 0.05$).

4.5 Hypothesis Testing Using Machine Learning Algorithms

To understand and predict the factors and their relationship in the theoretical model, the use of machine-learning classification algorithms was used in this paper. Many different methodologies such as decision trees, Bayesian networks, neural networks and if-then-else rules were used to gauge the relation between factors [52]. Weka (ver. 3.8.3) was applied to check the predictive model. Different classifiers such as J48, Logistic, OneR, BayesNet, LWL, and AdaBoostM1 were introduced for testing the predictive model [53, 54]. The outcomes seen in Table 6 conclude that the J48 had much better performance compared to many other classifiers in terms of forecasting speed & vividness (SV), perceived concentration (PC), perceived ease of use (PE), and perceived usefulness (PU) on the acceptance of audio-video material (AA). When we take into account the tenfold cross-validation, the prediction of AA by J48 was found to have an accuracy of 88.59% which is why, H1, H2, H3, and H4 are supported. The classifiers mentioned above depicted superior performance to other classifiers concerning precision (.886), TP rate (.885), and recall (.886).

Table 6. Predicting the AA by PU, PE, SV and PC.

Classifier	CCI[a] (%)	TP[b] rate	FP[c] rate	Precision	Recall	F-Measure
BayesNet	86.76	.867	.665	.868	.868	.869
Logistic	85.32	.853	.668	.853	.853	.854
LWL[d]	82.22	.822	.634	.822	.822	.830
AdaBoostM1	84.60	.846	.757	.846	.846	.851
OneR	87.22	.872	.718	.872	.872	.873
J48	88.59	.885	.787	.886	.886	.888

[a] CCI: Correctly Classified Instances.
[b]TP: True Positive.
[c]FP: False Positive.
[d]LWL: Locally Weighted Learning.

5 Discussion of Results

For the validation of the proposed model, a combination of PLS-SEM and machine learning algorithms is used by us, as we have mentioned previously that we used a complimentary multi-analytical method allowing it to contribute towards the IS literature since it is one of the few works where the use of machine learning algorithms is done for predicting the acceptance of audio-video material (AA) inside an educational environment. It is important that we mention that PLS-SEM can be used for the determination of dependent variables and the validation of mathematical models but expanding the current theory [55], likewise, it is possible for supervised machine learning algorithms (those which have a pre-defined dependent variable) can be used to predict dependent variable based on independent variables [52]. another interesting aspect regarding this research is that it uses multiple algorithms with differing methodologies as, "decision trees (J48), Bayesian networks, neural networks and if-then-else rules." in most of the instances J48 had better performance than most classifiers. It should also be noted that both continuous (numerical) and categorical variables were classified using the non-parametric decision tree, which separated the sample into homogeneous sub-samples on the basis of the most important independent variables [52]. The significant coefficients, contrarily, were evaluated using PLS-SEM (a nonparametric procedure) with substitutes of samples in order to obtain a large number of samples at random.

The previous studies have given us much insight into students' behavior in an online environment [10–14]. Nonetheless, only a few of the studies show the importance of audiovisual material in an educational environment [56–58]. The current study shows how important of a role audiovisual material has in an educational environment. The implementation of TAM helps in the identification of important factors that affect the acceptance behavior of students. The two constructs which are the perceived ease of use and perceived usefulness have an effective influence on the acceptance of audio-visual on the basis of the fact that all of the path coefficients were found to be statistically significant. We can conclude that the student's perceived ease of use and perceived usefulness directly affect the acceptance of audio-visual material. This arises because the students view audio-visual materials as resources that can be viewed anytime and repeatedly to aid in online learning. The students are in favor of getting benefits from offered educational resources as an alternative to traditional means of learning. The current results agree with results presented by [14, 39, 59] who explored students' behavior intention in terms of TAM two constructs, namely the perceived ease of use and perceived usefulness. According to these studies, students prefer to use such resources because the given material satisfies their needs in achieving their learning goals. Another more significant impact of this paper is that it has shown the importance of speed and vividness alongside perceived concentration which are important factors in determining the student's intention in using the audio-visual material. This study supports all previous studies in showing that vividness and speed are crucial in impacting the student's decision in using the audio-visual resources, such that the increase in the level of vividness increased perceived usefulness. This study mainly focuses on students at random rather than distinguishing them based on genders. This is a limitation of this study and for the future, it is better to conduct a study that has a gender distinction allowing us to understand the difference in intention and perception of using the available resources. Another limitation for this study comprises external factors like speed and vividness and perceived concentration, to combat these factors it would be appropriate for future studies to include such factors as content richness and satisfaction for the conceptual model. After we include these factors in future studies, we would be able to get more insights on technology acceptance as a whole. Lastly, this study only focuses on students from UAE, so to get a better grasp of student's perceptions from all around the world comparative studies must include students from different universities.

References

1. Salloum, S.A., Al-Emra, M., Habes, M.O., Alghizzawi, M.: Understanding the impact of social media practices on E-learning systems acceptance (2019). https://doi.org/10.1007/978-3-030-31129-2
2. Alghizzawi, M., Salloum, S.A., Habes, M.: The role of social media in tourism marketing in Jordan. Int. J. Inf. Technol. Lang. Stud. **2** (2018)
3. Alghizzawi, M., Ghani, M.A., Som, A.P.M., et al.: The impact of smartphone adoption on marketing therapeutic tourist sites in Jordan. Int. J. Eng. Technol. **7**, 91–96 (2018)
4. Habes, M., Salloum, S.A., Alghizzawi, M., Mhamdi, C.: The relation between social media and students' academic performance in Jordan: YouTube perspective. In: Hassanien, A.E., Shaalan, K., Tolba, M.F. (eds.) AISI 2019. AISC, vol. 1058, pp. 382–392. Springer, Cham (2020). https://doi.org/10.1007/978-3-030-31129-2_35
5. Rosa, F.O.: Analisis kemampuan siswa kelas X pada ranah kognitif, afektif dan psikomotorik. Omega J. Fis. Dan. Pendidik. Fis. **1**, 24–28 (2015)
6. Sudibyo, L.: Peranan dan Dampak Teknologi Informasi dalam Dunia Pendidikan di Indonesia. Widyatama **20** (2013)

7. Mutia, L., Gimin, G., Mahdum, M.: Development of blog-based audio visual learning media to improve student learning interests in money and banking topic. J. Educ. Sci. **4**, 436–448 (2020)
8. Kennedy, I.G., Latham, G., Jacinto, H.: Education Skills for 21st Century Teachers: Voices from a Global Online Educators' Forum. Springer, Cham (2016). https://doi.org/10.1007/978-3-319-22608-8
9. Miarso. Media Pengajaran. Sinar Barual Gensindo, Bandung (2007)
10. Liu, S.-H., Liao, H.-L., Pratt, J.A.: Impact of media richness and flow on e-learning technology acceptance. Comput. Educ. **52**, 599–607 (2009)
11. Lee, D.Y., Lehto, M.R.: User acceptance of YouTube for procedural learning: an extension of the technology acceptance model. Comput. Educ. **61**, 193–208 (2013). https://doi.org/10.1016/j.compedu.2012.10.001
12. Saeed, N., Sinnappan, S.: Effects of media richness on user acceptance of Web 2.0 technologies in higher education. In: Advanced Learning. IntechOpen (2009)
13. Sun, P.-C., Cheng, H.K.: The design of instructional multimedia in e-learning: a media richness theory-based approach. Comput. Educ. **49**, 662–676 (2007). https://doi.org/10.1016/j.compedu.2005.11.016
14. Chi-Yueh, H., Ci-Jhan, H., Hsiu-Hui, C.: Using technology acceptance model to explore the intention of internet users to use the audio and video fitness teaching. J. Eng. Appl. Sci. **12**, 4740–4744 (2017)
15. Kang, S.J., Lee, M.S.: Assessing of the audiovisual patient educational materials on diabetes care with PEMAT. Publ. Health Nurs. **36**, 379–387 (2019)
16. Sadjadi, S.O., Greenberg, C.S., Singer, E., et al.: The 2019 NIST audio-visual speaker recognition evaluation. In: Proceedings of the Speak Odyssey (submitted), Tokyo, Japan (2020)
17. Alshurideh, M.: Pharmaceutical promotion tools effect on physician's adoption of medicine prescribing: evidence from Jordan. Mod. Appl. Sci. **12** (2018)
18. Aburayya, A., Alshurideh, M., Al Marzouqi, A., et al.: An empirical examination of the effect of TQM practices on hospital service quality: an assessment study in UAE hospitals. Syst. Rev. Pharm. **11** (2020). https://doi.org/10.31838/srp.2020.9.51
19. Alhashmi, S.F.S., Salloum, S.A., Mhamdi, C.: Implementing artificial intelligence in the United Arab Emirates healthcare sector: an extended technology acceptance model. Int. J. Inf. Technol. Lang. Stud. **3** (2019)
20. Novak, T.P., Hoffman, D.L., Yung, Y.-F.: Modeling the structure of the flow experience among web users. In: INFORMS Marketing Science and the Internet Mini-Conference (1998)
21. Webster, J., Trevino, L.K., Ryan, L.: The dimensionality and correlates of flow in human-computer interactions. Comput. Human Behav. **9**, 411–426 (1993)
22. Adamo-Villani, N., Wilbur, R.B.: Effects of platform (immersive versus non-immersive) on usability and enjoyment of a virtual learning environment for deaf and hearing children. In: EGVE (Posters) (2008)
23. Huang, Y.-M., Huang, Y.-M., Huang, S.-H., Lin, Y.-T.: A ubiquitous English vocabulary learning system: evidence of active/passive attitudes vs. usefulness/ease-of-use. Comput. Educ. **58**, 273–282 (2012)
24. Larsen, T.J., Sørebø, A.M., Sørebø, Ø.: The role of task-technology fit as users' motivation to continue information system use. Comput. Human Behav. **25**, 778–784 (2009)
25. Davis, F.D.: Perceived usefulness, perceived ease of use, and user acceptance of information technology. MIS Q. **13**, 319–340 (1989). https://doi.org/10.2307/249008
26. Saeed, N., Yang, Y., Sinnappan, S.: Effect of media richness on user acceptance of blogs and podcasts. In: Proceedings of the Fifteenth Annual Conference on Innovation and Technology in Computer Science Education, pp. 137–141 (2010)

27. Nagy, J.T.: Evaluation of online video usage and learning satisfaction: an extension of the technology acceptance model. Int. Rev. Res. Open Distrib. Learn. **19** (2018). https://doi.org/10.19173/irrodl.v19i1.2886
28. McLean, G., Wilson, A.: Shopping in the digital world: examining customer engagement through augmented reality mobile applications. Comput. Human Behav. **101**, 210–224 (2019). https://doi.org/10.1016/j.chb.2019.07.002
29. Csikszentmihalyi, M., Csikszentmihalyi, I.: Beyond Boredom and Anxiety. Jossey-Bass, San Francisco (1975)
30. Csikszentmihalyi, M., Abuhamdeh, S., Nakamura, J.: Flow. In: Csikszentmihalyi, M. (ed.) Flow and the Foundations of Positive Psychology, pp. 227–238. Springer, Dordrecht (2014). https://doi.org/10.1007/978-94-017-9088-8_15
31. Csikszentmihalyi, M.: The flow experience and its significance for human psychology. In: Csikszentmihalyi, M., Csikszentmihalyi, I.S. (eds.) Optimal Experience: Psychological Studies of Flow in Consciousness, pp. 15–35. Cambridge University Press (1988). https://doi.org/10.1017/CBO9780511621956.002
32. Nakamura, J., Csikszentmihalyi, M.: The concept of flow. In: Flow and the Foundations of Positive Psychology, pp. 239–263. Springer, Dordrecht (2014). https://doi.org/10.1007/978-94-017-9088-8_16
33. Joo, Y.J., Joung, S., Kim, J.: Structural relationships among self-regulated learning, learning flow, satisfaction, and learning persistence in cyber universities. Interact. Learn. Environ. **22**, 752–770 (2014). https://doi.org/10.1080/10494820.2012.745421
34. Rodríguez-Ardura, I., Meseguer-Artola, A.: E-learning continuance: the impact of interactivity and the mediating role of imagery, presence and flow. Inf. Manag. **53**, 504–516 (2016). https://doi.org/10.1016/j.im.2015.11.005
35. Zhao, Y., Wang, A., Sun, Y.: Technological environment, virtual experience, and MOOC continuance: a stimulus–organism–response perspective. Comput. Educ. **144**, 103721 (2020). https://doi.org/10.1016/j.compedu.2019.103721
36. Steuer, J.: Defining virtual reality: dimensions determining telepresence. J. Commun. **42**, 73–93 (1992). https://doi.org/10.1111/j.1460-2466.1992.tb00812.x
37. Sukoco, B.M., Wu, W.-Y.: The effects of advergames on consumer telepresence and attitudes: a comparison of products with search and experience attributes. Expert Syst. Appl. **38**, 7396–7406 (2011). https://doi.org/10.1016/j.eswa.2010.12.085
38. Griffith, D.A., Gray, C.C.: The fallacy of the level playing field. J. Mark. Channels **9**, 87–102 (2002). https://doi.org/10.1300/J049v09n03_05
39. Hernandez, M.D.: A model of flow experience as determinant of positive attitudes toward online advergames. J. Promot. Manag. **17**, 315–326 (2011). https://doi.org/10.1080/10496491.2011.596761
40. Flavián, C., Gurrea, R., Orús, C.: The influence of online product presentation videos on persuasion and purchase channel preference: the role of imagery fluency and need for touch. Telemat. Inf. **34**, 1544–1556 (2017). https://doi.org/10.1016/j.tele.2017.07.002
41. Krejcie, R.V., Morgan, D.W.: Determining sample size for research activities. Educ. Psychol. Meas. **30**, 607–610 (1970)
42. Chuan, C.L., Penyelidikan, J.: Sample size estimation using Krejcie and Morgan and Cohen statistical power analysis: a comparison. J. Penyelid. IPBL **7**, 78–86 (2006)
43. Al-Emran, M., Salloum, S.A.: Students' attitudes towards the use of mobile technologies in e-evaluation. Int. J. Interact. Mob. Technol. **11**, 195–202 (2017)
44. Hair, J., Hollingsworth, C.L., Randolph, A.B., Chong, A.Y.L.: An updated and expanded assessment of PLS-SEM in information systems research. Ind. Manag. Data Syst. **117**, 442–458 (2017). https://doi.org/10.1108/IMDS-04-2016-0130
45. Nunnally, J.C., Bernstein, I.H.: Psychometric Theory (1994)

46. Kline, R.B.: Principles and Practice of Structural Equation Modeling. Guilford Publications (2015)
47. Hair, J.F., Ringle, C.M., Sarstedt, M.: PLS-SEM: indeed a silver bullet. J. Mark. Theory Pract. **19**, 139–152 (2011)
48. Henseler, J., Ringle, C.M., Sinkovics, R.R.: The use of partial least squares path modeling in international marketing. In: New Challenges to International Marketing, pp. 277–319. Emerald Group Publishing Limited (2009)
49. Al-Emran, M., Arpaci, I., Salloum, S.A.: An empirical examination of continuous intention to use m-learning: an integrated model. Educ. Inf. Technol. **25**(4), 2899–2918 (2020)
50. Salloum, S.A., Alhamad, A.Q.M., Al-Emran, M., et al.: Exploring students' acceptance of E-learning through the development of a comprehensive technology acceptance model. IEEE Access **7**, 128445–128462 (2019)
51. Chin, W.W.: The partial least squares approach to structural equation modeling. Mod. Methods Bus. Res. **295**, 295–336 (1998)
52. Arpaci, I.: A hybrid modeling approach for predicting the educational use of mobile cloud computing services in higher education. Comput. Human Behav. **90**, 181–187 (2019)
53. Frank, E., Hall, M., Holmes, G., et al.: Weka-a machine learning workbench for data mining. In: Maimon, O., Rokach, L. (eds.) Data Mining and Knowledge Discovery Handbook, pp. 1269–1277. Springer, Boston (2010). https://doi.org/10.1007/978-0-387-09823-4_66
54. Alomari, K.M., Al Hamad, A.Q., Salloum, S.: Prediction of the Digital Game Rating Systems Based on the ESRB
55. Alshurideh, M., Al Kurdi, B., Salloum, S.A., et al.: Predicting the actual use of m-learning systems: a comparative approach using PLS-SEM and machine learning algorithms. Interact. Learn. Environ. 1–15 (2020)
56. Akhtar, Z., Falk, T.H.: Audio-visual multimedia quality assessment: a comprehensive survey. IEEE Access **5**, 21090–21117 (2017)
57. Ongena, G., van de Wijngaert, L., Huizer, E.: Exploring determinants of early user acceptance for an audio-visual heritage archive service using the vignette method. Behav. Inf. Technol. **32**, 1216–1224 (2013)
58. Baehr, C.: Incorporating user appropriation, media richness, and collaborative knowledge sharing into blended e-learning training tutorial. IEEE Trans. Prof. Commun. **55**, 175–184 (2012)
59. Lai, J.-Y., Rushikesh Ulhas, K.: Understanding acceptance of dedicated e-textbook applications for learning: involving Taiwanese University students. Electron. Libr. **30**, 321–338 (2012)

Intellgenet Systems and Applications

Immunity of Signals Transmission Using Secured Unequal Error Protection Scheme with Various Packet Format

H. Kasban[1], Sabry Nassar[1](✉), and Mohsen A. M. M. El-Bendary[2]

[1] Nuclear Research Center, Egyptian Atomic Energy Authority, Cairo, Egypt
sabrynassar39@gmail.com

[2] Faculty of Industrial Education, Helwan University, Cairo, Egypt

Abstract. Low cost and self-configured communication make Wireless Sensor Network (WSN) is a promising technology for enhancing security in several nuclear applications. In this research paper, efficient techniques are proposed to immune audio signals transmitted over mobile communications channels. The various randomizing data tools are used and combined with error control mechanisms for enhancing the quality and performance of the transmitted signals over noisy channels. Various metrics of the error performance and received audio signals quality are used to investigate the suitability of the proposed audio scenarios of transmission. The Bit Error Rate (BER), Number of Lost Packets percentage (NLP), and Throughput (Th) are used as tools for measuring the error performance of the proposed techniques. While the Correlation coefficient (Cr) and Mean Square Error (MSE) are utilized for evaluating the quality of received audio signals. Reed Solomon codes are used for transmitted packet encoding. Various experiments with computer simulation are executed to examine the suggested scenarios with different velocities of the mobile terminal. The results of the experiments cleared the superiority of the presented scenarios for the transmission of audio signals.

Keywords: Audio signal · WSN · Data randomizing · Noisy channels

1 Introduction

With the continuous and accelerating advancements in technology, there has been a great interest in utilizing WSN in different fields of life. One of the most important uses of WSN is Homeland Security (HS), where a perimeter of critical and hazard areas is monitored and surveyed in a continuous manner using well-distributed audio sources [1]. The ease of installation, low cost, self-configured, very low cost of communication infrastructure and self-healing property make WSN a promising technology for enhancing homeland security in case of the military, nuclear, or other civilian applications [2, 3]. WSN is a wireless network that consists of multiple devices/nodes that are wirelessly connected to the controller and are intended to gather scalar data and monitor the object's special features. Advanced WSN is capable of handling multimedia signals as well as

© The Author(s), under exclusive license to Springer Nature Switzerland AG 2022
A. E. Hassanien et al. (Eds.): AMLTA 2022, LNDECT 113, pp. 267–277, 2022.
https://doi.org/10.1007/978-3-031-03918-8_24

the WMSN's scalar data. It has a variety of sensors and sends audio, video, and imagery. The functioning method of WMSN is based on audio, image capture, or video to detect any changes in a specified volume or in the surrounding environment characteristics. Scalar sensors, acoustic sensors, and other sensors detect this change.

The many applications of WSNs and WMSNs encountered some limits and obstacles, such as resource constraints, memory, power, processing, and so on. These wireless networks' service contexts are heterogeneous (various application fields, different surroundings, and changing communications conditions) (military, nuclear, industrial, civil, environmental monitoring). In various application scenarios, the WMSN's QoS criteria (high data rate, real-time necessity, data loss, sufficient throughput) are recognized as a difficulty. Data redundancy is a constraint in the WMSN applications, it is related directly to the sensor's deployment mechanism [4]. Also, security is an essential issue in the various applications of the WMSN, Authenticity, availability, confidentiality, freshness, integrity, and localization are some of the security needs, DOS, resource depletion, data theft, data alteration (pollution attack), and the injection of fake information are some examples of WSN attacks.

In this research paper, we aim to improve the efficiency with which audio signals are transmitted through a mobile wireless link by enhancing the immunity of these audio signals utilizing the randomized encoded packets. For this purpose; the interleaving techniques are used with error control schemes, and based on the chaotic Baker map, sophisticated chaotic randomization tools are employed for randomizing the encoded packets to enhance the quality of received acoustic signal and decrease the number of discarded packets on the mobile communication channel as evidenced by a number of computer simulation tests [5]. Different error control strategies are commonly employed to improve the performance of a problematic communication channel. Various error control strategies, different communication channel situations, and different packet lengths are discussed in this research study. Different scenarios are established for achieving better performance and more immune audio signals, so; several computer simulation tests were performed to analyze and evaluate the performance of the various scenarios presented [6, 7].

The following is how the rest of this research study is organized: The associated past work summary was discussed in Sect. 2. In Sect. 3, Security and Error Performance Improving Techniques over WMSN is discussed. Section 4 presents the proposed Model of immune Audio Signals Transmission. The simulation assumptions are mentioned in Sect. 4.1. In Sect. 4.2, the simulation results are shown. Section 5 puts the research paper to a conclusion.

2 Related Work Overview

There are a lot of recent works related to the security and performance of the audio transmission, an improved and secured voice collector is designed in [8], In order to encrypt audio transmissions, a 3DES method is paired with an ECC algorithm., better results related to encryption time, data integrity, data loss rate are obtained. Different chaotic encryption techniques are utilized to secure the transmission of audio signals over OFDM systems in [9, 10] where promising results are obtained for the suggested chaotic models with different measured performance parameters. A survey on the fundamental problems related to the privacy and security of 6G networks "next generation" is discussed in [11]. Wu et al. [12] review the research progress related to deep learning's technological implementation in wireless communication, in which security and privacy are considered one of the major aspects of the study. Physical layer security (PLS) has been identified as one of the most promising security technologies for protecting wireless network transmissions from eavesdroppers, so in [13] many issues related to PLS reviewed and discussed, such as cryptography techniques, the outline of MIMO communication, key-less PLS optimization approaches, their limitations, and constraints. Audio data performance for wireless VOIP (Voice over IP) is studied in [14], in which different wavelets are utilized for analyzing audio signals. Lonkar et al. [15] studied in Voice over LTE (VoLTE) application, the POLQA Mean Audio Performance Rating and the QoS Key Performance Induction (KPI) for video quality are studied, where the experimental analysis is made for different video quality analysis display formats. The statistics and distribution information for Uplink and Downlink Mean opinion score (MOS) is used to assess the audio quality of a VoLTE session.

In [16], in order to make the best use of bandwidth, high reliability of Bidirectional Selection Scheduling (BSS) scheme which can meet the high-quality demands of cloud-aware applications mobile fog computing (MFC), a novel burst error-correction for a cloud-aware MFC scheme is presented. Performance evaluation of a hybrid prefixing strategy implemented multicarrier cm-Wave wireless communication system on the transmission of an encrypted audio signal in a hostile flat fading channel was studied in [17], the simulation results demonstrated that the simulated system's repeat and accumulate channels were encoded when applying Cholesky decomposition-based signal retrieval, it is more resilient and effective. Myint et al. [18] introduce two state Markov-based 5G error model to represent the properties of the underlying error process in the 5G network, which can aid in a better understanding of the error process in 5G wireless communications, as well as the evaluation of error management solutions with less computing complexity and faster simulation timeframes. In [19], the contour-let transform and duffing oscillator are used in a new audio steganography technique to protect the transmission of spoken messages, The contributions of this work are; higher security level, greater hiding capacity, as well as maintaining the audio signal's transparency. Takabayashi et al. [20] propose a study on a physical layer (PHY) for a smart body area network (SmartBAN), as part of the Internet of medical things (IoMT) technology, these strategies are required for high-quality audio or video transmission, according to the findings. An efficient genetic algorithm (GA)-based routing algorithm for WMSNs is proposed in [21], where dynamic cluster formation, multipath routing, and cluster head selection are all included for multimedia data transfer.

As a branch of WMSN, an AWSN has many similar properties, but due to the implementation sometimes in harsh environment, and its vulnerability to a wide scale of malicious attacks, which can be categorized according to their impacts and action taken be malicious threat into two types of attacks; passive and active. The wireless links on the other hand are more susceptible to packet failure, so different techniques should be investigated to enforce the security, and improve the transmission performance of signals. Over the last few years, a number of packet recovery algorithms have been developed, which try to differentiate between packet corruption and packet loss. These methods attempt to recover corrupted packets at the receiver level, avoiding the requirement for packet retransmission. Packet reconstruction using forward error checksums (FEC) is one of the packet recovery strategies and other different techniques are discussed in [22]. The content verification of an encrypted image transmitted over AWGN wireless channel is discussed in [23], and an audio watermarking security enhancement technique for Bluetooth networks is presented in [24].

3 The Proposed Model of Immune Audio Signals Transmission

The proposed model of immune audio signals transmission has been presented in this section. Figure 1 shows the processing steps of audio signals transmission in the proposed model. This proposed model merges between the pseudo coding and error control schemes for enhancing audio signals quality, and improving error performance of the whole mobile wireless communication system. The pseudo schemes utilize the traditional block interleaving technique, and advanced chaotic tool. The randomizing process in the second tool is performed on the basis of a secret key for encrypting the transmitted packets by packet-by-packet behavior. As a result, the proposed model aims to improve error performance while also improving wireless link security.

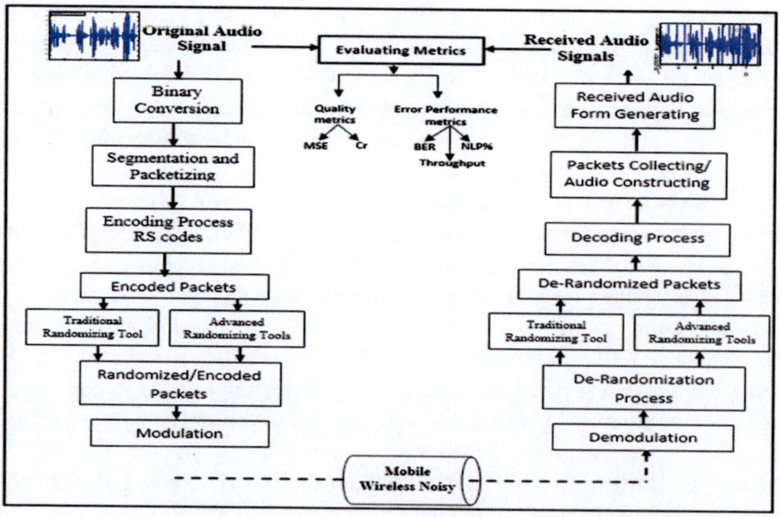

Fig. 1. Block diagram of the proposed audio signals immune model.

The simulation settings and various parameters are listed in Table 1. Where these parameters are selected to be closer to realistic and applicable conditions, where Jakes Model is widely used to simulate mobile wireless communication channel, BPSK modulation is used for its simplicity, and Chaotic Interleaving used as data randomizing for better security and data protection. The mobile wireless communication channel is simulated using the widely used Jakes model [25]. The error performance is measured using BER, NLP, and throughput while the quality of the received audio signals have been measured by the correlation coefficient, and MSE metrics [26]. In wireless communication, the balance between complexity and performance is always a major concern [27], in this research this issue is considered through introducing different audio signal transmission scenarios with different complexity over mobile wireless channel, where generally the system complexity is the amount of arithmetic and logic operation with the required memory, so there is a little increase in complexity in the proposed randomized error control techniques compared to the standard scenarios due to the randomized data tools included, but this little increase in the complexity considered nothing when compared to the performance improvement and received signal quality.

Table 1. Simulation setting parameters

Parameter	Simulation value
Packet Format	Un-coded Packets No error control protection Encoded Packets - Interleaved Encoded Packets
Packet Size	16384 Bits
Error control scheme	Reed-Solomon Codes
Modulation	BPSK modulation
Channel	Mobile Wireless Communications Channel
Channel Modeling	Jakes Model
Mobile Terminal Velocity	V = v1 = 2, v2 = 5, v3 = 20, v4 = 30 km/h
SNR	SNR = [0–35 dB]
Randomizing Tool	Block Interleaving based tool Chaotic Interleaving based tool
Secret Keys	Packet-by packet secret key (Optional)
Simulation tool	Matlab-Program 2016
Quality Metrics	Correlation Coefficient (Cr) Mean Square Error (MSE)
Error Performance	Bit Error Rate (BER) Throughput Number of Lost packets (NLP)

4 Computer Simulation Experiments

In this section, several computer simulations experiments have been carried out in order to evaluate the proposed model of immune audio transmission over mobile wireless link. In these experiments, various velocities of mobile terminals are tested. In the following, two groups of experiments have been devoted to testing audio quality and error performance for slow, and high mobility.

4.1 Slow Mobility with Different Transmission Scenarios

Quality metrics of the received audio signal are measured by the MSE and correlation coefficient as shown in Figs. 2 and 3. The audio signal is transmitted at different transmission scenarios, No interleaving - no FEC, no interleaving – FEC, packet block interleaving - FEC, and chaotic interleaving - FEC. these scenarios are performed for 2, and 5 km/h speed.

In these figures, the quality of received audio signal is decreased with increasing the speed, the degradation of audio signal quality due to the speed can be controlled by using FEC merged with randomized data tools. From the previous results, it's not enough to immune the transmitted audio signal over mobile channel. So the proposed model for longer packet (fourth scenario (FEC - Chaotic interleaving)) improves the quality of the received audio signal.

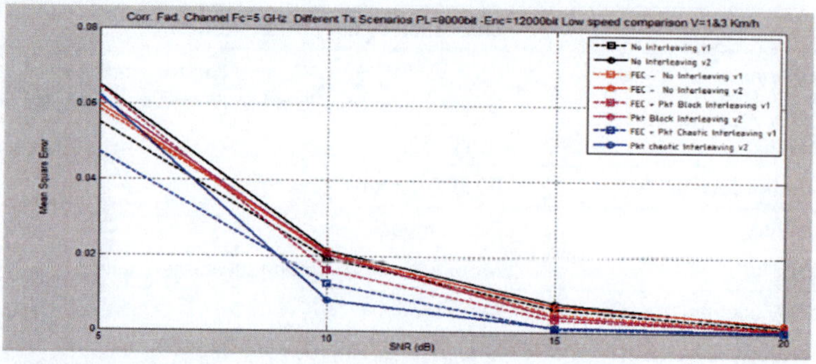

Fig. 2. MSE vs. SNR (dB) of the received audio signal with the various transmission scenarios over the mobile wireless channel ($v_1 = 2$ & $v_2 = 5$ km/h).

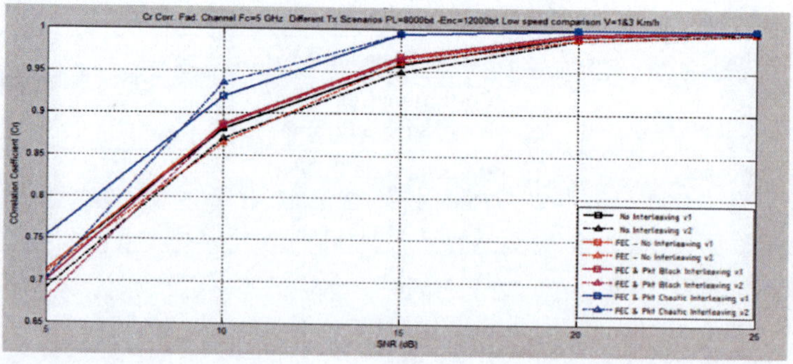

Fig. 3. Cr vs. SNR (dB) of received audio signal with the various transmission scenarios over mobile wireless channel ($v_1 = 2$ & $v_2 = 5$ km/h).

Throughput and BER vs. SNR are shown in Figs. 4 and 5 respectively. As shown from this experiment results, the error performance of the transmitted audio signals is improved using the proposed model due to employing the pseudo schemes with the traditional error control technique for long audio samples transmission over a mobile wireless channel.

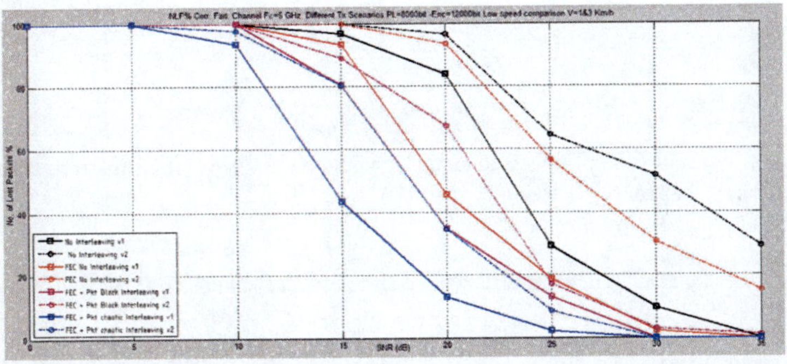

Fig. 4. NLP vs. SNR (dB) of Received audio signal with the various transmission scenarios over mobile wireless channel ($v_3 = 2$ & $v_4 = 5$ km/h).

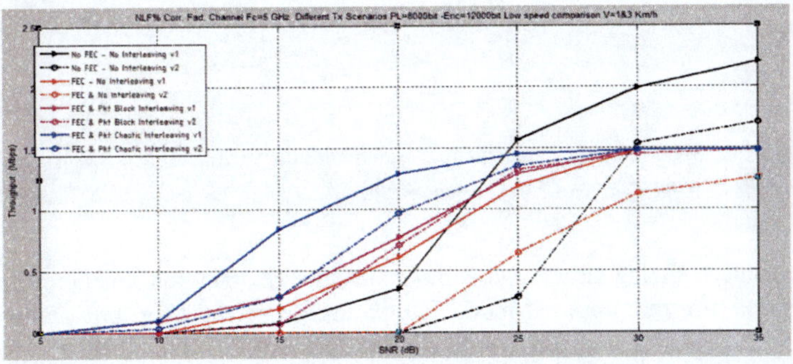

Fig. 5. Throughput vs. SNR (dB) of Received audio signal with the various transmission scenarios over mobile wireless channel ($v_3 = 2$ & $v_4 = 5$ km/h).

4.2 Higher Mobility with Different Transmission Scenarios

Testing and behavior checking of the proposed immune audio signal model have been studied in this section at higher mobile terminal velocities ($v_3 = 20$ & $v_4 = 30$ km/h). Quality of the received audio signals Cr and MSE metrics are given in Figs. 6 and 7, respectively.

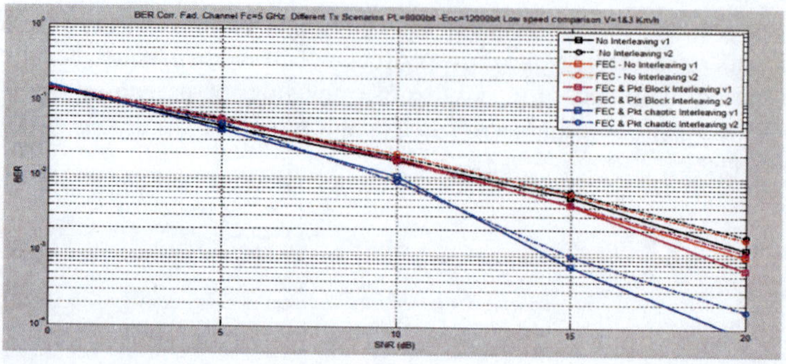

Fig. 6. BER vs. SNR (dB) of Received audio signal with the various transmission scenarios over mobile wireless channel ($v_1 = 2$ & $v_2 = 5$ km/h)

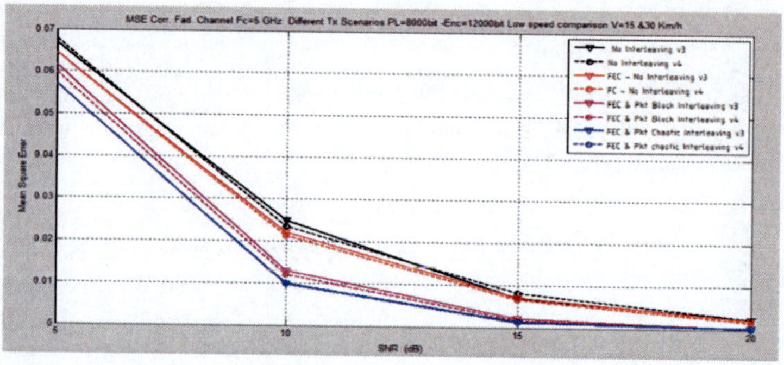

Fig. 7. MSE vs. SNR (dB) of Received audio signal with the various transmission scenarios over mobile wireless channel ($v_3 = 20$ & $v_4 = 30$ km/h).

The results showing more improvement than the previous experiment results in case of low speed. So, the proposed model can be used efficiently for long audio samples transmission over a high-speed mobile wireless channel.

The comparison of received audio signals quality (MSE and Cr) in case of slow and moderate mobile terminal speed has been shown in Figs. 8 and 9 for different audio transmission scenarios ($v_1, v_2, v_3,$ *and* v_4). It is clear in these figures, the best results are obtained with the proposed model. Traditionally, the quality of received audio signals at low speed is better than higher speed, with the proposed model this fact is changed, so the quality is improved at a higher speed due to utilizing the process of merging the pseudo code and error control scheme.

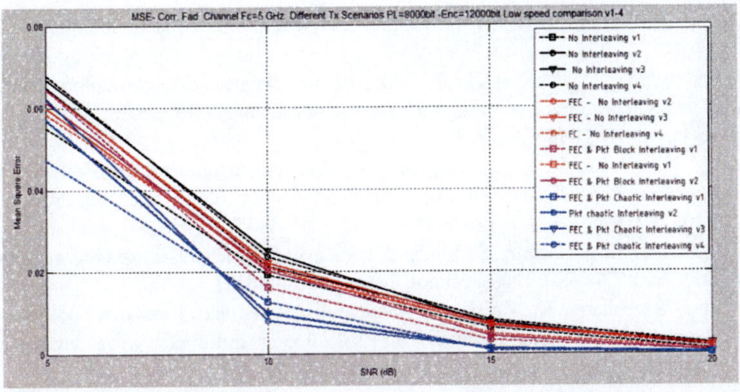

Fig. 8. MSE vs. SNR (dB) of Received audio signal with the various transmission scenarios over mobile wireless channel (v_1, v_2, v_3, v_4).

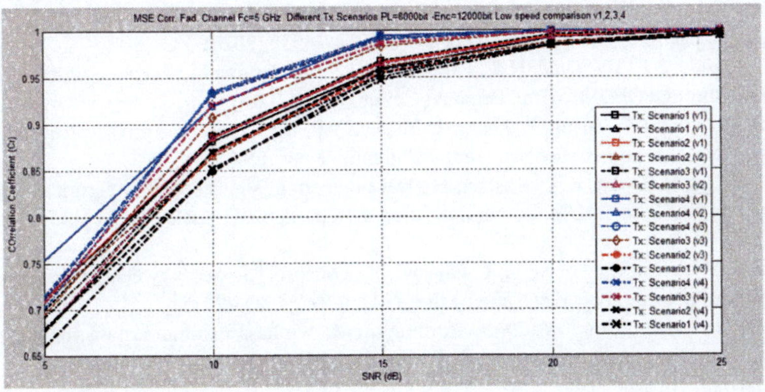

Fig. 9. Cr vs. SNR (dB) of Received audio signal with the various transmission scenarios over mobile wireless channel (v_1, v_2, v_3, v_4).

5 Conclusion

In this article, efficient model of immune audio signal transmission over mobile wireless channel for long packets has been proposed using the pseudo codes and FEC merging. The various velocities of the mobile terminal are tested using Jakes model for evaluating the proposed model. The mobility degrades the quality of audio signals, this issue can be solved by suitable experiments. Different scenarios are executed to evaluate the model, results of the experiments proved that the model can be used efficiently for long audio samples transmission over high-speed mobile wireless channel.

References

1. Grilo, A., Piotrowski, K., Langendoerfer, P., Casaca, A.: A wireless sensor network architecture for homeland security application. In: Ruiz, P.M., Garcia-Luna-Aceves, J.J. (eds.)

ADHOC-NOW 2009. LNCS, vol. 5793, pp. 397–402. Springer, Heidelberg (2009). https://doi.org/10.1007/978-3-642-04383-3_34
2. El-Bendary, M.A.M., Kasban, H., El-Tokhy, M.A.R.: Interleaved Reed-Solomon codes with code rate switching over wireless communications channels. Int. J. Inf. Technol. Comput. Sci. **16**(1), 10–18 (2014)
3. Kasban, H., Nassar, S., El-Bendary, M.A.M.: Medical images transmission over wireless multimedia sensor networks with high data rate. Analog Integr. Circ. Sig. Process **108**(1), 125–140 (2021)
4. Rajan, C., Geetha, K., Geetha, S.: Study of medical image transmission techniques in wireless networks. South Asian J. Eng. Technol. **2**(20), 43–50 (2016)
5. Hashima, S., Hatano, K., Kasban, H., Mohamed, E.M.: Wi-Fi assisted contextual multi-armed bandit for neighbor discovery and selection in millimeter wave device to device communications. Sensors **21**(2835), 1–19 (2021)
6. Ashraf, A., Zaghloul, A., Shaalan, A.A., Kasban, H.: Effect of fog and scintillation on performance of vertical free space optical link from earth to LEO satellite. Int. J. Satellite Commun. Netw. **39**, 3 (2021)
7. El-Bendary, M.A.M., Kasban, H., Haggag, A., El-Tokhy, M.A.R.: Investigating of nodes and personal authentications utilizing multimodal biometrics for medical application of WBANs security. Multim. Tools Appl. **79**(33–34), 24507–24535 (2020)
8. El-Zoghdy, S.F., El-Sayed, H.S., Faragallah, O.S.: Transmission of chaotic-based encrypted audio through OFDM. Wirel. Personal Commun. (2020)
9. Wang, M., Zhu, T., Zhang, T., Zhang, J., Yu, S., Zhou, W.: Security and privacy in 6G networks: new areas and new challenges. Digital Commun. Netw. (2020)
10. El-Tokhy, M.S., Ali, E.H., Kasban, H.: Development of signal recovery algorithm for overcoming PAPR in OFDMA communication system. Arab J. Nucl. Sci. Appl. **55**(1), 15–33 (2022)
11. KelechiIjemaru, G., Adeyanju, I., Olusuyi, K., Ofusori, T.J.: Security challenges of wireless communications networks: a survey. Int. J. Appl. Eng. Res. (2018)
12. Wu, H., Li, X., Deng, Y.: Deep learning-driven wireless communication for edge-cloud computing: opportunities and challenges. J. Cloud Comput. **9**(1), 1–14 (2020)
13. Sanenga, A., Mapunda, G.A., Jacob, T.M.L., Marata, L., Basutli, B., Chuma, J.M.: An overview of key technologies in physical layer security. Entropy **22**(11) (2020)
14. Singh, S., Gupta, A., Sohal, J.S.: Transmission of audio over LTE packet based wireless networks using wavelets. Wirel. Personal Commun. **112**(1), 541–553 (2020)
15. Lonkar, S.A., Reddy, K.T.V.: Analysis of audio and video quality of voice over LTE (VoLTE) call. Int. J. Inf. Technol. (2020)
16. Zhao, W., Dong, P., Guo, M., Zhang, Y., Chen, X.: BSS: a burst error-correction scheme of multipath transmission for mobile fog computing. Wirel. Commun. Mobile Comput. (2020)
17. Naznin, L., Ullah, S.E.: Secured audio signal transmission in hybrid prefixing scheme implemented multicarrier CmWave wireless communication system. Adv. Wirel. Commun. Netw. (2018)
18. Myint, S.H., Yu, K., Sato, T.: Modeling and analysis of error process in 5G wireless communication using two-state Markov chain. In: Special Section on Advances in Channel Coding for 5G and Beyond (2019)
19. Hameed, A.S.: A high secure speech transmission using audio steganography and duffing oscillator. Wirel. Personal Commun. **120**(1), 499–513 (2021)
20. Takabayashi, K., Tanaka, H., Sakakibara, K.: Toward an advanced human monitoring system based on a smart body area network for industry use. Electronics (2021)
21. Genta, A., Lobiyal, K., Abawajy, J.: Energy efficient multipath routing algorithm for wireless multimedia sensor network. Sensors (2019)

22. Khan, S.A., Moosa, M., Naeem, F., Alizai, M.H., Kim, J.: Protocols and mechanisms to recover failed packets in wireless networks: history and evolution. In: Special Section on Trends and Advances for Ambient Intelligence with Internet of Things (Iot) Systems, IEEE Access (2016)
23. Nassar, S.S., et al.: Content verification of encrypted images transmitted over wireless AWGN channels. Wirel. Personal Commun. **88**(3), 479–491 (2016)
24. El-Bendary, M.A.M., Haggag, A., Shawki, F., Abd-El-Samie, F.E.: Proposed approach for improving Bluetooth networks security through SVD audio watermarking. In: IEEE International Conference on Sciences of Electronics, Technologies of Information and Telecommunications, pp. 594–598 (2012)
25. El-Bendary, M.A.M., Kasban, H.: Efficient low computational complexity technique for burst error reduction in WiMAX networks. J. Comput. Sci. Inf. Technol. **5**(3), 351–356 (2015)
26. Kasban, H., Zahran, O., Arafa, H., El-Kordy, M., Elaraby, S.M.S., Abd El-Samie, F.E.: Quantitative and qualitative evaluation of gamma radiographic image enhancement. Int. J. Signal Process. Image Process. Pattern Recogn. **5**(2), 73–87 (2012)
27. Mohamed, E.M., Hashima, S., Hatano, K., Kasban, H., Rihan, M.: Millimeter-wave concurrent beamforming: a multi-player multi-armed bandit approach. Comput. Mater. Continua **65**(3), 1987–2007 (2020)

Check for updates

Overlapping Cell Segmentation with Depth Information

Tao Wang(✉)

School of Computer and Information Engineering, Harbin University of Commerce, Harbin 150028, China
wt@hrbcu.edu.cn

Abstract. With the advancement of artificial intelligence technology, analysis of cervical cells with the aid of the cervical cancer auxiliary diagnosis system can improve the accuracy and efficiency of cervical cancer screening. In order to better complete the overlapping cell segmentation in cervical cell samples, this paper proposes a segmentation algorithm for cytoplasm based on depth information. The segmentation algorithm based on depth information can make full use of the depth information in the stacked cervical cell samples, so as to better complete the overlapping cell segmentation. The experimental results show that, based on the ISBI2015 public data set, the cell segmentation algorithm based on depth information proposed in this paper is higher than other current methods in terms of positive predictive value, negative predictive value, precision rate and recall rate.

Keywords: Cervical cell · Cell segmentation · Depth information

1 Introduction

Cytoplasmic features have been shown to be critical for the identification of abnormal cell [1]. Accurate segmentation of cytoplasm is a core step in cervical cell segmentation. Once the cytoplasmic boundary is located, quantitative evaluation index values such as cell diameter and nuclei-cytoplasmic ratio can be calculated [2–7, 13]. Figure 1 shows the Cervical cytology samples with depth information and synthetic EDF images.

Phoulady et al. [8] used the depth information of cervical cell samples to segment overlapping cells within cell clusters for the first time, and the segmentation method won the ISBI2015 challenge. The method first uses an iterative threshold method to extract the nuclei in the EDF image. Cell clusters and background regions were then separated using a Gaussian mixture model. The core of the method is to segment overlapping cells using the depth information provided by multiple focal planes. By dividing different focal planes into region blocks, the similarity between each region block and different nuclei in the cell cluster is measured according to the proposed "nucleus-region block similarity" measure. According to their similarity, the region blocks are divided into the most similar nuclei, so as to achieve the purpose of coarse segmentation of overlapping cell boundaries.

© The Author(s), under exclusive license to Springer Nature Switzerland AG 2022
A. E. Hassanien et al. (Eds.): AMLTA 2022, LNDECT 113, pp. 278–287, 2022.
https://doi.org/10.1007/978-3-031-03918-8_25

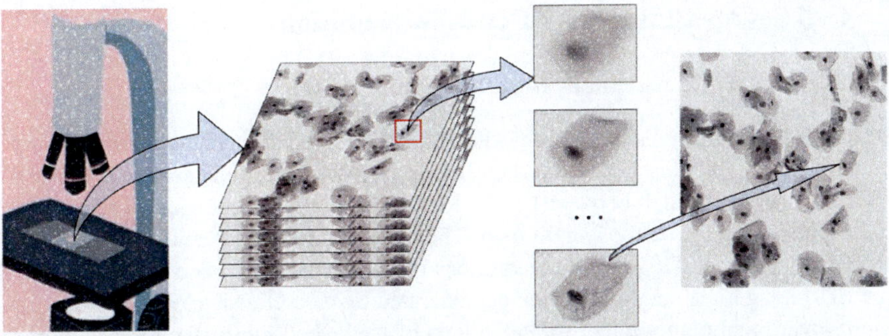

Fig. 1. Cervical cytology samples with depth information and synthetic EDF images

Literature [9] adds an overlap factor to the region block allocation function to improve the performance of literature [10]. By adjusting the overlap factor in the region block allocation function, the purpose of controlling the overlap rate between the roughly segmented cell regions can be achieved. With the addition of overlapping factors, the same area block can be assigned to different cells within the cell cluster. The rough segmentation boundary obtained by this method is closer to the real cell boundary. Literature [10] further works on the basis of literature [9]. The unreachable blocks in the coarse segmentation boundary are removed by the fine segmentation step of the region block. An unreachable block satisfies that the line connecting the block to the center of the cell block passes through the non-cellular area. Finally, the iterative smoothing method is used to complete the fine segmentation of the coarse segmentation boundary.

Most of the previous overlapping cell segmentation algorithms only utilize cell information on EDF images. Due to the serious occlusion between some cells in the EDF image, the cells cannot be identified from the cell cluster only by using the EDF image information. This leads to the unsatisfactory segmentation results of the cell segmentation algorithm in the cell cluster with a large overlap rate. The ISBI2015 challenge not only provides EDF images, but also provides stacked images with depth information in this field of view. Cervical cells exhibit different focal states in different focal planes in the stacked images. Therefore, it is more conducive to the extraction of cell contours by including the attributes of contour points and line segments containing depth information. The main contributions of the algorithm proposed in this paper are as follows: (1) Compared with other cell segmentation algorithms, the algorithm proposed in this paper does not use EDF images, but uses cervical cell sample images containing depth information; (2) Proposes a depth information-based contour point and line segment attributes, so that more accurate overlapping cytoplasmic segmentation boundaries can be obtained.

2 Cell Segmentation with Depth Information

2.1 Contour Point Attributes with Depth Information

It can be known from the literature [11] that the intensities of the contour points located on the same cell boundary have a certain similarity. Therefore, the literature [11] uses the similarity of the strength between the contour points to set the boundary weights which correspond to nodes on the weight map. The contour point attribute only considers the information on the EDF image in literature [11]. For the intensity of contour points, since the EDF image is a single-layer image generated by discrete wavelet transform of the multi-layer focal planes under the same field of view, the EDF image loses a lot of the original information of the multi-layer focal planes. In order to obtain better overlapping cell segmentation results, it is necessary to take full advantage of the depth information provided by the original stacked cervical cell image samples. As shown in Fig. 2(a), sampling locations C, D, E and F are taken on the multi-layer focal plane sample, where C and D are the boundary points of cell A, and E and F are the boundary points of cell B. Use the depth intensity vector $\overrightarrow{X} = (x_1, x_2, \cdots, x_M)$ to represent the intensity value of the sampling point at the same position in different focal planes.

As can be seen in Fig. 2(b), the depth intensity vectors of C and D sample points have relatively similar change curves with the change of the focal plane, and the depth intensity vectors of E and F sample points also have similar laws. However, the depth intensity vector curves of C, D sample points and E, F sample points are quite different. It can be concluded that the intensity values of the contour points of the same cell have similar characteristics of variation. Therefore, the depth information of the multi-layer focal plane is added on the basis of the mean value of the intensity window in [11]. The definition of the intensity mean window containing depth information is shown in formula (1).

$$M\_d = \frac{1}{N^2 \times M} \sum_{m=1}^{M} \sum_{(x_i, y_i) \in W} I_m(x_i, y_i) \tag{1}$$

Among them, N^2 represents the size of the intensity window; I_m represents the focal plane of the $m - th$ layer; the number M of focal planes is 20.

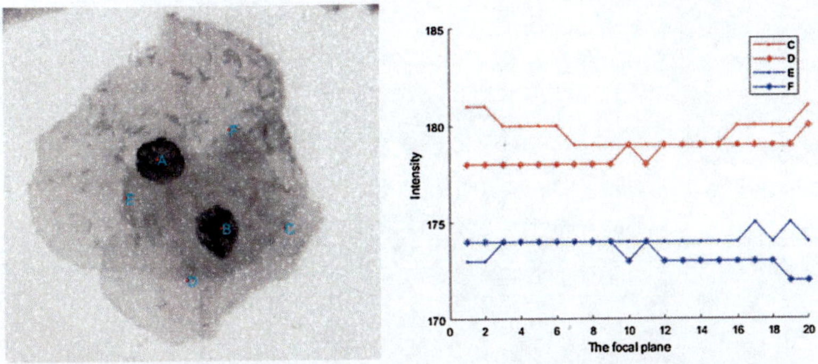

a) Sampling location b) Depth intensity distribution

Fig. 2. Comparison of the intensity values of different cell boundary points

Figure 3(a) shows the cells in the ISBI2015 dataset 001 sample, where the red line segment represents the radial direction, and "*" represents the candidate contour points. The mean value change curve of the intensity window obtained in the radial direction on the EDF image is shown in Fig. 3(b). Figure 3(c) shows the mean value change curve of the intensity window obtained by using the multi-layer focal plane containing depth information. Comparing Figs. 3(b) and (c), it can be seen that the candidate contour points on the boundary are more approximate in the Fig. 3(c) containing depth information. Therefore, the gradient-window mean value with depth information is defined as shown in Eq. (2).

$$W_b = |rg(x_i^\theta, y_i^\theta) - rg(x_j^{\theta+1}, y_j^{\theta+1})| \times |M\_d(x_i^\theta, y_i^\theta) - M\_d(x_j^{\theta+1}, y_j^{\theta+1})| \quad (2)$$

$rd(x_i, y_i)$ is the radial direction difference. The weights of contour points containing depth information are shown in formula (3).

$$W_p = \alpha W_b + \beta w_d \quad (3)$$

w_d is the distance weight between nodes.

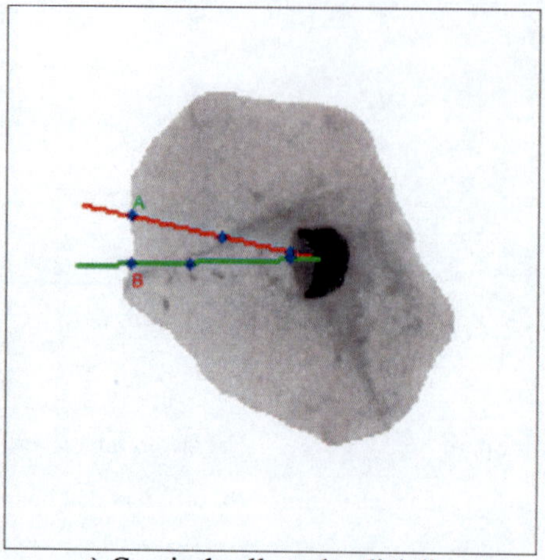

a) Cervical cells and radial direction

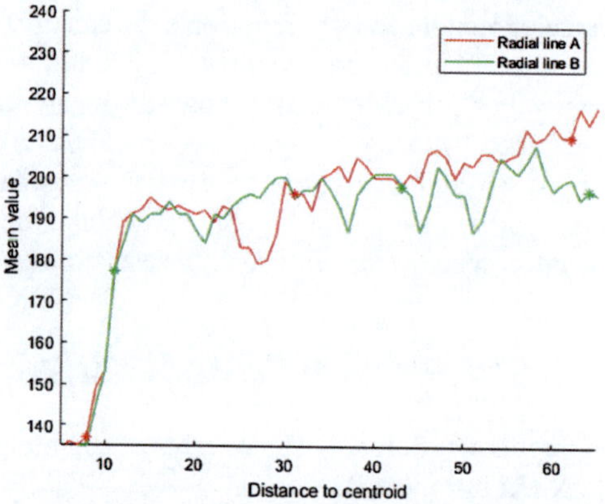

b) Mean change curve of intensity window

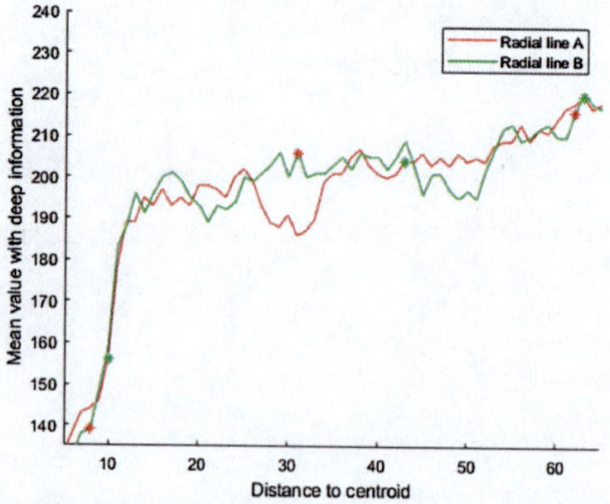

c) Mean change curve of intensity window with depth information

Fig. 3. Contrast of Mean change curve of intensity window

2.2 Contour Segment Attributes with Depth Information

Usually, the focal plane where the nucleus is located is the best focal plane of the cytoplasm corresponding to the nucleus. Therefore, it can be assumed that when the nucleus is in focus, the cytoplasm in its corresponding cell is also in focus, and vice versa. It can be seen that in cervical cell samples with multiple focal planes, the nucleus and its cytoplasm have similar focusing trends.

As the focal plane changes, the contrast in the nuclear and cytoplasmic regions changes accordingly. Area contrast is greatest when the area is in focus. In this paper, the regional variance is used to measure the focus state of the region. The contour point line segment attribute mentioned in the literature [11] contains the contour segment attribute based on the EDF image, and it does not contain the depth information of the cervical cell sample. In order to utilize the depth information of the samples, the consistency of the variance of the contour segment area and the cell area with the focal plane is added to the contour segment area attribute. Assuming the variances with the focal plane are V_{Nuc}^i and V_C^i for the nucleus area and the contour segment area, respectively. The difference in variance change between the contour segment area and the nucleus area is defined as formula (4).

$$d_v = \frac{1}{N}\sum_{i=1}^{N}\sqrt{(V_{Nuc}^i - V_C^i)^2} \tag{4}$$

In the formula, N represents the number of focal planes. In the ISBI2015 dataset, the number of focal planes is $N = 20$. The weight of the attribute of the contour segment region containing the depth information is defined as shown in formula (5).

$$W_{R_{bv}} = W_{R_b} \times d_v \tag{5}$$

Among them, W_{R_b} is the attribute weight of the defined contour segment region. The contour segment weight W_{RD} containing depth information is defined as formula (6).

$$W_{RD} = W_{dir} \times e^{-W_{R_{bv}}} \tag{6}$$

W_{dir} is the weight of the direction attribute of the contour segment. Combining the attributes of contour points with depth information and the attributes of contour segments containing depth information, the weights W_c between connected nodes in the weight graph are obtained as shown in formula (7).

$$W_c = W_P W_{RD} \tag{7}$$

After the weight graph is constructed, the Dijkstra dynamic programming algorithm is used to find the shortest path in the graph, and the shortest path in the graph corresponds to the rough segmentation contour in the cell image. Finally, the DRLSE level set model is used to finely segment the obtained rough segmentation boundaries.

3 Experimental Results and Discussion

3.1 Collection and Evaluation Method

The experimental samples in this paper are derived from the dataset published in the "Second Cervical Cell Image Segmentation Challenge" organized by the IEEE International Symposium on Biomedical Imaging. The dataset contains a total of 17 samples, each sample contains 20 focal planes at different depths under the same view, and each sample contains 1 EDF image. The images in the dataset are grayscale images with a resolution of 1024 * 1024. Each sample contains approximately 40 cells of varying overlap, texture, and contrast. Eight samples in the dataset are defined as the training set, which publishes manually annotated boundary images of the nucleus and cytoplasm. The remaining 9 samples are the test set, which only manually annotated the cytoplasmic boundaries.

The cervical cell segmentation evaluation methods used in the ISBI challenge include: pixel-based DSC, pixel-based true positive rate (TP_p), pixel-based false positive rate (FP_p) and object-based false negative rate (FN_o). If the DSC metric value of the overlapping area between the obtained cell segmentation area O_{seg} and the manually labeled (gold standard) cell area O_{GT} is higher than the specified threshold, the segmentation algorithm is said to correctly identify the cell. The definition of DSC is shown in formula (8). In the formula, $|\cdot|$ represents the number of pixels in the region. The ISBI Challenge specified threshold for correct cytoplasmic identification $DSC = 0.7$.

$$DSC = \frac{2|O_{GT} \cap O_{seg}|}{|O_{GT}| + |O_{seg}|} \tag{8}$$

This paper uses the geometric median metric GM of DSC and FN_o to measure the overall performance of the segmentation algorithm in terms of accuracy and recognition rate. The geometric median [10] is defined as Eq. (9).

$$GM = \sqrt{DSC(1 - FNRo)} \tag{9}$$

In this paper, Pre_{pix} and Rec_{pix} are also used to evaluate the results of cervical cell segmentation, as shown in formulas (10) and (11).

$$Pre_{pix} = \frac{Cd_p}{D_p} \tag{10}$$

$$Rec_{pix} = \frac{Cd_p}{Gt_p} \tag{11}$$

Cd_p represents the number of pixels that correctly detect the cell; D_p represents the number of pixels that detect the cell; Gt_p represents the number of pixels in the labeled image.

3.2 Results

In Table 1, based on the ISBI2015 challenge test set, the overlapping cervical cell segmentation algorithm proposed in the literature [11] and the cervical cell segmentation algorithm with depth information proposed in this paper are compared under the *DSC*, TP_p, FN_o and *GM* metric standards. Compared with the literature [11], the algorithm in this paper has been improved under the four metrics, among which the *DSC* metric has the largest improvement increased by 2.4%. Table 2 shows the statistical analysis results of pixel-based metrics of this paper and Lu's algorithm.

Table 1. Results of different algorithms based on ISBI2015 data set

Methods	*DSC*	TP_p	FN_o	*GM*
Literature [11]	0.882	0.851	0.318	0.774
Ours	0.891	0.871	0.305	0.784

Table 2. Statistical analysis based on pixel metrics

Algorithm	Pre_{pix}		Rec_{pix}		*DSC*	
Range	>0.8	>0.9	>0.8	>0.9	>0.8	>0.9
Lu [12]	96.4%	88.8%	**97.6%**	**78.8%**	**98.4%**	**93.2%**
Ours	**97.4%**	**90.4%**	94.1%	71.9%	98.2%	84.8%

3.3 Comparison and Analysis of Cell Segmentation Algorithms

It can be seen from Table 1 that the overlapping cell segmentation method based on depth information has great advantages over the traditional segmentation method based on EDF images. By utilizing the contour point attributes and contour segment attributes containing depth information, more comprehensive image information is applied to the weight graph. As shown in Fig. 4, the cell segmentation method with depth information is more effective for the segmentation of cell clusters with poor contrast.

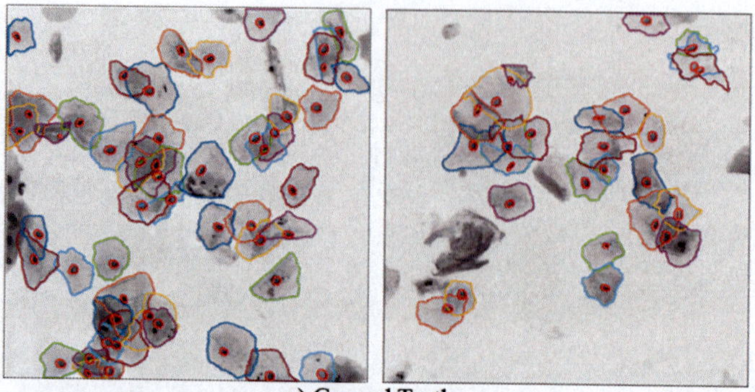

a) Ground Truth

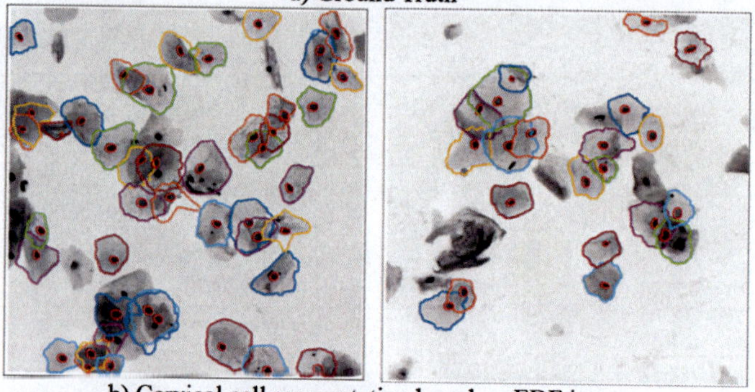

b) Cervical cell segmentation based on EDF image

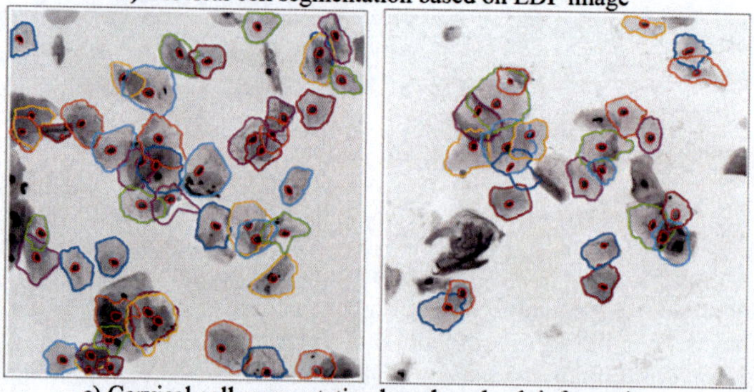

c) Cervical cell segmentation based on depth information

Fig. 4. Results of cervical cell segmentation

4 Conclusion

Overlapping cell segmentation of cervical cell images is a core step in computer-aided diagnosis systems. An attribute of contour points and contour segments between contour points based on depth information is proposed, which can obtain a more accurate rough segmentation boundary and use the level set algorithm to complete the fine segmentation of the boundary. The experimental results show that the proposed algorithm for the segmentation of cytoplasm of cervical cell samples has high segmentation accuracy.

Acknowledgement. This work was supported by the project of youth talented reserves funded by Harbin University of Commerce.

References

1. Plissiti, M.E., Nikou, C.: On the Importance of Nucleus Features in the Classification of Cervical Cells in Pap Smear Images. University of Ioannina (2012)
2. Yang-Mao, S.F., Chan, Y.K., Chu, Y.P.: Edge enhancement nucleus and cytoplast contour detector of cervical smear images. IEEE Trans. Syst. Man Cybern. B (Cybernetics) **38**(2), 353–366 (2008)
3. Harandi, N.M., Sadri, S., Moghaddam, N.A., et al.: An automated method for segmentation of epithelial cervical cells in images of ThinPrep. J. Med. Syst. **34**(6), 1043–1058 (2010)
4. Li, K., Lu, Z., Liu, W., et al.: Cytoplasm and nucleus segmentation in cervical smear images using Radiating GVF Snake. Pattern Recogn. **45**(4), 1255–1264 (2012)
5. Wu, H.S., Barba, J., Gil, J.: A parametric fitting algorithm for segmentation of cell images. IEEE Trans. Biomed. Eng. **45**(3), 400–407 (1998)
6. Tsai, M.H., Chan, Y.K., Lin, Z.Z., et al.: Nucleus and cytoplast contour detector of cervical smear image. Pattern Recogn. Lett. **29**(9), 1441–1453 (2008)
7. Chankong, T., Theera-Umpon, N., Auephanwiriyakul, S.: Automatic cervical cell segmentation and classification in Pap smears. Comput. Methods Prog. Biomed. **113**(2), 539–556 (2014)
8. Phoulady, H.A., Goldgof, D.B., Hall, L.O., et al.: An approach for overlapping cell segmentation in multi-layer cervical cell volumes. In: The Second Overlapping Cervical Cytology Image Segmentation Challenge-IEEE ISBI (2015)
9. Phoulady, H.A., Goldgof, D.B., Hall, L.O., et al.: A new approach to detect and segment overlapping cells in multi-layer cervical cell volume images. In: 2016 IEEE 13th International Symposium on Biomedical Imaging (ISBI), pp. 201–204. IEEE (2016)
10. Phoulady, H.A., Goldgof, D., Hall, L.O., et al.: A framework for nucleus and overlapping cytoplasm segmentation in cervical cytology extended depth of field and volume images. Comput. Med. Imaging Graph. **59**, 38–49 (2017)
11. Wang, T., Huang, J.J., Zheng, D.Q., He, Y.J.: Nucleus segmentation of cervical cytology images base d on depth information. IEEE Access. **8**, 75846–75859 (2020)
12. Lu, Z., Carneiro, G., Bradley, A.P.: An improved joint optimization of multiple level set functions for the segmentation of overlapping cervical cells. IEEE Trans. Image Process. **24**(4), 1261–1272 (2015)
13. Jinjie, H., Tao, W., Dequan, Z., Yongjun, H.: Nucleus segmentation of cervical cytology images based on multi-scale fuzzy clustering algorithm. Bioengineered **11**(1), 484–501 (2020)

Analysis of the China-Eurasian Economic Union Trade Potential Based on Trade Gravity Model

Shuying Lei(✉), Zilong Pan, and Chaoqun Niu

Harbin University of Commerce, Harbin, China
728970903@qq.com

Abstract. This article analyzes the current trade situation between China and the five countries of the Eurasian Economic Union. From the perspective of trade scale, the trade between China and Russia accounts for the largest share, followed by the trade between China and Kazakhstan. However, in terms of bilateral trade structure, the trade products are mostly labor-intensive products and primary products, the trade between China and the Eurasian Economic Union is still at a low level. Then, combined with panel data, using the trade gravity model, selecting the GDP, GNI, geographical distance, and borderline of the two trading countries as independent variables to measure the trade potential of China and the five countries of the Eurasian Economic Union. The trade potential of China has not yet been fully tapped, and there is a possibility of further development.

Keywords: Eurasian economic union · Trade gravity model · Trade potential

1 Preface

The Eurasian Economic Union (hereinafter referred to as EEU) is based on the Russia-Belarus-Kazakhstan Customs Union officially launched in 2010 by Russia, Belarus and Kazakhstan. On January 1, 2015, the signing of the Treaty on the Eurasian Economic Union marked the official establishment of the Eurasian Economic Union. Subsequently, Kyrgyzstan and Armenia also formally joined. The EEU currently has a population of about 182 million and a GDP of about 1.9 trillion US dollars. And the countries of the EEU strive to facilitate the free flow of goods, services, capital and labor, and implement coordinated economic policies [1].

In August 2015, Joint Statement on the Coordination and Cooperation of the Construction of the Silk Road Economic Belt and the Construction of the Eurasian Economic Union was issued by China and Russia. The statement is issued for strengthening and expanding financial cooperation between China and the EEU, and optimizing trade structure, improving investment environment, reducing financial risks during cooperation, as well as better connecting the The Belt and Road and the EEU [2]. In May 2018, Premier Li Keqiang of the State Council signed the economic and trade agreement between the people's Republic of China and the Eurasian Economic Union with the prime ministers of the member states of the EEU, which officially came into force in October, aiming to further reduce non-tariff trade barriers, improve the level of trade facilitation and create

© The Author(s), under exclusive license to Springer Nature Switzerland AG 2022
A. E. Hassanien et al. (Eds.): AMLTA 2022, LNDECT 113, pp. 288–297, 2022.
https://doi.org/10.1007/978-3-031-03918-8_26

a good environment for industrial development [3]. Therefore, it is of great practical significance to study the influencing factors and trade potential of bilateral trade between China and the five countries of the EEU.

2 Trade Status Between China and the Five Countries of the EEU

This part first explains the general trade situation of between China and the EEU, and then explains the current situation of bilateral trade between China and other countries.

2.1 General Situation

Among the five countries in the EEU, Russia and Kazakhstan are relatively rich in resource-intensive products such as mineral products, but weak in the domestic light industry and civilian industry. Belarus's industrial base is good, while agriculture and animal husbandry are also relatively developed. Kyrgyzstan's economy is mainly based on agriculture and animal husbandry, the industrial infrastructure of Kyrgyzstan is weak, and most of the exported products are raw materials. Armenia mainly exports gems and their semi-processed products, food, non-precious metals and their products, and minerals. Due to the scarcity of resource-intensive products such as mineral products in China, most of these resources are imported, while labor-intensive mechanical and electrical products and light industrial products with low technical content are exported [4]. Among them, the international trade with the countries of the EEU is mostly based on the import of mineral products, base metals and products; and the export of labor-intensive products such as mechanical and electrical products, manufactured products and light industrial products [5]. It can be seen that China and the countries in the EEU have formed a good complementarity in industry, which is conducive to the current trade. However, it is also found that the trade structure of the two sides is relatively simple, mainly primary products and labor-intensive products, but lack of cooperation in technology, environment, competition policy, finance, e-commerce, legal issues and other related aspects, which is not conducive to higher level of international trade [6].

At present, Sino-Russian bilateral trade accounts for the vast majority of China-EEU trade, generally between 65% and 85%. The proportion of China-Kazakhstan trade ranks second, generally between 14% and 25%. The proportion of bilateral trade between China and the other three countries is generally small, and the overall proportion is generally between 2% and 12%.

2.2 Sino-Russian Trade Situation

In the EEU, Russia has a population of 144 million and a GDP of about 1.66 trillion US dollars, accounting for 79.6% and 86.9% of the EEU population and GDP, and occupies a pivotal position in the alliance. While Russia inherited its political status after the disintegration of the Soviet Union, it also inherited its deformed industrial structure: excessive development of heavy industry and military industry, and lack of development of light industry and people's livelihood industry [7]. This makes its capital and technology-intensive products less competitive, and its export structure in international trade is relatively simple, mainly mineral products such as mineral fuels, mineral oil and its products, and resource-intensive products such as asphalt, which is also in line with Russia's own national factor endowments [8]. Mineral products are primary products, and compared with capital- and technology-intensive products, they have the disadvantage of less stable international prices. This simple international trade structure and the disadvantage of unstable primary product prices can easily be used by other countries to attack Russia's economy. For example, in 2014, the United States and Europe imposed economic sanctions on Russia over the Ukraine incident, they maliciously manipulated the international oil price, making it lower than Russia's normal oil extraction cost, which resulted in Russia's oil unable to export normally, and the domestic economy encountered serious difficulties.

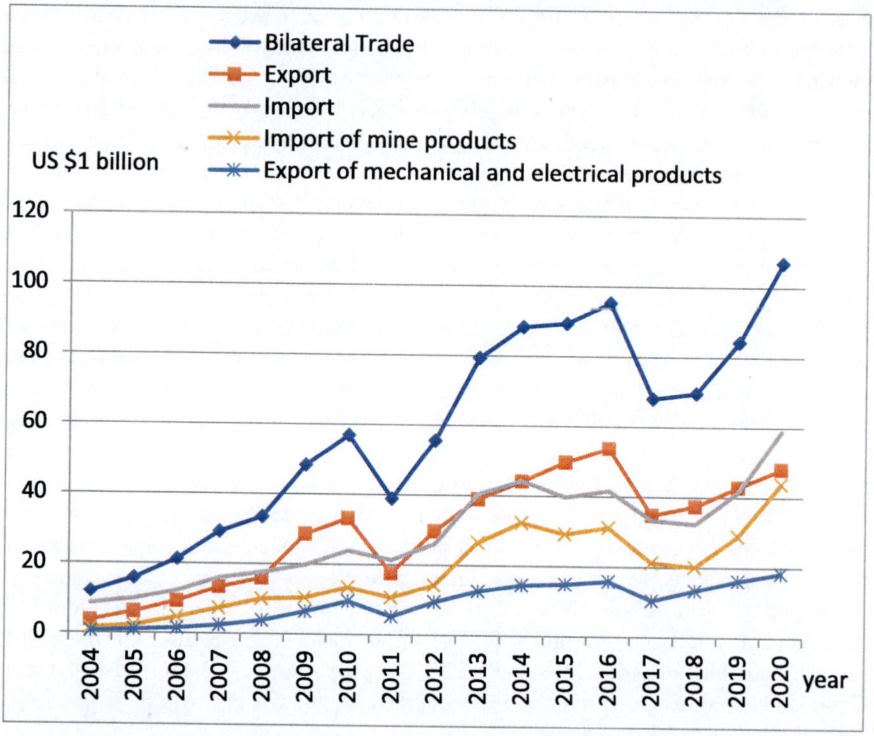

Fig. 1. Bilateral trade between China and Russia from 2004 to 2020 (US $1 billion). Data source: COMTRADE database (https://comtrade.un.org/data/), World Bank (https://data.worldbank.org.cn/)

As shown in Fig. 1, the bilateral trade volume between China and Russia showed an overall upward trend from 2004 to 2020, with US $11.927 billion reaching US $106.892 billion in 2020, with an annual growth rate of 14.69%. But there were two obvious declines, the sudden decline in bilateral trade volume in 2009 was mainly affected by the economic crisis, The decline in bilateral trade volume from 2016 to 2018 should be mainly due to Russia's economic sanctions from the United States and European countries, which forced the international oil price to fall and made Russian difficult to export oil. In 2020, China imported $42.71 billion in mineral products from Russia, an increase of 61.9%, accounting for 77.9% of China's total imports to Russia; The export of mechanical and electrical products reached US $26.45 billion, an increase of 3.9%, accounting for 50.7% of China's total exports to Russia, and the rest were mostly light industrial products. The bilateral trade volume increased from US $11.927 billion to US $106.892 billion, with an annual growth rate of 14.69%. China has become Russia's largest trading partner for nine consecutive years, but the actual trade volume is lower than the original expectations of the two governments, forcing the original plan to achieve US $200 billion in bilateral trade in 2020 to be postponed to 2024.

2.3 Sino- Kazakhstan Trade Situation

Kazakhstan has a population of about 15 million and a GDP of about 170.5 billion US dollars. It is second only to Russia in the alliance, and the per capita income level is high. It once reached 12090 US dollars in 2016. However, because it mainly depends on the export of mineral products, base metals and products and other resource intensive products, the domestic processing industry and light industry are relatively backward, and the main daily necessities depend on import.

Although the bilateral trade volume between Kazakhstan and China generally showed an upward trend, in 2016, due to the depreciation of the ruble and the decline of international oil prices, the Central Bank of Kazakhstan no longer maintained the tenge exchange rate level specified in the past, resulting in the sharp depreciation of the tenge of Kazakhstan, which had an adverse impact on China-Kazakhstan trade. From 2004 to 2020, the total bilateral trade volume increased from US $1955 million to US $19856 million, with an annual growth rate of 15.59%. In 2020, China's import of mineral products from Kazakhstan was US $4.069 billion, accounting for 47.7% of the total import to Kazakhstan; The export of mechanical and electrical products reached 2.88 billion US dollars, accounting for 24.9% of the total export to Kazakhstan, and the rest were mostly chemical products, base metals and products.

2.4 Trade Between China and Other Three Countries

The population and volume of Armenia, Kyrgyzstan and Belarus are significantly different from those of Russia and Kazakhstan. The bilateral trade volume between China and its three countries is also small, accounting for 2%–12% in China and the EEU.

As shown in Fig. 2, the bilateral trade volume between China and Kyrgyzstan increased rapidly from 2004 to 2010, reaching a peak of US $9.3 billion in 2010. However, affected by the financial crisis, the bilateral trade volume has not increased significantly since 2011. The bilateral trade volume between China and Armenia increased from US $8 million in 2004 to US $527 million, with an annual growth rate of 29.3%. The bilateral trade volume between China and Belarus increased from US $80 million in 2004 to US $1.716 billion, with an annual growth rate of 21.1%.

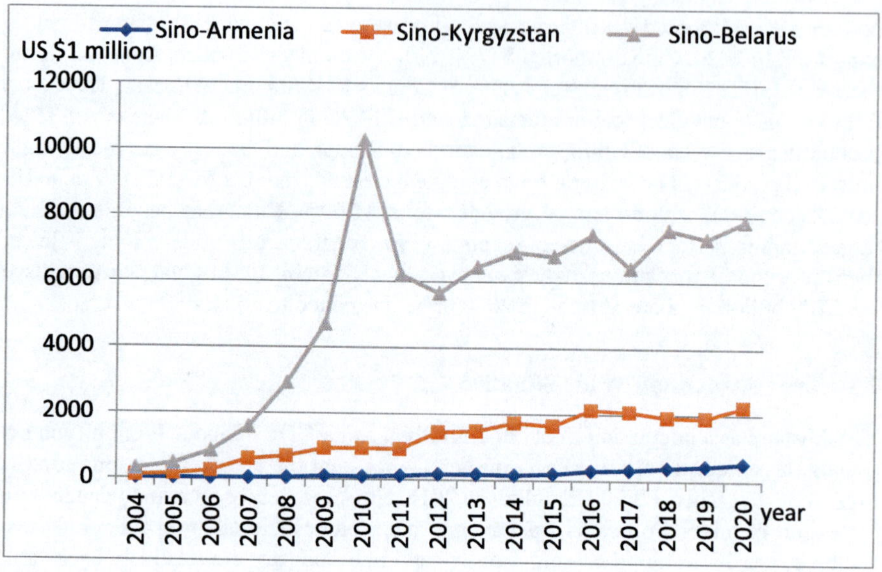

Fig. 2. Bilateral trade from 2004 to 2020 (USD). Data source: COMTRADE database (https://comtrade.un.org/data/), World Bank (https://data.worldbank.org.cn/)

3 Empirical Analysis of China-Eurasian Economic Union's Trade Potential

In this chapter, the trade potential of China and the five countries in EEU is measured respectively by trade gravity model using panel data.

3.1 Construction of Trade Gravity Model

Model. Based on the extended trade gravity model, this paper establishes the following model structure [8]:

$$\ln Trade = \beta_1 \ln GDP_i + \beta_2 GDP_j + \beta_3 Dis + \beta_4 pGNI_i + \beta_5 pGNI_j + \beta_6 Border + \mu$$

Trade represents the total bilateral trade volume; GDP_i and GDP_j represents the GDP of China and trading countries respectively; Dis is the geographical distance between China and trading countries, which is described as the distance between the capitals of the two countries; $pGNI_i$ and $pGNI_j$ represent the per capita national income of China and trading countries respectively; Border is a virtual variable, which means whether the two countries are bordered. If bordered, Border is set to 1, otherwise it is 0. Forecast: GDP_i, GDP_j, $pGNI_i$ and $pGNI_j$ and border are positively correlated with trade; Dis is negatively correlated with trade.

Data Selection. This paper selects the annual data of 17 years from 2004 to 2020. In addition to Russia, Belarus, Kazakhstan, Armenia, Kyrgyzstan and Uzbekistan and

Tajikistan who intend to join the alliance, the trading partner countries also include the United States, Japan, South Korea, Germany, Australia, the Netherlands, Singapore, Vietnam and India, Mexico, Britain, Brazil, Thailand and Malaysia are among the top trading partners of China, which can better reflect China's foreign trade situation. There are 374 groups of panel data in total. Trade data comes from COMTRADE database, GDP_i, GDP_j, $pGNI_i$ and $pGNI_j$ from the world bank database, the distance between China and the trading country comes from the absolute distance of map mapping, and whether China and the trading country are bordered, that is, the selected value of border comes from the map of China.

Model Estimation. Using Eviews to estimate the model, it is found that the model fitting of the mixed effect model is poor, mainly because the variables contained in the data, such as distance, have many repeated values, which are not suitable for the establishment of the mixed effect model; and the establishment of the fixed effect model loses higher degrees of freedom. Therefore, a random effect model is established in this paper, and the following results are obtained:

Table 1. Preliminary regression results of trade gravity model.

Variable	Coefficient	Std.	T Statistics	Confidence
C	−31.15	10.40	−3.00	0.0029
ln GDPi	1.69	0.49	3.43	0.0007
ln GDPj	0.72	0.11	6.72	0
ln pGNIi	−1.24	0.51	−2.46	0.0143
ln pGNIj	0.36	0.13	2.82	0.0051
ln Dis	−0.81	0.30	−2.72	0.0068
Border	0.81	0.45	1.79	0.074
R-squared	0.88			

Data source: COMTRADE database (https://comtrade.un.org/data/), World Bank (https://data.worldbank.org.cn/)

It can be seen from Table 1 that although the model fits well, the decisive coefficient reaches 0.88, the regression coefficient of each explanatory variable is significant, and at least passes the t-test under the confidence of 0.1, the regression coefficient of In $pGNI_i$ is negative, indicating that the rise of China's per capita national income is negatively related to the bilateral trade volume, which is inconsistent with the prediction. The reason for this phenomenon may be that country i represents only one country China, making the correlation between GDP_i and $pGNI_i$ is high, resulting in multicollinearity in the model. Using the correlation coefficient judgment method to test GDP_i and $pGNI_i$, it is found that the correlation coefficient of the two is 0.9987, and the correlation degree is very high, indicating that the model has serious multicollinearity.

Remove ln $pGNI_i$ from model (1), and the new model is as follows:

$$\text{lnTrade} = \beta_1 \text{lnGDP}_i + \beta_2 \text{GDP}_j + \beta_3 \text{Dis} + \beta_4 \text{pGNI}_j + \beta_5 \text{Border} + \mu \quad (2)$$

To estimate the model (2) and establish the random effect model, the following results are obtained:

Table 2. Regression results of adjusted trade gravity model.

Variable	Coefficient	Std.	T statistics	Cofidence
C	−6.71	3.09	−2.17	0.0304
ln GDPi	0.48	0.04	12.78	0
ln GDPj	0.76	0.11	7.10	0
ln Dis	−0.83	0.30	−2.78	0.0057
ln pGNIj	0.34	0.13	2.72	0.0068
Border	0.85	0.45	1.87	0.062
R-squared	0.88			

According to Table 2, it can be seen that the regression results are good, the regression coefficients of each explanatory variable are significant and the signs are consistent with the expected and actual trade conditions before the model was established, at least passing the t-test with a confidence level of 0.1; the coefficient of determination is 0.88, indicating the degree of fit of thc model good.

The Hausman test was performed on the model, and the P statistic value of the test result was $0.9874 > 0.05$, indicating that the random effect model was suitable to be established.

The model result is interpreted as follows:

The regression coefficient β_1 of $lnGDP_i$ is 0.48, indicating that of China's GDP increased by 1%, and the bilateral trade volume increased by 0.48%.

The regression coefficient β_2 of $lnGDP_j$ is 0.76, indicating that as GDP of the trading partner country increase by 1%, the bilateral trade volume increases by 0.76%. Comparing the magnitudes of β_1 and β_2, it can be seen that the increase in the GDP of the trading partner country promotes the bilateral trade volume more.

The regression coefficient β_3 of *lnDis* is −0.83, which indicates that the further the distance between the two countries is, the less conducive to the development of bilateral trade, and the impact on trade is greater than that of China's GDP and the GDP of trading partners.

The regression coefficient β_4 of $lnpGNI_j$ is 0.34, indicating that the per capita national income of the trading partner countries increased by 1%, and the bilateral trade volume increased by 0.34%.

The regression coefficient β_5 of the dummy variable Border is 0.85, which indicates that whether the two trading countries are adjacent to each other has a very significant impact on bilateral trade.

3.2 Measurement of Trade Potential Between China and the Five Countries of the EEU

The trade potential of China and the five countries of the EEU can be measured by the trade gravity model. The specific method is to calculate the ratio of the actual bilateral trade volume of the current year to the estimated volume of the trade gravity model. The lower the ratio, the greater the trade potential. The trade potential measurement results of bilateral trade data from 2004 to 2020 are shown in Table 3.

Table 3. Trade potential measurement results.

Year	Kazakhstan	Armenia	Belarus	Russia	Kyrgyzstan
2004	1.45	0.62	0.64	1.81	0.41
2005	1.83	0.36	0.79	1.78	0.50
2006	1.65	0.56	0.92	1.58	0.75
2007	1.72	0.63	1.67	1.51	1.00
2008	1.35	0.76	1.37	1.19	1.81
2009	1.49	0.98	1.26	1.13	2.03
2010	1.29	0.60	0.84	0.90	3.28
2011	1.06	1.01	0.85	0.75	1.81
2012	1.14	1.28	1.07	0.81	1.31
2013	0.99	1.11	0.94	0.80	1.13
2014	0.85	0.85	1.01	0.74	1.00
2015	0.77	0.99	0.78	0.66	0.83
2016	0.60	1.38	0.89	0.74	0.79
2017	0.44	1.62	1.11	0.76	0.70
2018	0.54	2.01	1.16	0.87	0.92
2019	0.65	2.02	0.97	0.88	0.77
2020	0.66	2.08	1.00	0.98	0.70

It can be seen from the table that the trade potential index of China and Kazakhstan was greater than 1.2 from 2004 to 2020, and the trade index decreased significantly after 2010, especially after the substantial depreciation of Kazakhstan tenge in 2016, indicating that there is great trade potential between China and Kazakhstan. The trade potential index of China and Armenia basically showed an upward trend from 2004 to 2020 and reached 2.08 by 2020, indicating that new trade growth factors need to be developed between China and Armenia. The trade potential index of China and Belarus fluctuated greatly from 2004 to 2010. The main reason may be that the bilateral trade volume and degree of cooperation were low at that time and were vulnerable to various factors. The trade potential index was relatively stable from 2010 to 2020, between 0.8 and 1.2, indicating that there is trade potential between Central Asia and waiting for further development. The trade potential index of China and Russia decreased significantly from 2004 to 2010 and remained between 0.7 and 1.0 from 2010 to 2020, indicating that although the bilateral trade between China and Russia has increased significantly in recent years, it still does not match the economic volume of the two countries, and the bilateral trade may be greatly improved. The trade potential index between China and Kyrgyzstan is low in recent years, which may be mainly due to the large gap in economic volume and per capita income between China and Kyrgyzstan in recent years, which is not conducive to the development of trade, and the trade between two sides still have great trade potential.

4 Conclusion

Properly handling China's one belt, one road initiative and the construction of EEU will help expand cooperation in investment, environment and technology and improve bilateral structure on the basis of original cooperation [9]. In the regression results of the trade gravity model, the GDP regression coefficient of trading partner countries is higher than that of China's GDP, indicating that the scale of China's foreign bilateral trade is more affected by the GDP of trading partners. Some studies believe that this situation is mainly because China's trade structure is still in the low stage, mainly labor-intensive or resource intensive products. In China's bilateral trade with the EEU, China mainly exports labor-intensive mechanical and electrical products and light industrial products, and the Union mainly exports resource intensive mineral products, base metals and other products. It is also concluded that the level of trade structure between China and the EEU is low and is in a disadvantageous position in international trade. At the same time, China has gradually realized the transformation from inter industry trade to intra industry trade, indicating that circumstance in China's international trade is gradually improving [10]. In addition, in the trade between China and the five countries of the EEU, Russia's trade with China and the other four countries's trade with China are obviously competitive in trade categories, which is likely to cause the conflict of strategic docking between China and Russia in the short term. Therefore, the connection between one belt, one road initiative and the EEU should be properly handled, and their mutual cooperation, mutual complement and mutual improvement should be mutually enhanced.

References

1. Lareen, A.Г., Matviev, B.A., Xiaohui, G.: How does Russia view the connection between the Eurasian Economic Union and the silk road economic belt. Euras. Econ. **2**, 18–26+125+127 (2016)
2. Armstrong, S.: Measuring trade and trade potential: a survey. In: Asian Pacific Economic Paper 368 (2007)
3. Battesse, G.E., Coelli, T.J.: Frontier production functions, technical efficiency and panel data: with application to paddy Famers in India. J. Prod. Anal. **3**(1), 153–169 (1992)
4. Yang, B., Zhuchang, T.: Co-building the silk road economic belt: a review of the academia of the Eurasian economic union. Euras. Econ. **3**, 102–124+126+128 (2019)
5. Xiang, Y.J., Zhang, J.P.: Obstacles and conflicts of strategic alignment of China-Russia regional economic cooperation. China Econ. Trade **2**(5), 33–38 (2016)

6. Zhang, Z.X., Hua, H.R., Lin.: Research on the trade relationship and potential between China and the West Asian countries along the "One Belt One Road". East China Econ. Manag. **12**, 13–19 (2019)
7. Wang, G.Z., Yanzhao, G.: Analysis of the economic and trade foundation and expected effect of the establishment of the China-GCC free trade zone. Indian Ocean Econ. Res. **5**, 99–141 (2017)
8. Junjiu, Q.J.: Research on China's export potential to APEC members and its influencing factors: An empirical test based on the trade gravity model. Asia Pac. Econ. **6**, 5–13 (2017)
9. Dong, B.C., Wanping, Y., Siyuan, N.: The impact of Sino-Japanese technical trade barriers on bilateral trade. J. Manag. **3**, 30–41 (2020)
10. Shuzhu, J., Xukun, Z.: The gravity model of ASEAN trade effect. Quantit. Tech. Econ. Res. **10**, 53–57 (2003)

Skip Truncation for Sentiment Analysis of Long Review Information Based on Grammatical Structures

Mengtao Sun[1](✉), Ibrahim A. Hameed[1], Hao Wang[2], and Mark Pasquine[3]

[1] Department of ICT and Natural Sciences, NTNU, 6009 Ålesund, Norway
{mengtao.sun,ibib}@ntnu.no
[2] Department of Computer Science, NTNU, 2815 Gjøvik, Norway
hawa@ntnu.no
[3] Department of International Business, NTNU, 6009 Ålesund, Norway
mapa@ntnu.no

Abstract. In reality, some emotional utterances are especially long, such as the critiques of a movie. Compared with short sentences, the sentiment of a longer paragraph is more difficult to be detected. Moreover, longer paragraphs are eased to overflow the GPU memory and sluggish to fit a large-parameterized network. Currently, the solutions include using a sentence-level summary; if not applicable, a long review is always directly truncated to an accepted length. However, the first solution will ignore the sentiment subtleties; The second solution will lose some important emotive expressions. This work puts forward a strategy to locate the potential sentiment information considering the grammatical structures and then proposes skip truncation to shorten the long review texts. This method can effectively delete irrelevant words without hitting the sentiment skeleton. The new skip truncation is compared against several baselines in binary and multiple sentiment perception. We conduct experiments with four types of standard deep neural networks. Experimental results verify that skip truncation can help reduce sentence length and improve performance in a large margin.

Keywords: Sentiment analysis · Deep learning · Long text · Grammatical structures

1 Introduction

Many reviews are especially long in the real-world applications of sentiment analysis [1], such as the dedicated reviews of movies, products, artworks, video games, and politics. Long paragraphs easily confuse the sentiment detection models because of too many emotionless words therein [2]. Moreover, they usually overflow the GPU memory and are sluggish to fit a deep network due to their oversize [3]. This problem can be well avoided for some NLP tasks, such as named entities recognition and machine translation because a document-level input can be divided into several sentence-level batches. However, sentiment perceptron is required to perceive the overall information

© The Author(s), under exclusive license to Springer Nature Switzerland AG 2022
A. E. Hassanien et al. (Eds.): AMLTA 2022, LNDECT 113, pp. 298–308, 2022.
https://doi.org/10.1007/978-3-031-03918-8_27

before predicting. Moreover, the emotion of each sentence in a document is diverse, so using sentence-level inputs will also lead to a great error. Consequently, it is important to find a solution for long review information sentiment analysis.

Currently, the solutions are still simple and insufficient [4]. The ideal way for sentiment analysis is to use a sentence-level input, but a summary of long text hardly comprises the emotional subtleties. More general practice is to truncate long paragraphs to a proper length. However, such an operation will lose much emotive content. On the one hand, feeling expressions may appear later in the reviews. Truncation may result in the texts seldom containing sentiment features. On the other hand, the sentiment reversals frequently appear in long texts, and the direct use of text truncation makes the model ignore this information. For a sentence example, "I do not want to say that the actor did a good job", by post-truncation, "I do not want to say" makes the sentence Negative, while by the pre-truncation, "The actor did a good job" makes the sentence Positive. Different snippets will influence the overall sentiment detection, and these errors may mislead the training of the sentiment analysis model. According to the perspective of human sentiment analysis, a more accurate judgment will be obtained when the texts contain many emotional expressions.

We assume that the length of input sentences for deep models generally does not exceed 200 words A larger input size often leads to memory overflow. In order to solve the abovementioned problems, this paper proposes a skip truncation. Through skip truncation, the information irrelevant to sentiment analysis can be reduced as much as possible. The truncated sentence maintains the original sentiment skeleton. This method is based on grammatical structure. Grammatical sentiment components can be a word, phrase, or clause that functions as an adjective or adverb to provide additional sentiment information. In this way, we can roughly capture the emotions in the early training stage. To summarize, this paper makes the following contributions to the sentiment analysis on long texts:

1. This paper proposes a new text truncation method for better sentiment analysis, called skip truncation.
2. This paper compares the performance of the skip truncation against different baselines in several typical deep neural networks.
3. This paper analyzes the performance of skip truncation in binary polarity classification and multiple polarity classification.

2 Related Works

Long text processing has always been a difficulty in deep learning training. To address the large scale of words, Pappagari et al. [5] use a hierarchical strategy to fit the Transformer [6] (a SOTA model in many NLP tasks) for document-level classification. They segmented the input into smaller chunks, fed them into the Transformer, then propagated each output through a single recurrent layer. He et al. [7] pointed out that document classification requires extracting high-level features from low-level word vectors. They proposed a new model incorporating the recurrent attention learning framework, focusing its attention on the discriminative parts. Wu et al. [8] proposed the hierarchical aspect-oriented framework for long document sentiment classification. Their model alleviated the challenge of the unstable sentiment information of the target aspect in long documents and the problem that the too-long input sequence can cause the model to forget previously learned information. Kaliamoorthi et al. [9] proposed a novel projection attention neural network that combines trainable projections with attention and convolutions. Their model is particularly effective on multiple large document text classification tasks.

More useful methods are based on the dictionary to address long texts sentiment analysis. First, Zhang et al. [10] extended the sentiment dictionary from degree adverb, network word, and negative word. The sentiment value of a micro-blog text is obtained by calculating the weight from the extended sentiment dictionary and Micro-blog texts on a topic are classified as positive, negative, and neutral. Okango et al. [11] employed a dictionary-based method for sentiment polarization from tweets related to coronavirus posted on Twitter and analyzed the effects and response in the pandemic. Yekrangi et al. [12] explored a wide related corpus using lexical resources and proposed a hybrid approach to building a lexicon specialized for financial markets sentiment analysis. Hossen et al. [13] proposed an improved lexicon-based model aiming at movie review data.

3 Materials and Method

This work proposes the use of skip truncation to construct a sentiment skeleton in order to reduce the scale of document content and amplify the subtle sentiment features. The overall workflow can be represented as follows: Raw Data Processing → Grammatical Components Tagging → Skip Truncating → Sentiment Detecting.

3.1 Raw Data

This study experimented with the idea of skip truncation by using Norwegian languages because the Norwegian language is more challengeable on text modeling due to its rich grammar rules, and it is often ignored in current sentiment analysis research [14]. Nevertheless, we believe the method can be applied to any language.

The experimental dataset is NoReC [14] (Full name: The Norwegian Review Corpus), proposed by University of Oslo. It is a Norwegian languages review dataset comprising a range of domains, including literature, movies, video games, restaurants, music, and theater, from eight rating websites in Norway. The release of the corpus consists of more than 35,000 reviews. Each review is rated with a numerical score on a scale of 1–6 and can be used for training and evaluating models for document-level sentiment analysis. These were defined by first sorting all reviews for each category by publishing date and then reserving the first 80% for training, the subsequent 10% for development, and the final 10% for testing.

The document-level reviews comprise averagely 426 words, which is lengthy for deep learning models [8, 14]. To cope with the long contents, the attributions of each document are summarized in a JSON file, containing the document ID, sentiment rating, and the manual excerpt of a document.

Data Cleaning. NoReC has already processed most of the noisy data. Here, we remove extra delimiters such as spaces and paragraph breaks. Then, we remove punctuations but keep periods, exclamations, and questions to identify sentence ending. We use word-level segmentation. In this way, the cleaned document is transformed into a sentence and can be represented as a long vector.

3.2 Sentiment on Grammatical Components

Next, the long sentences are annotated according to the linguistic structure, which is achieved by the spaCy toolkit. We use part-of-speech (POS) and syntactic dependency relation to label each token. POS and syntactic dependency relation can be regarded as the tags based on the word-level and the sentence-level grammars. We investigate each component that may be closely related to sentiment classification. Finally, according to empirical observations, the importance is summarized in Tables 1 and 2.

Table 1. Grammatical sentiment components based on POS tagging.

POS	Description	POS	Description
ADJ	Adjective	INTJ	Interjection
ADV	Adverb	AUX	Auxiliary
PART	Particle	PUNCT	Punctuation
VERB	Verb	O	All other tags in UPOS[1].

As a result, we divide each word into three categories, red, blue, and non-sentiment. Red marks will greatly impact sentiment classification; Blue marks are secondary grammatical sentiment components.

ADJ and ADV are the main research objects, such as 'det **store** huset' (the **big** house), '**nesten** ferdig' (**almost** finished). Some studies utilize ADJ and ADV to compute sentiment scores and construct sentiment dictionaries [10]. The words in PART tag cause an emotional reversal, such as 'Han liker **ikke** å spise is' (He **does not** like to eat ice cream). INTJ tag is mostly a phrase used for exclamations, such as '**bravo**'. AUX can enhance the expressive intensity, such as '**burde**' (should), '**må**' (must), etc.

Some words associated with PUNCT tags can enhance emotional expression. Moreover, some verbs have strong emotional tendencies, such as '**liker**' (like), but the VERB tag comprises a wide scope. Considering the actual situation of a document, the number of emotional expressing is limited. PUNCT and VERB are attributed to the secondary level. An elaborate specification of POS tags can be found in UPOS[1].

In sentence-level grammar, syntactic dependency relations describe the reliance between words. It points out the collocation between words in terms of their semantics. The auto-annotation has achieved 89% accuracy in the best Norwegian model. Similar to POS tagging, based on the syntactic dependencies, the importance of sentiment classification is depicted in Table 2.

[1] https://universaldependencies.org/u/pos/.

Table 2. Grammatical sentiment components based on syntactic dependency tagging.

Dependency	Description	Dependency	description
advmod	adverbial modifier	Root	root
amod	adjectival modifier	Punct	punctuation
aux	auxiliary	O	all other tags in UD[2]

Syntactic dependency relations are more complicated, but the labeling of grammatical sentiment components is equivalent to Table 1. For example, advmod is an adverb or adverbial phrase that modifies a predicate or modifier word, parallel to ADV in Table 1. An elaborate specification of syntactic dependency relations can be found in Universal Dependency[2].

3.3 Skip Truncation

Since the punctuations of ending the sentences are retained, we iteratively pass sentence-level input to obtain POS tags and syntactic dependency relations for every word in a document. After receiving all the grammars, we delete the non-sentiment words both in POS and syntactic dependency standard (See Table 1 and Table 2). Then, if the sequence length exceeds the threshold (200 in this work), we execute the following command and stop when the document length is acceptable:

1. Delete the blue words that appear once in POS or dependency,
2. Delete the blue words that appear both in POS and dependency,
3. Delete the red words that appear once in POS or dependency.

In this way, the document has been turned into sporadic words. Although the sequence is broken, the sentiment skeleton has been preserved. For documents whose length is 700, the truncated length is usually less than 100. If the file size is still more than the threshold, directly truncate the part exceeding the threshold. If the sentence length is less than the threshold, use zero padding after the sequence.

3.4 Sentiment Perceptron

We introduce a concise architecture for sentiment classification and explore how the neural networks behave in different truncations. The specification is shown as follows:

Embedding. In skip truncation, the context information is broken. We experimented with a pre-trained embedding and found that the pre-trained embedding can hardly get a satisfactory result, so it is suggested to train each model from scratch. To reduce overfitting, we apply dropout in the embedding layer.

[2] https://universaldependencies.org/u/dep/.

Hidden Neurons. After embedding, we verify four types of neurons for extracting sentiment features: Fully Connection units, convolutional units, Recurrent units, and Gated Recurrent units [15].

Classification. The classification part contains two layers of fully connection. The first layer furtherly extracts and integrates features, and the second layer is used for sentiment labeling.

4 Experimental Results

In experiments, we test ship truncation on binary sentiment polarity and multiple (six) sentiment polarity by different hidden neurons settings. This section first introduces the model settings, then illustrates the baselines, and finally presents the experimental results.

4.1 Model Settings

According to the architectures of sentiment perceptron, the details of parameters are introduced in Table 3. The model comprises one of the Hidden Neurons and one of the Classification settings each time, while the Embedding and Training settings are the same.

Table 3. Parameters of sentiment models.

Component	Value	Explanation
Embedding	Sentence length 200, embedding dimension 300, dropout rate 50%	Embeddings are trained from scratch
Hidden Neurons	Fully-Connection: unit 50 1D Convolution: unit 50, kernel size 3 bidirectional Recurrent: 50 units bidirectional Gated Recurrent: 50 units	for Fully-Connection Unit for Convolutional Unit for Recurrent Unit for Gated Recurrent Unit
Classification	2 Fully-Connection layers: unit 32 and unit 6 2 Fully-Connection layers: unit 32 and unit 2	for 6 sentiment polarities for 2 sentiment polarities
Training	Optimizer: Adam, Activation Function: ELU, Learning Rate: 0.001, Batch Size: 64, Epoch: 5	–

To evaluate the performance of our proposed model, we applied standard Precision (P), Recall (R), F1 Score between the model output and ground truth.

4.2 Baselines

We use three baseline truncations to prove the performance of the skip truncation. Truncated sentences have an identical fixed sentence length, 200, and all the experiments are performed in the same model setting.

Excerpt. Documents are manually summarized with sentence-level excerpts, and each excerpt is supplemented to 200 dimension using zero-padding in the end.

Brute Force I. Each document is selected 200 words from the beginning, using zero-padding to the end if the entire document is less than 200 words.

Brute Force II. Each document is selected 200 words from the end, using zero-padding to the beginning if the entire document is less than 200 words.

4.3 Results on Binary Sentiment Polarity

Binary sentiment polarity classification is more manageable due to its lower requirements on featurization. Therefore, we first integrate the original six sentiment labels to negative and positive. Labels 1, 2, 3 are negative and labels 4, 5, and 6 are positive. The performances of different neural networks against baselines are shown in Table 5, 6, 7, and 8.

Table 5. Results in fully connection (binary-polarity).

Fully connection	Precision	Recall	F1 Score
Excerpt	69.60	74.77	71.12
Brute force I	68.50	75.80	66.29
Brute force II	76.00	79.24	72.72
Skip-truncation	80.56	80.50	76.65

By fully connection, the F1 score of skip truncation reaches 76.65%, the Precision reaches 80.56%, and Recall reaches a rate of 80.50%. The F1 score, Precision, and Recall, is higher than the second about 4%, 4%, and 1%, respectively. On the other hand, the fully connection is not strong enough for featurization, which cannot display the full advantages of skip truncation.

Table 6. Results in convolutional unit (binary-polarity).

Convolutional unit	Precision	Recall	F1 Score
Excerpt	69.60	75.57	70.54
Brute force I	77.93	79.13	72.31
Brute force II	83.68	84.86	82.92
Skip-truncation	90.80	90.94	90.58

Convolutional units can better consider the relationship of neighbor words. The F1 score, Precision and Recall reaches 90.58%, 90.94% and 90.94%. Compared to corresponding metrics 83.68%, 84.86%, 82.92% in second place, the skip truncation shows stronger effectiveness.

The bidirectional recurrent architecture is especially suitable for text modeling because it considers the context from each time step. The performances by skip truncation manifest a further enhancement and achieve 93.15% F1 score, 93.13% Precision and 93.23% Recall.

The Gated recurrent unit is an improved version of the recurrent unit. The skip truncation keeps an equivalent good performance in a gated recurrent network. It produces a 93.83% F1 score, 93.83% Precision, and 93.92% Recall.

Overall, the Brute Force II method is better than Brute Force I because reviews include non-subjective descriptions in the front of the document, such as introducing the movie characters. Compared with baselines, skip truncation can improve the accuracy of binary sentiment polarity. It is especially effective for those high-capable networks.

Table 7. Results in recurrent unit (binary-polarity).

Recurrent unit	Precision	Recall	F1 Score
Excerpt	69.14	73.51	70.83
Brute force I	68.30	71.56	69.54
Brute force II	69.21	68.12	68.63
Skip-truncation	93.13	93.23	93.15

Table 8. Results in gated recurrent unit (binary-polarity).

Gated recurrent unit	Precision	Recall	F1 Score
Excerpt	70.39	73.05	71.52
Brute force I	74.01	75.69	74.71
Brute force II	80.51	81.65	80.93
Skip-truncation	93.83	93.92	93.83

It can break through the limitation of sentiment featurization and the most correctly perceive the document sentiment.

4.4 Results on Multiple Sentiment Polarity

In multiple sentiment polarity, the sentiment gradually changes from negative to positive from level one (negative) to level six (positive). The difficulty is perceiving the subtle sensation. The performances against baselines are shown in Tables 9, 10, 11 and 12.

Table 9. Main result in fully connection (multiple-polarity).

Fully connection	Precision	Recall	F1 score
Excerpt	33.01	37.61	33.66
Brute force I	31.42	36.47	32.02
Brute force II	35.07	39.91	35.83
Skip-truncation	51.77	50.92	47.57

Table 10. Main result in convolutional unit (multiple-polarity).

Convolutional unit	Precision	Recall	F1 Score
Excerpt	34.27	35.32	34.55
Brute force I	39.66	38.07	33.71
Brute force II	47.87	49.54	46.53
Skip-truncation	64.85	64.91	63.44

Table 11. Results in recurrent unit (multiple-polarity).

Recurrent unit	Precision	Recall	F1 Score
Excerpt	31.58	34.17	32.66
Brute force I	28.49	29.93	29.10
Brute force II	36.11	36.12	36.00
Skip-truncation	85.17	84.98	84.89

Experiments showed more significant differences in the multiple sentiment polarity because detailed expression changing will influence the prediction and the adjacent sentiment label is very hard to be discriminated. Merely by different truncation directions (see Brute Force I and II), the results showed a more than 10% gap in Precision, Recall, and F1 score. However, the baselines cannot sufficiently capture the fine-grained feelings. The best baseline got 50.11% Precision, 50.92% Recall, and 49.78% F1 score, while skip

Table 12. Results in gated recurrent unit (multiple-polarity).

Gated recurrent unit	Precision	Recall	F1 score
Excerpt	35.78	39.56	36.93
Brute force I	39.85	39.56	38.51
Brute force II	50.11	50.92	49.78
Skip-truncation	84.86	84.63	84.31

truncation improved them to 85.17%, 84.98%, and 84.89%. The results manifested that skip truncation can keep the correct sentiment features in a document and enable models to receive the complete sentiment skeleton. It can effectively reduce the interference of irrelevant words and amplify subtle information.

5 Conclusion

The document-level sentiment classification is difficult for deep learning training [1–8], while current truncations will reduce the expressive features and thus hinder the model capability. This paper proposed a new skip truncation method based on grammatical structures. The proposed truncation can reduce document length and enhance subtle sentiment detection. Using skip truncation for binary and multiple polarity got the best performances in different neural networks against baselines.

In future work, we will study more grammatical components towards sentiment information to keep the features extraction when truncation. We will also discuss more applications using the skip-truncation approach for text modeling.

References

1. Yin, Y., Song, Y., Zhang, M.: Document-level multi-aspect sentiment classification as machine comprehension. In: Proceedings of the 2017 Conference on Empirical Methods in Natural Language Processing, pp. 2044–2054. ACL, Copenhagen, Denmark (2017)
2. Rhanoui, M., Mikram, M., Yousfi, S., Barzali, S.: A CNN-BiLSTM model for document-level sentiment analysis. Mach. Learn. Knowl. Extract. **1**(3), 832–847 (2019)
3. Adhikari, A., Ram, A., Tang, R., Hamilton, W.L., Lin, J.: Exploring the limits of simple learners in knowledge distillation for document classification with DocBERT. In: Proceedings of the 5th Workshop on Representation Learning for NLP, pp. 72–77. ACL, Online Meeting (2020)
4. Fiok, K., Karwowski, W., Gutierrez, E., Davahli, M.R., Wilamowski, M., Ahram, T.: Revisiting text guide, a truncation method for long text classification. Appl. Sci. **11**(18), 8554 (2021)
5. Pappagari, R., Zelasko, P., Villalba, J., Carmiel, Y., Dehak, N.: Hierarchical transformers for long document classification. In: 2019 IEEE Automatic Speech Recognition and Understanding Workshop, pp. 838–844. IEEE, Singapore (2019)
6. Vaswani, A., et al.: Attention is all you need. In: Advances in neural information processing systems, pp. 5998–6008. Curran Associates, Long Beach, CA, USA (2017)

7. He, J., Wang, L., Liu, L., Feng, J., Wu, H.: Long document classification from local word glimpses via recurrent attention learning. IEEE Access **7**, 40707–40718 (2019)
8. Wu, Z., Gao, J., Li, Q., Guan, Z., Chen, Z.: Make aspect-based sentiment classification go further: step into the long-document-level. Appl. Intell. pp. 1–20 (2021)
9. Kaliamoorthi, P., Ravi, S., Kozareva, Z.: PRADO: projection attention networks for document classification on-device. In: Proceedings of the 2019 Conference on Empirical Methods in Natural Language Processing and the 9th International Joint Conference on Natural Language Processing, pp. 5012–5021. ACL, Hong Kong, China (2019)
10. Zhang, S., Wei, Z., Wang, Y., Liao, T.: Sentiment analysis of Chinese micro-blog text based on extended sentiment dictionary. Futur. Gener. Comput. Syst. **81**, 395–403 (2018)
11. Okango, E., Mwambi, H.: Dictionary based global twitter sentiment analysis of coronavirus (COVID-19) effects and response. Ann. Data Sci. (1/2022), 1–12 (2022)
12. Yekrangi, M., Abdolvand, N.: Financial markets sentiment analysis: developing a specialized Lexicon. J. Intell. Inf. Syst. **57**(1), 127–146 (2021)
13. Hossen, M.S., Dev, N.R.: An improved lexicon based model for efficient sentiment analysis on movie review data. Wireless Pers. Commun. **120**(1), 535–544 (2021). https://doi.org/10.1007/s11277-021-08474-4
14. Velldal, E., Øvrelid, L., Bergem, E.A., Stadsnes, C., Touileb, S. Jørgensen, F.: NoReC: the norwegian review corpus. In: Proceedings of the 11th edition of the Language Resources and Evaluation Conference, pp. 1–2. ACL, Miyazaki, Japan (2018)
15. Zhang, L., Wang, S., Liu, B.: Deep learning for sentiment analysis: a survey. wiley interdisciplinary reviews: Data Mining Knowl. Disc. **8**(4), e1253 (2018)

Check for updates

Improving the Power Quality of the Distribution System Based on the Dynamic Voltage Restorer

Alaa Yousef Dayoub(✉), Haitham Daghrour, and Nesmat Abo Tabak

Electrical Engineering Department, Tishreen University, Latakia, Syria
alaadayoub27@gmail.com

Abstract. In this research, work is done to improve the quality of electric power in the distribution system by eliminating harmonics to reduce voltage sag and swell by discussing different methods of voltage injection, where a completely new control technique has been proposed in order to achieve the conditions for quality of electric power through effective control of the capacitor supporting the DVR. To represent the load rays, single rays were used. In this research, the synchronous frame of reference theory was used to transform the three-phase voltages with rotating vectors into the fixed frame.

Keywords: Dynamic voltage restorer (DVR) · Voltage sag · Voltage swell · Voltage harmonics · Battery energy storage system (BESS) · PI controller

1 Introduction

The power quality in power systems is the most important source of concern, especially in power distribution systems, where it has been observed that the most power quality disturbances are voltage sag and voltage swelling in addition to harmonics and interruptions in the supply voltage. These disturbances occur due to the difference in the voltages of the supply voltage and the use of sensitive equipment and critical loads by the consumers of the distribution network. Given the importance of these disturbances and their great impact on the equipment and consequently the high cost as a result of the damage, the researchers resorted to using specialized power equipment to prevent the occurrence of these disturbances, which were divided into three types of The devices are as follows: power compensation devices connected in series with the power system, power compensation devices connected in parallel with the power system, and also parallel series compensation devices.

In this research paper, we will use a dynamic voltage restorer, a device that compensates energy through its connection in series with the power system, where it works to regulate voltage from disturbances that negatively affect power quality such as voltage sag, voltage swell and harmonics, where DVR is considered one of the most Energy efficient and effective in addressing energy quality problems that occur in the power distribution system compared to other energy devices designated for the same purpose in terms of small size, low economic cost and rapid dynamic response. The core of the DVR configuration is the control unit whose main task is to calculate the voltage that

© The Author(s), under exclusive license to Springer Nature Switzerland AG 2022
A. E. Hassanien et al. (Eds.): AMLTA 2022, LNDECT 113, pp. 309–322, 2022.
https://doi.org/10.1007/978-3-031-03918-8_28

must be added or removed in order to maintain a constant value of the load voltage. In this paper, a developed control unit has been designed using an integrative proportional controller (PI) with a PWM-type pulse generator. In the inverter, one of the most important components that are essential to the work of a dynamic voltage restorer is the phase-lock loop (PLL) and the dq0 transform.

This research paper looks at improving the quality of electrical power by improving the performance of DVR in the presence of three-phase faults during two different time periods.

2 DVR Operation Process

In the Fig. 1 a diagram of the DVR connected with the power supply system. Disruptions in the power system cause the power supply to be distorted and uneven. To keep the load voltage at a specified constant value and free from distortion and turbulence, Vinj the voltage injected into the system through the DVR, is included. Voltage *Vl* constant in its magnitude and undistorted. In Fig. 2 the various schematics of voltage injection are shown by a detailed phase diagram *VL*(pre-sag) is the voltage across the critical load prior to the voltage sag condition. A decrease in the value of the supply voltage occurs with the phase angle during the voltage sag, then the DVR starts injecting voltage to restore the load voltage *Vl* to its value before the sag occurred and maintains this value.

There are four different methods of voltage injection, all of which are based on phase angle [9].

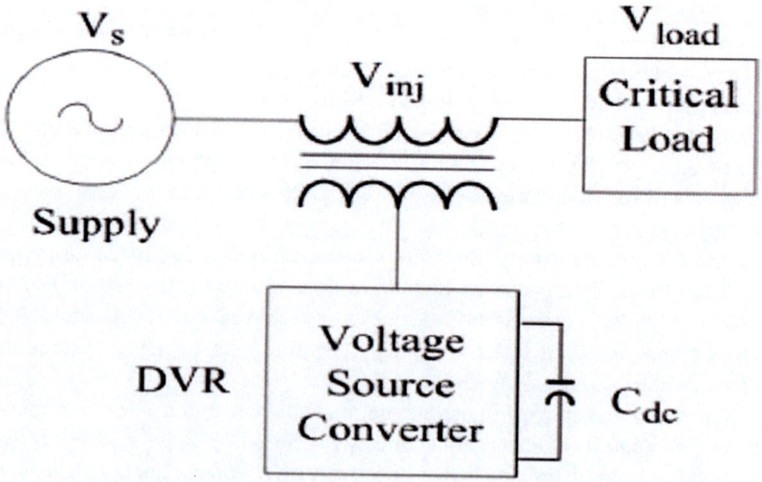

Fig. 1. DVR connection diagram with power supply system

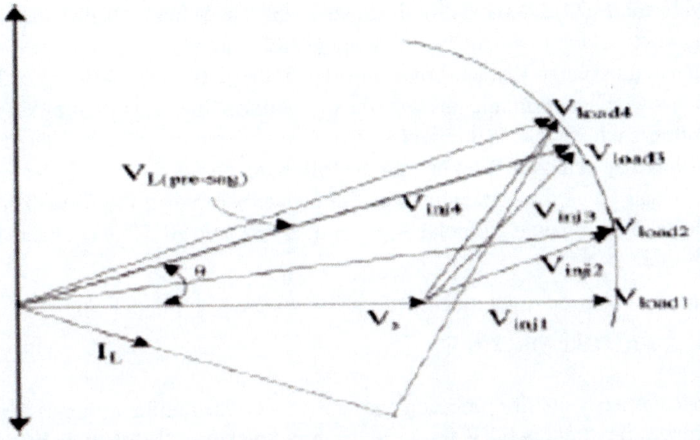

Fig. 2. Phase diagram of different voltage injection schemes using DVR

The value of the load voltage remains constant and the same as when Vinj2 is injected but becomes Vs at a small angle. In any case with Vinj1 the minimum converter rating remains to be met. In the case of Vinj3 it remains the same value for all phases than it was before the droop occurred and the angle between phases is ideal in this case [10]. The most suitable case for the support capacitor connected to the DVR is in the case that the injected voltage is squared with the current [7], which is suitable because it does not include any active power and this is what we find in Vinj4. The minimum rating of a converter is also achieved by Vinj1. In this scheme the DVR is operated with both a battery energy storage system (BESS) and also with a support capacitor.

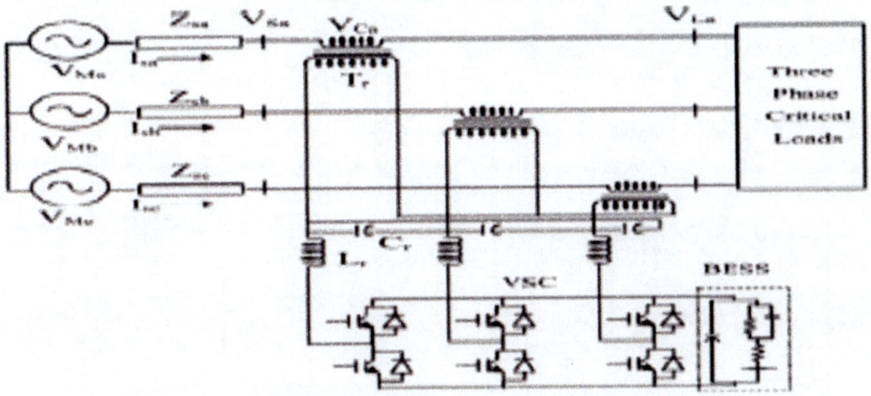

Fig. 3. DVR connection diagram with BESS

Figure 3 shows the schematic diagram of a DVR connected to a three-phase feed system in order to restore the sensitive load voltage connected to the feed system through three three-phase serial injection transformers. *VMa* is the supply voltage of phase A connected to the PCC and short circuit impedance *VSa*. When a disturbance occurs in the supply voltage, the DVR injects the voltage *VCa* into phase A in order to maintain the sensitive load voltage *VLa* at a constant value and free from any distortion. The DVR is a serial device that communicates with the distribution line through three single-phase *Tr* transformers whose task is to inject voltage. The injected voltage is filtered from the distortion wave by using the filter having the following components *Lr* is the inductance of the filter and Cr is the capacitance of the capacitor used in the filter. The inverter used consists of six insulated bipolar IGBT transistors, and BESS is connected to a DC busbar for that.

3 DVR Control System

The electrical energy quality problems that the power distribution system is exposed to can be addressed using the DVR through the actual power or the reactive power [7]. Its basic value is that the DVR does not need a battery to store energy in the DC bus of the VSC, meaning that the DVR is in a self-supporting state with the DC bus [3, 4], and the disturbed voltage is compensated by the DVR injecting actual power when the injected voltage is in phase with the current at it Batteries are required to store the energy needed to operate the VSC. The control techniques used in the DVR are required to take into account the voltage injection capacity requirements of the transformer and converter ratings and should also consider optimizing the battery storage size.

3.1 Controlling the DVR Using the Battery Energy Storage System

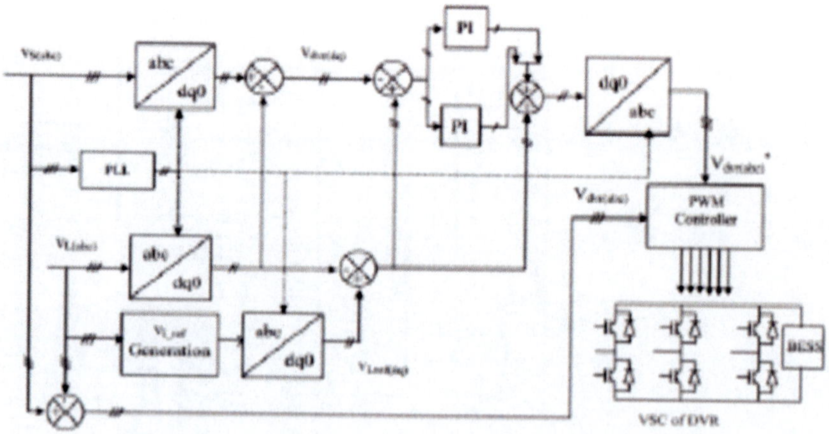

Fig. 4. DVR control circuit using SRF method to control BESS

In Fig. 4, the control part of the DVR is illustrated using the SRF theory used to determine the reference signal. The voltages are sensed at PCC(VS), and load voltage are sensed

(VL) and then the gate firing pulses of IGBTs are generated. From the unit vector the reference load voltage Vl^* is obtained [13]. Using the abc-dqo transformation that depends on the Park transformation with unit vectors (sinθ, cosθ) the load voltages (Vla, Vlb, Vlc) are converted to the rotating reference frame and then derived using a phase detection loop according to the following equations

$$\begin{bmatrix} vl_q \\ vl_d \\ vl_0 \end{bmatrix} = \frac{2}{3}\begin{bmatrix} \cos\theta & \cos\left(\theta - \frac{2\pi}{3}\right) & \cos\left(\theta + \frac{2\pi}{3}\right) \\ \sin\theta & \sin\left(\theta - \frac{2\pi}{3}\right) & \sin\left(\theta + \frac{2\pi}{3}\right) \\ \frac{1}{2} & \frac{1}{2} & \frac{1}{2} \end{bmatrix}\begin{bmatrix} vlaref \\ vlbref \\ vlcref \end{bmatrix} \tag{1}$$

Similar to what we did previously, the load voltages (Vla^*, Vlb^*, Vlc^*) and VS voltages of the PCC are transferred to the rotating reference frame, and then the DVR voltages are obtained according to the rotating reference frame as follows

$$VD_d = Vs_d - Vl_d \tag{2}$$

$$VD_q = VS_q - Vl_q \tag{3}$$

and then the reference DVR voltages are obtained according to the rotating frame as follows

$$VDd^* = Vsd^* - Vld \tag{4}$$

$$VDq^* = Vsq^* - Vlq \tag{5}$$

The error is regulated using the PI controller between the actual values of the DVR voltages and the reference values according to the rotating reference frame.

From Eq. (4), we take the value of VDd^* and from Eq. (5) we take the value of VDq^*, and for $VD0^*$, we consider its value as zero, then we get the values of the reference voltages for the DVR according to the abc frame, using the inverse Park transformations, as shown by the following relationship

$$\begin{bmatrix} V^*_{DVRa} \\ V^*_{DVRb} \\ V^*_{DVRc} \end{bmatrix} = \begin{bmatrix} \cos\theta & \sin\theta & 1 \\ \cos\left(\theta - \frac{2\pi}{3}\right) & \sin\left(\theta - \frac{2\pi}{3}\right) & 1 \\ \cos\left(\theta + \frac{2\pi}{3}\right) & \sin\left(\theta + \frac{2\pi}{3}\right) & 1 \end{bmatrix}\begin{bmatrix} VDd^* \\ VDq^* \\ VD0^* \end{bmatrix} \tag{6}$$

Note that the reference voltages ($V^*_{DVRa}, V^*_{DVRb}, V^*_{DVRc}$) and actual voltages ($V_{dvra}, V_{dvrb}, V_{dvrc}$) of the DVR are used to generate the gate pulses of the inverter used in the DVR via the PWM controller, which operates at a switching frequency of 10 kHz.

3.2 Self-supporting DVR Control (Has a Support Capacitor)

In Fig. 5, the detailed diagram of the capacitor supporting the DVR connected to the sensitive load is illustrated in addition to the control circuit used, which is based on the SRF theory that is illustrated in Fig. 6. The abc-dqo conversion is used to convert the voltages (VS) in the PCC to the rotating reference frame and by using low-pass filters (LPFs) the voltage harmonics are eliminated [13].

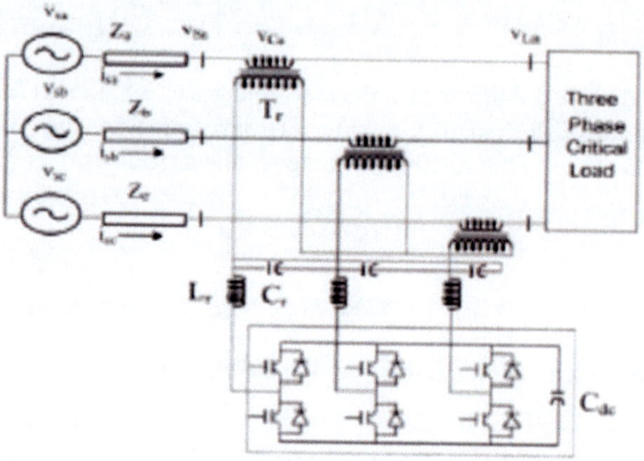

Fig. 5. Support capacitor diagram connected to the DVR system

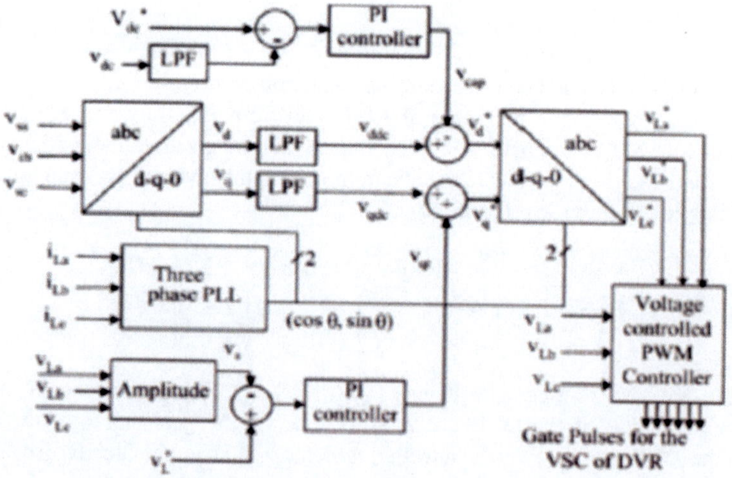

Fig. 6. DVR control circuit uses the SRF method to control the capacitor connected to it

In the following equations, we find, respectively, the components of the voltages on the d-axis and on the q-axis.

$$Vd = V_{ddc} + V_{dac} \tag{7}$$

$$V_q = V_{qdc} + V_{qac} \tag{8}$$

The compensation strategy depends in principle on compensating the voltage quality problems so that the load voltage remains at its specified value and free from any distortion.

The compensation strategy depends in principle on compensating the voltage quality problems so that the load voltage remains at its specified value and free from any distortion.

The PI controller is used at the DC busbar in the DVR in order to maintain the DC bus voltage of the self-supporting capacitor while the output is considered as a voltage ($V_{cap(n)}$) to counter its losses

$$V_{cap(n)} = V_{cap(n-1)} + K_{P1}\left(V_{de(n)} - V_{de(n-1)}\right) + K_{i1}V_{de(n)} \tag{9}$$

$V_{den(n)} = Vdc^* - V_{dc(n)}$ is the error between the reference voltage (Vdc^*) and DC voltage (V_{dc}) detected at the sampling moment (n^{th}).

K_{P1} and K_{i1} are the relative and integral gains of the DC voltage PI controller.

The reference load voltage (Vl) on the d-axis is expressed by the following equation

$$Vd^* = V_{ddc} - V_{cap(n)} \tag{10}$$

Another PI controller is used whose task is to adjust the value of the load voltage (Vl) to remain at the specified reference value for it ($V_l{}^*$).

In order to regulate the load voltage, the output of the PI controller is considered as the reactive component of the voltage V_{qr}.

The value of the load voltage (Vl) at the PCC is calculated by the value of the AC voltage.

$$(Vla, Vlb, Vlc) as Vl_l = (2/3)^2\left(Vla^2 + Vlb^2 + Vlc^2\right) \tag{11}$$

and then the PI controller is used to regulate the load voltage to match the required reference value according to the following

$$V_{qr(n)} = V_{qr(n-1)} + K_{P2}\left(V_{te(n)} - V_{te(n-1)}\right) + K_{i2}V_{te(n)} \tag{12}$$

Where $V_{te(n)} = V_l^* - V_{l(n)}$ indicates the error between the reference value (V_l^*), and the total value of the load voltage ($V_{l(n)}$) at the sampling moment (n^{th}) K_{P2} and K_{i2} are the relative and integral gains of the DC voltage PI controller. The reference load voltage is expressed on the quadrature axis according to the following equation.

$$V_q^* = V_{qdc} - V_{qr} \tag{13}$$

As in Eq. (6), using the inverse Park transform, the reference voltages for the load are obtained in the abc frame. The error value between the measured load voltage and the reference load voltage is more used in the control unit to generate the gate pulses of the DVR inverter.

4 Simulation and Modeling Results

The operating system that is being studied in this paper consists of a DVR connected with a three-phase feeding system and three-phase loads that are highly sensitive to power quality problems [11] and also as shown in Fig. 2 injection transformers connected in series with the system under the use of BESS Where the modeling and simulation were done in the MATLAB environment. Figure 2 shows the simulation model of Fig. 7 the operating system without the DVR control circuit, and therefore the simulation results will appear for the output voltage without any compensation for the voltage sag, and this is shown in Fig. 8 and without any compensation for the voltage swelling as shown in Fig. 9 Whereas Figs. 10 and 11 present the results of using SRF control theory in order to compensate for voltage sag and swell. In the same way, the capacitor-supported DVR system shown in Fig. 3 and the control system elements shown in Fig. 4 were simulated, where the results of this simulation appeared in Figs. 12 and 13 in terms of the performance of the studied system in compensating for the sagging and swelling of the voltage. The linear load with a power of 10 kVA and a power factor of $pf = 0.8$ is the equivalent load on which the results of using a DVR work system are being studied. The DVR reference voltages are obtained by sensing the voltages in the PCC and these voltages are (*Vsa*, *Vsb*, *Vsc*), as well as from the load voltages which are (*Vla*, *Vlb*, *Vlc*). In the DVR control circuit in this research, PWM was used as a pulse generation unit for IGBTs electronic switch gates located in the inverter of the DVR.

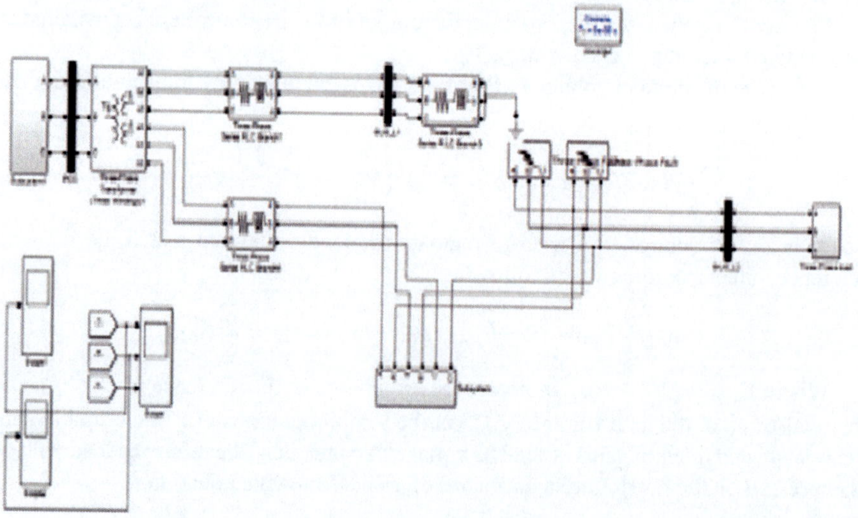

Fig. 7. Simulation model of DVR connected with BESS without control circuit

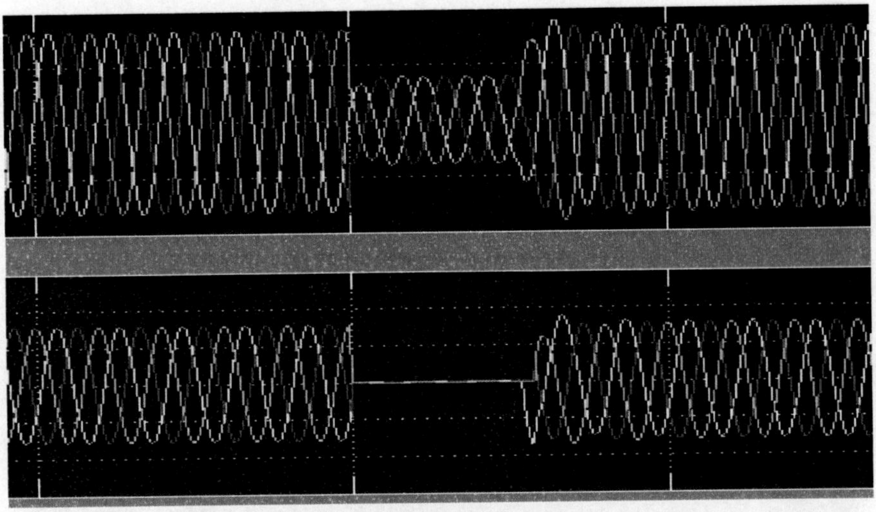

Fig. 8. The sag observed at PCC voltage and load voltage

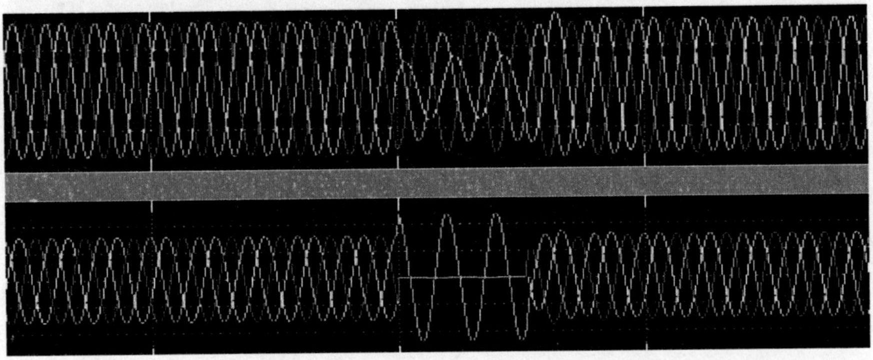

Fig. 9. Swelling observed at PCC voltage and load voltage

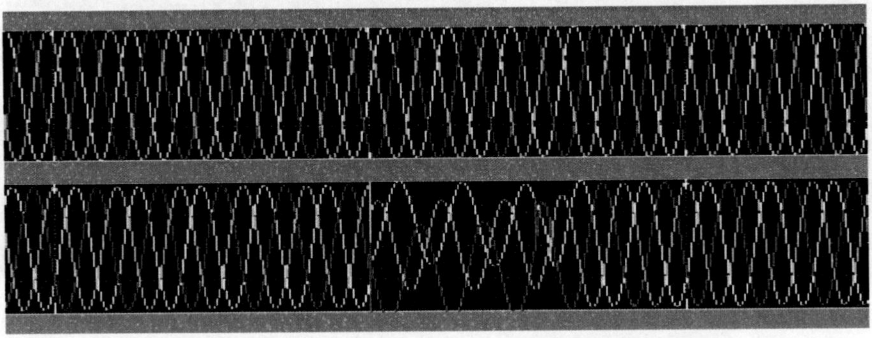

Fig. 10. Compensation for sag at PCC voltage and load voltage

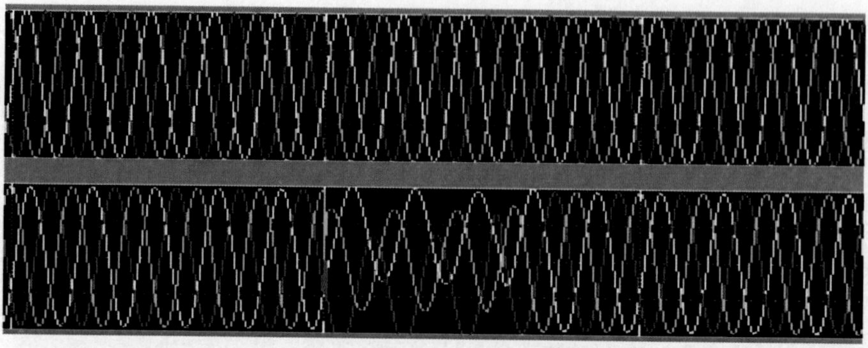

Fig. 11. Compensation for swelling at PCC voltage and load voltage

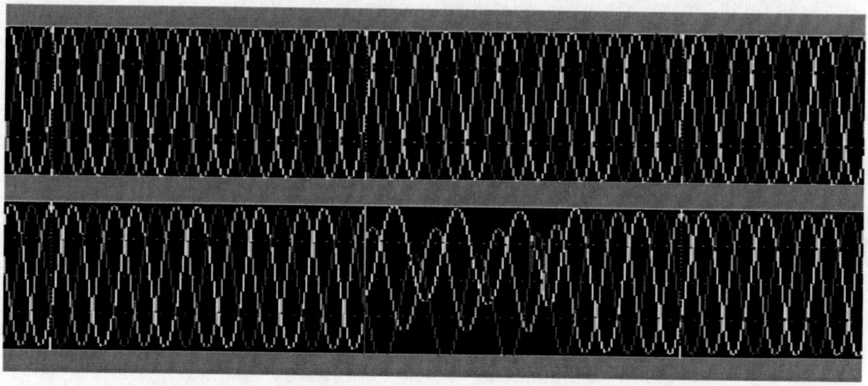

Fig. 12. PCC voltage and load voltage after sag compensation

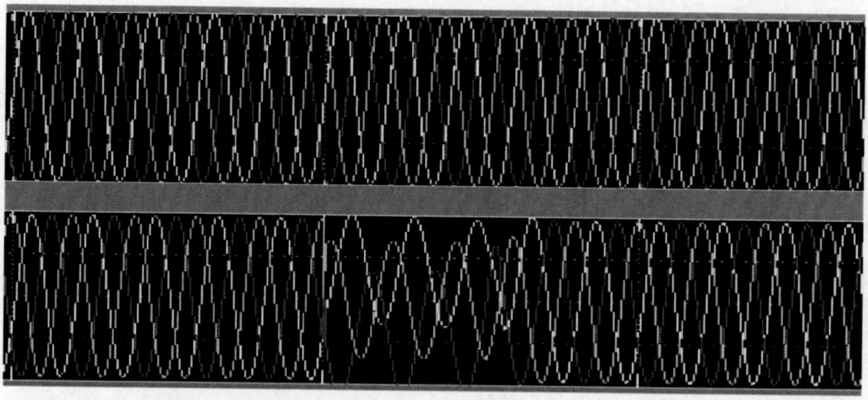

Fig. 13. PCC voltage and load voltage after swell compensation

5 DVR System Performance Evaluation

Authors in [1], the work of DVR and its performance for the task of solving the problems of various supply voltage disturbances such as voltage sag and voltage swell, Figs. 10, 11, 12 and 13 show the performance of the power system when conditions of sagging and swelling of the supply voltage occur and this is shown by the modeling section. Voltage sag was created for six cycles, from 0.2 s to 0.25 s, and voltage swell was carried out for one cycle, from 0.6 s to 0.65 s, thus we obtain a constant value in the load voltage under the conditions of voltage sag and swell. In Fig. 14 and 15 the total harmonic distortion in the supply voltage is illustrated in the presence of the support system by energy storage devices (batteries) and capacitors (Fig. 16).

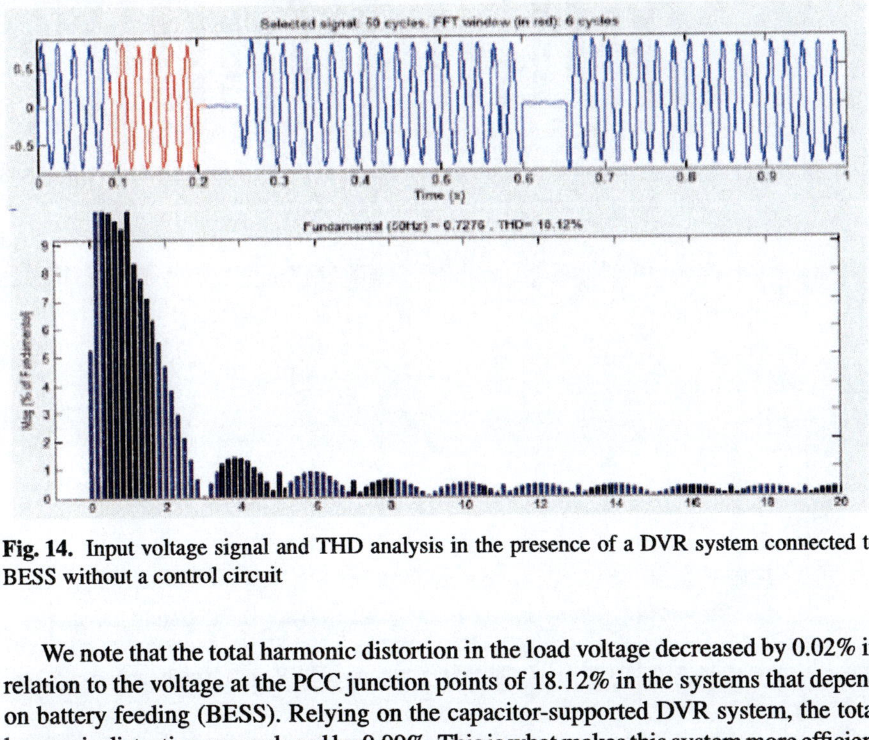

Fig. 14. Input voltage signal and THD analysis in the presence of a DVR system connected to BESS without a control circuit

We note that the total harmonic distortion in the load voltage decreased by 0.02% in relation to the voltage at the PCC junction points of 18.12% in the systems that depend on battery feeding (BESS). Relying on the capacitor-supported DVR system, the total harmonic distortion was reduced by 0.09%. This is what makes this system more efficient in performing its tasks compared to the system used in the reference [1]. In Table 1, we find the classifications of the four schemes of the DVR for each of the injection voltage, serial current and percent load capacity, and from this table and from Scheme 1 we find that the voltage injected into the Vinj1 phase through the phase diagram in Fig. 1, where we find that in this research paper the implementation of This scheme only and with improved efficiency. The DVR voltage is injected at a small angle of 30° in Scheme 2, the DVR voltage is injected at an angle of 45° in Scheme 3, while in Scheme 4 the DVR voltage is injected at an angle of 90°. The required extent of compensation when using Scheme 1 is significantly less compared to Scheme 1 (Table 2).

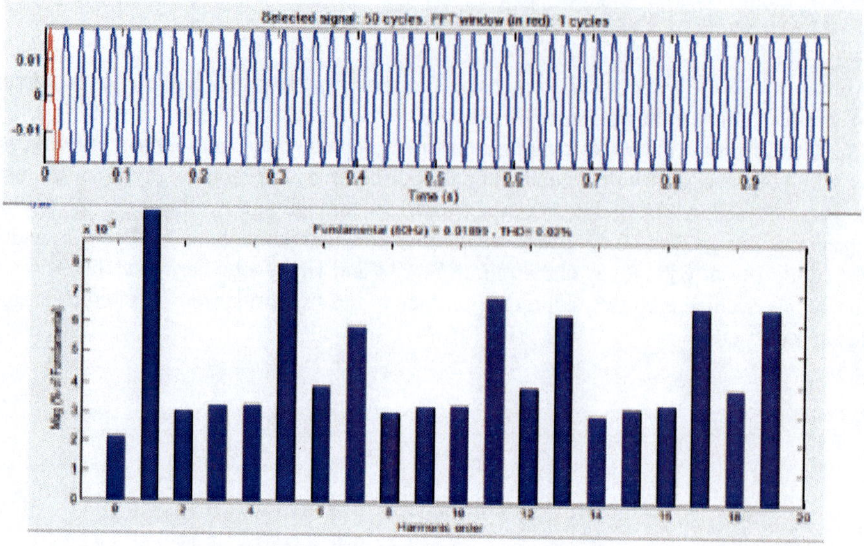

Fig. 15. Input voltage signal and THD analysis in presence of DVR system connected with BESS

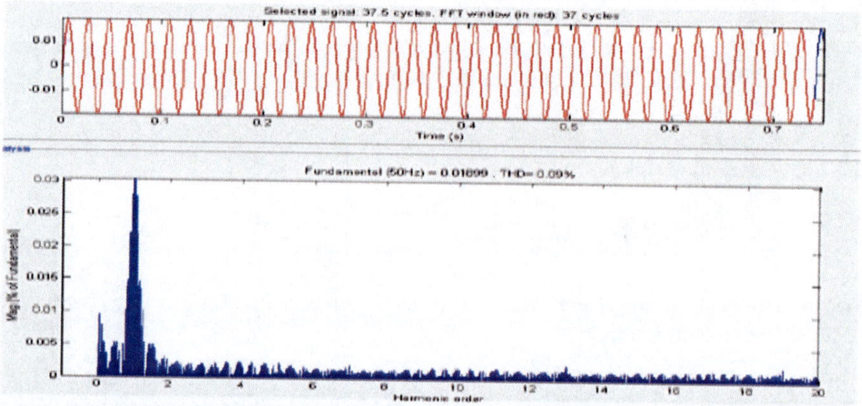

Fig. 16. Input voltage signal and THD analysis in presence of DVR system with support capacitor

Table 1. Voltage sag reduction by setting DVR rates

	Scheme-1	Scheme-2	Scheme-3	Scheme-4
Phase voltage	90	100	121	135
Phase current	13	13	13	13
VA per phase	1170	1300	1573	1755
kVA (% of load) load)	37.5%	41.67%	50.42%	56.25%

Table 2. Appendix

Supply voltage	415 V, 50 Hz
Sensitive loads	10-kVA 0.80-pf
Distribution line impedance	$Ls = 3.0$ mH, $Rs = 0.01\ \Omega$
Pass filter	$Cf = 10\ \mu$F, $Rf = 4.8\ \Omega$
Injection transformers	Three-phase, rated specification 10 kVA, 200 V/300 V
DVR with Battery Energy Storage System (BESS)	
DC voltage of the DVR	200 V
Ac seducer	0.005 H
PI controller gain on d-axis	
$kp = 0.63$	$kd = 0.0504$

6 Conclusion

The operation of the DVR with a new control technology is illustrated using different methods for the voltage-1 injection scheme from Table 1. Comparing the performance of the DVR with Scheme-1 of the one in [1] and the one made with the lower-rated VSC of BESS and the capacitor-supported DVR shows that the efficiency is higher in this paper compared to that in [1]. The reference load voltage was estimated using the unit vector method and achieved control of the DVR according to the controls in [1], with improved efficiency for both BESS and capacitor-supported DVR. The same SRF theory was used to estimate the DVR reference voltage as in [1]. It was concluded that in-phase voltage injection with PCC voltage results in minimal DVR rating but at a cost of DC bus power source along with better efficiency in mitigating power quality problems.

References

1. Jayaprakash, P., Singh, B., Kothari, D.P., Chandra, A., Al-Haddad, K.: Control of reduced-rating dynamic voltage restorer with a battery energy storage system (2008 and 2014)
2. Ghosh, A., Ledwich, G.: Power Quality Enhancement Using Custom Power Devices. Kluwer Academic Publishers, London (2002)
3. Vilathgamuwa, M., Perera, R., Choi, S., Tseng, K.: Control of energy optimized dynamic voltage restorer. In: Proceedings of the IEEE IECON 1999, vol. 2, pp. 873–878 (1999)
4. Nielsen, J.G., Blaabjerg, F., Mohan, N.: Control strategies for dynamic voltage restorer compensating voltage sags with phase jump. In: Proceedings of the IEEE APEC 2001, vol. 2, pp. 1267–1273 (2001)
5. Ghosh, A., Ledwich, G.: Compensation of distribution system voltage using DVR. IEEE Trans. Power Deliv. **17**(4), 1030–1036 (2002)
6. Aeloíza, E.C., Enjeti, P.N., Morán, L.A., Montero-Hernandez, O.C., Kim, S.: Analysis and design of a new voltage sag compensator for critical loads in electrical power distribution systems. IEEE Trans. Ind. Appl. **39**(4), 1143–1150 (2003)
7. Ghosh, A., Jindal, A.K., Joshi, A.: Design of a capacitor supported dynamic voltage restorer for unbalanced and distorted loads. IEEE Trans. Power Deliv. **19**(1), 405–413 (2004)

8. Ghosh, A.: Performance study of two different compensating devices in a custom power park. IEE Proc. Gener. Transm. Distrib. **152**(4), 521–528 (2005)
9. Nielsen, J.G., Blaabjerg, F.: A detailed comparison of system topologies for dynamic voltage restorers. IEEE Trans. Ind. Appl. **41**(5), 1272–1280 (2005)
10. Banaei, M.R., Hosseini, S.H., Khanmohamadi, S., Gharehpetian, G.B.: Verification of a new energy control strategy for dynamic voltage restorer by simulation. Simul. Model. Pract. Theory **14**(2), 112–125 (2006)
11. Jindal, A.K., Ghosh, A., Joshi, A.: Critical load bus voltage control using DVR under system frequency variation. Electr. Power Syst. Res. **78**(2), 255–263 (2008)
12. MahindaVilathgamuwa, D., Wijekoon, H.M., Choi, S.S.: A novel technique to compensate voltage sags in multiline distribution system—the interline dynamic voltage restorer. IEEE Trans. Ind. Electron. **54**(4), 1603–1611 (2007)
13. Chandra, A., Singh, B., Singh, B.N., Al-Haddad, K.: An improved control algorithm of shunt active filter for voltage regulation, harmonic elimination, power-factor correction, and balancing of nonlinear loads. IEEE Trans. Power Electron. **15**(3), 495–507 (2000)

Ecosystem of Health Care Software Engineering in 2050

Afrah Almansoori[1], Mohammed Alshamsi[2], and Said Salloum[3](✉)

[1] General Department of Forensic Science and Criminology, Dubai Police G.H.Q., Dubai, UAE
[2] Faculty of Engineering and IT, The British University in Dubai, Dubai, UAE
[3] School of Science, Engineering, and Environment, University of Salford, Salford, UK
S.A.S.Salloum@edu.salford.ac.uk

Abstract. We Emerging technologies towards digitalization and clinical technology are shifting health services to a different level of advancements. The intensification of trending technologies is enlightening patient aligned care, introducing new ways and means to improve health care at the same time cut actual cost of operation by enhancements towards automated information management, data and information repository, enabling instantaneous availability and accessibility of information to clinicians and last but not least technology that can proactively perform tasks consuming less space, time and cost. The purpose of the paper is to discuss the latest trends in clinical technology, tools that can enhance monitoring and diagnostic proficiencies, and healthcare needs that are growing apace as innovation them could play a key role in shaping the future of healthcare systems? A comprehensive analysis approach of healthcare policies and applications literature was used to identify the growing cost of healthcare and operations. The paper concludes that emerging technologies, hospitals and facilities architecture are key points or should we say pillars that could play vital role towards health care future. The paper will provide a summary of some of the technological trends that are making an impact in improving the quality of healthcare today and their contribution to health eco system in the future (2050). Why do we need these trends and technologies? Some of the hurdles, issues complications currently faced by the healthcare industry and how they can be resolved to shape the future of healthcare industry.

Keywords: Artificial Intelligence · Block chain · Future of health care · Health care eco system · Mixed Reality · Software engineering · Technology trends

1 Introduction

The mounting cost and cumulative demands of eminence health care services, remains a challenging and fascinating debate for almost all regions across the globe [1–8]. Settled countries have established better health care services and eco systems but still are fronting multifaceted challenges in supplying cost effective services without compromising the quality of health care [3, 4, 9, 10]; [11]. We find examples were regions and countries with all the limitations and challenges have endured rapid variations to meet the demands

© The Author(s), under exclusive license to Springer Nature Switzerland AG 2022
A. E. Hassanien et al. (Eds.): AMLTA 2022, LNDECT 113, pp. 323–336, 2022.
https://doi.org/10.1007/978-3-031-03918-8_29

and set standards towards the emergent enormousness and eminence of the health sector. The U.S. Centers for Medicare and Medicaid Services (CMS) testified that health costs have increased by an average of 4.3% globally in 2016 as compared to 2015 and is quite alarming as it is more than double the consumer price index (2.1%) in 2015 [12]. The proverb, "What goes up, must come down," doesn't seems to be likely applicable to the worldwide health care sector, with the aging and mounting population, countless occurrence of chronic diseases, exponential improvements in innovative, but costly, digital technologies, trends and other improvements continue to increase health care petition and expenditures [13].

2 Current Health Care Eco System

Experts and analyst's opinion are growing towards uncertainty and growing cost of service in health care institutes, the situation could be alarming as health authorities [14, 15], government and private sector won't be able to finance future health care costs. According to the 2016 survey of consumers for health care by Deloitte we are at the highest of discomfort due to the cumulative cost of health-related services. The urgency, demand of tailored medicine, exponential technologies, unsettling contenders, prolonged delivery sites and overhauled payment models is inoculating ambiguity towards healthcare economy across the globe, the growing demands and cost is resonating the industry and key stakeholders to plan on when and how to make the future moves maintain a balance between offering quality health care with latest trends and demands and also financially viable. Figure 1 [13] below show health care spending in USD (2017–2022).

Why it is essential to talk about the growing financial pressure as its effects are way too big, it would put pressure on consumers and most likely on health care financing units and delivery stakeholders. The biggest concern would be reductions on profitability and margin constraints that can damage and effect the future of the industry. Government and key stakeholder have a vital role when it comes to the future of health care services, government and private sector via partnerships and affiliations should work in collaboration to eliminate legal, ethical and financial hindrances faced, align health care costs to patient clinical necessities more effectively and create suitable incentives for health care industry to prosper.

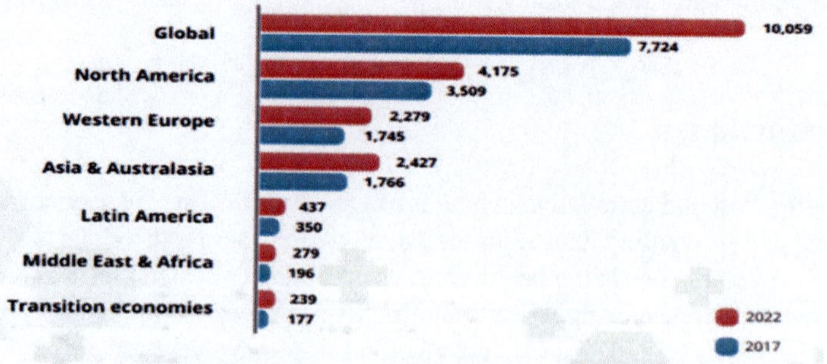

Fig. 1. Health care spending in USD [13].

3 Emerging Technologies in Health Care Industry

The IT industry has gone through tremendous amount of change, growth, restructuring and advancements in recent years and it doesn't stop here [16–21]. Every year new software machineries, tools like cross-platform application development [16, 22–27], Blockchain development [28], machine learning [29–32], cloud computing etc. is getting improved and enhanced, along with their progression demands of their utilizations and consumptions are increasing worldwide constantly. Before we go into detail discussion on all the latest trends evolving around today's health care industry, we should highlight a critical note that for any organization or industry to adopt [33, 34], consume, harness innovative steps and the latest trends, it should maintain operational integrity by modernizing legacy core systems, value for money and transforming the business technology while closing complying with or I should say zero compromise on the evolving cyber security and privacy risks [35].

Digital technology has revolutionized every industry and has disrupted the whole world. Today technology has become so ubiquitous and affordable not just cheap but easy to utilize, no learning curves required these days with resources and material available by various cradles and means to create business models and opportunities that never existed before in the past. Just for a while if we look around digital technology is everywhere around us we utilize it in daily basis for staying connected with friends and family, organize our daily routine and work [36], entertainment and education, not to forget it has expanded capabilities for business digitalizing government and public process and last but not least technology has brought huge breakthrough in health care via reengineering medical processes also introducing digital technology and eco systems which provide rapid and scalable information for faster actions and still there is no stopping it is still redefining cultural norms [35]. Figure 2 and Fig. 3 shows the technological trends by region in from 2019 to next 5 and 25 years along with different hurdles that the advanced technology are facing [12].

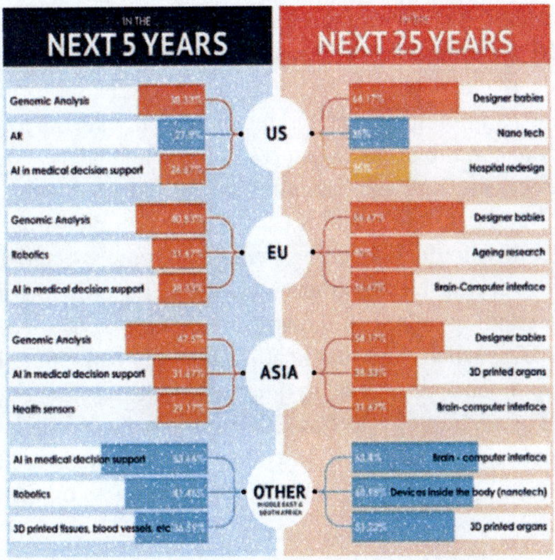

Fig. 2. Different hurdles facing advanced technology [12].

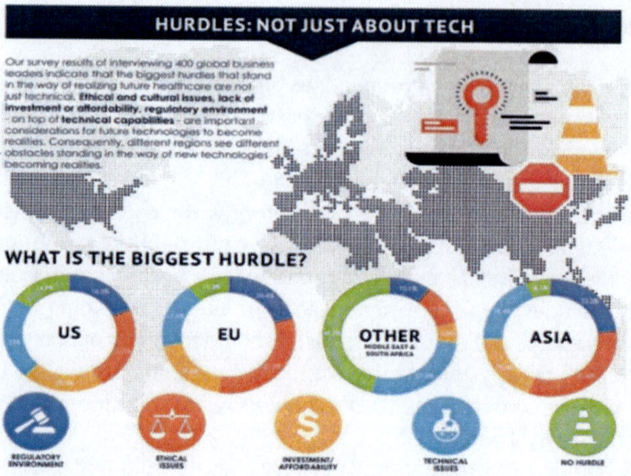

Fig. 3. Technological trends by region [12].

3.1 Mixed Reality

Mixed Reality (MR) has separate or I should say distributed definitions from experts where some call it as a possibility of both AR and VR in a same app, while others explain MR as continuum, the combination of real and virtual [37]. What we can conclude from the definition is that MR technology can provide real time of actual surrounding with a merge of integrated objects from virtual reality thus merge from both the physical and virtual worlds, hence opening way for new crossovers or should I say means of collaboration through gesture and voice commands which can allow miracles in the health care industry. Accenture Technology Labs' have demonstrated perfect examples to uplift the business standards by utilizing the technology opening new ways for digital businesses for almost all industries like consumer product goods, healthcare, oil and gas etc. Today several hardware companies are investing in developing product with the technology specially for utmost workforce utilization where a device utilizing enriched sensors can provide workers with a 360-degree setting for performing tasks in three environment-sensing technologies: infrared technology to map the physical atmosphere, infrared technology to capture or take gestures as input and lastly voice commands with natural language processing and voice recognition. I think the biggest break through the technology brought and can bring health care eco system do wonders in the future is remote experts (over the shoulder coaching), where no more specialist, experts would be asked to fly in order to access the situation saving travelling cost and time and interesting case study of MR technology that could revolutionize healthcare and predict healthcare future in 2050 via MR is when a leading healthcare client can remotely connect doctors with emergency respondents in the field to quickly offer care supervision and judgments [38]. Challenges: The technology today (2019) still it faces challenges with limited used cases, most applications, software, hardware and equipment are still like showcase and tricks, preservation uphold due to very limited awareness on the capabilities of the technology, lack of expertise for the health care industry and very less developers or resources available who can take care of all legal, ethical and technical issues [39].

3.2 Robotics

Mixed In the past 50 years from now medical facilities and hospitals have basically remained the same, although remarkable change, advancement and technology has been involved, adopted in medical industry and knowledge [40], which to some experts is no surprise as a medical facilities are overpoweringly complex structures, encompassing wide range of medical services and units, different complex departments like emergency rooms, operating/surgery theatres, labs and radiology centers, even food and housekeeping service for inpatients. These units since past long have been developed to work in shifts and coinciding manner forming a distended infrastructure that at times resist, seems impossible and way too costly to equip with the new innovation, technology latest IT systems. Experts suggest that instead of reaching to a deadlock and go for 100% reengineering of medical facilities infrastructure, business process and technology used it time to plan for future and push the accelerator towards future which demands heavy reliability on robots and digital technology. Robotic doesn't mean an actual robot would replace medical professionals and staff but physical and mental tasks can be automated utilizing combination of hardware and software which can leave immense amount of space in facilities and departments when less patients would be physically visiting and rest can avail services via technologies like telemedicine and remote health care.

Usage of technology and robotics is not new to health care, as 15 years ago the industry saw the first revolution via robotics with a successful tele-surgery procedure, when a surgeon in New York distantly (3,870 miles way) directed a three-armed surgical robot which was in France (Strasbourg), and performed surgery of a 68 years old patient diseased with gallbladder [12]. How do robots operate? How they help? What are the main components of robotics in health care? The answer to some common questions is that medical robotics are hardware and software with motion, self-sufficiency, astuteness etc. Motion comprises of actuated mechanism programmable for multi axis and are not fixed can be moved and relocated within an environment. Autonomy tells us that it's an automated process but still depends on human involvement to supervise the robot or mechanical device to perform tasks both in usual mode of maneuver and predominantly in case of a fault, issue or a problem. Intelligence is simply the effective utilization of knowledge or skills of an expert is used to enable an operator to do tasks with minimum knowledge/skill than a person would need to perform the same task on his own. There are several success cases and implementations that uplifted the medical industry standards like Balance Training Assist, Steerable Surgeons (also known as Nanobot), Robotic Systems for professional care (Nursing), Robotic rehabilitation treatment for therapy and not to forget robotic wheelchair, but these innovations are step forward for the future. According to research work and data collect by expert's robotic market and surgeries would grow by 2021, and in future 2050 several common tasks would be handled by robotics saving cost at the same time allow physicians to perform human skill tasks alongside robots for better outcome for public and industry [37].

Challenges: For robots to perform proactive measures in health care industry, the entire mechanism requires access to monitor and analyze huge amount of health data which is a big challenge at the moment due to regulatory complexity (like privacy, patient and doctor relationship, shared decision making etc.) and lack of data protection regulations [41]. Negative attitude has been found by patients towards robotics and trust is still not there [42].

3.3 Artificial Intelligence (AI)

Like robotics AI is also not new in the industry, several years back a true success pioneering work in AI was delivered which was an attempt to automate doctors and professionals work at Stanford University US, where utilizing a systematic coding process an algorithm was form which was able to diagnose infectious blood diseases, the automated process also had the capability to identify the cause of infection and even suggest a treatment course. The system was named as 'MYCIN', the core idea of the system was to pull professional and doctors from investing time performing observations, test, asking and listening to questions instead using machine intelligence to do these tasks, the notion could save immense number of doctors and professional's time. However unlike MYCIN several AI projects failed which wasn't much supportive for the technology (at least for health care industry) but with the passage of time and the advent of subterranean erudition algorithms in the current AI revitalization have improved to such an extent that machines could now even exceed productive output in wide range of tasks in the health care industry surpassing the performance of professionals, hence AI truly has transformative potential toward future of medicine and health care [12].

Babylon Health, in London is a perfect example supporting the above statement where an AI based system replaced medical professional in patient consultation which means that fields patients no more need to wait in queues for a General Practitioner doctor, the same AI solution is in pilot use by Britain's NHS (National Health Services) and with their current utilization of the system they predict that 85% of consultations could be replace by the smart system. The consequences are thoughtful of adopting such technologies, in future 2050 AI can prevent huge amount of spending and reduce cost of operations in facility at the same time more patients to be attended but on the other had has a rough side for professionals and doctors where machines and algorithms would take over half of their tasks there could be huge probability of facilities reducing manpower [12].

Challenges: Issues faced by AI is not much different from robotics in health care, despite AI can analyze and process data and predict disease and diagnosis it also creates ethical and social risks, not to forget the above mentioned rough side for professionals being isolated and can lead to fostering leaning i.e. medicine without doctors [41].

3.4 Internet of Things (IoT)

With all happening around in tech exhibitions and experts talking about we can easily narrate that the most hyped technological trend today and from the past ten years is the way internet has indulged in our life's by simply connecting various 'things' to operate in smart space with an intelligent interface to connect between real and virtual world [43–47]. 'Things' comprise of three main components: information, technology/robots/machines and people [37]. IoT has established a vision which has already kicked off and become popular in all industries that can be felt with the availability of online coffee machines and smart devices like the Amazon Echo or Google Home. Health industry has also seen fitness fanatics in the form of wearable tracking devices or watches that monitor heart-beat and measures the physical activity of the host. The aptitude or combination of a sensor into any object and connecting them to the internet, has benefits with no limits predominantly in healthcare. Patients save time by being examined remotely not just that but facilities will face lesser burdens on capacity directly improving the cost of operation, it doesn't end here doctors will be able to receive unremitting patient data, to monitor the patient's health without summoning him/her to the facility.

Perfect IoT cases in today's health care industry based on my preference are the real time bathroom scales that apprise central electronic health records over the air, or Kardia Mobile's USB-stick-sized device that takes EKG reading remotely in less than 30 s and worth mentioning remote sensor system that alerts based on configured threshold of a patient's recordings which professionals monitor and can dispatch help when required. The implementation has a unique and innovative method were the sensor is woven into a cloth or applied to the skin just like a tattoo. The most interesting part is that not just the impressive or innovative way of collecting the data, but the captured data is the real transformer of healthcare for the next ten years. Remote Patient monitoring and facility management, monitoring glucose level and inhaler via smart application, hearing aid and blood coagulation test are other revolutionary futuristic implementations including to the above mentioned given to health care industry via IoT which in future 2050 can help to congregate data unceasingly, ultimately paving way for early discoveries of deceases and medical happenings before time [48].

Challenges: Like AI and Robotics, IoT devices also generate tremendous amount of data, hence the biggest challenge in the future would be storage for the mountain of data produced by IoT devices and this data will be exchanged between various devices causing governance and security issues. Secondly the medical facilities infrastructure, which is a known issue for healthcare is mostly outdated and cannot support the implementation of such latest trends with huge amount of data flow. The security of IoT devices are also a concern and are vulnerable for hackers to infiltrate sensitive data [49].

3.5 Blockchain

The append only leger known as Blockchain is one of the greatest technology break through as with time it is proving to overcome almost all challenges in the past. It is a distributed transaction technology, any system built on its architecture won't require multiple levels of authentication to get access to on-demand chronologically arranged transparent and authentic data. Several proof of concepts and case studies have proven that the robust technology can drive all sorts of industries including health care improving quality of service with a relatively lower cost and undertaking all challenges faced like Fragmented Data, Timely access of patient records, Data security and governance, integration with legacy systems, incorporate the latest upcoming devices and technologies all this with cost effectiveness [50]. Figure 4 shows how a Blockchain health care eco system that can be established which was never possible before [50].

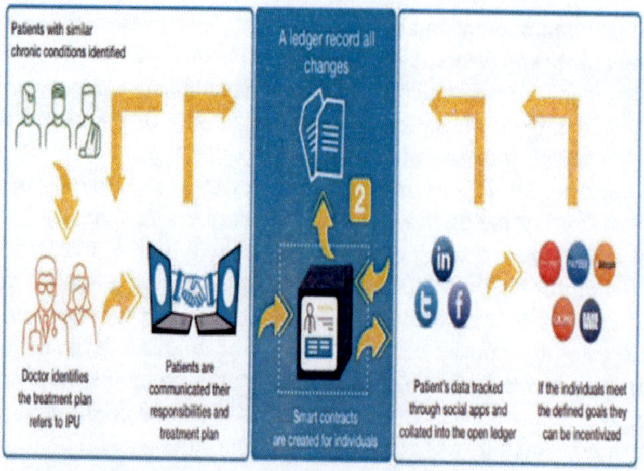

Fig. 4. Blockchain health care eco system [50].

Blockchain is the technology that can bring all the above mentioned technology to work as a single unit and can do fortune for the health care industry, In future 2050 solutions under the Blockchain technology layer can bring the signature of all records and data collected by whatever means on a single network and make it available for medical professional, pharmacies, drug industry and researches with an append only rule with all background and history to take action immediately and proactively. The current statues of health care records are disjoined but Blockchain is one technology that can allow safe, transparent share of sensitive information between authorities and facilities which was never been possible before several implementations and POC's demonstrate that blockchain technology embraces the aptitude to protect, unchallengeable, and enable easy access to medical records and has immense potential but also brings along with it complications and invites great challenges [51].

Challenges: As Blockchain is a social technology adopting it can't be more effective in isolation, hence a lot of work will be requiring to build consensus between major stakeholders to enable the impossible. Huge governance, regulations, data security and privacy is required to establish a Blockchain layer for all digital transactions which is an enormous challenge due to lack of vendors and experienced development team available to establish link between the technology and the legacy systems [52].

4 Literature Review

Healthcare industry and system are under great strain and with the passage of time they are facing greater complications due to increasing demand and spiraling cost of healthcare. Above discussed technologies has immense potential to overcome healthcare challenges, provide innovative and cost effective solutions, but all new technologies carry some adopting challenges, solution to each technology with 2050 future perspective will be discussed below:

4.1 Solution to MR Challenges

Mixed Reality in 2019 still is in its early stages. The challenges faced now can be overcome in future 2050 with mass production of units and price drop. Also legal and technical issues could be resolved with further POC in the healthcare industry, on the other hand medical facilities should take necessary steps by establishing special zones for the utilization of the technology at the same time create awareness of the implication of the technology and train professionals in parallel to adopt and led towards its perfection [39].

4.2 Solution to Robotics Challenges

Telemedicine is fastest moving trend for the healthcare industry and can play a major role for the future hospitals making them spacious, home alike on the other hand hospitals and facilities could focus on intensive care and robot-delivered surgeries. From literature review and evolving technologies expect most healthcare provisioning to be executed virtually rather in physical presence allowing quality, more efficient, convenient and cheaper access to healthcare. Robotic are the future for surgical procedures in medical fields in future, lot of research and work is in process to overcome the electromechanical restrictions of robotic systems, but till date 2019 work accomplished allows us to predict that in future the robotic system will be supported by much smaller instruments, overall less cost, shorter set-up time. In order to build public trust in for the future, robotics public engagement campaigns is a vital step that should be started as early as possible, training of healthcare professionals and public propagation towards positive robotic business cases could dissolve the negative impression and promote the recognition of robotic applications amongst healthcare staff and patients [42].

4.3 Solution to AI Challenges

Similarly, AI and machine leaning would require professionals to support the technology and perform special tasks in the future 2050, robots and machines would require human support for learning, adapting and mentoring based on data collect by AI and algorithms. Doctors will play a vital role in AI future by becoming part of the system supervising the progress of the machines and industry. Entire machine learning process would depend on medical professionals and doctors and they would be of great added value in the adoption of the technology in the future, hence techno specialist physicians would take over the medical landscape [53]. Also today and even in future patients will not prefer to communicate or respond to an algorithm/machine all the time, AI will be on role but to an extent, this answers the role of doctors and professional which would be vital in reviewing the diagnosis collected by AI, support and supervise patients to fulfil the treatment sequence remotely or physically coordinating and forming the interpersonal connections between the technology and the end users, which would be the most vital role for an effective health care eco system future by 2050 [37].

4.4 Solution to IoT Challenges

For IoT future in healthcare the only solution is incorporate security measures in the design of IoT devices, with proper risk assessment, appropriate POC before release, advance authentication mechanism between devices for data exchange with several security layer and device to device communication detailed logging and monitoring. It is known that any advance technology in earlier stage bring along with it new challenges, IoT devices and technology might get better and secure in the future but health care sector should analyze the devices before adopting them and take appropriate measure to secure them at the same time revamping the facilities infrastructure by future 2050 would be a must in order to utilize the advantage of all above mentioned technologies and incorporate the latest trends with legacy systems [49].

4.5 Solution to Blockchain Challenges

Blockchain technologies answer to all challenges in the future is can be achieved by collaboration, awareness and building consensus between participants, stakeholders and entities all security concerns and hurdles can be taken care off when everyone is a participating, the government, health care organization and all stakeholders who can play a valuable role as a health care provider should actively plans, share knowledge and experience to discuss progress, lessons learned, hurdles faced and work closely to set patterns for governance of a Blockchain health care eco system and identify unanswered questions [13].

5 Conclusion

Research is in process and towards all sectors of health care to comprehend and properly align each innovative idea and technology with medical aspects, with used cases and working examples to adopt and implementing applications and technology in the field. For technology like MR it is still in the early stage and yet to mature where MR devices are improving day by day. POC's and used cases are paving ways of interaction, adaptability and utilization also MR hardware is continuously evolving similar to the early evolution of smartphones which is a huge opportunity and experts promise huge capabilities and outcome, cost reduction, improved and enhanced services by utilizing MR technology [38].

Huge amount of data that is being collected and will be poised in the future will not just offer proactive medical support and services but great platform for researchers to expand the knowledge and improve the efficacy of drugs, medical care, symptoms also allow better understanding of development of cancer and other vital deceases. The introduction of Google Deep Mind, the machine learning arm is a great opportunity for future, as the search giant collaborating with Moor fields Eye Hospital can invite other experts of various industries to work close to accomplish the impossible. The idea here is to analyze huge amount of patient data for proactive detection of diabetes and age-related macular deterioration. Their technology is promises to analyze scans and data much more quickly than professional and doctors are doing at the moment, not to forget that the collected data can help to improve the treatment in the future 2050 [12]. For technologies like Robotics, AI and IoT, security and privacy of data that will be collect has the fear of leaking and misusage and has been found to be the biggest hurdle as of 2019 so for future 2050 software engineering giants, technology specialist and the entire medical industry should work towards to understand, resolve and take care of obstacles, required regulations and legal aspects involved should be highlighted at the earliest to enable the transition and adoption of the technology for a better future [53].

Infrastructure is the first and foremost domain that health care industry should start considering and proactively taking appropriate measures to tackle the challenges and updating the old infrastructure to state of art up to date infrastructure that can support hi-tech technologies in future 2050. It's a joint venture where technology giants, health care industry and medical professionals should deeply understand the challenges and build consensus around to enable effective measures and solutions for all stakeholders to develop, deploy and adapt technologies like IoT, AI, Robotics, MR and Blockchain that would make the impossible today possible by 2050.

Acknowledgment. This work is a part of a project undertaken at the British University in Dubai.

References

1. Alhashmi, S.F.S., Salloum, S.A., Abdallah, S.: Critical success factors for implementing artificial intelligence (AI) projects in Dubai Government United Arab Emirates (UAE) health sector: applying the extended technology acceptance model (TAM). In: Hassanien, A.E., Shaalan, K., Tolba, M.F. (eds.) AISI 2019. AISC, vol. 1058, pp. 393–405. Springer, Cham (2020). https://doi.org/10.1007/978-3-030-31129-2_36
2. Alshurideh, M., Al Kurdi, B., Abumari, A., Salloum, S.: Pharmaceutical promotion tools effect on physician's adoption of medicine prescribing: evidence from Jordan. Mod. Appl. Sci. **12**, 210–222 (2018)
3. Alhashmi, S.F.S., Salloum, S.A., Mhamdi, C.: Implementing artificial intelligence in the United Arab Emirates healthcare sector: an extended technology acceptance model. Int. J. Inf. Technol. Lang. Stud. **3**, 27–42 (2019)
4. Alhashmi, S., Alshurideh, M., Kurdi, B., Salloum, S.: A systematic review of the factors affecting the artificial intelligence implementation in the health care sector. In: Hassanien, A.-E., Azar, A.T., Gaber, T., Oliva, D., Tolba, F.M. (eds.) AICV 2020. AISC, vol. 1153, pp. 37–49. Springer, Cham (2020). https://doi.org/10.1007/978-3-030-44289-7_4
5. Alghizzawi, M., Habes, M., Salloum, S.A., et al.: The effect of social media usage on students' e-learning acceptance in higher education: a case study from the United Arab Emirates. Int. J. Inf. Technol. Lang. Stud. **3**, 13–26 (2019)
6. Habes, M., Salloum, S.A., Alghizzawi, M., Alshibly, M.S.: The role of modern media technology in improving collaborative learning of students in Jordanian universities. Int. J. Inf. Technol. Lang. Stud. **2**, 71–82 (2018)
7. Salloum, S.A., Al-Emra, M., Habes, M., Alghizzawi, M.: Understanding the impact of social media practices on E-learning systems acceptance (2019). https://doi.org/10.1007/978-3-030-31129-2
8. Alghizzawi, M., Salloum, S.A., Habes, M.: The role of social media in tourism marketing in Jordan. Int. J. Inf. Technol. Lang. Stud. **2**, 59–70 (2018)

9. AlShuweihi, M., Salloum, S., Shaalan, K.: Biomedical corpora and natural language processing on clinical text in languages other than English: a systematic review. In: Al-Emran, M., Shaalan, K., Hassanien, A.E. (eds.) Recent Advances in Intelligent Systems and Smart Applications. SSDC, vol. 295, pp. 491–509. Springer, Cham (2021). https://doi.org/10.1007/978-3-030-47411-9_27
10. Almansoori, A., AlShamsi, M., Salloum, S., Shaalan, K.: Critical review of knowledge management in healthcare. In: Al-Emran, M., Shaalan, K., Hassanien, A.E. (eds.) Recent Advances in Intelligent Systems and Smart Applications. SSDC, vol. 295, pp. 99–119. Springer, Cham (2021). https://doi.org/10.1007/978-3-030-47411-9_6
11. Mhamdi, C., Al-Emran, M., Salloum, S.: Text mining and analytics: a case study from news channels posts on Facebook. In: Shaalan, K., Hassanien, A.E., Tolba, F. (eds.) Intelligent Natural Language Processing: Trends and Applications. SCI, vol. 740, pp. 399–415. Springer, Cham (2018). https://doi.org/10.1007/978-3-319-67056-0_19
12. Economist T: The Future of Health care (2019)
13. Deloitte: 2019 Global Health care outlook (2019)
14. Aburayya, A., Alshurideh, M., Al Marzouqi, A., et al.: Critical success factors affecting the implementation of TQM in public hospitals: a case study in UAE Hospitals. Syst. Rev. Pharm. **11** (2020). https://doi.org/10.31838/srp.2020.10.39
15. Aburayya, A., Alshurideh, M., Alawadhi, D., et al.: An investigation of the effect of lean six sigma practices on healthcare service quality and patient satisfaction: testing the mediating role of service quality in Dubai primary healthcare sector. J. Adv. Res. Dyn. Control Syst. **12**, 56–72 (2020)
16. Salloum, S., Shaalan, K.: Adoption of E-book for university students. In: Hassanien, A.E., Tolba, M.F., Shaalan, K., Azar, A.T. (eds.) AISI 2018. AISC, vol. 845, pp. 481–494. Springer, Cham (2019). https://doi.org/10.1007/978-3-319-99010-1_44
17. Alnaser, A.S., Habes, M., Alghizzawi, M., Ali, S.: The Relation among Marketing ads, via Digital Media and mitigate (COVID-19) pandemic in Jordan The Relationship between Social Media and Academic Performance: Facebook Perspective View project Healthcare challenges during COVID-19 pandemic View project. DspaceUrbeUniversity (2020)
18. Salloum, S.A., Maqableh, W., Mhamdi, C., et al.: Studying the Social Media Adoption by university students in the United Arab Emirates. Int. J. Inf. Technol. Lang. Stud. **2**, 83–95 (2018)
19. Alshurideh, M., Al Kurdi, B., Salloum, S.A.: Examining the main mobile learning system drivers' effects: a mix empirical examination of both the expectation-confirmation model (ECM) and the technology acceptance model (TAM). In: Hassanien, A.E., Shaalan, K., Tolba, M.F. (eds.) AISI 2019. AISC, vol. 1058, pp. 406–417. Springer, Cham (2020). https://doi.org/10.1007/978-3-030-31129-2_37
20. Alghizzawi, M., Ghani, M.A., Som, A.P.M., et al.: The impact of smartphone adoption on marketing therapeutic tourist sites in Jordan. Int. J. Eng. Technol. **7**, 91–96 (2018)
21. Habes, M., Salloum, S.A., Alghizzawi, M., Mhamdi, C.: The relation between social media and students' academic performance in Jordan: YouTube perspective. In: Hassanien, A.E., Shaalan, K., Tolba, M.F. (eds.) AISI 2019. AISC, vol. 1058, pp. 382–392. Springer, Cham (2020). https://doi.org/10.1007/978-3-030-31129-2_35
22. AlGhanem, H., Shanaa, M., Salloum, S., Shaalan, K.: The role of KM in enhancing AI algorithms and systems. Adv. Sci. Technol. Eng. Syst. J. **5**, 388–396 (2020). https://doi.org/10.25046/aj050445
23. Yousuf, H., Salloum, S.: Survey Analysis: Enhancing the Security of Vectorization by Using word2vec and CryptDB
24. Mansoori, S., Salloum, S., Shaalan, K.: The impact of artificial intelligence and information technologies on the efficiency of knowledge management at modern organizations: a systematic review. In: Al-Emran, M., Shaalan, K., Hassanien, A.E. (eds.) Recent Advances in

Intelligent Systems and Smart Applications. SSDC, vol. 295, pp. 163–182. Springer, Cham (2021). https://doi.org/10.1007/978-3-030-47411-9_9
25. Yousuf, H., Zainal, A., Alshurideh, M., Salloum, S.: Artificial intelligence models in power system analysis. In: Hassanien, A.E., Bhatnagar, R., Darwish, A. (eds.) Artificial Intelligence for Sustainable Development: Theory, Practice and Future Applications, pp. 231–242. Springer, Cham (2021). https://doi.org/10.1007/978-3-030-51920-9_12
26. Salloum, S.A., Al-Emran, M., Abdallah, S., Shaalan, K.: Analyzing the Arab Gulf newspapers using text mining techniques. In: Hassanien, A.E., Shaalan, K., Gaber, T., Tolba, M.F. (eds.) AISI 2017. AISC, vol. 639, pp. 396–405. Springer, Cham (2018). https://doi.org/10.1007/978-3-319-64861-3_37
27. Salloum, S.A., Mhamdi, C., Al Kurdi, B., Shaalan, K.: Factors affecting the adoption and meaningful use of social media: a structural equation modeling approach. Int. J. Inf. Technol. Lang. Stud. **2**, 96–109 (2018)
28. AlShamsi, M., Salloum, S., Alshurideh, M., Abdallah, S.: Artificial intelligence and blockchain for transparency in governance. In: Hassanien, A.E., Bhatnagar, R., Darwish, A. (eds.) Artificial Intelligence for Sustainable Development: Theory, Practice and Future Applications. SCI, vol. 912, pp. 219–230. Springer, Cham (2021). https://doi.org/10.1007/978-3-030-51920-9_11
29. Alomari, K.M., Alhamad, A.Q., Mbaidin, H.O., Salloum, S.: Prediction of the digital game rating systems based on the ESRB. Opcion **35**, 1368–1393 (2019)
30. Salloum, S., Alshurideh, M., Elnagar, A., Shaalan, K.: Machine learning and deep learning techniques for cybersecurity: a review. In: Hassanien, A.-E., Azar, A.T., Gaber, T., Oliva, D., Tolba, F.M. (eds.) AICV 2020. AISC, vol. 1153, pp. 50–57. Springer, Cham (2020). https://doi.org/10.1007/978-3-030-44289-7_5
31. Salloum, S., Alshurideh, M., Elnagar, A., Shaalan, K.: Mining in educational data: review and future directions. In: Hassanien, A.-E., Azar, A.T., Gaber, T., Oliva, D., Tolba, F.M. (eds.) AICV 2020. AISC, vol. 1153, pp. 92–102. Springer, Cham (2020). https://doi.org/10.1007/978-3-030-44289-7_9
32. lshurideh, M., Kurdi, B., Salloum, S., Arpaci, I., Al-Emran, M.: Predicting the actual use of m-learning systems: a comparative approach using PLS-SEM and machine learning algorithms. Interact. Learn. Environ., 1–15 (2020). https://doi.org/10.1080/10494820.2020.1826982
33. Alshurideh, M., Salloum, S.A., Al Kurdi, B., et al.: Understanding the quality determinants that influence the intention to use the mobile learning platforms: a practical study. Int. J. Interact. Mob. Technol. **13** (2019). https://doi.org/10.3991/ijim.v13i11.10300
34. Salloum, S.A., Al-Emran, M., Khalaf, R., et al.: An innovative study of E-payment systems adoption in higher education: theoretical constructs and empirical analysis. Int. J. Interact. Mob. Technol. **13**, 68 (2019)
35. Deloitte: Tech Trends 2019 (2019)
36. Zainal, A., Yousuf, H., Salloum, S.: Dimensions of agility capabilities organizational competitiveness in sustaining. In: Hassanien, A.-E., Azar, A.T., Gaber, T., Oliva, D., Tolba, F.M. (eds.) AICV 2020. AISC, vol. 1153, pp. 762–772. Springer, Cham (2020). https://doi.org/10.1007/978-3-030-44289-7_71
37. Patel, A., Patel, R., Singh, N., Kazi, F.: Vitality of robotics in healthcare industry: an internet of things (IoT) perspective. In: Bhatt, C., Dey, N., Ashour, A.S. (eds.) Internet of Things and Big Data Technologies for Next Generation Healthcare. SBD, vol. 23, pp. 91–109. Springer, Cham (2017). https://doi.org/10.1007/978-3-319-49736-5_5
38. Accenture: Accenture Technology (2019)
39. Bilyk, V.: Augmented Reality Issues - What you need to know (2019)
40. Aburayya, A., Alshurideh, M., Al Marzouqi, A., et al.: An empirical examination of the effect of TQM practices on hospital service quality: an assessment study in UAE Hospitals

41. Dolic, Z., Castro Am, R.: Robots in healthcare: a solution or a problem? (2019)
42. Sheikh KCC-BS: Healthcare robotics a qualitative exploration of key challenges and future directions (2018)
43. Saeed Al-Maroof, R., Alhumaid, K., Salloum, S.: The continuous intention to use E-learning, from two different perspectives. Educ. Sci. **11**, 6 (2020)
44. Al-Maroof, R., Salloum, S.: An integrated model of continuous intention to use of google classroom. In: Al-Emran, M., Shaalan, K., Hassanien, A.E. (eds.) Recent Advances in Intelligent Systems and Smart Applications. SSDC, vol. 295, pp. 311–335. Springer, Cham (2021). https://doi.org/10.1007/978-3-030-47411-9_18
45. Al-Maroof, R., Arpaci, I., Al-Emran, M., Salloum, S., Shaalan, K.: Examining the acceptance of WhatsApp stickers through machine learning algorithms. In: Al-Emran, M., Shaalan, K., Hassanien, A.E. (eds.) Recent Advances in Intelligent Systems and Smart Applications. SSDC, vol. 295, pp. 209–221. Springer, Cham (2021). https://doi.org/10.1007/978-3-030-47411-9_12
46. Aburayya, A., Salloum, S.A.: The effects of subjective norm on the intention to use social media networks: an exploratory study using PLS-SEM and machine learning approach
47. Salloum, S., AlAhbabi, N., Habes, M., Aburayya, A., Akour, I.: Predicting the intention to use social media sites: a hybrid SEM - machine learning approach. In: Hassanien, A.-E., Chang, K.-C., Mincong, T. (eds.) Advanced Machine Learning Technologies and Applications: Proceedings of AMLTA 2021, pp. 324–334. Springer, Cham (2021). https://doi.org/10.1007/978-3-030-69717-4_32
48. Chowdary, R.: IoT in Healthcare: 20 Examples That'll Make You Feel Better (2019)
49. Matthews, K.: 5 Challenges Facing Health Care IoT in 2019 (2018)
50. Capgimini: Blockchain a Healthcare Industry View (2017)
51. Heston, T.: A case study in blockchain healthcare innovation (2017)
52. Eurocoinpay.io: Blockchain challenges in 2019 (2019)
53. Rushabh Shah, A.C.: Issues in Information System (2017)

Precision Education Approaches to Education Data Mining and Analytics: A Review

Abdulla M. Alsharhan[1] and Said Salloum[2(✉)]

[1] Faculty of Engineering and IT, The British University in Dubai, Dubai, UAE
[2] School of Science, Engineering, and Environment, University of Salford, Salford, UK
S.A.S.Salloum@edu.salford.ac.uk

Abstract. The COVID-19 pandemic has enabled a major digital transformation in education, which has resulted in an enormous amount of educational data and has allowed us to acquire rich insights on each student. Simultaneously, the development witnessed by the fields of artificial intelligence, machine learning, data mining, and data analytics can provide new opportunities to improve the current state of education. However, will recent developments allow for a transformation from one size fits all to a more personal learning experience tailored to each student? Precision education is an emerging field that has been borrowed from the field of precision medicine. This research aims to provide a systematic review of the current literature on precision education in the context of EDM. The results indicate that the field of precision education that started in 2016 is still relatively young. Most research has focused on introducing intervention methods, and prediction is the most frequently-used task in mining data related to precision education. Taiwan is most interested in this field. Several challenges have been identified, and future research opportunities in this field have been highlighted, the most prominent of which provides immediate personal feedback.

Keywords: Precision education · Educational Data Mining · EDM · Learning Analytics · LA · Big data · Knowledge discovery · Smart learning · Self-regulated learning · Personal education

1 Introduction

In February 2020, the World Health Organization (WHO) announced COVID-19 as a global pandemic. However, the world has seen unprecedented measures taken by governments worldwide, including cutting off 1.5 million students from physical attendance at schools and Universities [1–3]. These measures have helped in accelerating digital transformation practices in all fields [4]. The emergence of distance education was an adaptive response to the changing social norms and lifelong learning. Since then, a massive amount of educational data has been generated, which provides abundant opportunities for mining and harnessing big educational data [5, 6]. The uses of Learning Analytics (LA) and Educational Data Mining (EDM) [7] could provide potentially

© The Author(s), under exclusive license to Springer Nature Switzerland AG 2022
A. E. Hassanien et al. (Eds.): AMLTA 2022, LNDECT 113, pp. 337–356, 2022.
https://doi.org/10.1007/978-3-031-03918-8_30

beneficial information from large amounts of unstructured data. The educational insights generated from these data can improve online learning platforms, educational content, and learning activities, ultimately leading to more effective education [8, 9].

However, integrating big data in education can also highlight some emerging research gaps. If EDM and LA can extract all this information about each student, will it also allow a personalized learning experience? How to process big educational data to benefit students at the individual and micro-level? [10]. Some current researchers suggest that education is already moving from one size fits all to precision education and a more personalized learning experience ([11, 12]). The idea is similar to the emerging field of precision medicine. Big data can be used to determine patterns related to a specific group of patients, where preventive measures and treatment can be customized specifically for them [13]. Similarly, precision education can predict student performance and provide timed intervention to improve learning. Precision education is among the new educational models that consider personalized learning. Precision education brings new challenges to applying artificial intelligence [14], machine learning, and LA to improve teaching quality and overall learning performance [15, 16]. However, precision education is still in its infancy stage [17], as little is known about the relationship between EDM and precise education. This work attempts to provide the first step toward understanding EDM's role in the personalized learning experience, and identifying current trends, challenges, and possible research gaps. More specifically, this work aims to answer the following research questions:

- RQ1. What are the trends of articles published on precision education in the EDM context?
- RQ2. What are the main research purposes related to precision education in the context of EDM?
- RQ3. What are the main EDM tasks used in precision education?
- RQ4. What data mining tools have been used in precision?
- RQ5. What is the application of data mining in precise education?
- RQ6. What are some of the challenges faced by precision education?
- RQ7. What are the future directions for research in precision education using EDM?

The remainder of this paper is structured as follows. First, we outline the background and related work. This is followed by detailing the methodology, and discussing the results obtained. We then conclude and highlight some recommendations that merit future research.

2 Literature Review

2.1 Background

2.1.1 The Rise of Precision Care

In 2008, The President's Council of Advisors on Science and Technology identified precision medicine as a top priority for health care research, and defined it as the tailoring of medical treatment to the individual characteristics of each patient [18–20]. In 2015, the United States unveiled the Precision Medicine Initiative under the Obama administration [21], which was also synchronized with a similar effort in the United Kingdom to initiate Precision Medicine Catapult (PMC) initiative [22]. The basic concept behind this initiative is to individualize treatment and prevention based on genetics, environment, and lifestyle.

2.1.2 From Medicine to Other Fields

Many scholars believe that this approach should not be limited to biomedical diseases only. A personalized diagnostic and treatment approach would be ideal for educational purposes as well. For example, the precision education approach can help to classify and treat learning disabilities (LDs), and could be used in the everyday practice of education. Moreover, learning disabilities, which are a diverse group of mental disorders, are complex disorders with remarkable similarities to biomedical diseases. Biomedical diseases and LDs are characterized by a quantitative pattern of psychological, genetic, and environmental risk factors, which act independently and in an interactive manner. For both, treatment is complicated by individual differences in etiology and responses to treatment [11].

2.1.3 Personalized Education

The term ‘personalized education’ may refer to instructions that adapted teaching based on the need of each learner, and where the educational goals, methods, content, and sequence may change based on the learner’s needs. The learning activities are driven by the learner’s interests, and are often self-regulated [23]. In contrast, the term ‘precision education’ is relatively new. Figure 3 shows the number of published articles about precision education topics, which are relatively low, but growing rapidly.

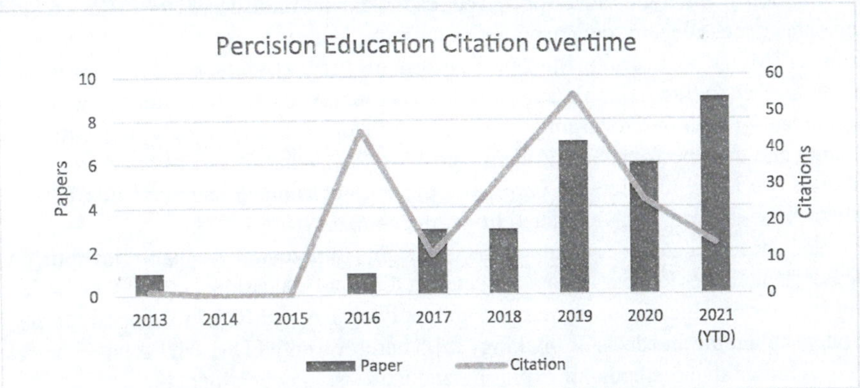

Fig. 1. Precision education citation overtime in Google Scholar

2.1.4 Precision Education

Precision education may refer to "engaging in efforts to acquire the right intervention in place for the right person for the right reason" [22]. Unlike the traditional "one size fits all" approach to teaching, precision education aims to provide individuals with personalized feedback and learning plans according to their learning profile. Precision education studies include four stages: diagnosis, prediction, treatment, and prevention [24] (Fig. 2).

Fig. 2. Precision education stages

In order to design effective learning in precision education, EDM and LA contributed not only as dashboards and intervention tools, but also as conceptual frameworks that guide research experiences [25].

2.2 Theoretical Framework

2.2.1 Educational and Technological Advancement

Artificial intelligence and machine learning technologies in education and psychology have led to significant developments in related fields such as educational intelligence, self-organized learning, and personalized education [25]. The recent breakthroughs in information technologies have allowed educators to have more access to big data, including data generated from social media, Open Course Ware (OCW), Massive Open Online Courses (MOOCs), Learning Management systems (LMSs), and sensors. Mobile devices alone are generating dynamic and complex personal records. Education has witnessed significant growth in the volume of data derived from students' interactions with technology and their personal and academic profiles [26]. What sets EDM apart from other fields is that it includes detailed particulars that are not available in other fields. Figure 1 shows how data are being extracted from educational systems [27].

Educational big data contains valuable insights to potentially enhance learning outcomes. Due to the richness of data collected in different educational settings, an increasing number of educational institutions are applying LA and EDM to support learners' strategic planning and decision-making [26]. Therefore, both EDM and LA have emerged to develop unique methods for exploring and harnessing educational big data.

2.2.2 Handling Big Data

Big data refers to the vast amount of data that is characterized by the four Vs: Volume (data size), Variety (data type), Velocity (data creation and transmission speed), and Veracity (data accuracy). With these four attributes of big data, it cannot be analyzed by traditional methods. On the other hand, machine learning has emerged as a subfield of

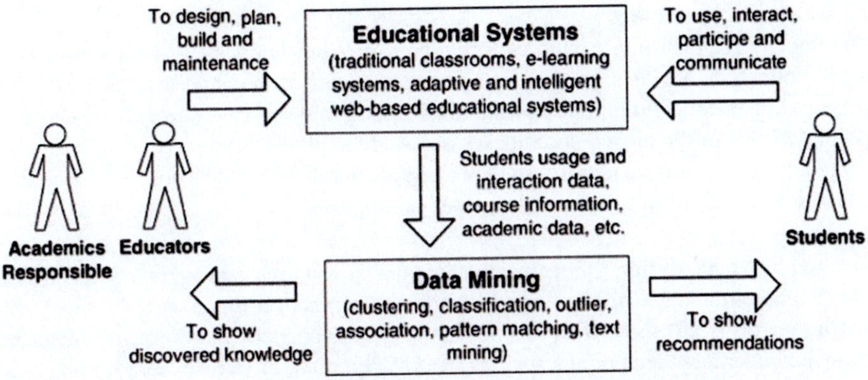

Fig. 3. Data mining application cycle in educational system [27].

artificial intelligence (AI). Machine learning focuses on building computer systems that can learn from big data and adapt to it without explicit programming [12].

2.2.3 Educational Data Mining

The educational data mining (EDM) community website[1] defines it as "an emerging discipline concerned with developing methods for exploring the unique and increasingly large-scale data that come from educational settings, and using those methods to understand students better, and the settings which they learn in". The application of EDM can identify learning patterns and students' behaviors, improve evolution methods, predict learning performance, and provide personalized support. It can also provide support to teachers by enhancing course planning, curriculum development, and teaching assistance. Moreover, it can offer teachers a live dashboard while students are working in real-time [28]. EDM usually applies the following methodologies:

- **Classification:** Classification is a training and testing method that classifies collected data together in predefined groups, and is helpful in predicting student performance, analyzing risks, monitoring students' systems, and error detection.
- **Clustering:** Similar to classification, but not in predefined discrete categories. This is useful when distinguishing between student learning styles.
- **Statistics:** Provide valuable insights for learning management systems to highlight extreme deviations from the mean.
- **Prediction:** A technique used to forecast success and dropout rate, in addition to designing the necessary intervention for student retention [29].
- **Relationship Mining:** Also known as Association Rule Mining: A tool used to identify the relationship between different variables for new knowledge discovery. This is useful to identify which variables are the most strongly associated with a single variable such as student's reasons for failure [30].

[1] www.educationaldatamining.org.

2.2.4 Learning Analytics

The Society for Learning Analytics Research (SoLAR)[2] defines LA as "the measurement, collection, analysis, and reporting of data about learners and their contexts, for purposes of understanding and optimizing learning and the environments in which it occurs". Some of the most common uses of LA are to predict student academic success and, more specifically, to identify students who are at risk of failing a course or dropping out of school. LA usually applies three methodologies:

- **Descriptive Analytics:** Offers insights into the past. It uses data aggregation and data extraction to understand trends and evaluative metrics over time.
- **Diagnostic Analysis:** Offers insights on why it happened. This form of advanced analytics features techniques such as data search, data discovery, data mining, and correlations to examine data or content to answer the question: "Why did this happen?"
- **Prescriptive Analysis:** Offers advice on potential outcomes by recommending one or more options using a combination of machine learning, algorithms, business rules, and computational modeling [31] (Fig. 4).

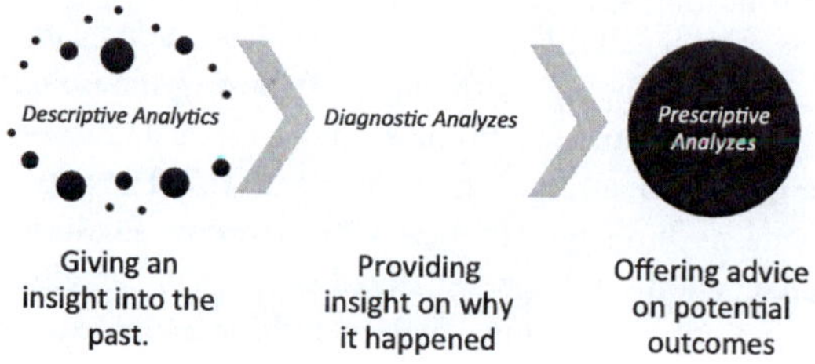

Fig. 4. LA methodologies

The use of LA and EDM could provide potentially beneficial information from large amounts of unstructured data. LA and EDM can enhance and improve the design of online learning platforms, educational materials and activities to ensure greater educational effectiveness and improve learning environments.

2.3 Related Work

EDM has attracted considerable attention from different parts of the world. Several systematic reviews on EDM were conducted in the last several years.

The work of [32] identified four major research trends in EDM empirical research: 1) Pedagogy-oriented issues from the student behavior perspective; 2) contextualization of the learning environment; 3) networked learning patterns and interactions; and 4) educational resources handling and recommendation.

The work of [33] focused on pedagogy-oriented issues by systematically reviewing EDM papers. The primary purpose was to identify EDM application areas and factors

[2] https://www.solaresearch.org/.

affecting student performance. The results indicate that no standard tools that could be used among educational institutes, which raised the need to standardize the EDM process. The work also identified research opportunities for designing a generalized framework and parameters that impact the learning process, which can eventually enhance the overall quality of education. Moreover, it highlighted classification as the most commonly used technique among researchers, which is also consistent with [34] findings.

Deep learning applications in EDM were the primary focus in both [34] and [35] reviews. The authors of [34] claimed to have compiled the first systematic literature review (SLR) of its kind. The objectives were to determine if deep learning has started to be applied in EDM, and the leading educational tasks in EDM that deep learning could address. The work indicates 2015 as the start of the application of deep learning applications on EDM, but with very little research.

The authors of [35] SLR aimed to identify the EDM tasks that have benefited from deep learning. The study indicates four out of thirteen tasks that have benefited from deep learning: 1) predicting students' performance, 2) detecting undesirable student behaviors, 3) generating recommendations, and 4) automatic evaluation. The remaining nine tasks represent research opportunities: 1) profiling and grouping students, 2) social network analysis, 3) providing reports, 4) creating alerts for stakeholders, 5) planning and scheduling, 6) creating courseware, 7) developing concept maps, 8) adaptive systems, and 9) scientific inquiries.

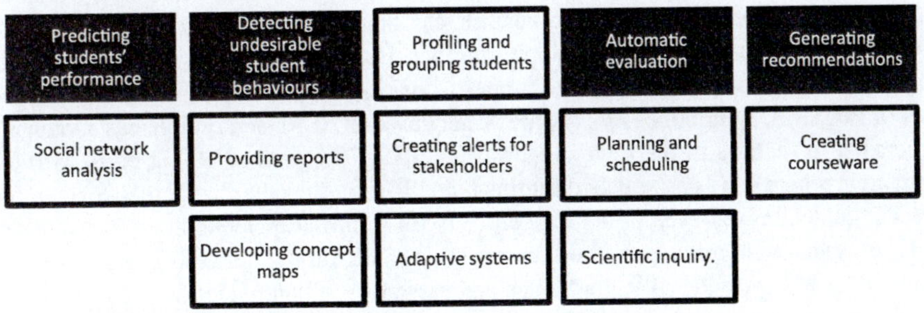

Fig. 5. EDM tasks that have benefited from deep learning (highlighted in blue)

The work of [36] compiled an SLR between 1983–2016, and found that most studies covered subject-specific research, but missed domain-specific EDM research such as EDM applications for improving organizational performance. The paper also highlighted research opportunities in several areas such as "learner annotation, the effect of classroom decoration to augment learning and teaching, implications of education affordability, the inclusion of semantic web in educationist usability, learner motivation, timetabling, examination scheduling, student profiling, and intelligent tutor systems" [36].

Another noteworthy SLR on using EDM in measuring self-regulated learning during the period of 2011–2019 was conducted by [37]. Their work highlighted an increasing interest in using EDM in self-regulated learning. Besides, time management was the most measured self-regulated learning behavior. Furthermore, a growing trend on linking new data with other self-reported measures to gain data triangulation was discussed.

The authors of [38] provided an SLR on the use of EDM in predicting university dropouts. The study identified six classification techniques: 1) decision trees, 2) K-nearest neighbors, 3) support vector machines, 4) Bayesian classification, 5) neural networks, and 6) logistic regression. The findings show that decision trees method was the most commonly used technique, followed by Bayesian classification and neural networks. Besides, it also identified 14 DM tools, and the most used ones are WEKA, followed by SPSS and R. Moreover, the study suggested open opportunities in determining model evaluation metrics and the level of reliability reached by the techniques used.

After analyzing the trends of EDM-related works in SLRs, some covered aspects of personalized learning and precision education such as prediction and self-regulated learning, but not as a whole concept. This opens a research opportunity to conduct an SLR specifically on precision education approaches to educational data mining and data analytics.

3 Method

An SLR is commonly used in fields that have rapid advancement, such as medical and information technology. SLR stands out from traditional methods in its abidance to a strict and explicit methodology [39]. One of the advantages of applying an SLR methodology is to capture the recent advancements and summarize the latest works on a specific research area. Moreover, an SLR can help identify research gaps in current research trends and future research opportunities. Besides, it provides the related framework and acts as a reference point to properly position new research [40]. The presented set of PRISMA[3] guidelines by [41] are widely used in SLRs and establishes a valuable checklist to include in the review. However, PRISMA seems to be biased to medical research, where it was established for this field [34]. On the other hand, there exist more recent guidelines to develop SLRs specifically for information technology. The work of [42] provides an eight-step method for writing an SLR for the field of IT. Moreover, the authors of [43] provide SLR guidelines dedicated to computer science research, which this work follows.

3.1 Identifying the Purpose

The purpose of this SLR is to identify the impact of EDM on precision education. More specifically the objectives of this work are:

1. To identify trends of articles published on precision education in the context of EDM.
2. To identify the primary research purposes related to precision education in the context of EDM.
3. To review the main EDM research objectives used in precision education.
4. To recommend the most EDM tasks in precision education.
5. To review the applications of data mining for precision education.

[3] Preferred Reporting Items for Systematic Reviews and Meta-Analyses.

6. To identify some of the challenges faced by EDM in precision education.
7. To highlight recommendations and possible research directions for precision education using EDM.

3.2 Selection Criteria

The reviewed literature must meet the criteria in Table 1 to be selected for inclusion.

Table 1. Inclusion and exclusion criteria

Inclusion criteria	Exclusion criteria
Articles published between 2016–2021	Articles published beyond 2016
English articles	Articles published in other languages
Articles that have full access	Articles that have limited access
Articles that are related to the computer science field	Articles that are not related to computer science
Must be related to EDM, big data or educational analytics and precise education	Papers related to precise education but not related to EDM, big data or learning analytics

3.3 Data Sources and Search Strategies

The articles in the literature were obtained via the following seven prominent databases related to the information technology field: WorldCat.org, IEEE, Emerald, ScienceDirect, Scopus, Google Scholar, Microsoft Academic, and ACM Digital Library.

The study was conducted in April 2021, using the following searching terms: "precision education" AND ("big data" OR "learning analytics" OR "data mining" OR "Machine Learning" OR "Deep Learning"). The period was set to be between 2016–2021. The results returned 204 results, as shown in Table 2 (Fig. 6).

Table 2. Initial search results

Online database	Results
Google Scholar	155
Microsoft Academic	19
Scopus	18
WorldCat	11
IEEE	2

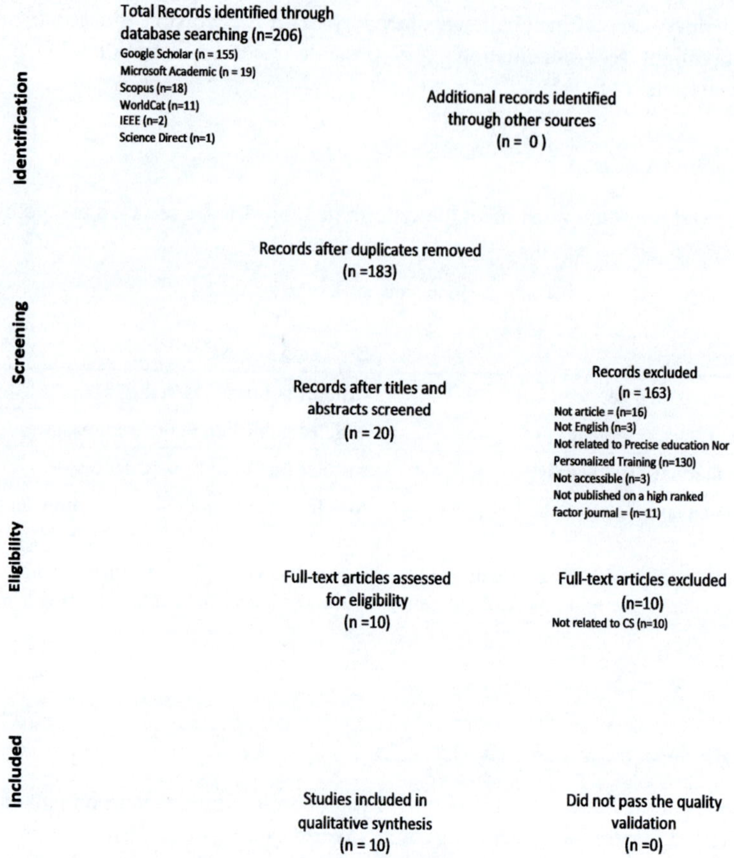

Fig. 6. Research PRISMA model

3.4 Quality Assessment

A significant step in an SLR is to appraise quality. This is also called screening for exclusion. Quality assessment criteria need to be cleared out to judge any work that will be excluded at this level for lacking quality [42]. Table 3 represents a list of several criteria used to judge the quality of the paper. The quality checklist is not meant to criticize the work of any researcher, but rather to ensure the papers have sufficient data for further analysis. The points assigned to each answer were 1 for every "Yes", 0 for every "No," and 0.5 for partly-answered questions.

The following table represents the results of the quality assessment questions for then 10 studies considered in this SLR. All filtered studies scored above 80%, indicating that they are qualified for further analysis (Table 4).

Table 3. Research questions

#	Question
1	Are the research objectives stated explicitly?
2	Are the methodologies undertaken detailed?
3	Does the paper provide clear findings?
4	Does the paper have novel contributions to add to the literature?
5	Was the paper published in a journal with a high impact factor (Q2-Q1)?

Table 4. Quality assessment score

Study	Q1	Q2	Q3	Q4	Q5	Total
S01	1	1	1	0.5	1	90%
S02	1	1	1	1	1	100%
S03	1	0	1	1	1	80%
S04	1	1	1	0.5	1	90%
S05	0.5	1	1	1	1	90%
S06	1	1	1	1	1	100%
S07	1	1	1	1	1	100%
S08	1	1	1	1	1	100%
S09	1	1	1	1	1	100%
S10	1	1	1	1	1	100%

3.5 Data Analysis and Coding Framework

The paper follows a data analysis system that uses a systematic methodology, following the use of data summarization and visualization. Since this SLR focuses on the relationship between precision education and EDM, ten papers fulfilled the selection criteria. Key data attributes were highlighted, such as main research purpose, EDM tasks, critical findings, EDM applications, EDM tools, and country.

4 Results

4.1 RQ1. What are the Trends of the Articles Published on Precision Education in the EDM Context?

Although the survey covered the last five years (2016–2021), there was no sign of articles issued in high impact factor publications before 2020. This suggests that precision education is a relatively new field in data and computer science, but it is snowballing. There was one study identified in 2020 focusing on dropout prediction [44], with 11 citations. The number increased to nine published studies as of May 2021, indicating that precision education is attracting more researchers to examine new research opportunities (Fig. 7).

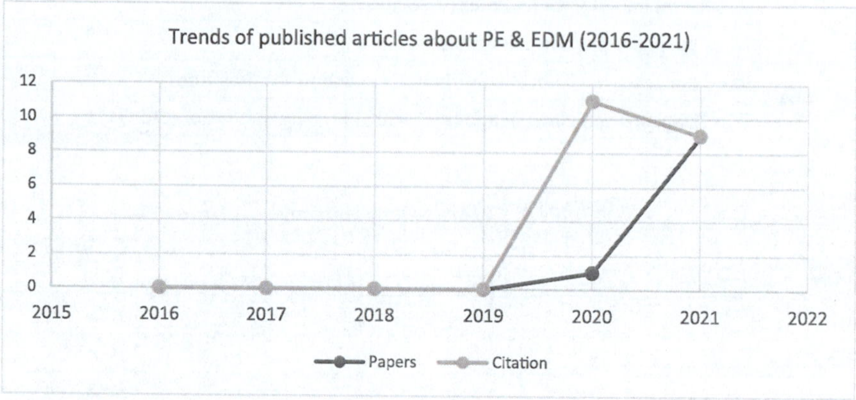

Fig. 7. Trends of published articles about precision education & EDM (2016–2021)

The following chart represents the research density by country. Most research concentration of precision education can be found in Far East countries. Taiwan has the most published research around precision education (n = 4), followed by Japan (n = 2). It can also be noticed that countries such as the United States, the United Kingdom, and China have not started research activities around this topic (Fig. 8).

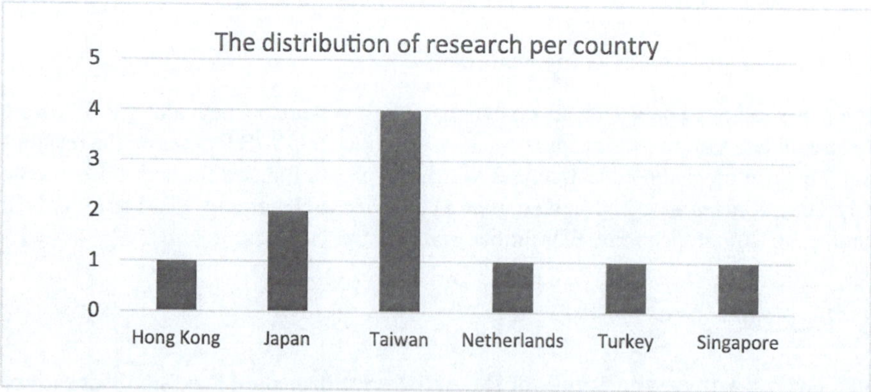

Fig. 8. The distribution of research per country

Most published topics around precision education seem to be dominated by one publication, “Education Technology & Society” (n = 9). In addition to one article published in the “International Journal of Educational Technology in Higher Education”. Both journals are classified with a high-ranking impact factor (Q1) from Scimago Journal and Country Rank (SJR). Table 5 presents the details of these publications.

Table 5. Top journals focusing on PE research

SJR	Country	Journal	Publisher	No. of papers
Q1	Taiwan	Educational Technology & Society	National Taiwan Normal University	9
Q1	Colombia	International Journal of Educational Technology in Higher Education	Springer	1

4.2 RQ2. What are the Main Research Purposes Related to Precision Education in the Context of EDM?

The analysis in the following chart indicates that most of the studies' objectives involve providing intervention (n = 4), followed by offering recommendation (n = 3). Two of the remaining studies provided literature review: one of them on personalized education in the context of language learning [45], while the second focused on the use of machine learning for precision education [46] (Fig. 9).

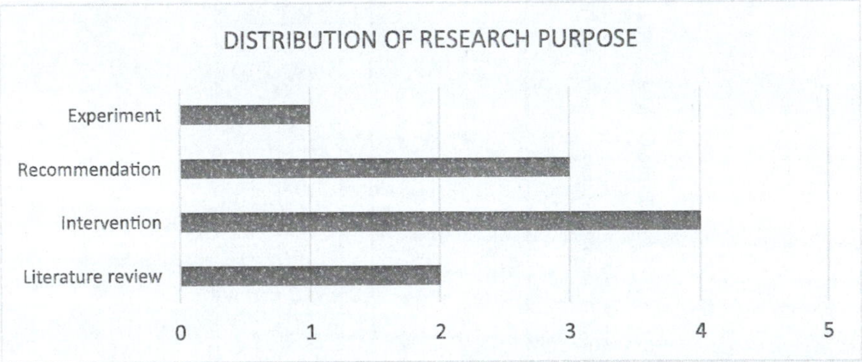

Fig. 9. Distribution of research purpose

4.3 RQ3. What are the Main EDM Tasks that are Used in Precision Education?

The following table presents the main DM tasks used in precision education in the surveyed papers. Most papers used a combination of several DM techniques. Predication was the most DM task used (n = 7), followed by association (n = 4), learning patterns (n = 4) and clustering (n = 4) (Table 6).

Table 6. Main DM tasks used in precision education

Source		Prediction	Association	Patterns	Clustering	Statistics	Classification	Regression	Social Network analysis	Text mining	Visualization	Decision-trees	Neural networks	Structural topic modeling	Mann-Kendall trend test
S01	[45]		x			x			x	x				x	x
S02	[24]	x	x	x	x										
S03	[47]	x						x			x				
S04	[46]	x													
S05	[48]	x		x		x									
S06	[25]	x	x	x	x										
S07	[44]	x				x		x					x		
S08	[10]				x		x		x	x	x				
S09	[26]		x	x	x										
S10	[49]	x					x					x			

4.4 RQ4. What Data Mining Tool Have Been Used in Precision Education?

Several data mining tools were used in PE. The most noticeable ones are R (n = 2) and BookRoll (n = 2). It was also noticed that some studies have combined several methods such as the work of [26] and [10] (Fig. 10).

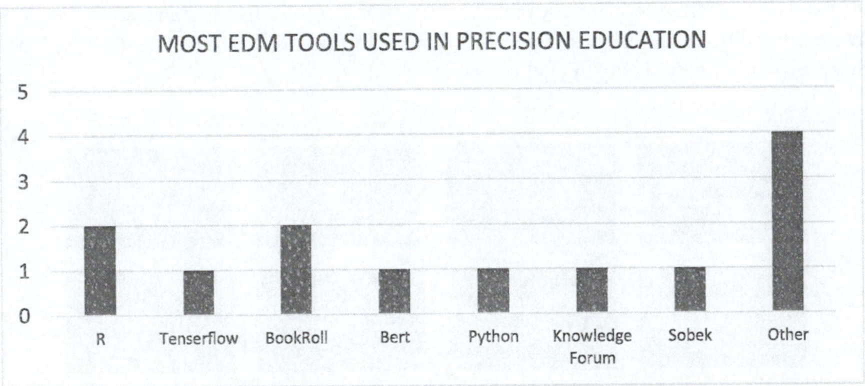

Fig. 10. Most EDM tools used in precision education

4.5 RQ5. What is the Application of Data Mining in Precise Education?

The following table shows the distribution of research applications and the different contexts for applying precision education (Table 7).

Table 7. Distribution of research purpose and their applications

	Educational task	Context
S01	Literature review	Personalized language learning
S02	Intervention	Self-regulated learning
S03	Recommendation	Symbiotic learning
S04	Literature review	Machine learning
S05	Recommendation	Dispositional LA
S06	Intervention	Assignment submission behavior
S07	Experiment	Dropout prediction
S08	Intervention	Idea identification
S09	Intervention	Ebook learning & Learning pattern
S10	Recommendation	Students' career planning

4.6 RQ6. What are Some of the Challenges Faced by Precision Educations?

Precision education is already a challenge in applying AI, ML techniques and LA to enhance overall teaching quality and learning performance [24–26, 45]. The work of [45] highlighted several issues faced by precision education, including data privacy policies, AI model training, content customization and instructor training, while the authors of [47] and [25] highlighted Generalizability. The work of [47] also mentioned that lack of internal validity, immediacy, transferability and interpretability are key challenges faced by precision education (Fig. 11).

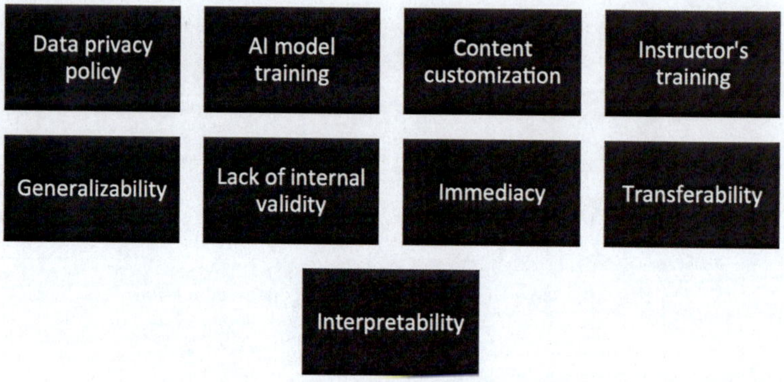

Fig. 11. Key obstacles faced by precision education

4.7 RQ7. What are Possible Future Directions for Research on Precision Education Using EDM?

Further directions for precision education research can potentially focus on immediate personalized feedback and assessment. Some recent work [45–47] focused on personalized recommendations, personalized context-aware ubiquitous learning systems, mobile chatbots, personalized content generation for game-based learning, AI for personalized diagnosis and adaptation, personalized LA dashboards, personalized practice in data-driven learning [45], in addition to advancing the development of intelligent agents based on rules and multimodal and multiple-source inputs [47]. Besides, future research could on focus on machine learning and neuroscience (e.g., EEG signals, eye-movement data) and lower education levels such as kindergarten, and elementary and secondary students [46]. Moreover, to overcome the generalizability issues faced by future precision, [25] suggest future research to be conducted in the context of open online courses. In addition, further work is suggested to incorporate more variables related to students' engagement, such as family, learning behavior aspects [44], self-regulation skills, and cognitive differences [25].

5 Discussion

After analyzing the primary results of this research, it can be said that the field of precision education is still vague, as it needs to have more definitive theoretical frameworks. Perhaps the work of [47] provides a first step toward achieving this goal. Some can argue that this field is either misunderstood by researchers or just another trendy interchangeable term for EDM and LA [17], as it includes many of the tools already used in the previous two areas such as prediction and diagnosis. One of the areas that researchers have referred to as a future research opportunity is the field of immediate personalized feedback, which seems to be in common with one of the serious gaming features (gaining immediate explicit feedback on performance) [50]. Gamification can allow a larger role to be given to the learner. With the right immediate feedback and recommendation system in place, precision education can be efficient. What should really make precision education stand out is offering an individual personal experience from the student's point of view, and not just within the four stages highlighted in Fig. 5. Unlike precision medicine, learners have a greater role in being active participants in the process. This cannot be initiate without considering the student being in the driving seat in many cases.

5.1 Limitations

The selection of the final papers was limited to publications with a high impact factor. This might lead to exclude some significant literature published in preprint repositories such as "arxiv.org" or recent conference proceedings that were not yet ranked or cited. The selection was also limited to open access papers, which might exclude papers from high impact computer science databases. There are also concerns that the results are very recent and mostly from one journal. The fields of "precision education" and "personalized education" are relatively new fields to computer science, or could be known by different terms or keywords.

6 Conclusion

In this paper, ten journal articles related precision education and EDM were systematically reviewed. It shows this field is relatively new, but accelerating rapidly. The study revealed several research opportunities in precision education. The results indicate that most research papers provide intervention, focus on prediction, which is in line with prior findings [46]. Moreover, the application of data mining in precision education was reviewed and several applications were identified. But there was no particular trend identified. Several challenges faced by EDM in precision education were identified. Precision education is already a challenge in applying AI, ML and LA to enhance teaching quality. Immediate personalized feedback and assessment were identified to be promising future directions for research on precision education. This review provides useful information for data scientists and educators on the current precision education applications, issues, and future research directions.

Acknowledgment. This study could not be possible without the guidance of Professor Sherief Abdallah. It is a part of a project done in the British University in Dubai. The authors declare no conflict of interest.

References

1. Akour, I., Alshurideh, M., Al Kurdi, B., et al.: Using machine learning algorithms to predict people's intention to use mobile learning platforms during the COVID-19 pandemic: machine learning approach. JMIR Med. Educ. **7** (2021). https://doi.org/10.2196/24032
2. Taryam, M., Alawadhi, D., Aburayya, A., et al.: Effectiveness of not quarantining passengers after having a negative COVID-19 PCR test at arrival to Dubai airports. Syst. Rev. Pharm. **11**, 1384–1395 (2020)
3. Alshurideh, M.T., Al Kurdi, B., AlHamad, A.Q., et al.: Factors affecting the use of smart mobile examination platforms by universities' postgraduate students during the COVID 19 pandemic: an empirical study. In: Informatics. Multidisciplinary Digital Publishing Institute, p. 32 (2021)
4. Kupchina, E.: Distance education during the Covid-19 pandemic. In: Proceedings of INTCESS 2021 8th International Conference on Education and Education of Social Sciences (2021)
5. Habes, M., Salloum, S., Alghizzawi, M., Mhamdi, C.: The relation between social media and students' academic performance in Jordan: YouTube perspective. In: Hassanien, A.E., Shaalan, K., Tolba, M.F. (eds.) AISI 2019. AISC, vol. 1058, pp. 382–392. Springer, Cham (2020). https://doi.org/10.1007/978-3-030-31129-2_35
6. Alghizzawi, M., Habes, M., Salloum, S.A.: The Relationship between digital media and marketing medical tourism destinations in Jordan: Facebook perspective (2020)
7. Salloum, S., Alshurideh, M., Elnagar, A., Shaalan, K.: Mining in educational data: review and future directions. In: Hassanien, A.-E., Azar, A.T., Gaber, T., Oliva, D., Tolba, F.M. (eds.) AICV 2020. AISC, vol. 1153, pp. 92–102. Springer, Cham (2020). https://doi.org/10.1007/978-3-030-44289-7_9
8. Salloum, S.A., Mhamdi, C., Al Kurdi, B., Shaalan, K.: Factors affecting the adoption and meaningful use of social media: a structural equation modeling approach. Int. J. Inf. Technol. Lang. Stud. **2**, 96–109 (2018)
9. Alghizzawi, M., Ghani, M.A., Som, A.P.M., et al.: The impact of smartphone adoption on marketing therapeutic tourist sites in Jordan. Int. J. Eng. Technol. **7**, 91–96 (2018)
10. Lee, A.V.Y.: Determining quality and distribution of ideas in online classroom talk using learning analytics and machine learning. Educ. Technol. Soc. **24**, 236–249 (2021)
11. Hart, S.A.: Precision education initiative: moving toward personalized education. Mind Brain Educ. **10**, 209–211 (2016). https://doi.org/10.1111/mbe.12109
12. Luan, H., Geczy, P., Lai, H., et al.: Challenges and future directions of big data and artificial intelligence in education. Front. Psychol. **11**, 580820 (2020). https://doi.org/10.3389/fpsyg.2020.580820
13. Salloum, S., Khan, R., Shaalan, K.: A survey of semantic analysis approaches. In: Hassanien, A.-E., Azar, A.T., Gaber, T., Oliva, D., Tolba, F.M. (eds.) AICV 2020. AISC, vol. 1153, pp. 61–70. Springer, Cham (2020). https://doi.org/10.1007/978-3-030-44289-7_6
14. Alhashmi, S.F.S., Salloum, S.A., Mhamdi, C.: implementing artificial intelligence in the United Arab Emirates healthcare sector: an extended technology acceptance model. Int. J. Inf. Technol. Lang. Stud. **3**, 27–42 (2019)
15. Aburayya, A., Alshurideh, M., Al Marzouqi, A., et al.: An empirical examination of the effect of TQM practices on hospital service quality: an assessment study in UAE hospitals. Syst. Rev. Pharm. **11**(9), 347–362 (2020)
16. Alshurideh, M., Al Kurdi, B., Salloum, S.A., et al.: Predicting the actual use of m-learning systems: a comparative approach using PLS-SEM and machine learning algorithms. Interact. Learn. Environ., 1–15 (2020)

17. Lian, A.-P.: Precision language education: a glimpse into a possible future. GEMA Online ® J. Lang. Stud. **17** (2017). https://doi.org/10.17576/gema-2017-1704-01
18. Priorities for Personalized Medicine. President's Council of Advisors on Science and Technology (U.S.) (2008)
19. Alshurideh, M., Al Kurdi, B., Abumari, A., Salloum, S.: Pharmaceutical promotion tools effect on physician's adoption of medicine prescribing: evidence from Jordan. Mod. Appl. Sci. **12**, 210–222 (2018)
20. Al-Maroof, R.S., Alhumaid, K., Akour, I., Salloum, S.: Factors that affect e-learning platforms after the spread of COVID-19: post acceptance study. Data **6**, 49 (2021)
21. The White House: FACT SHEET: President Obama's Precision Medicine Initiative | whitehouse.gov. The White House (2008)
22. Cook, C.R., Kilgus, S.P., Burns, M.K.: Advancing the science and practice of precision education to enhance student outcomes. J. Sch. Psychol. **66**, 4 (2018). https://doi.org/10.1016/j.jsp.2017.11.004
23. U.S. Department of Education: Reimagining the Role of Technology in Education (2017)
24. Yang, A., Chen, I., Flanagan, B., et al.: From human grading to machine grading. Educ. Technol. Soc. **24**, 164–175 (2021)
25. Kokoç, M., Akçapınar, G., Hasnine, M.N.: Unfolding students' online assignment submission behavioral patterns using temporal learning analytics. Educ. Technol. Soc. **24**, 223–235 (2021)
26. Yang, C.C.Y., Chen, I.Y.L., Ogata, H.: Toward precision education: educational data mining and learning analytics for identifying students' learning patterns with Ebook systems. Educ. Technol. Soc. **24**, 1176–3647 (2021)
27. Romero, C., Ventura, S.: Educational data mining: a survey from 1995 to 2005. Expert Syst. Appl. **33**, 135–146 (2007). https://doi.org/10.1016/j.eswa.2006.04.005
28. Baker, R.S.J.D., Yacef, K.: The state of educational data mining in 2009: a review and future visions. JEDM J. Educ. Data Min. **1**, 3–17 (2009)
29. Dahiya, V.: A survey on educational data mining educational data mining view project collaborative attacks in MANET view project a survey on educational data mining. Impact J. **6**, 23–30 (2018)
30. Peterson, B., Baker, P.S.J.D.: Data mining for education. Int. Encycl. Educ. **7**, 112–118 (2010)
31. Tsai, Y.-S.: What is learning analytics? Soc. Learn. Anal. Res. (2021)
32. Papamitsiou, Z., Economides, A.A.: Learning analytics and educational data mining in practice: a systematic literature review of empirical evidence. Educ. Technol. Soc. **17**, 49–64 (2014)
33. Narayan Singh, S., Khanna, L., Alam, M.: Educational data mining and its role in determining factors affecting students academic performance: a systematic review. In: Dynamic Resource Allocation in Cloud Environment View project Educational Data Mining and its Role in Determining Factors Affecting Students Academic Performance: A Systematic Review, pp. 1–7. ieeexplore.ieee.org (2016). https://doi.org/10.1109/IICIP.2016.7975354
34. Coelho, O.B., Silveira, I.: Deep learning applied to learning analytics and educational data mining: a systematic literature review. In: Brazilian Symposium on Computers in Education (Simpósio Brasileiro de Informática na Educação-SBIE), p. 143 (2017)
35. Hernández-Blanco, A., Herrera-Flores, B., Tomás, D., Navarro-Colorado, B.: A systematic review of deep learning approaches to educational data mining (2019). https://doi.org/10.1155/2019/1306039
36. Dutt, A., Ismail, M., Herawan, T.: A systematic review on educational data mining. IEEE Access **5**, 15991–16005 (2017)
37. Elsayed, A.A., Caeiro-Rodríguez, M., Mikic-Fonte, F.A., Llamas-Nistal, M.: Research in learning analytics and educational data mining to measure self-regulated learning: a systematic review. In: World Conference on Mobile and Contextual Learning, pp. 46–53 (2019)

38. Agrusti, F., Bonavolontà, G., Mezzini, M.: University dropout prediction through educational data mining techniques: a systematic review. J. E-Learn. Knowl. Soc. **15**, 161–182 (2019). https://doi.org/10.20368/1971-8829/1135017
39. Fink, A.: Conducting Research Literature Reviews: From the Internet to Paper. Sage Publications, Thousand Oaks (2019)
40. Kitchenham, B., Charters, S.: Guidelines for performing systematic literature reviews in software engineering (2007)
41. Moher, D., Liberati, A., Tetzlaff, J., Altman, D.G.: Preferred reporting items for systematic reviews and meta-analyses: the PRISMA statement. PLoS Med. **6**, e1000097 (2009). https://doi.org/10.1371/journal.pmed.1000097
42. Okoli, C.: A guide to conducting a standalone systematic literature review. Commun. Assoc. Inf. Syst. **37**, 879–910 (2015). https://doi.org/10.17705/1cais.03743
43. Weidt, N.F., da Silva, R., de Souza, L.: Systematic Literature Review in Computer Science - A Practical Guide (2016)
44. Tsai, S.-C., Chen, C.-H., Shiao, Y.-T., Ciou, J.-S., Wu, T.-N.: Precision education with statistical learning and deep learning: a case study in Taiwan. Int. J. Educ. Technol. High. Educ. **17**(1), 1–13 (2020). https://doi.org/10.1186/s41239-020-00186-2
45. Chen, X., Zou, D., Xie, H., Cheng, G.: Twenty years of personalized language learning: topic modeling and knowledge mapping. Educ. Technol. Soc. **24**, 205–222 (2021)
46. Luan, H., Tsai, C.-C.: A review of using machine learning approaches for precision education. Educ. Technol. Soc. **24**, 1176–3647 (2021)
47. Wu, J., Yang, C., Liao, C., et al.: Analytics 2.0 for precision education. Educ. Technol. Soc. **24**, 267–279 (2021)
48. Tempelaar, D., Rienties, B., Nguyen, Q.: The contribution of dispositional learning analytics to precision education. J. Item. Educ. Technol. Soc. **24**, 109–122 (2021)
49. Yang, T.-C.: Using an institutional research perspective to predict undergraduate students career decisions in the practice of precision education. Educ. Technol. Soc. **24**, 280–296 (2021)
50. Hull, D.C., Williams, G.A., Griffiths, M.D.: Video game characteristics, happiness and flow as predictors of addiction among video game players: a pilot study. J. Behav. Addict. **2**, 145–152 (2013). https://doi.org/10.1556/JBA.2.2013.005

The Impact of Strategic Orientation in Enhancing the Role of Social Responsibility Through Organizational Ambidexterity in Jordan: Machine Learning Method

Erfan Alawneh(✉) and Khaled Al-Zoubi

Faculty of Business and Management Sciences, Mutah University, Al Karak, Jordan
ealawneh2@gmail.com

Abstract. This study aimed to know the impact of the strategic orientation in enhancing the role of social responsibility through machine learning in the centers of Jordanian service ministries. The study population consisted of all the upper and middle departments employees in those ministries, which are (8) in number out of a total of (15) service ministries. The number of employees was (517). To achieve this, a descriptive and analytical method was used to measure this phenomenon through the questionnaire. (262) questionnaires were distributed, which constitute (50%) of the total sample size depending on the relative stratified sample, and after retrieving (254) of them, (4) non-conforming questionnaires were excluded. The rest (250) questionnaires were subjected to analysis by (SPSS) program, hypothesis testing through multiple linear regression, and path analysis (AMOS). The results showed that the strategic orientation in its dimensions impacted social responsibility in its combined dimensions in the Jordanian service ministries. At the same time, it did not affect the organizational ambidexterity in its dimensions. The study displayed that organizational ambidexterity has a mediating role in strategic orientation and social responsibility. The study recommended that the Jordanian service ministries adopt strategic thinking in planning processes through training and development, adopt or-generational ambidexterity by following accurate and effective standards in identifying needs, and pay attention to enhancing the role of social responsibility in those ministries by machine learning.

Keywords: Strategic orientation · Social responsibility · Organizational ambidexterity · Service ministries · Jordan

1 Introduction

Human resources are an essential asset for organizing institutional work and developing the performance of ministries and institutions in the public sector. They also play a key role in the organization's smooth running and achieving its goals. Thus, human

© The Author(s), under exclusive license to Springer Nature Switzerland AG 2022
A. E. Hassanien et al. (Eds.): AMLTA 2022, LNDECT 113, pp. 357–370, 2022.
https://doi.org/10.1007/978-3-031-03918-8_31

resources can become distinct, if they are managed effectively, and from here it is clear that the strategic orientation has a significant impact in enhancing the role of social responsibility, so that its importance increases in ministries and public sector institutions because of its direct effect on economic development and performance development, so The development of public sector institutions ensures the comprehensive growth of the national economy, especially in developing countries, and supports the focus on scarce resources and creative competencies in these institutions [1–3]. The present time requires the government, represented by its ministries, to create a new strategic orientation based on developing the efficiency of its strategies and procedures and the effectiveness of its pivotal and main role in the public sector, in order to ensure strategic consistency.so that the strategic orientation has the ability to see at the macro level more clearly, and those ministries are constantly working to achieve success to achieving continuous progress and development at the national level. In general, the strategy includes basic elements that determine its course, such as focusing, for example, on the long-term orientation of the government institution, and matching its activities with the available resources and its environment, in order to reduce threats and identify available opportunities. Senior management is the key factor for its future success [4]. Social responsibility is also one of the important topics that ministries and government institutions attach great importance to, and with the increasing role of the public sector and the large size of responsibilities entrusted to the government represented by its ministries and institutions, it has imposed a social role through which it contributes to achieving aspirations and societal goals. In order to achieve their success and sustainability, these institutions need to be ingenious and work to exploit the available opportunities, to adapt to their environment in light of the future circumstances and challenges they face [5]. The development and technological progress and the challenges faced by institutions and governments have led to the creation of strategic organizational orientations directly related to the social environment to re-subject it socially in light of the changes facing institutions or governments. These rapid changes in the organizational and institutional landscape force them to change, learn, and adapt in order to survive and grow and to promote better flexibility to the needs of the institution and the changing environment of which, organizational ambidexterity reduces cases of organizational collapse through the need to apply it in institutions to avoid threats that could affect the strategic orientation of ministries and institutions [6]. Therefore, this study came to examine the impact of the strategic orientation in enhancing the role of social responsibility through organizational ambidexterity in the service ministries in Jordan. Through machine learning, machine learning and artificial intelligence are therefore considered to be a tremendous ability to serve companies' businesses and help them achieve success by making informed decisions by enhancing human intelligence with powerful computer capabilities, accurate data analysis, and automating the required tasks [7, 8]. There is no doubt that machine learning-based innovation is poised to benefit from many potential applications in the strategic direction of institutions and organizations [9, 10].

2 Literature Review

The strategic orientation of the owners of family-owned companies tends to be product developers and innovators in their markets. According to its strategic orientation, the government needs to develop public policies aimed at developing greater CSR activities to create competitive advantages for managers and owners of family businesses so that they understand the importance of implementing responsible activities Social [11]. According to Adams, Freitas, and Fontana (2019) study [12], organizations with a common customer/technology orientation outperform organizations with a customer orientation or technology alone, as there is a positive impact of marketing management in promoting innovation success all orientations. Still, it is greater for organizations with a technology orientation so that the effect is moderate. For marketing management, the relationship between orientation and performance increases as more marketing mix elements are deployed simultaneously. Therefore, the study recommended increasing the levels of marketing because of its importance in determining the strategic orientation of organizations, especially service and marketing organizations. That strategic orientation has a great relationship with organizational savvy. The researcher recommends that the strategic orientation of companies be predicted about the dynamics of market competitiveness and the environment to effectively influence organizational ambidexterity [13]. Posch and Garaus's (2020) study [14] aimed to investigate the relationship between strategic orientation and organizational ambidexterity using a survey of 217 senior executives. The study sheds light on the importance of research on how executives use strategic orientation. The study results showed: Supporting the hypothesis that the positive or negative association of strategic orientation with organizational dexterity depends on other organizational factors. And that strategic orientation is only positively related to organizational ambidexterity when leaders' tendency toward innovation is unusually high, In addition to Ferreira, Coelho and Weersma, (2019) study [15], which aimed to examine the relationships between strategic orientation, innovation capacity, managerial capabilities and exploration and exploitation capabilities on competitive advantage and company performance. This study suggested that the role of exploration and exploitation capabilities in these relationships varies between the three dimensions of strategy orientation (cost-based leadership strategies, differentiation-based strategies, and product market scope) and performance. Modeling structural analysis was used to test hypotheses in a sample of (387) Portuguese small and medium-sized companies [17]. The empirical results indicate that innovation ability, managerial capabilities, and strategic orientation positively mediate the relationship between exploration and exploitation capabilities and performance, while strategic orientation affects competitive advantage and performance. The results of Mom *et al.*, (2019) study [18] showed that the downward effects of human resource practices affect the proficiency of the operational manager while not affecting senior managers due to the breadth of self-efficacy and intrinsic motivational guidance. In addition, the study results showed an ascending relationship between the operational manager and organizational skill and that it depends on human resource practices that enhance the company's chances of achieving its goals. The study recommends providing important new, multi-level insights about the effectiveness of strategic human resource systems in supporting individual and organizational ambidexterity. There are critical gaps in organizational dexterity systems, and knowledge computing can be directed to access information to improve performance in global strategic partnerships where it is necessary to work to exploit the large potential of cognitive computing, to interpret comprehensive data in a dynamic work environment, which can act as an enabler for organizational dexterity [19]. Social responsibility has been widely discussed and linked to an organizational performance by researchers. However, a significant research gap remains unexplored, and this measures the association between social responsibility and customer loyalty in a developing country context. Drawing on the resource-based viewpoint and stakeholder theory, Islam *et al.* (2021) study [20] develop the basic mechanism through which social responsibility affects customer loyalty by including company reputation, customer satisfaction, customer trust as intermediaries, and company capabilities, as a mediating variable. The questionnaire was used as a data collection tool within the five-year Likert scale, and (500) questionnaires were distributed to the study sample of communication users. The study found that CSR initiatives are significantly and positively related to the company's reputation and customer satisfaction and trust. Social responsibility greatly affects the organizational legitimacy of employees, as social responsibility does not have a positive impact on the social capital of employees directly, but through legitimacy and there is a positive impact of social capital on employee loyalty, so managers must make social responsibility procedures known to their employees and they must Involve them in these procedures [21]. According to Ferrell *et al.* (2019) study [22] aimed to demonstrate the role of social responsibility and work ethics in brand behavior. The descriptive analytical method was used in this study. To achieve this goal, a questionnaire was designed and distributed to the study sample consisting of (351) respondents and they reached results from Chief among them: Despite the importance of social responsibility attitudes, customers value business ethics as a key behavior in their perceptions of brand attitudes. The study recommended providing new insights related to customers' expectations, and their perceptions of social responsibility and work ethics behavior. According to Tuan, (2016) study [23], there is a positive relationship between organizational prowess and entrepreneurial orientation of companies, which were managed by the social responsibility of small and medium companies, where entrepreneurial orientation was also found as a strong indicator in achieving corporate social responsibility in the environment in which it operates. Therefore, the study recommended increasing the levels of corporate social responsibility and organizations in the multiple and changeable environments by searching for opportunities and exploring solutions through the organizational ambidexterity of organizations.

3 Theoretical Framework

In order to reach the goal of the study and to bridge the scientific gap, the following model was proposed, which shows the independent, dependent, and intermediate variables, where the previous studies were based on the preparation and development of the study model with its main variables, dimensions and main and subsidiary hypotheses [11, 19, 24] (Fig. 1).

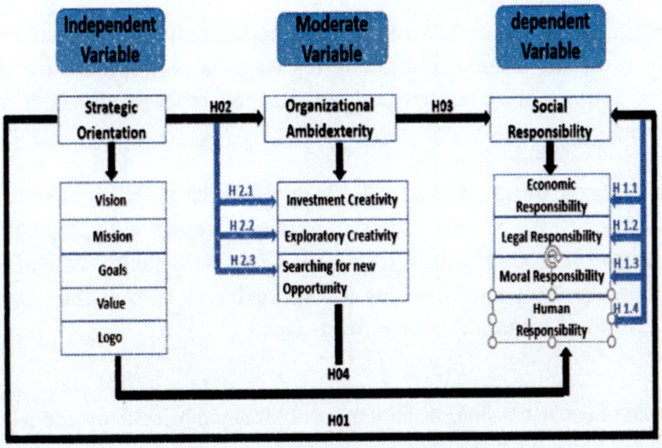

Fig. 1. Theoretical framework

4 Research Methodology

The study followed the descriptive-analytical approach, as this type is considered the most appropriate type to describe the phenomenon as accurately as it is in reality, based on the nature of the study and previous studies that dealt with the subject of the study [25]. The study population consisted of all the higher positions or departments (the minister, the secretary general, the assistant general secretary, the general managers, and directors of the directorates at the center of each ministry), and the middle departments at the centers. Of the ministries, directors of directorates in the governorates and regions are (517) employees. They work in (8) service ministries out of (15) service ministries in Jordan. The study sample was selected from the target population using the proportional stratified sampling method. The study population was estimated at (517) employees, of whom (50%) were selected, according to the stratified proportional method for each ministry that was surveyed. Hence, the number of sample members was (262) employees. The data collection process was carried out based on the questionnaire. The questionnaire consisted of two parts. The first dealt with the demographic information of the study sample members in terms of gender, marital status, educational qualification, years of experience, and finally, the job position. The second part included themes, dimensions, and paragraphs of the component study tool. From (57) paragraphs, all related to strategic orientation and its role in promoting social responsibility through organizational ambidexterity, they were built according to the five-point Likert scale (strongly agree, agree, neutral, disagree, strongly disagree).

5 Data Analysis

Progressive multiple linear regression analysis and path analysis were used to test the hypotheses of the study, to identify the impact of the strategic orientation with its combined dimensions (vision, mission, goals, values, logo), on social responsibility with its combined dimensions (economic responsibility, legal responsibility, moral responsibility, human responsibility). in the Jordanian service ministries, as well as in identifying the impact of the strategic orientation with its combined dimensions (vision, mission, goals, values, logo), on organizational ambidexterity with its combined dimensions (investment creativity, exploratory creativity, search for new opportunities), in the Jordanian service ministries.

The First Main Hypothesis (H01): "There is no statistically significant effect at the level of significance (0.05 ≥ α) for the strategic orientation with its combined dimensions (vision, mission, goals, values, logo), on social responsibility with its combined dimensions (economic responsibility, legal responsibility, responsibility ethical, human responsibility), in the Jordanian service ministries.

Table 1. Multiple linear regression of the impact of strategic orientation and its dimensions on social responsibility in its dimensions

Dimensions	B	Standard error	Beta	(T) value computed	The significance level for (T)
Vision	−0.464	0.496	−0.058	−0.935	0.351
The mission	0.649	0.467	0.087	1.390	0.166
Goals	0.923	0.471	0.123	1.960	0.051
Value	1.506	0.524	0.178	2.874	*0.004
Logo	2.710	0.535	0.313	5.070	*0.000

*Statistically significant at the significance level (0.05 ≥ α).

Table 1 shows that each of the dimensions of vision, mission, and goals did not have a statistically significant effect at the significance level (0.05 ≥ α) in social responsibility with its combined dimensions, in the Jordanian service ministries, where the T-test values for them, respectively (−0.935, 1.39, 1.96), which is less than the tabular (T) value of (2.35), while we find that both dimensions of values and the logo had a statistically significant effect at the significance level (0.05 ≥ α) in the strategic orientation with its combined dimensions in the ministries The Jordanian service, whose T-test values, respectively, amounted to (2.874, 5.07), which is higher than the tabular (T) value and thus confirms the existence of the effect of these dimensions.

First Sub-hypothesis (H1.1): To test the first sub-hypothesis, multiple linear regression analysis was used to identify the impact of strategic orientation on economic responsibility in Jordanian services ministries.

Table 2 shows that the dimensions of strategic orientation (vision, mission, goals, values, logo) did not have a statistically significant impact on economic responsibility in the Jordanian service ministries, and this is shown by the value of the significance level, which amounted to (0.224, 0.603, 0.984, 0.078, 0.575), respectively, which is greater than the significance level (α ≥ 0.05) and thus accepting the null hypothesis. The values of the correlation coefficient (R) between the dimensions of strategic orientation (vision,

Table 2. Multiple linear regression of the impact of strategic orientation in its dimensions on economic responsibility

Dimensions	R	B	Standard error	Beta	(T) value computed	Significance level for (T)
Constant		3.741	0.465		8.043	0.000
Vision	.149a	−0.209	0.092	−0.229	−0.266	0.224
The mission	.153c	−0.052	0.100	−0.050	−0.520	0.603
Goals	.152b	−0.002	0.079	−0.002	−0.020	0.984
Value	.192d	−0.161	0.091	−0.152	−1.768	0.078
Logo	.195e	−0.009	0.016	0.075	−0.562	0.575

*Statistically significant at the significance level (0.05 ≥ α).

mission, goals, values, logo) and economic responsibility were low. It is very negative and has a negative trend that appears through the value of the relationship trend (Beta), which came in succession (−0.266, −0.520, −0.020, −1.786, −0.562). The dimensions of the strategic orientation collectively explain the value of the variation in economic responsibility, and all these values confirm To accept the null hypothesis.

Second Sub-hypothesis (H1.2): To test the second sub-hypothesis, multiple linear regression analysis was used to identify the impact of strategic orientation on legal responsibility in Jordanian services ministries.

Table 3. Multiple linear regression of the impact of strategic orientation in its dimensions on legal responsibility

Dimensions	R	B	Standard error	Beta	(T) value computed	Significance level for (T)
Constant		1.141	0.417		2.736	0.007
Vision	.058a	0.098	0.054	0.116	2.818	0.030
The mission	.241c	0.198	0.061	0.207	3.275	0.001
Goals	.130b	0.130	0.054	0.153	2.389	0.018
Value	.322d	0.216	0.062	0.220	3.493	0.001
Logo	.322e	0.119	0.057	0.121	3.329	0.002

- Statistically significant at the significance level (0.05 ≥ α).

Table 3 shows that the dimensions of strategic orientation (vision, mission, goals, values, logo) have a statistically significant effect on legal responsibility in the Jordanian service ministries, and this is shown through the value of the significance level, which amounted to (0.030, 0.001, 0.018, 0.001)., 0.002), respectively, which is less than the significance level (α ≥ 0.05) and thus rejecting the null hypothesis, and the values of the correlation coefficient (R) between the dimensions of strategic orientation (vision, mission, goals, values, logo) and legal responsibility were high and positively oriented. It appears through the value of the orientation of the relationship (Beta), which came in a row (0.116, 0.207, 0.153, 0.220, 0.121). The dimensions of the strategic orientation collectively explain the value of the variance in legal responsibility. All these values confirm the rejection of the null hypothesis, the null hypothesis, and the acceptance of the alternative.

The Third Sub-Hypothesis (H1.3): To test the third sub-hypothesis, multiple linear regression analysis was used on the impact of strategic orientation on moral responsibility in the Jordanian services ministries.

Table 4. Multiple linear regression of the impact of strategic orientation on moral responsibility

Dimensions	R	B	Standard error	Beta	(T) value computed	Significance level for (T)
Constant		2.219	0.434		5.117	0.000
Vision	.141[a]	0.171	0.056	0.198	2.818	0.003
The mission	.163[c]	0.098	0.063	0.101	3.275	0.002
Goals	.141[b]	0.115	0.057	0.117	2.389	0.004
Value	.216[d]	0.138	0.064	0.139	3.493	0.032
Logo	.248[e]	0.117	0.060	0.127	3.329	0.001

- Statistically significant at the significance level ($0.05 \geq \alpha$).

From Table 4 it appears that the dimensions of strategic orientation (vision, mission, goals, values, logo) have a statistically significant effect on moral responsibility in the Jordanian service ministries, and this is shown through the value of the significance level, which amounted to (0.003, 0.002, 0.004, 0.032, 0.001), respectively, which is less than the significance level ($\alpha \geq 0.05$) and thus rejecting the null hypothesis. The values of the correlation coefficient (R) between the dimensions of strategic orientation (vision, mission, goals, values, logo) and economic responsibility were high and orientation. Positive shown by the value of the orientation of the relationship (Beta), which came in succession (0.198, 0.101, 0.117, 0.139, 0.127), as the dimensions of strategic orientation combined explain the value of the variance in moral responsibility, and all these values confirm the rejection of the null hypothesis and the acceptance of the alternative.

Fourth Sub-hypothesis (H1.4): To test it, multiple linear regression analysis was used to identify the impact of strategic orientation on human responsibility in Jordanian services ministries.

Table 5 shows that the dimensions of the strategic orientation (vision, mission, goals, values, logo) did not have a statistically significant impact on human responsibility in the Jordanian service ministries, and this is shown by the significance level value, which amounted to (0.483, 0.487)., 0.416, 0.384, 0.854), respectively, which is greater than the

Table 5. Multiple linear regression of the impact of strategic orientation in its dimensions on human responsibility

Dimensions	R	B	Standard error	Beta	(T) value computed	Significance level for (T)
Constant		1.540	0.459		3.355	0.001
Vision	.049[a]	−0.042	0.059	−0.044	0.702	0.483
The mission	.088[c]	−0.046	0.067	−0.043	0.696	0.487
Goals	.073[b]	−0.049	0.067	−0.051	0.815	0.416
Value	.170[d]	−0.394	0.068	−0.360	0.803	0.384
Logo	.095[e]	−0.012	0.063	−0.012	−0.184	0.854

- Statistically significant at the significance level (0.05 ≥ α).

significance level (α ≥ 0.05) and thus accepting the null hypothesis, as were the values of the correlation coefficient (R) between the dimensions of strategic orientation (vision, mission, goals, values, logo) and human responsibility Low and negative trend shown by the value of the relationship trend (Beta), which came in a row (−0.044, −0.043, −0.051, −0.360, −0.012), and the dimensions of the strategic orientation collectively explain the value of the variance in human responsibility, and all these values confirm On accepting the null hypothesis the null hypothesis.

Table 6 shows that both the investment innovation and exploratory innovation dimensions were not statistically significant at the significance level (0.05 ≥ α) in social responsibility in the Jordanian service ministries, where the T-test values for them, respectively, amounted to (−0.579)., 0.283), which is less than the tabular (T) value of (2.92), while we find that the dimension of searching for new opportunities had a statistically significant effect at the significance level (0.05 ≥ α) in social responsibility in the Jordanian service ministries, where the values of The (T) test has (−4,989), which is higher than the tabular (T) value and thus confirms the presence of the effect of this dimension.

Table 6. Statistically significant at the significance level (0.05 ≥ α)

Direct effect coefficient values		Indirect effect	Path	T	Indication level
Strategic orientation in organizational ambidexterity	0.565	0.268	Strategic Orientation → Organizational Dexterity	10.804	0.000
Organizational ambidexterity in social responsibility	0.498		Organizational ambidexterity → Social responsibility	11.907	0.000

The Fourth Main Hypothesis (H04): There is no statistically significant effect at the significance level (α 0.05) of the strategic orientation in social responsibility in the presence of organizational ambidexterity; as an intermediate variable; In the Jordanian services ministries. Table 7 the results of the path analysis to verify the direct and indirect effects of the strategic orientation in social responsibility through organizational ambidexterity in the Jordanian service ministries.

Table 7. The level of significance for predicting the values of strategic orientation through organizational ambidexterity in Jordanian services ministries

Statement	Chi^2	Gfi	Cfi	AGFI	RMSEA	Indication level
The impact of strategic orientation on social responsibility through organizational ambidexterity	15.44	0.897	0.991	0.819	0.013	0.000

- Statistically significant at the significance level (0.05 ≥ α).

6 Discussions

Tables 6 and 7 show that there is a statistically significant effect of strategic orientation on social responsibility through organizational ambidexterity in Jordanian ministries of service, such as the value of the chi-square (CH) (calculated) was (15.44) and the value of (GFI), which is the quality-fitness index (0.897), [9] which is close to the correct value (1), i.e., perfect fit, and the comparative fitness index value () (CFI (0.991), which is also close to the correct value, i.e. perfect fit, and the comparative fitness index value () Corrective Matching Quality (AGFI) () (0.819), which is also close to the correct one, that is, the corrected exact match, and the approximate root error (RMSEA) value reached (0.013), which is very close to zero, as all these values confirm that Good matching of the model and its validity for analysis according to the path analysis test. The study results showed that each of the dimensions of vision, mission, and goals did not have a statistically significant effect on social responsibility in its combined dimensions in the Jordanian service ministries. In contrast, the study results showed that both dimensions of values and the logo had a statistically significant effect on social responsibility with its combined dimensions in the ministries. This may be due to the unworkable goals adopted by the service ministries. It is known that the goals directly and indirectly affect and are affected by the vision and mission that the ministries will adopt. The lack of homogeneity and harmony in the goals, mission, and vision may become an obstacle for the ministry to influence social responsibility in all its dimensions. Concerning the dimensions of values and the logo and the impact they have on social responsibility in its dimensions, this is because all Jordanian ministries have a high level of values and adopt logos that are strongly linked to the practical reality of the Jordanian state, and this is consistent with the study of each of the Chine, (2017) [27]. The results of the study showed that each of the mission, goals, values, and logo dimensions did not affect organizational ambidexterity with its combined dimensions in the Jordanian service ministries. In contrast, the vision dimension had a statistically significant effect on organizational ambidexterity with its combined dimensions, and this may be attributed to the fact that many of the missions and goals The values and logos of ministries in Jordan are still in the process of building and basic organization and have not yet reached the point of creativity and innovation, which limits their ability to keep pace with the organizational ambidexterity represented and based mainly on creativity and innovation. Thinking, not thinking inside the box that limits the capabilities and energies of employees with bright visions. Concerning the visual dimension, which had an effective and statistically significant impact on organizational ambidexterity, this is because the vision does not specify limits, as it is an open field that allows employees and administrators to highlight all their energies, knowledge, expertise, and advanced skills, as the results of this study have partially agreed. Authors in [28] study confirmed a relationship between organizational ambidexterity and strategic thinking, which is one of the components of strategic orientation. The study results showed that both the investment innovation dimension and the exploratory innovation did not have a statistically significant effect on social responsibility in the Jordanian service ministries. In contrast, the search for new opportunities dimension had a statistically significant effect on social responsibility in the Jordanian service ministries, and this may be attributed to the fact that every employee looking for new opportunities must have a high level of knowledge and have a lot of motivation that will affect everyone around him, positively, in addition to that he considers himself responsible for providing the best services to everyone who deserves them. Also, the lack of impact of both investment creativity and exploratory creativity is due to this type of creativity. Therefore it will not have a noticeable impact on social responsibility. The lack of such creativity can be attributed to what is characterized by service ministries of following Outdated systems, in the promotion and appointment of employees, rely mostly on years of experience, ignoring all other aspects, which would be more important in determining the position or job position of the employee. The results of the study showed the existence of a statistically significant impact of the strategic orientation on social responsibility with the presence of organizational ambidexterity; as an intermediate variable; In the Jordanian service ministries, this may be attributed to the positive role of organizational ambidexterity in linking the strategic orientation and its dimensions, and social responsibility and its other dimensions. Organizational ambidexterity with its various dimensions plays a mediating role in reducing the gap between strategic orientations. Modern and constantly evolving and changing social responsibilities require a kind of creativity and innovation to keep pace with them and work to provide them. The results of this study agree with a study of Wegwu [13] and a study of Posch and Garaus [14], which showed that organizational ambidexterity plays a mediating role in strategic orientation and social responsibility, and also confirmed the existence of a positive relationship between organizational ambidexterity, and each of the trends Strategy and Social Responsibility.

7 Conclusions

The level of application of strategic orientation, organizational prowess, and social responsibility in the Jordanian service ministries has fallen within the medium level and each dimension of the strategic orientation. The results of the study showed that each of the dimensions of the vision, mission, and goals had no effect on social responsibility in its combined dimensions, in the Jordanian service ministries, and also the dimension of the mission, goals, values, and logo was not influential in organizational prowess with its combined dimensions, in the Jordanian service ministries. It was noted that both the investment innovation and exploratory innovation dimensions did not affect social responsibility in the Jordanian service ministries, while the search for new opportunities impacted social responsibility. Finally, it was found that there is an effect, which is that organizational ambidexterity plays a mediating role between strategic orientation and social responsibility as a mediating variable; in the Jordanian service ministries. With need to pay attention to new ideas of organizational ambidexterity that can be proposed in future studies that would encourage innovation and creativity and foresee the future in their application, Where more studies, research and development should be conducted on investment creativity and exploratory creativity because of their importance in social responsibility in the Jordanian service ministries, which the results of the current study showed a shortcoming in their application in the Jordanian service ministries.

References

1. Al-Shibly, M.S., Alghizzawi, M., Habes, M., Salloum, S.A.: The Impact of de-marketing in reducing jordanian youth consumption of energy drinks. In: Hassanien, A.E., Shaalan, K., Tolba, M.F. (eds.) AISI 2019. AISC, vol. 1058, pp. 427–437. Springer, Cham (2020). https://doi.org/10.1007/978-3-030-31129-2_39
2. Fulmer, I.S., Ployhart, R.E.: 'Our Most Important Asset' a multidisciplinary/multilevel review of human capital valuation for research and practice. J. Manage. **40**(1), 161–192 (2014)
3. Salloum, S.A., Al-Emran, M., Khalaf, R., Habes, M., Shaalan, K.: An innovative study of E-payment systems adoption in higher education: theoretical constructs and empirical analysis. Int. J. Interact. Mob. Technol. **13**(6) (2019)
4. Alawneh, E., Al-Zoubi, K.: The effect of strategic direction in enhancing the role of social responsibility through organizational prowess in jordan services ministries (2020)
5. Smith, A., Fressoli, M., Thomas, H.: Grassroots innovation movements: challenges and contributions. J. Clean. Prod. **63**, 114–124 (2014)
6. Alnawafleh, H., Alghizzawi, M., Habes, M.: The impact of introducing international brands on the development of Jordanian tourism. Int. J. Inf. Technol. Lang. Stud. **3**(2), 30–40 (2019)
7. Salloum, S.A., Al-Emran, M., Habes, M., Alghizzawi, M., Ghani, M.A., Shaalan, K.: What impacts the acceptance of E-learning through social media? An empirical study. In: Al-Emran, M., Shaalan, K. (eds.) Recent Advances in Technology Acceptance Models and Theories. SSDC, vol. 335, pp. 419–431. Springer, Cham (2021). https://doi.org/10.1007/978-3-030-64987-6_24
8. Habes, M., Elareshi, M., Ziani, A.: An empirical approach to understanding students' academic performance: YouTube for learning during the Covid-19 pandemic. Linguist. Antverp., 1518–1534 (2021)

9. Salloum, S.A., AlAhbabi, N.M.N., Habes, M., Aburayya, A., Akour, I.: Predicting the intention to use social media sites: a hybrid SEM - machine learning approach. In: Hassanien, A.-E., Chang, K.-C., Mincong, T. (eds.) AMLTA 2021. AISC, vol. 1339, pp. 324–334. Springer, Cham (2021). https://doi.org/10.1007/978-3-030-69717-4_32
10. Alshammari, R.: Arabic text categorization using machine learning approaches. Int. J. Adv. Comput. Sci. Appl. **9**(3), 226–230 (2018)
11. Aguilar, J.L.E., Maciel, A.S., Sánchez, J.L.Z., Hervert, M.deJ.P.: Strategic orientation of mexican family-owned businesses and its influence on corporate social responsibility practices. Organ. Mark. Emerg. Econ. **11**(1), 107–127 (2020)
12. Adams, P., Freitas, I.M.B., Fontana, R.: Strategic orientation, innovation performance and the moderating influence of marketing management. J. Bus. Res. **97**, 129–140 (2019)
13. Wegwu, M.E.: Strategic orientation and organizational ambidexterity practices of mobile communication firms in Port Harcourt. Int. J. Emerg. Trends Soc. Sci. **7**(1), 1–10 (2019)
14. Posch, A., Garaus, C.: Boon or curse? A contingent view on the relationship between strategic planning and organizational ambidexterity. Long Range Plann. **53**, 101878 (2019)
15. Ferreira, J.A.B., Coelho, A., Weersma, L.A.: The mediating effect of strategic orientation, innovation capabilities and managerial capabilities among exploration and exploitation, competitive advantage and firm's performance. Contaduría y Adm. **64**(SPE1), 0 (2019)
16. Al-Shakhanbeh, Z.M., Habes, M.: The relationship between the government's official facebook pages and healthcare awareness during Covid-19 in Jordan. In: Hassanien, A.-E., Elghamrawy, S.M., Zelinka, I. (eds.) Advances in Data Science and Intelligent Data Communication Technologies for COVID-19. SSDC, vol. 378, pp. 221–238. Springer, Cham (2022). https://doi.org/10.1007/978-3-030-77302-1_12
17. Puspaningrum, A.: Social media marketing and brand loyalty: the role of brand trust. J. Asian Financ. Econ. Bus. **7**(12), 951–958 (2020). https://doi.org/10.13106/JAFEB.2020.VOL7.NO12.951
18. Mom, T.J.M., Chang, Y.-Y., Cholakova, M., Jansen, J.J.P.: A multilevel integrated framework of firm HR practices, individual ambidexterity, and organizational ambidexterity. J. Manage. **45**(7), 3009–3034 (2019)
19. Kaur, S., Gupta, S., Singh, S.K., Perano, M.: Organizational ambidexterity through global strategic partnerships: a cognitive computing perspective. Technol. Forecast. Soc. Change **145**, 43–54 (2019)
20. Islam, T., et al.: The impact of corporate social responsibility on customer loyalty: The mediating role of corporate reputation, customer satisfaction, and trust. Sustain. Prod. Consum. **25**, 123–135 (2021)
21. Blanco-Gonzalez, A., Diéz-Martín, F., Cachón-Rodríguez, G., Prado-Román, C.: Contribution of social responsibility to the work involvement of employees. Corp. Soc. Responsib. Environ. Manag. **27**(6), 2588–2598 (2020)
22. Ferrell, O.C., Harrison, D.E., Ferrell, L., Hair, J.F.: Business ethics, corporate social responsibility, and brand attitudes: an exploratory study. J. Bus. Res. **95**, 491–501 (2019)
23. Tuan, L.T.: Organizational ambidexterity, entrepreneurial orientation, and I-deals: the moderating role of CSR. J. Bus. Ethics **135**(1), 145–159 (2016)
24. Claver-Cortés, E., Zaragoza-Sáez, P., Pertusa-Ortega, E.: Organizational structure features supporting knowledge management processes. J. Knowl. Manag. **11** (2007)
25. Alhumaid, K., Habes, M., Salloum, S.A.: Examining the factors influencing the mobile learning usage during COVID-19 Pandemic: an integrated SEM-ANN method. IEEE Access **9**, 102567–102578 (2021)
26. Elareshi, M., Habes, M., Ali, S., Ziani, A.: Using online platforms for political communication in bahrain election campaigns. Soc. Sci. Humanit. **29**(3), 2013–2031 (2021)

27. Chine, N.: The impact of strategic direction on the performance of business organizations, a case study: Naftal Company for the distribution and marketing of petroleum products, Algeria. Université Mohamed Khider-Biskra (2017)
28. Abu Zeid, A.: The role of strategic thinking in building organizational ingenuity: an applied study on Jordanian public shareholding companies. Jordanian J. Bus. Adm. **3**(15) (2019)

Three Mars Missions from Three Countries: Multilingual Sentiment Analysis Using VADER

Abdulla M. Alsharhan[3], Haroon R. Almansoori[1], Said Salloum[2](✉), and Khaled Shaalan[3]

[1] Faculty of Business Management, The British University in Dubai, Dubai, UAE
[2] School of Science, Engineering, and Environment, University of Salford, Salford, UK
S.A.S.Salloum@edu.salford.ac.uk
[3] Faculty of Engineering and IT, The British University in Dubai, Dubai, UAE

Abstract. In February 2021, the media around the world celebrated the success of three main space missions to Mars, Emirates hope mission from United Arab Emirates, followed by Tianwen-1, from China, and finally Mars 2020 mission from NASA. The aim of this research is to conduct a sentimental analysis of these tweets using a VADER-Multi, a method that is particularly aimed at multilingual social media texts. Major finding predicts that Tweets on Mars have positive polarity, while Emirates Hope mission, Tianwen-1 and NASA Mars2020 have neutral polarity that slightly shift toward positive. suggest VADER-Multi can generate good results on multilanguage but might not be very reliable on a large online corpus. Future work can be investigating different multilingual solutions, providing a web-based tool and investigating the root cause of the system overloaded with too many requests. This is study is among the first to analyze Mars space missions. It is also among the first to examine VADER multilingual tool (VADER-Multi).

Keywords: Multilingual sentiment analysis · Multilanguage · VADER · VADER-Multi · Mars mission · Emirates Mars Hope Mission · Mars2020 · Tianwen-1 · NASA persevere · Natural Language Processing

1 Introduction

The United Arab Emirates' Mars Mission is an uncrewed space exploration mission to Mars. The Hope orbiter was launched on 19 July 2020 and reached Mars on 9 February 2021. The successful mission arrival on Mars' orbit created a lot of media publicity, part of the generated coverage was from international media and experts. The local news reported #ArabsToMars topped the trends with 2.7 billion engagements around the world and more than 56,000 tweets and comments [1].

By the end of February 2021, the media around the world celebrated the success of three main space missions to Mars, Emirates Hope mission from United Arab Emirates, followed by Tianwen-1, from China, and finally Mars 2020 mission from NASA. These achievements have been recognized on a wide scale as a milestone for mankind, however, there are some voices that still doubts the high cost associated with these missions, and if the resources could have been used more wisely.

© The Author(s), under exclusive license to Springer Nature Switzerland AG 2022
A. E. Hassanien et al. (Eds.): AMLTA 2022, LNDECT 113, pp. 371–387, 2022.
https://doi.org/10.1007/978-3-031-03918-8_32

The mankind support for these missions is crucial to scientific advancement in space exploration, without it; these missions will not have necessary resources to keep going, but how the policy makers can be certain on people readiness for embracing these shifts in resource expenditure?

With a wealth of data from users around the world through social media, in addition to powerful algorithms that allow automating the data in many ways, including the sentiment feeling of an entire society. Sentiment analysis tools can give valuable insights to decision makers, starting by measuring the popularity of certain policies up to product development and companies' service reputation.

This study will explore sentiment analysis to decode the public sentiment feeling toward mars mission based on Twitter data collected around Emirates Hope orbiter, China's Tianwen-1 and NASA's Mars2020 missions. However, since these missions were launched by three different nations who speak multilanguage, it's important to have a tool that can analyze multilingual feelings. This study will try to understand the kind of feelings associated with these three Mars missions. More specifically, to answer the following research questions:

- RQ01. What is the general feeling associated with the three Mars missions?
- RQ02. What is impact of VADER-Multi Analysis on the three Mars missions' Twitter accounts?

2 Literature Review

2.1 Background

Sentiment analysis is a technique that calculate the sentiment score of any phrase using Natural Language Processing (NLP) data collecting and cleaning methodologies. The sentiment score can determine the overall feeling of the phrase from the writing mode whether its positive, negative, or neutral [2]. However, the way people express themselves changes based on their individual differences, momentary mood and cultural context. All of which could affect the way people chose to express themselves, there are also significant differences across cultural over the use of the power of expressing various feelings.

There are many common obstacles are faced when using social media sentiment analysis. Some of the most common ones are (1) The use of slang language and spelling mistakes, (2) expressing more than one opinion in one single post, (3) Modifiers and negators (strengthening/weakening or inverting the sentiment), (4) ambiguous words (words that can have more than one meaning), (5) Emojis, symbols and emoticons, (6) Sarcasm and irony [3]. There are many methods and tools for sentiment analysis. For example, TextBlob is one of Python libraries that provides a simple Application Programming Interface (API) that carries various NLP tasks. One of them is sentiment analysis [4].

Most studies on sentiment analysis have addressed a single language, which usually English, however, there are roughly 6,500 languages are spoken in the world today [5]. With the rapid growth of the internet usage, users around the world express their feeling in all sorts of languages. Analyzing sentiment in one language may risk losing essential feelings written in foreign languages in the process. Besides, some countries are home for many nationalities speaking different languages [6]. For example, over 8 million from over 200 countries call the UAE home speaking all sort of languages. Therefore, a multilingual framework is necessary to analyze data from different languages [7] (Fig. 1).

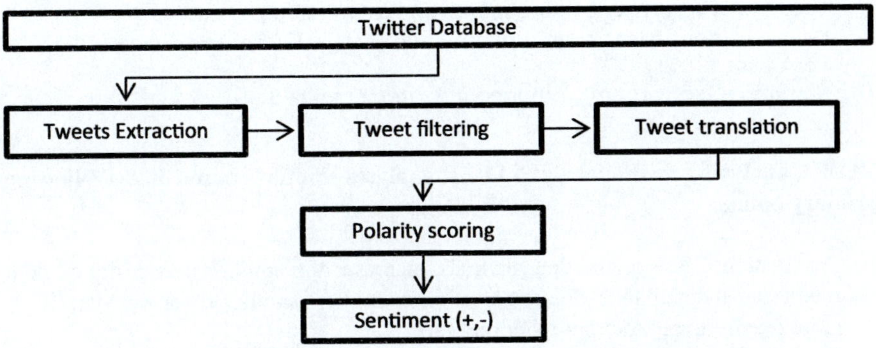

Fig. 1. A proposed multilingual sentiment analysis framework for Twitter

In 2014, Hutto & Gilbert have introduced Valence Aware Dictionary and sentiment Reasoner. (VADER), a parsimonious rule-based model for sentiment analysis of social media texts. VADER model achieved (96%) accuracy and was able to outperform individual human rate classification accuracy which was at (84%). VADER is also known for its capability in consuming less resources compared to machine learning models, as it does not require a considerable amount of training data. The VADER methodology is not impacted heavily by the speed-performance trade-off [8], besides, VADER is also fully open-sourced under the MIT License. VADER is also available on different programming language including python, Java, JavaScript, PHP, Scala, C#, Rust, Go and R [9]. Some experts believe VADER Sentiment Analysis works better for texts from social media. It is based on lexicons of sentiment-related words. an expert from [10] stated that Twitter sentiment analysis using VADER was significantly better compared to textBlob [10]. Nevertheless, sentiment analysis methods are evolving constantly and more recent versions of both VADER and Textblob might have different results.

VADER scoring methodology: VADER output the sentiment value of any given statement by calculating the polarity score.

```
analyser.polarity_scores("I never woke up feeling so
ecstatic and relieved right now Launched on july 2020
and just landed earlier this mo")

{'compound': 0.8039, 'neg': 0.051, 'neu': 0.63, 'pos':
0.319}
```

From the score above, it can be stated that the given statement is 32% positive, 51% negative, 63% neutral. Overall, the statement falls under the positive category.

The compound score is a metric that has been scaled between −1 as the most negative value and 1 as the most positive value. The following shows how the VADER methodology works [9].

```
Positive sentiment: compound score >= 0.05
Neutral sentiment: (compound score > -0.05) and (compound
score < 0.05)
Negative sentiment: (compound score <= -0.0
```

VADER analyzing methodology: VADER analyses emotions based on the following primary points:

1. Punctuation: The exclamation mark (!), increases the emotional intensity without modifying the semantic orientation. For example: "Another rover on Mars!!!" is more intense than "Another rover on Mars."
2. Capitalization: Writing using ALL-CAPS to stress an emotional word in the context of other lowercase words, increases the sentiment intensity without affecting the semantic orientation. For example: "Another rover on MARS." Increase the feeling intensity than "Another rover on Mars."
3. Degree modifiers: (intensifiers, booster words, or degree adverbs) affect emotion degree by increasing or decreasing the intensity. For example: "Mars is extremely cold." is more intense than "Mars is cold.", whereas "The Mars is somewhat cold." decreases the intensity.
4. Conjunctions: Polarity change due to Conjunctions such as "but". They signal a shift in emotion polarity, where the sentiment of the following text is dominant. For example: "Mars is cold, but I am still going anyway." has mixed feelings, where the last half of the statement is dictating the overall rating.
5. Preceding Tri-gram: detecting Polarity Negation by analyzing the sequence of three items preceding a sentiment-laden lexical feature. This way can dedicate up to 90% of the cases where negation change the polarity of the statement. For example, a negated phrase would be "Mars isn't really that faraway" [8].

2.2 Related Work

Most research on sentiment analysis has focused on modifying VADER to adapt to one single language. The research by [11] provides an adapted version of VADER to support Bengali language without the need for an intermediate translator, such as translator from Google, Python or MyMemory. Although The experiments indicate a significant improved result in detecting feelings compered to translated text from popular translators, the method was not made available on an online repository for further investigating or testing on a different language.

Despite this interest in Multilingual sentiment analysis (MSA), only few studies have explored multilingual analysis in the context of social media. The first article [12] from Brazil has focused on evaluating the current efforts to do specific sentiment analysis, this research has evaluated twenty one sentiment analysis methods in English language, and compared them with two language-specific methods. A detailed quantitative analysis for a multi-language approach was provided based on nine language-specific datasets. Later on, in the same year, the same group of researchers in addition to others have introduced a multilingual benchmarking system for sentiment analysis based on sentence-level named iFeel 2.0. The proposed solution is a web-based model that conducts nineteen sentence-level analysis and simplifies the process by allowing users to label a dataset with all of them. The introduced solution provides an easy-to-use comparative tool of new methods with basic approaches. And it can also be a useful comparative sentiment analysis tool for the researchers [13]. The following Table 1 shows a comparative result of the sixteen different methods using iFeel 2.0 based on the input "I'm so happy we finally have another rover on".

Table 1. A comparative result of the 16 different methods using iFEEL

Your input: *"I'm so happy we finally have another rover on"*			
Method name	Status	Method score	Polarity
OPINIONLEXICON	Completed	1	Positive
SENTISTRENGTH	Completed	0.25	Positive
HAPPINESSINDEX	Completed	0.8	Positive
SANN	Completed	1	Positive
EMOTICONSDS	Completed	1	Positive
SENTIMENT140	Completed	13.71	Positive
STANFORD	Completed	1	Positive
AFINN	Completed	3	Positive
MPQA	Completed	1	Negative
NRCHASHTAG	Completed	17.75	Positive
EMOLEX	Completed	1	Positive
EMOTICONS	Completed	0	Neutral
PANAST	Completed	1	Positive
SASA	Completed	0	Neutral
SENTIWORDNET	Completed	0.7452080246055627	Positive
VADER	Completed	0.6115	Positive

3 Methodology

The methodology was adapted from [9] and (Loria 2020 and explained by [2] and [14], using a rule-based system that is particularly aimed at social media texts using Python. The Python library that was used is called VADER, in addition to Textblob.

Since the analyzed data are taken from Arabic and possibly Chinese tweets as well; an updated multilingual version of VADER is used (VADER-Multi 3.2.2.1). This version incorporates Google Translate API through Python library. This tool requires an active online connection to work and detect to behave like the original version [15].

Live samples were collected from twitter API using Twitter developer account by Tweepy library in python. Tweets were searched from the major mission official accounts:

(@HopeMarsMission", "@NASAPeersevere", "@CNSA_en").

The returned tweets were filtered, cleaned, and analyzed for their overall sentiment value. Twitter library will be used and various codes were applied to download the tweets. Next, the dictionary will be converted using a Panda data frame to plot the result as a bar graph. The implementation details are stored in a GitHub repository:

https://github.com/isharhan/3Mars_missions.

3.1 Design of Experiments

VADER vs Textblob

After applying a generic sentiment analysis to Twitter data using Textblob package in Python. The polarity value for each tweet on the related search term has been calculated, then it was plotted in a histogram to determine the general feeling towards the query. First the accuracy of Textblob package was evaluated and compared with VADER against actual sentiment. The initial result of random tweets that both tools (textblob and VADER) have fairly similar accuracy at 80% compared to the actual emotion (Table 2). However, Textblob version used in this experiment is only able to analyze text in English, therefore, the following analysis will cover 1000 tweets in English from each of the three missions.

Table 2. Accuracy comparative table between Textblob and VADER

	Textblob	VADER	Actual
Claim UAE has reserved jobs for those born on 9 Feb the day when HOPE Probe arrived on the orbit of Mars	Neutral 0.0	Positive 0.5622	Neutral
Most trending hashtagArabsToMars Proud moment for UAEHope Probe draws over 56000 tweets and comments	Positive 0.65	Positive 0.5209	Positive
Amazed and wowed by the recent achievements obtained by the Emirates After the successful landing of the UAEs H	Positive 0.375	Positive 0.891	Positive
The landmarks of our country Emirates are the beauty and decoration of hope	Neutral 0.0	Positive 0.7717	Positive
Mars is looking as beautiful as ever UAE Emirates science NASA space	Positive 0.85	Positive 0.5994	Positive
Accuracy	80%	80%	100%

VADER vs VADER-Multi

To test the new function the following experimental design was used. After successfully installing VADER in Python in Google Collaboratory notebook; the following statement was used using VADER polarity score method, which calculates the score for a given sentence.

```
analyser.polarity_scores("I'm so happy we finally have
another rover on Mars")
```

The output of that function is:

```
{'compound': 0.6115, 'neg': 0.0, 'neu': 0.693, 'pos':
0.307}
```

Since the first statement was explicitly positive, another negative statement was tested.

```
analyser.polarity_scores("Why wasting money on Mars mis-
sion when you don't have money for water and electric-
ity")
```

The return output of the second statement has returned negative, which suggests the basic sentiment analysis is working.

```
{'compound': -0.4019, 'neg': 0.162, 'neu': 0.838, 'pos':
0.0}
```

The original VADER scoring methodologies do not handle text in different languages apart from English. To confirm this, the original VADER was tested using a following statement were used:

```
#In Arabic
analyser.polarity_scores(" ناسا الأمريكية تهنئ الإمارات بمناسبة وصول مسبار الأمل
إلى المريخ")
#In Chinese
analyser.polarity_scores("我很高兴我们终于有了另一辆流浪者")
#In Hindi
analyser.polarity_scores("जब आप पानी और बबजलीके बलएपैसा नहीं देते
तो मंगल बमशनपर पैसा क्यों बबाादकरना")
```

The original VADER output has returned the same output for all statements in every language, which confirm the previous statement.

```
{'compound': 0.0, 'neg': 0.0, 'neu': 1.0, 'pos': 0.0}
{'compound': 0.0, 'neg': 0.0, 'neu': 1.0, 'pos': 0.0}
{'compound': 0.0, 'neg': 0.0, 'neu': 1.0, 'pos': 0.0}
```

VADER-Multi is handling language differently, to test the new method, the original VADER was uninstalled, and replaced with VADER-Multi 3.2.2.1. Once running the same input using the previous three statements under the new model, the following polarity scores have returned:

```
#In Arabic
{'compound': 0.4588, 'neg': 0.0, 'neu': 0.7, 'pos': 0.3}
#In Chinese
{'compound': -0.362, 'neg': 0.164, 'neu': 0.76, 'pos':
0.076}
#In Hindi
'compound': 0.4404, 'neg': 0.0, 'neu': 0.805, 'pos':
0.195}
```

Afterwards, Tweepy was installed and used from Python library to access the Twitter API functions. In addition, some additional libraries were imported for data collection and data preprocessing like Pandas, NumPy, Matplotlib, Time and Re.

Since this experiment is using real-time data; a Twitter developer account is essential to continue this experiment. The necessary credentials for authentication were entered to extract the tweets from Twitter. The tweets were pulled from the main three Mars missions accounts using the following input:

```
comp_searches = ("@HopeMarsMission", "@NASAPersevere",
"@CNSA_en")
```

After defining the accounts that we need to extract tweets from; 1000 tweets were extracted from each account. The tweets were stored in a sentiment dictionary with the user date. Table 3 shows the extracted array in a data frame from different languages.

To extract useful insights from the tweets, the data needed to be cleaned from its unnecessary symbols, such as retweets, username handles, URL links, which make it easier to analyze them afterwards (Fig. 2).

Table 3. Extracting the array in a data frame from different languages

	User	text	Date
0	@HopeMarsMission	RT @HHShkMohd: Another successful mission has ...	Thu Feb 18 20:57:41 +0000 2021
1	@HopeMarsMission	RT @HHShkMohd: انضمت اليوم بعثة جديدة من كوكب ...	Thu Feb 18 20:57:39 +0000 2021
2	@HopeMarsMission	Inspired by @NASA's work since we were childre...	Thu Feb 18 20:57:11 +0000 2021
3	@HopeMarsMission	RT @HHShkMohd: نهنيء جمهورية الصين الشعبية بإن...	Wed Feb 17 05:19:55 +0000 2021
4	@HopeMarsMission	RT @MohamedBinZayed: The transmission of the H...	Sun Feb 14 09:37:55 +0000 2021
5	@HopeMarsMission	RT @MohamedBinZayed: إرسال أول صورة للمريخ بعد...	Sun Feb 14 09:37:53 +0000 2021
6	@HopeMarsMission	RT @HHShkMohd: من ارتفاع ٢٥ ألف كم عن سطح الكو...	Sun Feb 14 09:37:45 +0000 2021
7	@HopeMarsMission	RT @DXBMediaOffice: عمران شرف، مدير مشروع #الإ...	Fri Feb 12 07:08:09 +0000 2021
8	@HopeMarsMission	RT @DXBMediaOffice: Omran Sharaf, Project Dire...	Fri Feb 12 07:07:58 +0000 2021
9	@HopeMarsMission	RT @DXBMediaOffice: .@SarahAmiri1, Minister of...	Fri Feb 12 07:07:53 +0000 2021

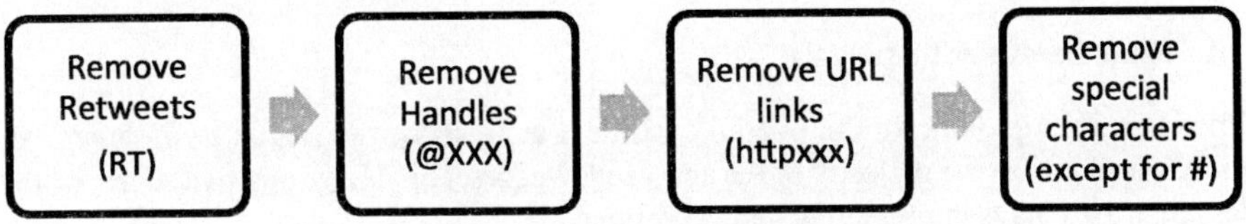

Fig. 2. Data cleaning and filtering process

The following data shows a sample of the tweets after being cleaned and filtered:

```
0      Another successful mission has just reached M...
1      . . . انضمت اليوم بعثة جديدة من كوكب الأرض للكوكب ا
2     Inspired by 's work since we were children, we...
3      نهنيء جمهورية الصين الشعبية بإنجازها العظيم و . . .
4      The transmission of the Hope Probe's first im...
```

Afterwards the scores for tweets needs to be extracted, the polarity score was used to find all positive, negative, and neutral scores for each tweet and have been plotted in Table 4.

Finally, the results are combined in a pivot table of all the mean compound scores of all account tweets where the result can be analyzed.

Table 4. The associated scores with the respective tweets for the account @HOPEMARSMISSION

	text	Compound	Positive	Negative	Neutral
0	Another successful mission has just reached M...	0.9403	0.547	0.0	0.453
1	انضمت اليوم بعثة جديدة من كوكب الأرض للكوكب ا...	0.8689	0.266	0.0	0.734
2	Inspired by 's work since we were children, we...	0.7351	0.256	0.0	0.744
3	نهنىء جمهورية الصين الشعبية بإنجازها العظيم و...	0.9403	0.344	0.0	0.656
4	The transmission of the Hope Probe's first im...	0.4404	0.121	0.0	0.879

4 Results

The following section will try to answer the first research question (RQ01) "What is the general feeling associated with the three Mars missions?"

4.1 Mars Sentiment Analysis

The chart in Fig. 3 shows the overall sentiment analysis of the search term "mars" is mostly positive, as the majority of the analyzed tweets seem to have positive score, while less than 10% have expressed negative feelings.

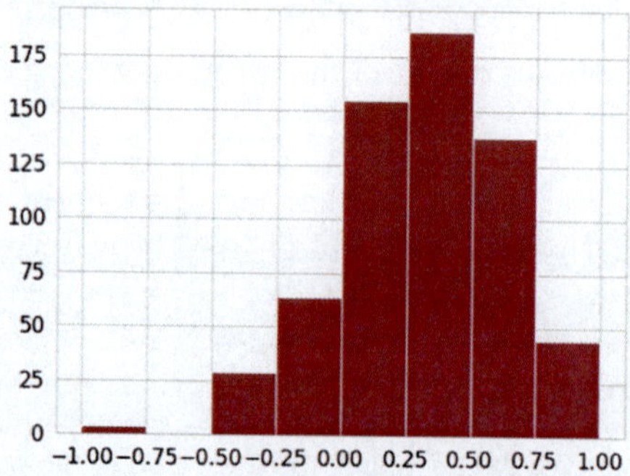

Fig. 3. Sentiments from Tweets on The Mars

Emirates Hope Mars Sentiment Analysis: The chart in Fig. 4 of Emirates hope mission shows a compound positive emotion, with no signs of any negative feelings detected from the analyzed tweets. However, it can be noticed that 50% of the positive polarity is closer to the neutral feeling. Which suggest more effort is needed to shift the polarity to the positive side.

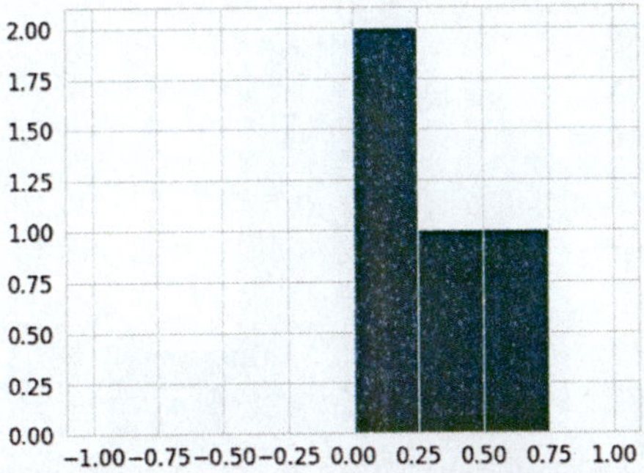

Fig. 4. Sentiments from Tweets on Emirates Hope Mission

Tianwen-1 Sentiment Analysis: Most of tweets mentioning Tianwen-1 have a compound positive feeling (Fig. 5), with polarity around neutral feelings. There are some negative feelings dedicated, but they are less than the total positive feelings in general. Which is good distribution of emotions.

Mars2020 Sentiment Analysis: The distribution of sentiment around the search term "Mars2020" is 66% positive with 44% slightly negative tone. (Fig. 6), The polarity of sentiment seems to be around neutral feelings. Which suggest a fair number of contradictory opinions around the search term "Mars2020".

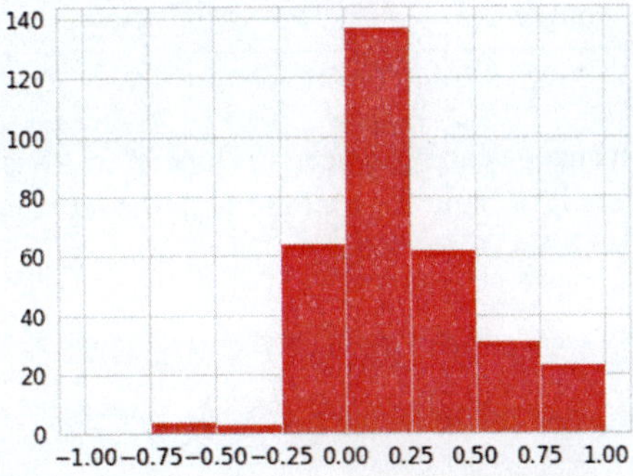

Fig. 5. Sentiments from Tweets on Tianwen-1

4.2 The Impact of MSA on Mars Missions' Tweets

The Impact of MSA on Positive Tweets: The following section will try to answer the second research question (RQ02) "What is the impact of VADER-Multi Analysis on the three Mars missions' Twitter accounts?". The following Table 5 shows that impact of using VADER multilingual (VADER-Multi) tool vs the normal VADER. The result shows the new multilingual tool has affected the final result by including results from

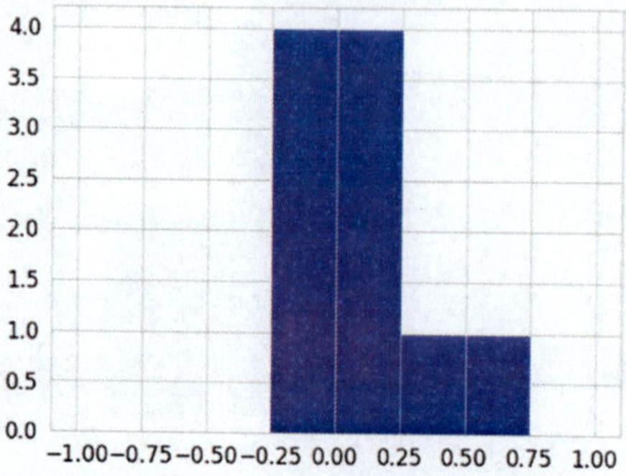

Fig. 6. Sentiments from Tweets on Mars2020

foreign language (mostly Arabic). The score of the CNSA (15%) has not been affected, while emirates hope and NASA Persevere have witnessed some changes (from 9% to 10%).

Table 5. Three Mars mission accounts Positive score

User	Positive score using VADER	Positive score using VADER-Multi
@CNSA_en	0.15125	0.15125
@HopeMarsMission	0.06259	0.16547
@NASAPersevere	0.09262	0.09635

The most significant change can be noticed in Hope Mars Mission account sentiment (from 6% to 17%), where it became the account with the most positive tweets (Fig. 7) only after analyzing it with VADER-mulit.

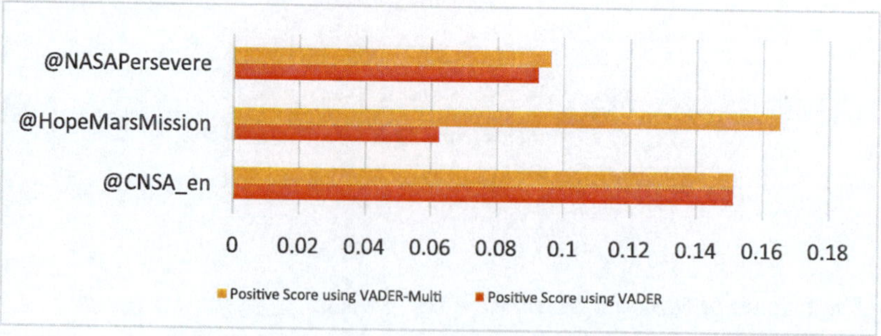

Fig. 7. Three Mars mission accounts positive score

The Impact of MSA on Mars Missions' Compound Tweets: After analyzing the result in Table 6, we can notice that CNSA (41%) has the highest compound score among the three analyzed accounts under the normal VADER tool, however, when applying the multilingual tool of VADER-mulit, Hope Mars Mission (45%) has topped the list again. NASA Persevere has been impacted and increased slightly by VADER-Multi (from 18% to 19%).

Table 6. Three Mars mission accounts compound score

User	Compound score VADER	Compound score VADER-Multi
@CNSA_en	0.406475	0.406475
@HopeMarsMission	0.143513	0.456652
@NASAPersevere	0.177393	0.186416

Table 7. Three Mars mission accounts negative score

User	Negative score VADER	Negative score VADER-Multi
@CNSA_en	00.00000	00.00000
@HopeMarsMission	00.00310	0.00376
@NASAPersevere	0.01211	0.01211

Figure 8 shows Hope Mars Mission has the highest compound, which indicates most tweets posted by this account was positive, followed by CNSA than NASA Persevere mission (19%).

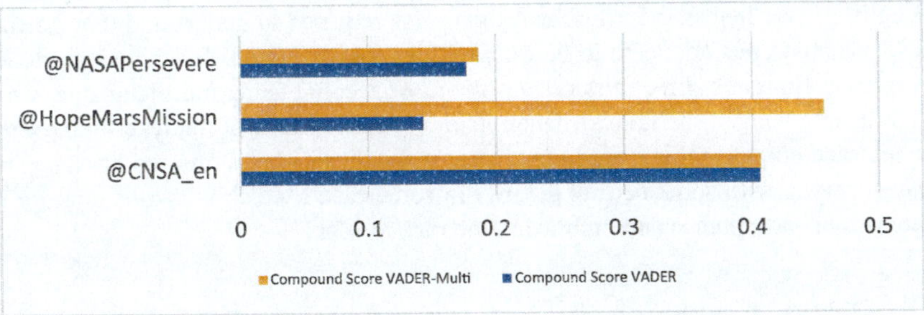

Fig. 8. Three Mars mission accounts compound score

The Impact of MSA on Mars Missions' Negative Tweets: By looking at the most negative results, Table 7 can shows the account who posted the maximum number of negative accounts. Multilingual sentiment analysis doesn't seem to impact the negative score much, as most of the three analyzed accounts have a positive tone in general.

Although the accounts from CNSA (0%) and NASA (1%) did not have any impact, VADER-Multi tool was still able to detect slight change in negativity score in Hope

Mars mission account (from 0.03% to 0.04%), however, this did not change the overall ranking. And both VADER and VADER-Multi provided almost similar result in this scenario (Fig. 9).

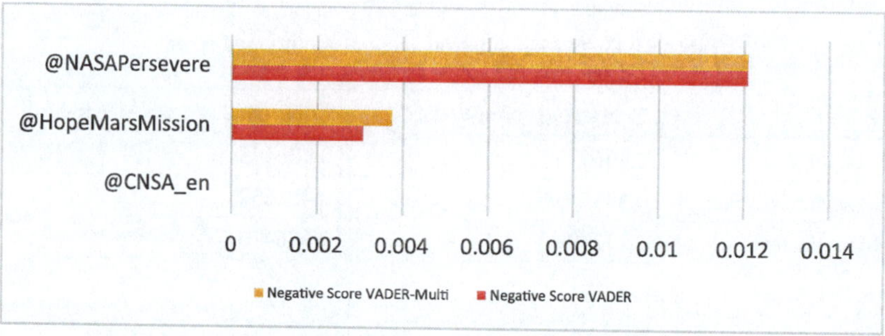

Fig. 9. Three Mars mission accounts negative score

5 Discussion

The previous results shows that all three mars Twitter account missions have a good positive compound score. Posts from Emirates Hope Mission is predicted to be the most positive, while posts from NASA Persevere here are predicted to be the most negative. One of the interesting findings is the returned result of CNSA to be the least negative on with a result equal to = 0.0. a possible reason for this that CNSA do not use Twitter to address their nation. However, they use a positive tone to address the outside world only in English, nevertheless, a further investigation is required to understand this, besides, CNSA account does not seem to be impacted by VADER-Multi for that same reason. In Contrast, Emirates Hope mission was the most account using multilanguage, where a significant change was noticed when using VADER-Multi to analyze their tweets. The returned output of this experiment in general indicates Mars mission has a positive polarity among with some neutral polarity in some cases, which is a good sign for the three nations' acceptance and embracing the race to Mars.

Table 8. Translating capitalization to different languages in Google Translate

	Normal	Capitalization
English	Mars is awesome!	Mars is AWESOME!
Spanish	¡Marte es asombroso!	¡Marte es IMPRESIONANTE!
Arabic	المريخ رائع !	المريخ رائع !
Chinese	火星真棒！	火星真棒！

VADER analyses emotion based on the five-primary style of writing: Punctuation, Capitalization, Degree modifiers and Conjunctions, while most might of these styles are preserved when translated to other languages, some of them are not preserved. For example, Capitalization is not applicable when translating text from Arabic to English (Table 8), which means VADER-Multi is using only four out five (80%) of its capabilities when dealing with Arabic corpus. In addition, when repeating the experiment using different parameters, the system returns "Http Error" on many occasions. One reason for that is believed to be the dependance on an active online connection to retrieve the tweets and translate them via Google Translate, which makes it one of the downsides of this method.

```
HTTPError: HTTP Error 429: Too Many Requests
```

Many experts believe this type of error is not an error, but it's a request from the server to stop spamming requests. It seems using this method may generate a lot of requests. Some suggest making sure everything is setup properly to receive a "Retry After" header along with the error.

Future work can be developing a mitigation solution for that error to be able to perform multilanguage tools that bypass these types of obstacles.

The work [13, 16–26] concluded that translating an input from a foreign language into English can achieve better results than existing language specific efforts. These studies could have been further investigated by including Arabic. The complexity of Arabic is not observed by its structure, but also by its slang and different dialects.

6 Conclusion

In this study, we have tried to describe the theoretical framework of multilingual sentiment analysis, highlighted some challenges, and review VADER tools as one of the potential tools to be used for multilingual analysis. A real live-data case study was used to test this methodology, VADER-Multi and Textblob was used to examine the sentiment associated with success of three main space mission to Mars from three countries. The aim of this research was to gather sentiment analysis around the three Mars mission, in addition, determine if VADER-Multi was a reliable tool to analyze multilingual social media texts. The study was limited by time, constrains, limited expertise, and one methodology using VADER-Multi. Future work can be investigating different multilingual solutions such as EvoMSA. In addition, providing a web-based tool that is easier to use and investigating the root cause of the system overloaded with too many requests. This study hopes to contribute to multilingual sentiment analysis and space exploration, as it's believed to be among the first to examine Mars missions' sentiment analysis, and among the first to review VADER multilingual tool (VADER-Multi).

Acknowledgment. This is a part of a project conducted at the British University of Dubai. Special thanks to (Hutto & Gilbert 2014), (Loria 2020), (Band 2020) and (Morrissey et al. 2020), Python Software Foundation, Google Collaboratory and Twitter for providing the tools and methods to make this experiment possible. The author(s) declares no form, sources of any funding used to conduct this research and no type of conflict of interest regarding the research and the aim of this article.

References

1. Sircar, N.: UAE Mars mission: Success of Hope Probe overwhelms world media - News | Khaleej Times (2021)
2. Band, A.: Twitter Sentiment Analysis using Vader & Tweepy | Python in Plain English (2020)
3. Vesterberg, P.: Sentiment analysis and the moon landing | Towards Data Science, Medium (2019)
4. Loria, S.: textblob Documentation (2020)
5. Klappenbach, A.: The 12 Most Spoken Languages in the World – Busuu Blog, Busuu (2021)
6. MOE: Languages, Population and Religion, Ministry of Economy (2020)
7. Dashtipour, K., et al.: Multilingual sentiment analysis: state of the art and independent comparison of techniques. Cogn. Comput. **8**(4), 757–771 (2016). https://doi.org/10.1007/s12559-016-9415-7
8. Swarnkar, N.: VADER Sentiment Analysis in Algorithmic Trading. QuantInsti.Com (2020)
9. Hutto, C.J., Gilbert, E.: VADER: a parsimonious rule-based model for sentiment analysis of social media text. In: Proceedings of the Eighth International AAAI Conference on Weblogs and Social Media, pp. 216–255 (2014)
10. Stack Overflow: Sentiment Analysis in Python: Textblob vs Vader? Stack Overflow (2017)
11. Amin, A., Hossain, I., Akther, A., Alam, K.M.: Bengali VADER: a sentiment analysis approach using modified VADER. In: 2nd International Conference on Electrical, Computer and Communication Engineering, ECCE 2019. Institute of Electrical and Electronics Engineers Inc. (2019). https://doi.org/10.1109/ECACE.2019.8679144
12. Araújo, M., Pereira, A., Reis, J., Benevenuto, F.: An evaluation of machine translation for multilingual sentence-level sentiment analysis. In: Proceedings of the ACM Symposium on Applied Computing, pp. 1140–1145. Association for Computing Machinery, New York (2016). https://doi.org/10.1145/2851613.2851817
13. Araújo, M., et al.: iFeel 2.0: a multilingual benchmarking system for sentence-level sentiment analysis (2016)
14. Morrissey, M., Wasser, L., Diaz, J., Palomino, J.: Analyze The Sentiment of Tweets From Twitter Data and Tweepy in Python | Earth Data Science - Earth Lab, Earth Lab (2020)
15. Hutto, C.J.: vader-multi · PyPI, Python Software Foundation (2019)
16. Araújo, M., Pereira, A., Benevenuto, F.: A comparative study of machine translation for multilingual sentence-level sentiment analysis. Inf. Sci. **512**, 1078–1102 (2020). https://doi.org/10.1016/j.ins.2019.10.031
17. Galeshchuk, S., Analytics, G., Jourdan, J., Qiu, J.: Sentiment analysis for multilingual corpora. In: The 7th Workshop on Balto-Slavic Natural Language Processing, pp. 120–125 (2019)
18. Salloum, S.A., Al-Emran, M., Shaalan, K.: A survey of lexical functional grammar in the Arabic context. Int. J. Comput. Netw. Technol. **4**(03) (2016)
19. Salloum, S.A., Khan, R., Shaalan, K.: A survey of semantic analysis approaches. In: Hassanien, A.-E., Azar, A.T., Gaber, T., Oliva, D., Tolba, F.M. (eds.) AICV 2020. AISC, vol. 1153, pp. 61–70. Springer, Cham (2020). https://doi.org/10.1007/978-3-030-44289-7_6
20. Wahdan, K.A., Hantoobi, S., Salloum, S.A., Shaalan, K.: A systematic review of text classification research based on deep learning models in Arabic language. Int. J. Electr. Comput. Eng **10**(6), 6629–6643 (2020)
21. Yousuf, H., Lahzi, M., Salloum, S., Shaalan, K.: Systematic review on fully homomorphic encryption scheme and its application. In: Al-Emran, M., Shaalan, K., Hassanien, A.E. (eds.) Recent Advances in Intelligent Systems and Smart Applications. SSDC, vol. 295, pp. 537–551. Springer, Cham (2021). https://doi.org/10.1007/978-3-030-47411-9_29
22. Alshamsi, A., Bayari, R., Salloum, S.: Sentiment analysis in English texts. Adv. Sci. Technol. Eng. Syst. J. **5**(6), 1683–1689 (2020)

23. AlShuweihi, M., Salloum, S., Shaalan, K.: Biomedical corpora and natural language processing on clinical text in languages other than English: a systematic review. In: Al-Emran, M., Shaalan, K., Hassanien, A.E. (eds.) Recent Advances in Intelligent Systems and Smart Applications. SSDC, vol. 295, pp. 491–509. Springer, Cham (2021). https://doi.org/10.1007/978-3-030-47411-9_27
24. Yousuf, H., Salloum, S.: Survey analysis: enhancing the security of vectorization by using word2vec and CryptDB. Adv. Sci. Technol. Eng. Syst. J. **5**(4), 374–380 (2020)
25. Elnagar, A., Yagi, S.M., Nassif, A.B., Shahin, I., Salloum, S.A.: Systematic literature review of dialectal Arabic: identification and detection. IEEE Access **9**, 31010–31042 (2021)
26. Al-Sarayrah, W., Al-Aiad, A., Habes, M., Elareshi, M., Salloum, S.A.: Improving the deaf and hard of hearing internet accessibility: JSL, text-into-sign language translator for Arabic. In: Advanced Machine Learning Technologies and Applications: Proceedings of AMLTA 2021, p. 456 (2021)

Applying the Uses and Gratifications Theory to College Major Choice Using Social Networks Online Video

Mohammed Habes[1(✉)], Mohd Hashem Salous[1], and Marcelle Issa Al Jwaniat[2]

[1] Faculty of Mass Communication, Radio and TV Department, Yarmouk University, Irbid, Jordan
{mohammad.habes,mohadsalous}@yu.edu.jo

[2] Faculty of Mass Communication, Department of Journalism, Yarmouk University, Irbid, Jordan
marcelle.jwaniat@yu.edu.jo

Abstract. This study aims to review studies regarding the impact the of online video on students' choice of academic major in general. The researcher used Uses and Gratifications Theory as conceptual background of the study. The researcher employed experimental approach to measure the uses and gratifications to student's major college choices, using online video. Results showed that students gratify certain needs while searching for college-related information through online videos. These needs involved: Participation in discussions, interactions with other students, sharing academic information and to understand the academic concepts. Moreover, it also plays a key role to motivate the students, boosting their interest in watching videos on YouTube, Facebook and use these more frequently than other social network sites. Besides, Video Content (video effect, sound effect, music, live videos, and animation) are important factors that influence their decisions for the selection of college majors. Therefore, the findings of the surveyed studies will assist the scholars in pursuing their future studies.

Keywords: Social networks · Uses and gratifications theory · Online video · Student · College major · Video content

1 Introduction

People from all age groups are using social networks technology, due to the exponential growth of new media in recent years. A large proportion of users of social media technology are of the young generation today and it facilitated their every aspect of life including social, personal and others [1–3]. This is because, during the past 10 years, the role and place of social media have prominently changed in young people's lives. Technological revolution and ease of access due to technology development are the leading causes of increased social media integration. Mobiles phones, laptops, tablets and other devices along-with faster internet offering Wi-Fi and 4G technology are the major reasons behind it [4]. Young people prefer using social media for all the leading

© The Author(s), under exclusive license to Springer Nature Switzerland AG 2022
A. E. Hassanien et al. (Eds.): AMLTA 2022, LNDECT 113, pp. 388–400, 2022.
https://doi.org/10.1007/978-3-031-03918-8_33

purposes including communication, education, information and entertainment. Facilitating interaction, communication and education are the major advantages of these Social Networking today [5]. Also affirmed by [6] as they consider it as a beneficial shift from traditional media to new media offering several advantages. According to the researchers, social media usage has altered their lifestyle and provided them with ease of access to information worldwide. They can communicate and interact with people in other parts of the world and they can even make the discussion about prevalent issues. Briskly increasing number of social networking sites tend to perform different functions that are even helpful for the young generation's social development. Thus, social media usage facilitates the overall needs and requirements of the young generation today [7].

Similarly, transferring information through social media is one of the major benefactions offered by these online platforms. Users are autonomous to choose what they want to see and this is how they experience the content of their choice. Especially choosing the suitable field of study at the university level is prominent in this regard. The way to make relevant choices are made and students are better able to make decisions on their own [8]. Due to this, considering online videos to seek information about the relevant field of study is highly practised today. Moreover, online video platforms for personal information and education gain has changed greatly during the last few years. Technology is rapidly improving and even students can save these videos, consider them again and again and make the relevant decisions. Feasibility to search, record and reconsider them all are the reasons, students today are well capable of brainstorming about their study choices [9]. Even in students' opinion, making their decisions on their own and brainstorming is feasible due to online advisory platforms. This helps them to actively choose a field of interest and further to study well as they make study choices according to their aptitude. In this regard, online videos particularly advisory content helps them in three dimensions: They can actively search and evaluate the information of their choice (visualized benefits), they can make their study and career decisions on their own (cognitive benefits) and, they are capable of performing well in their educational arena as their choices represented their aptitude and an active consent (psychological benefit) [10]. Indeed, online videos are an enriched source of educational guidance, information and learning today. They are serving as a powerful tool of accessing and facilitating the students for many aspirations. They serve as a fast and efficient source of getting information that not only guides the students but also benefits them regarding their educational matters [11]. By keeping in view the importance of online videos, the focus of this study is the integration of the social networks online video and the uses and gratifications theory to conduct empirical research to determine the effect of online video on students' selection of their college major. As the Uses and Gratification Theory states the users' reasons for using social networks, where the audience are active selectors of the media. Here individuals are not passive recipients, but actively select the content based on their specific needs [12]. Furthermore, the motivations to use social media include searching for a companion, relaxing, learning, escapism, interaction, education and others [13]. In this regard, the needs of the user or audience in using the media content determine the content consumed. Each user has his/her own need, but it is witnessed that leisure and information are the foremost reasons for users to use online video Galan et al. [14]. The online resources from learning material to career information, every information is available in online videos. Students to seek information, help and guidance are the benefactions offered by these resources [15]. This need of information as described by Uses and Gratifications Theory and [14, 16–18], individual needs and academic needs for media use in selecting academic disciplines, and the influence of the online video in the educational process is crucial today. Likewise, it is more convenient for the students today to stay online and search for the guidance they need. They gather information, make favourable educational decisions which further helps to study well and make achievements [19]. This is because online videos are potentially compatible to alter the attitudes and decisions of the young students. These online platforms are according to the propositions made by Uses and Gratifications theory [20] which describes an active selection of content to gratify certain needs from media. Especially from new media, students gratify their informational and educational needs besides communication. Students today tend to prefer social media especially online video content as a convenient source of information which facilitates them regarding their educational needs [20]. Therefore, the researcher scrutinized different aspects that affirm a strong association between informational videos available online and the decision to choose their college majors. First, the researchers discussed the role and importance of social media for the students particularly, helping them to make relevant educational decisions. The researcher reviewed the available literature that validated the phenomenon and, in the end, made the conclusion accordingly.

2 Literature Review

Here the researcher will review the relevant literature which will further validate the propositions of this study as "reviewing the literature is an essential step in conducting research studies" [21, 22]. Katz, Blumler, and Gurevitch [23] postulated the needs of individuals for using media including traditional television (video) generally into five major classes: the need for social interaction, to ease tension, to fulfil influential or emotional needs, to address cognitive needs and for social integration. Likewise, the extent to which individuals tends to use new media is highly correlated with their experiences especially gratifications offered by online networking platforms. Increasing motivations to use social networking sites indicate their increased preference by the users. This concept of usefulness affirms the postulations given by uses and gratification theory. [24] Moreover, these uses and gratifications have changed among the young users particularly students of the new media and their motives modified through means of social communication to become a platform for the exchange of experiences, information and acquaintance [25–28]. In this regard, social media have altered the way audiences interact with the creator of information, and how information is discussed and shared among users [25]. To strongly affirm the importance of social media for the students, [29] witnessed the advantages of learning and information gathering that online videos offer. According to the results of their study, the advantages of the online videos include classroom interactions, enhanced reasoning, management and thinking skills. Another study to analyse the benefactions of social media videos content was conducted by [30]. The researchers surveyed the teachers and results unveiled that instructors also consider the usefulness of exposure to learning strategies, grades, and degrees by online video benefitting the students on every level. [31, 32] also examined the use of online platforms by undergraduate college students. Findings revealed that students at undergraduate level tend to prefer social media usage for positive constructive aspirations. They use online platforms for information gathering regarding journalistic purposes which further helped to elect journalism as their major subject in college. While College Major Choice, by using Social Networks Online Video is typically associated with lectures on topics like biology and literature, there's also a regular series on computer science that looks into everything from the history of computing to, yes, artificial intelligence and machine learning. [33] This using Social Networks Online Videoon ML helps to demystify a typical ML pipeline and includes a solid overview of some of the more common machine learning algorithms, [34] like support-vector machines to College Major Choice by using Social Networks Online Video.

2.1 Uses and Gratifications of Student to Major Academic Choices by Online Video

Alhumaid et al. [12] highlighted the propositions according to the uses and gratifications perspective, that motivate students to use online videos and gratify their informational and educational needs. These needs may include: keeping a follow up on official university platforms, searching for college-related information, to participate in discussions with other students, to share academic information, to understand the academic concepts, for collaborative learning, to exchange information, to find out about new undergraduate majors, video conferencing, experience interaction with teachers, get academic information such as the academic majors, to participate in an online activity, for social utility, and others [14, 36, 38, 39, 40–42]. Table 1 gives a brief overview of the uses and gratifications identified through different studies by using social media that are capable of influencing the decisions made by students regarding the selection of major subjects.

The motivations regarded mainly as uses and gratifications, behind students' selection of college major when using online videos by social media include freedom of education which student decide and receive from these sources. Similarly, ease of access and usage convenience, advantages of deep exploration, digital ad spending, trust, or the voice of teachers [46]. Furthermore, Raikos and Waidyasekara [44] also witnessed a high-level consideration, evaluation and deep exploration by students when they select college majors and consider online videos as a source of advice and guidance. While [47] affirmed the use of social media for information gathering purposes which further leads to improving academic performance by the students. Another study to validate the use of social media for information and educational purposes was conducted by [48]. The researchers surveyed teenage girls from different schools and colleges in Pakistan. Findings revealed that teenager use social media mainly for information gathering purposes. Moreover, the results also witnessed the positive impacts of social media on the students' academic performance. Aloraini [39] examined the needs gratified by social media and found that these are purely experimental needs. Marketing, consumer trust, relationship management, and interpersonal attraction were cited as reasons for students to use social media online video.

Table 1. Students' needs and gratification obtained from using online video.

Reference	Uses and gratifications factors
[36, 43]	Entertainment, Seeking Information, Interaction, self-expression like using, convenience & advantage, knowledge, social needs internal satisfaction, experience, opinions, comments discussion, learn new things and to keep pace with the modern world
[36]	Interaction with group member, opportunity to discuss, exchange of information, facilitate interaction, Collaborative learning, Needs personal relationships, Intention create groups and chat, improve my research skills, collaborative learning, Participation, Knowledge
[14]	Influencing the opinion, learning, continues the academic activities, interaction with students and gather information. university information & news, event and activity. Freedom of Education and Student Life (pictures, videos, accommodation, Entertainment, job opportunities)
[39, 44]	Scientific achievements, information exchange, behaviour, values, discussion with groups, Effects Values and needs, learning and making new decisions
[40, 42]	Knowledge sharing, reinforces motivation, cooperative behaviour, communication enhancement and information sharing Entertainment, opinion interpersonal attraction, participate, increased awareness, Social attendance, Education and training
[38]	Knowledge, experience, deep exploration, integration of information high-level thinking, Engagement by students, Needs, information. Users, Developing opinion, awareness, Behaviour, Flow information, Discussion. Collaborative learning, interactive with peers, Interactive with teachers, academic performance, Continuous communicating

2.2 Motivations of Student to Major Academic Choices, by Online Video

"Most studies have pointed out that certain needs and gratifications are the influencing factors that motivate social media usage among students. Besides communication, education and information gathering are also among the most prominent reasons [13]. In addition to the impacts of media, the public's expectations of social media are mainly to guide, inform and benefit them [14, 21, 22] in their studies found that the motivations for the student to use YouTube or video by Facebook, for instance, are to chat with other users, broadcast live videos, share and receive information etc. Additionally, according to the previous studies, the factors that motivate students to use and obtain gratification from online videos include keeping a follow up on university's social media pages and websites, searching for college-related information, to participate in discussions, engagement in the online learning environment and sharing academic information, [21, 49] Similarly, in order to understand educational concepts, for collaborative learning, exchange of information, searching about new undergraduate majors, experiencing interaction with teachers, to get academic information, enhance new academic skills, for scientific consulting, convenience, entertainment, engender positive thinking are some highlighted gratifications offered by online videos [50]. Furthermore, [3, 22] reported that aspirations like information seeking from published ads, finding interesting information about academic majors, as well as understanding the instructions of the courses provided, and obtaining more information about the university faculties are the major benefactions offered by social media video content. The Motivations regarding online videos search to select major college choices, are briefly outlined in Table 2.

Table 2. The motivation of users in using online video social networks

Reference	Motivations
[50]	Interaction, environment for learning, interaction with peers, new academic skills, Chat, academic motivation, Sharing Information, positive constructive thinking and gaining educational guidance
[30]	Published educational ads, and gaining interesting information about academic majors, as well as understanding the instructions of the courses provided, communication with the friends and feel part of the Facebook community
[51]	Social media platforms are used for educational and informational purposes. Influence students" perceptions and help them to make effective decisions
[3]	Convenience, entertainment, accessibility, social interaction, personal communication, positive thinking, improved interaction, and social responsibility
[52]	Enhance new academic skills and scientific consulting

Table 2 affirms different important factors regarding the increased usage of social media among students. These factors involve: the most motivating social media channels (YouTube, Facebook) and the effect of new media have on students' selection of college majors, ease of use, internet accessibility, knowledge of academic majors, continuous consumption of information, an environment for learning, creating perceptions, audio and video, private information. According to. Ali et al. [53], the different types of media content especially videos influence attitude, thinking, knowledge, and emotions and work as sources of persuasion for the students (chatting, commenting, posting, and sharing). There are other similar factors for measuring the attitudinal change concerning online videos and motivating the students to select certain majors. Some were taken into account while others such as blogging and microblogging tools are still under consideration by the researchers [52]. For [51] social media platforms are used for educational and information purposes by the students. They help the students to make an important educational decision which also affects their academic performance positively.

2.3 Video Content with Uses and Gratifications

Table 3 further affirms the influence of social media platforms on the decision making process of the young generation, particularly while making educational decisions. According to [54], social media platforms i.e. blogs, websites, personal networks, YouTube videos, live chats, comments, pictures, and videos deeply influence

Table 3. Social Media content and its influence on academic decisions & performance

References	Academic decisions and performance
[54]	Modified ways of communication, interaction and information-seeking behaviour among the students
[29, 52]	Influencing the learning patterns and information-seeking behaviour, modified the educational performance of the students
[55]	Social media content helps students to make effective decisions regarding their educational matters. It also helps them to ameliorate their academic performance
[56]	Social media content provides efficiency and self-sufficiency to the students to gain autonomy regarding their educational matter and independently make better decisions
[57]	High-quality content on social media tends to attract students. Students take it under consideration and get influenced by it and make decisions regarding the field of their interest

our behaviour and perceptions. These have changed the way they have transformed how humans as a whole, communicate with each other. Moreover, studies conducted by Cardoso et al. [29]; Nwangwa et al. [52] regarding videos and tutorials revealed that they mainly influence the learning process and educational decisions made by the college students. Furthermore, [55] examined the impacts of social media content on learning behaviours of the students in Pakistan. Researchers found that social media content helps students to make influential educational decisions. These decisions help them to gather academic help and support. According to [56] social networking has an immense influence on individuals' overall decision-making behaviour. In their study, the researchers analysed the use of social media for educational purposes and its impacts on making education concerning matters. Results revealed that it makes students self-efficient and also provides them with the autonomy to make their decisions on their own. Social Networking usage is briskly preferred in both public and private level institutions. It has improved the ways of gathering information, making decisions and also provided autonomy regarding the educational matters to the students [57]. Social Media content contains texts, video, images, sound effects, image effects, the use of high-quality content, innovative design, animation, special video production programs which indicate the power of the video content influencing students in their selection of the college major [50, 58].

3 Discussion

According to [59], social media platforms have audio and video content that helps the student to read, watch and listen to the programs of their choice. Especially, online video sharing sites are prominent in this regard as they have an enormous number of relevant videos for the students. This trend of social media usage is rapidly increasing day by day. The students prefer them for entertainment and educational aspirations. They also encourage community platforms and discussion forums where students can discuss different topics and get guidance. Students get engaged in conversation about different issues and make relevant conclusions [60]. Even [61] considers respective social media content as a source of consultation and guidance for the students at every level. This expanding usage is benefiting the students as social media platforms are providing educational support and enabling able to educational and career decisions on their own. Therefore, the discussion on the use, preference and impacts of online vides content has validated the phenomenon under the Uses and Gratification Perspective of Media. Social networking sites tend to offer content that facilitates the connection, information and education. Many studies affirm how these social media gratify the information and education needs of the studies. This is because students prefer social media to gratify their needs and get benefitted from them. These illuminating qualities sparked more opportunities for the others to actively engage, get information and make the decisions on their own [62]. Social media platforms are benefiting the students for communication, interaction, teamwork, learning new skills and gaining information. Also, an increasing number of social networking sites are distinctly serving these purposes on a broader level. Majority of the students use them to gratify their education and information gathering needs [63]. Moreover, students tend to prefer social networking as they offer them to experience important content. Here the researcher's proposition is the facilities provided by online videos that besides gratifying the needs, also provide educational guidance and information to the students. This further helps them to make decisions regarding subject selection. For example, the YouTube platform provides online videos where besides entertainment, information and education are available. Tutorials and even videos from different educational institutions are also available. These videos are divided into different categories mainly depending on their themes. YouTube provides categories like blogs, education, entertainment and others and the people choose what they want to watch [64]. Likewise, the continued growth of social networking sites is directly correlated with the academic performance of the students. This an important example of social media popularity among the students these days. They are focused on their educational matters and make their decisions accordingly [56]. Even these social networking sites work as a pathway of communication between students, teachers and educational institution. This shows how deeply social media integrated into students' life and everyday educational matters [65].

4 Conclusion and Recommendations

This study highlighted the importance of social media-based video content that help and support the students to gather relevant educational guidance and support. The researcher made arguments in the light of gathered studies that witnessed social media content as helpful, influential, offering differs gratification and attracting the students. Also affirmed by [13, 66] as they highlighted common services offered by online video material i.e., sharing the opinion, continuous learning, follow events, communicate with friends & knowledge that gratify the needs of the individuals. This is because online videos have greatly altered the information gathering ways [29, 52, 54, 67–69]. The type of video content offered by social media affects the students' perceptions about their education, aptitude and later help them to make decisions on their own. The same propositions are given by Uses and Gratification Theory as it is stated that "the users' reasons behind using online video, where the media and public audience are active recipients of the media message". According to this notion, individuals are not passive recipients of the media, but actively consume the video content based on their specific needs. Thus, the reasons behind student's selection of college major when using online video by the social network; searching for college-related information, to participate in discussions and interactions with other students, sharing academic information, to understand the academic concepts, for collaborative learning, to exchange information, to find out about new undergraduate majors all are as similar as described by Uses and Gratifications theory. Additionally, YouTube and Facebook are the most preferred social networking sites. They deeply influence the choice of academic majors and offer ease of use, internet accessibility, knowledge of college majors, continuous consumption of information, an environment for learning and creating the perceptions. Finally, the researcher found online video content as a most influential factor on students' choice of college major: (choosing texts, high-quality production, video effect, and sound effect, the use of high-quality cameras, music, live videos, and animation). Now that research has highlighted the online videos as a major source of influence on students' academic matter, more studies are required to examine the students' psychological and social motivation through exposure to online videos. Likewise, a new strategy should also be adopted to understand the successful social networks. Future studies should include an analysis of the uses and gratifications of students and the environment in which online content is used to address the problems faced by students who do not have enough skills to search for college majors. Studies should also continue to examine the types of video content on social networks which students consider favourable regarding making their educational decisions. This will have a stronger relevance with the uses and gratifications perspective of mass media.

References

1. Salloum, S., Al-Emran, M., Khalaf, R., Habes, M., Shaalan, K.: An innovative study of E-payment systems adoption in higher education: theoretical constructs and empirical analysis. Int. J. Interact. Mob. Technol. (iJIM) **13**(06), 68 (2019). https://doi.org/10.3991/ijim.v13i06.9875
2. Ali, M., Raja, R.A.I.B., Nuri Al Amin, B.E., Iftikhar, B.: Determining the academic use of social media with technology acceptance models. NFC IEFR J. Eng. Sci. Res. **5** (2017)
3. Aljaad, N.H.M.: The role of social communication tools in education from the Saudi female students' perceptions. Int. Educ. Stud. **9**(8), 194 (2016). https://doi.org/10.5539/ies.v9n8p194
4. Mason, M.: The use of the internet and social media by young people. ResearchGate, March 2017
5. Neelamalar, M., Chitra, P.: New media and society: a study on the impact of social networking sites on Indian youth. Media (2009)
6. Rani, P.U., Padmalosani: Impact of social media on youth. Int. J. Innov. Technol. Explor. Eng. **8**(11), 786–787 (2019). https://doi.org/10.35940/ijitee.K1138.09811S19. Special Issue
7. Christofferson, J.P.: How is social networking sites effecting teen's social and emotional development: a systemic review (2016)
8. Rajadell, M., Garriga-Garzón, F.: Educational videos: after the why, the how. Intang. Cap. **13**(5), 902–922 (2017). https://doi.org/10.3926/ic.1042

9. Buchner, J.: How to create educational videos: so gelingen lernvideos. R&E Source **12**, 1–10 (2018)
10. Kosterelioglu, I.: Student views on learning environments enriched by video clips. Univers. J. Educ. Res. **4**(2), 359–369 (2016). https://doi.org/10.13189/ujer.2016.040207
11. O. of E. Technology: Letter from the Director (2017)
12. Alhumaid, K., Habes, M., Salloum, S.A.: Examining the factors influencing the mobile learning usage during COVID-19 Pandemic: an integrated SEM-ANN method. IEEE Access **9**, 102567–102578 (2021)
13. Al-Shakhanbeh, Z.M., Habes, M.: The relationship between the government's official Facebook pages and healthcare awareness during Covid-19 in Jordan. In: Hassanien, A.-E., Elghamrawy, S.M., Zelinka, I. (eds.) Advances in Data Science and Intelligent Data Communication Technologies for COVID-19. SSDC, vol. 378, pp. 221–238. Springer, Cham (2022). https://doi.org/10.1007/978-3-030-77302-1_12
14. Galan, M., Lawley, M., Clements, M.: Social media's use in postgraduate students' decision-making journey: an exploratory study. J. Mark. High. Educ. **25**(2), 287–312 (2015). https://doi.org/10.1080/08841241.2015.1083512
15. Lebeničnik, M., Pitt, I., Starčič, A.I.: Use of online learning resources in the development of learning environments at the intersection of formal and informal learning: the student as autonomous designer. Cent. Educ. Policy Stud. J. **5**(2), 95–113 (2015)
16. Nwazor, J.C., Godwin-Maduike, C.C.: Social media and academic performance of business education students in South-East Nigeria. J. Educ. Pract. **6**(25), 127–134 (2015)
17. Al-Rahmi, W., Othman, M.S., Yusuf, L.M.: The role of social media for collaborative learning to improve academic performance of students and researchers in Malaysian higher education. Int. Rev. Res. Open Distance Learn. **16**(4), 177–204 (2015). https://doi.org/10.5901/mjss.2015.v6n4s1p193
18. Tayseer, M., Zoghieb, F., Alcheikh, I., Awadallah, M.N.S.: Social Network: academic and social impact on college students. In: ASEE 2014 International Conference (2014). http://www.asee.org/documents/zones/zone1/2014/Student/PDFs/185.pdf
19. International Baccalaureate Organization: Online learning, teaching and education continuity planning for schools (2020)
20. Ebersole, S.: Uses and gratifications of the web among students. J. Comput. Commun. **6**(1), 0 (2006). https://doi.org/10.1111/j.1083-6101.2000.tb00111.x
21. Elareshi, M., Habes, M., Ali, S., Ziani, A.: Using online platforms for political communication in Bahrain election campaigns. Soc. Sci. Humanit. **29**(3), 2013–2031 (2021)
22. Salloum, S.A., AlAhbabi, N.M.N., Habes, M., Aburayya, A., Akour, I.: Predicting the intention to use social media sites: a hybrid SEM - machine learning approach. In: Hassanien, A.-E., Chang, K.-C., Mincong, T. (eds.) AMLTA 2021. AISC, vol. 1339, pp. 324–334. Springer, Cham (2021). https://doi.org/10.1007/978-3-030-69717-4_32
23. Katz, E., Blumler, J.G., Gurevitch, M.: Uses and gratifications research. Public Opin. Q. **37**(4), 509–523 (1973)
24. Habes, M., et al.: Digital media and students' AP improvement: an empirical investigation of social TV. In: International Conference on Innovation and Intelligence for Informatics, Computing and Technologies (2020)
25. Cha, J., Chan-Olmsted, S.M.: Substitutability between online video platforms and television. J. Mass Commun. Q. **89**(2), 261–278 (2012)
26. Salloum, S., Shaalan, K.: Factors affecting students' acceptance of e-learning system in higher education using UTAUT and structural equation modeling approaches. In: Hassanien, A.E., Tolba, M.F., Shaalan, K., Azar, A.T. (eds.) Proceedings of the International Conference on Advanced Intelligent Systems and Informatics 2018, pp. 469–480. Springer, Cham (2019). https://doi.org/10.1007/978-3-319-99010-1_43

27. Mhamdi, C., Al-Emran, M., Salloum, S.A.: Text mining and analytics: a case study from news channels posts on Facebook. In: Shaalan, K., Hassanien, A.E., Tolba, F. (eds.) Intelligent Natural Language Processing: Trends and Applications. SCI, vol. 740, pp. 399–415. Springer, Cham (2018). https://doi.org/10.1007/978-3-319-67056-0_19
28. Salloum, S.A., Al-Emran, M., Shaalan, K., Tarhini, A.: Factors affecting the E-learning acceptance: a case study from UAE. Educ. Inf. Technol. **24**(1), 509–530 (2018). https://doi.org/10.1007/s10639-018-9786-3
29. Cardoso, A.F., Moreli, L., Braga, F.T.M.M., Vasques, C.I., Santos, C.B., Carvalho, E.C.: Effect of a video on developing skills in undergraduate nursing students for the management of totally implantable central venous access ports. Nurse Educ. Today **32**(6), 709–713 (2012). https://doi.org/10.1016/j.nedt.2011.09.012
30. Junco, R., Heiberger, G., Loken, E.: The effect of Twitter on college student engagement and grades. J. Comput. Assist. Learn. **27**(2), 119–132 (2011). https://doi.org/10.1111/j.1365-2729.2010.00387.x
31. Ebersole, S.: Uses and gratifications of the web among students. J. Comput. Commun. **6**, JCMC614 (2000)
32. Lin, C.A.: Perceived gratifications of online media service use among potential users. Telemat. Informatics **19**(1), 3–19 (2002)
33. Al-Sarayrah, W., Al-Aiad, A., Habes, M., Elareshi, M., Salloum, S.A.: Improving the deaf and hard of hearing internet accessibility: JSL, text-into-sign language translator for Arabic. In: Hassanien, A.-E., Chang, K.-C., Mincong, T. (eds.) AMLTA 2021. AISC, vol. 1339, pp. 456–468. Springer, Cham (2021). https://doi.org/10.1007/978-3-030-69717-4_43
34. Tsao, S.F., Chen, H., Tisseverasinghe, T., Yang, Y., Li, L., Butt, Z.A.: What social media told us in the time of COVID-19: a scoping review. Lancet Digit. Heal. **3**(3), e175–e194 (2021). https://doi.org/10.1016/S2589-7500(20)30315-0
35. Abu-Shanab, E., Al-Tarawneh, H.: The influence of social networks on high school students' performance. Int. J. Web-Based Learn. Teach. Technol. **10**(2), 49–59 (2015)
36. Wang, T.-L.: The usage behaviors, motivations and gratifications of using user-generated media: the case study of Taiwan's YouTube. Adv. J. Commun. **2**(04), 137 (2014)
37. Wang, X., Goh, D.H.-L.: Video game acceptance: a meta-analysis of the extended technology acceptance model. Cyberpsychol. Behav. Soc. Netw. **20**(11), 662–671 (2017)
38. Simuforosa, M.: The impact of modern technology on the educational attainment of adolescents. Int. J. Educ. Res. **1**(9), 1–8 (2013)
39. Aloraini, S.: The impact of using multimedia on students' academic achievement in the College of Education at King Saud University. J. King Saud Univ. Lang. Transl. **24**(2), 75–82 (2012). https://doi.org/10.1016/j.jksult.2012.05.002
40. Ngai, E.W.T., Moon, K.K., Lam, S.S., Chin, E.S.K., Tao, S.S.C.: Social media models, technologies, and applications: an academic review and case study. Ind. Manag. Data Syst. **115**(5), 769–802 (2015)
41. van Langenberghe, A.N.K., Calderon, S.D.P.: A cross-platform study of brand recall in online video advertising. BI Norwegian Business School (2017)
42. Park, D.Y., Goering, E.M.: The health-related uses and gratifications of YouTube: motive, cognitive involvement, online activity, and sense of empowerment. J. Consum. Health Internet **20**(1–2), 52–70 (2016)
43. Phua, J., Jin, S.V., Kim, J.J.: Uses and gratifications of social networking sites for bridging and bonding social capital: a comparison of Facebook, Twitter, Instagram, and Snapchat. Comput. Human Behav. **72**, 115–122 (2017)
44. Raikos, A., Waidyasekara, P.: How useful is YouTube in learning heart anatomy? Anat. Sci. Educ. **7**(1), 12–18 (2014)

45. Habes, M., Alghizzawi, M., Khalaf, R., Salloum, S.A., Ghani, M.A.: The relationship between social media and academic performance: Facebook perspective. Int. J. Inf. Technol. Lang. Stud. **2**(1) (2018)
46. Owusu-Acheaw, M., Larson, A.G.: Digital natives and social media behaviors. J. Educ. Pract. (2015)
47. Alhammad, K.L., Habes, M., Al Olaimat, F., Haddad, I.: Attitudes of students of the faculty of mass communication at Yarmouk university towards using video platforms in distance education. Rev. Int. Geogr. Educ. Online **11**(5), 1041–1052 (2021)
48. Ali, S.: Social media usage among teenage girls in Rawalpindi and Islamabad. Glob. Media J. **16**(31), 1 (2018)
49. Ziani, A.-K., Elareshi, M., Habes, M., Tahat, K., Ali, S.: Digital media usage among Arab journalists during Covid-19 outbreak. In: Abdalmuttaleb, M.A., Al-Sartawi, M., Razzaque, A., Kamal, M.M. (eds.) Artificial Intelligence Systems and the Internet of Things in the Digital Era: Proceedings of EAMMIS 2021, pp. 116–129. Springer, Cham (2021). https://doi.org/10.1007/978-3-030-77246-8_12
50. Tahat, K.M., Al-Sarayrah, W., Salloum, S.A., Habes, M., Ali, S.: The influence of YouTube videos on the learning experience of disabled people during the COVID-19 outbreak. In: Hassanien, A.-E., Elghamrawy, S.M., Zelinka, I. (eds.) Advances in Data Science and Intelligent Data Communication Technologies for COVID-19. SSDC, vol. 378, pp. 239–252. Springer, Cham (2022). https://doi.org/10.1007/978-3-030-77302-1_13
51. Peter, O.: Social media and academic performance of library and information science undergraduates, 203–229 (2015). https://doi.org/10.4018/978-1-4666-7415-8.ch011
52. Nwangwa, K.C.K., Yonlonfoun, E., Omotere, T.: Undergraduates and their use of social media: assessing influence on research skills. Univers. J. Educ. Res. **2**(6), 446–453 (2014). https://doi.org/10.13189/ujer.2014.020602
53. Ali, S., Habes, M., Youssef, E., Alodwan, M.: A cross-sectional analysis of digital library acceptance, & dependency during Covid-19. Int. J. Comput. Digit. Syst. (2021)
54. Kushin, M.J., Yamamoto, M.: Did social media really matter? College students' use of online media and political decision making in the 2008 election. Mass Commun. Soc. **13**(5), 608–630 (2010). https://doi.org/10.1080/15205436.2010.516863
55. Abbas, J., Aman, J., Nurunnabi, M., Bano, S.: The impact of social media on learning behavior for sustainable education: evidence of students from selected universities in Pakistan. Sustain **11**(6) (2019). https://doi.org/10.3390/su11061683
56. Boahene, K.O., Fang, J., Sampong, F.: Social media usage and tertiary students' academic performance: examining the influences of academic self-efficacy and innovation characteristics. Sustainability **11**(8), 2431 (2019)
57. Al-Mukhaini, E.M., Al-Qayoudhi, W.S., Al-Badi, A.H.: Adoption of social networking in education: a study of the use of social networks by higher education students in Oman. J. Int. Educ. Res. **10**(2), 143–154 (2014). https://doi.org/10.19030/jier.v10i2.8516
58. Khan, M.L.: Social media engagement: what motivates user participation and consumption on YouTube? Comput. Human Behav. **66**, 236–247 (2017). https://doi.org/10.1016/j.chb.2016.09.024
59. Hussain, I.: A study to evaluate the social media trends among university students. Procedia. Soc. Behav. Sci. **64**, 639–645 (2012). https://doi.org/10.1016/j.sbspro.2012.11.075
60. Mingle, J., Adams, M.: A comparative analysis of social media usage and academic performance in public and private senior high schools, vol. 7, pp. 13–22 (2018)
61. Mullen, P.R., Griffith, C., Greene, J.H., Lambie, G.W.: Social media and professional school counselors: ethical and legal considerations. J. Sch. Couns. **12**(8) (2013)
62. Tajuddin, M., Azlili, N.: Social media usage among university students: a study on selfie and its impacts. Global J. Bus. Soc. Sci. Rev. **1**, 126–134 (2015)

63. Zamri, N.S., Zaihan, N.A., Samat, M.F.: A review on social media usage among students. Konvensyen Kebangs. Pemimpin Pelajar **14**(15) (2018)
64. Tezci, E., Icen, M.: High school students' social media usage habits. J. Educ. Pract. **8**(27), 99–108 (2018)
65. Oueder, M., Abousaber, I.: A Study on the Impact of Social Media Usage on Student Academic Performance: University of Tabuk an Example (2018)
66. Salloum, S., Al-Emran, M., Habes, M., Alghizzawi, M., Ghani, M., Shaalan, K.: What impacts the acceptance of E-learning through social media? An empirical study. In: Al-Emran, M., Shaalan, K. (eds.) Recent Advances in Technology Acceptance Models and Theories. SSDC, vol. 335, pp. 419–431. Springer, Cham (2021). https://doi.org/10.1007/978-3-030-64987-6_24
67. Yu-Liang, T., Tai, Y., Tseng, T.-H., Shin-Ping, T.: Innovative use of mobile video conferencing in face-to-face collaborative science learning: the case of reflection in optics. J. Educ. Technol. Soc. **21**(3), 74 (2018)
68. Cofield, J.L.: An assessment of streaming video in web-based instruction (2002)
69. Hoag, A., Grant, A.E., Carpenter, S.: Impact of media on major choice: survey of communication undergraduates. NACADA J. **37**(1), 5–14 (2017)

Determinants of Unemployment in the MENA Region: New Evidence Using Dynamic Heterogeneous Panel Analysis

Qusai Mohammad Qasim Alabed[1](✉), Fathin Faizah Said[1], Zulkefly Abdul Karim[1], Mohd Azlan Shah Zaidi[1], and Mohammad Mansour[2]

[1] Faculty of Economic and Management, Department of Economic, Universiti Kebangsaan Malaysia, Kuala Lumpur, Malaysia
qmqa89@yahoo.com, fatin@ukm.edu, {myzak1972,azlan}@ukm.edu.my
[2] Tayem Alharbi, Faculty of Social Science, University of Jordan, Amman, Jordan

Abstract. Unemployment is a crucial issue in the economy around the world. The Middle East and North Africa (MENA) region experience higher rates of unemployment than other regions. For that reason, this study extends the determinants of unemployment rates in the MENA region for a panel of 20 countries for the period (1996–2017), using a dynamic heterogeneous panel method in modelling the determinants of unemployment. The empirical analysis carried out using pooled mean group PMG approach of Pesaran, Shin and Smith, (1999). The findings proved that unemployment in the MENA region is influenced by economic determinants such as the level of trade, gross domestic product, inflation rate, and investment. Population growth is found to be one of the significant demographic elements affecting unemployment rate. The results also support the negative impact of institutional quality factors on unemployment. Therefore, our findings suggest that effective fiscal, monetary, and trade policies implementation taking into account shift unemployment to the equilibrium point to bring unemployment rates down in these countries from the policymaking perspective. Besides, government policies should pay attention to young people by increasing investments to create more job opportunities. Moreover, improvements in the institutional environment should not be ignored to get good institutional quality factors.

Keywords: MENA region · Unemployment · Dynamic heterogeneous panel · Pooled Mean Group

1 Introduction

Global development in all fields reflected significant challenges for organizations and states and created a highly competitive climate, and economic returns became very important for countries. Recently, many economies suffered from unemployment, such as the European Union at 7.6%, OECD countries at 6.8%, South Africa at 17.3%, and Latin America at 8.1% in 2017. It seems that employment growth does not rise even

© The Author(s), under exclusive license to Springer Nature Switzerland AG 2022
A. E. Hassanien et al. (Eds.): AMLTA 2022, LNDECT 113, pp. 401–411, 2022.
https://doi.org/10.1007/978-3-031-03918-8_34

though economic growth has increased. Because unemployment rates are high in most countries, the main question is how to create jobs to increase employment growth, as affected by economic policies [2]. Thus, unemployment stills a problematic issue at the global level due to its ramifications on national economies, social developments, and political stability. Despite significant economic growth in the world, unemployment rates remain high. Thus, youth in the labor market in the MENA region face difficulties in raising education levels, mainly for women, accompanied by a lack of demand for skilled people in the service sector [3]. The MENA region has generated a significant challenge in terms of job creation. Creating jobs become the main challenge for governments in this region because of the unemployment crises' consequences on the stability of these countries, both politically and economically. Additionally, the deterioration of the investment situation in most of the MENA region is one of the crucial reasons for the lack of job opportunities. In contrast, foreign direct investment reached 49.909 billion dollars in 2016. This environment has been unattractive for investors because of the high taxes imposed and political instability in many countries in the MENA region [4]. Recently, economists and researchers have focused on the economic quality factors in the business sector and their effectiveness [5]. A study by [6] pointed out that institution factors should be part of the production possibility curve in all economies that want to develop their countries. Moreover [7], the importance of institutions come true since they figure the behavior of economic agents and stakeholders. Therefore, the role of institutional quality in the labor market very little investigated in the literature. Empirical studies associate the high quality of institutions with investment, industrial growth, and firm performance [8]. Growth, in turn, creates jobs and raises well-being through increasing the labor demand [9], Noticeably low unemployment countries such as the USA 4.3%, Singapore 3.6%, Germany 3.7%, and Japan 2.8% associate with high institutions index with 84.6, 88.95, 89.53, and 82.68 respectively in 2017. In contrast, high unemployment countries such as Egypt 11.8%, Iran 12.1%, and Tajikistan 10.7% associate with low institutions index values of 22.62, 21.67, and 31.43, respectively.

2 Literature Review

Unemployment is a problem of any economy, whether developed or developing. Thus, debates have been taking place with policymakers and economists to investigate the determinants of unemployment, which could help avoid the reflections of high unemployment rates on the stability of these countries economically and politically. Therefore, economists and researchers have agreed on the importance of studying the determinants of unemployment as a problematic topic in most countries and regions. In the literature, numerous determinants have been characterized as key elements affecting unemployment, such as aggregate demand, trade policies, education, wage, demographic changes, etc. However, a literature review about the determinants of unemployment will be shown in this section and classified into three main groups: economic, demographical, and institutional determinants. On the economic front, [10] argued that aggregate demand is commonly considered the most significant determinant of unemployment. The economic crisis leads to a decrease in aggregate demand, which increases unemployment rates. For instance, during the recession case, employers try to decrease the number of workers; hence, the unemployment rate will increase. This refers to the significant effects of the economic situation on unemployment, whereas an economic downturn decreases the labor demand and increases unemployment rates. [11] Confirmed the negative nexus between aggregate demand and unemployment. Theoretically, by increasing the unemployment rate by 1%, the GDP of a particular country will decline by about 2% than its potential GDP. This relationship is expected to be negative between these two variables. Moreover, Phillips (1958) represents one of the appropriate explanations for knowing the relationship between inflation and the lack of job opportunities. It establishes a negative nexus between these two elements. An unemployment rate of a country rises as inflation decrease and the opposite right. Moreover, [12] pointed out that trade enhances the overall output of labor, which decreases unemployment due to creating more job opportunities. Similarly, [13] argue in their search-unemployment model that removes trade restrictions or barriers such as duties and surcharges and nontariff barriers (trade liberalization) decrease unemployment as a result of the development of the gross product. Moreover, relying on the efficient theory of demand, employment is a function of income. There are two essential factors in the case of effective demand, the first is aggregate demand, and the second is aggregate supply. This confirms that employment influenced by the total level of demand determined by consumption and investment. Therefore, job opportunities can be raised by raising the rate of consumption and investment [14–17]. Concerning the demographic side, demographic changes such as a change in the ageing body of the people, child mortality, fertility rates influence the labor market structure [18]. According to [19], the population is recognized as a significant factor affecting unemployment. A large number of young people may help to raise the unemployment rate. This is because young people may have the same qualifications as adult's people. Consequently, the growth of people directly affects unemployment. For example, countries with a high unemployment rate, such as the MENA region and Sub-Saharan Africa are associated with a high population growth rate. Finally, since the 1990s, various aspects of the role of institutions in economic life have been discussed in the literature, such as the role of institutions in attracting new investments and developing the economic sectors, which reduces poverty and income allocation. The term "institutions" has been defined by many economists in different ways. [20] Describes an institution as a collection of laws and constraints that help structure human interaction. Both formal and informal constraints are present in this set of laws and limitations. According to [21], institutions provide the right environment for doing business, which provides protection for the economic sectors and boosts financial growth. [20] Makes a persuasive case for institutions by stressing their role in decreasing transaction costs, which allows for greater creativity and productivity. These two variables are contributing to the country's long-term growth. From the same point of view, [22] clarifies that total factor productivity is improved by better institutional quality. According to an [23] report, strong institutions can develop foreign investment, which helps boost the country's overall growth. Starting with [24] to investigate the main determinants of unemployment in China, ARDL's results show a negative link between the GDP and the availability of job opportunities. In contrast, the inflation rate and foreign direct investment have been linked positively and significantly to unemployment. For a panel of eight European countries, [25] emphasized based on a survey of 2000 employers that scholarly outputs do not match with markets' needs for essential skills. Recently, youth unemployment has been the central issue in many industrial economies. [26] Studied unemployment between older and younger persons in OECD. They showed the difficulties facing older individuals to find, while problems facing youth are related to the business cycle. The study emphasizes the role of a country's growth rate as the primary determinant of youth unemployment. These findings confirmed with the study of [27], where they examined several factors affecting unemployment among youth in a group of 31 OECD countries. Empirical results confirm the adverse effects of growth, inflation, and saving on youth unemployment, whereas labor productivity has a supportive effect on job availability, these results also coincide with those of [28–33] who found that GDP, inflation, government expenditure, and exchange rate are key determinants of unemployment. Moreover, a favorable link between high population growth and unemployment can be detected because a rise in population leads to a rise in the demand for new jobs. Many researchers, such as [34].

3 Methodology and Data Used

3.1 Data Descriptions

The panel data with observation for 20 countries of the MENA region see Table A1 with annual data for the period (1996–2017) for a total of 22 years. The set of variables consists of seven variables. The dependent variable includes unemployment rates, and the independent variables are gross domestic product growth, inflation rate, trade openness, investment level, population, and institutions index. The institution's index variable here is used to examine the effects of institutional quality on unemployment since effective institutions are beneficial to the productivity of labor and development [35–37]. The institutional quality index is achieved by averaging the six dimensions: political stability, the rule of law, voice and accountability, regulatory quality, control of corruption, and government effectiveness. The definitions of variables and the source of data are described in Table A2. The descriptive statistics of the variables used to observe the distribution of the variable is presented in Table A3. The main measures of our variables, including the mean, standard deviation, minimum value, maximum value, and the number of observations, As discussed earlier, the MENA region is characterized by high levels of unemployment, as observed from the high rates of unemployment displayed in Table A3. For instance, the average rate of unemployment in the MENA area is 9.392, ranging from 0.340 to 31.222. This wide disparity in the rates of unemployment is due to our sample, including 20 countries. Where the UN is the dependent variable and refers to the unemployment rate. The set of explanatory variables that are supposed to affect unemployment include GDP refers to the growth rate of real GDP, INF denotes the inflation, TRADE refers to trade as a percentage of GDP, INV is investment level, POP is the population, and II represents the institutional quality index. ε_{it} Is the error term, indicates to countries, and t represents time. The choice of the study period in the baseline model is based on the availability of data. In this regard, to get a superior perception of the relationship of explanatory variables with the dependent one, the pooled mean group (PMG) of the [1] method will be employed. [38] Pointed out that the advantages of pooling data include extra variability, fewer co-linearity, a higher level of independence of expression and efficiency. Thus, PMG is an approach that makes the long-run parameters similar, but the short-run parameters and variances errors are different among groups. According to [39],

4 Empirical Findings

Panel data analysis required a panel unit root test to data and must be conducted before the co-integration method can be applied. There is a possibility of inconsistent estimation if the variables exceed I (1); thus, all variables should be stationary at level I (0) or first difference I (1). Different unit root tests are used in the current study include [40, 41] test, [42] test. Due to the drawbacks of both the LLC and IPS structures, [40] recommend using a non-parametric Fisher-type test based on a combination of the p-values of the test-statistics for a unit root of every cross-sectional unit. Therefore, as reported in Table 1, the overall evidence indicates that the included variables are integrated of order one, except for GDP and INF combined of order zero in the MENA region, thus supporting the use of the PMG estimator. Next, the current study analyzed the relationship between the unemployment rate proxies in the panel co-integration using [43] test to see whether the long-run relationship does exist or not. Table 1 illustrates the panel co-integration method results for the MENA region. The results rejected the null hypothesis of no co-integration (−1.3314). Therefore, estimated using a PMG estimator, which will provide reliable inference about the long-run and short-run determinants of unemployment rate for all the MENA countries.

Table 1. Pedroni panel co-integration results

	ADF t-statistic	*p*-value
MENA region	−1.3314*	0.0915

The results of the PMG estimation using ARDL is (1, 1, 1, 1, 1, 1, 1) for the MENA region; as shown in Table 2, the value of the coefficient of the error-correction terms (EC) is −0.0632, which is significant at 1% level and within the stable unit circle. This means that a long-run relationship exists among the variables in the estimation model. In other words, this confirms that the variables in the estimation model are co-integrated and providing reliable interpretation to the long-run estimates. Moreover, the error-correction term indicates the speed of adjustment of disequilibrium. In this case, −0.0632 implies that equilibrium is reached in 15.82 years. As shown in Table 2, the long-run coefficients of GDP growth, inflation rate INF, trade openness TRADE, institutional quality index II are negative and significant at 1% level, investment level INV is negative and significant at 10% level. Meanwhile, total population POP is positive and significant at 1% level. This implies that all determinants have a significant impact on the unemployment rate in the MENA region. Therefore, regarding the economic determinants of unemployment, the study found that 1% rise in GDP growth rate contributes to decreasing unemployment UN by 0.104%; this finding is consistent with economic theories. As indicated by [11] an increase in the country's income will reduce the unemployment level. Several empirical studies support these results, such as [11, 44] Moreover, a 1% increase in inflation rate INF leads to a fall in unemployment UN by 0.065%, these results in line with the theoretical and empirical reviews. [45] illustrated a trade-off relationship between unemployment and inflation; he argued that the inflation rate increase as results of high growth rates leads to more job opportunities. [46, 47] revealed the same results. Moreover, a 1% increase in trade openness TRADE leads to a decrease in unemployment UN by 0.954%. The outcomes are consistent with economic theories, whereas the higher share of trade as a percentage of GDP means more employment opportunities created in the trade linkages sectors. These findings revealed by many empirical studies in different countries, such as [48, 49] Furthermore, a 1% increase in investment INV contributes to decreasing unemployment UN by 0.001%; the Keynesian school argued that increasing investment brings more to economic growth and then encourages the employment of people. These results are consistent with [14, 50] studies. Furthermore, concerning the institutional quality determinants of unemployment, the study revealed that a 1% rise in institutional quality index II leads to a decrease in unemployment UN by 0.435%, as discussed earlier in the theoretical and empirical section that improving institutional quality could lead to decrease unemployment. As pointed by [20], institutions reduce transaction costs and open up the way for innovations and productivity. Moreover, [22] explain that a better quality of institution raises the total factor productivity. Therefore, good institutions needed for labour and human capital advancement. Numerous empirical studies showed that high institutional quality reduces unemployment rates, such as [51, 52], Finally, for the demographic determinants of unemployment. The study found a 1% increase in the total population POP leads to an increasing unemployment UN to 0.511%. The higher population will lead to higher competition to find a limited job opportunity, thus a higher unemployment rate, these outcomes in line with studies of [53, 54]. Unemployment tends to increase when the population increases. Concerning the policy implication, the government should enhance early education towards future job prospects, particularly in the digital economy (e.g. software developer, smart factory specialist, and maintenance, repair, and overhaul [MRO]). This will lead the population to have their comparative advantage, becoming high-skilled workers, thus penetrating the global market. Moreover, Table 2 presents the results of regression obtained from the mean group (MG). Higher standard errors and speed of adjustment have resulted from the restriction of typical long-run coefficients of the mean group MG. This outcome anticipated as the mean group MG protocol is less strict and thus theoretically inefficient [1]. The [55] test used to determine the slop of homogeneity. The basic idea of the Hausman test is to compare two sets of estimation methods, one consistently based on both the null and the alternative hypothesis, and the other consistently based on the null hypothesis only. In favor of the alternative hypothesis, a large difference between the two estimation techniques is taken as evidence. The main difference between the two estimators is that the pooled mean group PMG approach enforces a homogeneity restriction on the long-run association among variables while the MG approach does not. The homogeneity restrictions enforced by the theory can be verified empirically via the Hausman test [1] Based on Table 2 (0.8868 for the MENA region), the Hausman test cannot reject the null hypothesis (greater than 0.1) of the homogeneity restriction on the regressors in the long run, which means that pooled mean group PMG is a more efficient technique than mean group MG. The dynamic fixed effect (DFE) estimator, similar to the pooled mean group PMG estimator, restricts the coefficients of the co-integrating vector to be equivalent overall panels. The speed of the adjustment coefficient and the short-run coefficients are also limited in the dynamic fixed effect DFE model argued that the dynamic fixed effect DFE models are subject to a simultaneous equation bias from the endogeneity between the error term and the lagged dependent variable. Hausman test also used for determining the endogeneity, the outcomes in Table 2 show that the simultaneous equation bias is negligible for these data, Therefore, the dynamic fixed effect model favored over the mean group MG model.

Table 2. Estimation results of long-run and short-run coefficients (MENA region)

Variable	Mean Group (MG)		Pooled Mean Group (PMG)		Dynamic Fixed Effect (DFE)	
	Coefficient	Significance value	Coefficient	Significance value	Coefficient	Significance value
Long-run coefficients						
GDP	−0.0250***	0.0032	−0.1044***	0.0000	−0.0164	0.2391
INF	−0.0213	0.4362	−0.0652***	0.0001	−0.0114	0.1417
TRADE	−0.6789	0.4083	−0.9537***	0.0003	−0.3863**	0.0305
INV	−0.01719	0.9443	−0.0011*	0.0891	0.0588	0.6421
II	−1.1203*	0.0531	−0.4346***	0.0058	0.0432	0.8644
POP	1.2221	0.2106	0.5112***	0.0000	−1.0953**	0.0120
Error correction coefficient	−0.6398***	0.0000	−0.0632***	0.0017	−0.0851***	0.0000
No. of observation	420		420		420	
Hausman statistics (PMG is efficient estimation than MG) Prob > $chi^2 = 0.8868$						
Hausman statistics (DFE is efficient estimation than MG) Prob > $chi^2 = 1.0000$						
Note: *** Indicates significance at the 1% level; ** indicates significance at the 5% level, and * indicates significance at the 10% level.						

5 Conclusions

Unemployment is a problem in most countries, whether developed or developing. The current study examines the dynamic heterogeneity of unemployment rates in the MENA region. Using annual data collected from various sources for a panel of 20 countries from 1996 to 2017, we applied a pooled mean group PMG/ARDL method to investigate the effects of economic, demographic, and institutional quality factors on unemployment rate in the MENA region. Our sample split into two subgroups, and it includes countries that consider oil to be their primary source of income or countries that do not have oil. The main results reveal the existence of a long-run relationship between the level of unemployment in the MENA region and its determinants that include GDP growth, inflation rate, domestic investment, trade openness, demographic factors, and institutional quality. The results indicate that GDP growth, inflation rate, domestic investment, trade openness, and institutional quality all negatively affect unemployment rates. Moreover, the study results reveal that demographic factors have positive and significant effects on unemployment rates. In addition to that, the error correction factor is negative and significant, explaining the speed of adjustment of disequilibrium and finally, our results are robust by splitting the full sample into two subgroups, namely, oil and non-oil exporting countries. The overall results of this study are consistent with standard economic theories and many empirical works. The above findings imply that economic environment and demographic changes have an essential role in clarifying unemployment in the MENA region. Moreover, institutional quality significantly explained changes in unemployment in the MENA region, which means that a country with good institutional quality is associated with low unemployment rates. Therefore, the study's findings are important in policy implementation to bring unemployment rates down. According to our model, policymakers in the MENA region should adopt effective monetary and fiscal policies to motivate economic growth by decreasing inflation rates, representing a significant influence on investments both domestic and foreign. Moreover, policymakers have to pay attention to trade policies to improve the capacity of exports and international competitiveness to increase the production capacity for countries. All the above economic recommendations tend to more job opportunities, which leads to lower rates of unemployment. Regarding institutional quality, policymakers should eliminate corruption, enhance government performance, and maintain country stability, an effective role of law, account table bureaucracy to reform institutional policies as strong institutions enhance the business environment and human capital progress. Hence, improving institutional quality factors should be a vital recommendation for policymakers in the MENA region.

References

1. Pesaran, M.H., Shin, Y., Smith, R.P.: Pooled mean group estimation of dynamic heterogeneous panels. J. Am. Stat. Assoc. **94**(446), 621–634 (1999)
2. Ebaidalla, E.M.: Determinants of youth unemployment in OIC member states: a dynamic panel data analysis. J. Econ. Coop. Dev. **37**(2), 81 (2016)
3. Chigunta, F., Chisup, N., Elder, S.: Labour market transitions of young women and men in sub-Saharan Africa. Work. Publication Ser. ILO, no. 6, p. 90 (2016)
4. IMF International Monetary Fund: MENAP Regional Economic Outlook, pp. 203–214, October 2016
5. AlghzawI, M., Alghizzawi, M., Tarabieh, S.: Consumer impulsive buying behavior of services sector in Jordan. J. NX-A Multidiscip. Peer Rev. J. **6**(7), 227–237 (2020)
6. Frey, B.S., Stutzer, A.: Happiness and Economics: How the Economy and Institutions Affect Human Well-Being. Princeton University Press, Princeton (2010)

7. Sabir, S., Zahid, K.: Macroeconomic policies and business cycle: the role of institutions in selected SAARC countries. Pak. Dev. Rev. **4**, 147–158 (2012)
8. Commander, S., Svejnar, J.: Business environment, exports, ownership, and firm performance. Rev. Econ. Stat. **93**(1), 309–337 (2011)
9. Padalino, S., Vivarelli, M.: The employment intensity of economic growth in the G-7 countries. Int'l Lab. Rev. **136**, 191 (1997)
10. Ebaidalla, E.M.: Do ICTs reduce youth unemployment in MENA countries? J. Econ. Coop. Dev. **38**(4), 95–122 (2017)
11. Ruxandra, P.: The specifics of Okun's law in the Romanian economy between 2007 and 2013. Ann. Ser. **1**, 50–53 (2015)
12. Dutt, P., Mitra, D., Ranjan, P.: International trade and unemployment: theory and cross-national evidence. J. Int. Econ. **78**(1), 32–44 (2009)
13. Felbermayr, G., Prat, J., Schmerer, H.-J.: Globalization and labor market outcomes: wage bargaining, search frictions, and firm heterogeneity. J. Econ. Theory **146**(1), 39–73 (2011)
14. Anowor, O.F., Uwakwe, Q.C., Chikwendu, N.F.: How investment does affect unemployment in a developing economy. Sumerianz J. Econ. Financ. **2**(7), 82–88 (2019)
15. Alnawafleh, H., Alghizzawi, M., Habes, M.: The impact of introducing international brands on the development of Jordanian tourism. Int. J. Inf. Technol. Lang. Stud. **3**(2), 30–40 (2019)
16. Elareshi, M., Habes, M., Ali, S., Ziani, A.: Using online platforms for political communication in bahrain election campaigns. Soc. Sci. Humanit. **29**(3), 2013–2031 (2021)
17. Al-Shakhanbeh, Z.M., Habes, M.: The relationship between the government's official Facebook pages and healthcare awareness during Covid-19 in Jordan. In: Hassanien, A.-E., Elghamrawy, S.M., Zelinka, I. (eds.) Advances in Data Science and Intelligent Data Communication Technologies for COVID-19. SSDC, vol. 378, pp. 221–238. Springer, Cham (2022). https://doi.org/10.1007/978-3-030-77302-1_12
18. Salvador, R.G., Leiner-Killinger, N.: An analysis of youth unemployment in the euro area. ECB Occas. Pap., no. 89 (2008)
19. Arslan, M., Zaman, R.: Unemployment and Its determinants: a study of Pakistan economy (1999–2010). J. Econ. Sustain. Dev. **5**(13), 20–24 (2014)
20. North, D.C.: Institutions, Institutional Change and Economic Performance. Cambridge University Press, Cambridge (1990)
21. Mansoor, A., Quillin, B.: Migration and remittances: Eastern Europe and the former Soviet Union. The World Bank (2006)
22. Olson, M., Sarna, N., Swamy, A.V.: Governance and growth: a simple hypothesis explaining cross-country differences in productivity growth. Public Choice **102**(3), 341–364 (2000)
23. IMF international monetary: Making the Global Economy Work for All (2003)
24. Şahin, D.: Determinants of unemployment: empirical analysis for China. Akad. Sos. Araştırmalar Derg. **4**(22), 50–58 (2016)
25. Mckinsey, Brodherson, M., Heller, J., Perrey, J., Remley, D.: Creativity's Bottom Line : How Winning Companies Turn Creativity into Business Value and Growth. Digital McKinsey (2017)
26. Axelrad, H., Malul, M., Luski, I.: Unemployment among younger and older individuals: does conventional data about unemployment tell us the whole story? J. Labour Mark. Res. **52**(1), 1–12 (2018). https://doi.org/10.1186/s12651-018-0237-9
27. Bayrak, R., Tatli, H.: The determinants of youth unemployment: a panel data analysis of OECD countries. Eur. J. Comp. Econ. **15**(2), 231–248 (2018)
28. Chowdhury, M., Hossain, M.: Determinants of unemployment in Bangladesh: a case study. Dev. Ctry. Stud. **4**(3) (2014)
29. Khumalo, Z.Z., Eita, J.H.: Determinants of unemployment in Swaziland. J. Appl. Sci. **15**(9), 1190 (2015)

30. Ramli, S.F., Firdaus, M., Uzair, H., Khairi, M., Zharif, A.: Prediction of the unemployment rate in Malaysia. Int. J. Mod. Trends Soc. Sci. **1**, 38–44 (2018)
31. Al-Shibly, M.S., Alghizzawi, M., Habes, M., Salloum, S.A.: The impact of de-marketing in reducing jordanian youth consumption of energy drinks. In: Hassanien, A.E., Shaalan, K., Tolba, M.F. (eds.) AISI 2019. AISC, vol. 1058, pp. 427–437. Springer, Cham (2020). https://doi.org/10.1007/978-3-030-31129-2_39
32. Salloum, S., Al-Emran, M., Khalaf, R., Habes, M., Shaalan, K.: An innovative study of E-payment systems adoption in higher education: theoretical constructs and empirical analysis. Int. J. Interact. Mob. Technol. (iJIM) **13**(06), 68 (2019). https://doi.org/10.3991/ijim.v13i06.9875
33. Elbasir, M., Elareshi, M., Habes, M.: The influence of trust, security and reliability of multimedia payment on the adoption of EPS in Libya. Multicult. Educ. **6**(5), 53–68 (2020)
34. Yanan, W., Wise, E.: Research article determinants of the rate of unemployment in Nigeria. Int. J. Inf. Res. Rev. **04**(01), 3593–3595 (2017)
35. Acemoglu, D., Johnson, S., Robinson, J.A.: Institutions as a fundamental cause of long-run growth. Handb. Econ. Growth **1**, 385–472 (2005)
36. Salloum, S., Al-Emran, M., Habes, M., Alghizzawi, M., Abd, M., Ghani, K.: What impacts the acceptance of E-learning through social media? An empirical study. In: Al-Emran, M., Shaalan, K. (eds.) Recent Advances in Technology Acceptance Models and Theories, pp. 419–431. Springer International Publishing, Cham (2021). https://doi.org/10.1007/978-3-030-64987-6_24
37. Al-Sarayrah, W., Al-Aiad, A., Habes, M., Elareshi, M., Salloum, S.A.: Improving the deaf and hard of hearing internet accessibility: JSL, text-into-sign language translator for Arabic. In: Hassanien, A.-E., Chang, K.-C., Mincong, T. (eds.) AMLTA 2021. AISC, vol. 1339, pp. 456–468. Springer, Cham (2021). https://doi.org/10.1007/978-3-030-69717-4_43
38. Goswami, G.G., Junayed, S.H.: Pooled mean group estimation of the bilateral trade balance equation: USA vis-à-vis her trading partners. Int. Rev. Appl. Econ. **20**(4), 515–526 (2006)
39. Demetriades, P., Hook Law, S.: Finance, institutions and economic development. Int. J. Financ. Econ. **11**(3), 245–260 (2006)
40. Maddala, G.S., Wu, S.: A comparative study of unit root tests with panel data and a new simple test. Oxf. Bull. Econ. Stat. **61**(S1), 631–652 (1999)
41. Levin, A., Lin, C.-F., Chu, C.-S.J.: Unit root tests in panel data: asymptotic and finite-sample properties. J. Econom. **108**(1), 1–24 (2002)
42. Im, K.S., Pesaran, M.H., Shin, Y.: Testing for unit roots in heterogeneous panels. J. Econom. **115**(1), 53–74 (2003)
43. Pedroni, P.: Critical values for cointegration tests in heterogeneous panels with multiple regressors. Oxf. Bull. Econ. Stat. **61**(S1), 653–670 (1999)
44. Anyanwu, J.C.: Characteristics and macroeconomic determinants of youth employment in Africa. African Dev. Rev. **25**(2), 107–129 (2013)
45. Philips, A.W.: The relation between unemployment and the rate of change of money wage rates in the United Kingdom 1861–1957. Economica **25**(100), 283–299 (1958)
46. Dritsaki, C., Dritsaki, M.: Phillips curve inflation and unemployment: an empirical research for Greece. Int. J. Comput. Econ. Econom. **3**(1–2), 27–42 (2013)
47. Shaari, M.S., Abdullah, D.N.C., Razali, R., Saleh, M.L.A.-H.M.: Empirical analysis on the existence of the Phillips curve. In: MATEC Web of Conferences, vol. 150, p. 5063 (2018)
48. Gozgor, G.: The impact of trade openness on the unemployment rate in G7 countries. J. Int. Trade Econ. Dev. **23**(7), 1018–1037 (2014)
49. Dalmar, M.S.: Factors affecting unemployment in Somalia. J. Econ. Sustain. Dev. **8**(2) (2017). ISSN 2222–1700 (Paper) ISSN 2222–2855 (Online)
50. Irpan, H.M., Saad, R.M., Nor, A.H.S.M., Ibrahim, N.: Impact of foreign direct investment on the unemployment rate in Malaysia. J. Phys. Conf. Ser. **710**(1), 12028 (2016)

51. Baccaro, L., Rei, D.: Institutional determinants of unemployment in OECD countries: does the deregulatory view hold water? Int. Organ. **61**, 527–569 (2007)
52. Ederveen, S., Thissen, L.: Can labour market institutions explain high unemployment rates in the new EU member states? Empirica **34**(4), 299–317 (2007)
53. Adekola, P.O., Allen, A.A., Olawole-Isaac, A., Akanbi, M.A., Adewumi, O.: Unemployment in Nigeria; a challenge of demographic change? Int. J. Sci. Res. Multidiscip. Stud. ISROSET **2**(5), 1–9 (2016)
54. Maijama'a, R., Musa, K.S., Yakubu, M., Mohammed, N.: Impact of population growth on unemployment in Nigeria: dynamic OLS approach. J. Econ. Sustain. Dev. **10**(22), 79–89 (2019)
55. Hausman, J.A.: Specification tests in econometrics. Econ. J. Econ. Soc., 1251–1271 (1978)

The Relationship Between Digital Transformation and Quality of UAE Government Services Through Machine Learning

Rashed Abdulla AlDhaheri[1](✉), Ibrahim Fahad Sulaiman[1], and Haleima Abdulla Al Matrooshi[2]

[1] Faculty of Economics and Management Sciences, University Sains Islam Malaysia, Negeri Sembilan, Nilai, Malaysia
p808@hotmail.com, ibrahimfahad@usim.edu.my

[2] Ghazali Shafie Graduate School of Government - Universiti Utara Malaysia, Changlun, Malaysia

Abstract. Digital transformation has become necessary for institutions and bodies seeking to develop and improve their services to beneficiaries, and this study aims to discover The Relationship Between Digital Transformation using machine learning and Quality of UAE Government Services and requirements for digital transformation in the UAE, according to the literature, digital transformation has become necessary for UAE institutions and bodies that seek to develop and improve their services to beneficiaries, and there is a trend towards digital transformation in UAE government institutions. It is possible to take advantage of new digital technologies to create innovative and high-quality services in them, and thus enhance the services provided, which calls for research on the current and expected effects of digital transformation and how to benefit from it as well. Further discussions and implications are also presented in the study.

Keywords: Digital transformation · UAE Government · Quality services · Machine learning

1 Introduction

The current era is witnessing many rapid and successive developments and changes, as a result of the information and communication revolution that invaded the lives of peoples, which had profound effects that led to what is known as the age of information and knowledge, in which information became a basic resource and completely different from the traditional and old ways of services, communication, information and smart transformation. This made government and service institutions face many challenges and difficulties in how to keep pace with them and how to adapt to them [1, 2]. The digital transformation of government services has become a necessity for various institutions, organizations and entities that seek to develop, improve and innovate services

© The Author(s), under exclusive license to Springer Nature Switzerland AG 2022
A. E. Hassanien et al. (Eds.): AMLTA 2022, LNDECT 113, pp. 412–421, 2022.
https://doi.org/10.1007/978-3-031-03918-8_35

for the public within the institution [3]. The digital transformation system is also a comprehensive and integrated program that affects the way and style of the institution and its work internally and externally through the services provided and achieves competitiveness and distinction in the quality of services provided to it. It achieves flexibility in providing service and completing transactions faster and more smoothly [4]. As a result, governments and service institutions dealt with the digital revolution as a future method of work and an innovative way of working with the public, providing them with services, and providing digital innovations and smart solutions to customers [5]. This transformation has become necessary in light of the massive spread of technologies, smart devices, the Internet, and the availability of all digital services to users. As a result, the government of the United Arab Emirates has put digital transformation steps in government services and service institutions as one of the most important priorities for the next stage, starting in the year 2002 when His Highness Sheikh Mohammed bin Rashid, Vice President of the UAE, Prime Minister and Ruler of Dubai launched the e-government project, where the government set transformation plans Digital and building digital strategies in services as an affliction so that the United Arab Emirates is at the forefront of developed countries in digital transformation and technology through the digital infrastructure in the country that allows work to provide diverse and distinct content and service in government work that covers all areas of life and service that the individual needs [6]. In a manner that contributes to the transfer and exchange of knowledge and services worldwide and the importance of this in raising the level of efficiency and work of the government apparatus, excellence in providing services to customers, reaching global competitive indicators and levels in the government services sector, service and technical development sector, and anticipating the digital future [7, 8]. Thus, this study discusses the issue of digital transformation, which is the main focus in the work of institutions and organizations in the current era in light of the information technology revolution and smart transformation in various areas of life with the aim of reaching a set of proposals and results that it is hoped will contribute to the development of theoretical and practical literature in the digitization and quality of government work in the United Arab Emirates.

2 Literature Review

Identifying the levels of quality of government work and levels of raising the efficiency and quality of government services provided through the transition from traditional governments to digital governments and smart governments, as digital transformation has become a basic criterion in evaluating the performance of government work and Baalbek plays a key role In promoting the digital economy in private, governmental and service institutions, digital transformation using machine learning has also become an essential criterion in the process of achieving global indicators for governments and reaching strategic goals through government digital maturity standards [9]. The UAE Ministry of Justice is one of the government institutions that has developed a comprehensive strategy in the process of digital transformation to upgrade and develop judicial facilities and the role of the judiciary and develop the litigation and pleading system, in addition to providing general information to the public and services remotely, through websites and smart services in order to ensure the quality of services provided, low costs. and speed in completing it [10–12]. The UAE government has set standards and indicators to measure levels of government digital maturity in the quality of services provided to the public through six main indicators and axes to measure quality levels in government institutions: leadership, strategy, governance, emerging technologies, technology, and regulating legislation [13, 14] which are standards aimed at achieving government directions, upgrading the capabilities of workers, and work procedures to leadership ranks, providing services that meet the aspirations of stakeholders and leaders, and fully digitizing government work and supporting the UAE vision 2017 [15–17].

3 Digital Transformation of Governments

The digital transformation of governments is no longer an optional issue, especially for institutions and bodies that deal directly with the public, and that seek to develop and improve their services and facilitate their access to citizens [18]. The concept of digital transformation goes beyond the use of technological applications, to become an approach and work method that contributes to providing better services for government institutions [19]. Add to the concept of digital transformation means the integration of digital technology into all areas of work, and it is a cultural change, which requires institutions to constantly challenge the status quo, and constantly experiment [20]. As well as the transformation of businesses or governments by making radical changes, including work, procedures and processes, and the transformation may include the process of changing the product, or the way the service is provided entirely; In line with modern changes, the transformation may be strategic, in line with digital development, as it enters into all functions of the organization, from sales to supply, information technology, and the entire value chain [21–23]. Building excellence requires the participation of all functional, supervisory and administrative activities; To produce a natural interaction, stimulate radical variables that create a continuous voluntary movement, and generate a kind of correct polarization, which constitutes a transitional leap that leads to the integration of many institutional sectors [24]. Transformation - as an abstract concept - calls for seeing the basic concepts of reform. On the one hand, decision-makers and writers refer to the word “transformation” physically as: the process of changing the form without changing the content, and on the other hand, transformation is intended in practice: improving the efficiency and effectiveness of the public service in light of the needs of individuals [19]. The previous concept is consistent with what was indicated by (Mansour, 2016) that the process of digital transformation or digitization: the process of obtaining and managing electronic text collections, by converting information sources available on traditional storage media to an electronic image, and thus traditional content becomes digitized content, which can be View it through computer applications [10]. He also agreed with [25, 26] about his vision of digital transformation; He notes that digital transformation is forcing universities to rethink basic assumptions about books and lectures [27].

Berger et al. (2019) [20] adds that digital transformation or digitization is a key factor for the university’s change, not to change the existing, but to provide a new field full of potentials that help success. The meaning of digitization is not only limited to technological tools; But the obligation to think about how to control the mechanisms and administrative processes, and the skills of the individual and how to apply them. Gunaratna (2016) [28] point out that the concept of "digital transformation" or "digitization" involves technical and cultural transformation, and is reflected in all areas, and enhances and identifies methods, methods and opportunities, and that avoiding that transformation seems impossible and thus transformation In the way governments work, so that the monotonous work is reduced, and the time for thinking about development increases; Digital transformation is the acceleration of the daily way of working, so that the great development of technology is exploited to serve your customers better and faster. Digital transformation is also increased efficiency in the workflow, so that errors are reduced and productivity increases [6]. Digital transformation is to increase the number of team members without hiring, simply put: digital transformation for governments is to harness technology to work for people [20].

4 Digital Transformation in Government Services in the UAE

In the context of the pioneering process and positive interaction pursued by the UAE, and its keenness to spread knowledge at the regional and Arab levels, the Telecommunications Regulatory Authority (TRA) issued the Arabic version of the United Nations e-Government Study 2018, and the study provides an opportunity for governments to view global and advanced experiences, reports, studies and standards (formerly the Division of Public Administration, Development Management), of the United Nations Department of Economic and Administrative Affairs [29, 30]. As for the UAE's ranking in the e-government development index, according to the e-government development study, issued in 2018 by the United Nations Department of Economic and Social Affairs, the UAE has made a qualitative leap in the overall index of e-government development from the twenty-ninth place in In 2016, it reached the twenty-first place in 2018, advancing by eight places, to become one of the twenty-five leading countries in this indicator [31]. The index also monitors the level of progress in the digital transformation of global governments [4]. The e-government development index consists of three sub-indices, including: the e-services index, the human capital index, and the telecom infrastructure readiness index [32]. All state institutions, each within its sphere of competence, impose order and the rule of law through the administrative and judicial bodies on the one hand, and social awareness and the responsibility of civil society to maintain social security on the other. Among the important state institutions that bear the burden of providing security are the police agencies, where the police departments that carry out the task of the state's right to provide security and safety and achieve an integrated security system work [33]. The United Arab Emirates launched the strategic plan for the fourth strategic cycle 2017–2021, which aimed to achieve the UAE Vision 2021, delight employees and customers, achieve leadership in digital services, provide legal and judicial services in an innovative way, and develop pioneering legislation that guarantees the rule of law and the protection of freedoms and rights. All of this is in compliance with the directives of the wise leadership, and in fulfillment of its ambition to make the UAE one of the best countries in the world [34, 35]. The plan aims to impose the best legislation and laws that keep pace with global trends and internal changes, and in line with the traditions of the state, and achieve the needs and aspirations of citizens towards the future, and to attract, train, retain and motivate the best human cadres, especially judicial ones, that work within an innovative environment on the application of laws and legislation, besides providing innovative government and justice services to all client groups, based on efficiency, quality and effectiveness, and spreading legal culture and information to all segments of society, through innovative and multiple communication channels, and building local and international strategic partnerships that contribute to government cooperation and exchange of experiences [36, 37].

5 Quality of Service for Customers in Government Institutions

This development in the field of service quality helps service providers increase the level of service efficiency and improve its quality, after the service sector - as the sector that deals with service as an intangible and difficult to measure activity - was characterized by low performance and low ability to improve it. Because what cannot be measured cannot be managed. Today it is clear that the services are undergoing the process of great development. Perhaps this great experience in managing and improving service quality is the appropriate and good basis for the subsequent development of electronic quality, which is essentially a digital electronic service [38]. Electronic quality is the last - and perhaps the finest - form of evolving quality concerns, areas, policies, and efforts. It is evident that the development since the nineties has taken two overlapping trends, namely: the trend of development in the quality of knowledge and information, and the trend of quality and electronic services [39]. The Internet was able to achieve effective focus on the customer, and interaction with him everywhere, and at the right time; Instead of preparing in advance the production of goods or services and placing them in warehouses according to the inventory plan, the Internet can interact automatically with the customer; To determine what he really wants to achieve high clienteles. Also, through application software, it is possible to verify and scrutinize the previous choices of customers, to better respond to customer needs and preferences [3, 40]. The subject of digital quality is undoubtedly a new one; For this, it needs more methodological effort in order to clarify the concept; It still means many things to different people; Some see that it means the element of predictability (predictability) of the customer and consistency (consistency) in its presentation by the company. Others see it as achieving efficient mobility on the network, increasing the volume of data provided, while maintaining a consistent behavior of the characteristics, and others believe that it means the different layers of data service provided by network sources to the higher service layers at the expenses of the lower layers (higher service at a lower cost), and some see It is an attempt to appropriate the allocation of network resources to the characteristics of particular data flows [41]. Add to given that the new goals of government performance based on information technology necessarily require new standards [10]. It has now become established that the success of the performance of government services cannot be judged through spending control, and the commitment that spending remains within or less than the amounts allocated in the budget [42]. Therefore, the Government Accounting Standards Board and other international bodies suggested using the "value criterion" as a means of evaluating performance in light of the new technological environment for government performance, and among those services is the support and support of leaders in public organizations for the transformation process, while working to maintain technical and administrative expertise and competencies to reduce resistance during the digital transformation process [43]. Develop plans and policies to ensure the safe implementation of the digital transformation process, and divide this process into detailed steps to clarify the role and importance of each process in the digital transformation process, with an emphasis on familiarity with all operational steps [44]. Also, design mechanisms that enable citizens to know how to benefit from digital transformation in public services. Training and preparing workers in service organizations - oriented to the process of digital transformation - on new business and procedures, and encouraging them to be creative and innovative in that, while making it clear to them that they are partners in success [45, 46]. In addition to developing the infrastructure, material and technology that allows the digital transformation process as a tool for transferring data and information, without which the digital transformation process will become an implementable trend, and the constructive participation of all state media institutions, including and others, to encourage the trend of the digital transformation process [3]. Also, building information security systems to maintain the confidentiality of information, and not to breach it while providing backup copies of that information to preserve it from loss and damage, and finally issuing legislation and laws that explicitly stipulate the direction of the digital transformation process as a general trend of the state and a target for all public and governmental organizations [7]. Digital transformation at the enterprise level is a major project that calls for complex organizational, social, political, financial and technological measures. Countries that have succeeded in carving their way through the digital age can and by employing artificial intelligence and machine learning algorithms in government work However, in its experience, it offers what it offers to regional governments seeking to integrate AI applications into their government activities across different sectors and over their broader government services [14].

6 Discussion

Digital transformation using ML provides huge opportunities for government and private institutions in various aspects, the most important of which are: achieving the objectives of the institutions, and reaching their strategic vision, with less wasted capabilities at the present time or at a time before digital transformation. Digital transformation will help organizations improve their industrial path, and use their materials in optimum and higher efficiency. The digital transformation using ML will also provide greater opportunities after opening the dialogue between the public and private sectors, and the partnership between them, in cooperation with all ministries. As awareness of the inevitability of this shift, and working collectively, contribute mainly to the growth and prosperity of these sectors, which will positively reflect on the progress of countries, to be more flexible and aware at work, and abler to predict and plan for the future. Thus, e-government in the UAE contributes to enhancing the role of digital transformation using ML in improving government services. This result, we talked about previously, and how it contributed to the transformation in the quality of services in the UAE. Thus, this result also agreed with the study [3, 33, 47, 48] as researcher Lewis & Molyneux, (2018) stated in her letter that in the past five years, the UAE has made great leaps in digital transformation, starting with the state leadership's announcement of the launch of the smart government in May 2013, and what followed. From rapid developments, most government services have been made available to the public of customers, through digital channels, and by smart means, based on anticipation, interaction and continuous development, taking into account the principle of customer centrality. The results of the study also indicate that the application of digital transformation using ML levels in the UAE was - through a comprehensive examination of the digital government services in the UAE government and - helps to improve quality, and digital services in the ministry save time and effort, and digital services contribute to improving the performance of the ministry Continuous, which includes all work within government, and gives sufficient time for benchmarking with respect to improvement activities, to a significant degree. The results of the study also showed that applying the levels of government services through digital transformation systems in the government and making comparisons helped improve the quality of digital services and integrate data and applications with other government services, and that digital services in the government may suffer from penetration and electronic and cyber-attacks. The public, laws, regulations, reports and bulletins subject to change are an obstacle to the development of the digital transformation of the Ministry, and the decision-making process on the results of consultations and public participation is an obstacle to the transformation in the quality of services provided to the public.

7 Conclusion

The UAE realized at an early stage the importance of the transition to the digital environment, which prompted its government to adopt an integrated strategy; To achieve the leadership's vision and directives in accomplishing this transformation, within specific time frames and goals, including various work sectors, and this indicates the employees' awareness of the need for continuous development towards digital transformation, and their understanding of its major strategies; For a successful transformation in the Arab world and the world at large. The importance of giving data the importance it deserves, so that data in the current era is a crucial element in understanding patterns and foreseeing behaviors, and then developing solutions based on deep learning and artificial intelligence. It is possible to take advantage of new digital technologies to create innovative and quality services in them, and thus enhance the services provided, which It calls for a search for the current and expected effects of digital transformation and how to benefit from it, as well as a search for the prominent role of the dimensions of digital transformation in creating an advanced service map. By committing to embracing and developing new digital technologies to enhance the performance of government services, in addition to searching for challenges and obstacles that prevent the implementation of digitization. Knowing the factors that affect the level of adoption of digital government services using ML, and workers in the field of digital transformation who are interested in knowing global trends in this field.

References

1. Lewis, S.C., Molyneux, L.: A decade of research on social media and journalism: assumptions, blind spots, and a way forward. Media Commun. **6**(4), 11–23 (2018). News and Participation through and beyond Proprietary. https://doi.org/10.17645/mac.v6i4.1562
2. Ziyadin, S., Suieubayeva, S., Utegenova, A.: Digital transformation in business. In: Ashmarina, S.I., Vochozka, M., Mantulenko, V.V. (eds.) ISCDTE 2019. LNNS, vol. 84, pp. 408–415. Springer, Cham (2020). https://doi.org/10.1007/978-3-030-27015-5_49
3. Alhumaid, K., Habes, M., Salloum, S.A.: Examining the factors influencing the mobile learning usage during COVID-19 pandemic: an Integrated SEM-ANN Method. IEEE Access (2021)
4. Al-Shakhanbeh, Z.M., Habes, M.: The relationship between the government's official Facebook pages and healthcare awareness during Covid-19 in Jordan. In: Hassanien, A.-E., Elghamrawy, S.M., Zelinka, I. (eds.) Advances in Data Science and Intelligent Data Communication Technologies for COVID-19. SSDC, vol. 378, pp. 221–238. Springer, Cham (2022). https://doi.org/10.1007/978-3-030-77302-1_12
5. Elareshi, M., Habes, M., Ali, S., Ziani, A.: Using online platforms for political communication in bahrain election campaigns. Soc. Sci. Humanit. **29**(3), 2013–2031 (2021)
6. Alhammad, K.L., Habes, M., Al Olaimat, F., Haddad, I.: Attitudes of students of the faculty of mass communication at Yarmouk University towards using video platforms in distance education. Rev. Int. Geogr. Educ. Online **11**(5), 1041–1052 (2021)
7. Belias, D., Koustelios, A.: Transformational leadership and job satisfaction in the banking sector: a review. Int. Rev. Manag. Mark. **4**(3), 187–200 (2014)
8. Belias, D.: Leadership and job satisfaction: a review. Eur. Sci. J. **10**(8), 24–46 (2014)
9. Sedik, A., et al.: Deploying machine and deep learning models for efficient data-augmented detection of COVID-19 infections. Viruses **12**(7), 769 (2020)
10. Pei Lin, C., Xian, J., Li, B., Huang, H.: Transformational leadership and employees' thriving at work: the mediating roles of challenge-hindrance stressors. Front. Psychol. **11**, 1–19 (2020). https://doi.org/10.3389/fpsyg.2020.01400
11. Almarzooqi, A.: Towards an artificial intelligence (AI)-driven government in the United Arab Emirates (UAE): a framework for transforming and augmenting leadership capabilities. ProQuest Diss. Theses **204** (2019)
12. Salloum, S.A., AlAhbabi, N.M.N., Habes, M., Aburayya, A., Akour, I.: Predicting the intention to use social media sites: a hybrid SEM-machine learning approach. In: Advanced Machine Learning Technologies and Applications: Proceedings of AMLTA 2021, pp. 324–334 (2021)
13. Yıldız, I.G., Şimşek, Ö.F.: Different pathways from transformational leadership to job satisfaction. Nonprofit Manag. Leadersh. **27**(1), 59–77 (2016). https://doi.org/10.1002/nml.21229
14. Ziani, A.-K., Elareshi, M., Habes, M., Tahat, K.M., Ali, S.: Digital media usage among Arab Journalists during Covid-19 outbreak. In: European, Asian, Middle Eastern, North African Conference on Management & Information Systems, pp. 116–129 (2021)
15. Madichie, N.O.: Professional sports: a new 'services' consumption mantra in the United Arab Emirates (UAE). Mark. Rev. **9**(4), 301–318 (2009). https://doi.org/10.1362/146934709x479890
16. Al-Swidi, A.K., Nawawi, M.K.M., Al-Hosam, A.: Is the relationship between employees' psychological empowerment and employees' job satisfaction contingent on the transformational leadership? A study on the Yemeni Islamic banks. Asian Soc. Sci. **8**(10), 130 (2012)
17. Alnawafleh, H., Alghizzawi, M., Habes, M.: The impact of introducing international brands on the development of Jordanian tourism. Int. J. Inf. Technol. Lang. Stud. **3**(2) (2019)

18. Mandell, B., Pherwani, S.: Relationship between emotional intelligence and transformational leadership style: a gender comparison. J. Bus. Psychol. **17**(3), 387–404 (2003)
19. Long, C.S., Yusof, W.M.M., Kowang, T.O., Heng, L.H.: The impact of transformational leadership style on job satisfaction. World Appl. Sci. J. **29**(1), 117–124 (2014). https://doi.org/10.5829/idosi.wasj.2014.29.01.1521
20. Berger, R., Czakert, J.P., Leuteritz, J.P., Leiva, D.: How and when do leaders influence employees' well-being? Moderated mediation models for job demands and resources. Front. Psychol. **10**(December), 1–15 (2019). https://doi.org/10.3389/fpsyg.2019.02788
21. Sujata, J., Mukul, P., Hasandeep, K.: Role of smart communication technologies for smart retailing. Int. J. Innov. Technol. Explor. Eng. **8**(6, 4), 1015–1020 (2019). https://doi.org/10.35940/ijitee.F1209.0486S419
22. Alghizzawi, M., Habes, M., Salloum, S.A.: The relationship between digital media and marketing medical tourism destinations in Jordan: Facebook perspective. In: Hassanien, A.E., Shaalan, K., Tolba, M.F. (eds.) AISI 2019. AISC, vol. 1058, pp. 438–448. Springer, Cham (2020). https://doi.org/10.1007/978-3-030-31129-2_40
23. Al-Shibly, M.S., Alghizzawi, M., Habes, M., Salloum, S.A.: The impact of de-marketing in reducing Jordanian youth consumption of energy drinks. In: Hassanien, A.E., Shaalan, K., Tolba, M.F. (eds.) AISI 2019. AISC, vol. 1058, pp. 427–437. Springer, Cham (2020). https://doi.org/10.1007/978-3-030-31129-2_39
24. Habes, M., Alghizzawi, M., Ali, S., SalihAlnaser, A., Salloum, S.A.: The relation among marketing ads, via digital media and mitigate (COVID-19) pandemic in Jordan. Int. J. Adv. Sci. **29**(7), 2326–12348 (2020)
25. Habes, M., Salloum, S.A., Alghizzawi, M., Mhamdi, C.: The relation between social media and students' academic performance in Jordan: YouTube perspective. In: Hassanien, A.E., Shaalan, K., Tolba, M.F. (eds.) AISI 2019. AISC, vol. 1058, pp. 382–392. Springer, Cham (2020). https://doi.org/10.1007/978-3-030-31129-2_35
26. Salloum, S.A., Al-Emran, M., Khalaf, R., Habes, M., Shaalan, K.: An Innovative study of E-payment systems adoption in higher education: theoretical constructs and empirical analysis. Int. J. Interact. Mob. Technol. **13**(6) (2019)
27. Mirkamali, S.M., Thani, F.N., Alami, F.: Examining the role of transformational leadership and job satisfaction in the organizational learning of an automotive manufacturing company. Proc. Soc. Behav. Sci. **29**, 139–148 (2011). https://doi.org/10.1016/j.sbspro.2011.11.218
28. Gunaratna, R.: Global terrorism in 2016 LOBAL. Rev. UNISCI **39**, 133–138 (2016)
29. Salloum, S.A., AlAhbabi, N.M.N., Habes, M., Aburayya, A., Akour, I.: Predicting the intention to use social media sites: a hybrid SEM - machine learning approach. In: Hassanien, A.-E., Chang, K.-C., Mincong, T. (eds.) AMLTA 2021. AISC, vol. 1339, pp. 324–334. Springer, Cham (2021). https://doi.org/10.1007/978-3-030-69717-4_32
30. King'Oina, J.O.: Factors influencing value added tax compliance among the construction firms in Kisumu County, Kenya. pp. 1–62 (2016)
31. Sokolovska, O., Sokolovskyi, D.: Market and government failures related to the introduction of tax incentives regime. Econ. si Sociol. Rev. Teor. **4**, 17–26 (2015)
32. Habes, M., Elareshi, M., Ziani, A.: An empirical approach to understanding students' academic performance: YouTube for learning during the Covid-19 pandemic. Linguist. Antverp. 1518–1534 (2021)
33. Choi, S.L., Goh, C.F., Adam, M.B.H., Tan, O.K.: Transformational leadership, empowerment, and job satisfaction: the mediating role of employee empowerment. Hum. Resour. Health **14**(1), 73 (2016)
34. Abelha, D.M., Carneiro, P.C.D.C., Cavazotte, F.D.S.C.N.: Transformational leadership and job satisfaction: Assessing the influence of organizational contextual factors and individual characteristics. Rev. Bras. Gest. Negocios **20**(4), 516–532 (2018). https://doi.org/10.7819/rbgn.v0i0.3949

35. Salloum, S.A., Al-Emran, M., Habes, M., Alghizzawi, M., Ghani, M.A., Shaalan, K.: What impacts the acceptance of e-learning through social media? An empirical study. Recent Adv. Technol. Accept. Model. Theor. 419–431 (2021)
36. Tahat, K.M., Al-Sarayrah, W., Salloum, S.A., Habes, M., Ali, S.: The influence of YouTube videos on the learning experience of disabled people during the COVID-19 outbreak. In: Hassanien, A.-E., Elghamrawy, S.M., Zelinka, I. (eds.) Advances in Data Science and Intelligent Data Communication Technologies for COVID-19. SSDC, vol. 378, pp. 239–252. Springer, Cham (2022). https://doi.org/10.1007/978-3-030-77302-1_13
37. Ali, S., Habes, M., Youssef, E., Alodwan, M.: A cross-sectional analysis of digital library acceptance, & dependency during Covid-19. Int. J. Comput. Digit. Syst. (2021)
38. Unes, B.V.J., de Castro Camioto, F., Guerreiro, É.D.R.: Relevant factors for customer loyalty in the banking sector. Gest. e Prod. **26**(2) (2019). https://doi.org/10.1590/0104-530X2828-19
39. Lasrado, F.: Legacy of Excellence: The Case of the United Arab Emirates (UAE), pp. 37–56. Springer, Cham (2018)
40. Cheng, X., Dale, C., Liu, J.: Statistics and social network of YouTube videos. IEEE Int. Work. Qual. Serv. IWQoS, 229–238 (2008). https://doi.org/10.1109/IWQOS.2008.32
41. Lee, B.C., Yoon, J.O., Lee, I.: Learners' acceptance of e-learning in South Korea: theories and results. Comput. Educ. **53**(4), 1320–1329 (2009). https://doi.org/10.1016/j.compedu.2009.06.014
42. Hay, I.: Transformational leadership: characteristics and criticisms. E-J. Organ. Learn. Leadersh. **5**(2), 2–19 (2006)
43. Oeij, P.R.A., de Looze, M.P., Ten Have, K., van Rhijn, J.W., Kuijt-Evers, L.F.M.: Developing the organization's productivity strategy in various sectors of industry. Int. J. Product. Perform. Manag. **61**(1), 93–109 (2011). https://doi.org/10.1108/17410401211187525
44. Karaca, H., Kapucu, N., Van Wart, M.: Examining the role of transformational leadership in emergency management: the case of FEMA (2013). https://doi.org/10.1002/rhc3.10
45. Ferreira, G.B., Borges, S.: Media and misinformation in times of COVID-19: how people informed themselves in the days following the Portuguese declaration of the state of emergency. Journal. Media **1**(1), 108–121 (2020). https://doi.org/10.3390/journalmedia1010008
46. Habes, M., Alghizzawi, M., Salloum, S.A., Mhamdi, C.: Effects of Facebook personal news sharing on building social capital in Jordanian universities. In: Al-Emran, M., Shaalan, K., Hassanien, A.E. (eds.) Recent Advances in Intelligent Systems and Smart Applications. SSDC, vol. 295, pp. 653–670. Springer, Cham (2021). https://doi.org/10.1007/978-3-030-47411-9_35
47. Hussein, Z.: Leading to Intention: the role of attitude in relation to technology acceptance model in E-learning. Proc. Comput. Sci. **105**, 159–164 (2017). . https://doi.org/10.1016/j.procs.2017.01.196
48. Elareshi, M., Habes, M., Ziani, A.-K.: New media users' awareness of online inflammatory and mobilisation methods for radical and extreme activities. Ilkogr. Online **20**(5), 5567–5576 (2021)

Key Factors Determining the Expected Benefit of Customers When Using Artificial Intelligence

Abdulsadek Hassan[1(✉)], Mahmoud Gamal Sayed Abd Elrahman[2], Faheema Abdulla Mohamed[1], Sumaya Asgher Ali[1,2], and Nader Mohammed Sediq Abdulkhaleq[1,2]

[1] Ahlia University- Kingdom of Bahrain, Manama, Bahrain
dr.sadek1964@yahoo.com
[2] Faculty of Mass Communication, Radio and Television Department, Beni-Suef University, Beni Suef, Egypt

Abstract. The study aims to identify the relationship of Artificial Intelligence (AI) with the field of e-commerce; The relationship of AI with the e-commerce sector is about broad technologies such as data analysis and machine learning to make better and smarter actions to improve the experience of today's digital shoppers. Given the importance of the mechanisms of artificial intelligence (AI) and the Internet of Things (IoT) in improving the e-commerce process, it was necessary to prepare this article to learn about the images of the use of AI techniques and how to use them to provide an easier and smarter shopping experience through the online Store.

Keywords: Artificial intelligence · E-commerce · Marketing

1 Introduction

The trend of consumers to e-commerce is increasing at a very acceleration, which leads to a clear growth in the size of the e-commerce market, as it is expected that 1.9 billion people will make electronic purchases by the end of 2021, and more than 2 billion people by the year 208. The high demand prompted companies to think of new ways to reach more audiences [8]. Artificial Intelligence can help merchants get better forecasts about their sales, better support their customers, and retarget customers who have left them. When launching the first online Store, the last thing you'll think of is that one day you'll have to work alongside bots! Well, that day has come in the form of AI and machine learning. AI will have a huge impact on the e-commerce sector, just as e-stores have affected all of our lives, as data shows that among all sectors, e-commerce is a fertile environment for AI investment [4].

These methods include personalized email marketing, voice search, and image search configuration; And a lot of these methods rely mainly on automation and artificial intelligence, which affects very much e-commerce now, and in the future in particular [11]. AI will ensure that customers have an easier, smoother, convenient, private, and faster shopping experience than ever before. Rather than being afraid of bots, merchants will

© The Author(s), under exclusive license to Springer Nature Switzerland AG 2022
A. E. Hassanien et al. (Eds.): AMLTA 2022, LNDECT 113, pp. 422–431, 2022.
https://doi.org/10.1007/978-3-031-03918-8_36

seek to work with them. Of course, for merchants in the e-commerce sector and e-store design, it's not about the technology itself, but what matters primarily are turnovers and sales [19].

Artificial intelligence is a valuable tool to support new marketing features, such as personalization, predictive offers, and voice search. These tools require analyzing data packets to extract valuable results with all the promising opportunities that artificial intelligence technologies present in marketing; knowing how Which has obtained the most significant impact is essential for marketers at this time [22].

Marketers must consider the capabilities of artificial intelligence before adopting them, as effective marketing of digital commerce requires analyzing huge amounts of data and identifying the relationships between these data to make decisions and take the most appropriate approach. Many digital commerce companies and brands have successfully implemented this for several years [7].

2 Reasons for Using Artificial Intelligence in E-commerce

2.1 Reduce Cart Abandonment

Emails sent to a consumer who does not purchase after some products are in the cart have an open rate of 45%, resulting from AI-powered email marketing automation [5]. Abandoning a purchase after placing items in the shopping cart often indicates an error or problem before completing the purchase; automation and email marketing help reduce this problem [4]. Many reasons may prevent a user from completing an asset, such as high shipping costs and the need to create an account before purchasing; Understanding these reasons is necessary to overcome the problem and facilitate the purchase process for the user; So automation is an excellent way to get customers back, while also collecting information that can help you prevent more customers from dropping out of their shopping cart, for example through an email with a survey [12].

2.2 Facilitate Voice Search

A study by ComScore predicted that by 2020, 50% of all internet searches would be done by voice. And with the advent of devices like Alexa, Echo, and Apple devices that use Siri and Google Home, customers can search for products using their voice. As a result, companies have to ensure that their products are found through voice search [6].

Companies should start improving their websites for voice search. For example, many companies can now use machine learning by letting customers shop with Alexa on their websites. Customers are looking for more convenience in their online shopping process, and voice search provides customers with this convenience by searching for items without the need for a laptop or phone, making the shopping experience more efficient [16].

2.3 Enhance Your Targeting of a More Specific Audience

AI eliminates the need for guesswork and assumptions when it comes to targeting the right consumer; Instead of adopting the concept of "one size fits all" and creating ads with broad and ineffective targeting, companies can now target customers according to their buying behavior and the nature of their interaction digitally thanks to artificial intelligence [11].

Digital marketing automation and artificial intelligence tools make it easier to collect target audience data and create dynamic ads that take this data into account. Accordingly, these ads are published on platforms and channels appropriate to the nature of the content, making the target person more likely to see the ad. Retargeting allows companies to target customers who have already interacted with the brand in one way or another [7].

2.4 Improve Search Results

A marketer can create excellent and attractive content, but he cannot achieve marketing goals and increase sales if customers cannot find it. Many customers search for products using search engines and search within the Store itself. It is worth noting that more than 40% of e-commerce website visits come through natural search results for the Google search engine; This is what makes the site's configuration appear in search engines vital and crucial to the Store's success or not [13].

AI-powered tools can help marketers drive more traffic to their sites in a way that encourages a smooth flow of buyers through your e-commerce store by analyzing the site and making sure you choose the right keywords, organizing content perfectly and without errors such as site performance analysis tools, and filter related products, etc. [10].

For many years, artificial intelligence has been associated with robotic devices and the shape of machines that are expected to exist in the future. But today, it is impossible to have a conversation about the future of digital marketing without mentioning artificial intelligence as a major part of this future. Thus, marketers must constantly think about how to exploit the capabilities of artificial intelligence to make more effective marketing decisions to achieve better successes and enhance the reach of the target audience [9].

3 Artificial Intelligence in 2020

AI is a broad category, and not every AI feature can serve every retail business. This article will cover the latest AI tools in e-commerce, what to expect in the coming years, and how entrepreneurs can make AI work for their specific needs through research and expert interviews. Here are the AI tools to watch in 2020.

(A) **Enhanced chatbots**

So far, many e-commerce customers have used some form of a chatbot, a website bot that acts as a customer service representative. The demand for easy access to 14/7 customer service has led many companies to implement chatbots. Customer service is an important matter, so we find many who request it and need a quick response [14]. Therefore, growing companies should find a way to reduce the pressure on customer service. For example, chatbots are an alternative to automated phone systems. Chatbot features are useful if the company receives the same questions repeatedly, as the bot can refer customers to simple answers to common questions [16].

(B) **Voice recognition**

Voice commerce allows customers to make purchases online using a device with a voice assistant, such as a smartphone, Amazon Echo, or Google Home. AI voice robots are beginning to understand what people are saying [6]. The technology is now good enough to recognize what a person is saying in addition to being able to form a voice. Some tools can even translate languages. But we're not at the stage where we can recognize every word [17].

(C) **Augmented reality**

Augmented Reality (AR) is a modern and ever-evolving e-commerce tool to look out for in 2021. On certain devices, customers can virtually try on accessories and clothing. Potential car buyers can use augmented reality apps to switch between different color options [16].

(D) **Visual search**

Through visual search, customers can upload product images and start searching based on visual algorithms. This technology has improved over the past few years and is becoming more widely available. (Google) Lens, for example, allows users to tap and hold on to an image, which starts a product search in online stores. Google and iOS have recently started offering visual search [19].

(E) **Product Recommendations**

One of the popular AI e-commerce options, the product recommendation tool, suggests new products to customers based on demographics, location data, past individual behavior, and other metrics [16]. These recommendations may appear on the e-commerce site itself, or through targeted social media ads. Product recommendations usually don't require a lot of data to trigger, and while performance varies, they can be useful. As AI collects more of this data over time, self-learning mechanisms improve recommendations [2].

(G) **The use of artificial intelligence in e-commerce marketing**

Artificial intelligence and the management of marketing campaigns in the world of e-commerce over the Internet has increased significantly in recent years, relying on the data provided only, due to the company's interest in presenting a selection and variety of consumer behavior, and also interest in presenting their data and interests [19].

(H) **The impact of artificial intelligence on e-commerce**

- After the advent of artificial intelligence, all e-commerce companies know the appropriate strategy to market their products [8].
- Artificial intelligence has made social networking sites, such as Facebook, Twitter, and Instagram, which facilitate the company's marketing campaigns [2].
- The artificial intelligence program aims to facilitate these campaigns and the ways and means of managing them, and all this is done without any human intervention [3].
- Chatbots will get smarter by activating artificial intelligence software [1].
- All commercial markets and e-commerce companies currently use artificial intelligence in chatting [5].

- Chatbots experience has been implemented, allowing all users to communicate with all brands easily and conveniently [12].
- This is done by using natural language, which is suitable for chatting with users [15].
- Chatbots have become better than traditional customer services, especially in matters related to marketing and sales [20].
- Thanks to the artificial intelligence program, this makes it easier for the company to achieve more profits [14].

(I) **Using artificial intelligence to write content**

Recently, artificial intelligence has become the leader in various areas of life and is no longer confined to e-commerce sites. It was discovered that a large amount of content available on news sites was written through artificial intelligence [15]. In the case of comparing these writings with the writings of the human element, it can be noted that artificial intelligence robots cannot be underestimated, due to their ability to write many product specifications, on websites at the lowest cost, with ease, quality, and high efficiency [15].

4 Benefits of Using Artificial Intelligence Techniques in E-commerce

4.1 It Saves Time and Energy

Since automation uses artificial intelligence to perform activities; Tasks such as marketing, communication, fulfillment, shipping, and tracking can be done by AI without stress [21]. Business owners no longer need to waste time collecting, entering, to arrange, and to analyze data as it was easy to create and implement all this by automation software [4]. The energy needed to take care of customers can be manually channeled into creative thinking about designing or creating more products that can generate more revenue for the business [18]. Also, the energy needed to move up and down trying to reach customers can be used to create engaging content and ads on the automation software [5]. This, in turn, will free up more time to focus on business development [29].

4.2 Improves Customer Satisfaction

Business automation software allows customers to access products faster,

It allows them to take action faster, and more. Chatbots, for example, will answer their questions [7].

Business automation software makes it easy for business owners to meet the needs of their customers without stress because customers are just a click away [15].

As there is no hindrance to the movement on the automation software, i.e., customers can submit their complaints, recommendations and requests at any time of the day.

Customer complaints can be easily resolved without complicated legal procedures as everyone is just a click away [15].

4.3 Software can Make Business Owners Better Serve Their Customers in the Following Ways

- Make it easy for customers to quickly access the information or products they are looking for [8].
- Allow customers to see what service/products are available, and the quantity ordered or delivered [15].
- Terms and conditions of service are easily accessible and visible to customers
- Clients enjoy independence as the software can easily access [15].
- Order status, create quotations, receive automation notification related to their products, etc. [14].

4.4 Optimizing Operations and Business Intelligence

Business intelligence (BI) is software and services that take raw data. It turns them into actionable and relevant insights that companies can use to strengthen their business positions and decisions [14]. This tool is automation software that helps with data collection and analysis [8]. This helps business owners make the right decision and expand and build a more robust future for the business [2]. Automation software reduces the chances of business owners engaging in repetitive activities such as attending to customers over and over again [8]. Since customers can now help themselves, business owners can now focus on how to create engaging ads and reach their potential customers, which will translate into more orders in the long run, and business productivity will be increased automatically also [18]. They will have time to strategize how to retain existing customers and build stronger relationships between employers and workers [20].

4.5 Reduces the Error

Fatigue often leads to error.

Why deal with inventory management, drop shipping, order fulfillment, customer support, digital marketing, and data collection when they can be fully automated and executed? [6]. Automation software reduces the chances of errors in orders and deliveries as customers can easily track their orders are well selected from the software, so there is no miscommunication or interpretation of orders [12]. We have seen many examples – including vehicle carriers, logistics, and others- who have harnessed the power of automation software to ensure safe and adequate delivery to customers without a hitch [1]. Customers have complete information on when and how orders will be delivered; there is no need to visit the job site [11] endlessly. The software can handle multiple tasks simultaneously without mixing things up, leading to customer happiness and business improvement [7].

4.6 Improves Marketing

Email marketing, social media marketing, digital marketing, affiliate marketing, and many other tools are proven tools that help in automating e-commerce business ideas [5]. Automation software makes marketing easier and helps target ads to the right customers [8]. Unlike the previous methods of advertising like sharing flyers, TV or radio broadcasts, giving free samples, etc., which don't know if potential customers are being reached or not [18]. The automation software helps sort out leads and make sure you can reach them at any time of the day. It also helps collect information from potential customers, which can later be used to generate more productivity [8].

5 How Will AI Change E-commerce?

AI supports every business to grow and explore new horizons. According to Tractica's forecast, AI is expected to generate $118.6 billion from AI by 208 [16]. The increasing tendency and constant competition in the new trend technologies lead to the improvement of the e-commerce industry at a growing pace [12]. AI is transforming traditional ways of managing different business tasks in smarter ways, from local vendors and manufacturers to giant retailers and digital startups [1].

AI Helps Every Business
Transformations brought by AI in improving the shopping experience for customers and business owners [15].

Artificial Intelligence and the Virtual Assistant
AI makes it possible to help customers virtually with the help of chatbots [4]. While browsing e-commerce sites, you will find the message pop up from smart bots that can solve the puzzle piece for the customers [6]. The main advantage is that you do not need to employ multiple resources to serve customers who speak various languages [18]. Unlike the old chatbots, integrated chatbot geniuses can communicate with customers just like humans [2]. Based on customer requirements and inquiries through machine learning, chatbots can guide them through every shopping step [12]. From product selection to the checkout process, chat sites assist customers at every stage [8]. In addition to chatbots, voice assistance can be useful to end users. It makes user communication easy and fast and prevents users from typing in their queries. On the one hand, voice search saves you time and adds a personal touch [2]. On the other hand, Voice Assistance allows your customers to assist them with bots that answer their queries by speaking to them rather than displaying them on the screen [14].

Advanced Visual Search
Visual search is one of the great features that everyone craves. The reason is very simple, we mostly know the product we want, but sometimes it is confusing. Getting the exact result you want is impossible [9]. Thanks to the direct product search option from the AI, you can scan the product you want to search. Your search can bring up hundreds of variations of the product you were looking for [13]. The method is to scan the product, and for the product, you will find all the information displayed along with the available

quantity and available variants [15]. E-commerce giants like Amazon are already starting to use the visual search option. Later, this feature can be added to the Store [14].

Personalized Product Recommendations

Customer satisfaction is the ultimate goal of every business [5]. Therefore, each of your user requirements should be taken care of [22]. When talking about physical shops, it becomes very crowded on holidays or during peak hours [4]. The store owner and staff can understand the customers' shopping behavior and show the products according to their needs, but considering each customer's choice, even when he/she is a first-time visitor; It will be a little tricky [8]. E-commerce allows every visitor's choice to be observed, and AI helps provide them with a personalized experience [13]. One customer search displays 100 products, and with step-by-step education, you will find a list of recommended products in the Products You May Like section. The list of products is solely based on your research and customers who have already liked these products [1]. You will also find products that you can buy together, it's like a profitable deal that can entice customers to buy the product [12].

Manage your Inventory Based on Expected Sales

The sales forecast feature is an asset for companies. Based on previous purchases, user demographics, customer behavior, time of the season, and various other factors, AI helps you predict your future sales [5].

Based on this analysis, you can manage your inventory and stock up on products that have more sales chances in the meantime [18]. Sales forecasts can help inform marketing strategies and reduce costs when they are not valuable [10].

Artificial Intelligence and Customer Relationship Management (CRM)

Unlike in the old days of managing your customer database via spreadsheets, now CRM software has taken the perfect tool for it [17]. When CRM makes it easier for you to manage data and sales for B2B and B2C customers separately, it also makes it easier to communicate with potential customers [15].

Furthermore, AI makes it easier for you to collect data from social media, helps in evaluating potential leads, completes data entry, and enables data analysis to support marketing and sales strategies [14].

In the end, the above benefits of AI are the vital ones we rely on. There is a plethora of opportunities that you can discover when it takes to develop your e-commerce business with more customization for your customers [15].

6 Conclusion

The study showed that Artificial intelligence applications have become used in all fields. It is expected shortly that these applications will draw the features of e-commerce. In the era of the ever-changing digital market, in which merchants expect to receive a view of shoppers' activities through multiple "omnichannel" marketing channels, a feature that links the activities of customers during the moment, whether they are online or offline, and therefore it is very important to be able to respond in a way that A quick. Facilitate business dealings between the customer and the Store. It also shows Artificial intelligence provides easy data analysis. Ultimately, these insights will be useless if marketers cannot make decisions based on them. Artificial intelligence enables algorithmic solutions systems to automatically monitor and record customer behavior instead of manually tuning systems that may miss these basic insights. The study concluded that more and more institutions are now investing in AI to address barriers to product growth, such as cutting operational costs, making a path for more accurate product matching, and getting legitimate product reviews. Although AI technology is still far from perfect, e-commerce companies are constantly improving their tools to keep up with market demand. Collaboration between companies also paves the way for sharing their AI competencies, creating more complex AI tools. Also, artificial intelligence's role in our daily lives may not be noticed, but it is very important. From platforms like Facebook and Snapchat to online shopping apps, we can be sure that AI technology will dominate the digital age as an engine of exponential growth.

References

1. Ameen, N., Tarhini, A., Reppel, A., Anand, A.: Customer experiences in the age of artificial intelligence. Comput. Hum. Behav. **114**, 106548 (2020)
2. Aoki, N.: An experimental study of public trust in AI chatbots in the public sector. Gov. Inf. Q. **37**(4), 101490 (2020)
3. Ballestar, M.T., Grau-Carles, P., Sainz, J.: Predicting customer quality in ecommerce social networks: a machine learning approach. RMS **13**(3), 589–603 (2019)
4. Canhoto, A.I., Clear, F.: Artificial intelligence and machine learning as business tools: a framework for diagnosing value destruction potential. Bus. Horiz. **63**(2), 183–193 (2020)
5. Chung, M., Ko, E., Joung, H., Kim, S.J.: Chatbot e-service and customer satisfaction regarding luxury brands. J. Bus. Res. **117**(1), 587–595 (2020)
6. Ciechanowski, L., Przegalinska, A., Magnuski, M., Gloor, P.: In the shades of the uncanny valley: an experimental study of human–chatbot interaction. Futur. Gener. Comput. Syst. **92**(1), 539–548 (2019)
7. Colicev, A., Kumar, A., O'Connor, P.: Modeling the relationship between firm and user-generated content and the stages of the marketing funnel. Int. J. Res. Mark. **36**(1), 100–116 (2019)
8. Dabija, D.C., Bejan, B.M., Tipi, N.: Generation X versus millennials communication behaviour on social media when purchasing food versus tourist services. Econ. Manag. **21**(1), 191–205 (2018)
9. Dospinescu, O., Anastasiei, B., Dospinescu, N.: Key factors determining the expected benefit of customers when using bank cards: an analysis on millennials and generation Z in Romania. Symmetry **11**(12), 1449–1469 (2019)
10. Enache, M.C.: E-commerce trends. Annals of the University Dunarea de Jos of Galati: Fascicle: I. Econ. Appl. Inform. **14**(2), 67–71 (2018)
11. Fernandes, T., Oliveira, E.: Understanding consumers' acceptance of automated technologies in service encounters: drivers of digital voice assistants adoption. J. Bus. Res. **122**(1), 180–191 (2020)
12. Koehn, D., Lessmann, S., Schaal, M.: Predicting online shopping behaviour from clickstream data using deep learning. Expert Syst. Appl. **50**, 113342 (2020)

13. Massaro, A., Vitti, V., Lisco, P., Galiano, A., Savino, N.: A business intelligence platform Implemented in a big data system embedding data mining: a case of study. Int. J. Data Min. Knowl. Manag. Process **9**(1), 1–20 (2019)
14. Micu, A., Micu, A.E., Geru, M., Căpățînă, A., Muntean, M.C.: The impact of artificial intelligence use on the e-Commerce in Romania. Amfiteatru Economic **6**(56), 137–154 (2021)
15. Moriset, B.: e-Business and e-Commerce. Int. Encyclopedia Hum. Geogr. 1–10 (2020)
16. Nichifor, E., Trifan, A., Nechifor, E.M.: Artificial intelligence in electronic commerce: basic chatbots and the consumer journey. Amfiteatru Economic **6**(56), 87–101 (2021)
17. Pantelimon, F.-V., Georgescu, T.-M., Posedaru, B.-Ş: The impact of mobile e-Commerce on GDP: a comparative analysis between Romania and Germany and how Covid-19 influences the e-Commerce activity worldwide. Inform. Econ. **14**(2), 27–41 (2020)
18. Sheehan, B., Jin, H.S., Gottlieb, U.: Customer service chatbots: anthropomorphism and adoption. J. Bus. Res. **115**(1), 14 (2020)
19. Soni, V.D.: Emerging roles of artificial intelligence in e-commerce. Int. J. Trend Sci. Res. Dev. **4**(5), 26–28 (2020)
20. Vanneschi, L., Horn, D.M., Castelli, M., Popovič, A.: An artificial intelligence system for predicting customer default in e-commerce. Expert Syst. Appl. **104**, 1–21 (2018)
21. Wang, Q., Cai, R., Zhao, M.: E-commerce brand marketing based on FPGA and machine learning. Microprocess. Microsyst. 103446 (2020)
22. Xueming, L., Tong, S., Fang, Z., Qu, Z.: Machines versus humans: the impact of AI chatbot disclosure on customer purchases. Mark. Sci. **38**(6), 937–947 (2019)

Check for updates

Examining Factors Affecting Job Employment in Egyptian Market

Lamiaa Mostafa(✉)

Business Information System Department, ArabAcademy for Science and Technology and Maritime Transport, Alexandria, Egypt
Lamiaa.mostafa31@aast.edu, Lamiaa.mostafa31@gmail.com

Abstract. Higher education value is measured through the employment rate of the graduates. Employment rate also affect the government performance rate. Many factors affect hiring decision such as Academic Performance (AP), Personality (PE), Communication Skills (CK), Leadership and Motivation (LM), Technical Skills (TS), Team work and Problem solving (TP), University Reputation (UR), Continuous Learning (CL), Career Marketing Demand (CMD), and Job Search Tool (JST). In this paper, factors that affect graduate employability (GE) are examined. Primary data collected from a sample of 350 employers through a structured questionnaire working as the hiring managers. Exploratory factor analysis was used and structural equation modeling with AMOS (24) to test the hypothesized relationship between each independent variable and the dependent one. The research provides a model that can be used by educational institutions to assist graduates in finding a job and creating their career.Results showed that TS,AP,TP,UR,CL,CMD positively significantly affect Graduate employment and PE,CK,LM, and JST affect Graduate employment but insignificantly. AP, TP, UR, CL and CMD have positive significant impact with values respectively: 0.461, 0.582, 0.497, 0.422, 0.509, and 0.437. While PE, CK, LM, JST has a positive impact on GE but insignificantly.

Keywords: Employement · University graduate · SEM · Covid · Egypt

1 Introduction

The world economy is affected by the pandemic. Connection between graduates and market is formed using the internet tools. Less employment rate is rising in different countries. Graduate employability is an emerging topic, which is the focus of higher education institutions and government.

© The Author(s), under exclusive license to Springer Nature Switzerland AG 2022
A. E. Hassanien et al. (Eds.): AMLTA 2022, LNDECT 113, pp. 432–444, 2022.
https://doi.org/10.1007/978-3-031-03918-8_37

Graduates should adapt new skills and technologies while Universities must enhance their relationship with the markets and develop curriculums to cope with the technologies development.

Many factors affect hiring decision such as Academic Performance (AP), Personality (PE), Communication Skills (CK), Leadership and Motivation (LM), Technical Skills (TS), Teamwork and Problem solving (TP), University Reputation (UR), Continuous Learning (CL), Career Marketing Demand (CMD), and Job Search Tool (JST).

The paper will examine the effect of the previous factors on graduate employment through statistical method. Factors are extracted from previous researchers will be examined using structured equation modeling. A research model created based on the variables selected from previous research, the model would help career development institutions and educational academic advisors' staff to assist students in university to find their strengths and weakness regarding the skills that can be used in working field.

The rest of this paper will be organized as follows. Section 2 summaries literature review on higher education, career development, factors affecting job employment. Section 3 discusses research methodology though hypothesis formulation and data collection. Section 4 discusses the results of descriptive statistics; reliability analysis, factor analysis and Sect. 5 provide the conclusion and the future work.

2 Literature Review

The quick development of job market stress graduates to enhance their capacities to match this quick development. Career learning and consultation can make a big difference in graduate life. Higher education curriculum should be designed and developed to be aligned with Egyptian job market, Technological field needs developing skills and continuous learning [1]. Educational institution must provide their graduates with the required skills to meet the job market requirements and employer's needs. Strengthen the English language and the computer skills provide competitive advantage for seeking job vacancies in the future [2]. Secondary education should focus on engaging students in the market career and students must learn more about the different careers in the market. When they understand their future career, they will be able to enhance the required skills and communication skills in certain areas. Career guidance an education can help students make decisions and choices in the different fields [3, 4]. Educatoinal institutions should focus on graduates career developement to enhance their employability skills, such as communication, leadership, teamwork, problem-solving, and technical skills [5].

In the same context, researchers in [6] provide a model of career choices while investigating the mediating role of career exploration. They also consider social support and career self-efficacy predictors among students at University Putra Malaysia [6]. Career process is affected by socioeconomic, personality, communication skills, educational, cultural background, and family variables [7]. Based on [8] career exploration has a significant positive relationship with career self-efficacy, social support, and career choice.

Other researchers like [9, 10] tested the graduate unemployment in China country. Graduates from universities with high reputation are more successful in finding a suitable job than other graduates. Arts and social science or agriculture graduates find jobs faster than graduates with law and science degrees Female graduates work earlier than male graduates do. Engineering and business graduates find jobs more easily than law and science graduates. 60% of the respondents' reason for unemployment was family concern and decided not to find a job. Business graduates find a job easily based on their course performance and the necessary theoretical and technical skills they gained in school helped them to find job [10, 11].

COVID-19 in 2020 complicated the situation of the employment of college graduates. Graduates feel pressure of employment and epidemic. Employment force focus on college graduates. Researchers in [12] tried to understand the employment characteristics of college graduates and the changes of college graduates' employment expectations under the epidemic situation. 2, 000 data questionnaireswere filled. Results covered the problem of employment of college graduates and provided future solutions [12]. The following table will summarize the previous research (Table 1).

Based on the previous table, the hypothesis is formulated. It was noticed that the variables Technical Skills [13, 14], Performance [14, 15], personality [14, 16], communication skills-leadership [14, 15, 17, 18], teamwork [15, 19], university reputation [22], job search tool [23], continuous learning [20] and career marketing demand [22, 23] are used in previous researchers due to their importance in employment process.

Table 1. Summary of literature review

Paper	Technical skills	Academic performance	Personality	Communication skills	Leadership	Team work and Problem solving	University reputation	Continuous learning	Career marketing demand	Job search tool	Culture	Attitude
13	X											
14	X	X	X	X	X							
15		X		X	X	X						
16			X								X	X
17				X	X							
18	X			X	X							
19						X						
20							X					
22		X						X	X			
23									X	X		
This paper	X	X	X	X	X	X	X	X	X	X		

3 Research Methodology

3.1 Hypothesis Formulation

The following subsections will define each variable and the suggested hypothesis based on the previous work. Figure 1 represents the research model.

1. **Technical Skills (TS):** Graduate Employment (GE) is the focus of each university. University must develop the student capabilities to meet job market needs [13]. Technical support and information technology skills influence the employment of graduates and give the student the competitive edge to be employed [13, 14].

   Hypothesis 1 (H1): TS can positively and significantly affect GE

2. **Academic Performance (AP):** Student grade point average (GPA)is a good reflector for his academic performance. Most companies consider GPA while hiring student in a job [15].

   Hypothesis 2 (H2): AP can positively and significantly affect GE

3. **Personality (PE):** Personality (PE) defined as distinctive manner the combines thinking, feeling, and behaving. Researchers in [16], conducted a study to examine PE and perceived employability while they concluded that personality affect employment positively [16].

   Hypothesis 3 (H3): PE can positively and significantly affect GE

4. **Communication Skills (CK):** Commutation skills (CS) are the ability to communicate with others effectively while sending and receiving the intended meaning. A study in Europe was analyzing the effect of communication skills [17], they concluded that soft skills involving communication skills, interpersonal skills and problem-solving skills increase the employability of the students/graduates.

   Hypothesis 4 (H4): CK can positively and significantly affect GE

5. **Leadership and Motivation (LM)***:* Leadership skill is the ability to lead a group of team members while motivation is the ability to encourage self and others to reach a specific goal. An extensive literature survey that focuses on India market was developed in [18] while providing a future recommendation and discussing the importance of leadership and motivation in the employment.

   Hypothesis 5 (H5): LM can positively and significantly affect GE

6. **Team work and Problem Solving (TP):** Teamwork and problem solving are essential skills to get a job and sustain job longer time. Researchers in [19] investigated

the impact of soft skills on Bangladesh business graduates, they emphasized the importance of problem solving and teamwork skills.

Hypothesis 6 (H6): TP can positively and significantly affect GE

7. **University Reputation (UR):** Large university reputation is premium for employment. Researchers in [20] concluded that candidates graduated from well-valued universities are 40% more likely to receive a job acceptance.

Hypothesis 7 (H7): UR can positively and significantly affect GE

8. **Continuous Learning (CL):** University student should work on developing the skills using different courses and certificates. Continuous learning will provide the student with the new market skills and follow the market change continuously. In [21] a job matching model which matches the candidate personal and educational details with Information and Communication Technology National Competency framework (ICT-NCF) with the ability to enhance candidate learning path that will also develop his career and provide him with the suitable employment job title.

Hypothesis 8 (H8): CL can positively and significantly affect GE

9. **Career Marketing Demand (CMD):** There is a big demand to connect university curriculum with the job market demand. Researchers in [22], developed an AI curriculum to be connected to the market and include both universities and companies' information.

Hypothesis (H9): CMD can positively and significantly affect GE

10. **Job Search Tool (JST):** Job candidates search for new career opportunities through job portals and hiring agencies such as: wuzzuf.net, indeed.com, egycareers.com, forasna.com. LinkedIn Talent Solution offered job recruiter facilities to design qualified applicants' best work prospects [23]. Job search tools assist candidate to be employed.

Hypothesis 10 (H10): JST can positively and significantly affect GE

3.2 Data Collection

A questionnaire was conducted, involving 350 human resources employees. The survey involves ten independent variables and one dependent variables, four demographic information. A 30 questions are placed in the questionnaire, and the Likert-type five level scales are employed. The options for each question using points 1 to 5 to represent "extremely disagree," "disagree," "neutral," "agree," and "extremely agree," respectively. In this study, two statistical software: SPSS25.0 and AMOS23.0 were used. Descriptive analysis, reliability test was done. The structural equation model (SEM) was established, and the confirmatory factor analysis and hypothesis verification analysis was conducted.

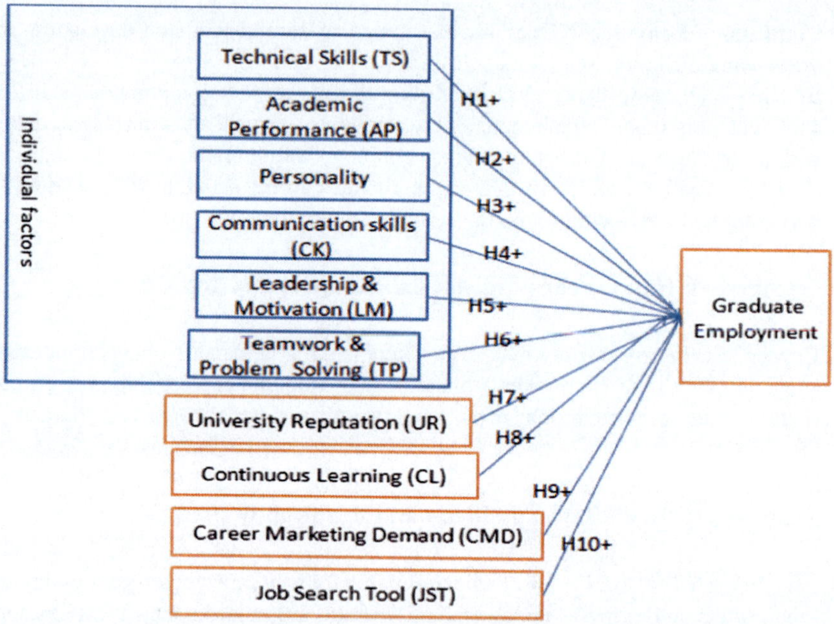

Fig. 1. Research model

4 Discussion of Results

4.1 Descriptive Statistics

A survey strategy was employed using an online questionnaire. The target population was human resources employees. An anonymous questionnaire was developed in English using Google Forms. 350 participants were invited by email, and Facebook Messenger to participate in this survey. The Data collection lasted for 30 days. Of the 370 responses received, 350 responses (effective rate is 90.9%) were considered valid for further analysis after verifying incomplete questionnaire and data. Detailed descriptive statistics of respondents' characteristics are shown in Table 2.

Table 2. Respondents' profile

Attribute	Category	Frequency	Percent
Age	25–36	109	31.2%
	Over 36	241	68.8%
	Total	350	–
Gender	Male	155	44.2%
	Female	195	55.8%
	Total	350	–
Education level	Masters	269	76.8%
	Bachelor	81	23.2%
	Total	350	–
Years of experience	5–15	109	31.2%
	More than 15	241	68.8%
	Total	350	–

4.2 Reliability Analysis

26 items relevant to eleven constructs of the proposed research model were adopted from existing literature and refined based on the specific topic of this study. Cronbach's alpha coefficient was employed to determine the reliability of the questionnaire. Based on [24], Cronbach's alpha coefficient showed is equal or greater than 0.7, which is similar to constructs constraints. All items in the survey were measured using a five-point Likert scale ranging from (1) strongly agree to (5) strongly disagree as shown in Table 3.

Table 3. Reliability analysis

Construct	Cronbach's alpha	Items
Technical skills	0.78	3
Academic performance	0.81	3
Personality	0.80	3
Communication skills	0.97	3
Leadership-motivation		3
Team work and Problem solving	0.96	3
University reputation	0.97	1
Continuous learning	0.88	1

(continued)

Table 3. (*continued*)

Construct	Cronbach's alpha	Items
Career marketing demand	0.98	2
Job search tool	0.87	3
Graduate employment	0.91	1
Total		26

4.3 Exploratory Factor Analysis (EFA)

Four assumptions were used to assets Exploratory factor [25]: Kaisers–Mayesolkin measure greater than 0.5; the minimum value for each factor; considering the sample size, factor loading of 0.50. After examining the pattern matrix of EFA, it was found that all the items had factor loadings greater than 0.50 as shown on Table 4.

Table 4. EFA

Latent variable	Item	Factor load	CR	AVE
Technical skills	TS1	0.875	0.848	0.701
	TS2	0.78		
	TS3	0.661		
Academic performance	AP1	0.818	0.791	0.711
	AP2	0.721		
	AP3	0.829		
Personality	PE1	0.708	0.82	0.604
	PE2	0.829		
	PE3	0.789		
Communication skills	CK1	0.854	0.881	0.782
	CK2	0.779		
	CK3	0.881		
Leadership and motivation	LM1	0.832	0.886	0.677
	LM2	0.886		
	LM3	0.678		
Team work and Problem solving	TP1	0.829	0.845	0.645

(*continued*)

Table 4. *(continued)*

Latent variable	Item	Factor load	CR	AVE
	TP2	0.738		
	TP3	0.779		
University reputation	UR1	0.857	0.894	0.737
Continuous learning	CL	0.811	0.745	0.695
Career marketing demand	CMD1	0.831	0.788	0.718
	CMD2	0.878		
Job search tool	JST1	0.775	0.856	0.768
	JST2	0.911		
	JST3	0.882		
Graduate employment	GE	0.887	0.891	0.801

4.4 Conformity Factor Analysis (CFA)

Confirmatory factor analysis (CFA) is a statistical technique used to verify the factor structure of the observed variables. The relative Chi-Square for this model was 4.034 that is smaller than 5.0 as recommended by [26]. The comparative fit index (CFI) is 0.955 that greater than threshold recommended by [27]. The root mean residual (RMR) value was found to be0.061, which is less than 0.08 defined by [28].

Goodness of fit index (GFI) of the model is 0.912 which is more than the recommended value of 0.90 suggested [13]. The adjusted goodness of fit index (AGFI) was found to be 0.882 which matches the threshold recommended by [29]. The root mean square error of approximation (RMSEA) is 0.073, which is also less than the suggested fit of [30]. Finally, the standardized means square residual (SRMR) is 0.069 which is less than 0.08 recommend by [30]. Table 5 defines the confirmatory factor analysis model fit.

Table 5. Confirmatory factor analysis model fit

Model fitting index	Value	Level of acceptance	Reference
Chi-square/df	4.034	<5.0	[26]
Comparative fit index (CFI)	0.955	>0.90	[27]
Root mean residual (RMR)	0.061	<0.08	[28]
Goodness of fit index (GFI)	0.912	>0.90	[13]
Adjusted goodness of fit index (AGFI)	0.882	>0.85	[29]
Root mean square error of approximation (RMSEA)	0.073	<0.08	[30]
Standardizedmeans square residual (SRMR)	0.069	<0.08	[30]

Structural equation modeling was utilized to identify the significant relationships among ten different employability of AAST university graduates. The path diagram showed that TS, AP, TP, UR, CL and CMD have positive significant impact with values respectively: 0.461, 0.582, 0.497, 0.422, 0.509, and 0.437. While PE, CK, LM, JST has a positive impact on GE but insignificantly. Table 6 represents path coefficient and hypothesis support.

Table 6. Path coffiecient and hypothesis support

Path	Hypothesis	Standard cofficient	Significance	Support
GE←TS	H1	0.461	√	Supported
GE←AP	H2	0.582	√	Supported
GE←PE	H3	0.193	x	Not supported
GE←CK	H4	0.186	x	Not supported
GE←LM	H5	0.121	x	Not supported
GE←TP	H6	0.497	√	Supported
GE←UR	H7	0.442	√	Supported
GE←CL	H8	0.509	√	Supported
GE←CMD	H9	0.437	√	Supported
GE←JST	H10	0.141	x	Not supported

Results had shown a similarity of hypothesis including Academic performance, technical skills, teamwork and problem solving supporting according to previous findings of [14, 15, 16, 17, and 22]. However, results did not agree on Communication skills variable and personality variable with the results on previous research [14, 31] since the differences in the culture that affects personality, job market demand that affect communication skills and the quality of education.

5 Conclusion and Future Work

Job Marketers required specific capabilities of graduates. Establishing relationship between industry and university is essential to ensure the perfect match between job and graduates Career consulting provide a good connection point to help graduates manage their skills in the learning path in the future. Understanding the requirements of local and global market is the main concern of higher education. Universities should provide enhanced curriculum based on the job demand of the market. Limitation of the research represented as follows: data collection was limited to 350 in which data can be expanded to select a bigger dataset. Egyptian market is the only tested market while a comparative study including other countries can be examined. Research may expand to include gulf countries. Examined factors are limited to 10 factors; Factors like culture and attitude can be examined on future.

References

1. Cho, H.: The role of career counselling in job and career development. Pertanika J. Soc. Sci. Humanit. **25**(S), 29–38 (2017)
2. Balingbing, A.: Tracer study of BS in information technology (BSIT) graduates of Camarines Sur Polytechnic Colleges, Nabua, Camarines Sur from 2004 to 2010. Asia Pac. J. Multidiscip. Res. (2014)
3. Career Industry Council of Australia: Career education: Why is it important? Career Industry Council of Australia (2020). https://pssfw.myskills.gov.au/media/1754/career-education-why-is-it-important.pdf
4. Arshad, M., Tahir, A., Khan, M., Basit, A.: Impact of career counseling and vocational guidance on employment in TVET sector. Int. J. Hum. Resour. Stud. **8**(1), 274–289 (2018)
5. Puad, M., Desa, H.: Dissecting perceptions of new graduates on work orientation and self-confidence in employability skills training program. Univ. J. Educ. Res. **8**(1), 70–75 (2020)
6. Yusran, N., Puad, M., Omar, M.: Role of career exploration in influencing career choice among pre-university student. Soc. Sci. Humanit. (2021)
7. Mohd Rasdi, R., Ahrari, S.: The applicability of social cognitive career theory in predicting life satisfaction of university students: a meta-analytic path analysis. Plos One **15**(8), 1–22 (2020). https://doi.org/10.1371/journal.pone.0237838
8. Chan, C.: Factors affecting career goals of Taiwanese college athletes from perspective of social cognitive career theory. J. Career Dev. **47**(2), 193–206 (2020)
9. Jun, K., Fan, J.: Factors affecting job opportunities for university graduates in China–the evidence from university graduates in Beijing. Res. World Econ. (2011)
10. Michael, G., Albino, M., & Albino, F.: Employability of information technology education graduates of president Ramon Magsaysay State University, Castillejos, Zambales from 2015–2020. **2**(3), 440–443 (2021)
11. Nadi, A.: An Investigation of Factors Related to Graduate Student Success. University of Calgary, Calgary, AB (2019)
12. Kong, H., Chen, X.: Analysis of Multiple impacts of COVID-19 outbreak and employment of college graduats–based on the questionniareosurvey and anlaysis of fresh graduates in Uangtze River Delta Region. World Sci. Res. J. (2011)
13. Joreskog, K., Sorbom, D.: LISREL 8: Users' Reference Guide. Scientific Software International, IL (1993)
14. Hossain, M., Alam, M., Alamgir, M., Salat, A.: Factors affecting business graduates' employability–empirical evidence using partial least squares (PLS). Educ. Train. **62**(3), 292–310 (2020)
15. Fenta, H., Asnakew, Z.S., Debele, P.K., Nigatu, S.T., Muhaba, A.M.: Analysis of supply sidefactors influencing employability of new graduates: a tracer study of Bahir Dar University graduates. J. Teach. Learn. Grad. Employab. **10**(2), 67–85 (2019)
16. Neneh, B.N.: An empirical study of personality traits, job market appraisal and self-perceived employability in an uncertain market. High. Educ. Skills Work-Based Learn. **10**(1), 255–274 (2019)
17. Succi, C., Canovi, M.: Soft skills to enhance graduate employability: comparing students and employers' perceptions. Stud. High. Educ. **45**(9), 1834–1847 (2020)
18. Subbu, N., Rajasekaran, V.: Employability skills: a review. IUP J. Soft Skills **12**(1), 29–37 (2018)
19. Nusrat, M., Sultana, N.: Soft skills for sustainable employment of business graduates of Bangladesh. High. Educ. Skills Work Based Learn. **9**(3), 264–278 (2019)
20. Nogales, R., Córdova, P., Urquidi, M.: The impact of university reputation on employment opportunities: experimental evidence from Bolivia. Econ. Labour Relat. Rev. **31**(4), 524–542 (2020)

21. Mostafa, L., Elbarawy, A.: Enhance job candidate learning path using gamification. In: ICCTA18 Proceedings of the 28th International Conference on Computer Theory and Applications, Alexandria, Egypt (2018)
22. Yaojie, L., Wang, X., Xin, D.: An inquiry into AI university curriculum and market demand: facts, fits, and future trends. In: SIGMIS-CPR, 2019, 20–22 June 2019, Nashville, TN, USA (2019)
23. Mostafa, L., Beshir, S.: Job candidate rank approach using machine learning techniques. In: AMLTA 2021 (2021)
24. Kannan, V., Tan, K.: Just in time, total quality management, and supply chainmanagement: understanding their linkages and impact on business performance. Omega **33**(2), 153–162 (2015)
25. Hosain, M., Ameen, M., Mustafi, A., Parvin, T.: Factors affecting the employability of private university graduates: an exploratory study on Bangladeshi employers. Res. Rev. (2021)
26. Marsh, H., Hocevar, D.: Application of confirmatory factor analysis to the study of selfconcept: first-and higher order factor models and their invariance across groups. Psychol. Bull. **97**(3), 562–582 (1985)
27. Bentler, P.: Comparative fit indexes in structural models. Psychol. Bull. **107**(2), 238–246 (1990)
28. Hu, L., Bentler, P.: Fit indices in covariance structure modeling: sensitivity to underparameterized model misspecification. Psychol. Methods **3**(4), 424–453 (1998)
29. Anderson, J., Gerbing, D.: The effect of sampling error on convergence, improper solutions, and goodness-of-fit indices for maximum likelihood confirmatory factor analysis. Psychometrika **49**(2), 155–173 (1984)
30. Browne, M.W., Cudeck, R.: Alternative ways of assessing model fit. Sage Focus. Edit. **154**, 126–136 (1993)
31. Nazron, M., Lim, B., Nga, J.: Soft skills, attributes and graduate employability: a case in university Malaysia Sabah. Malays. J. Bus. Econ. **4**(2), 65–76 (2017)

Intrinsic Interference Reduction: A Channel Estimation Approach for FBMC-OQAM Systems

Shaimaa E. Elghetany(✉), Saly Hassaneen, Islam E. Shaalan, and Heba Y. M. Soliman

Electrical Engineering Department, Faculty of Engineering, Port Said University, Port Said 42524, Egypt

{shaimaa.elsedeek,saly_hasanen,is_shaalan,hebayms}@eng.psu.edu.eg

Abstract. A viable alternative to orthogonal frequency division multiplexing (OFDM) is filter bank multicarrier with offset quadrature amplitude modulation (FBMC-OQAM). Because of what is known as intrinsic interference (II), channel estimation is a major challenge in the FBMC-OQAM system.

A proposed FBMC-OQAM channel estimation technique is provided in this study to reduce the effect of the II term. According to theoretical research, the proposed method enhances bit error rate (BER) performance while also reducing system complexity. It even reduces the system's power consumption and enhancing the system data rate.

Keywords: FBMC-OQAM · OFDM · AP · ZF · LS · II · IIR

1 Introduction

Filter bank multicarrier with offset quadrature amplitude modulation (FBMC-OQAM) first introduced by Chang [1] and developed by Saltzberg [1, 2] has been appeared to replace orthogonal frequency division multiplexing (OFDM). Degrading of spectrum efficiency due to cyclic prefix (CP) and high out of band emission (OOBE) due to the slowly decaying behavior in the frequency domain are of the major drawbacks of OFDM that FBMC beats [2, 13]. Despite FBMC overcomes the previously mentioned drawbacks, more cost is expected due to the change of orthogonality condition from complex field conditions to less strict real field conditions. As a result, channel estimation becomes more challenging in FBMC-OQAM due to the appearance of the intrinsic interference term.

Channel estimation techniques can be classified as pilot-based channel estimation, decision-directed, and blind techniques. In this paper we will focus on pilot-based techniques. Preamble-based channel estimation (PBCE) comes from replacing pilots at the beginning of each frame. PBCE could be classified into pair of pilots (POP) [3], interference approximation method (IAM) [4], and the interference cancellation method (ICM) [5, 6]. These methods are suitable for time-invariant channel with expected performance degradation for high noise levels.

© The Author(s), under exclusive license to Springer Nature Switzerland AG 2022
A. E. Hassanien et al. (Eds.): AMLTA 2022, LNDECT 113, pp. 445–455, 2022.
https://doi.org/10.1007/978-3-031-03918-8_38

Scattered pilots; auxiliary pilot (AP) method [7] or coding method [8], are other types of pilot-based channel estimation techniques. In order to estimate channel coefficients without wasting the data rate, as in the AP method and in a simpler way than the Coding method, this paper proposes a channel estimation method based on mitigating the intrinsic interference by an iterative process.

The remainder of this paper is organized as follows: FBMC-OQAM system is represented in Sect. 2, followed by Sect. 3 where types of scattered pilots channel estimation techniques will be explained. In Sect. 4, the proposed method is introduced and the results will be discussed in Sect. 5. Finally, conclusions are summarized up in Sect. 6.

2 FBMC-OQAM System

The block diagram of FBMC-OQAM [9] is presented in Fig. 1. Where it differs from the OFDM block diagram in; cp block is removed, well-localized filters are used and OQAM-preprocessing and OQAM-postprocessing blocks are inserted in transmitter and receiver respectively. The transmitted signal can be expressed as:

$$x(t) = \sum_k \sum_n s_{\mathrm{k,n}} P_k(t - \mathrm{n}\tau) \tag{1}$$

where $s_{\mathrm{k,n}}$ is the real valued transmitted symbol at k^{th} subcarrier and n^{th} time instant, τ is the duration for one FBMC-OQAM symbol which equal to one-half the duration for one OFDM symbol ($\tau = \mathrm{T}/2$, where T is one OFDM symbol time) and $P_k(t - \mathrm{n}\tau)$ is the filter function [9] which can expressed as:

$$P_k(t) = \mathrm{p}(t) e^{j2\pi \mathrm{fkt}} e^{j(\mathrm{k+n})\frac{\pi}{2}} \tag{2}$$

where p(t) is the prototype filter which is designed to satisfy the real orthogonality condition [10] expressed as

$$\Re\left(\sum_m p_{\mathrm{k,n}}[m] p_{\mathrm{k}',\mathrm{n}'}[m]\right) = \delta_{\mathrm{k,k}'} \delta_{\mathrm{n,n}'} \tag{3}$$

where $\delta_{\mathrm{k,k'}}$ is known as Kroncker delta function.

Merging (2) with (1), it follows that

$$x(t) = \sum_k \sum_n s_{\mathrm{k,n}} p(t - \mathrm{n}\tau) e^{j2\pi \mathrm{fkt}} e^{-\mathrm{j}\pi \mathrm{kn}} e^{j(\mathrm{k+n})\frac{\pi}{2}} \tag{4}$$

putting $(k+n)\frac{\pi}{2} - kn\pi = \phi_{k,n}$ this leads to:

$$x(t) = \sum_k \sum_k s_{\mathrm{k,n}} p(t - \mathrm{n}\tau) e^{j2\pi \mathrm{fkt}} e^{\mathrm{j}\phi \mathrm{k,n}} \tag{5}$$

Considering the channel coefficient $h_{k,n}$ and noise contribuation $\gamma_{k,n}$at k^{th} subcarrier, n^{th} time instant, the received signal after passing through the channel can be expressed as:

$$r(t) = \sum_k \sum_n s_{\mathrm{k,n}} p(t - \mathrm{n}\tau) h_{\mathrm{k,n}} e^{j2\pi \mathrm{fkt}} e^{\mathrm{j}\phi_{\mathrm{k,n}}} + \gamma_{\mathrm{k,n}} \tag{6}$$

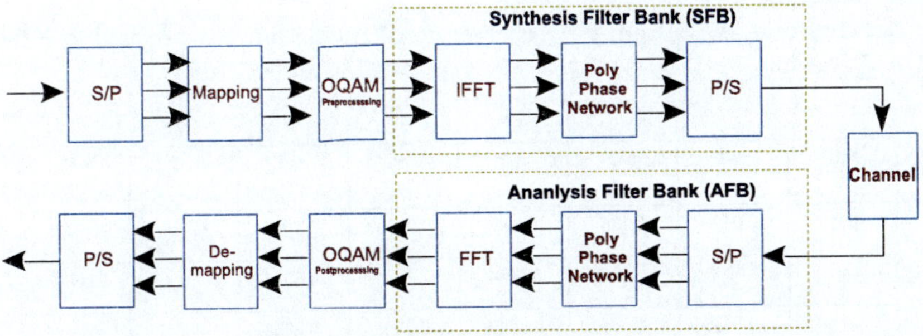

Fig. 1. FBMC-OQAM block diagram.

The demodulated signal at k' th subcarrier, n' th time instant results from the projection between the received signal and the bank of filters at the receiver side [11]:

$$y_{k',n'} = \langle r(t), p_{k',n'} \rangle$$
$$\left(\sum_{k'} \sum_{n'} s_{k,n} p(t - n\tau) p(t - n'\tau) h_{k,n} e^{j2\pi fkt} e^{-j\pi kn} e^{j(k+n)\frac{\pi}{2}} e^{-j2\pi fk't} e^{j\pi k'n'} e^{-j(k'+n')\frac{\pi}{2}} \right) \quad (7)$$
$$+ \langle \gamma_{k,n'}, p_{k',n'} \rangle$$

as the energy of the impulse response is localized around the considered symbol [12], Eq. (7) becomes

$$y_{k',n'} = s_{k',n'} h_{k',n'} + \left[\sum_{(k,n) \neq (k',n')} s_{k,n} p(t - n\tau) P(t - n'\tau) h_{k,n} e^{j\phi_{k,n}} e^{-j\phi_{k',n'}} \right] + \Upsilon_{k',n'} \quad (8)$$

we assume that the intrinsic interference term depends only on this tied set (denoted by Ω). Moreover, assuming that the channel is constant at least over this summation area, we can write as in [11, 12]

$$y_{k',n'} = s_{k',n'} h_{k',n'} + h_{k',n'} I_{k',n'} + \Upsilon_{k',n'} \quad (9)$$

$I_{k',n'}$ is the intrinsic interference caused by the neighbors of the symbol represented as:

$$I_{\Omega} = \sum_{\substack{(p,q) \in \Omega \\ (p,q) \neq (k',n')}} s_{p,q} \langle P_{p,q}, P_{k',n'} \rangle = \sum_{w} s_w \alpha_w \quad (10)$$

where α donates the intrinsic interference coefficient given by Fig. 2 [19]. As the filters are designed to be well localized in time domain and frequency domain, so the first order neighbors have the most effect on $I_{k',n'}$ (i.e.,$k',n' \in \{-1,0,1\}$,$(k',n') \neq (0,0)$) [14] this leads to $I_{k',n'}$ is a pure imaginary part. Equation (9) becomes

$$y_{k',n'} = s_{k',n'} h_{k',n'} + jh_{k',n'} I_{k',n'} + \Upsilon_{k',n'} \quad (11)$$

So, recovering $s_{k',n'}$ from $y_{k',n'}$ is a very challenging problems related to FBMC-OQAM system.

(f,t)	n-3	n-2	n-1	n	n+1	n+2	n+3
k-1	-0.0429	0.125	-0.2058	0.2393	-0.2058	0.125	-0.0429
k	0.0668	0	0.5644	1	-0.5644	0	-0.0668
k+1	-0.0429	-0.125	-0.2058	-0.2393	-0.2058	-0.125	-0.0429

(f,t)	n-3	n-2	n-1	n	n+1	n+2	n+3
k-1	0.0429	0.125	0.2058	0.2393	0.2058	0.125	0.0429
k	-0.0668	0	-0.5644	1	0.5644	0	0.0668
k+1	0.0429	-0.125	0.2058	-0.2393	0.2058	-0.125	0.0429

Fig. 2. The intrinsic interference coefficient when PHYDYAS filter is used, upper table for even k and lower table for odd k [19].

3 Scattered Pilots Channel Estimation Techniques

In order to track the time varying and frequency selective channel characteristics, the pilot tones are inserted along both the time and frequency axis with given periods. Transmitting a single or a pair of pilots on the time-frequency lattice, surrounded by data symbols, is a suitable choice for a fast fading channels.

3.1 AP Method

In order to cancel the intrinsic interference term at the pilot position, the AP method relies on sacrificing a symbol where a signal with an opposite effect to the interference is added. For example, to cancel the effect of first-order neighbors around the pilot shown in Fig. 3, the data at AP can be driven as:

$$s_{\text{aux}} = \alpha_{\text{aux}}^{-1} \sum\nolimits_{i=1}^{7} s_i \alpha_i \tag{12}$$

The channel response could then be obtained by least square channel estimation. The AP method has a bad effect on the data rate and power efficiency of the system [16, 17].

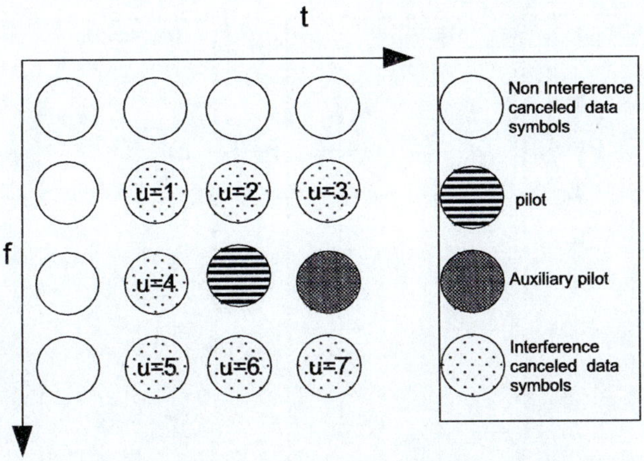

Fig. 3. Pilot structure of eight data symbols cancels AP.

3.2 Coding Method

Choosing to stay safe away from AP's cons as in [8], we can achieve this by using N time-frequency positions to spread N-1 data symbols around the pilot. The data symbols are precoded by the matrix c so that the data symbols after mapping can be expressed as:

$$\mathrm{x} = \mathrm{cs} \tag{13}$$

matrix c is chosen to fulfill two conditions:

- 1-C must be orthogonal to the interference which allows us to cancel the interference

$$\alpha^T \mathrm{c} = 0 \tag{14}$$

- 2-C must be orthogonal matrix, the design of the coding matrix is discussed in detail in [16, 18, 19].

4 Proposed Method

The Intrinsic Interference Reduction (IIR) approach, as shown in Fig. 4, allows us to reduce the effect of intrinsic interference and improve the accuracy of channel estimation over least square (LS) channel estimation.

Assume the output symbols at the equalization output are correct if a zero forcing (ZF) equalizer is employed. After demodulation, the receiver performs an LS estimation at the pilot location, followed by an interpolation process to estimate the channel coefficient at the other symbols, and finally the received symbols enter our IIR loop.

The received symbols are equalized in the first iteration, then intrinsic interference terms are estimated using (12) and the table in Fig. 2, and the interference term is

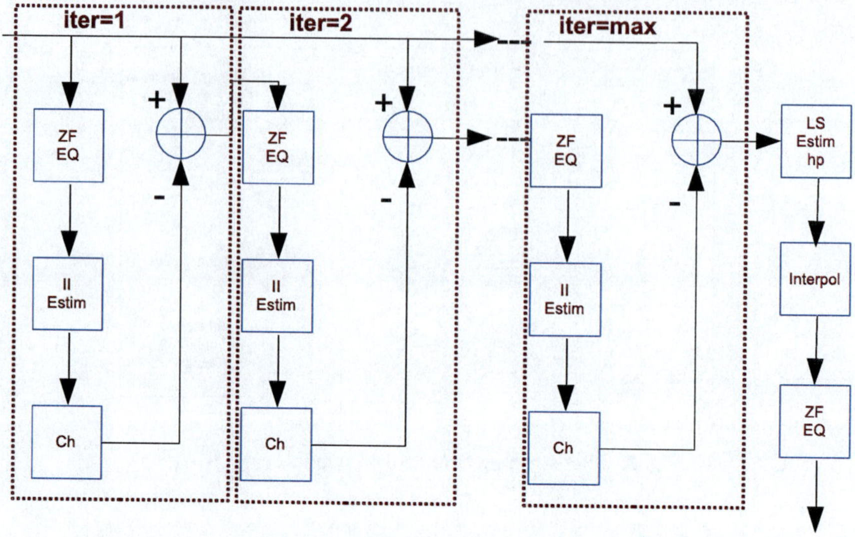

Fig. 4. Block diagram of IIR receiver.

weighted by multiplying these terms by the channel coefficient. We can reduce the errors in received symbols by subtracting the estimated interference from them.

This process can be repeated until we reach an acceptable performance at which the channel estimation could be performed more accurately. Unlike AP method IIR doesn't waste any symbols, so it can save our data rate.

5 Simulation Results

Our simulation results are based on using PHYDYAS [19] filter with overlapping factor = 4, 1024 QAM modulation, 24 subcarriers, we assume a doubly flat channel model, and the number of data symbols whose interference canceled by AP, Coding and IIR equals to eight symbols.

Figure 5 shows the simulated result curves of bit error rate (BER) with a signal to noise ratio (SNR). It can be seen that at BER of $10^{-0.9}$, AP and Coding have the same performance, OFDM has about 1 dB better performance than the other previous FBMC methods.

Using LS in FBMC results in a very bad performance about 9.5 dB lower compared to OFDM, introducing the IIR method to FBMC enhances the performance at which it surpasses OFDM by about 1 dB.

In order to show the effect of an increasing number of iterations in the IIR method, let's look at Fig. 6, increasing the iteration from 2 to 5 or from 5 to 8 each results in 1 dB improvement in performance. Increasing iterations more than 8 have no performance enhancement, but it increases the complexity and wasting the processing time of the system.

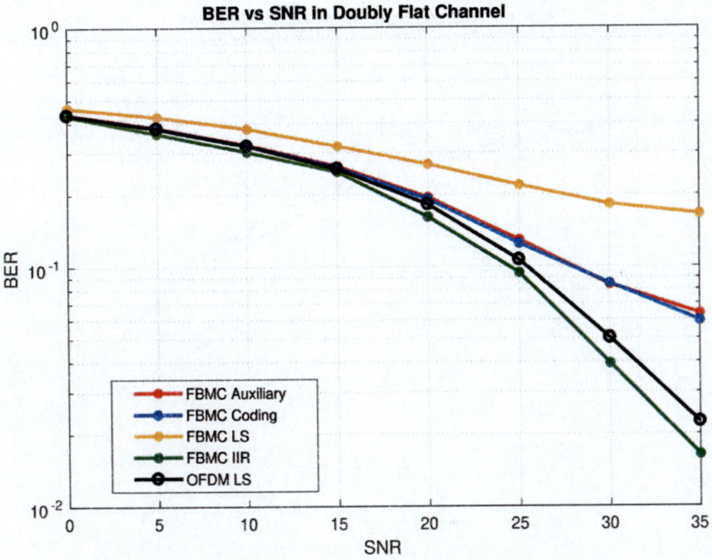

Fig. 5. The BER performance of different channel estimation methods.

Figure 7 presents the pilot structure used in the IIR method, where there is no need to waste the system data rate by using many non-data-symbols, in addition to the pilot symbols, as in AP method presented in Fig. 8.

The number of real multiplications needed for calculating the channel coefficient for one pilot is performed to evaluate different FBMC/OQAM CE approaches in terms of computational complexity. Table 1 shows how many processes are needed from each side of the communication system.

Table 1. The number of additional multiplication at each FBMC/OQAM CE method

	AP	Coding	IIR
Transmitter side	N_s	$N_s(N_s - 1)$	–
Receiver side	–	$N_s(N_s - 1)$	$R(N_s + 2)$

Where N_s is the number of interference canceled neighbors and R is the number of iterations in IIR method.

Figure 9 is derived using the parameters used to produce the result given in Fig. 5 and 6, where $N_s = 28$.

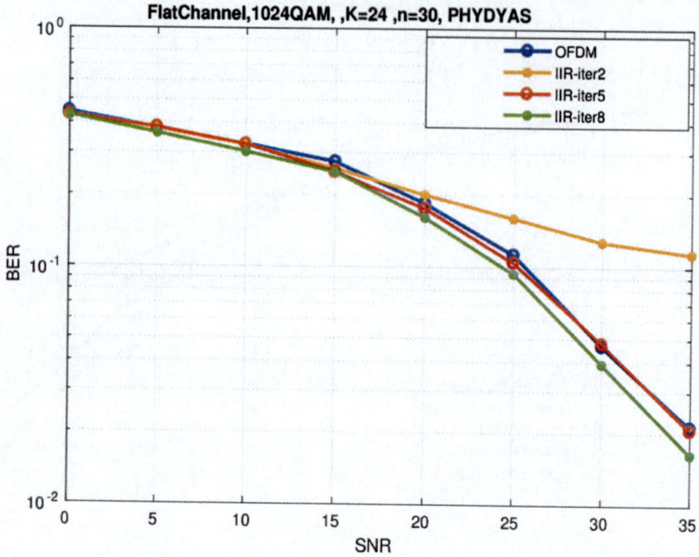

Fig. 6. The BER performance of the IIC method with a different number of iterations.

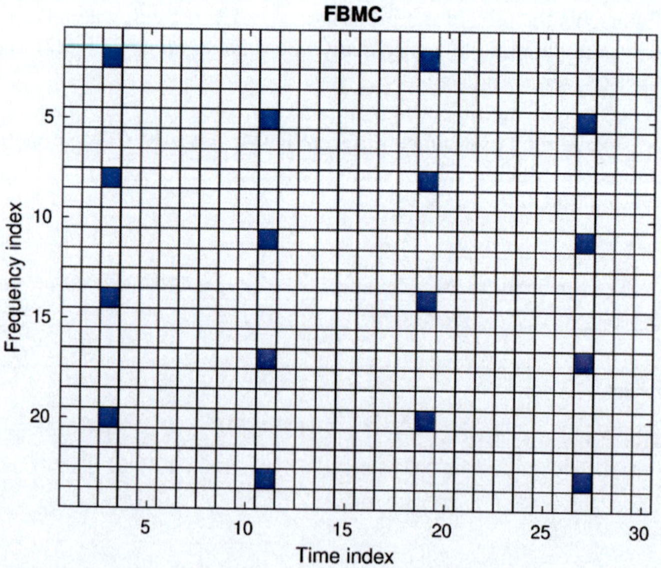

Fig. 7. Frame structure used in the proposed approach, (blue = pilot, white = data).

The peak to average power ratio (PAPR) is a common difficulty for FBMC, just as it is for OFDM. As can be observed in Fig. 10, changing the type of channel estimation method has no effect on the problem, implying that a significant effort should be taken to eliminate it.

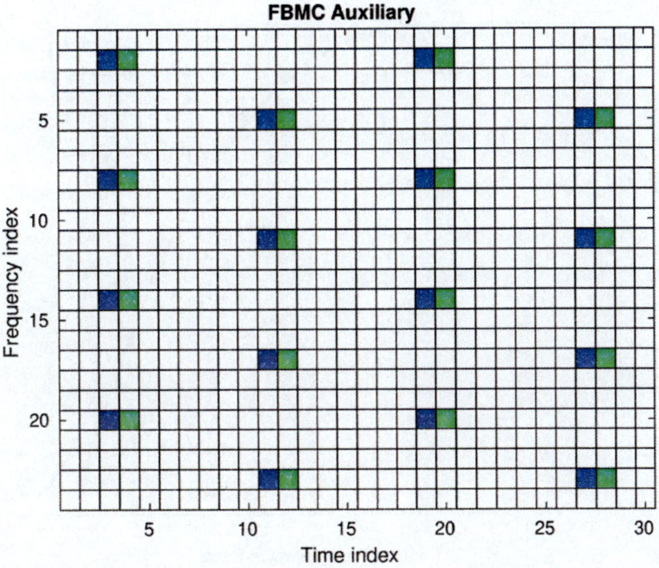

Fig. 8. Frame structure used in the proposed approach, (blue = pilot, green = AP, white = data).

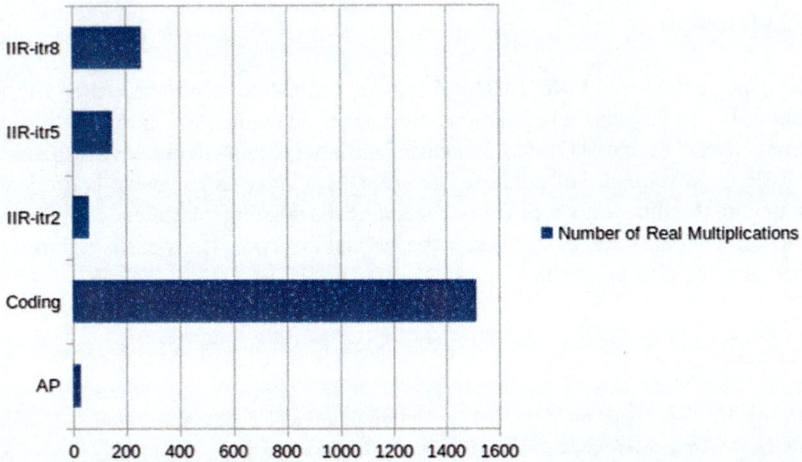

Fig. 9. The number of multiplications obtained using various FBMC/OQAM CE techniques.

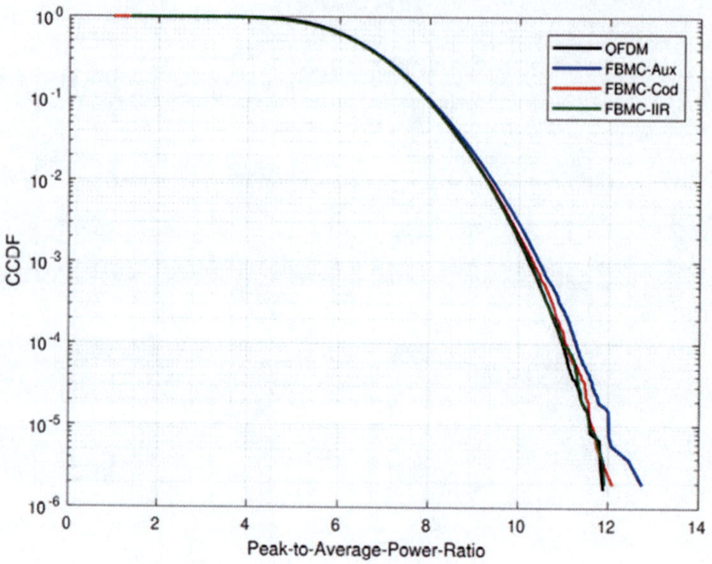

Fig. 10. PAPR performance of different methods.

6 Conclusion

In this paper, a proposed FBMC-OQAM channel estimation technique, called IIR, was proposed. Under flat channel conditions, the suggested technique improves the system BER until it reaches that of OFDM. Increasing the inner II cancellation loop also resulted in a BER improvement that outperformed OFDM. Unlike other methods of channel estimates, this methodology is based on cancelling the inherent interference term with no loss in power or data rates, as opposed to the AP method, and it is a more straightforward method than the Coding method.

References

1. Chang, R.: High-speed multichannel data transmission with bandlimited orthogonal signals. Bell Syst. Tech. J. **45**, 1775–1796 (1966)
2. Saltzberg, B.: Performance of an efficient parallel data transmission system. IEEE Trans. Commun. Technol. **15**(6), 805–811 (1967)
3. Lélé, C., Javaudin, J.-P., Legouable, R., Skrzypczak, A., Siohan, P.: Channel estimation methods for preamble-based OFDM/OQAM modulations. Eur. Trans. Telecommun. **19**(7), 741–750 (2010)
4. Du, J., Signell, S.: Novel preamble-based channel estimation for OFDM/OQAM systems. In: Proceedings of the IEEE International Conference on Communications, ICC2009, pp. 4135–4140, IEEE, Dresden, Germany, June 2009
5. Kofidis, E., Katselis, D., Rontogiannis, A., Theodoridis, S.: Preamble-based channel estimation in OFDM/OQAM systems: a review. Signal Process. **93**(7), 2038–2054 (2013)

6. Eleftherios, K., et al.: FBMC channel estimation techniques. In: Orthogonal Waveforms and Filter Banks for Future Communication Systems, ch. 11, sec. 3, pp. 263–268. Elsevier, London, UK (2017)
7. Javaudin, J., Lacroix, D., Rouxel, A.: Pilot-aided channel estimation for OFDM/OQAM. In: Proceedings of the 57th IEEE Semiannual Vehicular Technology Conference. VTC 2003, pp. 1581–1585, Jeju, South Korea (2003)
8. Lélé, C., Legouable, R., Siohan, P.: Channel estimation with scattered pilots in OFDM/OQAM. In: Proceedings of the IEEE 9th Workshop on Signal Processing Advances in Wireless Communications (SPAWC 2008), pp. 286–290, Recife, Brazil, July 2008
9. Farhang-Boroujeny, B.: OFDM versus filter bank multicarrier. IEEE Signal Process. Mag. **28**(3), 92–112 (2011). https://doi.org/10.1109/MSP.2011.940267
10. Siohan, P., Siclet, C., Lacaille, N.: Analysis and design of OFDM/OQAM systems based on filterbank theory. IEEE Trans. Signal Process. **50**(5), 1170–1183 (2002)
11. Zakaria, R., Le Ruyet, D.: A novel filter-bank multicarrier scheme to mitigate the intrinsic interference: application to MIMO systems. IEEE Trans. Wirel. Commun. **11**(3), 1112–1123 (2012). https://doi.org/10.1109/TWC.2012.012412.11-0607
12. Katselis, D., Kofidis, E., Rontogiannis, A., Theodoridis, S.: Preamble- based channel estimation for CP-OFDM and OFDM/OQAM systems: a comparative study. IEEE Trans. Signal Process. **58**(5), 2911–2916 (2010)
13. Vaezi, M., Ding, Z., Poor, H.V. (eds.): Multiple Access Techniques for 5G Wireless Networks and Beyond. Springer, Cham (2019). https://doi.org/10.1007/978-3-319-92090-0
14. Jérôme, L., et al.: Equalization and demodulation in the receiver (single antenna). In: ICT –211887, PHYDYAS, July 2008
15. Nissel, R.: Filter bank multicarrier modulation for future wireless systems, Ph.D. dissertation, Abbrev. Dept., TU Wien. Univ., Vienna (2017)
16. Xue, L., Chen, H., Zhang, K., Qiu, S.: Iterative auxiliary pilots channel estimation for OFDM/OQAM systems. In: 2017 3rd IEEE International Conference on Computer and Communications (ICCC), Chengdu, pp. 136–140 (2017). https://doi.org/10.1109/CompComm.2017.8322528
17. Cui, W., Qu, D., Jiang, T., Farhang-Boroujeny, B.: Coded auxiliary pilots for channel estimation in FBMC-OQAM systems. IEEE Trans. Veh. Technol. **65**(5), 2936–2946 (2016). https://doi.org/10.1109/TVT.2015.2448659
18. Choi, J., Oh, Y., Lee, H., Seo, J.: Pilot-aided channel estimation utilizing intrinsic interference for FBMC/OQAM systems. IEEE Trans. Broadcast. **63**(4), 644–655 (2017). https://doi.org/10.1109/TBC.2017.2711143
19. Bellanger, M., et al.: FBMC physical layer: a primer, PHYDYAS, Technical report (2010). http://www.ict-phydyas.org

Impact of Using Different Color Spaces on the Image Segmentation

Dena A. Abdelsadek[1](✉), Maryam N. Al-Berry[2], Hala M. Ebied[2], and Mosab Hassaan[1]

[1] Faculty of Science, Benha University, Banha, Egypt
{dena.abdelsadek,mosab.hassaan}@fsc.bu.edu.eg
[2] Faculty of Computer and Information Sciences, Ain Shams University, Cairo, Egypt
{marym_nabil,Halam}@cis.asu.edu.eg

Abstract. Image segmentation is considered one of the most difficult challenges in image processing. Recently many advanced applications have emerged in this field. Color images provide more information and more reliable in segmentation than grayscale images. In this paper, the color spaces RGB, YCbCr, XYZ, and HSV are compared using four different methods of image segmentation. These methods are k-means, Fuzzy C-means, Region growing, and Graph Cut. The main objective of image segmentation is to simplify and change the image to something more meaningful and easier to analyze. In this study, we used single-color space components. In addition to this, we vote between the three components of every color space in the segmented image to get the best image segmentation result. Different RGB color images from Berkeley databases are used. The accuracy of the image segmentation is measured using the peak signal-to-noise ratio (PSNR) and mean square error (MSE). The experimental results show that the voting between color components achieved good segmentation accuracy.

Keywords: Color image segmentation · Color spaces · K-mean · Fuzzy C-mean · Region growing

1 Introduction

Image Segmentation is an important task in image processing, that dividing an image into parts, objects, or different regions based on pixels. Each set of pixels in a region has similar properties such as color, intensity or texture, etc. [1, 2].

Image segmentation continues to be widely used in advanced applications in many fields like machine vision, face recognition, traffic control systems, medical imaging, and others [3]. The image represents a finite set of digital values called pixels. Images may be grayscale or color images. So, image segmentation may be either gray scale image segmentation or color image segmentation. Color images as provide more information and enhance the processing of analysis images than grayscale images [4].

© The Author(s), under exclusive license to Springer Nature Switzerland AG 2022
A. E. Hassanien et al. (Eds.): AMLTA 2022, LNDECT 113, pp. 456–471, 2022.
https://doi.org/10.1007/978-3-031-03918-8_39

The pixel's of digital color images can be represented in different color space models for different color image segmentation techniques [5]. A color space represents visual information that describes the color spectrum as a multidimensional model which most color spaces have three dimensions and the basic digital color image is given in RGB coordinate [4, 6].

Image segmentation has an important influence on image analysis and understanding [7]. There are different methodologies for the image segmentation process, these methodologies can be organized into different groups: Histogram, Thresholding, Clustering, Region-based, Edge-based, Model-based, Fuzzing approaches, and Neural Network-based [7, 8].

In this paper, different image segmentation methods are used to study the impact of changing the color space models on image segmentation accuracy. K-means, FCM, Region Growing, and Graph Cut methods are used in this paper. In experimentl results, we used the RGB, YCbCr, XYZ, HSV color spaces. The image segmentation methods apply to the single color component of the RGB, YCbCr, XYZ, HSV color spaces. We also vote on the segmented image of the three components of every color space. The performance is measured using the peak signal-to-noise ratio (PSNR) and mean square error (MSE).

The rest of the paper is organized as follows. Section 2 presents the related work. Section 3 describes the color spaces used in our experiments. Section 4 describes the image segmentation methods. Section 5 shows the experimental results. Section 6 concludes the results of the paper.

2 Related Work

It is an important and interesting problem to understand and analyze the image for color image segmentation, but not found the specific color space can be best for every image segmentation problem [9]. In this section, we survey several image segmentation methods based on different color spaces. In recent years, a lot of researchers received attention to image segmentation and the relation between color spaces and segmentation.

Jurio et al. [10] presented a comparative study between different color spaces in cluster-based image segmentation. The comparison carried out between four color spaces HSV, YUV, CMY, and RGB. The authors concluded that CMY is the best color space for most cases. Burney et al. [11] analyzed the algorithm of k-means clustering for image segmentation methods. This study proposed how to work on the k-means cluster algorithm for some color spaces (RGB and LAB). This work concluded that the LAB color space achieved better than RGB color space. Khattab et al. [12] presented a comparative study using different color spaces to evaluate the performance of color image segmentation using the automatic Grab Cut method. The author used RGB, HSV, XYZ, CMY, and YUV color spaces. The experiments result show that RGB color space is the best color.

Mythili et al. [13] k-means clustering and Effective robust Kernclized fuzzy C-means(ERKFCM) for color image segmentation based on different color spaces such as RGB,HSV,YIQ,and XYZ by using the PSNR and MSE to evaluate the performance that gave the combine segmentation of various color spaces give more accurate segment of signal color.

Shih et al. [14] presented an efficient segmentation algorithm for the color image using automati seed region growing. The authors firstly convert input RGB color space to YCbCr color. Second, automatically select initial seeds. Third, the color image is segmented into regions. Finally, region merged This method carried out good results used k-means clustering and Effective Robust Kernelized fuzzy C-means (ERKFCM) for color image segmentation based on different color spaces such as RGB, HSV, YIQ, and XYZ. The authors used the PSNR and MSE to evaluate the performance. The experiments result show that combined segmentation of various color spaces archived more accurate segment than using signal color space.than existing algorithms in the paper.

Wang et al. [15] applied a comparison of different color spaces for image segmentation using Graph Cut. The tested colors, RGB, HSV, L*U*V, and L*a*b show that the color space L*a*b archived higher or similar quality as all other methods, but RGB worse than segmentation-based on any other tested color space.

Busin et al. [16] analyzed the use of color spaces and image segmentation. They described the choice of color space significantly depending on the kind of image to be segmented and the segmentation method. The evaluation was used for the selection that was based on spectral color analysis.

3 Color Spaces

A color space is an abstract mathematical model describing the way colors can be represented as tuples numbers (e.g. triples in RGB or quadruples in CMYK). However, there are a lot of different color spaces, each with its properties, advantages, limitations, and areas of application [10, 17].

In this study, we are comparing four color spaces in image segmentation are RGB (Red, Green, Blue), HSV (Hue, Saturation, Value), YCbCr (Luminance (Y), Chrominance (Cb, Cr)), and XYZ (Luminance (Y), Chromaticity coordinates (X, Z)).

3.1 RGB

The RGB color space is all possible colors that can be made from three colorants red, green, and blue. The RGB model considers the most natural space and is commonly used in image processing, computer graphics, and multimedia systems. So, all color spaces can be derived from the RGB information, so that we are going to use this color space as in [9].

3.2 HSV

The HSV (Hue, Saturation, Value) color space is very close to the RGB color space in which humans create and perceive colors in the sense of color. HSV color space attempts to characterize colors according to their hue (H), light objects in saturation (S), and intensity (value, or lightness).

Note that, hue represents the color degree of the image, light objects in saturation is the amount of the color, and intensity is amount of light (it allows the distinction between a dark color and a light color). The relationship between HSV and RGB can be described as follows.

$$H = \begin{cases} 0. & \text{if } max = min \\ \left(60^\circ \times \frac{G-B}{max} + 360^\circ\right) mod\ 360^\circ. & \text{if } max = R \\ 60^\circ \times \frac{B-R}{min} + 120^\circ. & \text{if } max = G \\ 60^\circ \times \frac{\pi - min}{max - min} + 240^\circ. & \text{if } max = B \end{cases} \tag{1}$$

$$S = \begin{cases} 0. & \text{if } max = 0 \\ \frac{max-min}{max} = 1 - \frac{min}{max}. & \text{otherwise} \end{cases} \tag{2}$$

$$V = max \tag{3}$$

3.3 YCbCr

The YCbCr is a technique of color spaces known as transmission primaries. It is used for digital video and photography systems. It belongs to the family of television transmission color spaces. The family includes others such as YUV and YIQ [18]. The YCbCr separates RGB into Luminance (Luma Y) and (chroma [Cb and Cr]). (Luma Y) is the brightness that occurs using black and white gray shades. And Chrominance (chroma [Cb and Cr]), is the color information in a signal mainly concentrated on “YCbCr” either in a red or blue signal [19]. The following equations show to convert from RGB to YCbCr.

$$\begin{bmatrix} Y \\ Cb \\ Cr \end{bmatrix} = \begin{bmatrix} 0{\cdot}2989 & 0{\cdot}5866 & 0{\cdot}1145 \\ -0{\cdot}1688 & -0{\cdot}3312 & 0{\cdot}5000 \\ 0{\cdot}5000 & -0{\cdot}4184 & -0{\cdot}0816 \end{bmatrix} \begin{bmatrix} R \\ G \\ B \end{bmatrix} \tag{4}$$

3.4 XYZ

The XYZ is the first color space created by the international commission on illumination in 1931 so that XYZ called CIE XYZ [20]. The model can be described as the luminance component Y and the chromaticity information. The value of X, Y, Z are computed using a linear transformation from RGB color coordinates, as shown in the following equations:

$$\begin{bmatrix} X \\ Y \\ Z \end{bmatrix} = \begin{bmatrix} 0{\cdot}4125453 & 0{\cdot}357580 & 0{\cdot}180423 \\ 0{\cdot}212671 & 0{\cdot}715160 & 0{\cdot}072169 \\ 0{\cdot}019334 & 0{\cdot}119193 & 0{\cdot}950227 \end{bmatrix} \begin{bmatrix} R \\ G \\ B \end{bmatrix} \tag{5}$$

4 Segmentation Techniques

This section reviews the K-Means, Fuzzy C-Means, Region Growing, and Graph Cut in the following subsections:

4.1 K-Means Clustering Segmentation

K-Means algorithm is the most popular partitioning-based clustering technique. K-Means algorithm is an unsupervised clustering algorithm that is used to divide an image into K clusters [21]. The aim of K-Means Clustering analysis is to group data in such a way that similar objects are in one cluster and objects of different clusters are dissimilar. Partitional clustering (K-Means) is faster than hierarchical clustering but gives different results with different initial chosen centroids. Let us consider an image with a resolution of $(x \times y)$ that has to be clustered into K number of clusters. Let $p(x, y)$ be input pixels to be clustered for the image and c_k be the cluster centers of the color image. The procedure of the K-Means algorithm is given as follows [22]:

Algorithm 1. k-means method
1. Select randomly k pixels as the initial cluster centroids.
2. For each pixel of the image, $p(x.y)$ do.
3. Calculate the Euclidean distance $d = \|p(x.y) - c_k\|$.
4. Group the pixels based on the min distance.
5. Update centroids by calculate the mean of the pixels in the same group.
6. Go to step 2, until no moving of objects between different groups.

4.2 Fuzzy C-Means Clustering Segmentation

The Fuzzy C-Means (FCM) is an extension of K-Means clustering which allows one pixel to belong to two or more clusters [23], unlike K-Means Clustering which each pixel is a set of membership levels are associated. This method was developed by (Dunn in 1973) and improved by (Bezdek in 1981) [24]. Fuzzy is frequently used in pattern recognition and several applications, such as medical imaging and security systems. The procedure of the FCM algorithm is given as follows [25]:

Algorithm 2. Fuzzy C-Means method

1. Receive the image in the form of data matrix X.
2. Fix the number of clusters c, $(2 \leq c \leq n)$ *where* n is the image length.
3. Assume the partition matrix U.
4. Calculate the cluster centers V_i, $i = 1, 2, \ldots c$ using the following equation
$$V_i = \frac{\sum_{j=1}^{n} (U\,ij)^m x_j}{\sum_{j=0}^{n} (U\,ij)^m}, \qquad \text{for } i = 1,2,\ldots c \text{ and } j = 1,2,\ldots n$$
5. Calculate the Euclidean distance matrix, d, using the following equation
$$d_{ij} = \|X_j - V_i\|.$$
6. Compute the cost or objective function. Stop if either it is below a certain tolerance value or its improvement over the previous iteration is below a certain threshold
$$J(U, c_1, \ldots \ldots, c_c) = \sum_{i=1}^{c} \sum_{j=1}^{n} u_{ij}{}^m d_{ij}{}^2.$$
7. Compute a new U using the next equation. Go to step 2
$$U_{ij} = \sum_{k=1}^{c} \left\{ \frac{\|X_j - V_i\|}{\|X_j - V_k\|} \right\}.$$

4.3 Region Growing Segmentation

Region growing (RG) is a simple region-based image segmentation method. It is also classified as a pixel-based image segmentation method since it involves the selection of the initial seed point (pixel). It starts with a pixel and goes on adding the pixels based on similarity properties (gray level, textures, color), to the region. This is repeated until all pixels belong to some region. Connected regions have multiple criteria at the same time such as the advantages of the region growing. It gives a good result with less noise but gives inefficient segmentation when the image is noisy or has intensity variations [7, 26]. The procedure of the Region Growing algorithm is given as follows:

Algorithm 3. Region Growing method

1. Select seed point P_{seed} in the color image.
2. Check the neighber pixels of P_{seed} and add them to the region if they are similar to the P_{seed}.
3. Repeat step 2 for each of the newly added pixels; STOP if no more pixels can be added.

4.4 Graph Cut Segmentation

The Graph Cut is used as the fundamental object/background segmentation method [27]. A graph cut is a process of partitioning a directed or undirected graph into disjoint sets. In the graph model, each pixel is considered as a node and connected to its neighbor nodes through edges. Edges between pixel nodes are called n-links (pixel similarity), additionally, there are two terminal nodes: S (source) and T (sink). These nodes represent object and background. Each pixel node has two edges connected to S and T, which are called t-links. All links between two-pixel nodes i and j are assigned to weight w_{ij}. A cut is a set of edges in the graph that completely separate the foreground this cut is calculated using Eq. (6)

$$c(A, B) = \sum\nolimits_{i \in A, j \in B} W_{ij}, \tag{6}$$

where A and B correspond to the two disjoint sets of nodes of the result of bipartitioning. This cut is determined using the min-cut/max-flow algorithm [28]. Figure 1, shows a sample example of the graph cut process for a 3 × 3 image [29].

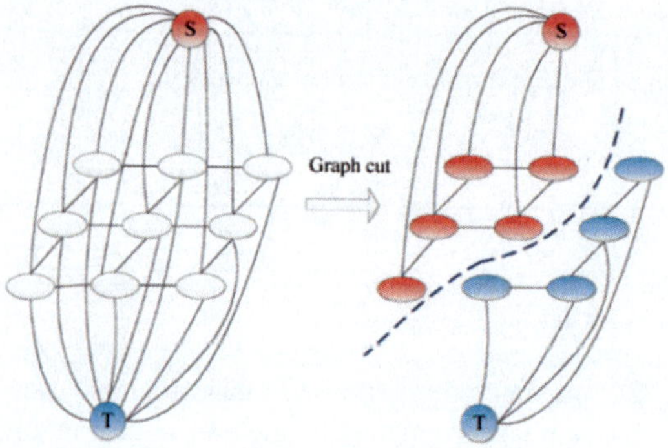

Fig. 1. An example of the graph cut process for a 3 × 3 image.

5 Experimental Results

5.1 Dataset

In this paper, we have applied the four algorithms described above on 150 images from Berkeley Segmentation Database (BSD300) [30]. The images are 481 × 321 pixels. For each color image, a set of benchmark segmentation results (called ground truth) is available.

5.2 Accuracy Criteria

The accuracy is measured by the following equation [31]:

$$Accuracy(\%) = \frac{TP + TN}{TP + TN + FP + FN} * 100. \tag{7}$$

This method computes the accuracy depending on the use of a confusion matrix. True Positive (TP) is the number of object pixels correctly identified as an object, True

Negative (*TN*) is the number of background pixels correctly identified as background, False Positive (*FP*) is background pixels incorrectly identified as an object, False Negative (*FN*) is object pixels incorrectly identified as back ground. To objectively evaluate the segmentation results we used two different measures which are MSE (Mean Square Error) [32] and PSNR (Peak to Signal Noise Ratio) [33].

5.3 Results

In this study, we evaluated four different segmentation methods, namely K-Means, FCM, Region Growing, and Graph Cut. We used these segmentation methods with different color space models, including RGB, HSV, YCbCr, and XYZ. Each color space model was applied on a single color component. In addition, we vote between the segmented three components for every color space in one vector. The final segmentation results are obtained for all 150 images from Berkley Segmentation Database. Figure 2 shows some examples of images that were used to perform the experiments presented in this study.

Fig. 2. Some samples of images of Berkely database.

For a quantitative comparison, the first experiment was carried out to measure the accuracy of the four segmentation methods (K-Means, FCM, Region Growing, and Graph Cut) with the $R.G.$ and B components individually besides the vote between three segmented components in one vector. Table 1 presented PSNR, MSE, and segmentation accuracy for four algorithms using the $R.G.$ *and* B components and voting between the three components. In this experiment, the B component achieved better performance than the G and R components. Voting RGB also provided good results compared to the G and R components. The PSNR has the highest value with the Graph Cut and FCM algorithms. MSE has the lowest value with the the Graph Cut and FCM algorithm. Also, they have the best segmentation accuracy (see Fig. 4).

Figure 3 shows a sample example of the original image of the three components ($R.G.$ and B) and ground-truth and Fig. 4 presents results of the k-means segmentation, FCM segmentation, Region Growing segmentation, and Graph Cut segmentation.

The second experiment was performed using the HSV color space. The results are shown in Table 2 where PSNR, MSE, and segmentation accuracy for the four algorithms using the H, S, and V components and vote between the three segmented components (voting HSV).

Table 1. Average of PSNR, MSE, and accuracy of k-means, FCM, Region Growing, and Graph Cut.

Color component		R	G	B	voting RGB
K-Means	PSNR	0.3609	0.3704	0.3765	0.3805
	MSE	0.1375	0.1321	0.1127	0.1112
	Accuracy	84.026	83.385	84.650	85.370
FCM	PSNR	0.3703	0.3753	0.3801	0.3858
	MSE	0.1423	0.1220	0.1077	0.1040
	Accuracy	83.884	84.178	86.847	85.130
Region Growing	PSNR	0.3680	0.3730	0.3736	0.3731
	MSE	0.1524	0.1331	0.1177	0.1191
	Accuracy	82.766	83.969	84.205	84.075
Graph Cut	PSNR	0.3714	0.3853	0.3917	0.3939
	MSE	0.1217	0.1120	0.1066	0.1120
	Accuracy	84.249	85.390	86.028	86.077

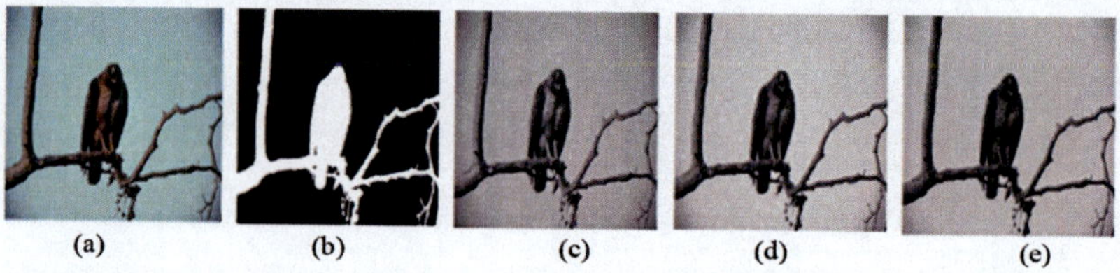

Fig. 3. (a) Original image, (b) Ground truth, (c) Red component, (d) Green component, and (e) Blue component.

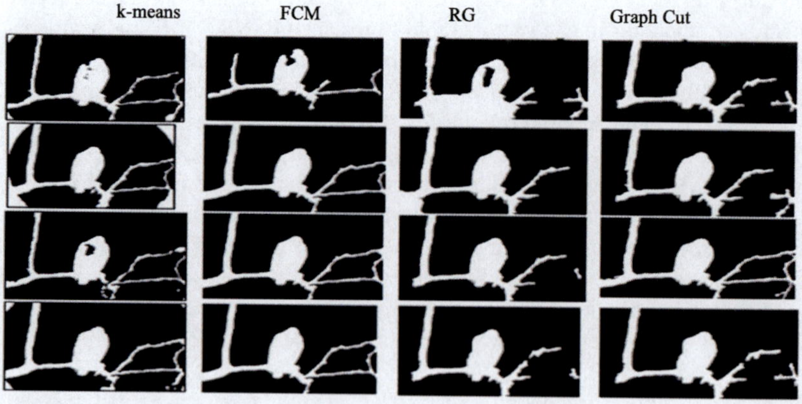

Fig. 4. First row for red (*R*) component segmentation, second row for green (*G*) component segmentation, third row for blue (*B*) component segmentation, and fourth row for voting RGB.

Table 2. Average of PSNR, MSE, and accuracy of k-means, FCM, Region Growing, and Graph Cut for the component $H.S.V.$ and voting HSV.

Color component		H	S	V	voting HSV
K-Means	PSNR	0.3574	0.3670	0.3961	0.3865
	MSE	0.2622	0.1764	0.1193	0.1214
	Accuracy	79.054	84.569	87.538	87.338
FCM	PSNR	0.3505	0.3565	0.3842	0.3625
	MSE	0.1918	0.1514	0.1400	0.1504
	Accuracy	79.060	82.054	85.361	85.776
Region Growing	PSNR	0.3190	0.3562	0.3809	0.3780
	MSE	0.2139	0.1596	0.1164	0.1368
	Accuracy	77.456	82.532	86.541	85.081
Graph Cut	PSNR	0.3669	0.3799	0.3752	0.3613
	MSE	0.1758	0.1491	0.1276	0.1355
	Accuracy	80.567	81.355	85.396	84.783

It is clear that voting HSV is not the best as V component is giving good results in some cases, while the H component is the worest in the most cases. The PSNR has the highest value with the Graph Cut algorithm. MSE has the lowest value with the the Graph Cut algorithm. Also, has the color space that was converted from the RGB color space model. It shows the ground truth of the image.

Figure 6 shows the results of the k-means segmentation, FCM segmentation, Region Growing segmentation, and Graph Cut segmentation using HSV color space.

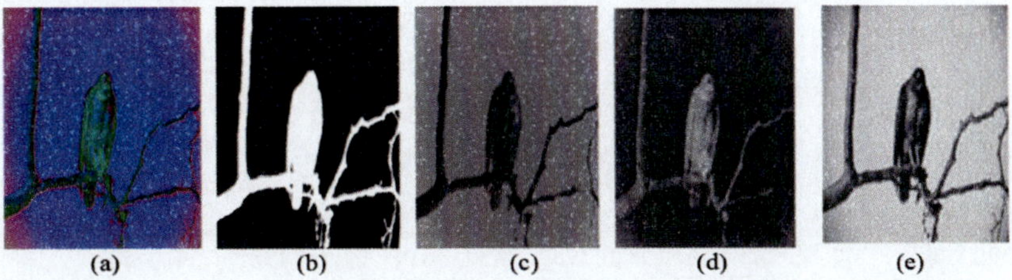

Fig. 5. (a) HSV image, (b) Ground truth, (c) H component, (d) S component, and (e) V component.

The third set of experiments was performed using the YCbCr color space with the segmentation methods presented in this study. Table 3 shows the PSNR, MSE, and segmentation accuracy for the four algorithms using the Y. Cb. and Cr components and vote between the three segmented components (voting YCbCr).

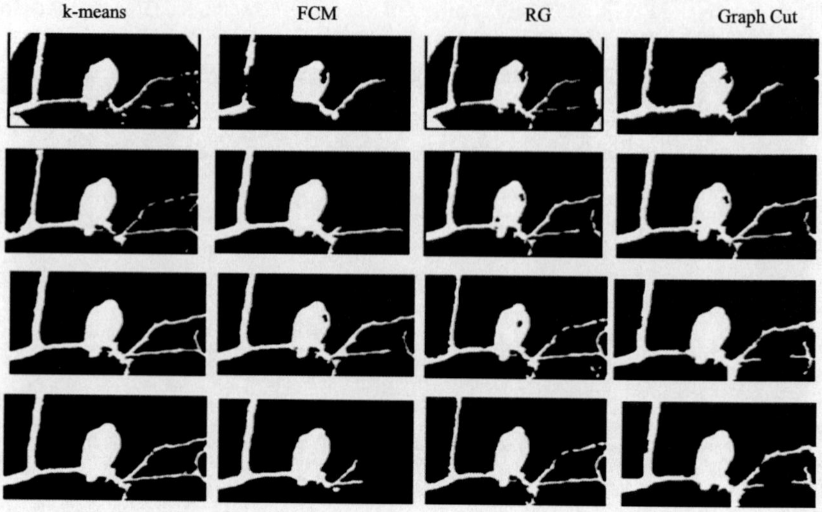

Fig. 6. First row for *H* component segmentation, second row for *S* component segmentation, third row for *V* component segmentation, and fourth row for voting (HSV).

Table 3. Average of PSNR, MSE and accuracy of k-means, FCM, Region Growing, and Graph Cut for the component $Y.Cb.Cr$ and voting YCbCr.

Color component		Y	Cb	Cr	voting YCbCr
K-Means	PSNR	0.4168	0.3540	0.3734	0.3686
	MSE	0.0629	0.1368	0.1129	0.1089
	Accuracy	87.091	84.703	84.518	85.441
FCM	PSNR	0.4094	0.3396	0.3434	0.3677
	MSE	0.0779	0.1566	0.1671	0.1001
	Accuracy	86.103	80.476	80.890	85.273
Region Growing	PSNR	0.3742	0.3054	0.3398	0.3489
	MSE	0.1184	0.2540	0.1829	0.1547
	Accuracy	85.815	75.055	80.072	82.027
Graph Cut	PSNR	0.3982	0.3414	0.3438	0.3811
	MSE	0.0990	0.1452	0.1530	0.1078
	Accuracy	87.821	82.830	80.750	84.104

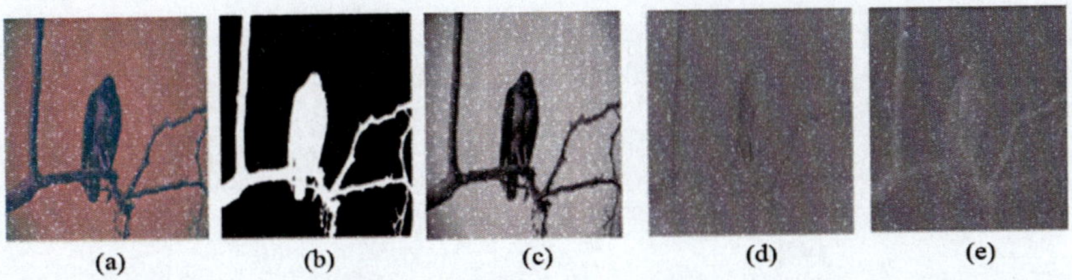

Fig. 7. (a) YCbCr image, (b) Ground truth, (c) *Y* component, (d) *Cb* component, and (e) *Cr* component.

We observed that the Y component has good results in most cases. In addition, the voting YCbCr had good results. The Graph Cut and the K-Means have the best performance accuracy Fig. 7 shows a sample example of the *Y*, *Cb*, and *Cr* components of YCbCr color space that was converted from the RGB color space model. It also shows ground truth of the image. Figure 8 shows the results of the segmentation.

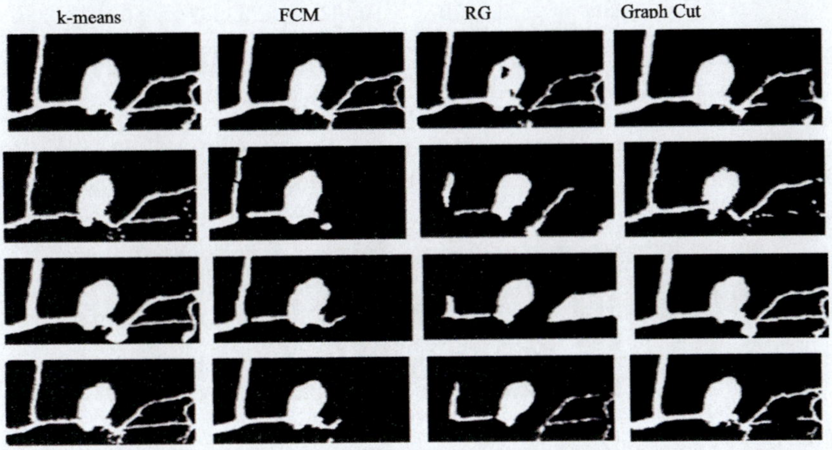

Fig. 8. First row for *Y* component segmentation, second row for *Cb* component segmentation, third row for *Cr* component segmentation, and fourth row for voting YCbCr.

The last experiment was carried out to measure the accuracy of segmentation using the XYZ color space with the four segmentation methods used in this paper. Table 4 shows PSNR, MSE, and segmentation accuracy for four algorithms using single individual components of color space XYZ and voting XYZ. The results show the *X* and *Y* component gives better results in most experiments. the votingXYZ between the three segmented components achieved good results. Graph Cut achived the highest PSNR and highest performance accuracy in addition to the lowest MSE.

Table 4. Average of PSNR, MSE and accuracy of k-means, FCM, Region Growing, and Graph Cut based for Segmentation for the X. Y. Z and voting XYZ.

Color component		X	Y	Z	voting XYZ
K-Means	PSNR	0.3678	0.3890	0.3646	0.3730
	MSE	0.1184	0.1009	0.1329	0.1042
	Accuracy	84.236	85.899	81.731	84.813
FCM	PSNR	0.3742	0.3760	0.3781	0.3720
	MSE	0.1151	0.1066	0.0910	0.1091
	Accuracy	84.313	84.312	84.033	85.996
Region Growing	PSNR	0.3625	0.3760	0.3521	0.3720
	MSE	0.1223	0.0928	0.1162	0.1140
	Accuracy	83.959	84.164	81.802	83.084
Graph Cut	PSNR	0.3727	0.3752	0.3766	0.3849
	MSE	0.1107	0.1426	0.1177	0.1122
	Accuracy	85.033	84.821	83.538	86.007

Figure 9 shows three component $X.Y.$ *and* Z of the three components and ground truth of original image. Figure 10 shows the results of the segmentation algorithms.

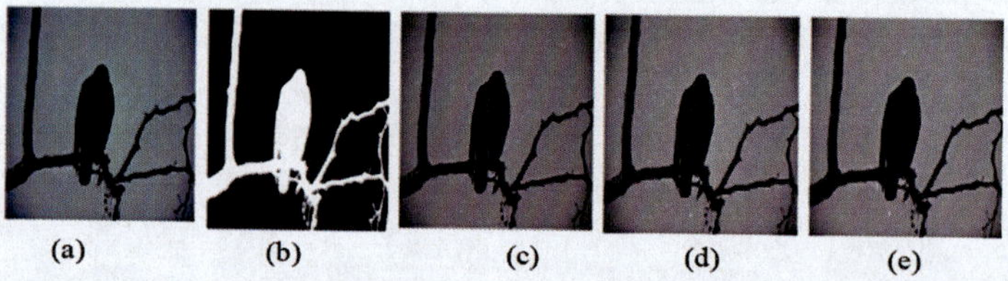

Fig. 9. (a) XYZ image, (b) Ground truth, (c) X component, (d) Y component, and (e) Z component

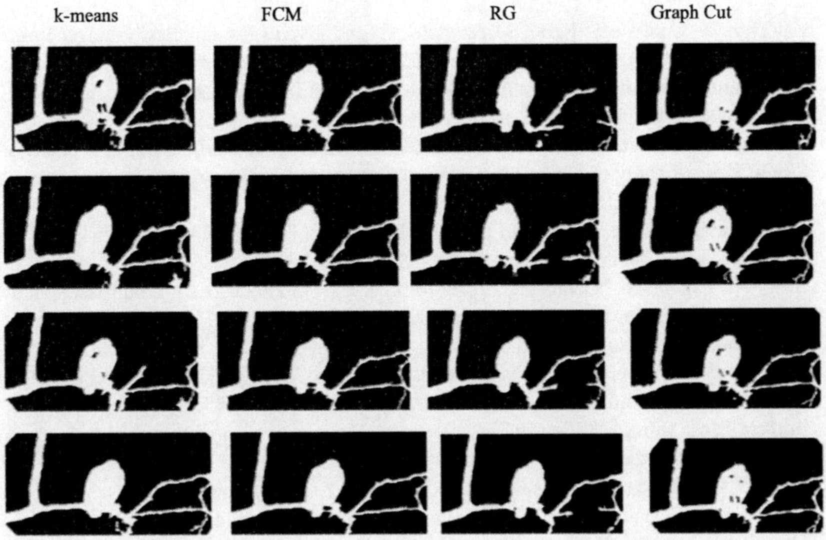

Fig. 10. First row for X component segmentation, second row for Y component segmentation, third row for Z component segmentation, and fourth row for voting XYZ.

6 Conclusion

In this study, we have focused on different color spaces for color image segmentation. Different color spaces carry out a different amount of information. Four different color spaces were examined. The results were obtained using the single components of RGB, HSV, YCbCr, and XYZ color spaces. We also voted between the segmented single components based on the K-Means, Fuzzy C-Means, Region Growing, and Graph Cut segmentation techniques. A comparative study of color spaces based on segmentation methods shows that the results of all color spaces are close to each other and the V and Y components have the high performance compared to the other signal components. Also, we found that the voting of the segmented color components of all color spaces has the best performance.

References

1. Jeevitha, K., Iyswariya, A., RamKumar, V., Basha, S.M., Kumar, V.P.: A review on various segmentation techniques in image processing. Eur. J. Mol. Clin. Med. **7**(4), 1342–1348 (2020)
2. Manoharan, S.: Performance analysis of clustering based image segmentation techniques. J. Innov. Image Process. (JIIP) **2**(01), 14–24 (2020)
3. Zheng, X., Lei, Q., Yao, R., Gong, Y., Yin, Q.: Image segmentation based on adaptive K-means algorithm. EURASIP J. Image Video Process. **1**, 1–10 (2018)
4. Bora, D.J., Gupta, A., Khan, F.A.: Comparing the performance of L*A*B* and HSV color spaces with respect to color image segmentation. Int. J. Emerg. Technol. Adv. Eng. **5**(2) (2015)
5. Agrawal, S., Xaxa, D.K.: Survey on image segmentation techniques and color models. Int. J. Comput. Sci. Inf. Technol. **5**(3), 3025–3030 (2014)

6. Zaldivar, D., Cuevas, E., Perez, M.A.: Color spaces advantages and disadvantages in image color clustering segmentation. In: Hassanien, A., Oliva, D. (eds.) Advances in Soft Computing and Machine Learning in Image Processing. SCI, vol. 730, pp. 3–22. Springer, Cham (2018). https://doi.org/10.1007/978-3-319-63754-9_1
7. Zaitoun, N.M., Aqel, M.J.: Survey on image segmentation. Techniques **65**, 797–806 (2015)
8. Singh, K.K.: A study of image segmentation algorithms for different types of images. Int. J. Comput. Sci. Issues (IJCSI) **7**(5), 414 (2010)
9. Garcia-Lamont, F., Cervantes, J., López, A., Rodriguez, L.: Segmentation of images by color features a survey. Neurocomputing **292**, 1–27 (2018)
10. Jurio, A., Pagola, M., Galar, M., Lopez-Molina, C., Paternain, D.: A comparison study of different color spaces in clustering based image segmentation. In: Hüllermeier, E., Kruse, R., Hoffmann, F. (eds.) Information Processing and Management of Uncertainty in Knowledge-Based Systems. Applications. IPMU 2010. CCIS, vol. 81, pp. 532–541. Springer, Berlin, Heidelberg (2010). https://doi.org/10.1007/978-3-642-14058-7_55
11. Burney, S.A., Tariq, H.: K-means cluster analysis for image segmentation. Int. J. Comput. Appl. **96**(4), 1–8 (2014)
12. Khattab, D., Ebied, H.M., Hussein, A.S., Tolba, M.F.: Color image segmentation based on different color space models using automatic grab cut. Sci. World J. (2014)
13. Mythili, C., Kavitha, V.: Color image segmentation using ERKFCM. Int. J. Comput. Appl. **41**, 21–28 (2012)
14. Shih, F.Y., Cheng, S.: Automatic seeded region growing for color image segmentation. Image Vis. Comput. **23**(10), 877–886 (2005)
15. Wang, X., Hänsch, R., Ma, L., Hellwich, O.: Comparison of different color spaces for image segmentation using graph-cut. In: International Conference on Computer Vision Theory and Applications (VISAPP), pp. 301–308. IEEE (2014)
16. Busin, L., Vandenbroucke, N., Macaire, L.: Advances in imaging and electron physics. **151**, 65–168 (2008)
17. Pm, N., Chezian, R.: Various colour spaces and colour space conversion algorithms. J. Glob. Res. Comput. Sci. **4**, 44–48 (2013)
18. Gopinathan, S., Gayathri, M.S.: A study on image enhancement techniques using YCbCr color space methods. Int. J. Adv. Eng. Res. Sci. **3**(8), 236818 (2016)
19. Ibraheem, N.A., Hasan, M.M., Khan, R.Z., Mishra, P.K.: Understanding color models: a review. ARPN J. Sci. Technol. **2**(3), 265–275 (2012)
20. Plataniotis, K., Anastasios, N.: Venetsanopoulos, Color image Processing and Applications. Springer Science & Business Media, Heidelberg (2013)
21. Jipkate, B.R., Gohokar, V.V.: A comparative analysis of fuzzy c-means clustering and k means clustering algorithms. Int. J. Comput. Eng. Res. **2**(3), 737–739 (2012)
22. Dhanachandra, N., Chanu, Y.J.: Image segmentation method using k-means clustering algorithm for color image. Adv. Res. Electr. Electron. Eng. **2**(11), 68–72 (2015)
23. Li, X., Lu, X., Tian, J., Gao, P., Kong, H., Xu, G.: Application of fuzzy c-means clustering in data analysis of metabolomics. Anal. Chem. **81**(11), 4468–4475 (2009)
24. Bezdek, J.C., Ehrlich, R., Full, W.: FCM: the fuzzy c-means clustering algorithm. Comput. Geosci. **10**(2–3), 191–203 (1984)
25. Ganesan, P., Rajini, V., Sathish, B.S., Kalist, V., Basha, S.K.: Satellite image segmentation based on YCbCr color space. Indian J. Sci. Technol. **8**(1), 35 (2015)
26. Hore, S., et al.: An integrated interactive technique for image segmentation using stack based seeded region growing and thresholdin. Int. J. Electr. Comput. Eng. **6**(6), 2088–8708 (2016)
27. Boykov, Y., Veksler, O.: Graph cuts in vision and graphics: theories and applications. In: Paragios, N., Chen, Y., Faugeras, O. (eds.) Handbook of Mathematical Models in Computer Vision, pp. 79–96. Springer, Boston, MA (2006). https://doi.org/10.1007/0-387-28831-7_5

28. Boykov, Y., Kolmogorov, V.: An experimental comparison of min-cut/max-flow algorithms for energy minimization in vision. IEEE Trans. Pattern Anal. Mach. Intell. **26**(9), 1124–1137 (2004)
29. Zheng, Q., Li, W., Hu, W., Wu, G.: An interactive image segmentation algorithm based on graph cut. Procedia Eng. **29**, 1420–1424 (2012)
30. Martin, D.R., Fowlkes, C.C., Tal, D., Malik, J.: A database of human segmented natural images and its application to evaluating segmentation algorithms and measuring ecological statistics. In: Proceedings Eighth IEEE International Conference on Computer Vision. ICCV 2001, vol. 2, no. 2, pp. 416–423 (2001)
31. Prabha, D.S., Kumar, J.S.: Performance evaluation of image segmentation using objective methods. Indian J. Sci. Technol. **9**(8), 1–8 (2016)
32. Wang, Z., Bovik, A.C.: Mean squared error: love it or leave it? A new look at signal fidelity measures. IEEE Signal Process. Mag. **26**(1), 98–117 (2009)
33. Moreno, J., Jaime, B., Saucedo, S.: Towards no-reference of peak signal to noise ratio. Int. J. Adv. Comput. Sci. Appl. (IJACSA) **4**(1) (2013)

The Relationship Between Functional Empowerment and Creative Behavior of Workers During the COVID-19 Pandemic in the UAE

Sultan Obaid AlZaabi[1](✉) and Hussein Mohammed Abu Al-Rejal[2]

[1] School of Government, University Utara Malaysia, 06010 Sintok, Kedah, Malaysia
sultan_obaid@gsgsg.uum.edu.my

[2] Knowledge Science Research Lab, School of Technology Management and Logistics, College of Business, University Utara Malaysia, 06010 Sintok, Kedah, Malaysia
abualrejal@uum.edu.my

Abstract. Employees' functional Empowerment has always been a crucial topic for management sciences researchers. It involves several characteristics that help the workforce to manage their tasks through creative ideas and problem-solving behavior. However, the current pandemic has adversely affected the organizations leading the employees to take their functional Empowerment under consideration. The current study also examined the impact of functional Empowerment on the creative behavior of workers during the Covid19 pandemic in the United Arab Emirates. We used the research reviewed approach to validate the proposed argumentation. Thus, we found a set of deficiencies in the administrative and functional Empowerment of workers. Besides, functional Empowerment was strongly associated with improved creative behavior of the workforce individuals. Thus, we conclude that, if Emirati organizations focus more on creating, accelerating, and strengthening the functional Empowerment of the employees, they will exhibit a strong creative behavior to cope with Covid-19. Further, we discussed the study contributions and recommendations for future research accordingly.

Keywords: Organizational management · Functional empowerment · Workforce wellbeing · Covid-19 · Creative behavior

1 Introduction

Governments are fighting battles on several fronts during the rapid spread of the new Coronavirus worldwide. Many countries have taken series of measures to prevent the transmission of infection, by restricting travel, closing schools and companies, and instructing citizens to stay at home [1–6]. However, it is notable that, the current pandemic has raised several social, economic, political, and other barriers leading to halting the organizational functions at every level [7–11]. Notably, today, when technological advancements are booming and creating ease of access with the ease of use for their users, coping with the healthcare crisis, is not problematic anymore [12–14]. Similarly,

© The Author(s), under exclusive license to Springer Nature Switzerland AG 2022
A. E. Hassanien et al. (Eds.): AMLTA 2022, LNDECT 113, pp. 472–486, 2022.
https://doi.org/10.1007/978-3-031-03918-8_40

the recent developments in the era of management and the new concepts and practices causing functional Empowerment are some of the modern administrative methods. We consider functional Empowerment as enabling the employees to exercise full authority in their field of work. It is also one of the basic building blocks in reaching excellence and creativity in the work environment [15]. Business organizations and institutions widely adopt functional empowerment, to exchange trust between workers and management and also to involve workers in the decision-making process. Organizations consider functional Empowerment as building channels of communication and relations to improve their ability to transcend organizational boundaries between workers and management [16]. As a result, employees take a position to take responsibility for decision-making and participate with senior management in determining the future of the organization through their creative efforts [17]. In this context, functional Empowerment is one of the most important elements of organizational success, that distinguishes them from others as it is the driving force for the functional and creative behavior patterns of various workers. Functional Empowerment gives an impetus to the employee to improve their work performance towards achieving the organizational goals and increasing productivity levels. Especially, during the current era of information technology and the development of management concepts and smart organizations [18, 19]. According to Diab et al. (2017), functional Empowerment is the main driver in unleashing creative energies, which is reflected in the employee's job satisfaction and motivation towards productivity. Today, the functional Empowerment of workers is crucial to the success of organizations due to its characteristics that give employees complete freedom to choose the required task management techniques. Employees enjoy the full authority to make decisions related to the work without the interference of any external parties [21]. Employees' importance in the decision-making process is related to the quality of their work and enhance the ability of employees to perform the tasks required of them in an optimal manner [22]. Additionally, functional Empowerment helps in promoting concepts related to self-reliance during crises and sort them out effectively [23]. Thus, enhancing employees' ability by highlighting their roles to influence the processes of selecting appropriate alternatives, taking the appropriate decision on issues, and the nature of their work boosts their morale and confidence as well [24, 25]. We propose the United Arab Emirates-based organizations must adopt creative behavior organizations to achieve the organizational excellence and desired goals particularly, during the Covid-19 pandemic. As noted by Bani-Melhem et al. (2021) [26], creative behavior aspires to distinguish institutions and motivate employees towards creative behavior and innovation at work. Therefore, this study examined the impact of job empowerment. on the creative behavior of workers during the Covid-19 pandemic in the United Arab Emirates. We aimed to develop suitable strategies to propose and modify the administrative and functional empowerment system f in the UAE federal institutions in a more effective way to achieve the desired strategic goals [27].

2 Literature Review

As mentioned earlier, the emergence of the Covid-raised economic and social repercussions by halting the professional arenas both at national and international levels. The outbreak of Covid-19 imposed a package of precautionary measures such as isolation, quarantine, social distancing, travel ban, and complete closure of all state institutions. As a result, the sudden closure of schools, universities, companies, factories, entertainment venues, and tourism companies negatively affected the economies of all countries of the world and entered the global system into a state of socio-economic stagnation [28]. This massive closure of institutions also affected the United Arab Emirates as it is one of the most dominant global economies facing severe financial challenges [29–31]. Likewise, the changes in the administrative business environment and the organizations' quest to apply unified, global foundations for quality and institutional excellence, and openness to the world, have prompted to possess a sustainable competitive advantage and continue to achieve the level of achievement during the pandemic. All these indicate creativity and Empowerment in the work environment as one of the elements of survival in our time [32, 33]. Functional empowerment includes the essential elements to achieve administrative creativity in all administrative organizations. Management scholars agree that contemporary organizations live in complex conditions, that further demand creative behavior, teamwork, participation in decision-making, and contribution to solving problems and confronting crises [34]. To cope with problems, it is important to promote functional empowerment, accelerate human resources, activate the capabilities, energies, and talents for continuous improvement in the organization [35]. The importance of creativity is evident through the human resource in the business environment, especially during the current competitive environments [36]. The significance of functional Empowerment can be determined by Henry Ford's quote, as he argued "*Take everything from me and leave the individuals to me, and I guarantee you that I will come back strong again*" [37]. Ford's argument highlights the role and contribution of Human resources as an engine for the organization as a whole further presuming that creativity as a personal behavior comes from the mind, indicating it as a strong mental cognitive process [38]. Further validated by [39], as that stated that creativity is a basic need for organizations to help them excel in performance and maintain competitive strategy. Contemporary institutions have realized this need and established administrative units under the auspices of creativity and innovators. The importance of creativity increased when the decision-makers realize that there is a discrepancy between the actual performance and the desired performance. As a result, creativity calls the management to search for new methods to accelerate the performance as here creativity appears as the main tool that ensures organizational development and success [40]. The importance of creativity from the researcher's point of view lies in the individual's ability to utilize the current resources to propose new problem-solving techniques. It must be emphasized that creativity in contemporary organizations is no longer a matter taken or granted, rather it is indispensable to thrive and survive in rapidly changing business environments [26]. Therefore, institutions lacking creativity also lack future visions and success as creativity is that the main driver of the intellectual and scientific capabilities of human resources [20, 22]. There is a correlation and impact of strategic intelligence or machine learning on the creative behavior of employees, as the more dimensions of intelligence are available at a high level, the greater the effectiveness and efficiency of strategic leadership and the levels of job Empowerment for workers in UAE government institutions.

2.1 The Importance of Empowering Employees

Empowerment is defined as an applied concept that can provide benefits to organizations and create a sense of belonging and pride in employees. Empowerment enhances the relationship between organizations and employees [41]. This relationship constitutes the cornerstone for the success and adoption of developed methods within institutions [42]. The importance of Empowerment lies in the fact that empowering workers is a new way to change and develop the organization itself, making it more flexible, able to learn, adapt to different environmental variables, and keep in touch with scientific and technological developments. Empowerment gives employees satisfaction by giving them wide authority in carrying out their tasks [43]. Empowerment also creates an entrepreneurial spirit in the organization and raises the willingness of employees to work with customers with high enthusiasm [44]. Empowerment also accelerates the process of administrative functions, contributes to enhanced unused human energies based on significant individual differences [45]. These differences can be social, physical, psychological, educational, and political [46]. The process of Empowerment helps officials and administrators identify creative and talented workers, and better invest their time in important administrative matters [47]. Also, empowering employees makes the organization recognize its needs, and reduce costs in line with market changes [34]. The importance of empowering workers increases in the organizations to cope with the market competition, reduce administrative levels efforts, and reduce the task burden on senior management with routine matters especially during the Covid-19 outbreak [39]. Previous studies have dealt with job Empowerment and creative behavior in an extensive way. However, the current research is focused on the discovery of the important positive role of job Empowerment and creative behavior of workers and the relationship between them from the point of view of workers in government agencies and institutions during the COVIED19 pandemic [8, 26, 46]. Table 1 provides a summary of studies witnessing the benefits of employees' empowerment.

2.2 Employees Empowerment with Creative Behavior

Creativity in institutions depends on a set of basic pillars that must be available in the workplace environment. One of the most important of these pillars is the climate of functional Empowerment that prevails within the work environment, in terms of working conditions and workers within the institution. Creativity allows the members to achieve their ambitions and help them to utilize their capabilities, participate in decision-making, to satisfy their desires, use the right planning method, free from the traditional tendency, and follow excellence [48]. Creative behaviors reflect finding unique solutions that lead towards change. Empowering workers stimulates the creative energies of workers. Independence for workers makes them feel less restricted than others, which helps to address the impact of the dimensions of functional Empowerment on the creative behavior of

Table 1. Summary of Benefits of Employees Empowerment

Study	Research purpose	Method	Findings
[16]	Studying the impact of administrative Empowerment on the employee performance management process at Jadara University	The study followed the quantitative approach, with a total of (56) male and female employees	The level of practice of administrative Empowerment and the process of employee performance management at Gadara University was moderate. It was found that there is a significant impact of administrative Empowerment on employees' performance management in employee performance expectations, supervising the progress of employee performance, and evaluating employee performance
[17]	Detecting patterns of administrative Empowerment among information and communication technology (ICT) coordinators during them in schools	Qualitative approach	Emphasize all participants of personal Empowerment through their new role. Findings expressed self-efficacy as a sense of responsibility, intrinsic motivation to work, and the ability to lead change and overcome resistance
[19]	Highlighting how to turn obstacles into opportunities so that MoE staff can continue to work remotely during critical circumstances	Qualitative approach	There are visions about activating remote work in response to sudden incidents by highlighting the measurements and factors that contribute to the readiness of entities. Creativity influences the value and authenticity of the Ministry of Education's teleworking practices and their impact on the wider community

(*continued*)

Table 1. *(continued)*

Study	Research purpose	Method	Findings
[20]	Investigating the effect of Empowerment on organizational creativity among health managers in the Jordanian Ministry of Health	Descriptive analytical method	There is an effect of enabling management on organizational creativity. Besides, enabling management individually positively affects organizational creativity. However, training, motivation, and delegation of authority have the most decreasing effect

employees. These dimensions mainly involve self-motivation, the delegation of authority, training, and participation in decision-making [49]. The impact of self-motivation on creative behavior and the distinguished administrative organizations is based on an organizational environment that proposes creative trends and executes them effectively [50, 51]. Moreover, incentives work as the main ingredient in creative organizations and provide a basic foundation for the existence and development of creativity. Today, incentives have an important impact on creativity, but if the relationship between incentives and performance related to creativity weakens, it will adversely affect the creativity of employees [23]. Previous studies firmly believe that distinguished administrative organizations are based on their adoption of an organizational environment that gives legitimacy to employees' creativity. These organizations trust the functional values and practices based on employees' creativity that further enhances their work performance in a better possible manner [22].

2.3 The Effect of Decentralization of Authority on the Creative Behavior

Successful institutions enable their managers to seek to bear major responsibilities and tasks creatively. These organizations believe in the decentralization of powers and authority, indicating the lower-level employees as equally capable of participating in the decision-making functions. Centralized power, on the other hand, deprives the lower-level employees of decision-making contributions that can further affect their confidence, creativity, and job performance [23, 52]. Therefore, the decentralized authority leads to the development of creative and innovative abilities, skills, also and developing these capabilities through their tentative credibility and administrative functions. Decentralization of Authority also motivates and reduces any potential risks to accomplish the assigned tasks in a better manner. As a result, it boosts employees' morale and increases their self-confidence to show their creativity [53]. Thus, decentralization of authority provides an opportunity to exercise tasks with f freedom, flexibility, authority, and a sense of responsibility, that further improves increasing performance effectiveness, and sharpens enthusiasm [15, 54] (Table 2).

Table 2. Summary of employees empowerment, creative behavior

Study	Research purpose	Method	Findings
[21]	To empirically investigate the impact of transformational management on followers' inventiveness and organizational innovation	Qualitative approach	Transformational leadership has a significant positive relationship with both followers' creativity and organizational innovation. Furthermore, a significant positive relationship was also found between followers' creativity and organizational innovation
[22]	Empowerment and its relationship to functional and creative performance from the perspective of balanced scorecards for a youth care specialist	Descriptive method	The skills and methods of administrative creativity are among the necessities required in the management of the organization and among workers in institutions, especially during crises
[23]	Detecting the impact of job Empowerment on the creative behavior of workers in the Tourism Authority	Quantitative method, n = 65 questionnaires	The level of job Empowerment in its dimensions (delegation of authority, participation in decision-making, self-motivation, work environment, and confidence-building) among workers in the hospitality sector was high, and the level of impact of job Empowerment in its various dimensions (delegation of authority, participation in decision-making, self-motivation, environment Work and building confidence) in the creative behavior of workers in the hospitality sector positively
[24]	Figuring out and rating the numerous antecedents to evaluate overall performance through workers' proposal schemes (ESSs) UAE based organizations	Descriptive method	The findings showed that management guide control style, personal attributes, and company approach are the maximum essential elements that impact overall performance through ESS

(continued)

Table 2. (*continued*)

Study	Research purpose	Method	Findings
[26]	Determining the impact of administrative Empowerment represented by power-sharing, information exchange, work teams, and expansion of powers in achieving competitiveness in five-star hotels in the United Arab Emirates	Quantitative method	The competitiveness level in five-star hotels came to a high degree, and there was a statistically significant effect of the areas of administrative Empowerment in the competitiveness of five-star hotels in the UAE. Also, there were statistically significant differences between the respondents' perceptions of the employee Empowerment level based on gender and academic rank. However, there were no statistically significant differences in the respondents' perceptions about employee Empowerment level based on experience and job title. in five-star hotels in the United Arab Emirates

Thus, administrative Empowerment has many advantages and characteristics that are of great importance to ensure the functioning of the administrative process in a perfect manner [21]. On the other hand, during the current pandemic, which necessitated them to work to maintain their stability and growth, organizations turned towards the development and promotion of creative behavior to accomplish their tasks, achieve innovation and effectiveness in performance during the COVID 19 pandemic in the United Arab Emirates [40].

3 Factors of Creative Behavior of Employees in UAE

The government sector in the United Arab Emirates is the main pillar that maintains public security and protects lives, property, and honor. The strategic plan of the United Arab Emirates aims to ensure that the security, justice, and safety sector keeps pace with the socio-economic growth Establishing security and achieving justice and safety for society and individuals are the primary goals of the Emirati government [19]. In this context, the current study has exceptional importance in looking for the relationship between functional Empowerment and creative behavior, which contributes significantly to the performance of Emirati institutions [8, 55]. The issue of functional Empowerment and creative behavior is receiving increasing attention from all business organizations in the United Arab Emirates. This increased interest is due to the strive to attain a high ability to meet the challenges and modern developments witnessed in the current century. According to Goher et al. (2021) [15], the government of the United Arab Emirates exercises its functions in an organized and planned manner and under various administrative directives that enhance the creative behavior of employees. Also, the local government considers the human element as constituting the most important element in all organizations, and it is the main and main driver of organizations [56]. The government administration always seeks to employ those individuals having strong work experience, with distinguished qualifications as these people are creative along with greater working skills. For this purpose, Human Resource Managers of government institutions actively seek the desired employees through different resources [26]. Despite participation in the decision-making process has many advantages, bureaucratic organizations do not involve employees in the decision-making process as they consider skills and qualifications as important components regarding the decision-making process [22]. During the decision-making process, Emirati organizations acknowledge every individual and equally encourage everyone to share their opinion. As a result, the climate of opinion resonates with organizational creativity and compliance with the norms among employees. It is notable that, a successful group makes efforts to fulfill the assigned tasks and creatively solve the problems. Thus creativity depends on the extent to which stakeholders, managers, and leaders boost the morale and confidence of the workforce to participate in the decision-making process [21]. The main objective of effective participation is to develop and stimulate creative ideas and suggestions, by ensuring employees' participation in the decision-making operations. Consequently, creative suggestions and ideas flow in a way that contributes to task completion and organizational development [21]. Thus, we conclude that employees' participation in making various administrative decisions by expressing their opinions and suggestions is of greater significance. Employees' participation in organizational functions and decision-making processes are better able to serve their organizational interests as they are actively proposing the creative idea to accelerate their work performance [57, 58].

4 Creative Behavior of Workers During the Covid-19 in UAE

Covid-19 crisis has posed an emotional challenge for many people as it has changed their everyday life in unprecedented ways. For government institutions, there is no longer an option to conduct business as usual [24]. Thus, governmental institutions developed and implemented several plans to support their employees following the most conservative guidelines for making changes to the general policy of governmental institutions [1]. Currently, some government institutions are actively working to benchmark their efforts that are also complied by other institutions to determine appropriate policies and to support their employees [34]. In this regard, government sector institutions are distinguished as the local government ensured employees' welfare, improvement and strengthened by motivating them to share their creativity to cope with the challenges raised by Covid-19 [59]. As a result, employees from all levels have a strong sense of responsibility for addressing the pandemic and working to mitigate its effects. From the beginning of Covid-19, the government administration is constantly working to manage the crisis. Their efforts are based on evaluating the effectiveness of the measures taken at each stage and the extent to which they achieve the desired goals. Emirati government administration is carefully working to access the next stage of organizational functionality with deliberate and conscious steps. These efforts are also enhancing employees' confidence which is further boosting the morale of public members to participate in the government's crisis management efforts. Consequently, both employees and civilians are more aware of the importance of obligating the precautionary instructions as they are active partners in managing the pandemic [15, 53]. Emirati government also proposed some more models to clarify the preventive measures against Covid-19 that are designed by the World Health Organization [29]. These measures are equally imposed on workers' safety and healthcare measures. Currently, the UAE is recovering from the Covid-19 crisis with a transparent strategy, based on dialogue with all ministries, institutions, and national bodies. The Emirati government aims to develop a system to cope with the emergencies in future, that may ensure the continuity of business and services, under the latest internationally recognized standards [27].

5 Discussion

Findings revealed the existing set of deficiencies in the administrative efforts regarding the functional Empowerment of workers. These results are compatible with the study conducted by Alhashmi et al. (2020) [40] as it also indicated the lack of institutional efforts to support the functional Empowerment of workers i.e., providing a technology-based environment that allows workers to access information promptly, effective incentive systems to encourage employees to improve their performance, and the lack of training systems aimed at developing and improving the competencies of employees. In a similar context, the study conducted by Alikaj et al. (2020) [22] indicated that superiors show less interest in boosting the self-confidence of their workforce subordinates as they still believe in conventional ways of organizational management. As a result, hierarchal authority centralizes power that further leads to unequal power relations in an organization. Moreover, there is a gap between the practical and the theoretical side of administrative empowerment, which negatively affects the institutional performance of institutions and organizations. This gap also affects the procedures and methods that organizations rely on to develop the creative behavior of their employees [20]. Thus, creative behavior through functional Empowerment during the Corona crisis in the UAE government institutions was due to workers' critical ability and crisis management skills to sustain their work despite the organizational closure and remote working [28, 60]. Also affirmed by Alhashmi et al. (2020) [40] and Wren (2021) [59] as they emphasized raising the capabilities of workers to achieve equal opportunities, invest the available resources, and also encourages workers to present unique ideas. Previous studies conducted by Valeriani et al. (2020) [37] and Yarberry & Sims (2021) [39] indicated the impacts of functional Empowerment on the behavior of employees, which was through the decentralization of authority. Consequently, this decentralization leads to leads to the development of creative and innovative abilities, creative problem-solving skills, and the ability to propose new ideas and encourage creativity. Furthermore, training generates a desire to be more creative so that they may deal effectively with the problems. Thus, we conclude the impact of administrative Empowerment factors and also on individuals as positive and productive to enhance the employees' creative working skills. Especially, in the United Arab Emirates where the government is comparatively more focused to develop the creative skills of their employees during the current healthcare crisis. For this purpose, both Human Resource Managers are resorting to different tactics including innovation technology to facilitate the workforce as much as they can. These efforts are consistent with the ideas innovatively AlMarar et al. (2021) [19], Arnout (2020) [32], Avidov-Ungar & Hanin-Itzak (2019) [17] Siswanti & Muafi (2020) [29] as they also indicated that functional Empowerment and creative behavior help workers to participate in decision-making process. This active participation leads to raise morale, boosts self-confidence, and also motivates them to utilize their creative abilities. Finally, the results showed that the employees' Empowerment levels were consistent with their sense of belonging, job conformity as the more they feel creative, the more they tend to perform creatively Therefore, active participation in decision-making is a greater motivation for the workforce individuals regarding their creative abilities that they are ought to show through their work performance [20, 26, 61, 62].

6 Conclusion and Future Work

We have discussed the impact of the relationship between functional Empowerment and the creative behavior of workers during the COVID-19 pandemic in the UAE. The importance of functional Empowerment and the creative behavior of workers during the ongoing crisis cannot be denied. Enhancing career Empowerment in the UAE government institutions by providing workers with knowledge and multiple skills, especially compatibility, conflict resolution, leadership, and confidence-building, through training and development is the need of the day. Creativity and excellence activate participation in decision-making among workers in the UAE government institution. As a result, this participation raises their morale and increases the self-confidence among workers in the UAE government institutions, which leads them to show their skills and creative behavior. Therefore, in the light of current findings, the researchers emphasize the functional Empowerment of the workforce in the United Arab Emirates. For this purpose, their involvement in the decision-making process is important to boost their confidence and improve their creative thinking. We recommend studies in the future to examine the impacts of functional empowerment, especially in the Emirati and Gulf-based institutions, to empirically examine the relevant impacts effectively.

References

1. Malkawi, E., Bawaneh, A.K., Bawa'aneh, M.S.: Campus off, education on: UAEU students' satisfaction and attitudes towards e-learning and virtual classes during COVID-19 pandemic. Contemp. Educ. Technol. **13**(1), 1–14 (2021)
2. Al-Maroof, R.S., Alhumaid, K., Akour, I., Salloum, S.: Factors that affect e-learning platforms after the spread of COVID-19: post acceptance study. Data **6**(5), 49 (2021)

3. AlQudah, A.A., Salloum, S.A., Shaalan, K.: The role of technology acceptance in healthcare to mitigate COVID-19 outbreak. In: Arpaci, I., Al-Emran, M., A. Al-Sharafi, M., Marques, G. (eds.) Emerging Technologies During the Era of COVID-19 Pandemic. SSDC, vol. 348, pp. 223–244. Springer, Cham (2021). https://doi.org/10.1007/978-3-030-67716-9_14
4. Gaid, M.L., Salloum, S.A.: Explore the relationship between COVID-19 testing rates with the number of cases. In: Hassanien, A.E., et al. (eds.) AICV 2021. AISC, vol. 1377, pp. 33–45. Springer, Cham (2021). https://doi.org/10.1007/978-3-030-76346-6_4
5. Tahat, K.M., Al-Sarayrah, W., Salloum, S.A., Habes, M., Ali, S.: The Influence of YouTube videos on the learning experience of disabled people during the COVID-19 outbreak. In: Hassanien, A.-E., Elghamrawy, S.M., Zelinka, I. (eds.) Advances in Data Science and Intelligent Data Communication Technologies for COVID-19. SSDC, vol. 378, pp. 239–252. Springer, Cham (2022). https://doi.org/10.1007/978-3-030-77302-1_13
6. Taryam, M., et al.: Factors affecting the uptake of COVID-19 vaccine among Dubai Airport's professionals. South East. Eur. J. Public Health (2022)
7. Youssef, E.: Role of social service institutions on social empowerment of women at the United Arab Emirates: a field analysis study. Multicult. Educ. **6**(4), 99–111 (2020)
8. Habes, M., Alghizzawi, M., Ali, S., SalihAlnaser, A., Salloum, S.A.: The relation among marketing ads, via digital media and mitigate (COVID-19) pandemic in Jordan. Int. J. Adv. Sci. **29**(7), 2326–12348 (2020)
9. Abu Al-Rejal, H.M.E., Udin, Z.M., Hassan, M.G., Sharif, K.I.M., Al-Rahmi, W.M., Al-kumaim, N.H.: Green information technology adoption antecedence: a conceptual framework. In: Saeed, F., Mohammed, F., Gazem, N. (eds.) IRICT 2019. AISC, vol. 1073, pp. 1098–1108. Springer, Cham (2020). https://doi.org/10.1007/978-3-030-33582-3_103
10. Yousuf, H., Zainal, A.Y., Salloum, S.A.: System of systems as a solution to mitigate the spread of Covid-19. In: Hassanien, A.-E., Elghamrawy, S.M., Zelinka, I. (eds.) Advances in Data Science and Intelligent Data Communication Technologies for COVID-19. SSDC, vol. 378, pp. 77–91. Springer, Cham (2022). https://doi.org/10.1007/978-3-030-77302-1_5
11. Elnagar, A., Afyouni, I., Shahin, I., Nassif, A.B., Salloum, S.A.: The empirical study of e-learning post-acceptance after the spread of COVID-19: a multi-analytical approach based hybrid SEM-ANN. arXiv Prepr arXiv:2112.01293 (2021)
12. Khalid, A., Ali, S.: COVID-19 and its challenges for the healthcare system in Pakistan. Asian Bioeth. Rev. **12**(4), 551–564 (2020). https://doi.org/10.1007/s41649-020-00139-x
13. Almatrooshi, F., Alhammadi, S., Salloum, S.A., Shaalan, K.: Case study: the implications of knowledge management tools on the process of overcoming COVID-19. In: Al-Emran, M., Al-Sharafi, M.A., Al-Kabi, M.N., Shaalan, K. (eds.) ICETIS 2021. LNCS, vol. 322. Springer, Cham. https://doi.org/10.1007/978-3-030-85990-9_49
14. Ahmed, D., Salloum, S.A., Shaalan, K.: Sentiment analysis of Arabic COVID-19 tweets. In: Al-Emran, M., Al-Sharafi, M.A., Al-Kabi, M.N., Shaalan, K. (eds.) ICETIS 2021. LNCS, vol. 322. Springer, Cham. https://doi.org/10.1007/978-3-030-85990-9_50
15. Goher, G., Masrom, M., Amrin, A., Abd Rahim, N.: Disruptive technologies for labor market information system implementation enhancement in the UAE: a conceptual perspective. Int. J. Adv. Comput. Sci. Appl. **12**, 370–379 (2021)
16. Al-bdareen, R.: The impact of the administrative empowerment on the employees performance management process. Int. J. Asian Soc. Sci. **10**(4), 193–206 (2020)
17. Avidov-Ungar, O., Hanin-Itzak, L.: Sense of empowerment among school ICT coordinators: personal, subject-area and leadership empowerment. Technol. Knowl. Learn. **24**(3), 401–417 (2019)
18. Saoula, O., Fareed, M., Abd Hamid, R., Al-Rejal, H.M.E.A., Ismail, S.A.: The moderating role of job embeddedness on the effect of organisational justice and organisational learning culture on turnover intention: a conceptual review. Humanit. Soc. Sci. Rev. **7**(2), 563–571 (2019)

19. AlMarar, R., Alkhodari, A., Atia, D.Y., Mohamed, S., Shubair, R.M.: Working remotely during the COVID-19 pandemic: efforts of the Ministry of Education in the United Arab Emirates. TIJAB (Int. J. Appl. Bus.) **5**(1), 1–17 (2021)
20. Diab, S.M., Kamal, H.M., Mansour, A.I., Fawzy, R.M., Azab, B.S.: Clinical significance of Matrilin-3 gene polymorphism in Egyptian patients with primary knee osteoarthritis. Eur. J. Rheumatol. **4**(3), 200 (2017)
21. Al Harbi, J.A., Alarifi, S., Mosbah, A.: Transformation leadership and creativity: effects of employees pyschological empowerment and intrinsic motivation. Pers. Rev. (2019)
22. Alikaj, A., Ning, W., Wu, B.: Proactive personality and creative behavior: examining the role of thriving at work and high-involvement HR practices. J. Bus. Psychol. **36**(5), 857–869 (2020). https://doi.org/10.1007/s10869-020-09704-5
23. Teng, C., Hu, C., Chang, J.: Triggering creative self-efficacy to increase employee innovation behavior in the hospitality workplace. J. Creat. Behav. **54**(4), 912–925 (2020)
24. Mehrajunnisa, M., Jabeen, F.: Antecedents to employee suggestion schemes: the study of UAE. TQM J. (2020)
25. Abu al-Rejal, H.M., Mohamed Udin, Z., Yusoff, R.Z.: The role of supply chain management toward performance of manufacturing companies in Yemen. Int. J. Supply Chain Manag. **7**(6), 92–99 (2018)
26. Bani-Melhem, S., Zeffane, R., Abukhait, R., Shamsudin, F.M.: Empowerment as a pivotal deterrent to employee silence: evidence from the UAE hotel sector. Hum. Perform. **34**(2), 107–125 (2021)
27. Al Khayyal, A.O., Alshurideh, M., Al Kurdi, B., Salloum, S.A.: Women empowerment in UAE: a systematic review. In: Hassanien, A.E., Slowik, A., Snášel, V., El-Deeb, H., Tolba, F.M. (eds.) AISI 2020. AISC, vol. 1261, pp. 742–755. Springer, Cham (2021). https://doi.org/10.1007/978-3-030-58669-0_66
28. Habes, M., et al.: E-learning acceptance during the Covid-19 outbreak: a cross-sectional Study. In: Musleh Al-Sartawi, A.M.A., Razzaque, A., Kamal, M.M. (eds.) EAMMIS 2021. LNNS, vol. 239, pp. 65–77. Springer, Cham (2021). https://doi.org/10.1007/978-3-030-77246-8_7
29. Siswanti, Y., Muafi, M.: Empowering leadership and individual creativity: the mediation role of psychological empowerment in facing covid-19 pandemic. J. Asian Financ. Econ. Bus. **7**(11), 809–816 (2020)
30. Salloum, S.A., AlAhbabi, N.M.N., Habes, M., Aburayya, A., Akour, I.: Predicting the intention to use social media sites: a hybrid SEM - machine learning approach. In: Hassanien, A.-E., Chang, K.-C., Mincong, T. (eds.) AMLTA 2021. AISC, vol. 1339, pp. 324–334. Springer, Cham (2021). https://doi.org/10.1007/978-3-030-69717-4_32
31. Alhumaid, K., Habes, M., Salloum, S.A.: Examining the factors influencing the mobile learning usage during COVID-19 pandemic: an integrated SEM-ANN method. IEEE Access **9**, 102567–102578 (2021)
32. Arnout, B.A.: Predicting psychological service providers' empowerment in the light of the COVID-19 pandemic outbreak: a structural equation modelling analysis. Couns. Psychother. Res. **20**(3), 406–418 (2020)
33. Habes, M., Alghizzawi, M., Salloum, S.A., Mhamdi, C.: Effects of Facebook personal news sharing on building social capital in Jordanian Universities. In: Al-Emran, M., Shaalan, K., Hassanien, A.E. (eds.) Recent Advances in Intelligent Systems and Smart Applications. SSDC, vol. 295, pp. 653–670. Springer, Cham (2021). https://doi.org/10.1007/978-3-030-47411-9_35
34. Nuzul, A.A., Witjaksono, A.D., Kistyanto, A.: Empowering leadership and employee creativity: the mediating role of psychological empowerment. SEISENSE J. Manag. **3**(6), 14–25 (2020)

35. Capurso, M., Dennis, J.L., Salmi, L.P., Parrino, C., Mazzeschi, C.: Empowering children through school re-entry activities after the COVID-19 pandemic. Contin. Educ. **1**(1) (2020)
36. Falicov, C., Niño, A., D'Urso, S.: Expanding possibilities: flexibility and solidarity with under-resourced immigrant families during the COVID-19 pandemic. Fam. Process **59**(3), 865–882 (2020)
37. Valeriani, G., Sarajlic Vukovic, I., Lindegaard, T., Felizia, R., Mollica, R., Andersson, G.: Addressing healthcare gaps in sweden during the COVID-19 outbreak: on community outreach and empowering ethnic minority groups in a digitalized context. Healthcare **8**(4), 445 (2020)
38. Grützmacher, L.S., Schermuly, C.C.: Psychologically empowered during the job search: how empowering counseling affects job search process quality. Consult. Psychol. J. Pract. Res. **73**, 251 (2021)
39. Yarberry, S., Sims, C.: The impact of COVID-19-prompted virtual/remote work environments on employees' career development: social learning theory, belongingness, and self-empowerment. Adv. Dev. Hum. Resour. **23**(3), 237–252 (2021). https://doi.org/10.1177/15234223211017850
40. Alhashmi, S.F.S., Alshurideh, M., Al Kurdi, B., Salloum, S.A.: A systematic review of the factors affecting the artificial intelligence implementation in the health care sector. In: Hassanien, A.-E., Azar, A.T., Gaber, T., Oliva, D., Tolba, F.M. (eds.) AICV 2020. AISC, vol. 1153, pp. 37–49. Springer, Cham (2020). https://doi.org/10.1007/978-3-030-44289-7_4
41. Rahmasari, O.A., Febriandika, N.R.: The application of Wadiah contract on Islamic banking savings products through branchless banking (conformity analysis on Fatwa DSN-MUI and POJK), vol. 101, no. Iconies 2018, pp. 285–289 (2019)
42. Þorkelsson, J.: The impact of communication factors on job satisfaction among icelandic employees in the public sector (2018). BSc Psychol.
43. Voorveld, H.A.M.: Brand communication in social media: a research agenda. J. Advert. **48**(1), 14–26 (2019)
44. Mangold, P.-M.: Was sind Schwaben und wie erkennt man sie? (2009)
45. Osborne, S., Hammoud, M.S.: Effective employee engagement in the workplace. Int. J. Appl. Manag. Technol. **16**(1), 4 (2017)
46. Mohammadi, K.: Health media and management of the corona Covid-19 crisis in light of the spread of fake news on social media sites. J. Soc. Empower. **2**(3), 34–56 (2020)
47. Yowell, C.M., Gordon, E.W.: Youth empowerment and human service institutions. J. Negro Educ., 19–29 (1996)
48. Chung, N., Lee, H., Lee, S.J., Koo, C.: The influence of tourism website on tourists' behavior to determine destination selection: a case study of creative economy in Korea. Technol. Forecast. Soc. Change **96**, 130–143 (2015)
49. Habes, M.: The influence of YouTube videos on ELA during the COVID-19 outbreaks in Jordan, December, 2020
50. Habes, M., Ali, S., Salloum, S.A., Elareshi, M., Ziani, A.-K., Manama, B.: Digital media and students' AP improvement: an empirical investigation of social TV
51. Abbas, J., Aman, J., Nurunnabi, M., Bano, S.: The impact of social media on learning behavior for sustainable education: evidence of students from selected universities in Pakistan. Sustainability **11**(6), 1683 (2019)
52. Al-Shibly, M.S., Alghizzawi, M., Habes, M., Salloum, S.A.: The impact of de-marketing in reducing Jordanian youth consumption of energy drinks. In: Hassanien, A.E., Shaalan, K., Tolba, M.F. (eds.) AISI 2019. AISC, vol. 1058, pp. 427–437. Springer, Cham (2020). https://doi.org/10.1007/978-3-030-31129-2_39

53. Cherian, J., Gaikar, V., Paul, R., Pech, R.: Corporate culture and its impact on employees' attitude, performance, productivity, and behavior: an investigative analysis from selected organizations of the United Arab Emirates (UAE). J. Open Innov. Technol. Mark. Complex. **7**(1), 45 (2021)
54. Elareshi, M., Habes, M., Ali, S., Ziani, A.: Using online platforms for political communication in Bahrain election campaigns. Soc. Sci. Humanit. **29**(3), 2013–2031 (2021)
55. Mohamed, M.S., Khalifa, G.S.A., Nusari, M., Ameen, A., Al-Shibami, A.H., Abu-Elhassan, A.E.: Effect of organizational excellence and employee performance on organizational productivity within healthcare sector in the UAE. J. Eng. Appl. Sci. **13**(15), 6199–6210 (2018)
56. Tabash, M.I., Yahya, A.T., Akhtar, A.: Financial performance comparison of Islamic and conventional banks in the United Arab Emirates (UAE). In: International Conference on Advances in Business and Law, vol. 1, no. 1, pp. 477–494 (2017)
57. Almarzooqi, A.: Towards an Artificial Intelligence (AI)-driven government in the United Arab Emirates (UAE): a framework for transforming and augmenting leadership capabilities. ProQuest Diss. Theses, p. 204 (2019)
58. Hanaysha, J.R.M., Khalid, K., Kamariah Nik Mat, N., Sarassina, F., Yahya Bin Ab Rahman, M., Sazali Bin Zakaria, A.: Transformational leadership and job satisfaction. Am. J. Econ. **2**(4), 145–148 (2012)
59. Wren, S.: The expansion of e-learning in the UAE: implications and opportunities in the post-COVID-19 era. In: Online Teaching and Learning in Higher Education during COVID-19, Routledge, pp. 229–240 (2021)
60. Ali, S., Qamar, A., Derindag, O.F., Habes, M., Youssef, E.: The mediating role of gender in ICT acceptance & its impacts on students' academic performance during Covid-19. Int. J. Adv. Trends Comput. Sci. Eng. **10**(2) (2021)
61. Ayub, N.: Ijarah: 'the concept, development and progress of Islamic finance leasing in Ireland's aviation sector'. Dublin Business School (2016)
62. Tahat, D.N., Tahat, K.M., Habes, M.: Jordanian newspapers coverage of cyberbullying during COVID 19 pandemic. PalArch's J. Archaeol. Egypt/Egyptology **17**(7), 15390–15403 (2020)

The Role of Strategic Leadership to Achieving Institutional Excellence for Emirati Federal Institutions

Mubarak Alnuaimi(✉)

Faculty of Business Muamalat, University Science Islamic Malaysia, Nilai, Malaysia
mubarak@raudah.usim.edu.my

Abstract. Despite a long history of management research regarding leadership, scholars have a mutual consensus about the potential of strategic leadership. Especially, for achieving institutional excellence, strategic leadership plays a significant role. Thus, we also analyzed the role of strategic leadership in achieving institutional excellence in the UAE scenario. However, we also anticipated the role of five primary tactics that a strategic leader keeps under consideration to attain organizational excellence. We gathered data from one prominent federal institution of the United Arab Emirates and employed close-ended, self-administered questionnaires. Findings showed that strategic leadership is significantly linked to institutional excellence in the UAE. Besides, Institutional Excellence, Strategic Organizational processes, Leading & Developing People, Culture & Value Systems, Distinct Competencies, and Effective Networking Competencies are the basic tactics that help a strategic leader to attain institutional excellence. Thus, we discussed the results and made the relevant conclusions.

Keywords: Institutional excellence · Strategic leadership · Strategic organizational process · United Arab Emirates · Culture & value system

1 Introduction

Organizational excellence is one of the most investigated topics in management sciences, having different dimensions, and methods to achieve it. We define organizational excellence as the growth and improvement of an organization in different aspects. As a result, all the stakeholders, workforce, and other concerned individuals attain maximum social and economic benefits on both individual and collective levels [1, 2]. Shirvani & Iranban (2013) described organizational excellence as a strong managerial infrastructure that may cause improvement by depending upon primary concepts and principles paying attention to the quality management system. Institutional excellence also provides an instrument to determine the capacity of systems and self-assessments to determine all the organizational activities for performance improvement [2, 4, 5]. The dynamicity of business arenas demands every organization to achieve institutional excellence. If an organization wants to attain and sustain excellence, it should keep all the factors that affect an institution under consideration [6]. In this context, business development,

© The Author(s), under exclusive license to Springer Nature Switzerland AG 2022
A. E. Hassanien et al. (Eds.): AMLTA 2022, LNDECT 113, pp. 487–501, 2022.
https://doi.org/10.1007/978-3-031-03918-8_41

quality management, and customers" satisfaction are the core factors that need strong consideration to achieve organizational excellence [2]. As noted by Al-Jedaiah & Albdareen (2020), accomplishing organizational excellence demands learning and achieving new tactics, that can further help to achieve the sustainable development goals faster than the other competitors. [1, 5]. The sustainability of an organization depends on different factors such as policies, their strategic implementation, and their practicality regarding execution. Similarly, organizational excellence is a world-class consideration for business organizations in the United Arab Emirates. As Emirati organizations are focusing not only to generate a greater revenue also the focus is towards providing socio-economic benefits at both individual and collective levels without any discrimination [8]. Talking specifically about the public sector, federal institutions in the United Arab Emirates, are popular worldwide for their organizational growth, balance, achievement, and prosperity. These Emirati organizational federal institutions are focused on improving their cooperation and morale, quality acceptance as a strategic concern, the value gained from their self-assessment, and an increased pace of improvement [9]. Many researchers attribute these increased considerations as a result of strategic leadership (See Cebula et al. 2012; O'Regan & Lehmann 2008; Owusu-Boadi 2019; L. Singh & Panda 2015). As noted by Shobaki & Naser (2017), there is a mutual consensus about the role of leadership in achieving organizational excellence. Especially, in the Emirati organization, this role is comparatively stronger and prominent in their administrative departments, that further highlights how strategic leaders are working effectively by obligating their responsibilities to attain the designated organizational goals. This research is also focused on examining the role of strategic leadership in achieving organizational excellence. However, it is notable that, studies to examine the extent to which strategic leaders are using components of strategic leadership in achieving institutional excellence are scarce. Particularly, in the Emirati scenario where federal institutions are efficiently working, the reason behind their success is yet to be explored. Therefore, we structured our article as the first section discusses the importance of institutional excellence and the role of strategic leadership. In the second section, we discussed the factors that affect strategic leaders' behavior to attain organizational excellence. In the third section, we discussed the most suitable research methods for our research. The fourth section consists of statistical analysis, and the fifth and final chapter discussed the results accordingly.

2 Development of Conceptual Model

2.1 Strategic Leadership and Institutional Excellence

While the role of manager qualities in organizational outcomes has long been debated in the field of management, academicians are frequently limited by data availability and computational power [15]. To address these issues, academics are beginning to apply recent advances in computer science to management study. This symposium will focus on new measures in strategic leadership research that employ big data and machine learning techniques. As leadership is ought to have core abilities to influence and guide others, strategic leadership is one of the leadership styles having these abilities in the best possible manner. A strategic leader helps to make strong organizational decisions, motivate the workforce, and enhance the prospects for both short-term and long-term organizational advantages [16]. In this regard, Rahman et al. (2018) suggest that organizational leaders have to develop strategic leadership qualities, as they should be strategic, direct, and general. As leaders are not only evaluated on their current tasks, also their task management, risk management, teamwork, and other relevant capabilities are equally important for their followers. This means that strategic leaders are having high-performing teams, a thinking organization, and a workforce that is highly committed to their task completion. To further affirm this, examined the role of strategic leadership in achieving organizational excellence in the colleges of Karbala University, Iraq. Data gathered from the structured questionnaires indicated that strategic leadership helps the college worker (including instructors) to seek motivation and guidance from the leadership. As a result, employees exhibit strongly cooperative and constructive behavior that helps them to seek new techniques to work effectively and develop their performance. Thus, Mann et al. (2012) attributed strategic leadership as directly providing a pathway to organizational excellence improving the working condition, and quality of the organizational workforce.

H1: There is a significant relationship between strategic leadership and institutional excellence in the Emirati federal institutions.

2.2 Role of Development Strategies and Organizational Processes in Achieving Institutional Excellence

Strategic planning of the organizational functions provides a strong baseline to attain organizational efficiency and excellence. Developing effective organizational strategies and their practical implementation are the core pathways to achieve organizational excellence. Here the role of the workforce is of greater significance as a strategic leader is subjected to motivate the workforce to accelerate the performance [7]. For example, the administration in the Palestinian Universities pays special consideration to develop strong organizational strategies that are both short-term and long-term goal-oriented. Despite some individuals consider these strategies to be less effective and require more improvement, the role of leadership in achieving organizational excellence is prominent [18]. Similarly, Owusu-Boadi (2019) also cited an example of the potential relationship strategic leadership in using several development strategies resulting in attaining institutional excellence. As noted that, successful leaders often make certain improvements in their practical strategies. However, these strategies are based on long-terms advantages and institutional excellence. Here Cebula et al. (2012) some characteristics of a strategic leader, that can materialize the goal of achieving institutional excellence. i.e., result-oriented, optimistic, purpose drive, competent, upright, and trustworthy. These characteristics are further directly attributed to strategic leadership as providing a direct link to achieve organizational excellence.

H2: Development strategies and organizational processes significantly mediate the relationship between strategic leadership and institutional excellence in the Emirati federal institutions.

2.3 Role of Leading and Developing People in Achieving Institutional Excellence

For example, multinational organizations such as AT&T, GM, IBM, and Sears take special transformational initiatives to cope with the gigantic organizational challenges. For this purpose, their leadership adopted strong strategic policies, that helped their workforce to perform well, leading to organizational stability and excellence at both national and international levels [19]. Further Lyons & In (2016) also highlighted the importance of developing the workforce to achieve organizational excellence. As noted that, effective leadership is comparatively more focused on supporting and developing their workforce individuals. For this purpose, strategic leaders introduce all the effective policies and their practical execution that are aimed at incorporating all the activities for workforce learning and development. [21] highlighted the role of leadership in increasing employees' development opportunities to achieve organizational excellence. The researchers cited an example of Bulgarian private business enterprises. As noted that, leadership can strongly affect the employees" motivation and organizational conformation through focusing more on workforce development and wellbeing programs. Here the role of managers and leaders is of greater pre-eminence due to their distinguished position in proposing, designing, and implementing new policies to achieve institutional excellence. Hence, attaining and sustaining organizational excellence directly relies upon the organizational workforce. If leadership knows the capabilities, and actively searches for employee development opportunities, that can positively affect the organizational performance as well [22].

H3: Leading and developing people significantly mediate the relationship between strategic leadership and institutional excellence in the Emirati federal institutions.

2.4 Role of Developing Culture and Value System in Achieving Institutional Excellence

According to Rahman et al. (2018), the changing business environments and technology incorporation in business organizations, are two major factors that demand employees development. From the structural perspective, organizations are based on several individuals that support, inspire and motivate the subordinates. This hierarchal structure also supports the leadership to take their workforce and group under consideration. As a result, direct contact is the bridge between leadership and employees, which further guides them to adapt to the organizational culture and values in the best possible manner. Under the previous argumentation, Al-Jedaiah & Albdareen (2020) also linked organizational conformity as an achievable yet requiring leadership role. As argued that, organizations decide and design the way affiliated individuals should act and interact. These decisions are based on organizational culture and norms, that are part of an organization and also provide a pathway to achieve the organizational goals and excellence. A strategic leadership not only notifies the workforce about the organizational culture but also helps them to adopt the norms and follow the organizational culture, that further results in attaining organizational conformity. Asif & Basit (2020) also validated the role of strategic leadership in achieving organizational excellence through developing organizational culture and value system. As stated that, strategic leaders belong to executive leaders or top leaders. Risk management, adopting change, obligating the higher moral values, and reinforcing organizational culture adoption are some basic characteristics that not only distinguish strategic leadership but also help to achieve organizational excellence.

H4: Developing culture and value system significantly mediate the relationship between strategic leadership and institutional excellence in the Emirati federal institutions.

2.5 Role of Developing Distinct Organizational Competencies in Achieving Institutional Excellence

The distinguished combination of measurable and observable skills, experiences, attributes, and abilities helps to accelerate the workforce performance that further leads to organizational success. The concept of organizational competencies demands developing the same capabilities that further result in achieving organizational excellence [24]. But Ellström & Kock (2012) questioned why and how an organization attains excellence? The answer can be determined by the fact that excellence is a result of individual and collaborative efforts that not only adds financial benefits, also helps individuals to build their confidence through a successful task management process.

Norzailan et al. (2015) consider strategic leadership as developing distinct competencies that are aimed at achieving organizational excellence. As noted that, strategic leadership deals with all the issues that are concerned with organizational well-being. However, strategic leadership differs from developing merely supervisory skills, where leaders only guide the workforce regarding task completion. Strategic leadership, on the other hand, guides, motivates and also helps to work for the employees' well-being, which helps them to improve their performance and achieve organizational excellence [27].

H5: Developing distinct organizational competencies significantly mediate the relationship between strategic leadership and institutional excellence in the Emirati federal institutions.

2.6 Role of Developing Effective Networking Competencies in Achieving Institutional Excellence

Today's fastest-changing technological environment accelerates one's desire to rely more on internet technology. As a result, individuals use digital platforms to communicate, particularly for business and other professional purposes. Similarly, organizations have also integrated Information Communication Technology to keep pace with the fastest growing business competencies [28]. To cope with the changing digital environments and nature of human needs, both formal and informal organizations are giving special attention to introducing technological advancements that can further add to their professional competencies. Digital technology not only helps to work faster but also provides data management and communicative sustainability to an organization. As the idea is, the more people stay connected with each other, the more they stay updated about the new professional trends and challenges [29]. To further broaden the concept of networking competencies, Kriger & Zhovtobryukh (2013) examined the role of strategic leadership in attaining institutional excellence through digital technology. Data gathered from the quantitative method revealed that strategic leaders are focused on integrating technology to further enhance the professional competency of their organizations. In this context, they not only support technology Acceptance but also motivate their workforce to actively learn tactical technology usage to maximize organizational excellence. Even in crises, when organizational stakeholders look for sustainable solutions, technology implementation and usage accelerated by strategic leadership provide practical solutions [31].

H6: Developing effective networking competencies significantly mediate the relationship between strategic leadership and institutional excellence in the Emirati federal institutions (Fig. 1).

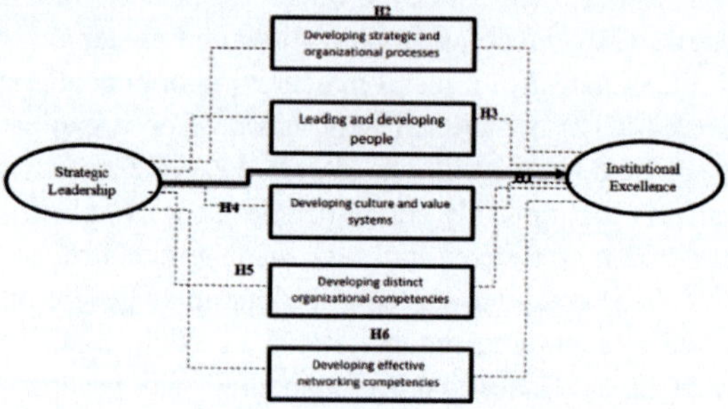

Fig. 1. Conceptual model of current study

3 Research Methodology

We executed cross-sectional design in this research and used structure, close-ended questionnaires designed on Five-Point Likert Scale for data gathering purposes as suggested by [32, 33]. We designed the questionnaire by selecting measurements scales from the different, authenticated resources approved by previous studies [34–37]. Table 1 below summarizes the sources of measurements scales:

Table 1. Sources of measurements scales

Scale	Source
Strategic leadership	[38]
Role of strategic leadership	[39, 40]
Institutional excellence	[6, 41]

3.1 Convergent Validity

After data gathering, we carefully manipulated and coded the data to further perform the statistical analysis by using Statistical Package for Social Sciences and AMOS.

Here we conducted both descriptive and inferential data analysis, involving frequency calculation and Structural Equation Modelling according to the study requirements [42]. In this regard, we first determined the reliability and validity of the research model, to affirm the internal consistency and status of our propositions. For this purpose, we conducted three different analyses as suggested by [1], including, Convergent Validity analysis, Construct Reliability analysis, and Discriminant Validity analysis. Table 2 provides the summary of internal validity examination through the relevant techniques. We can observe the Intercoder Reliability assessment of the research items is indicating the relevant values as ranging from .736 to .872, and the values of Composite Reliability are ranging from .729 to 1.35. Here we found that all the values of both assessments are greater than the threshold value of 0.7. Likewise, all the Factor Loading values are ranging from .709 to .936 and Average Variance Extracted Values are ranging from .787 to .919. We also found that the AVE and Factor Loading values are greater than the threshold value of 0.5. Thus, we conclude that the Convergent validity of the research model is successfully established.

Table 2. Summary of convergent validity analysis

Study variables	Items	FL-value	AVE	CA	CR
Strategic leadership	SLP1	767	.801	.872	.775
	SLP2	.765			
	SLP3	.872			
Institutional excellence	IEE1	.862	.910	.816	.924
	IEE2	.935			
	IEE3	.935			
Strategic organizational pocess	SOP1	.926	.919	.736	.931
	SOP2	.931			
	SOP3	.888			
Leading & developing people	LDP1	.798	.880	.820	1.35
	LDP2	.886			
	LDP3	.772			
Culture & value system	CVS1	.771	.787	.849	.752
	CVS2	.811			
	CVS3	.781			
Distinct competencies	DCS1	.807	.827	.768	.818
	DCS2	.784			

(*continued*)

Table 2. (*continued*)

Study variables	Items	FL-value	AVE	CA	CR
	DSC3	.892			
Effective networking competencies	ENC1	.709	.772	.741	.729
	ENC2	.755			
	ENC3	.853			

3.2 Discriminant Validity of Research Model

We further assessed the discriminant validity of the research model as suggested by [44]. For this purpose, we first conducted the Heterotrait-Monotrait Ratio Scale analysis (See Table 3a). The careful calculation of average values and applying the HTMT formula through MS Excel 2016, we found the HTMT value at .791 is lower than the designated value of 0.85, indicating that the HTMT criterion is successfully fulfilled. Further, we conducted Fornell-Larcker Criterion assessment (See Table 3b) and found that, the square of all the Average Variance Extracted values are greater than the mentioned correlations values, indicating that the Fornell-Larcker Criterion is also is fulfilled. Thus, we found that the discriminant validity of the research model is fully established [5, 35].

Table 3. (a) Heterotrait-Monotrait Ratio Scale (b) Fornell-Larcker Criterion

a

	SLP	IEE	SOP	LDP	CVS	DCS	ENC
SLP							
IEE	.191						
SOP	.271	.955					
LDP	.490	.216	.250				
CVS	.523	.198	.230	.744			
DCS	.241	.316	.280	.383	.704		
ENC	.021	.612	.573	.028	.113	.544	

b

	SLP	IEE	SOP	LDP	CVS	DCS	ENC
SLP	.641						
IEE	.133	.828					
SOP	.165	.855	.844				
LDP	.512	.164	.837	.774			
CVS	.570	.221	.202	.702	.619		
DCS	.457	.472	.261	.549	.518	.683	
ENC	.045	.780	.488	.058	.101	.362	.595

4 Data Analysis and Findings

4.1 Sampling and Demographics of Participants

As the study is focused on Federal Institutions in the United Arab Emirates, the population involves all the employees currently working in the relevant institutions. However, according to research criteria and certain limitations, we randomly selected the Ministry of Human Resources and Emiratization for the data gathering aspirations. Thus, despite major criticism on convenience sampling [5, 45], we selected the relevant sampling method due to its greater suitability for the current research. We visited the head office of the Ministry of Human Resources and Emiratization in both Dubai and Abu Dhabi and obtained data from the $n = 350$ employees. Participants were also provided with the informed consent and the autonomy to withdraw from the process whenever they want [46]. After data gathering, we calculated the response rate which was 86.0% as 14% ($n =$ 49) questionnaires were incompletely or wrongly filled. Thus, the frequency calculation of demographical data showed that a majority of respondents (59.2%) were females and 48.8% were males. Similarly, the age calculation of respondents indicated that 21.8% of participants were 26 to 30 years old, followed by the group of individuals who were 31 to 35 years old, 13.0% of respondents were from 18 to 25 years old, 9.3% were 36 to 40 years old, 6.0% were 41 to 45 years old, and only 5.0% of respondents were 46 years or above. Moreover, the descriptive analysis of qualification indicated that 23.0% of participants were graduated individuals, 21.3% were holding Master/Post-graduation degrees, 17.3% of individuals were having professional diplomas or certifications, and only 9.0% of respondents were holding under graduation degrees. Lastly, we found that 63.2% of participants were from the branch office situated in Abu Dhabi and 36.8% were from the branch office in Dubai. We also conducted a One-Way Analysis of Variance to examine the impacts of demographical variables on participants' responses as the study has one independent variable [47]. We can observe that the significance values of the demographical variables are ranging from .000 to .069, indicating that the relevant variables are controlled and have no difference in the responses based on participants' demographical data. Table 4 summarizes the results of demographics calculation and Analysis of Variance:

Table 4. Demographics of respondents & analysis of variance

Variables	Constructs	f	%	Mean	SD	Levene Statistics	P-value
Gender	Male	122	30.5%	1.59	.492	5.159	.000
	Female	179	59.2%				
Age	18–25	**52**	13.0%	2.85	1.415	2.394	.028
	26–30	87	21.8%				
	31–35	81	20.3%				
	36–40	37	9.3%				
	41–45	24	6.0%				
	46 or Above	20	5.0%				
Qualification	Under graduation	36	9.0%	2.84	1.181	1.974	.069
	Graduation	92	23.0%				
	Masters	85	21.3%				

(continued)

Table 4. (*continued*)

Variables	Constructs	f	%	Mean	SD	Levene Statistics	P-value
	Diploma/Certification	69	17.3%				
Branch	Dubai	110	36.8%	1.6346	.48236	40.246	.000
	Abu Dhabi	191	63.2%				

4.2 Coefficients of Determination R^2

We conducted Coefficients of Determination R^2 as it helps to determine the predictive power of the research model concerning the structural equation modeling [48, 49]. As visible in Table 5, the R^2 values are ranging from .630 to .836, we found that the research model has strong predicting potential in the current study.

Table 5. Coefficients of determination R^2

S/R	Variables	R^2
1.	Institutional excellence	.836
2.	Strategic organizational process	.681
3.	Leading & developing people	.742
4.	Culture & value system	.799
5.	Distinct competencies	.813
6.	Effective networking competencies	.630

4.3 Hypotheses Testing: Path Analysis, Regression, and Mediation Analysis

Further, we assessed the proposed relationship between study variables. First, we conducted a simple Linear regression analysis between strategic leadership and Institutional Excellence. With the path value at and significance value at $p \geq .001$, we found a strong significant relationship between strategic leadership and Institutional Excellence. Likewise, we conducted the mediation analysis to analyze the impact of meditating variables on the proposed relationship between strategic leadership and Institutional Excellence. In this regard, we first found that the path value of the mediating effect of Strategic Organizational Process is strongly significant ($p = 1.149$) along with the indirect effects value of .254 and significance value of $p \geq .000$, indicating that the **H2** of the study is validated. Moreover, regarding the **H3** of the study, the significance value of $p \geq 0.015$ and indirect effects value also validated the relevant research postulation. The H4 of the study is also validated with the significance value of $p \geq .038$ *and the value of the indirect effect at .068*. Furthermore, the mediating role of Distinct Competencies is also affirmed with the significance value at $p \geq .001$ and indirect effect value at .066. Finally, with the significance level at $p \geq .071$ and indirect effects value at .032, we also validated the mediating role of Effective Networking Competencies in the relationship between strategic leadership and Institutional Excellence. Table 6 summarizes the outputs of hypotheses testing including path analysis, significance values, and mediation analysis:

Table 6. Path analysis, regression analysis, and mediation analysis

Hyp.	*Relationship*	*Path*	*t*	*f*	*Sign*
H1	**SLP>IEE**	.452	19.785	11.279	.001
Hyp.	*Relationship*		*Path*	*Indirect Effect*	*Sign*
H2	SLP>SOP>IEE		1.149	.254	.000
H3	SLP>LDP>IEE		.366	.076	.015
H4	SLP>CVS>IEE		.421	.068	.038
H5	SLP>DCS>IEE		.533	.066	0.001
H6	SLP>ENC>IEE		.419	.032	0.071

According to Năstase (2010) current organizational trends and challenges place a higher pressure on organizational capabilities to meet the requirements and keep pace with the fastest-changing organizational environments. For this purpose, managerial competencies are of greater importance, as they bear the responsibility to bring and implement positive and constructive tactics, that may ensure collective wellbeing and growth. As a result, institutional excellence is the primary objective that provides a pathway to avail the collect organizational goals [50]. Similarly, the current study no only provided an overview of the factors that help the strategic leadership to attain the organizational goals, also described their mediating role at the individual level. Especially, in the United Arab Emirates where federal organizations are successful entities, their organizational excellence is the secret of the country's social and economic stability [9]. Additionally, these results also provided the role of strategic leadership in achieving organizational excellence in Emirati federal institutions like the Ministry of Human Resources and Emiratization [51]. We found that the strategic leadership of the Ministry of Human Resources and Emiratization gives special consideration to strategic organizational process, professional development of their workforce, having a distinguished organizational cultural and value system, possess distinct competencies, and having modern technology as a part of their functional system. As a result, today Emirati federal organizations are the center of attention providing equal opportunities to their employment and working effectively for the common wellbeing and progress [8]. These findings are also consistent with the previously conducted studies, witnessing strategic leadership as providing a direct pathway to organizational excellence worldwide [2, 10, 12, 23, 26, 38, 52]. Thus, we recommend more studies that may witness the role of strategic leadership in attaining organizational excellence, especially in the Emirati corporate sector, where organizational excellence is a primary goal for the organizational management and leaders [42, 54].

5 Conclusion

The results show that strategic leadership is closely related to organizational excellence in the United Arab Emirates. Besides, organizational excellence, strategic organizational processes, leadership and development of people, culture and value systems, distinct competencies, and effective networking competencies are the primary tactics that help a strategic leader achieve organizational excellence. The strategic leadership in the Emirati organization, this role is comparatively stronger and prominent in their administrative departments, that further highlights how strategic leaders are working effectively by obligating their responsibilities to attain the designated organizational goals., that further highlights how strategic leaders are working effectively by obligating their responsibilities to attain the designated organizational goals. In the Emirati scenario where federal institutions are efficiently working, the reason behind their success is yet to be explored.

References

1. Salloum, S.A., et al.: What impacts the acceptance of e-learning through social media? an empirical study. Recent Adv. Technol. Accept. Model. Theor. 419–431 (2021)
2. Rahman, N.R.A., et al.: Impact of strategic leadership on organizational performance, strategic orientation and operational strategy. Manag. Sci. Lett. **8**(12), 1387–1398 (2018). https://doi.org/10.5267/j.msl.2018.9.006
3. Shirvani, A., Iranban, S.J.: Organizational excellence performance and human force productivity promotion: a case study in south zagros oil and gas production company. Iran 2(3), 3010–3015 (2013)
4. Alhumaid, K., Habes, M., Salloum, S.A.: Examining the factors influencing the mobile learning usage during COVID-19 Pandemic: an integrated SEM-ANN method. IEEE Access (2021)
5. Salloum, S.A., AlAhbabi, N.M.N., Habes, M., Aburayya, A., Akour, I.: Predicting the intention to use social media sites: a hybrid sem-machine learning approach. Adv. Mach. Learn. Technol. Appl. Proc. AMLTA **2021**, 324–334 (2021)
6. Abu Naser, S., Al Shobaki, M.: Organizational excellence and the extent of its clarity in the palestinian universities from the perspective of academic staff. Int. J. Inf. Technol. Electr. Eng. **6**(2), 47–59 (2017)
7. Al-Jedaiah, M.N., Albdareen, R.: The effect of strategic human resources management (SHRM) on organizational excellence. Probl. Perspect. Manag. **18**(4), 49–58 (2020). https://doi.org/10.21511/ppm.18(4).2020.05
8. Dubai, R.I.: Organizational Excellence (2000)
9. Al Ameri, S.: Organisational Excellence in Cultural-Social Development Organisations, **70** (2011)

10. Owusu-Boadi, B.Y.: The Role of Strategic Leadership in the Profitability of Large Organizations. ProQuest Diss. Theses, p. 163 (2019)
11. Cebula, N., Craig, E., Eggers, J., Fajardo, M.D., Gray, J., Lantz, T.: Achieving performance excellence: the influence of leadership on organizational performance. Static. Nicic. Gov (2012)
12. Singh, L., Panda, B.: Impact of organisational culture on strategic leadership development with special reference to Nalco. Int. J. Res. Dev. - A Manag. Rev. **4**(1), 133–142 (2015)
13. O'Regan, N., Lehmann, U.: The impact of strategy, leadership and culture on organisational performance: a case study of an SME. Int. J. Process Manag. Benchmark. **2**(4), 303–322 (2008). https://doi.org/10.1504/IJPMB.2008.021790
14. Al Shobaki, M.J., Naser, S.S.A.: Learning Organizations and their role in achieving organizational excellence in the palestinian universities. Int. J. Digit. Publ. Technol. **1**(2), 40–85 (2017)
15. Salloum, S.A., AlAhbabi, N.M.N., Habes, M., Aburayya, A., Akour, I.: Predicting the intention to use social media sites: a hybrid SEM - machine learning approach. In: Hassanien, A.-E., Chang, K.-C., Mincong, T. (eds.) AMLTA 2021. AISC, vol. 1339, pp. 324–334. Springer, Cham (2021). https://doi.org/10.1007/978-3-030-69717-4_32
16. Olwan, A.A.: Strategic leadership competencies and its impact on achieving organizational excellence via the mediating role of organizational commitment: a case study in national center for security and crisis management (NCSCM). J. Soc. Sci. **8**(1), 106–138 (2019). https://doi.org/10.25255/jss.2019.8.1.106.138
17. Mann, R., Mohammad, M., Agustin, M.T.A.: UNDERSTANDING an Awareness Guidebook Guide Book for SMEs. pp. 1–13 (2012)
18. Samy, S.: Organizational excellence in Palestinian universities of gaza strip. Int. J. Inf. Technol. Electr. Eng. **6**(4), 20–30 (2017)
19. Pinar, M.: Investigating the Impact of Organizational Excellence and Leadership on Achieving Business Performance : An Exploratory Study of Turkish Firms 1 College of Business Administration Valparaiso University Valparaiso , IN 46383 , USA Girard , Tulay Penn State," 73, 29–45 (2008)
20. Lyons, R., In: Strategic Human Resource Development Impact On Organizational Performance: Does Shrd Matter? North Dakota State University (2016)
21. Tzvetana, S., Ivaylo, I.: Employee engagement factor for organizational excellence. Int. J. Bus. Econ. Sci. Appl. Res. (2017)
22. Singh, A.: Sustaining organizational excellence through talent management: an empirical study. Int. Rev. Bus. Soc. Sci. **1**(10), 16–23 (2012)
23. Asif, A., Basit, A.: Exploring strategic leadership in organizations: a literature review. Strateg. Leadersh. Organ. 212 GMR **5**(2), 211–230 (2020)
24. Puteh, F., Kaliannan, M., Alam, N.: Employee core competencies and organizational excellence. In: Proc. Australas. Conf. Bus. Soc. Sci. 2015, Sydney (in Partnersh. with J. Dev. Areas), vol. 2, no. 1998, pp. 331–340 (2015)
25. Ellström, P.E., Kock, H.: Competence development in the workplace: concepts, strategies and effects. Int. Perspect. Compet. Dev. Dev. Ski. Capab. **9**(1), 34–54 (2012). https://doi.org/10.4324/9780203523032
26. Norzailan, Z., Yusof, S.M., Othman, R.: Developing strategic leadership competencies. J. Adv. Manag. Sci. **4**(1), 66–71 (2015). https://doi.org/10.12720/joams.4.1.66-71
27. Năstase, M.: Developing a strategic leadership approach within the organizations. Rev. Manag. Comp. Int. **11**(3), 454–460 (2010)
28. De Janasz, S.C., Forret, M.L.: Learning the art of networking: a critical skill for enhancing social capital and career success. J. Manag. Educ. **32**(5), 629–650 (2008). https://doi.org/10.1177/1052562907307637

29. De Klerk, S.: The importance of networking as a management skill. South African J. Bus. Manag. **41**(1), 37–49 (2010). https://doi.org/10.4102/sajbm.v41i1.512
30. Kriger, M., Zhovtobryukh, Y.: Rethinking strategic leadership: stars, clans, teams and networks. J. Strateg. Manag. **6**(4), 411–432 (2013). https://doi.org/10.1108/JSMA-09-2012-0051
31. Johansson, C., Bäck, E.: Strategic Leadership Communication for Crisis Network Coordination, pp. 1–20 (2017)
32. Habes, M., Ali, S., Salloum, S.A., Elareshi, M., Ziani, A.-K., Manama, B.: Digital Media and Students' AP Improvement: An Empirical Investigation of Social TV. Int. Conf. Innov. Intell. Informatics, Comput. Technol. Progr., (2020)
33. Habes, M., Alghizzawi, M., Salloum, S.A., Mhamdi, C.: Effects of Facebook personal news sharing on building social capital in Jordanian universities. In: Al-Emran, M., Shaalan, K., Hassanien, A.E. (eds.) Recent Advances in Intelligent Systems and Smart Applications. SSDC, vol. 295, pp. 653–670. Springer, Cham (2021). https://doi.org/10.1007/978-3-030-47411-9_35
34. Elareshi, M., Habes, M., Ziani, A.-K.: New media users' awareness of online inflammatory and mobilisation methods for radical and extreme activities. Ilkogr. Online **20**(5), 5567–5576 (2021)
35. Habes, M., et al.: E-Learning acceptance during the covid-19 outbreak: a cross-sectional study. In: European, Asian, Middle Eastern, North African Conference on Management & Information Systems, pp. 65–77 (2021)
36. Al-Sarayrah, W., Al-Aiad, A., Habes, M., Elareshi, M., Salloum, S.A.: Improving the deaf and hard of hearing internet accessibility: JSL, text-into-sign language translator for arabic. In: Hassanien, A.-E., Chang, K.-C., Mincong, T. (eds.) AMLTA 2021. AISC, vol. 1339, pp. 456–468. Springer, Cham (2021). https://doi.org/10.1007/978-3-030-69717-4_43
37. Elareshi, M., Habes, M., Ali, S., Ziani, A.: Using online platforms for political communication in bahrain election campaigns. Soc. Sci. Humanit. **29**(3), 2013–2031 (2021)
38. Gupta, M.: Strategic leadership: an effective tool for sustainable growth. SIBM Pune Res. J., **XV**(0), 2249–1880 (2018)
39. Davies, B.J., Davies, B.: Strategic leadership. Sch. Leadersh. Manag. **24**(1), 29–38 (2004). https://doi.org/10.1080/1363243042000172804
40. Al-Shibly, M.S., Alghizzawi, M., Habes, M., Salloum, S.A.: The Impact of de-marketing in reducing jordanian youth consumption of energy drinks. In: Hassanien, A.E., Shaalan, K., Tolba, M.F. (eds.) AISI 2019. AISC, vol. 1058, pp. 427–437. Springer, Cham (2020). https://doi.org/10.1007/978-3-030-31129-2_39
41. Sharqawi, A.W.: Strategic planning: the basis of excellence, entrepreneurship and institutional creativity. Arab Bus. Adm. Assoc. **186**(28), 21–22 (2020)
42. Al-Skaf, S., Youssef, E., Habes, M., Alhumaid, K., Salloum, S.A.: The acceptance of social media sites: an empirical study using PLS-SEM and approaches. Adv. Mach. Learn. Technol. Appl. Proc. AMLTA **2021**, 548–558 (2021)
43. Mat Nawi, F.A., Abdul Malek A.T., Muhammad Faizal, S., Wan Masnieza, W.M.: A review on the internal consistency of a scale: the empirical example of the influence of human capital investment on malcom baldridge quality principles in tvet institutions. Asian People J., **3**(1), 19–29, https://doi.org/10.37231/apj.2020.3.1.121 (2020)
44. Henseler, J., Ringle, C.M., Sarstedt, M.: A new criterion for assessing discriminant validity in variance-based structural equation modeling. J. Acad. Mark. Sci. **43**(1), 115–135 (2014). https://doi.org/10.1007/s11747-014-0403-8
45. Etikan, I.: Comparison of convenience sampling and purposive sampling. Am. J. Theor. Appl. Stat. **5**(1), 1 (2016). https://doi.org/10.11648/j.ajtas.20160501.11
46. Roache, R.: Why is informed consent important? J. Med. Ethics **40**(7), 435–436 (2014). https://doi.org/10.1136/medethics-2014-102264

47. Scofield, T.L.: ANOVA **2**, 1–11 (2018)
48. Ziani, A.-K., Elareshi, M., Habes, M., Tahat, K.M., Ali, S.: Digital media usage among arab journalists during covid-19 outbreak. In: European, Asian, Middle Eastern, North African Conference on Management & Information Systems, pp. 116–129 (2021)
49. Dufour, J.-M.: Coefficients of Determination ∗ no. March 1983 (2011)
50. Ringrose, D.: 10 Benefits of Implementing an Organizational Excellence Model. pp. 10–13 (2011)
51. Lasrado, F.: Legacy of Excellence: The Case of the United Arab Emirates (UAE), pp. 37–56. Springer, Cham (2018)
52. Suzana, R.: The relationship of strategic leadership characteristics, gender issues and the transformational leadership among institutions of higher learning in Malaysia. In: Acad. Bus. Res. Inst. Conf. - Las Vegas 2010 Conf. Proceeding, vol. LV2010, pp. 1–14 (2010)
53. Elbasir, M, Elareshi, M., Habes, M.: The influence of trust, security and reliability of multimedia payment on the adoption of EPS in Libya. Multicult. Educ., **6**(5) (2020)
54. Lasrado, F., Uzbeck, C.: The excellence quest: a study of business excellence award-winning organizations in UAE. Benchmarking **24**(3), 716–734 (2017). https://doi.org/10.1108/BIJ-06-2016-0098

An Extended Modeling Approach for Marine/Deep-Sea Observatory

Charbel Geryes Aoun[1,2](✉), Loic Lagadec[2], and Mohammad Habes[3]

[1] Information Technology, Institut Catholique d'Arts et Métiers (ICAM), Toulouse, France
charbel.aoun@icam.fr

[2] Information Technology, École Nationale Supérieure de Techniques Avancées (ENSTA), UMR CNRS 6285 Lab-STICC 2, Bretagne, France
loic.lagadec@ensta-bretagne.fr

[3] Mass Communication, Yarmouk University, Irbid, Jordan
mohammad.habes@yu.edu.jo

Abstract. A Sensor Network (S.N.) is responsible for performing two main activities: (1) observation/measurement, which means accumulating data collected at each sensor node; (2) transferring the collected data to processing centers (e.g., Smart Sensors, Smart Fusion Servers) within the S.N. The infrastructure of Marine/Deep-sea Observatory is an Underwater Sensor Networks (UW-SN) to perform collaborative monitoring tasks over a given ocean/sea area. This observation should consider the environmental constraints since it may require specific logical and physical components. The physical ones could be specific tools, materials, and devices such as marine cables, servers, etc.). As for the logical ones, specific algorithms could validate the implementation phase early, such as validating the allowable entered bandwidth ranges of underwater acoustic channels. This paper presents our approach in extending the modeling languages to include new domain-specific concepts and constraints. Thus, we propose an extended meta-model that is used to generate a new design tool that contains the new constraints. We illustrate our proposal with an example from the Marine Observatory (MO) domain on object localization with several acoustics sensors. Additionally, we generate the corresponding simulation code for a standard network simulator using our self-developed domain-specific model compiler. Our approach helps to reduce the complexity and time of the design activity of a Marine Observatory. It provides a way to share the different viewpoints of the designers in the MO domain and obtain simulation results to estimate the network capabilities. The major improvement is to provide an early validation step via models and a simulation approach to consolidate the system design.

Keywords: Smart sensors · ArchiMate · Sensor networks · Domain-specific modeling language · Enterprise architecture

1 Introduction

Applications dedicated to monitoring different environments depend on extensively on deploying a network of sensors to collect scientific data. A sensor network comprises

© The Author(s), under exclusive license to Springer Nature Switzerland AG 2022
A. E. Hassanien et al. (Eds.): AMLTA 2022, LNDECT 113, pp. 502–514, 2022.
https://doi.org/10.1007/978-3-031-03918-8_42

specialized sensors with a communication infrastructure designed to monitor and record terms and scientific data at various locations. Gathered data is relayed to dedicated workstations in real-time for analysis to become useful information. Marine Observatory (MO) applications are not different in concept, though they differ in technology [1]. They provide new opportunities for sea surveys, such as a continuous observation of the sea [1]. The MO uses Underwater Acoustic Sensor Networks (UASNs) to operate underwater. These sensors collect data collaboratively and send it to the dedicated workstations. To achieve their objective, sensors make full use of the Sensor Networks' (S.N.) main advantages. These advantages include but are not limited to the autonomous nature of the S.N. and the operation of S.N. as a distributed information system. Autonomy is based on the ability of the S.N. to adapt to the characteristics of the ocean environment. At the same time, the operation of the S.N. as a distributed information system allows it to process queries regarding any service requirements. To be able to satisfy all of these features above, the sensor network has to be complex [2–5].

Although it can be applied to any MO project, we chose the Marine e-Data Observatory Network (MeDON) as our research scope [6]. As with any MO project, the MeDON project allows the continuous observation of the sea. MeDON uses different communication protocols (REST, SOAP, and proprietary ones) to connect its different components (Hydrophones, Fusion Servers, Object Localization Algorithms) [6, 5]. It seems obvious by now that the implementation of such distributed system is considered complex [2–5]. This complexity originates from different sources. However, in this paper we focus on two of them: (1) the architecture of a distributed system by itself is a complex system since it comprises many heterogeneous components (underwater sensor network and the rest of the information system); (2) deployment of such a system requires high-level value accuracy of several properties of these components.

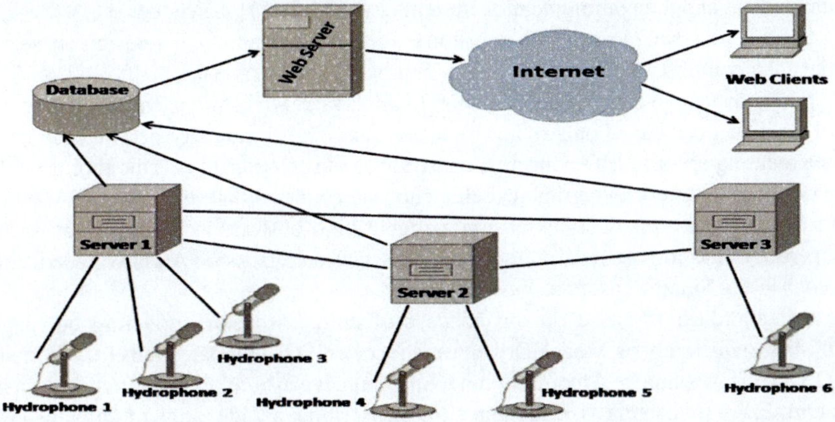

Fig. 1. Structure of MeDON - an example: N = 6, Y = 3

Our scope in MeDON project is in the design of the Smart Sensor Network and the system to localize the underwater objects. Designing an information system such as the one of MeDON may generate errors while performing the design activity. These errors will negatively affect the deployment activity by resulting in inappropriate and inapplicable designs of Sensor Networks. According to [6, 5], the complexity of the design is a result of (1) the different domains of experience (business process modeling, Information system modeling, and the underlying infrastructure modeling) that are required from the designer(s) to be able to model and describe such systems; (2) distributed software structure of MeDON Information System (Fig. 1) since each component (e.g. Data Fusion Server, Smart Sensor, etc.) is responsible for performing set of specific tasks.

Our global objective is to help MO designers to reduce the complexity of the design activity. The deployment of a set of sensors (Sensor Network) is a costly operation due to: the necessary equipment such as specific boats, marine cables, Sensors (Hydrophones), Data Fusion Servers, experts in diving, etc. In addition, we cannot ignore that the deployment operation is risky because any errors that may occur at the design level can negatively affect the deployment phase in terms of performance and quality of service (QoS) provided. In the DSO/MO context, the available bandwidth of underwater acoustic channels is crucial for the deployment of S.N. to ensure and provide high QoS. This available bandwidth is limited and dramatically depends on transmission range and depth [2, 3]. For this purpose, during the design level we have to take into consideration the following constraints of the bandwidth ranges [2]: for long-range communication in deep water, the available bandwidth ranges from 500 Hz to 10 kHz; for medium-range communication in shallow water, the available bandwidth ranges from 10 to 100 kHz; and for short-range communication in deep water, the available bandwidth ranges from 100 to 500 kHz.

About our scope in MeDON (localization of underwater moving objects), lack of suitable bandwidth ranges during the communication between the physical components (e.g., Smart Sensors, Fusion Servers) results in complexity and errors during the deployment phase. Thus, an integration between the information system (logical and physical components) and the communication system (e.g. I.M.S.) [7] is needed.

For this purpose, our research question is: how do we improve the design phase and reduce the complexity of the deployment and maintenance phase?

In this paper, we propose an extension to ArchiMO. ArchiMO is a modeling design tool we developed based on ArchiMate that reduces the complexity of MO designs by implementing specific MO concepts, constraints, and relationships. This tool provides the designer with reusable graphical elements and concepts that respect ArchiMate [8] and the MO concepts. This extension brings a high abstract level constraint on the properties of the communication tools (e.g., association relation in ArchiMO) dedicated to connecting Smart Sensors to Fusion Servers.

Our approach is based on the concept of domain-specific modeling languages (DSMLs), which rely on Model-Driven Engineers (M.D.E.) fundamentals [9]. To model MO systems, we choose ArchiMate modeling language as it relies on Enterprise Architecture (E.A.) framework [10, 11] that allows describing a wide range of domains [12]. We use meta-models to generate the tools that belong to different development activities using Eclipse Modeling Framework (E.M.F.) [13].

ArchiMate is proper to model systems from the I.T. domain [8]. Our proposal extends the ArchiMate meta-model (abstract and concrete syntax) to add a new abstract constraint of MO to ArchiMO. We add specific constraints to the grammar of the design tool according to the extended MO meta-model. On one hand, a main feature of (E.A.) frameworks is sharing multiple viewpoints [12]. This reduces complexity of one view to a manageable size. E.A. frameworks introduce interoperability issues between views and their dedicated software [12]. On the other hand, our proposed constraint is extensible, where the developers may extend it and add new sub-constraints, concepts, and standards according to the progress and needs in MO domain.

Linking our extended MO meta-model to the IP Multimedia Subsystem (IMS) one (proposed previously in [14] helps to integrate the different smart sensors of the sensor network to the rest of the information system through the core network [7]. We apply our design model to a model compiler to generate simulation code directly in the NS-3 network simulator [15].

The paper content is organized as follows. In Sect. 2, we present the related work connected to the added constraint on the design tools. Section 3 presents MO project. In Sect. 4, we present MDE fundamentals, DSMLs, ArchiMate, and our proposal constraint for the MO/MeDON. Section 5 explains the proposed constraint's abstract syntax, concrete syntax, and semantics. In Sect. 6, we present the newly added communication constraint and how it is generated with the MO design tool and the simulation approach. In Sect. 7, we conclude and discuss our future work.

2 Related Work

In this section, we present the related work in relation to the possibility of having high-level abstract constraints for an association relation that connects two physical components (e.g., Smart Sensor and Data Fusion) that are added to a design tool.

In relation with the concept of Architectural Description Languages (ADLs) [16–19] and their design tools; we are interested in the following concerns that we shall specify and analyze in this section: (C1) error prevention at the design level by invoking language structure or syntax of languages; (C2) multiple viewpoints[1] that are represented in the architectural description [20]; (C3) extensibility of design tool; (C4) diversity of components; (C5) testing/execution platform.

According to the preventing errors concern, the extended design tool early prevents errors that may be made by the designer during design activity, rather than correcting them afterward. This error prevention approach is available in [21–23]. Like in our approach, it's avoided by invoking the abstract syntax (our proposed constraints) where we have defined and added our specific concepts, constraints and relations.

Concerning the multiple viewpoints concern, the extended design tool provides different viewpoints for the designers according to their specialties and domains of experience. In [21–23], the design tool provides only one view-point in order to fit software development tasks. This design tool does not provide the ability to share the design between different designers. This is due to the non-existent architectural framework which generates a design tool to be fully compliant with the above concern [24]. Our approach considers this issue thanks to the different layers of EA standard that separates between perspectives.

[1] Viewpoint: is a work product establishing the conventions for the construction, interpretation and use of architecture views to frame specific system concerns.

Regarding the extensibility concern, the extension of a meta-model allows the extension of a design tool by adding new concepts and constraints to it [21, 22]. It's realized in our approach by extending the ArchiMate meta-model by new constraints, then generating a new design tool that contains our newly created constraints in the toolbox like in [7, 12].

Concerning the heterogeneity concern, the existence of different components and communications that are related to different contexts and activities.

We are facing this heterogeneity in the software components and models in [21–23]. In our approach, the diversity of components appears in our MO model, which contains many Smart Sensors connected to a large number of Data Fusion servers.

According to the execution test platform concern, we can find an integration between two different platforms to provide an automatic execution test of a given complex model like in [25]. Also, we can find some platforms where the designer is not able to test and verify his models or instances on an executable platform like in [21–23]. However, based on [7], our approach provides us the ability to validate the created models on an executable platform implemented in the same framework where the creation of models occurs (see Sect. 6). For example, Smart Sensors and Fusion Servers can exchange messages using (IMS).

3 Marine Observatories

Underwater Sensor networks that aim to environmental data acquisition will play an essential role in the development of future large data acquisition systems [6, 26, 27]. They allow the data to be exchanged and treated between the different devices (servers, sensors). We can have software components on all these devices to treat and store the data. MO is the Marine e-Data Observatory Network (MeDON) (Refer to Fig. 1).

The designer should include N acoustic sensors connected to the Y fusion servers in this context, as shown in (Fig. 1). These servers treat the hydrophones' acoustic data and then diffuse them on the network. Servers store their data on the same database. The Database server provides the treated data to the web server where the configuration of a web application is done. Thus, the web server diffuses the information detected by the hydrophones, such the voice of the dolphin to the web clients through a graphical interface.

4 Model-Driven Engineering (MDE) and Domain-Specific Modeling Languages (DSML)

MDE [20] is "a software development method which focuses on creating and exploiting domain models. It allows the exploitation of models to simulate, estimate, understand, communicate, and produce code". MDE helps to manage complexity thanks to the modeling concept and model transformations. Modeling helps to describe the design in a highly abstract way and model transformation helps to have a generated design tool.

Domain-Specific Modeling Languages (DSMLs) [28] enable designers from different domains and backgrounds to participate in software development tasks and to specify their own needs using domain concepts. A DSML [29] is they are comprised of three components: abstract syntax, concrete syntax, and semantics. The abstract syntax defines modeling concepts and their relationships. There are several kinds of concrete syntaxes: visual, XML-based, textual, etc. [30]. The concrete syntax is associated with a set of rules that define the abstract syntax's representation. Semantics describe a model's meaning and are related to the abstract syntax. They are well-formed rules for the model and are used to constrain the concrete syntax [29].

In general, errors caught during the design cycle are much less time-consuming to identify and correct than those found during testing. To avoid errors in the design activity, we have implemented constraints that are defined in the abstract syntax of the language (meta-model) (Fig. 2) [31]. The concrete syntax that is associated with these added constraints can be implemented in the design tool such as 'ArchiMO' tool in our context. This tool is generated relying on Eclipse-EMF (tool generation concept thanks to model transformations).

Modeling languages are used to describe a system with a high level of abstraction (e.g. UML 2.0) [30]. For MeDON/MO, and in relation with our objectives, we describe distributed systems. UML is not enough to cover our needs, as it has only one layer that contains all of the design concepts, and these concepts are too general [32]. Thus, we selected ArchiMate modeling language that meets UML in some concepts, but that can describe the systems from

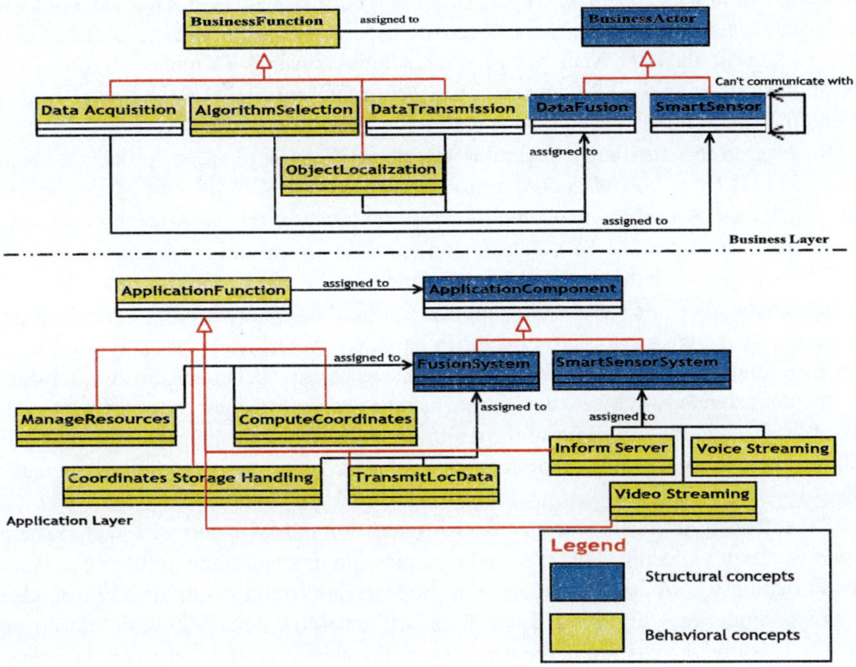

Fig. 2. Extending business and application layers of ArchiMate: proposal of MO Meta-Model

IT domain and share multiple viewpoints during the design relies on the TOGAF framework [20]. Additionally, as of January 2018, the latest version of the NATO Architecture Framework (NAF v4) can be created using The Open Group's ArchiMate meta-model [24]. NAFv4 is a standard for developing architectures.

ArchiMate relies on Enterprise Architecture (EA) framework [11, 20]. It decomposes the system design into three layers: business, application, and technology. In our approach, we present these layers in the following way:

1. Business layer: specifies the end-user functions and actors. It describes the service activities as perceived by the end-user and the flow between them;
2. Application layer: specifies the functions and software components of the service. It describes the capability of the system under study and the way of performing its tasks;
3. Technology layer: specifies the functions, topology, hardware elements, and signaling protocols of the underlying platform. It describes the execution platform that offers functions to be used by the functions of the application layer.

5 Contribution

In general, a meta-model of DSL represents the concepts/operations and constraints that belong to the domain specificity (MO in our case). Our previous work has extended the ArchiMate meta-model in order to take into consideration the domain specificity of MO [31]. As a result, the ArchiMate meta-model includes concepts, elements, relations, and constraints related to the MO domain. This section presents our contribution to further extend this meta-model, through its concrete syntax and design tools.

Relying on the distributed fusion architecture (DFA) in [33], our proposed meta-model in [31] (Fig. 2), and according to our context MeDON [6], our MO extended meta-model and ArchiMO generated design tool already accommodates the following components: Smart Sensor (SS), Data Fusion (DF) and several other components (Fig. 2).

In this paper, we extend a built-in association relation in Archi by adding the constraints of the available bandwidth ranges as properties. Therefore, by extending the abstract and concrete syntax and semantic of ArchiMO, we added to the association relation the following constraints: once the designer select this association from the ArchiMO palette, our extended ArchiMO design tool ask the designer to enter the appropriate values of the bandwidth according to the depth of water (Fig. 3). Otherwise, the designer will not be able to establish a communication between any two physical components (Fig. 4).

The extended ArchiMO tool considers different domains of experience, each domain expert works in a specific layer (Business, application or technology) as the model created in section VI. We implement our new proposed constraint in our already extended MO meta-model (see Fig. 2), and our already it generated ArchiMO design tool (see Fig. 6). This implementation is the grammar of the new proposed constraint. Our contribution tackles the issues presented in 2 by: (C1) enhancing the design process by minimizing syntax and relation errors; (C2) providing three layers according to each domain specificity by relying on Enterprise Architecture; (C3) extend MO meta-model to take into account additional constraints by generating ArchiMO that includes these constraints; (C4) deploy several physical components (sensors and servers), and logical components such acquisition/localization algorithms by creating an MO model that contains this variety of components.

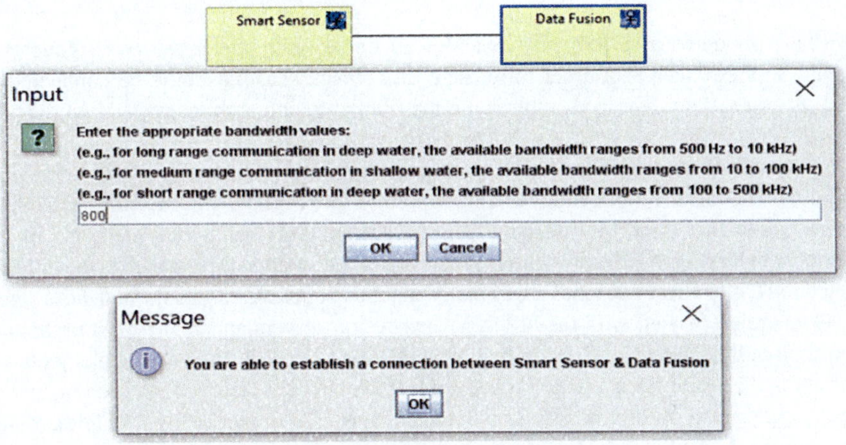

Fig. 3. Appropriate values of bandwidth

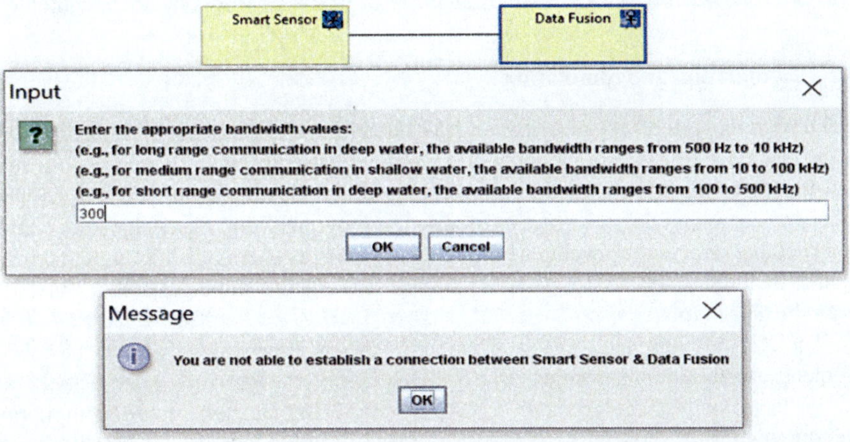

Fig. 4. Inappropriate values of bandwidth

6 Object Localization Case Study

To validate our proposed tool, we use it to model the application of Object Localization using the different new elements proposed in the meta-model (Fig. 2). Then we apply the design model to a model compiler (Fig. 6) that we have developed to perform some error checks and automatically generate simulation code for NS-3. This simulation code runs in the NS-3 tool, a standard and classical simulator in the networking domain.

6.1 Design Model

We have modeled a system that localizes an underwater object using our generated design tool ArchiMO. In order to localize this object, sensors should be connected to data fusion servers. We have applied the distributed fusion architecture (DFA) [33] for this design.

The design model is composed of three views regarding the layers of ArchiMate (Fig. 5): Business, Application, and Technology. In (Fig. 5), we present parts of the large model that ArchiMO designs. The model contains behavioral elements in the business layer (Fig. 5). It shows the first activity of the smart sensor, which is the dolphin detection1, etc. These activities are assigned to their proper smart sensors, and these smart sensors are associated with the different data fusion servers and smart sensors that are required in the DFA [33]. Concerning the application layer, the behavioral elements are the compute coordinates function triggered by the resources reservation function, and so on. ArchiMate allows the association between layers, as shown in Fig. 5. For example, the Inform A Application Function aims to inform the fusion server A of detecting an object through the smart sensor1A. An extensive series of functions are associated with the technology layer (e.g. *send to*) to execute this application function. The *sendto* function forwards/sends a message of type SIP or Diameter from one node to another.

6.2 Compilation and Simulation

The design tool ArchiMO generates an XMI file to represent the graphical design. This helps to conduct the design model to other tools. We use the XMI file to input our self-developed domain-specific model compiler to generate the simulation code (Fig. 6). This hides the complexity of constructing simulation programs from the designer and saves considerable time in the development process. The code generator needs both the meta-model, including the abstract syntax of DSML for MO, and the input model generated from the design tool.

The XPAND template in (Fig. 6) contains the mapping rules between the model elements and their representations in NS-3 [15].

We have run the generated code in NS-3 (version 3.13), and the results of compilation and running results show the code is error-prone. Traces and logs (e.g., PCAP files) were generated to analyze the simulation outputs.

Figure 7 shows the system design architecture that NS-3 generates for the mentioned design model. NS-3 generated a hardware representation (nodes, interfaces, wires) for the design model elements. The blue-colored stream represents a message that is exchanged between two nodes at a given point in time. This confirms that the behavioral elements were mapped as expected.

We have used our approach in different application domains and network simulators (Video Conferencing system [14, 15], and MO context). The common design concept

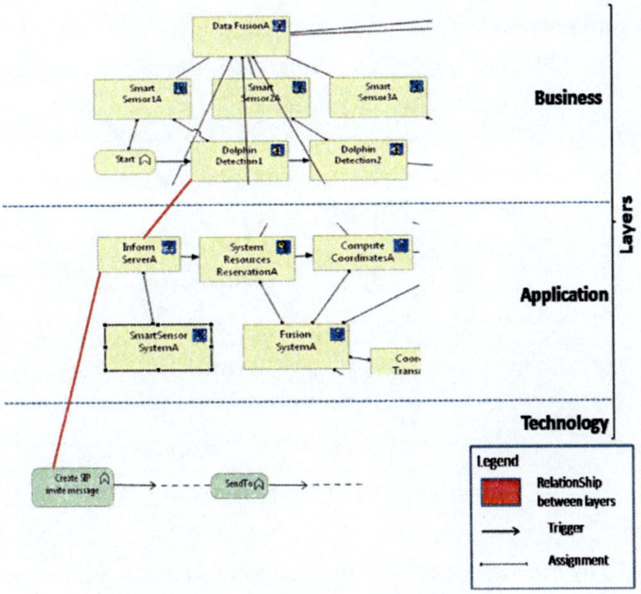

Fig. 5. Object localization underwater

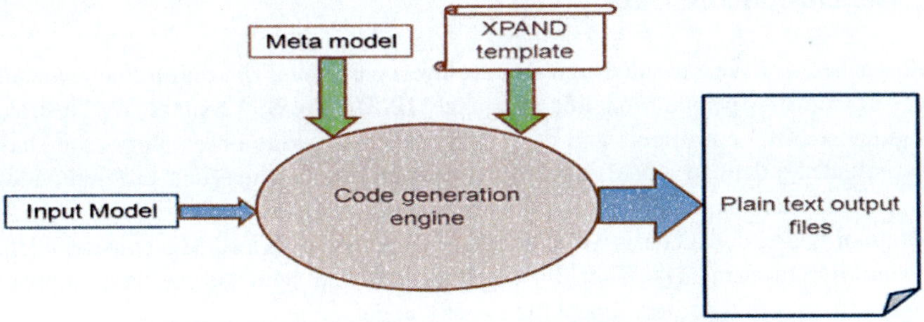

Fig. 6. The code generator workflow in XPAND language

between all these use cases is the underlying platform (IMS) that represents the Platform Specific Model (PSM) [30].

In other words, considering using one tool (e.g. NS-3), we could change the application domain relying on ArchiMate and our extensions (DSMLs) by fixing the underlying platform that is represented in the technology layer. This confirms that our extended design tool (ArchiMO) creates models that follow the same meta-model and domain-specific concepts/constraints.

Our testing approach demonstrates to tackle the issue presented in 2 by: (C5) provides the designer the ability to test and validate his MO created model on an executable platform included in the same framework where the designer creates this model well.

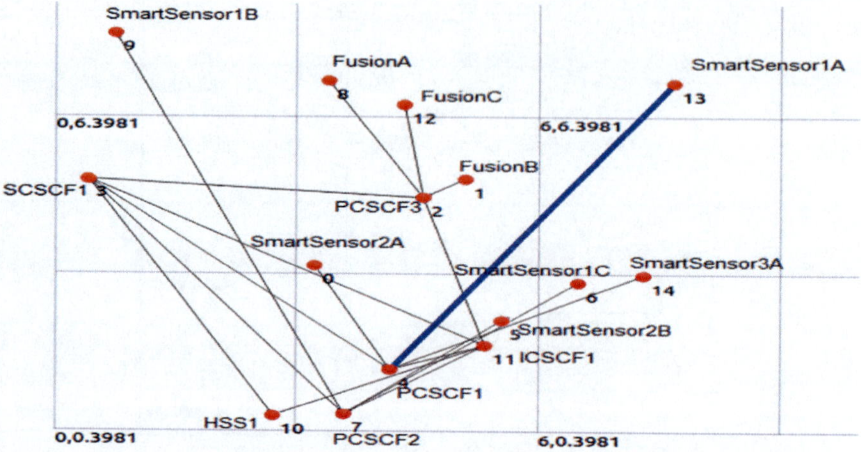

Fig. 7. Snapshot from the animation through NetAnim tool after running NS-3 simulation

This is provided by generating the simulation code of this model then executing it using NS-3.

7 Conclusion and Future Work

In this paper, we have presented high abstract level constraints. This constraints are extensions of Domain-Specific Modeling Language (DSML) for MO context. We illustrated the proposed MO constraints and design tool using a marine observatory case study. We presented a defined model for marine observatories showing their different views: business, application, and technology. These models are designed using our extended version of the ArchiMO design tool containing the new proposed MO constraints that rely on MDE fundamentals. Then, the resulting consistent model is simulated using the NS-3 network simulator to validate the system model.

We rely on a standard and open tool (Archi) that we extend through developing the modeling language and Java implementations. Our extended ArchiMO protects the designers from making design errors earlier than the other design activities and the code generation step. Another advantage is the extensibility of our proposed meta-model/tool. The developers may extend it and add new concepts and standards according to the progress in MO domain. ArchiMO provides the added MO concepts (e.g. Smart Sensor, Data Fusion, etc.) in different applications, activities, models, or instances.

ArchiMO reduces the time of the design activity by having the specific elements/concepts and constraints in the palette of this tool. Additionally, we conserve the standard constraints in the abstract syntax (meta-model) of ArchiMate since the newly added concepts inherit concepts from standard ArchiMate concepts.

On the other side, representing and meta-modeling the domain knowledge is a hard job that needs experience and a high level of accuracy, especially when setting the grammar of the DSML according to the meta-model constraints.

As perspectives, we will extend our meta-model to satisfy and cover the possible required operations, concepts and activities in the context of DSO/MO.

References

1. Zein, O.K., Champeau, J., Kerjean, D., Auffret, Y.: Smart sensor meta- model for deep sea observatory. In: OCEANS 2009 - EUROPE, pp. 1–6 (2009)
2. Yang, G., Dai, L., Wei, Z.: Challenges, threats, security issues and new trends of underwater wireless sensor networks. Sensors **18**(11), 3907 (2018)
3. Vihman, L., Kruusmaa, M., Raik, J.: Overview of fault tolerant techniques in underwater sensor networks. arXiv preprint arXiv:1910.00889 (2019)
4. Fattah, S., Gani, A., Ahmedy, I., Idris, M.Y.I., Targio Hashem, I.A.: A survey on underwater wireless sensor networks: requirements, taxonomy, recent advances, and open research challenges. Sensors **20**(18), 5393 (2020)
5. Schneider, J.-P., Champeau, J., Kerjean, D.: Domain-specific modelling applied to integration of smart sensors into an information system. In: International Conference on Enterprise Information Systems (ICEIS 2011), Lille, France (2011)
6. MeDON - Acoustic Data. https://keep.eu/projects/7945/Marine-e-Data-Observatory-Ne-EN/
7. Chiprianov, V., Alloush, I., Kermarrec, Y., Rouvrais, S.: Telecommunications service creation: towards extensions for enterprise architecture modeling languages. In: 6th International Conference on Software and Data Technologies (ICSOFT), Seville, Spain, vol. 1, pp. 23–29 (2011)
8. The Open Group, ArchiMate 1.0 Specification. http://www.opengroup.org/subjectareas/enterprise/archimate
9. Pérez-Medina, J.-L., Dupuy-Chessa, S., Front, A.: A survey of model driven engineering tools for user interface design. In: Winckler, M., Johnson, H., Palanque, P. (eds.) TAMODIA 2007. LNCS, vol. 4849, pp. 84–97. Springer, Heidelberg (2007). https://doi.org/10.1007/978-3-540-77222-4_8
10. Noran, O.: An analysis of the Zachman framework for enterprise architecture from the GERAM perspective. Ann. Rev. Control **27**(2), 163–183 (2003)
11. Quartel, D., Engelsmanb, W., Jonkersb, H., van Sinderenc, M.: A goal-oriented requirements modelling language for enterprise architecture. In: IEEE International Enterprise Distributed Object Computing Conference, 2009, EDOC 2009, pp. 3–13. IEEE (2009)
12. Chiprianov, V., Kermarrec, Y., Rouvrais, S.: Extending enterprise architecture modeling languages: application to telecommunications service creation. In: The 27th Symposium on Applied Computing, Trento, pp. 21–24. ACM (2012)
13. Eclipse Modeling FrameWork. http://www.eclipse.org/modeling/emf/
14. Alloush, I., Chiprianov, V., Kermarrec, Y., Rouvrais, S.: Linking telecom service high-level abstract models to simulators based on model transformations: the IMS case study. In: Szabó, R., Vidács, A. (eds.) EUNICE 2012. LNCS, vol. 7479, pp. 100–111. Springer, Heidelberg (2012). https://doi.org/10.1007/978-3-642-32808-4_10
15. Alloush, I., Kermarrec, Y., Rouvrais, S.: A generalized model transformation approach to link design models to network simulators: NS-3 case study. In: International Conference on Simulation and Modeling Methodologies, Technologies and Applications (SIMULTECH 2013), pp. 337–344. SciTePress Digital Library (2013)
16. Jazayeri, B., Schwichtenberg, S., Küster, J., Zimmermann, O., Engels, G.: Modeling and analyzing architectural diversity of open platforms. In: Dustdar, S., Yu, E., Salinesi, C., Rieu, D., Pant, V. (eds.) CAiSE 2020. LNCS, vol. 12127, pp. 36–53. Springer, Cham (2020). https://doi.org/10.1007/978-3-030-49435-3_3

17. Crnkovic, I., Sentilles, S., Feljan, A., Chaudron, M.: A classification framework for software component models. IEEE Trans. Softw. Eng. **37**, 593–615 (2011)
18. El Hachem, J., Pang, Z.Y., Chiprianov, V., Babar, A., Aniorte, P.: Model driven software security architecture of systems-of-systems. In: 2016 23rd Asia-Pacific Software Engineering Conference (APSEC), pp. 89– 96 (2016)
19. Medvidovic, N., Taylor, R.: A classification and comparison framework for software architecture description languages. IEEE Trans. Softw. Eng. **26**, 70–93 (2000)
20. Chiprianov, V.: Collaborative construction of telecommunications services. An enterprise architecture and model driven engineering method. Ph.D. thesis, Telecom Bretagne, France (2012)
21. Touraille, L., Traoré´e, M.K., Hill, D.R.C.: A model-driven software environment for modeling, simulation and analysis of complex systems. In: Proceedings of the 2011 Symposium on Theory of Modeling & Simulation: DEVS Integrative M&S Symposium. TMS-DEVS 2011, San Diego, CA, USA, pp. 229–237 (2011)
22. Achilleos, A., Yang, K., Georgalas, N.: Context modelling and a context-aware framework for pervasive service creation: a model-driven approach. Pervasive Mob. Comput. **6**(2), 281–296 (2010)
23. Bakker, J.-L., Jain, R.: Next generation service creation using XML scripting languages, vol. 4, pp. 2001–20074 (2002)
24. NATO Architecture Framework. https://www.nato.int/
25. Brumbulli, M., Gaudin, E., Teodorov, C.: Automatic verification of BPMN models. In: 10th European Congress on Embedded Real Time Software and Systems (ERTS 2020), Toulouse, France (2020). https://hal.archives-ouvertes.fr/hal-02441878
26. Sorribas, J., Barba, A., Trullols, E., Del Rio, J., Manuel, A., de la Muela, M.: Marine sensor networks and ocean observatories. A policy based management approach. In: 2008 The Third International Multi-Conference on Computing in the Global Information Tech nology, ICCGI 2008, pp. 143–147 (2008)
27. NEPTUNE - Ocean Networks Canada. https://www.oceannetworks.ca/
28. Zekai Demirezen, M.M.T., Bryant, B.R.: DSML design space analysis. In: UAB, Birmingham, AL 35294, USA (2011)
29. Cho, H., Gray, J., Syriani, E.: Creating visual domain-specific modeling languages from end-user demonstration. In: 2012 ICSE Workshop on Modeling in Software Engineering (MISE), pp. 22–28 (2012)
30. Kurtev, I., Bézivin, J., Jouault, F., Valduriez, P.: Model-based DSL frameworks. In: Companion to the 21st ACM SIGPLAN Symposium on Object-oriented Programming Systems, Languages, and Applications, OOPSLA 2006, pp. 602–616. ACM, New York (2006)
31. Aoun, C.: An enterprise architecture and model driven engineering based approach for sensor networks. Ph.D. thesis, ENSTA Bretagne (2018)
32. Sommerville, I.: Sofware Engineering, 9th edn. Pearson, London (2011)
33. Liggins, M.E., Hall, D., Llinas, J.: Multisensor Data Fusion, Theory and Practice, p. 849. Taylor & Francis Group, LLC (2009)

Internet of Things and Smart Cities

Internet of Vehicles and Intelligent Routing: A Survey-Based Study

Abeer Hassan(✉), Radwa Attia, and Rawya Rizk

Electrical Engineering Department, Port Said University, Port Fuad, Egypt
abeerhassan648@gmail.com, {radwa_yousef,r.rizk}@eng.psu.edu.eg

Abstract. Nowadays, vehicles are used daily by more and more people. With the rapid development in Intelligent Transportation Systems (ITS), modern vehicles are expected to be involved such as Vehicular Ad hoc NETworks (VANETs). VANETs is a type of Mobile Ad hoc NETwork (MANET) with a highly dynamic network structure; for dealing with the fast mobility of vehicles. However, the expansion of the network scale and the need for real-time information processing have led to turning real VANETs into an automotive Internet of Vehicle (IoV) for achieving an effective and smart future ITS. The main objective of this paper is to introduce a solid analysis of the most significant IoV routing proposals. A summary of features of existing approaches is presented. This survey concludes with further points for investigation.

Keywords: ITS · VANETs · MANETs · IoV · Optimization techniques · Routing

1 Introduction

Vehicular Ad hoc NETworks (VANETs) [1] are a special type of Mobile Ad hoc NETworks (MANETs) [2]. In VANETs, the vehicles are able to act as routing nodes that exchange information, which may be varied depending on different applications such as video and security notifications [3]. The communication in VANETs uses Dedicated Short-Range Communications (DSRC) and Wireless Access in Vehicular Environment (WAVE) standards [4], it consists of two main types; Vehicle-to-Vehicle (V2V) communication which uses On-Board Units (OBUs) [4] for taking place between vehicles, and Vehicle-to-Infrastructure (V2I) communication that allows the vehicle to communicate with the roadside infrastructure through equipped Road-Side Units (RSUs) [4]. However, numerous challenges impact the achievements of VANETs such as [5]; dynamic topology, high mobility, less scalability, signal losses, limited bandwidth, connection, and routing; that in turn led to a growing demand for turning real VANETs into the Internet of Vehicles (IoV) [6] for achieving an effective and smart future transportation system. The structure of the IoV network is illustrated in Fig. 1, IoV can be seen as a group of VANETs cooperating to extend the VANETs scale, structure, and applications; to put more emphasis on information interaction among vehicles, humans, and roadside units [6].

© The Author(s), under exclusive license to Springer Nature Switzerland AG 2022
A. E. Hassanien et al. (Eds.): AMLTA 2022, LNDECT 113, pp. 517–531, 2022.
https://doi.org/10.1007/978-3-031-03918-8_43

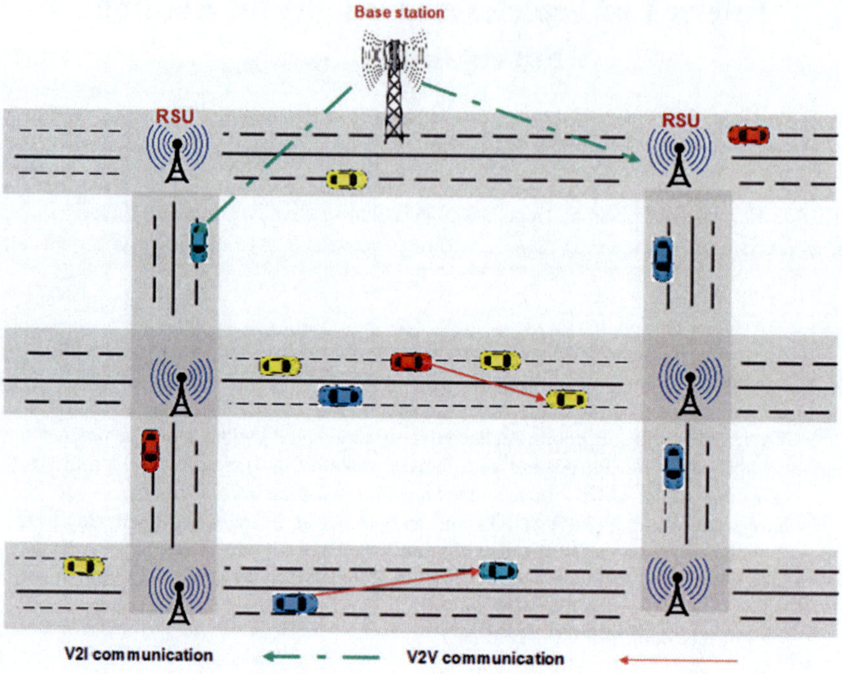

Fig. 1. Structure of IoV network.

Routing is a crucial issue in IoV, due to the high mobility and changes in the network topology; new types of routing protocols are required to be performed by mobile and unreliable vehicles. The traditional protocols for IoV can be grouped into four general categories [7]; Topology-based, position-based, multicast-based, and broadcast-based routing. However, these approaches are less efficient and less realistic for IoV in terms of Packet Delivery Ratio (PDR), End-to-End Delay (E2ED), Normalized Routing Overhead (NRO), and routing Quality-of-Service (QoS) metrics [8]. These limitations lead to poor packet transmissions quality and in turn, result in vehicle breakdown services. Alternative routing algorithms [8], based on a wide range of bio-inspired processes such as evolution, swarm intelligence, and the immune system, have emerged over the past decade to ensure efficient routing and improve IoV services.

This paper provides a survey of using bio-inspired routing protocols to achieve efficient routing in IoV. The remainder of the paper is organized as follows; Sect. 2 describes the architecture and main characteristics of IoV. An overview of the functionality for different communication proposals in IoV is provided in Sect. 3. Section 4 introduces a review of heuristic bio-inspired routing protocols and their evaluation approaches. Finally, Sect. 5 concludes the paper and defines topics for future research.

2 Motivation from VANETs to IoV: Architecture and Characteristics

VANETs lack the processing power to handle large amounts of data; they are best suited for short-term applications or small-scale services [5]. Nowadays, vehicles need to interact with infrastructure, the Internet, and people; to deliver a variety of modern applications, which led to the motivation from VANETs to the IoV, which follows the Internet of Things (IoT) paradigm [9].

2.1 Architecture of IoV

The IoV architecture is discussed in many proposals [6]. As shown in Fig. 2, it can be basically divided into three levels; the perception layer, the network layer, and the application layer.

- **Perception layer:** This is the first level that contains all the sensors within the vehicle to gather environmental data and detect specific events of interest such as driving patterns and vehicle situations, environmental conditions, among others.
- **Network layer:** The communication layer which supports different wireless communication modes such as V2V, V2I, and Vehicle-to-Sensor (V2S) to ensure seamless connectivity to emerging networks such as GSM, Wi-Fi, LTE, Bluetooth, and IEEE 802.15 [10].
- **Application layer:** It is responsible for storing, analyzing, processing, and making decisions among others regarding various risk situations such as traffic congestion, and dangerous road conditions.

2.2 Characteristics of IoV

- Many aspects of automotive networking are influencing the development of IoV technology. Some features may be difficult to design, while others may offer benefits [7].
- **Highly Dynamic Topology:** Vehicles may move at high speeds, forcing the network topology to change frequently.
- **Variable Network Density:** Mostly depends on traffic density, which can be very high in the cities center, or very low, as in rural areas.
- **Geographical Communication:** In IoV, nodes are defined by the geographical position where packets need to be forwarded; in safe-driving applications.
- **Predictable Mobility:** Vehicles do not move randomly. It is confined by road architecture, resulting in mobility predictability.
- **Sufficient Energy and Storage**: Vehicles are not energy-constrained; they are equipped with ample battery resources, computing, and storage capability.
- **Different Types of Communication**: Vehicular networks are separated into two environments; highways and urban. Highways have a relatively simple environment, whereas urban environments are far more complicated, with obstructions and no direct dimensional movement.

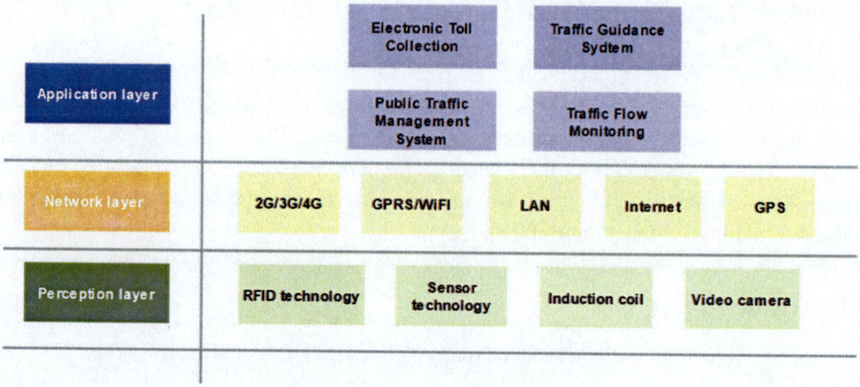

Fig. 2. Architecture layers of IoV.

3 Routing Protocols in IoV

Inter-vehicular communication protocols in VANETs play an essential role for IoV; because they enable various levels of interaction among vehicles, humans, and roadside units. They can provide alternate routes efficiently and quickly if a problem arises with the current route. Routing is a key study area in IoV, which is the network-layer protocol that aims to improve throughput while minimizing packet loss and overhead. The main IoV communication protocols are introduced in Table 1.

3.1 Topology-Based Routing

The main principle of topology-based routing considers topological links between nodes along the source-destination path in order to determine routes. This protocol is categorized to [11]:

Proactive (Table-Driven). Aims to find paths in advance for all source and destination pairs in each node and store them in the form of tables, so at any time, the path is ready to be used without any delay for the discovery process. However, with rapid topology changes in VANETs, proactive routing may generate many control packets and hence cause more overhead [12].

Reactive (On-demand). Necessary paths are discovered only when they are required, in order to minimize routing overhead [12].

3.2 Position-Based Routing

All nodes recognize their own location and neighbor nodes location using pointing devices like a Global Positioning System (GPS) [13]. The data packets are forwarded to the destination without knowledge of the network topology, where a relay node is chosen as the closest to the destination [14].

3.3 Multicast Routing Protocols

Cluster-Based Routing. In order to ensure scalability in the VANETs, cluster-based routing is used. This category is based on creating a virtual partial network infrastructure called a cluster; the cluster has one cluster head that is responsible for intra and inter-cluster control management. After clustering and head election, the source node is decided to select the vehicle that plays a gateway role. In [15], various clustering algorithms are analyzed in view of VANETs.

Geo-cast Routing. It is a type of location-based multicast routing [11] that aims to send a packet from a source node to all other nodes within a defined geographical region; known as the Zone of Relevance (ZoR). Geo-cast routing in urban VANETs routing protocol [16] is a GPS-based inter-vehicle communication protocol used for alarm message dissemination among vehicles in a highway in risk situations.

3.4 Broadcast Routing

In this protocol, each node can broadcast messages to all other nodes. Broadcast is a common routing mechanism in IoV for sharing traffic, weather, emergency, and road condition information among cars and other applications [11].

Table 1. Comparison of various IoV protocols

Broadcast protocols	Geo-cast protocols	Cluster based protocols	Position based protocols	Reactive protocols	Proactive protocols	Protocols
Flooding routing	Zone routing	Intra- and inter-cluster forwarding	Based on destination's position	On demand route discovery	Maintain routes to all destinations	Forwarding method
Highway	Highway	Urban	Highway	Urban	Urban	Environment
– High scalability	– Low overhead	– Reliable packet transmission	– Good performance and high stability	– Beaconless – Link failure maintenance	– Consume lower bandwidth and latency	Advantages
– High delay especially in large scale networks	– High delay	– High overhead	– The need of GPS and in turn deadlock	– High latency	– High overhead	Disadvantages
BROADCOMM [17]	Streetcast [16]	COIN, HCB [17]	GPSR, GSR [14]	DSR, AODV [12]	OLSR [12]	Examples

4 Heuristic Bio-inspired Routing Protocols Literature

Routing is a key study area in IoV [18]. This main challenge requires an optimization technique such as bio-inspired algorithms; to facilitate automatic tuning of parameters configuration in the IoV environment. Bio-inspired methods are more efficient for large-scale IoV because; the behavior of species during the discovery of the food is very similar to the identification of IoV communication routes [8, 19], also, because of the low complexity of bio-inspired protocols in performing the computational problems. Bio-inspired algorithms are typically divided into three main classes, as shown in Fig. 3. The IoV routing optimization problems give rise to a number of challenges [8]:

- **Network Scalability:** With the large network size and the frequent IoV topology changes, the performance is quickly degraded, especially when more vehicles want to communicate at the same time.
- **Computational Complexity:** In such large-scale networks, computational costs like; run time and the number of resources utilized to solve routing problems, are very high, hence bio-inspired approaches can be successfully used to provide optimal pathways with low complexity.
- **Adaptability:** Self-organizing and making non-user-based routing is an important challenge. Bio-inspired approaches enable the deployment of adaptive routing solutions better than traditional techniques.
- **Network Robustness:** This is the ability to provide secure paths between source and destination against network failures, disruptions, and attack threats.
- **Quality of Services:** QoS implies ensuring the successful delivery of transmitted messages with a minimum number of dropped packets, with high bandwidth.

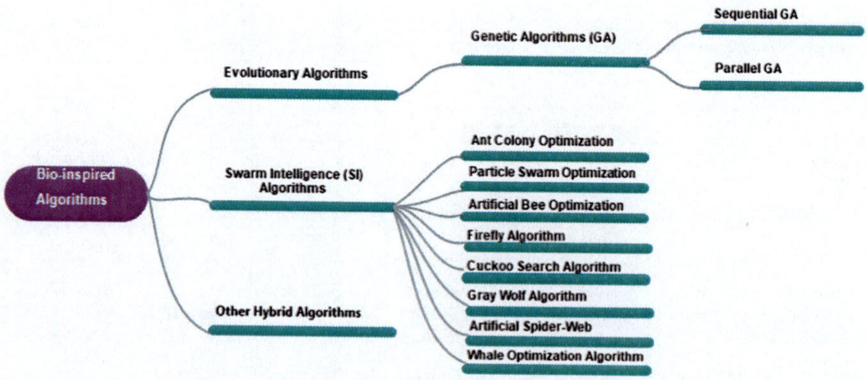

Fig. 3. Taxonomy of bio-inspired algorithms for IoV.

4.1 Evolutionary Algorithms

The evolution of species inspires Genetic Algorithm (GA) [19] that is an optimization approach based on a population of chromosomes to generate another population during iteration. GAs can be further investigated in wireless networks to find the best routing strategy; however, it has some disadvantages such as; the computation time and the excessive growth of its individuals [8]. In [20], the author presented the Adaptive Weighted Clustering Protocol (AWCP), as an optimized clustering protocol that takes into account the highway number, vehicle direction, position, speed, and the number of neighboring cars to improve network topology stability. In [21], the author proposed a Software-Defined network-based Geographic Routing (SDGR) protocol, where an Improved Genetic Method (IGM) is used to create an optimal forwarding path that identifies the shortest path based on shorter road length and higher vehicles density.

GA can also be applied in a parallel manner that allows reaching high-quality results in an acceptable execution time. In [22], the author proposed an energy-aware OLSR routing to reduce the power consumption of OLSR [12] protocol in IoV by using a parallel GA algorithm, this protocol makes OLSR compatible with large realistic IoV scenarios, experimental analysis proved substantial reductions in power consumption and significantly enhance the network overload, while the modified protocol only suffered abounded degradation in the packet delivery metric. The drawback of using parallel GA is unaffectedness to cover huge search space frequently, even with the help of local search compared to other hybrid approaches.

4.2 Swarm Intelligence

Ant Colony Optimization (ACO). It is inspired by the foraging behavior of ants. In ACO, artificial ants communicate with each other indirectly through the pheromones that deposit in their path, once a subsequent ant follows the path, it lays down new pheromone over the path, then depending on the amount of pheromone previously deposited, the best paths are selected and the others are ignored [19]. Many modifications are made on ACO to solve the route optimization for existing dynamic routing problems [23]. In [24], the author presented an adaptive QoS-based routing protocol for IoV (ARQV) with ACO, which can determine the QoS route intersection through which data packets can be transferred from source to destination on a periodic basis. However, this protocol misses two important features self-organization and the ability to adapt to the failure of RSUs. In [25], an ACO-based delay-sensitive routing protocol is proposed which utilizes pheromone information of transmission delay and heuristic information of vehicles. The authors in [26] introduced a reactive route setup technique, which relies on ACO to choose the shortest path from source to destination, based on connectivity, delay, and packet delivery ratio, However, increasing network size can increase the network congestion, which generates a very important overhead.

Particle Swarm Optimization (PSO) is inspired by the social behavior of bird flocking. PSO is a swarm of particles that fly around in a multidimensional search space. During flying, each particle adjusts its movement based on its best prior position as well as the swarm's best previous position. Thus, the PSO system combines local and global search methods, attempting to balance exploration and exploitation [27]. Many researchers have used the ACO and PSO algorithms for local and global optimization; respectively [23]. In [28], the author proposed PSO based Routing method for VANETs (PSOR) that considers several objective criteria of vehicles' location, distance, and vehicle speed to determine the next forwarding vehicle. But the density parameter is not considered. The author in [29] presents clustering V2V routing based on PSO in IoV, which consists of three components; cluster creation, route particle coding, and routing within or between clusters. This protocol can improve the stability of the network; however, it has the limitation of tuning network stability with delay.

Artificial Bee Colony (ABC). It is an optimization algorithm that mimics the food foraging behavior of bee colonies. Some bees (called scouts) explore the region in search of food, if they are discovered; they return to the hive performing a waggle dance to inform their mates of their finding, some bees (called foragers) are recruited to exploit this discovery [8]. ABC algorithm uses few control parameters, and has fast convergence; so many researchers use it in optimization problems for IoV [30–32]. In [33], a QoS-based routing protocol is proposed; it uses fuzzy logic to determine a feasible path from multiple possible ones discovered by the ABC algorithm, where the path must meet requirements such as bandwidth, latency, jitter, and link expiration time, however this protocol may overhead the network with control packets.

4.3 Other Bio-inspired Approaches

Firefly Algorithm (FA). It is an optimization algorithm based on the flashing behavior of fireflies, which acts as a signal system to attract other fireflies. It is able to solve multi-dimensional problems with fast convergence, and it can be used for both global and local search problems [27]. In [34], the author proposed a Reputation-based Weighted Clustering Protocol (RWCP) for stabilizing the IoV topology using FA algorithm, taking into account the vehicles' direction, position, velocity, and other parameters, however, comparison with IoV clustering is not considered. In [35], the author concentrates on the QoS multicast routing problem by using FA with the Levy distribution algorithm to prevent the local optima convergence.

Cuckoo Search Algorithm (CS). It is a meta-heuristic algorithm inspired by the cuckoo bird's method of existence; these birds are known as "Brood parasites". Cuckoos eject one of the host bird's eggs and lay eggs that more closely resemble the host's eggs. If the host bird identifies the different eggs, it either throws that eggs away from its nest or leaves its nest and builds a new one. The advantage of CS over other optimization algorithms is its simplicity of using fewer control parameters [27]. In [36], the authors proposed a clustering technique that uses velocity and distance of vehicles to create a stable cluster structure, CS algorithm is triggered to select the super cluster-head while achieving optimum distance, minimum delay, and high network lifetime. In [37], the author developed a NetCLEVER protocol to support the intelligent routing of the data packets using broadcast communication in order to avoid the broadcast storm problem, in this approach, CS technique is used by considering the impact of a road intersection and traffic lights on link stability.

Gray Wolf Algorithm (GWO). It is a population-based algorithm that uses the grey wolf social intelligence to pick the best prey for the hunt. In particular, the three best candidate solutions α (the first one), β (the second one), γ (the third one) are randomly generated respecting the constraints, after this; other possible solutions are generated according to these three solutions and adjusted their positions accordingly [38]. In [39], a GWO clustering protocol is proposed to solve the vehicle clustering problem, the proposed algorithm reduces the transmission cost of the entire network by effectively reducing the number of clusters, but it did not consider the bandwidth metric that affects the network efficiency. In [40], a social-based routing scheme is proposed based on the GWO algorithm to diminish the social network services and enhance the throughput.

Whale Optimization Algorithm (WOA). It is inspired by the bubble net feeding behavior of humpback whales in the ocean. In [41], authors proposed an enhanced WOA, they use Adaptive Weighted Clustering Protocol (AWCP) for grouping the network topology. In [42], Optimal Adaptive Data Dissemination Protocol (OADDP) for vehicle road safety is proposed; it uses the WOA for clustering and predictor-based decision-making algorithm for control overhead messages reduction.

4.4 Hybrid Approaches

A routing algorithm using the ACO and Evolving Graph (EG) model is proposed in [43], ACO is used to discover the ideal route, but has stagnation behavior and slow convergence in the high mobility network, thus the EG model is used to find the optimal route with better QoS support. In [44], the author combined GA and ACO algorithms to stabilize the cluster in IoV, they use GA for cluster formation and ACO to optimize the number of clusters. However, this approach uses static control weights that do not give more analysis results for different communication settings. In [45], a Modified Cognitive Tree Routing Protocol (MCTRP) is proposed, it incorporates a routing protocol with the cognitive radio technology for efficient channel assignment, it promises lower overheads with effective channel utilization than the OLSR [12]; however, this type of tree-based solutions is no more appropriate for IoV networks due to traffic changes. In [46], an enhanced hybrid routing protocol is proposed, it employs fuzzy logic and CS approaches to find the most stable path between the source and the destination taking into consideration, the route lifetime, route reliability, and average available buffer.

Considering the above-related works, most of the proposals have a number of drawbacks as shown in Table 2. In [32], Advanced Greedy Hybrid Bio-Inspired (AGHBI) routing protocol is proposed to improve the performance of IoV. AGHBI makes use of a modified hybrid routing scheme with the help of ABC to select the highest QoS route and keep the route with minimum overflow.

Table 2. Evolutionary & Swarm intelligence algorithms for IoV routing (Satisfied: ✔, Un-satisfied: ⊠, Un-defined: N/D)

Ref.	Implementation approach	Routing protocols	Mobility model	Simulated area (Km^2)	Performance Metrics		
					Delay	PDR	Overhead
[20]	Use GA to balance latency and cost in software-defined IoV	Geographic	Urban	1	✔	✔	⊠
[21]	Use GA to enhance the network topology stability	Clustering	Highway	4	✔	N/D	✔
[22]	Energy-aware OLSR routing using a PGA algorithm	Proactive	Urban	360	N/D	N/D	✔
[24]	Use ACO with intersection routing to satisfy QoS requirements	Intersection	Urban	5	N/D	✔	✔
[25]	Use ACO to utilize transmission delay and heuristic information of vehicles	Hybrid	Highway	1000 m $*$ 25 m	✔	✔	✔
[26]	Use reactive ACO method to establish the stable route between intersections	Reactive & Intersection	Urban	5 × 5	✔	✔	⊠
[28]	PSO tuned OLSR protocol	Proactive		1.4 × 1.2	✔	✔	✔

(*continued*)

Table 2. *(continued)*

Ref.	Implementation approach	Routing protocols	Mobility model	Simulated area (Km^2)	Performance Metrics		
					Delay	PDR	Overhead
[29]	Efficient Clustering V2V Routing Based on PSO	Clustering	Urban	360 m^2	✔	✔	✔
[32]	Use ABC with a greedy method to select the highest QoS route with minimum overflow	Geographic & Intersection	Urban	4000 × 4000 m^2	✔	✔	☒
[33]	Use ABC and fuzzy logic approaches for discovering routes complying with QoS criteria	Unicast	Highway	1	✔	✔	✔
[34]	RWCP that optimize cluster heads using FFA algorithm	Clustering	Urban	750 m^2	✔	✔	☒
[35]	QoS multicast routing problem using FFA with the Levy distribution	Multicast	N/D	N/D	✔	✔	☒
[36]	Clustering technique to create stable cluster structure using CS	Clustering	Highway	3 × 4	✔	N/D	N/D
[37]	Adaptive intelligent safety message routing protocol based on CS algorithm	Broadcast	Highway	800 m × 750 m	✔	N/D	N/D
[39]	A GWO clustering protocol that solves the clustering problem in heterogeneous IoV	Clustering	Highway	400 × 400 m	N/D	N/D	✔

(continued)

Table 2. (*continued*)

Ref.	Implementation approach	Routing protocols	Mobility model	Simulated area (Km^2)	Performance Metrics		
					Delay	PDR	Overhead
[40]	Social-based routing scheme for IoV based on GWO algorithm	Unicast	Random	1	✔	✔	✔
[41]	AWCP protocol based on an enhanced whale optimization algorithm	Clustering	Highway	1 × 1	✔	✔	✔
[42]	Uses the WOA for clustering and predictor-based (PDM) algorithm for control overhead messages reduction	Clustering	Urban	3 × 3	✔	✔	✔
[43]	Use ACO and EG to discover the ideal route in IoV	Unicast	Highway	10	✔	✔	✔
[44]	Use ACO and GA to improve the performance in IoV	Clustering	N/D	N/D	✔	✔	✔
[45]	A modified cognitive tree routing protocol based on genetic WOA	Unicast	N/D	1	✔	N/D	✔
[46]	Intelligently employed fuzzy and cuckoo approach for finding the most stable path	Hybrid	Urban	7 × 7	✔	✔	☒

5 Conclusion

The development of an intelligent transportation system that supports both safe driving and comfort application has received much attention from the automotive industry and government agencies. IoV offers communication services among vehicles or with roadside infrastructure. However, routing in IoV gives rise to a number of challenges because of frequent network topology changes. This paper provides a survey on the most significant IoV routing proposals by describing their functionality, characteristics, and drawbacks. The survey shows that bio-inspired methods are more efficient for large-scale IoV, which disseminate data packets with low complexity and improve QoS routing performances metrics such as delay, packet delivery ratio, routing overhead, and bandwidth. On the horizon, there is still a lot of work to be done. From a technical point of view, achieving IoV still presents many challenges to be solved related to devices, protocols, applications, and services. The investigations concerning the routing and handoff decision are still below expectations.

References

1. Silva, F.A., Boukerche, A., Silva, T.R., Ruiz, L.B., Cerqueira, E., Loureiro, A.A.: Vehicular networks: a new challenge for content-delivery-based applications. ACM Comput. Surv. **49**(1), 1–29 (2016)
2. Conti, M., Giordano, S.: Mobile ad hoc networking: milestones, challenges, and new research directions. IEEE Commun. Mag. **52**(1), 85–96 (2014)
3. Al-Sultan, S., Al-Doori, M.M.: A comprehensive survey on vehicular Ad Hoc network. J. Netw. Comput. Appl. **37**, 380–392 (2014)
4. Wu, W., Yang, Z., Li, K.: Internet of vehicles and applications. In: Internet of Things, pp. 299–317 (2016)
5. Karagiannis, G., et al.: Vehicular networking: a survey and tutorial on requirements, architectures, challenges, standards and solutions. IEEE Commun. Surv. Tutorials **13**(4), 584–616 (2011)
6. Contreras-Castillo, J., Zeadally, S., Guerrero-Ibañez, J.A.: Internet of vehicles: architecture, protocols, and security. IEEE Internet Things J. **5**(5), 3701–3709 (2017)
7. Senouci, O., Aliouat, Z., Harous, S.: A review of routing protocols in internet of vehicles and their challenges. Sens. Rev. **39**, 58–70 (2019)
8. Bitam, S., Mellouk, A.: Bio-inspired Routing Protocols for Vehicular Ad Hoc Networks. Wiley, New York (2014)
9. Nanda, A., Puthal, D., Rodrigues, J.J., Kozlov, S.A.: Internet of autonomous vehicles communications security: overview, issues, and directions. IEEE Wireless Commun. **26**(4), 60–65 (2019)
10. Abo Hashish, S.M., Rizk, R.Y., Zaki, F.W.: Energy efficiency optimization for relay deployment in multi-user LTE-advanced networks. Wireless Pers. Commun. **108**(1), 297–323 (2019)
11. Sharef, B.T., Alsaqour, R.A., Ismail, M.: Vehicular communication ad hoc routing protocols: a survey. J. Netw. Comput. Appl. **40**, 363–396 (2014)
12. Ali, T.E., al Dulaimi, L.A.K., Majeed, Y.E.: Review and performance comparison of VANET protocols: AODV, DSR, OLSR, DYMO, DSDV & ZRP. In: 2016 Al-Sadeq International Conference on Multidisciplinary in IT and Communication Science and Applications (AIC-MITCSA), pp. 1–6 (2016)

13. Kumar, S., Verma, A.K.: Position based routing protocols in VANET: a survey. Wireless Pers. Commun. **83**(4), 2747–2772 (2015)
14. Yang, X., Li, M., Qian, Z., Di, T.: Improvement of GPSR protocol in vehicular ad hoc network. IEEE Access **6**, 39515–39524 (2018)
15. Bali, R.S., Kumar, N., Rodrigues, J.J.: Clustering in vehicular ad hoc networks: taxonomy, challenges and solutions. Veh. Commun. **1**(3), 134–152 (2014)
16. Yi, C.W., Chuang, Y.T., Yeh, H.H., Tseng, Y.C., Liu, P.C.: Streetcast: an urban broadcast protocol for vehicular ad-hoc networks. In: 2010 IEEE 71st Vehicular Technology Conference, pp. 1–5. IEEE, Taiwan (2010)
17. Devangavi, A.D., Gupta, R.: Routing protocols in VANET—a survey. In: 2017 International Conference on Smart Technologies for Smart Nation (SmartTechCon), pp. 163–167. IEEE, India (2017)
18. Cheng, J., Cheng, J., Zhou, M., Liu, F., Gao, S., Liu, C.: Routing in internet of vehicles: a review. IEEE Trans. Intell. Transp. Syst. **16**(5), 2339–2352 (2015)
19. Hajlaoui, R., Guyennet, H., Moulahi, T.: A survey on heuristic-based routing methods in vehicular ad-hoc network: technical challenges and future trends. IEEE Sens. J. **16**(17), 6782–6792 (2016)
20. Hadded, M., Zagrouba, R., Laouiti, A., Muhlethaler, P., Saidane, L.A.: A multi-objective genetic algorithm-based adaptive weighted clustering protocol in VANET. In: 2015 IEEE Congress on Evolutionary Computation, pp. 994–1002. IEEE, Japan (2015)
21. Lin, C.C., Chin, H.H., Chen, W.B.: Balancing latency and cost in software-defined vehicular networks using genetic algorithm. J. Netw. Comput. Appl. **116**, 35–41 (2018)
22. Toutouh, J., Nesmachnow, S., Alba, E.: Fast energy-aware OLSR routing in VANETs by means of a parallel evolutionary algorithm. Clust. Comput. **16**(3), 435–450 (2013)
23. Jindal, V., Bedi, P.: An improved hybrid ant particle optimization (IHAPO) algorithm for reducing travel time in VANETs. Appl. Soft Comput. **64**, 526–535 (2018)
24. Li, G., Boukhatem, L., Wu, J.: Adaptive quality-of-service-based routing for vehicular ad hoc networks with ant colony optimization. IEEE Trans. Veh. Technol. **66**(4), 3249–3264 (2016)
25. Ding, Z., Ren, P., Du, Q.: Ant colony optimization based delay-sensitive routing protocol in vehicular ad hoc networks. In: Li, B., Yang, M., Yuan, H., Yan, Z. (eds.) IoTaaS 2018. LNICSSITE, vol. 271, pp. 138–148. Springer, Cham (2019). https://doi.org/10.1007/978-3-030-14657-3_15
26. Srivastava, A., Prakash, A., Tripathi, R.: An adaptive intersection selection mechanism using Ant Colony optimization for efficient data dissemination in urban VANET. Peer-to-Peer Netw. Appl. **13**(5), 1375–1393 (2020). https://doi.org/10.1007/s12083-020-00892-8
27. Masegosa, A.D., Osaba, E., Angarita-Zapata, J.S., Laña, I., Ser, J.D.: Nature-inspired meta-heuristics for optimizing information dissemination in vehicular networks. In: Proceedings of the Genetic and Evolutionary Computation Conference Companion, pp. 1312–1320 (2019)
28. Yelure, B., Sonavane, S.: Particle swarm optimization based routing method for vehicular ad-hoc network. In: 2020 International Conference on Communication and Signal Processing (ICCSP), pp. 1573–1578. IEEE, India (2020)
29. Bao, X., Li, H., Zhao, G., Chang, L., Zhou, J., Li, Y.: Efficient clustering V2V routing based on PSO in VANETs. Measurement **152**, 107306 (2020)
30. Hashem, W., Nashaat, H., Rizk, R.: Honey bee based load balancing in cloud computing. KSII Trans. Internet Inf. Syst. (TIIS) **11**(12), 5694–5711 (2017)
31. Gamal, M., Rizk, R., Mahdi, H., Elnaghi, B.E.: Osmotic bio-inspired load balancing algorithm in cloud computing. IEEE Access **7**, 42735–42744 (2019)
32. Attia, R., Hassaan, A., Rizk, R.: Advanced greedy hybrid bio-inspired routing protocol to improve IoV. IEEE Access **9**, 131260–131272 (2021)

33. Fekair, M.E.A., Lakas, A., Korichi, A., Lagraa, N.: An efficient fuzzy logic-based and bio-inspired QoS-compliant routing scheme for VANET. Int. J. Embedded Syst. **11**(1), 46–59 (2019)
34. Joshua, C.J., Duraisamy, R., Varadarajan, V.: A reputation based weighted clustering protocol in VANET: a multi-objective firefly approach. Mobile Netw. Appl. **24**(4), 1199–1209 (2019)
35. Elhoseny, M.: Intelligent firefly-based algorithm with Levy distribution (FF-L) for multicast routing in vehicular communications. Expert Syst. Appl. **140**, 112889 (2020)
36. Malathi, A., Sreenath, N.: An efficient clustering algorithm for VANET. Int. J. Appl. Eng. Res. **12**(9), 2000–2005 (2017)
37. Purkait, R., Tripathi, S.: Network condition and application-based data adaptive intelligent message routing in vehicular network. Int. J. Commun. Syst. **31**(4), e3483 (2018)
38. Farshin, A., Sharifian, S.: A chaotic grey wolf controller allocator for software defined mobile network (SDMN) for 5th generation of cloud-based cellular systems (5G). Comput. Commun. **108**, 94–109 (2017)
39. Fahad, M., et al.: Grey wolf optimization based clustering algorithm for vehicular ad-hoc networks. Comput. Electr. Eng. **70**, 853–870 (2018)
40. Sharma, S., Kad, S.: Enhancing social based routing approach using grey Wolf optimization in vehicular ADHOC networks. Int. J. Comput. Appl. **975**, 0975–8887 (2019)
41. Kittusamy, V., Elhoseny, M., Kathiresan, S.: An enhanced whale optimization algorithm for vehicular communication networks. Int. J. Commun. Syst. e3953 (2019)
42. Dwivedy, B., Bhola, A.K.: Improved data dissemination protocol for VANET using whale optimization algorithm. In: Solanki, V.K., Hoang, M.K., Lu, Z., Pattnaik, P.K. (eds.) Intelligent Computing in Engineering. AISC, vol. 1125, pp. 153–161. Springer, Singapore (2020). https://doi.org/10.1007/978-981-15-2780-7_19
43. Wang, X., Liu, C., Wang, Y., Huang, C.: Application of ant colony optimized routing algorithm based on evolving graph model in VANETs. In: 2014 International Symposium on Wireless Personal Multimedia Communications (WPMC), pp. 265–270. IEEE, NSW (2014)
44. Goswami, V., Verma, S.K., Singh, V.: A novel hybrid GA-ACO based clustering algorithm for VANET. In: 2017 3rd International Conference on Advances in Computing, Communication & Automation (ICACCA), pp. 1–6. IEEE, India (2017)
45. Mohanakrishnan, U., Ramakrishnan, B.: MCTRP: an energy efficient tree routing protocol for vehicular ad hoc network using genetic whale optimization algorithm. Wirel. Pers. Commun. **110**(1), 185–206 (2020)
46. Yahiabadi, S.R., Barekatain, B., Raahemifar, K.: TIHOO: an enhanced hybrid routing protocol in vehicular ad-hoc networks. EURASIP J. Wirel. Commun. Netw. **2019**, 1–19 (2019)

Location Privacy-Preserving of Vehicular Ad-Hoc Network in Smart Cities

Yasmin Alkady[1(✉)] and Rawya Rizk[2]

[1] Faculty of Information Technology and Computer Sciences, Sinai University, Ismailia, Egypt
yassmin.hosny@su.edu.eg

[2] Electrical Engineering Department, Port Said University, Port Said, Egypt
r.rizk@eng.psu.edu.eg

Abstract. With rapid growth of technology involved in smart city networks, Nowadays, A navigation in Vehicular Ad-hoc Network (VANET) with the help of Location-Based Services (LBSs) is used to provide communications between nearby vehicles and between vehicles and fixed infrastructure on the roadside and to provide road safety, driving comfort, and infotainment. However, the privacy issues of using LBS in VANET are still challenging today. However, many LBSs run-on third-party cloud infrastructures. So, privacy is still in a huge problem. In this paper, an efficient LBS Query scheme called Privacy Preserving of Fully Homomorphic Encryption over Advanced Encryption Standard is proposed. In an encrypted manner, the LBSP's data are outsourced to the cloud server and a registered user can get accurate LBS query results without revealing the information of location to the LBSP and cloud server provider. A real traffic scenario in Ismailia city, Egypt has been applied in the simulation. The results showed that the proposed scheme is better than the existing schemes in terms of response time, accuracy rate, and processing time.

Keywords: Advanced Encryption Standard · Cloud computing · Fully Homomorphic Encryption · Location Based Service · Privacy Preserving · Vehicular Ad-hoc Network

1 Introduction

The world is focusing on evolution of smart cities and new techniques that emerge from innovations in information technology. Security and expectations of privacy are the main challenges. Social systems and public venues are on their path to full connectivity known as the Internet of Things (IoT) as an example Vehicular Ad-hoc Network (VANET) system. VANET system is very important nowadays to reduce significant time and gas wasted every day as a result of traffic congestion and slow traffic [1]. Much attention is paid by governments to better manage traffic using VANET system.

VANET allows cooperative driving, approaching vehicle data to increase vehicular safety, reducing congestion in roads, and providing access to Location Based Services (LBSs) to request certain query of destination [2]. So that in VANET system, there is a small hardware device installed on a vehicle. The device can determine current

© The Author(s), under exclusive license to Springer Nature Switzerland AG 2022
A. E. Hassanien et al. (Eds.): AMLTA 2022, LNDECT 113, pp. 532–543, 2022.
https://doi.org/10.1007/978-3-031-03918-8_44

location of vehicle and then find the shortest path to the required destination based on receiving Global Positioning System (GPS) signals [3]. VANET system has some entities as vehicle, On-Board Unit (OBU), Road Side Units (RSUs), Registration authority (RA), LBS Provider (LBSP), and Cloud Server Provider (CSP). Revealing the physical location of vehicles by LBSP to introduce a service can present a violation for user's location privacy. It is a great concern over how to protect the user's privacy from the ability of data sniffing or revealing.

Vulnerabilities are the main problem confronting VANET system. Location tracking of vehicles increases these vulnerabilities as using location information of vehicles fraudulently because most crimes occur due to misuse of location information [4, 5]. Un-trusted RSU leads to disaster, because it is the core that supports the way of communication between any entities. Un-trusted LBSP leads to big problems according to profile sensitive information like real identity and precise location to find the nearest path to requested location.

Then, this research proposes a scheme to preserve privacy for LBS query in VANET system which is called Privacy Preserving Fully Homomorphic Encryption over Advanced Encryption Standard (PP-FHE-AES) scheme. It improves FHE by AES to avoid noise concatenated with encrypted message which leads to communication overhead. Using AES helps proposed scheme to eliminate defects of noise from using only FHE.

The sections of this paper are organized as follows: Sect. 2 presents the related work. Section 3 presents the preliminaries. Section 4 covers the proposed system. Section 5 shows performance of the proposed scheme. Finally, Sect. 6 concludes the work.

2 Related Work

The most related schemes that tend to mask the real users' identities in VANET system are presented as follows:

K-Anonymous region [6] depends on the idea of K-Anonymous region formation, in which the location will be sent as a group of k users not by the real user. So, his exact location will still be unrevealed. However, this scheme has a huge drawback as the user must hold on till the number of k users is in place to be able to send his query. So, it cannot be applied in low user density areas and the waiting time also lowers the service quality due to the delays that happen. This means it is difficult to apply this scheme in real-time services. Another point is the need of a Trusted Third Party (TTP) which can be an obvious bottleneck that is when compromised, all location data is on the verge to be leaked as TTP has all the real locations.

The Updating Pseudonyms Scheme UPS [7] presented the idea of using temporary identities (pseudonyms) that have no relation to the real identity of vehicle or user. Users update their identities with every query and this confuse attackers. However, since some applications need long term communications, which could be interrupted by continuously changing pseudonyms and these communications are hard to be reestablished, thus limiting the implementation of this concept. Also, as RSUs deliver pseudonyms to vehicle users, the traffic amount is elevated resulting in bandwidth reduction that can affect other applications.

Robust Location Privacy Scheme RLPS [8] and Location Privacy Preserving Scheme LPPS [9] introduced the concept of forming a group allowing the user to access LBS anonymously where only a node called Group Leader will communicate with LBS instead of all the other group members to mask the source of the request. The vehicle user sends his request in anonymous way through the Group Leader which is responsible for many tasks rather than just forwarding messages such as signature, verification process, encryption, and decryption functions. So, Group Leader can easily be a performance bottleneck. If the Group Leader left the group before receiving response from the server because of vehicles rapid motion, the communication will be so difficult to maintain. Being a single node for communication, Group leader can also be a single point of failure in accessing services.

Because there are certain criteria that need to be addressed like authenticity, confidentiality, and data integrity, so position privacy realization is envisioned by Homomorphic concept. To hide the user identity, POSTER [10] and TopK-FHE [11] schemes has been used to respond to requests without unmasking the request information. So that the computational results related to ciphertext and plaintext are the same, so that the cryptosystems using homomorphic concept are promising from the location privacy application point of view. The main disadvantage of these schemes is the requirement for more exponential computations, and high overhead production because of the noise concatenating with the ciphertext.

In brief, all the mentioned schemes have huge privacy issues that may give attackers a chance to get into the network to profile the users and learn about the real identities to track their destinations. In the FHE over AES proposed scheme, solutions for the previously mentioned problems are addressed.

3 Preliminaries

This section presents some preliminaries that are used to build the proposed FHE over AES.

3.1 Advanced Encryption Standard (AES)

The Advanced Encryption Standard (AES) algorithm is a symmetric Encryption algorithm that have three block ciphers, AES-128, AES-192 and AES-256 [12]. The proposed scheme is based on the AES-128 Encryption. Each 128-data bits block along with the 128-bit cipher key are processed through a 4×4 state matrix and key matrix respectively. AES-128 encryption consists of 10 rounds [13]. Also AES is widely deployed and used in security-ware applications, so it is relevant to implement it. The circuit of AES has a regular (and quite "algebraic") structure, which is subjected to optimizations as well as parallelism. Also, AES is often used as a benchmark for implementations of protocols for Secure Multi-party Computation (SMC), for example [14–17].

3.2 Homomorphic Cryptography

The homomorphic cryptography objective is to assure the data privacy in communication, storage, or in use through processes with mechanisms and techniques similar to conventional cryptography, but with adding the abilities to perform computation operations over encrypted data, and searching an encrypted data. FHE is used to perform any arbitrary computation over encrypted data. It has been introduced by Rivest, Adleman and Dertouzos in 1978 [18].

3.3 Fully Homomorphic Over Advanced Encryption Standard (FHE Over AES)

FHE-AES relies on matrix form operations which are characterized by light computations. It uses symmetric small size keys to making it suitable for several data centric applications. It gains its security from being too hard to factorize a large integer, which is the base of many public key cryptosystems. As shown in Fig. 1, FHE over AES scheme is used for designing an efficient and practically affordable FHE that uses AES symmetric algorithm.

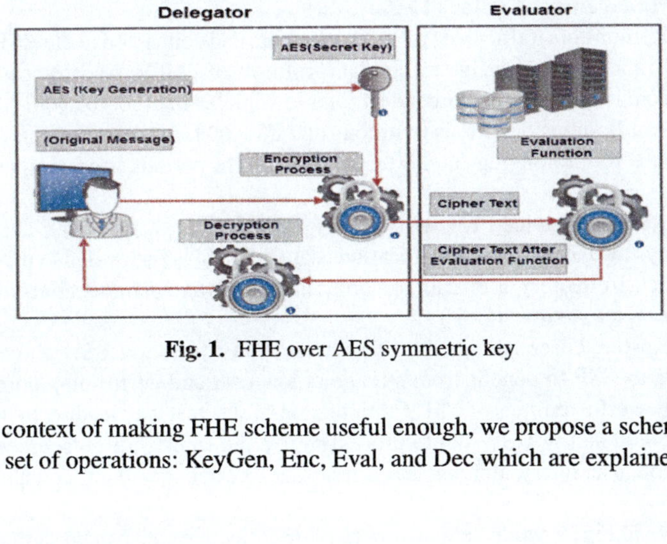

Fig. 1. FHE over AES symmetric key

In the context of making FHE scheme useful enough, we propose a scheme with the following set of operations: KeyGen, Enc, Eval, and Dec which are explained in details in [14].

4 Research Scheme

In this section, the focus is on the way LBS provide efficient and accurate service for users relying on VANET system [19, 20] while achieving privacy of location data.

4.1 System Design

In our system model, we mainly focus on how the LBSP offers accurate and efficient LBS to VANET [21–23] system users. Specifically, the system consists of seven entities: OBU, RSUs, RA, Traffic Management Center (TMC), LBS user, LBSP, and CSP.

- **Vehicles**
  Connecting with each other and with the RSUs located along the road. They can store the credentials of cryptography and can run the algorithms of cryptography.

Usually equipped with GPS receivers to determine its location. They use the location information to decide if the segments provided by RSUs are in its routes or not.

- **On Board Unit (OBU)**
  A chip placed on the top of each vehicle to allow the vehicle to communicate with other vehicles and with the infrastructure [24].
- **Road Side Units (RSUs)**
  RSUs are access points that are placed alongside the roads through which vehicles can communicate with different applications' servers. RSUs receive encrypted routes from vehicles passing in their segments as a representative for Traffic Management Center (TMC). RSUs are interconnected via wired networks and connected with the TMC via wired cables, 4G, or WiMax as a requirement for fast communication technology.
- **Registration Authorities (RA)**
  The provider for important services such as authorization and authentication for LBSP and vehicles.
- **Traffic Management Center (TMC)**
  Each TMC monitors traffic in its relevant segments and connected to the RSUs in these segments, and able to receive traffic data from other TMCs. As it processes traffic information, it sends its recommended notes to vehicles in order to avoid congestions and slow-traffic areas. TMCs also can control traffic lights in their segments to improve traffic status by prolonging or shortening green light periods according to the need [25].
- **Location Based Service Provider (LBSP)**
  LBSP plays the role of recording locations data that RSUs forwarded to process them with the data from other information sources like TMCs and manufacturers of vehicles.
- **Cloud Server Provider (CSP)**
  After a registered user submits the query in encrypted form to LBSP who outsources these data to CSP to benefit from storage at low cost and performing computations with the powerful abilities of CSP. Computation of shortest way leading to destination of choice is done by CSP. While processing overall query, all outsourced location information, user query, and precise current user location are all unknown to the CSP.

As shown in Fig. 4, while vehicles are moving collective data can be shared between them as well as through RSUs to servers. Units or parties can be interconnected in permanent way in this environment that gathers traffic managers and permitting external server's access.

4.2 System Processing

A huge amount of LBS data is constructed as a category set and a location data set. A CATEGORY denotes the general name of location data sets. Each location data set is a four-parts {Pseudo Identification (PID), Destination title (T_d), Destination coordination (X_d,Y_d), Description of destination(D_d)} where d indicates destination location and belongs to Destination category (C_d) in a category set. In this paper query size is assumed 30 KB, and FHE is processed over AES to encrypt LBS data. The proposed PP-FHE-AES scheme enables the LBSP to give exactly the same query service over encrypted location data as the plaintext environment without being revealed. There are three user's specifications in VANET system as shown in Fig. 2: New vehicle user (preregistered user), Registered user (authorized user), and Revoked user (un authorized user).

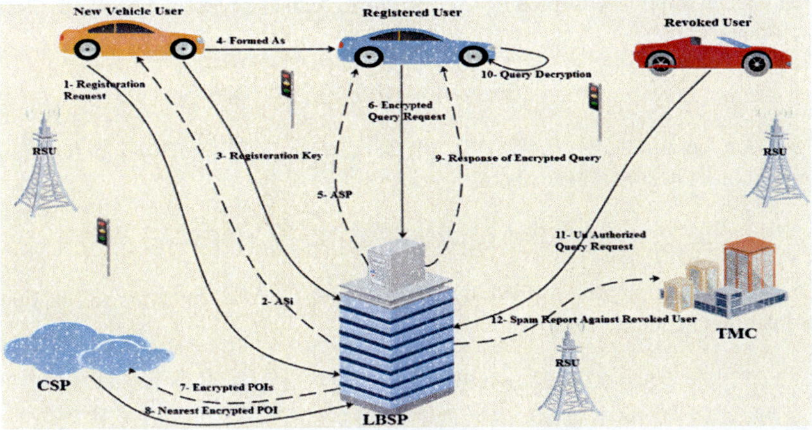

Fig. 2. Types of vehicle users

The structure of all processes at PP-FHE-AES refer that RA Generates public security parameter (λ) for each vehicle. Then Vehicle user (VU_i) sends registration request to LBSP, the VU_i receives λ as input from corresponding RSU. After that, RA publishes the public generator parameter g for LBSP which sends Attribute-set (AS_i) to VU_i to fill it with needed information. When VU_i perform randomly selection for value of $X \in Z_n^*$ to perform a secure cryptographic Message Digits5 hashing function (MD_5). Then VUi performs a process of key generation to compute secret key (SK) in order to encrypt query as formed in (1). The key format is structured as a matrix of 4×4 bytes in the set of all integers (Z_n^*).

$$\text{keyGen}(\lambda) \rightarrow \text{SK} \tag{1}$$

Then VU_i sends X to LBSP through RSU via secure channel. In order to LBSP selects a random value $D_i \in Z_n^*$ and computes $g^{X/Di}$. Then LBSP sends them to VU_i through RSU via secure channel, VUi selects $r_i \in Z_n^*$ randomly. Then registration key ($regK_i$) is computed by (2) and sends it to LBSP to store it to list of registration keys (KList).

$$\text{regK}_i = g^{X/D_i} \times g^{r_i} \tag{2}$$

After that, new vehicle user is formed as registered user. When VU_i sends query request to LBSP who checks authorization of the query by checks the regKi that listed at KList or not. If the request is authorized, LBSP will send Access Structure Policy (ASP) to VU_i. LBSP negotiates permissions with VU_i to get access to groups of data records. If the request is not authorized, LBSP will revoke the request.

Then VU_i encrypts all location information of query request using AES as formed by encrypted part 1 as E1, encrypted part 2 as E2, encrypted part 3 as E3, and encrypted part 4 as E4.

E1 Can be implicitly formed by (3) to perform AES Encryption function to title and description attributes.

$$E1 = \text{Encryption AES } (T_d, D_d, SK) \tag{3}$$

E2 Can be implicitly formed by (4) to perform AES Encryption function to coordination of destination attribute.

$$E2 = \text{Encryption AES}(M_4(X_d, Y_d), SK) \tag{4}$$

E3 Can be implicitly formed by (5) to perform AES Encryption function to coordination of user.

$$E3 = \text{Encryption AES } (M_4(X_i, Y_i), SK) \tag{5}$$

Then concatenating each encrypted parts to be formed by (6)

$$EQ1 = \text{Encryption AES}(M_4(X_i, Y_i), SK) \tag{6}$$

LBSP performs homomorphic evaluation function to detect all the ways leading to required destination which refer to required Place of Interest (POIs). It can be implicitly formed by (7)

$$EQ2 = \text{Evaluation } (f, EQ1) \tag{7}$$

LBSP outsourced list of encrypted POIs to CSP to detect the closest POIs to the user location. It can be implicitly formed by (8). So that the powerful CSP searches over encrypted outsourced location data and output encrypted nearest target location by using haversine formula [26].

$$EQ3 = Q(\text{closest POI}) \tag{8}$$

Then sends the result to LBSP, who formalizes the encrypted query result to be formed as LBS data then sends it to VU_i that decrypts the result with own secret key of AES.

5 Performance Evaluation

In this section, PP-FHE-AES scheme performance will be evaluated from three perspectives; VU, LBSP, and, CSP.

5.1 Simulation Tools

The software and hardware configurations of VU_i and LBSP side are performed on a 64-bit Ubuntu 12.04 LTS system with an Intel Core i7 processor and 16 GB RAM. The CSP side is a virtual machine with Intel processor, 64 GB memory, and VMware VSphere.

The Network Simulation tools (NS2) [27] is implemented since it is the most widely used network simulator. A Simulation of Urban Mobility SUMO [28] is a well-designed traffic simulation and also implemented to successfully simulation of the proposed system to develop road topology with its vehicle, road, junction, and traffic light built with parameters to define vehicles traffic direction and speed. CloudSim [29] is a toolkit supports modeling of cloud environment and it used for PP-FHE-AES.

5.2 Performance Metrics

These metrics are the main scale that is used to perform comparisons between the schemes: K-anonymous Region [6], UPS [7], RLPS [8], LLPS [9], POSTER [10], TopK-FHE [11], and the proposed PP-FHE-AES scheme. During simulation, a communication channel with the support of IEEE 802.11 is assumed. The total number of moving vehicles is from 10 to 500 vehicles in a random walkway with a maximum speed of 120 km/h. The real map used is based on the coordinates of Ismailia city, Egypt. As for the size of the query request, it is fixed at 30 kbytes. The movement model for vehicles is Map-Route-Movement.

5.2.1 Responding Time

The definition meant by Time of responding is the time between the end of the query request and the response beginning. Figure 3 shows that simulation of architectures has been done for varying query requests number (20, 50, 200, and 500) per second while keeping the sizes of query request and query response to be fixed. It can also be clearly seen that when the number of requests per second increases, there is a rapid-rate increase in the response time for all schemes, but the PP-FHE-AES scheme achieved the least response time.

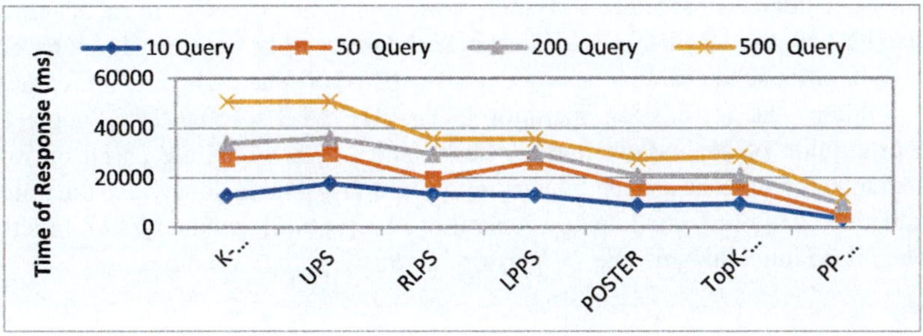

Fig. 3. Responding time

5.2.2 Accuracy Rating

Accuracy rate can be expressed as the number of requested queries divided by the response queries number. So, the best performance is obtained when the accuracy rate value is equal to or near to 1. The worst case is assumed to happen in a VANET simulation containing 500 vehicles, and all of them send query requests at the same time leading to formation of a bottleneck. The results as shown in Fig. 4, the performance drop for the proposed scheme was approximately less than 10% during the occurrence of bottleneck which is considered a very little and an acceptable variation. In other times with variant requests number, the accuracy rate of the proposed scheme value can always reach higher than 91%. While in the best case (only 20 requests), the accuracy can reach approximately 100%.

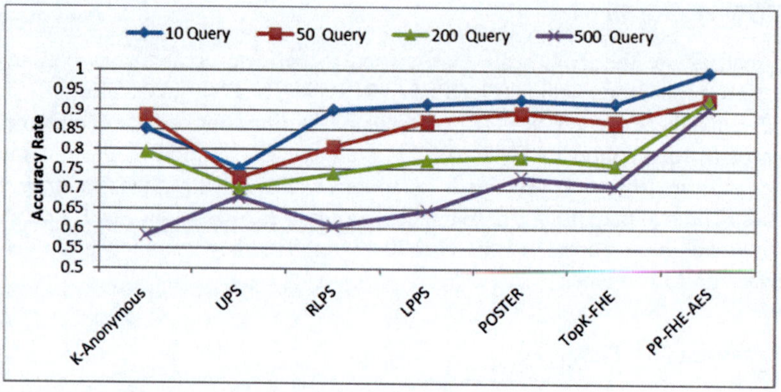

Fig. 4. Accuracy rate

5.2.3 Processing Time

The relative processing time for the operations of Generation, Encryption, Evaluation, and Decryption is shown in Fig. 5. The two schemes K-anonymous region and UPS has been excluded from this test cryptosystem is not supported in these schemes. It can also be seen that in PP-FHE-AES scheme the processing time for all operations is less than in other schemes, that is why the PP-FHE-AES scheme is more efficient than others although the verification operation in PP-FHE-AES scheme takes longer time. No consumption of evaluation time for RLPS and LLPS which are based on hybrid cryptography (symmetric and asymmetric cryptography) that do not involve the function of evaluation. They only include key generation, encryption, and decryption functions. And they need much longer time for performing this cryptographic type.

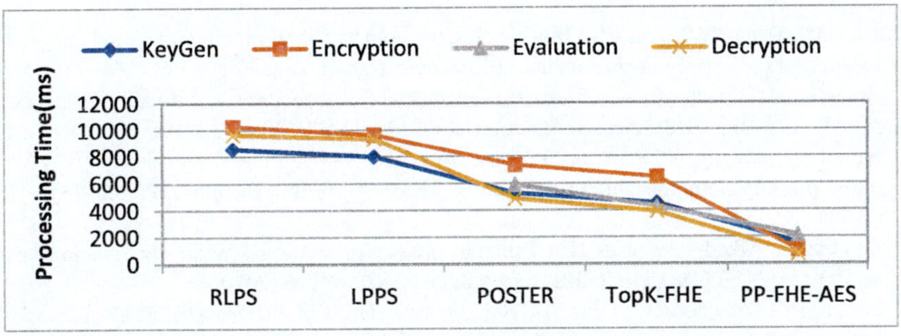

Fig. 5. Processing time

6 Conclusion

In this research, the threats emerging on location privacy in VANET system because of the illegitimate vehicle tracking according to their broadcasts, and the possible threats on privacy based on identifying the LBS applications that a vehicle access. A scheme called PP-FHE-AES is proposed as a solution that is based on FHE technique over AES symmetric cryptography in order for preventing the noise associated with data, and then LBS data is outsourced to cloud in a privacy preserving manner and identity privacy is protected. PP-FHE-AES scheme gives CSP the ability to perform the computations necessary to detect the shortest way to the destination of choice on encrypted data, thus keeping service data confidential from RSUs, LBSP, and CSP.

The simple model was designed for studying LBS usage in VANET system, so a setup with real traffic scenario of Ismailia city, Egypt with variant node densities has been created to help in the analysis of LBS usage in VANET performance metrics (Time of responding, Accuracy Rate, Processing Time). The mentioned scenario has been applied and evaluated via NS-2 network simulator and SUMO traffic simulator. The results of analysis showed that PP-FHE-AES scheme performance is better than other related work schemes in terms of real time and dynamic environment. Where the results of analysis showed that PP-FHE-AES scheme achieved the least response time. The accuracy rate percentage reached values as high as 100% in some cases. The results of analysis showed that in PP-FHE-AES scheme the processing time for all operations is less than in other schemes, that is why the PP-FHE-AES scheme is more efficient than others although the verification operation in PP-FHE-AES scheme takes longer time. The focus of the ongoing and upcoming work is on the application of this security scheme in V2V communication such as in cooperative driving in order to reduce congestions of traffic and increasing the safety for vehicles.

References

1. Rasheed, A., Gillani, S., Ajmal, S., Qayyum, A.: Vehicular Ad Hoc Network (VANET): a survey, challenges, and applications. In: Laouiti, A., Qayyum, A., Mohamad Saad, M.N. (eds.) Vehicular Ad-Hoc Networks for Smart Cities. AISC, vol. 548, pp. 39–51. Springer, Singapore (2017). https://doi.org/10.1007/978-981-10-3503-6_4

2. Lu, E.C., Tseng, V.S., Yu, P.S.: Mining cluster-based temporal mobile sequential patterns in location-based service environments. IEEE Trans. Knowl. Data Eng. **23**(6), 914–927 (2011)
3. Braasch, M., Dempster, A.: Tutorial: GPS receiver architectures, front-end and baseband signal processing. IEEE Aerosp. Electron. Syst. Mag. **34**(2), 20–37 (2019)
4. Farouk, F., Alkady, Y., Rizk, R.: Privacy preserving location based services query scheme based on fully homomorphic encryption. J. Theor. Appl. Inf. Technol. **99**(16), 4098–4121 (2021)
5. Farouk, F., Alkady, Y., Rizk, R.: Efficient privacy-preserving scheme for location based services in VANET system. IEEE Access **8**(1), 60101–60116 (2020)
6. Shokriy, R., Troncosoz, C., Diazz, C., Freudigery, J., Hubaux, J.P.: Unraveling an old cloak: k-anonymity for location privacy. In: Proceedings of the 9th Annual ACM Workshop on Privacy in the Electronic Society (WPES 2010), New York, NY, USA, 4 October 2010, pp. 115–118 (2010)
7. Buttyán, L., Holczer, T., Vajda, I.: On the effectiveness of changing pseudonyms to provide location privacy in VANETs. In: Stajano, F., Meadows, C., Capkun, S., Moore, T. (eds.) ESAS 2007. LNCS, vol. 4572, pp. 129–141. Springer, Heidelberg (2007). https://doi.org/10.1007/978-3-540-73275-4_10
8. Sampigethaya, K., Li, M., Huang, L., Poovendran, R.: AMOEBA: robust location privacy scheme for VANET. IEEE J. Sel. Areas Comm. **25**(8), 1569–1589 (2007)
9. Xue, X., Ding, J.: LPA: a new location-based privacy-preserving authentication protocol in VANET. Security Comm. Netw. **5**(1), 69–78 (2012)
10. Hu, P., Zhu, S.: POSTER: location privacy using homomorphic encryption. In: Deng, R., Weng, J., Ren, K., Yegneswaran, V. (eds.) SecureComm 2016. LNICSSITE, vol. 198, pp. 758–761. Springer, Cham (2017). https://doi.org/10.1007/978-3-319-59608-2_45
11. Hur, M., Lee, Y.: Privacy preserving top-k location-based service with fully homomorphic encryption. J. Korea Soc. Simul. **24**(4), 153–161 (2015)
12. Ghewari, P.B., Patil, M.J.K., Chougule, A.B.: Efficient hardware design and implementation of AES cryptosystem. Int. J. Eng. Sci. Technol. **2**, 213–219 (2010)
13. Alkady, Y., Farouk, F., Rizk, R.: Fully homomorphic encryption with AES in cloud computing security. In: Hassanien, A.E., Tolba, M.F., Shaalan, K., Azar, A.T. (eds.) AISI 2018. AISC, vol. 845, pp. 370–382. Springer, Cham (2019). https://doi.org/10.1007/978-3-319-99010-1_34
14. Abdel-Kader, R.F., El-sherif, S., Rizk, R.Y.: Two-stage cryptography scheme for secure distributed data storage in cloud computing. Int. J. Electr. Comput. Eng. (IJECE) **10**(3), 3295–3306 (2020)
15. Gentry, C., Halevi, S., Smart, N.P.: Homomorphic evaluation of the AES circuit. In: Safavi-Naini, R., Canetti, R. (eds.) CRYPTO 2012. LNCS, vol. 7417, pp. 850–867. Springer, Heidelberg (2012). https://doi.org/10.1007/978-3-642-32009-5_49
16. Pinkas, B., Schneider, T., Smart, N.P., Williams, S.C.: Secure two-party computation is practical. In: Matsui, M. (ed.) ASIACRYPT 2009. LNCS, vol. 5912, pp. 250–267. Springer, Heidelberg (2009). https://doi.org/10.1007/978-3-642-10366-7_15
17. Alkady, Y., Habib, M.I., Rizk, R.: A new security protocol using hybrid cryptography algorithms. In: Proceedings of 9th International Computer Engineering Conference (ICENCO), Cairo, Egypt, 29–30 December 2013, pp. 109–115 (2013)
18. Rivest, R., Adleman, L., Dertouzos, M.: On data banks and privacy homomorphisms. In: Foundations of Secure Computation, pp. 169–179. Academia Press (1978)
19. Zhankaziev, S., Gavrilyuk, M., Morozov, D., Zabudsky, A.: Scientific and methodological approaches to the development of a feasibility study for intelligent transportation systems. Transp. Res. Procedia **36**, 841–847 (2018)
20. Qu, F., Wu, Z., Wang, F., Cho, W.: A security and privacy review of VANETs. IEEE Trans. Intell. Transp. Syst. **16**(6), 2985–2996 (2015)

21. Yao, X., Zhang, X., Ning, H., Li, P.: Using trust model to ensure reliable data acquisition in VANETs. Ad Hoc Netw. **55**, 107–118 (2017)
22. Pan, J., Cui, J., Wei, L.: Secure data sharing scheme for VANETs based on edge computing. J. Wirel. Commun. Netw. **2019** (2019)
23. Lu, Z., Qu, G., Liu, Z.: A survey on recent advances in vehicular network security, trust, and privacy. IEEE Trans. Intell. Transp. Syst. **20**(2), 760–776 (2019)
24. Zhang, X., Chen, X.: Data security sharing and storage based on a consortium blockchain in a vehicular ad-hoc network. IEEE Access **7**, 58241–58254 (2019)
25. McKenney, D., White, T.: Distributed and adaptive traffic signal control within a realistic traffic simulation. Eng. Appl. Artif. Intell. **26**(1), 574–583 (2013)
26. Chopde, N.R., Nichat, M.K.: Landmark based shortest path detection by using A* and Haversine formula. Int. J. Innov. Res. Comput. Commun. Eng. (IJIRCCE) **1**(2), 2320–9801 (2013)
27. NS-2 2008 the Network Simulator – ns 2. http://www.isi.edu/nsnam/n
28. SUMO– Simulation of Urban Mobility. http://sumo.sourceforge.net/
29. CloudSim. http://www.cloudbus.org/cloudsim/

Post-pandemic Education Strategy: Framework for Artificial Intelligence-Empowered Education in Engineering (AIEd-Eng) for Lifelong Learning

Naglaa A. Megahed, Rehab F. Abdel-Kader(✉), and Heba Y. Soliman

Port Said University, Port Said, Egypt
{naglaaali257,rehabfarouk,hebayms}@eng.psu.edu.eg

Abstract. The unprecedented rate of technological advancements led by the artificial intelligence (AI) revolution is transforming teaching and learning. The rapidly dynamic specifics and the distinct nature of engineering education transcend the incorporation of core technical skills to accommodate other essential skills, including collaboration, creativity and innovation, critical thinking, and problem-solving skills. AI-empowered education (AIEd) systems offer customized and personalized experiences for different learning groups, teachers, and decision-makers, thus supporting lifelong learning. This study systematically analyzes the body of scientific literature on the field of AIEd in Engineering (AIEd-Eng), which has undergone significant developments over the last decades. Based on the results, the study proposes a framework and strategy for directing future research initiatives in AIEd-Eng. In addition, the study intends to assess the influence of AI on different educational processes, including instruction, learning, management, and decision-making practices, in engineering education.

Keywords: Artificial intelligence · Education · Engineering · Conceptual framework · Learning · Post-pandemic

1 Introduction

The COVID-19 pandemic has caused a significant shift in attention toward online education in the previous years. Consequently, the use of digital technology has influenced not only the general methodology of teaching and learning in higher education but also the practical hands-on teaching and learning strategies essential to engineering education. Engineering has been considered one of the most important academic degrees in the 21st century owing to its facilitation of essential technological literacy, which is vital in today's world. Scholars have predicted that engineering will play a pivotal role in overcoming some of the most significant challenges prevalent in society. Most of these challenges—including renewable energy, medical engineering, bio-informatics, sustainable development, and access to clean water—are multidisciplinary. Four major aspects characterize the future of engineering education. First, engineering education is an active

© The Author(s), under exclusive license to Springer Nature Switzerland AG 2022
A. E. Hassanien et al. (Eds.): AMLTA 2022, LNDECT 113, pp. 544–556, 2022.
https://doi.org/10.1007/978-3-031-03918-8_45

learning process requiring high levels of student involvement in the classroom and laboratory. Second, it requires individuals to possess various cognitive and problem-solving skills and competencies. Third, it requires several transferable competencies, including teamwork and communication skills. Finally, it is a practice-based experience based on theories [1–6].

Throughout the existing ongoing pandemic, witnessing the potential of artificial intelligence (AI) for education is consequential with the vast amount of data being collected through online learning management systems (LMSs) and massive online open courses [1, 2]. The key objectives of education systems are to maximize student satisfaction, improve academic performance, and enhance the ongoing learning process. The COVID-19 pandemic is an unprecedented crisis that has led to the closure of the majority of face-to-face activities in educational institutions in most countries in an effort to contain COVID-19 and alleviate its impact [3, 4]. As such, e-learning solutions and smart learning systems have been crucial to facilitating the availability of education during the pandemic, especially during the remote working era. The rapid evolution of information and communication technology has played a significant role in accelerating the adoption of AI and adaptive learning technologies in educational systems. AI-empowered education (AIEd) has become one of the currently emerging fields in novel educational technology and an essential strategy for attaining competitive advantages in the education services market. As they continue to cope with the impacts of COVID-19, the majority of institutions of higher education are incorporating AIEd-Eng to render educational experiences safer, more efficient, and more adaptable to changing environments. Research indicates a growth rate of 43% in the use of AI in education in the period from 2018 to 2021. Moreover, scholars predict a substantial growth in AI applications for teaching and learning in the future [5, 6].

Notably, the role of AI in higher learning is to develop human cognitive skills, enhance the educational process, and exceed to produce procedures for content delivery, control, and assessment. The application fields of AIEd include profiling and prediction; evaluation and assessment; and development of adaptive and personalized systems [7]. However, analysis of the recent literature revealed that relatively less research has been conducted on this area in relation to engineering education. To fill this research gap, we developed an implementation strategy called AI-empowered Education in Engineering (AIEd-Eng), which is composed of four stages, namely, processes, communication models, scenarios, and applications. The proposed strategy targets engineering students in the post-pandemic era and is based on reviving the advent, development, and future trends of AIEd-Eng. To accomplish this objective, Sect. 2 presents the background of AI, whereas Sect. 3 provides a review of the related literature on AIEd-Eng. Section 4 describes the main processes, models, and scenarios involved in AIEd-Eng. Section 5 discusses the recent applications, prospective trends, and challenges of AIEd. Lastly, Sect. 6 summarizes and concludes.

2 Background of AI

Engineering education is extremely connected to highly dynamic industrial and environmental demands, which are dominated and sustained by robust information systems, modern technological advancements, and the adoption of AI-empowered technologies. Progress in the field of AI has been diversely reshaping nearly all aspects of society and life [8, 9]. In essence, AI is a framework that enables computers to mimic human cognitive capabilities, automatically learn rules, and detect patterns through experience (data) instead of explicit predefined rules. As a division of AI, machine learning (ML) focuses less on simulating intelligence but more on adding special capabilities to computers to supplement human intelligence and perform tasks beyond human abilities. Specifically, deep learning (DL) is an ML algorithm that utilizes powerful, multi-layer neural networks empowered by massive training data to solve challenging problems. ML algorithms can be classified into one of three learning methodologies, namely, supervised, unsupervised, and reinforcement learning [10, 11]. These learning models can serve various applications by performing various ML operations, such as detection, classification, regression, and clustering.

Several efforts are underway to define the future impact of AI technologies on society. The expanding use of AI technologies is expected in every aspect of modern life due to digital transformations in public safety, governance, manufacturing, logistics, transportation, business, finance, tourism, entertainment, medicine, bioscience, and education. Today, significant investments are made in AI-empowered technologies, which have become a conventional aspect of daily life. For example, virtual assistants, such as Amazon's Echo, Google Assistant, and Apple's Siri are not only improving lives but also making the world more accessible, especially for people with disabilities, special needs, or learning difficulties [12, 13].

3 Evolution in AIEd-Eng Research

The field of AIEd has undergone significant developments over the last years. Recent advancements in AI research along with the evolution in Internet technology and the communication sector have changed the education system paradigm. These new technologies are constantly changing the models of teaching and learning and transforming the interaction between teachers/educators and students, which, thus, induces new teaching methods and alters the traditional classroom settings. AIEd solutions are becoming prevalent in daily life through solutions, such as AI chatbots, smart student learning systems, and virtual labs [14–16]. Furthermore, AI can play a significant role in strategic or institutional decision-making in educational systems. This aspect involves the incorporation of learning analytics, which involves big data algorithms, statistics, and AI techniques to collect data about past and current activities within the system and analyze the collected data to make predictions and eventually propose strategies for prospective students [17–20]. In general, AIEd characterizes the development of computer models that can perform cognitive, learning, and problem-solving tasks, which are similar to those performed by humans [21, 22]. However, engineering education exhibits distinctive characteristics that are more concerned with mastering special knowledge and advanced practical skills specifically tailored to the field. Engineering education has been reshaped to become increasingly learner-centered. At the same time, however, such a shift has weakened the dominance of the role of instructors in the learning process [23–25]. Therefore, AI tools are applied to assist students in obtaining an advanced understanding of the knowledge and skills of interest. AI techniques support engineering education through five models of learning and management, namely, intelligent instructor model (IIM), intelligent student model (ISM), intelligent learning model (ILM), exploratory learning model (ELM), and policy-making adviser (PMA) [23, 24].

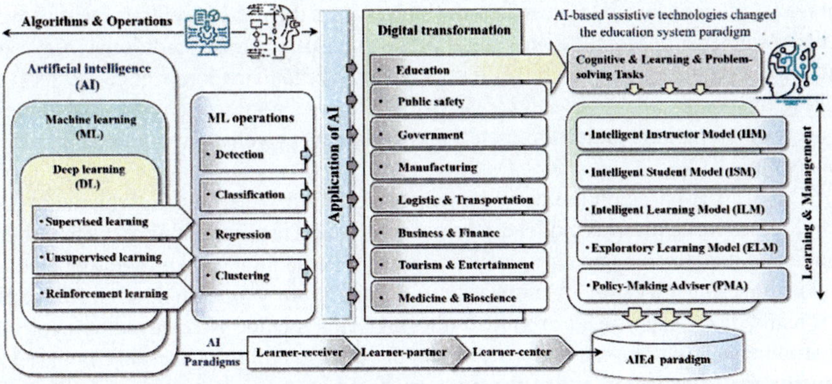

Fig. 1. The conceptual framework for artificial intelligence, machine learning, and deep learning from broad and narrow perspectives.

3.1 AIEd-Eng Models: From Learning to Management

Based on the literature [23, 24], this section identifies the main components that can be amalgamated to a conceptual framework for the application of AI, ML, and DL in engineering education. From the perspective of educational applications, several concepts and related models of AI in the proposed framework can be used to enhance the abovementioned learning and management processes (Fig. 1). The subsequent text provides detailed descriptions of these five models.

The IIM has been developed to improve engineering education in conventional engineering classes, where the number of students being directly educated by one instructor influences their knowledge and comprehension of the subject. A few students with special needs may require additional attention from the instructor, which includes direct individual tutorials to facilitate efficient and effective learning. This model is not limited to providing learners with platforms that contain supporting instructional materials but includes adaptation according to their capabilities [26]. Every learner is distinctive in terms of ability and knowledge, whereas advanced students are generally less dependent on the instructor and instructional materials. This model takes advantage of the AI algorithms to obtain learning assistance tools. The AI algorithms can represent knowledge as a set of production rules, detect the behavioral patterns of students, and provide automatic comments or instructions [23, 24].

The ISM aims to engage learners in the context of assisting others in understanding complicated topics. Conversely, this method may improve higher-order thinking skills and knowledge levels, which includes the smart clustering of students based on their performance within the educational system and on their interests. AI models and methodologies can gain information and experience through human interactions. The learning capacity of AI models can assist in the creation of more advanced systems in the future [23, 27, 28].

The ILM is a critical problem from the standpoint of constructivism and student-centered learning. Gadgets can assist students in collecting, analyzing, and visualizing data in an efficient and effective manner, which enables them to concentrate on improving skills or higher-order thinking instead of performing basic tasks. Traditional mindtools, such as concept-mapping tools, facilitate learners in organizing knowledge by passively linking the connections between ideas. Through this idea-mapping process, the intelligent concept-mapping tool recommends tips to learners as well as evaluates the produced concept maps [28].

The ELM is based on the development of exploratory learning environments [24]. Scholars proposed that this model can be used to address the limitations of the previously mentioned models in engineering education. ELM concentrates on providing personalized engineering education by emphasizing the opportunity to learn through free exploration and discovery instead of guided tutoring. This method is achieved by exploring environments, simulations, and virtual practices. This notion of learning is based on actively engaging learners thru the concept that learning patterns can be effectively transferred to dissimilar situations through meta-reflection.

Lastly, the PMA utilizes AI approaches to gain predictive information, which can be used in the formation of policies or regulations. As a result, developing this model for educational policy development is viable and practicable. Policymakers may better comprehend trends and challenges in educational settings from the macro and micro perspectives with the use of AI technology, which can help in the development and assessment of successful educational policies [23].

Based on these five models and AI-empowered technologies, the future of the engineering education paradigm will be reshaped. Future trends of AIEd-Eng will exhibit a migration from conventional AI paradigms such as learner–receiver and learner–partner to one that is learner-centered, to prepare the new generation of students and engineers who are capable of facing upcoming challenges, especially during pandemics.

4 AIEd-Eng Strategy, Main Processes, and Applications

The proposed AIEd-Eng strategy is a framework that integrates AIEd processes, models, scenarios, and applications (Fig. 2). The strategy explains the different AI practices that are actively being implemented or investigated by different stakeholders in engineering education.

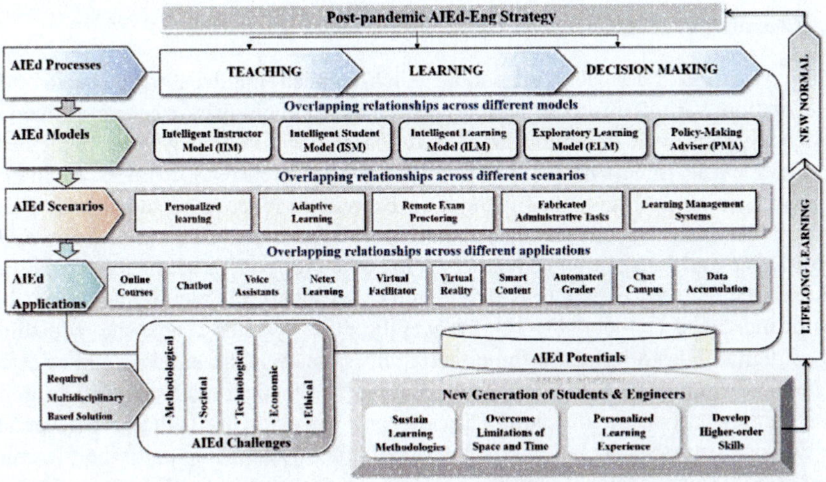

Fig. 2. Components of the proposed AIEd-Eng strategy

According to the e-learning industry, a significant increase has been noted in the application of AI in modern learning systems. In the past three years, nearly 47% of learning management products incorporate AI capabilities. Although AI-empowered solutions have been present in the field of education technology, the industry has been reluctant to embrace them. Alternatively, the COVID-19 pandemic drastically changed the scene and pushed educators to depend on technology for distance and virtual instruction. Currently, technology signifies the new normal for education according to 86% of educators [3, 4, 6, 25, 29]. The present and near future of AIEd-Eng research in terms of the combination of various roles and models of educational processes holds the potential to improve learning and teaching by facilitating the evolution of the education industry. The rapid advancement of computing technologies has facilitated the implementation of AIEd technologies or application programs in educational settings to facilitate learning, teaching, or decision-making processes [23, 29, 30].

4.1 Teaching Process

One of the key areas that perceived the massive impact of AI systems is instruction or teaching. AI has enabled the creation and deployment of systems that employ effective pedagogical tools. The recent literature describes various applications of AI as a pedagogical tool or an instructional platform. Furthermore, such systems may provide simulation-based instruction that incorporates various forms of technology, such as virtual tours or laboratories to demonstrate concepts or practical teaching resources [29, 31].

4.2 Learning Process

Learning is an integral aspect of education, which has been significantly influenced by the AI revolution. Various efforts in modern educational technology have been motivated by the demand to provide tailored instruction to learners within large groups. Implementing intelligent learning systems, incorporating the preferences of learners, evaluating personal learning data, and other methods have rendered adaptive/personalized learning feasible. The main objective of the AI-empowered learning systems is to improve the performance and satisfaction of learners. This objective can be attained by providing personalized learning experiences that identify and satisfy the requirements and abilities of individual learners [27, 32]. Frequently, these systems denote the methodical use of digital data to influence important choices, such as the efficient and effective clustering/distribution of students within projects or practical assignments. Currently, these learning proponents are taking a broader view and claim that colleges must encourage the social, emotional, and physical development of students to tailor their learning experience to specific talents, abilities, preferences, backgrounds, and experiences [33]. Blended strategies for instruction combine face-to-face teaching, technology-assisted training, and student-to-student cooperation to capitalize on the interests of each student and promote deep learning. Examples of such applications are chatbots, voice assistants, Netex, and many others.

Chatbots are low-cost computer programs that mimic human discussions through textual or speech interactions and provide responses to repetitive or frequently asked questions. Moreover, they deliver learning content with recognized inspiration in its interesting, social, and personal aspects when required [34, 35]. Typical examples of chatbots include Talkbots, chatterbots, conversational agents, and artificial conversational entities. Learners may employ chatbots to assist in the recall, review, and retention of acquired information. Voice assistants, such as Amazon's Alexa, Apple's Siri, Microsoft's Cortana, are other AI applications that instructors are effectively using in the classroom instead of conventional printed instructions or complex web-based resources. These voice assistants enable learners to interact with instructional materials without the need for the commitment of the teacher [36]. For example, the Netex learning platform facilitates professors in building, managing, and revising digital content in a single location. It plays an important role in personalized learning, which stimulates high-performance students by enabling microlearning, skills mapping, and providing content recommendations. In this context, the virtual teaching and learning environment can create a flexible professional environment for instructors and learners. As an evaluation tool, AI can be used to grade papers and exams and free up the teacher's time [29].

4.3 Decision-Making Process

AI can play a significant role in strategic or institutional decision-making in AIEd-Eng. With the help of AI technologies, policymakers can precisely understand the trends and difficulties in the educational settings from the macro and micro perspectives. In this manner, AI can eventually help them build and predict future educational policies [23, 37]. This aspect involves learning analytics that involves big data algorithms, statistics, and ML. The AI-empowered predictive analytics models collect data about past and current activities within the system. Models can then assess the effectiveness of the education program and determine whether students are acquiring the required competencies. Moreover, the system can monitor problems in student selection, dropout, and group behavior tendencies and analyze the collected data to make predictions and, eventually, redirect strategies for prospective students. LMSs use software applications to organize, track, report, automate, and deliver educational courses, training programs, and learning and development programs. A tremendous increase in the use of LMSs has been noted for detecting training and learning gaps due to the emphasis on remote learning during the COVID-19 pandemic [38]. LMSs are mainly used for online learning. However, they may also be used for purposes, such as serving as a repository for electronic content (e.g., asynchronous and synchronous courses). LMSs may enhance classroom management for instructor-led instruction or a flipped classroom in higher education. Moreover, modern LMSs use intelligent algorithms to produce automatic course suggestions based on the ability profile of a user and extract information from learning materials to improve the accuracy of recommendations [36, 39, 40].

5 Prospective Trends and Challenges

Innovations in AI lead to new potentials and challenges for teaching and learning in engineering education, especially in the post-COVID-19 era. AIEd-enabled learning systems can sustain various learning methodologies, overcome the limitations of space and time, provide learners with personalized learning experiences, and assist learners in developing higher-order skills. At present, various AIEd-empowered solutions are used in engineering education. However, AI continues to pose many challenges that limit the expectations of specialists [41–44].

Table 1 presents the results of the strengths, weaknesses, opportunities, and challenges (SWOC) analysis of AIEd-Eng, which were obtained through a short survey of its methodological, technological, societal, economic, and ethical aspects. This study may guide various stakeholders in the education sector. Furthermore, it may contribute to the perspective development of knowledge that identifies and discusses the different channels through which AI influences education in general and engineering education in particular in the face of future pandemics. However, predicting the future of higher education in the post-COVID-19 era is difficult, especially with the global increase in the acceptance of e-learning and computer-supported learning platforms. Undoubtedly, however, the current pandemic expedited the transition in the adoption and development of ongoing educational systems with emerging technologies. Consequently, online, blended, and remote learning are becoming a necessity with the aid of AI-empowered technologies [56, 57, 59].

Table 1. Strengths, weaknesses, opportunities, and challenges (SWOC) analysis of AIEd-Eng. Adopted from [2, 16, 17, 23–29, 33, 40, 45–59].

	Strengths	Weaknesses	Opportunities	Challenges
Methodological	• Continuity during pandemics • Applicability for all stages in education • Efficient use of the time of learners/teachers • Personalized progress monitoring, evaluation, and review • Efficient inclusion of students with special abilities or needs	• Lack of accessibility to certain virtual laboratories • Lack of soft skills • Limited research results concerning engineering education	• Creativity and innovation • Metacognitive skills: critical thinking and problem-solving • Virtual teaching • Chatbots • Virtual assistant/tutors • Global classrooms • Learner profiling • Exceptional query resolution	• Lack of training data for AI models • Management of massive amounts of teaching and learning data
Technological	• Radical transformation in engineering education • Automated grading systems • Innovative pedagogical approach	• Real possibility of augmenting teachers	• Global movement toward digital transformation • IoT-based teaching-learning activities • Increased software flexibility and reduced cost	• Technological actions are required to detect students' behaviors during learning processes • Impossibility of building a completely secured system • Threats of cyberattacks
Societal	• Students benefit from effective learning • Forecasting future learning outcomes and problems • Decrease physical contact during COVID-19	• The multidisciplinary nature of AIEd presents a unique challenge for researchers with different backgrounds • Lack of a mutual understanding of what constitutes successful learning and how to measure outcomes • Poor communication and student motivation • Increased need for teachers to adapt to the post-pandemic era to fully utilize the benefits of AIEd	• Re-definition of the role of the teacher • Relatively more effective and enjoyable mode of education • Effective assessment of learning • Safety during pandemics and disasters	• Digital illiteracy • Low acceptance rate • Cultural awareness and competence • Inclusion and equity in AI education • Lack of diversity in the AI industry • Scarcity of talented AI developers • Importance of reflecting the values incised in AI technologies
Economical	• Relatively cheaper and more effective mode of education	• Hidden costs of AI systems and unstable infrastructure • Perceived threat to teachers and instructors	• Changing strategic and decision-making policies accordingly	• Cost of re-engineering current systems to accommodate standards

(*continued*)

Table 1. (*continued*)

	Strengths	Weaknesses	Opportunities	Challenges
Ethical	• Ensuring rapid knowledge acquisition and improvement in learning outcomes	• Data privacy and security • Academic dishonesty due to rise of technology-enhanced learning • Privacy, surveillance, autonomy, prejudice, and discrimination issues	• Lack of solutions for online course integrity • Proctored assignments and examinations are a typical component of many schemes that aim to regulate academic dishonesty • AI-based grading focuses on activities and student engagement instead of grading	• Ethics in AI needs to be embedded in the entire development pipeline

6 Conclusions

Looking into the impacts of COVID-19 and its application to foster the digital transformation phenomenon, engineering education needs to adopt AIEd-Eng as an essential strategy. This study reviewed recent emerging theoretical and technological advances of AI in the educational setting as well as the processes that contribute to the education of a new generation of students and engineers with the help and advantage of AI technologies. The adoption of AIEd-Eng is a recent terminology that should be addressed in post-COVID-19 education because it may facilitate and enhance the effectiveness of engineering educational processes. Consequently, it will influence overall quality and efficiency to render educational experiences safer and more adaptable to the changing environment. Furthermore, this paper serves to raise awareness of the opportunities and challenges that accompany AI for pedagogical adaptation and initiate a dialogue. In this context, the education community will need to be re-educated to work under the new normal. Although several communities promote the broad adoption of AI technologies in education, other groups remain more resistant to technology, skeptical of its opportunities, or worried about its consequences. The findings can be used as a reference for future investigations regarding the application of AI, ML, and DL technologies in AIEd-Eng. The proposed strategy and framework provide a theoretical approach that integrates AIEd processes, scenarios, and applications that utilize state-of-the-art technologies and computational models. Evidently, integrating knowledge from different research disciplines into the future of the AIEd-Eng paradigm is crucial for validating the theoretical proposal within an empirical framework.

References

1. Reeves, S.M., Crippen, K.J.: Virtual laboratories in undergraduate science and engineering courses: a systematic review, 2009–2019. J. Sci. Educ. Technol. **30**(1), 16–30 (2021)
2. Chaudhry, M., Kazim, E.: Artificial Intelligence in education (Aied) a high-level academic and industry note 2021. Available at SSRN 3833583, 24 April 2021
3. Cepal, N.: Education in the time of COVID-19

4. Megahed, N., Ghoneim, E.: E-learning ecosystem metaphor: building sustainable education for the post-Covid-19 era. Int. J. Learn. Technol. (2022, in press)
5. Radanliev, P., De Roure, D.: Emergent role of artificial intelligence in higher education, May 2021
6. Thomas, S.: Future ready learning: Reimagining the role of technology in education. In: 2016 National Education Technology Plan. Office of Educational Technology, US Department of Education, January 2016
7. Zawacki-Richter, O., Marín, V.I., Bond, M., Gouverneur, F.: Systematic review of research on artificial intelligence applications in higher education–where are the educators? Int. J. Educ. Technol. High. Educ. **16**(1), 1–27 (2019)
8. Owoc, M.L., Sawicka, A., Weichbroth, P.: Artificial intelligence technologies in education: benefits, challenges and strategies of implementation. In: Owoc, M.L., Pondel, M. (eds.) IFIP International Workshop on Artificial Intelligence for Knowledge Management, pp. 37–58. Springer, Cham (2019). https://doi.org/10.1007/978-3-030-85001-2_4
9. Woschank, M., Rauch, E., Zsifkovits, H.: A review of further directions for artificial intelligence, machine learning, and deep learning in smart logistics. Sustainability **12**(9), 3760 (2020)
10. Shalev-Shwartz, S., Ben-David, S.: Understanding Machine Learning: From Theory to Algorithms. Cambridge University Press, Cambridge (2014)
11. Saxe, A., Nelli, S., Summerfield, C.: If deep learning is the answer, what is the question? Nat. Rev. Neurosci. **22**(1), 55–67 (2021)
12. Cantú-Ortiz, F.J., Galeano Sánchez, N., Garrido, L., Terashima-Marin, H., Brena, R.F.: An artificial intelligence educational strategy for the digital transformation. Int. J. Interact. Des. Manuf. (IJIDeM) **14**(4), 1195–1209 (2020). https://doi.org/10.1007/s12008-020-00702-8
13. Aljowaysir, N., Ozdemir, T.O., Kim, T.: Differentiated learning patterns with mixed reality. In: 2019 IEEE Games, Entertainment, Media Conference (GEM), pp. 1–4. IEEE, 18 June 2019
14. Peña-López, I.: Innovating education and educating for innovation. The Power of Digital Technologies and Skills (2016)
15. Selwyn, N.: Rethinking education in the digital age. In: Digital Sociology, pp. 197–212. Palgrave Macmillan, London (2013)
16. Tuomi, I.: The impact of artificial intelligence on learning, teaching, and education. Publications Office of the European Union, Luxembourg (2018)
17. Zhuang, Y.T., Wu, F., Chen, C., Pan, Y.H.: Challenges and opportunities: from big data to knowledge in AI 2.0. Front. Inf. Technol. Electron. Eng. **18**(1), 3–14 (2017)
18. Sarker, I.H.: Data science and analytics: an overview from data-driven smart computing, decision-making and applications perspective (2021)
19. Sarker, I.H.: Machine learning: Algorithms, real-world applications and research directions. SN Comput. Sci. **2**(3), 1–21 (2021)
20. Roll, I., Wylie, R.: Evolution and revolution in artificial intelligence in education. Int. J. Artif. Intell. Educ. **26**(2), 582–599 (2016)
21. Baker, T., Smith, L.: Education rebooted? Exploring the future of artificial intelligence in schools and colleges. Nesta Foundation (2019)
22. Chen, X., Xie, H., Zou, D., Hwang, G.: Application and theory gaps during the rise of artificial intelligence in education. Comput. Educ. Artif. Intell. **1**, 100002 (2020)
23. Hwang, G., Xie, H., Wah, B.W., Gašević, D.: Vision, challenges, roles and research issues of Artificial Intelligence in Education, vol. 1, p. 100001 (2020)
24. Ouyang, F., Jiao, P., Alavi, A.H.: Artificial intelligence-based smart engineering education. In: Sensors and Smart Structures Technologies for Civil, Mechanical, and Aerospace Systems 2020, vol. 11379, p. 113790C. International Society for Optics and Photonics, 23 April 2020

25. Megahed, N., Hassan, A.: A blended learning strategy: re-imagining the post-Covid-19 architectural education. Int. J. Architect. Res. ArchNet-IJAR (2021, in press)
26. Noh, N.M., Ahmad, A., Halim, S.A., Ali, A.M.: Intelligent tutoring system using rule-based and case-based: a comparison. Proc. Soc. Behav. Sci. **10**(67), 454–463 (2012)
27. Al-Tarabily, M.M., Abdel-Kader, R.F., Azeem, G.A., Marie, M.I.: Optimizing dynamic multi-agent performance in E-learning environment. IEEE access. **15**(6), 35631–35645 (2018)
28. Ahmad, S.F., Rahmat, M., Mubarik, M.S., Alam, M.M., Hyder, S.I.: Artificial intelligence and its role in education. Sustainability. **13**(22), 12902 (2021)
29. Chen, L., Chen, P., Lin, Z.: Artificial intelligence in education: a review. IEEE Access. **17**(8), 75264–75278 (2020)
30. Baker, M.J.: The roles of models in artificial intelligence and education research: a perspective view. J. Artif. Intell. Educ. **11**, 122–143 (2000)
31. Timms, M.J.: Letting artificial intelligence in education out of the box: educational cobots and smart classrooms. Int. J. Artif. Intell. Educ. **26**(2), 701–712 (2016)
32. Bulger, M.: Personalized learning: the conversations we're not having. Data Soc. **22**(1), 1–29 (2016)
33. Xie, H., Chu, H.C., Hwang, G.J., Wang, C.C.: Trends and development in technology-enhanced adaptive/personalized learning: a systematic review of journal publications from 2007 to 2017. Comput. Educ. **140**, 103599 (2019)
34. Winkler, R., Söllner, M.: Unleashing the potential of chatbots in education: a state-of-the-art analysis (2018)
35. Vanichvasin, P.: Chatbot development as a digital learning tool to increase students' research knowledge. Int. Educ. Stud. **14**(2), 44–53 (2021)
36. Sáiz-Manzanares, M.C., Marticorena-Sánchez, R., Ochoa-Orihuel, J.: Effectiveness of using voice assistants in learning: a study at the time of COVID-19. Int. J. Environ. Res. Public Health **17**(15), 5618 (2020)
37. Tsai, Y.S., Poquet, O., Gašević, D., Dawson, S., Pardo, A.: Complexity leadership in learning analytics: drivers, challenges and opportunities. Br. J. Edu. Technol. **50**(6), 2839–2854 (2019)
38. Yeh, Y.J., Lai, S.Q., Ho, C.T.: Knowledge management enablers: a case study. Ind. Manag. Data Syst. **106**(6), 793–810 (2006)
39. Bradley, V.M.: Learning management system (LMS) use with online instruction. Int. J. Technol. Educ. **4**(1), 68–92 (2021)
40. Louhab, F.E., Bahnasse, A., Bensalah, F., Khiat, A., Khiat, Y., Talea, M.: Novel approach for adaptive flipped classroom based on learning management system. Educ. Inf. Technol. **25**(2), 755–773 (2019). https://doi.org/10.1007/s10639-019-09994-0
41. Keerthiwansha, N.B.: Artificial intelligence education (AIEd) in English as a second language (ESL) classroom in Sri Lanka. Artif. Intell. **6**(1), 31–36 (2018)
42. Hagendorff, T., Wezel, K.: 15 challenges for AI: or what AI (currently) can't do. AI Soc. **35**(2), 355–365 (2019). https://doi.org/10.1007/s00146-019-00886-y
43. Bird, E., Fox-Skelly, J., Jenner, N., Larbey, R., Weitkamp, E., Winfield, A.: The ethics of artificial intelligence: issues and initiatives. European Parliamentary Research Service, Technical report PE, March 2020
44. Tremblay, K., Lalancette, D., Roseveare, D.: Assessment of higher education learning outcomes: feasibility study report, Volume 1–Design and Implementation. Organisation for Economic Co-operation and Development, Paris, France (2012)
45. Guan, C., Mou, J., Jiang, Z.: Artificial intelligence innovation in education: a twenty-year data-driven historical analysis. Int. J. Innovat. Stud. **4**(4), 134–147 (2020)
46. Woolf, B.P., Lane, H.C., Chaudhri, V.K., Kolodner, J.L.: AI grand challenges for education. AI Mag. **34**(4), 66–84 (2013)
47. Schellen, E., Wykowska, A.: Intentional mindset toward robots—open questions and methodological challenges. Front. Robot. AI **11**(5), 139 (2019)

48. Dwivedi, Y.K., et al.: Artificial Intelligence (AI): multidisciplinary perspectives on emerging challenges, opportunities, and agenda for research, practice, and policy. Int. J. Inf. Manage. **57**, 101994 (2021)
49. Davenport, T., Guha, A., Grewal, D., Bressgott, T.: How artificial intelligence will change the future of marketing. J. Acad. Mark. Sci. **48**(1), 24–42 (2019). https://doi.org/10.1007/s11747-019-00696-0
50. Leslie, D.: Understanding artificial intelligence ethics and safety: a guide for the responsible design and implementation of AI systems in the public sector. Available at SSRN 3403301, 10 Jun 2019
51. Atoum, Y., Chen, L., Liu, A.X., Hsu, S.D., Liu, X.: Automated online exam proctoring. IEEE Trans. Multimedia **19**(7), 1609–1624 (2017)
52. Davis, A.B., Rand, R., Seay, R.: Remote proctoring: the effect of proctoring on grades. In: Advances in Accounting Education: Teaching and Curriculum Innovations. Emerald Group Publishing Limited, 11 Jan 2016
53. Olt, M.R.: Ethics and distance education: Strategies for minimizing academic dishonesty in online assessment. Online J. Dist. Learn. Admin. **5**(3), 1–7 (2002)
54. Torresen, J.: A review of future and ethical perspectives of robotics and AI. Frontiers Robot. AI. **15**(4), 75 (2018)
55. McCoy, D.: Domain models, student models, and assessment methods: three areas in need of standards for adaptive instruction. In: The Adaptative Instructional System (AIS) Standards Workshop of the 14th International Conference of the Intelligent Tutoring Systems (ITS) Conference, Montreal, Quebec, Canada, June 2018
56. Megahed, N., Ghoneim, E.: Blended learning: the new normal for post-Covid-19 pedagogy. IJMBL **14**(1) (2022, in press)
57. Händel, M., Stephan, M., Gläser-Zikuda, M., Kopp, B., Bedenlier, S., Ziegler, A.: Digital readiness and its effects on higher education students' socio-emotional perceptions in the context of the COVID-19 pandemic. J. Res. Technol. Educ. **9**, 1–3 (2020)
58. Megahed, N.A.: Augmented Reality based-learning assistant for architectural education. Int. J. Adv. Educ. Res. **1**(1), 35–50 (2014)
59. Zawacki-Richter, O.: The current state and impact of Covid-19 on digital higher education in Germany. Hum. Behav. Emerg. Technol **3**(1), 218–226 (2021)

An Intelligent Algorithmic Approach for Data Collection in a Smart Warehouse Testbed

Ngoc-Bich Le[1,2], Duc-Canh Nguyen[3], Xuan-Hung Nguyen[3], Manh-Kha Kieu[4,5], Vu-Anh-Tram Nguyen[4], Tran-Thuy-Duong Ninh[4], Minh-Dang-Khoa Phan[6], Narayan C. Debnath[7], and Ngoc-Huan Le[3](✉)

[1] School of Biomedical Engineering, International University, Ho Chi Minh City, Vietnam
lnbich@hcmiu.edu.vn

[2] Vietnam National University Ho Chi Minh City, Ho Chi Minh City, Vietnam

[3] Mechanical and Mechatronics Department, Eastern International University, Thu Dau Mot, Binh Duong, Vietnam
{canh.nguyen,hung.nguyenxuan,huan.le}@eiu.edu.vn

[4] Becamex Business School, Eastern International University, Thu Dau Mot, Binh Duong, Vietnam
{kha.kieu,tram.nguyen,duong.ninh}@eiu.edu.vn

[5] School of Business and Management, RMIT University Vietnam, Ho Chi Minh City, Vietnam

[6] EIU Fablab, Eastern International University, Thu Dau Mot, Binh Duong, Vietnam
khoa.phan@eiu.edu.vn

[7] School of Computing and Information Technology, Eastern International University, Thu Dau Mot, Binh Duong, Vietnam
narayan.debnath@eiu.edu.vn

Abstract. In this paper, a smart warehouse testbed was designed and simulated. The system serves two purposes. It simulates a real-world industrial setting and emphasizes the significance of experimental research to assess the feasibility of new ideas in real-world situations. The suggested simulation model has five rack systems with a total capacity of 1000 pallets. An AGV moves in the middle of each rack system, picking up and returning pallets to both sides in the direction of movement. A circulating conveyor system is used in the proposed system to enhance transaction count data to enable intelligent algorithms such as machine learning. Accordingly, each pallet (attached with one RFID tag) is managed by a management code when stored in the warehouse. After taking out, that pallet can be cleared of the old management code, assigned a new code, and returned to the input conveyor of the system. One of the biggest challenges of the proposed system is ensuring the continuous operation at the inlets of the AS/RS AGV's pallet receiving conveyors. The feasibility was demonstrated by conducting the simulation to build a polynomial regression function from which the value of Tseparator can be attained according to the given set of values (Vconveyor and Tstoring). In addition, when the set of values (Vconveyor, Tstoring, and Tseparator) is known, the time required to store the entire inventory may also be calculated.

Keywords: Intelligent algorithm · Artificial intelligence · Machine learning · IoT · Logistics · AS/RS · Simulation modelling · Testbed

© The Author(s), under exclusive license to Springer Nature Switzerland AG 2022
A. E. Hassanien et al. (Eds.): AMLTA 2022, LNDECT 113, pp. 557–566, 2022.
https://doi.org/10.1007/978-3-031-03918-8_46

1 Introduction

In recent years, Vietnam's logistics sector has grown at a breakneck pace. As a result, warehouses and warehousing services in Vietnam are critical to improving logistical competitiveness [1]. Simple storage, processing, value-adding, and piece-picking are all limits in warehouse services [2]. Applying new technologies to warehouse operations, such as IoT, deep learning, and so on, is one of the problems and impediments to the development of warehousing services.

For the distribution and manufacture of commodities, automated storage and retrieval systems have been widely deployed. Racks and cranes are running in aisles between the racks to receive and store goods in automated storage and retrieval systems (AS/RS). AS/RSs are used to store and retrieve items (e.g., raw materials or (semi-) completed products) from storage [3]. As a result, the AS/RS lowers labor costs while enhancing productivity.

System configuration [4], transit time estimation [5], storage assignment [6], dwell point location [7], and request sequencing [8] are the current AS/RS system difficulties being explored and developed. Simulations have long been utilized in logistics operations as a decision-making aid. Many research studies [4, 9] have concentrated on simulation modelling for large-scale AS/RS operations.

Building a smart warehouse serving as a testbed is essential in the 4.0 era. Testbeds are fundamental platforms for the rigorous and replicable testing of ideas, computational tools, new technologies, and systems [10]. According to the authors, the concept of testbed has facilitated the acquisition of practical and theoretical understanding [11]. In most universities, students are expected to cope with a large quantity of theory; nevertheless, testbeds allow students to work on and solve difficulties - such as algorithm programming control or augmented reality use - while designing and materializing the testbed. With the advancement of artificial intelligence, machine learning, and other related technologies, an ideal smart warehouse testbed should support checking the optimization of sorting algorithms, the optimization of AGV transportation algorithms, and so on, to save operation time, transportation costs, and energy costs. As a result, this study presents a testbed that can address the following research questions (RQs) using a simulation model.

RQ1 – Can the proposed model be easily upgraded based on the existing warehouse systems of small and medium-sized companies?
RQ2 – Can it generate various transactions to provide data for optimization problems through artificial intelligence and machine learning approach?
RQ3 – Is the flexibility in storing and returning goods optimal?

For RQ1, the study surveyed service warehouses and warehouses at each company in provinces with large industrial zones such as Binh Duong, Dong Nai, Long An, and Ho Chi Minh City, Vietnam. Almost all of the companies that were surveyed have the same arrangement, including a typical aisle for forklifts and racks on both sides.

For RQ2, the study proposed a pallet circulation mechanism. When a request to depart the warehouse is received, that pallet will be cleared of the old management code and reassigned with a new management code and then stored in the warehouse by the connection between the output conveyor system and the input conveyor.

For RQ3, a Simio commercial package simulation model was conducted to determine the time to deliver pallets to the input conveyor system, so no pallet blockage occurs according to the correlation between conveyor speed and storage velocity. In addition, the storage time is also determined.

The paper has solved all the technical problems to build a warehouse testbed starting from building the warehouse structure, operating the AGVs, managing pallet information, and the circulating conveyor system. Moreover, the study simulated the system with 1000 pallets, five racks to determine the optimal parameters of the value set (Tseparator, Vconveyor, Tstoring) so that the system does not suffer from bottlenecks at the top of input conveyors.

2 Introduction

Detailed solution approaches are described in this section including basic requirements and the methods used.

2.1 Basic Requirements

The smart warehouse system was developed to realize two primary purposes. First, the system is utilized as a testbed center for regional demands. Second, it is used for research purposes, especially for logistic and AI algorithms. These demands require continuous operation and numerous data collecting. Consequently, it is necessary to recycle and circulate the pallets. This circulation requires solving two main obstacles: the first is the mechanical hardware system to transfer the pallets from the output back to the input, and the second is to delete the old management code and reassign the new management code to the pallets.

2.2 Solution Approaches

To solve the problem of pallet circulation, the following solutions were considered (1) manually pick and place; (2) AGV mobile robots; (3) crane system; and (4) conveyor system. Using the decision matrix with the evaluation criteria described in Table 1 gave a superior evaluation result of the conveyor system solution. Consequently, the conveyor system approach was designated.

The conveyor system solution has the following features: (1) continuous circulation of pallets between output and input; (2) the inlet (green conveyor in Fig. 1 (a)) and outlet (red conveyor in Fig. 1 (a)) conveyor systems were structured in two different levels to enhance the system's flexibility; (3) Pneumatic cylinder pushers were installed at the perpendicular turns to transfer the pallet; (4) The inlet conveyor of each AS/RS AGV acts as a buffer for queuing pallets; (5) Prior to the AGV's output conveyors, sensors and pneumatic cylinders are arranged to block coming pallets on the main conveyor when needed to avoid collisions. With the above structure and features, the conveyor system allows the integration of automation solutions such as arranging stations to read, write

Table 1. Decision matrix for circulation solution.

Method	Criteria			Total
	Cost	Auto. compatible	Cont. operation	
Manual pick and place	4	1	1	6
AGV mobile robots	1	4	3	8
Crane system	2	2	2	6
Conveyor system	**3**	**3**	**4**	**10**

and erase management codes for pallets as well as ensuring continuous operation. These requirements can be well met if AGV mobile robots are used. Nevertheless, to meet the continuous operation requirements, many AGV mobile robots are necessary. As a result, the cost would significantly levitate. Recharging AGV mobile robots after each operating cycle also contributes to the multiplication of robot quantity.

To solve the pallets' assigned management code writing and erasing issues, the following solutions have been considered, including (1) manual operation, (2) bar codes, and (3) RFID. Applying the decision matrix with similar evaluation criteria gives the top evaluation result of the RFID solution (Table 2).

Table 2. Decision matrix for pallets' assigned management code writing and erasing solution.

Method	Criteria			Total
	Cost	Auto. compatible	Cont. operation	
Manual operation	3	1	1	5
Barcode	2	2	2	6
RFID	**1**	**3**	**3**	**7**

With its intrinsic features, the RFID solution provides a straightforward approach for management code writing and erasing. Every pallet was assigned with an attached RFID tag for coding purposes. However, this task is quite challenging when using manual or barcode methods. Both of these methods require detaching the old printed code from the pallet and attaching the new one. Furthermore, reading codes to redirect pallets into the respective AS/RS AGV's conveyors are carried out by arranging the corresponding RFID reading stations. This problem can be solved similarly with the barcode method. It is still challenging with the manual operation approach because it is necessary to arrange the discrete sensors along the conveyor to trace and map all the pallets on the conveyor system. Consequently, RFID is considered the best solution in this current application.

3 System Design

3.1 Description of the System

The first Fig. 1(a) shows a novel approach to pallet circulation for data collecting in a smart warehouse testbed that we suggest in this research. The presented system is used for both storage and retrieval operations. The automated guide vehicles (AGVs) in this system can travel inside storage racks. AGVs are stacker cranes with telescopic forks that are used to process storage and retrieval tasks, as shown in Fig. 1(b).

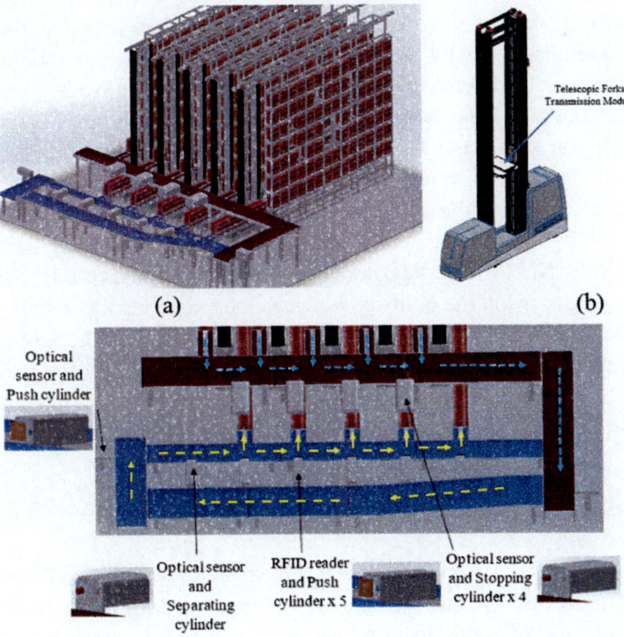

Fig. 1. (a) The smart warehouse testbed in 3D view; (b) AGV; (c) Pallet circulation mechanism.

The conveyor system circulates pallets by using horizontal conveyors to provide pallets to the system, save pallets when they leave the system, and circulate pallets out of the system to return to the system's feeding area. There are two vertical conveyors in front of each rack system to collect pallets from the horizontal conveyor before AGV picks them up and brings them into storage and transfers pallets from AGV to the horizontal conveyor. The horizontal conveyor system also serves as a buffer, allowing the testbed system to run automatically when the ratio of pallets entering and exiting the system is near 1. When enormous amounts of data are needed for AI and profound learning challenges, this is one of the system's advantages to improve transactions.

A sensor and a lever on the horizontal conveyor system, just before the pallet reaches the vertical conveyor position, adjust the minimum time to load pallets into the storage area (Tseparator). Tseparator is extremely important in avoiding pallet bottlenecks before transitioning from horizontal to vertical conveyors. Figure 1(c) shows an RFID reader on the horizontal conveyor directly in front of each vertical conveyor that reads the pallet's RFID code to determine where the pallet is stored.

3.2 Description of the Operation of the System

The proposed testbed operation can be summarized as follows: (1) When an order is received from a customer, the warehouse management system (WMS) software runs an optimization algorithm to determine the optimal position to arrange the pallets; (2) Then the corresponding number of pallets (each pallet is attached with 1 RFID tag) will be assigned management codes by the RFID writer that placed on the horizontal conveyor system; (3) The pallets move to the separator unit, in turn, to move into the storage area at specific intervals; (4) Next, when the pallet arrives in front of each rack, the RFID reader will scan the RFID tag to determine which rack the pallet is stored. The process repeats until all pallets are stored in the warehouse; (5) When it is necessary to take out the pallets according to the customer's order, the WMS software will determine the order to take the pallets. The AGVs will perform the task of bringing the pallets to the vertical conveyors. Next, the vertical conveyors will transfer the pallets to the horizontal conveyor. In order to avoid the collision between the existing pallet and the pallet about to enter the conveyor, before the intersection of the vertical and horizontal conveyors, there is a sensor and a lever that blocks the moving pallets until the pallets from the vertical conveyor are already inside the horizontal conveyor; (6) After that, the output pallets will be stored on the horizontal conveyor until a new order arrives.

4 Analysis of the Smart Warehouse Testbed

In the proposed system, it is most important to calculate the values (Tseparator, Vconveyor and Tstoring) to operate continuously without getting interrupted when transferring pallets from horizontal to vertical conveyors. The Tseparator is the minimum interval for a new pallet on the horizontal conveyor to enter the storage area; Vconveyor is the speed of the horizontal and vertical conveyors (we assume these two speeds are the same); and Tstoring is the travel time of the AGV from the time of receiving the pallet, taking it to storage and moving back to the receiving position. Simio simulation software was deployed to create the system's simulation model.

4.1 Simulation Model for Searching (Tseparator, Vconveyor, and Tstoring)

The simulation model assumes that Tstoring is a constant (average value of arrangements at one rack) in the considered cases. Determining the ideal sorting time is indeed an open problem that is still being worked on. Similarly, for real-world systems, the conveyor speed varies widely in the range of [0.1–3] m/s. This is also the parameter from which the simulation results are derived.

Determining Tseparator based on the change of the data set (Vconveyor and Tstoring) is a multivariate problem. Therefore, the experiment was conducted by fixing the value pair (Tstoring and Vconveyor) and continuously changing the Tseparator value to avoid pallet blockage on the vertical conveyor. During this experiment, five rack systems that can store up to 1000 pallets were utilized. At each appropriate Tseparator value, the minimum time required to load 1000 pallets into the entire stock was recorded.

The Fig. 2 below is the pairs of values (Tstoring and Vconveyor) applied in the simulation experiments to find Tseparator and working hours, respectively.

Vconveyor (m/s) / Tstoring (s)	**7**	**10**	**13**	**18**	**25**	**35**	**50**	**70**
0.1								
0.2								
0.3								
0.5				Tseparator (s)?				
0.8				Working Hours?				
1.0								
1.2								
1.5								
2.0								
3.0								

Fig. 2. Applied simulation experiment data set (Vconveyor and Tstoring).

4.2 Details of the Simulation Model

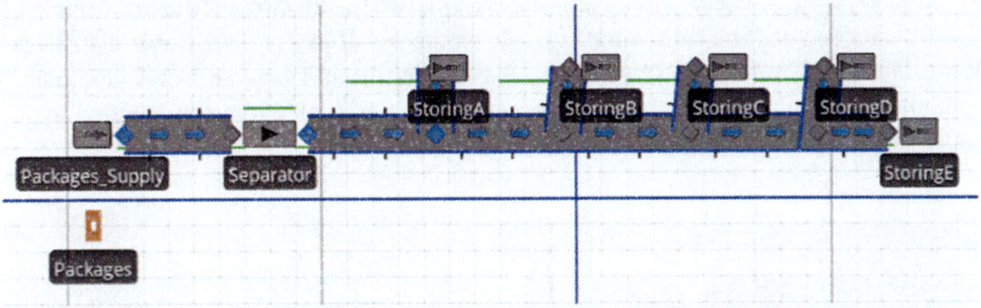

Fig. 3. Vertical conveyors, horizontal conveyor and separator unit in the simulation model.

The simulation model for the proposed design is presented in this section. Figure 3 depicts screenshots from an animated model of vertical conveyors, a horizontal conveyor, and a separator unit.

In the simulation model, the following constraints were setup: (1) The maximum number of pallets that can be accommodated on the vertical conveyors is 3. The size of each vertical conveyor is 2.5 m long × 2 m wide; (2) Dimensions of the horizontal conveyor are 20 m long × 2 m wide; (3) The size of each pallet is 0.63 m long × 1.2 m wide × 0.37 m high.

5 Results and Discussion

The simulation results are presented in this section. At each combination (Tstoring and Vconveyor), when the optimal value of Tseparator is reached, the number of pallets distributed in each rack system (A, B, C, D, and E) were saved. As shown in Fig. 4, the distribution of pallets is random and fluctuates around the expected value at each rack system (200 pallets).

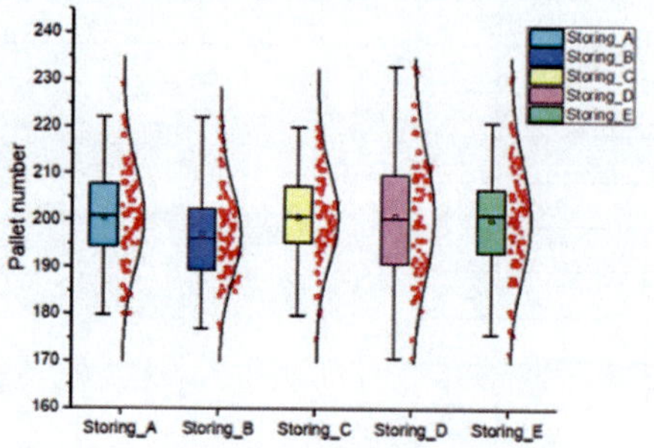

Fig. 4. The number of pallets was distributed into five racks in 80 simulated combinations.

Polynomial regression with response surface analysis is a complex statistical method that has grown in popularity in multisource feedback studies. Researchers can use this method to see how combinations of two predictor variables affect an outcome variable, which is especially useful when the disparity between the two predictor variables is a significant factor. The following Fig. 5(a) is the result of the Tseparator distribution according to the remaining two values. It is evident that Tseparator changes linearly without any singularity. This helps the generated polynomial regression function to give high accuracy prediction results.

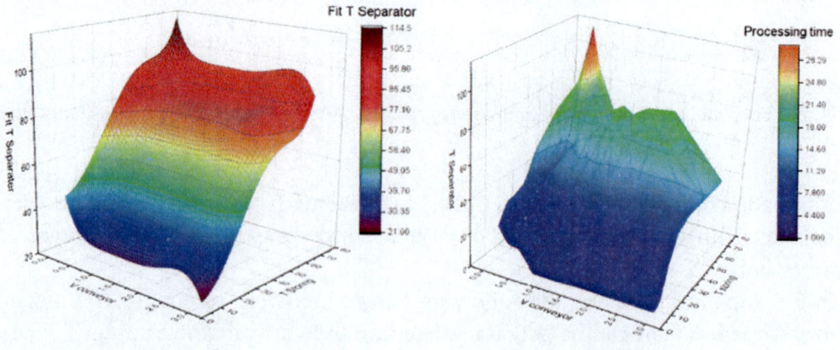

a) b)

Fig. 5. (a) Graph of regression results of Tseparator based on (Tstoring and Vconveyor); (b) Graph of regression results of working time based on (Tseparator, Tstoring and Vconveyor).

Equation 1 is a polynomial surface regression function for searching Tseparator.

$$z = z_0 + A_1x + A_2x^2 + A_3x^3 + A_4x^4 + A_5x^5 + B_1y + B_2y^2 + B_3y^3 + B_4y^4 + B_5y^5 \quad (1)$$

Table 3. Regression results.

	Value	Standard error	t-Value	Prob > (t)	Dependency
z0	21.63063	26.02391	0.83118	0.40874	0.99852
A1	−43.15849	53.93387	−0.80021	0.42633	0.99982
A2	12.46406	129.25855	0.09643	0.92346	0.99999
A3	16.51231	126.05197	0.131	0.89616	1
A4	10.15704	52.26997	−0.19432	0.8465	1
A5	1.51109	7.57091	0.19959	0.84239	1
B1	0.35862	5.99022	0.05987	0.95243	0.99998
B2	−0.00331	0.48842	−0.00678	0.99461	1
B3	0.00243	0.01735	0.14027	0.88886	1
B4	−6.5260E-5	2.73477E-4	−0.23863	0.8121	1
B5	4.74016E-7	1.55581E-6	0.30468	0.76153	1

The obtained coefficients of Eq. 1 are listed in Table 3.

Figure 5(b) demonstrates the time distribution to arrange total stock according to the set of 3 basic parameters (Tseparator, Tstoring and Vconveyor).

6 Conclusion and Future Work

In this study, a smart warehouse testbed was proposed. The testbed can help to eliminate the difficulty of establishing a testing environment for new hardware and software. Because of the similarities in hardware design, the proposed testbed system can be readily updated from current warehouses. Pallet circulation is one of the testbed's strengths; as it can complete transactions fast and with little human interaction, allowing it to collect significant amounts of data for optimization challenges using AI and deep learning.

Through simulation results, the optimal determination of the value set (Tseparator, Vconveyor and Tstoring) has been conducted so that the storage time is the fastest without pallet blockage on the conveyor. Mathematical models have been built based on experimental results. Accordingly, Tseparator based on a given set of values (V conveyor and T storing) can be estimated. Similarly, the storage time to arrange total stock (1000 pallets) based on a given set of values (Tseparator, Vconveyor and Tstoring) can be assessed. Further study will be conducted to introduce a whole testbed system with mathematical functions that can be applied to any number of rack systems and pallets to be stored in the near future.

Acknowledgement. This research is financially supported by Eastern International University, Binh Duong Province, Vietnam.

References

1. Blancas, L.C.: Rapid growth, Challenges and opportunities in Vietnam's logistics limited connectivity (2019). http://vietnamsupplychain.com/assets/files/530ef941689c9done_2_Blancas_Vietnam_Logistics_Challenges.pdf. Accessed June 2019
2. Arnold, J., Arvis, J.F., Mustra, M.A., Horton, B., Carruthers, R., Ojala, L.: Trade and Transport Facilitation Assessment: A Practical Toolkit for Country Implementation (2010). https://openknowledge.worldbank.org/handle/10986/2490. Accessed 01 June 2010
3. Roodbergen, K.J., Vis, I.F.A.: A survey of literature on automated storage and retrieval systems. Eur. J. Oper. Res. **194**(2), 343–362 (2009). https://doi.org/10.1016/j.ejor.2008.01.038
4. Jerman, B., Ekren, B.Y., Küçükyasar, M., Lerher, T.: Simulation-based performance analysis for a novel AVS/RS technology with movable lifts. Appl. Sci. **11**(5), 2283 (2021). https://doi.org/10.3390/app11052283
5. Shi, H.-y., Lu, X., Li, D.-w.: Travel time analysis of the single and dual command of AS/RS. In: 2017 29th Chinese Control And Decision Conference (CCDC), Chongqing, China, pp. 3407–3413. IEEE (2017). https://doi.org/10.1109/CCDC.2017.7979095
6. Hu, H., Jia, X., He, Q., Fu, S., Liu, K.: Deep reinforcement learning based AGVs real-time scheduling with mixed rule for flexible shop floor in industry 4.0. Comput. Ind. Eng. **149** (2020). https://doi.org/10.1016/j.cie.2020.106749
7. Li, M.P., et al.: Simulation analysis of a deep reinforcement learning approach for task selection by autonomous material handling vehicles. In: 2018 Winter Simulation Conference (WSC), Gothenburg, Sweden, pp. 1073–1083. IEEE (2018). https://doi.org/10.1109/WSC.2018.8632448
8. Zhou, T., Tang, D., Zhu, H., Zhang, Z.: Multi-agent reinforcement learning for online scheduling in smart factories. Robot. Comput. Integr. Manuf. **72** (2021). https://doi.org/10.1016/j.rcim.2021.102202
9. Gaku, R., Takakuwa, S.: Simulation modeling of shuttle vehicle-type mini-load AS/RS systems for E-commerce industry of Japan. In: 2017 Winter Simulation Conference (WSC), Las Vegas, NV, USA, pp. 1–10. IEEE (2017). https://doi.org/10.1109/WSC.2017.8248036
10. Boynton, P.: Summary Report — Measurement Challenges and Opportunities for Developing Smart Grid Testbeds, Maryland. (2015). https://dss-lab.github.io/pub/misc/SG-Testbed-Workshop-Report-FINAL-12-8-2014.pdf. Accessed 05 July 2021
11. Kaczmarczyk, V., Baštán, O., Bradáč, Z., Arm, J.: An industry 4.0 testbed (Self-Acting barman): principles and design. IFAC-Papersonline **51**(6), 263–270 (2018). https://doi.org/10.1016/j.ifacol.2018.07.164

Fog, Edge, and Cloud Computing

Mobility-Aware Task Offloading Enhancement in Fog Computing Networks

Heba Raouf[1](✉), Rania Abdallah[2], Heba Y. M. Soliman[2], and Rawya Rizk[2]

[1] Arab Academy for Science Technology and Maritime Transport, Port Said, Egypt
heba.raouf@eng.psu.edu.eg

[2] Electrical Engineering Department, Port Said University, Port Said 42526, Egypt
{eng_rania99,hebayms,r.rizk}@eng.psu.edu.eg

Abstract. The evolution of mobile networks from 1G to 5G, as well as the development of mobile devices and new applications are the reasons to develop Quality of Services (QoS). The fog computing concept aims to improve QoS for high-sensitivity applications in particular. Due to the great distance of propagation, these apps are unable to deal with the cloud, which may result in further delays. Also, the mobility of users must be taken into account because it results in services migration. The paper enhances migration performance by instead of dealing with the cloud when the user moves from one fog to another, the user can deal with cooperative fog servers along the direction of motion. The proposed model investigates the system cost in terms of CPU cycles and time delay when offloading and in case of local computing. The proposed model consists of three layers; cloud, fog computing node (FCN) and user equipment (UE) layer. A large number of users move with different speeds around base stations which require different resources. FCN layer has many fogs with available computing resources and connected in backhaul. However, mobility may cause UE cannot complete the required task and in this case, the cloud will serve the user through the fog layer in between. This migration concept will cost the system more energy and latency. The proposed model will be more efficient in migrating than using the cloud since it will save the user revenue, which is a function of the time delay and energy required to send and execute tasks.

Keywords: Cloud computing · Fog computing · MEC servers · Mobility · Service migration

1 Introduction

Cloud computing provides capabilities such as scalability, on-demand resource allocation, reduced management efforts, and easy applications and services provisioning. It is widely used however it still has some drawbacks. Basically the connectivity between the cloud and the end devices is set over the internet and is not convenient to latency sensitive applications. In addition, data generated from billions of sensors, known as big data, cannot be transmitted and processed in the cloud. Furthermore, some Internet of

© The Author(s), under exclusive license to Springer Nature Switzerland AG 2022
A. E. Hassanien et al. (Eds.): AMLTA 2022, LNDECT 113, pp. 569–580, 2022.
https://doi.org/10.1007/978-3-031-03918-8_47

Things (IoT) applications need to be processed faster than the cloud's current capability [1].

Fog computing is an emerging paradigm that exploits computation, communication, and storage capabilities at the edge of the network. Unlike cloud computing, fog computing can support delay sensitive service requests from farthest users with reduced energy consumption and low traffic congestion. Basically, fog networks can be considered as offloading to core computation and storage. Fog nodes in fog computing decide to either process the services using its available resource or send to the cloud server. Thus, fog computing helps to achieve efficient resource utilization and higher performance considering the delay, bandwidth, and energy consumption [2].

Figure 1 gives a diagram about fog model. It has three layers: The cloud layer, the fog layer, and the IoT/end users layer. The fog layer can be formed by one or more fog domains, controlled by the same or different providers. Each of these fog domains is formed by the fog nodes that can include edge routers, switches, gateways, access points, smartphones, etc. Fog layer works under provisioning of cloud, thus it cannot operate in a standalone mode. This pays attention to the interactions between the fog and the cloud. Moreover, fog has an n-tier architecture, offering more flexibility to the system [3, 4].

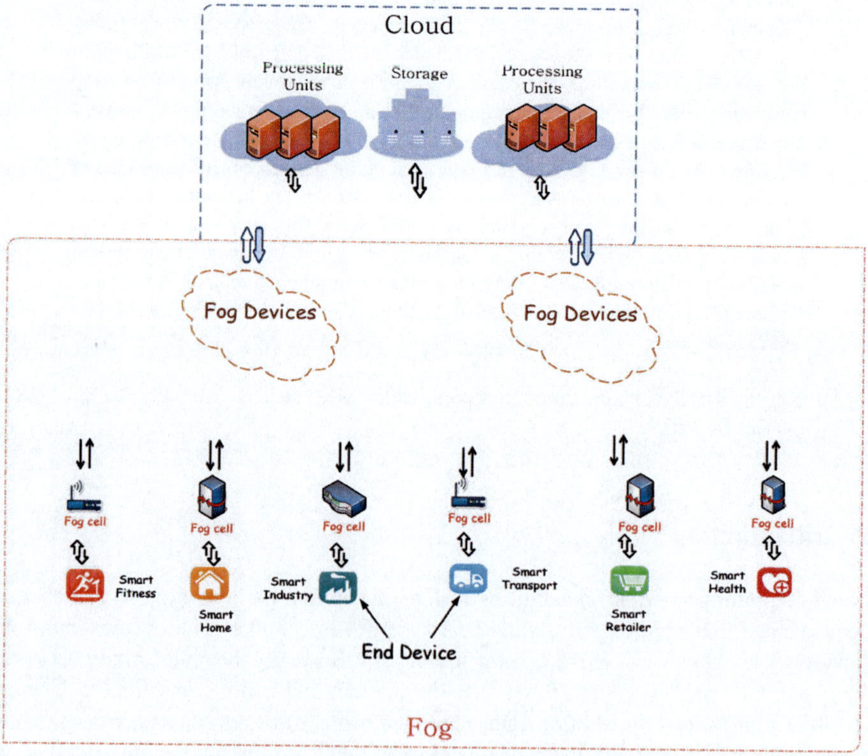

Fig. 1. Fog computing architecture.

Compared to cloud which is centralized computing paradigm, fog computing is a distributed computing paradigm where processing is done at the edge of the network with the interaction of the cloud infrastructure. It enables a computing facility for IoT environments or other latency sensitive applications. It is a virtualized platform that provides computing, storage, and networking services between IoT devices and traditional cloud computing Data Centers (DCs). The term fog computing or edge computing means that rather than connecting and working from a centralized cloud, fog systems operate on network ends. Fog location can be near devices with processing capability like telecommunication network and it will connect to devices anywhere in factory, vehicles or homes [5].

In this paper, a proposed model for enhancing migration performance is presented. When the user moves from one fog to another, the user deals with cooperative fog servers along the direction of motion. The proposed model investigates the system cost in terms of CPU cycles and time delay when offloading and in case of local computing.

The rest of this paper is organized as follows. In Sect. 2, computing paradigms, mobile edge computing and vehicular edge computing, are described. In Sect. 3, related work is presented. In Sect. 4, the proposed model is demonstrated. In Sect. 5, the problem is formulated and validated the results. Section 6 draws the conclusion and future work.

2 Computing Paradigms

2.1 Mobile Edge Computing (MEC)

Figure 2 shows MEC servers. They can be placed on Base Station (BS) tower of existing cellular network. This server provides storage and processing abilities to users near edge. They can work separated from cloud [5]. MEC provides an Information Technology (IT) service environment and cloud computing capabilities at the edge of the mobile network and near mobile subscribers [6, 7]. This concept enhances computation and avoids congestion and system failure. It provides low latency services because it is near end user which makes it suitable to low latency services. Edge distributed devices utilize low-level signaling for information sharing. MEC receives information from edge devices within the Local Access Network (LAN). It needs four components; User Equipment (UE), network operator (manages operation at BS, core mobile network and MEC server), internet provider (maintain internet connection routers) and application service provider (maintain application services in DC).

2.2 Vehicular Edge Computing (VEC)

VEC network combines MEC concept and vehicular network. Mobility is the driving factor for emerging VEC concept [8]. In VEC, MEC servers or Road Side Units (RSU) are deployed along the road during mobility. This makes offloading easier to vehicles to compute their tasks by offloading to these servers instead of using cloud. MEC servers provides communication and computation resources for passing vehicles. Offloading cause communication overhead such as power and bandwidth and computation resources (CPU cycles needed to process the task). Offloading is useful when it occurs if the user

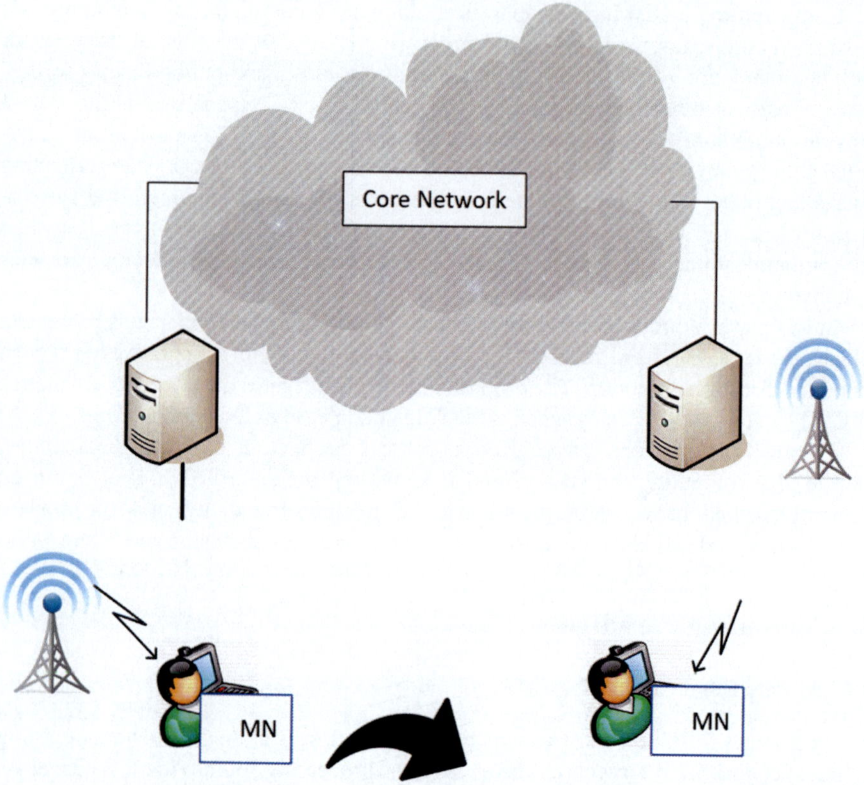

Fig. 2. MEC model.

inside the coverage area of the MEC server [9, 10]. However, due to mobility user can receive computing results in another server in another location along direction of motion. This concept is known as migration that user may receive result in another RSU to avoid service discontinuity. So, user or vehicular user need to know when to offload task and to which MEC server can offload task [11]. Migration may fail if user migrate task to wrong fog, crowded one or it may has not enough resource to compute task.

3 Related Work

The development of smart phones and emerging applications is the reason for searching new techniques for processing data. Also, IoT devices that generate big data need huge computing which is not available in local devices. Offloading helped in these issues to save time and battery life of devices by computing the task in fogs or cloud. So, resources must be distributed according to the need and type of task. Also, MEC concept attracts attention due to supporting vehicular users as well as stable users. This is done efficiently by jointly optimizing offloading decision in presence of mobility and resources allocation. Many papers studied the issue and we will present some of them in this section.

In [12] the problem of resources allocation and task offloading is studied to enhance the user revenue when offloading to MEC servers. The author measured user revenue in terms of energy consumed and time completion of computing task and formulated the problem into MINLP. In [13], the problem is formulated by jointly optimizing computation resources allocation and task offloading to minimize service delay and solve it by alternating direction method of multipliers (ADMM) method and difference of convex functions (D.C.) programming. Also, a location based offloading scheme is proposed as a function of computing overhead. Besides, joint offloading and task scheduling in cooperative MEC system problem is formulated in [14] as MINLP problem to solve task delay and resource consumption.

Additionally, heuristic mobility aware offloading algorithm (HMAOA) is proposed in [15] to obtain approximate offloading decision. [16] formulated the problem of offloading decision and task scheduling as MINLP and proposed partheno genetic algorithm (PGA) and heuristic rules to achieve an approximate optimal solution. In [17], a collaborative computation offloading and resource allocation optimization (CCORAO) scheme is studied to solve the problem of offloading decision and resource allocation in MEC enabled vehicular network in which offloading decisions are taken through game-theoretic approach.

[18] investigates the energy minimization of offloading considering constraints of service delay. Optimization algorithm is developed that is artificial fish swarm algorithm (AFSA) to solve energy optimization problem. Service migration also has been studied in cellular network to support real time services in [19]; the model is QoS aware scheme based on handover a model for minimizing overhead problem for MEC enabled vehicular network. [20] developed a non-convex problem to minimize the overhead which is induced by mobility of user and resulting computing offloading.

Performance of cooperated and independent MEC servers is studied in [21] in conjunction with location based offloading scheme to enhance system computation and communication overhead. In [22], three layers VEC model is proposed and the authors focused on wireless resource allocation because it cause system degradation. Also mobility is considered to improve resources management. In [23], system cost is defined by delay and communication and computing cost due to mobility and the author developed matching based task offloading and resource allocation algorithm.

4 Proposed Model

The model consists of three layers; cloud, fog computing node FCN and user equipment UE layer. Large number of users move with different speed around base stations which requires different resources. FCN layer has many fogs with available computing resources and connected in backhaul. However, mobility may cause UE cannot complete required task. In this case, cloud will serve the user through fog layer in between. This migration concept will cost the system more energy and latency. The paper investigates two models; local computing model and fog computing model. Mobility of user is formulated here by exponential function of sojourn time [24].

4.1 Fog Computing Architecture

UEs are distributed randomly around FCNs. $N_f = \{1, 2, \ldots, \mathrm{F}\}$ is the set of FCNs. $N_u = \{1, 2, \ldots, \mathrm{U}\}$ is the set of UEs. Each user has only one task Z_u defined by 3 parameters $\{D_u, f_u, T_u^{max}\}$. D_u is the data size of task. f_u is the required computing resources (CPU cycles) to process task $f_u = \varepsilon D_u$. T_u^{max} is the maximum latency allowed to perform task. Let $\mathrm{S} = \{s_u, u \in N_u\}$ is the set of offloading decisions. $s_u = \{0, 1\}$. $s_u = 0$ means UE will not offload the task Z_u and compute locally at UE. $s_u = 1$ means UE will offload the task Z_u to FCN. The system uses Orthogonal Frequency Division Multiple Access OFDMA as a multiple access technique in uplink. Each FCN can connect to many users simultaneously, but UE can access one FCN. Each user is assumed to occupy one sub-channel in uplink to avoid interference. $\mathrm{A} = \{a_{uf}, \text{ for } u \in N_u \text{ f } \in N_f\}$ is the set of all FCNs selection variables. $a_{uf} = 1$ means that UE decided to offload the task. $\sum_{u \in N_u} a_{uf} \le K$; For K is the number of sub-channels of each FCN. So, fog can manage K UEs simultaneously. $\sum_{f \in N_f} a_{uf} \le 1$ defines that each UE can access only one FCN. $U_f = \{u \in N_u, a_{uf} = 1\}$ is the set of UEs will offload their tasks. Revenue of UE is defined by the amount of revenue the user will gain when offloading a task compared to performing it locally. Revenue is function of energy consumed and computing delay.

4.2 Local Computing Model

Here $s_u = 0$. The task execution needs energy E_u^{local} and time T_u^{local}. We define local computing capacity of UE c_u^{local} as CPU cycles, we can define local execution time as [25]:

$$T_u^{local} = \frac{f_u}{c_u^{local}} \tag{1}$$

For f_u is the computing resource or CPU frequency (CPU cycles/bit). The energy consumption is:

$$E_u^{local} = kc^2 = k\left(c_u^{local}\right)^2 f_u \tag{2}$$

E_u^{local} is the energy consumption. K is capacitance of CPU. C is required computing resource. Now the total overhead incurred by performing task locally is:

$$q_u^{local} = \lambda_u^E E_u^{local} + \lambda_u^T T_u^{local} \tag{3}$$

For λ_u^E, λ_u^T are weighting factor depend on task type and its requirements $\lambda_u^E + \lambda_u^T = 1 \, for \, \lambda_u^E = \{0, 1\} \,\&\, \lambda_u^T = \{0, 1\}$.

4.3 Fog Computing Model

Here $s_u = 1$. So, the total processing delay T_u^F consists of transmission time from UE to FCN on uplink (t_{uf}^{up}), execution time needed for processing on fog (t_{uf}^{exe}) and transmission time from fog back to UE but this factor is ignored here. Transmission time t_{uf}^{up} is given by:

$$t_{uf}^{up} = \frac{D_u}{r_{uf}}, \quad u \in N_u \;\&\, f \in N_f \tag{4}$$

For r_{uf} is the transmission rate. By Shannon theorem, transmission rate is [26]:

$$r_{uf} = w log_2(1 + SINR_{uf}) \tag{5}$$

w is the transmission bandwidth, $SINR_{uf}$ is the signal interference noise ratio of wireless channel between UE and FCN. Let $P = \{p_u, u \in N_u\}$ is the maximum power the users send their tasks and g_{uf} is the sub-channel gain from UE to FCN. Consequently, $G = \{g_{uf} \forall u \in N_u \,\&\, \mathrm{f} \in N_f\}$ the set of all sub-channels gains for all users accessing FCN. So

$$SINR_{uf} = \frac{p_u \quad g_{uf}}{\sigma^2 + \sum_{i \in N_{F_{/f}}} \sum_{j \in u_i} a_{ij} p_i g_{ij}} \tag{6}$$

σ^2 is the noise variance, $\sum_{i \in N_{f/\{f\}}} \sum_{j \in u_i} a_{ij} p_i g_{ij}$ is the accumulated interference from adjacent sub-channels. Execution time t_{uf}^{exe} depends on computation capacity and CPU cycles/second available on fog. c^F is the total computing capacity of each fog and Computing resources applied to users is c_{uf}^F which cannot exceed c^F. So, execution time is

$$t_{uf}^{exe} = \frac{f_u}{c_{uf}^F} \tag{7}$$

Also, total computing time on FCN:

$$T_{uf}^F = t_{uf}^{up} + t_{uf}^{exe} \tag{8}$$

Total energy consumption:

$$E_{uf}^{up} = p_u t_{uf}^{up} \tag{9}$$

Finally total overhead incurred or offloading cost:

$$q_{uf}^F\left(c_{uf}^F\right) = \lambda_{uf}^E E_{uf}^{up} + \lambda_{uf}^T T_{uf}^F \tag{10}$$

4.4 Fog Computing Model with Mobility

Mobility of users is described by sojourn time t_{uf}^s and modeled by exponential distribution as following PDF:

$$f_{\tau_{uf}} = \frac{1}{\tau_{uf}} e^{\frac{-t}{\tau_{uf}}}, \quad t \geq 0, \quad \forall u \in N_u, f \in N_f \tag{11}$$

τ_{uf} is the average sojourn time of UE u in FCN f. Because each user has different mobility trace, τ_{uf} has different value between them. So, the sojourn time of each user are assumed independently identical distributed i.d.d (Gaussian i.d.d). We have two cases describe the offloading to fog based on mobility of UE and relation between processing time and sojourn time.

Case 1: $\mathbf{t^s_{uf} > T^F_{uf}}$ (offloading + no result migration)

In this case sojourn time t^s_{uf} of UE u in a FCN f coverage is greater than processing time T^F_{uf}. This means that computing results will be sent to UE before leaving the FCN. There is no need for results migration and probability of this case denoted by $P_{\tau_{uf}}(t^s_{uf} > T^F_{uf})$. Finally, we can define total cost of UE u on FCN f is given as:

$$q^{F,1}_{uf} = q^F_{uf}\left(c^F_{uf}\right), \forall u \in N_u, f \in N_f \tag{12}$$

Case 2: $\mathbf{t^s_{uf} < T^F_{uf}}$(offloading + result migration)

UE sojourn time is lower than processing time, so task Z_u will be offloaded to FCN f and processing results will be migrated to UE via cloud. Probability in this case is denoted by $P_{\tau_{uf}}(t^s_{uf} < T^F_{uf})$. Migration will result in energy consumption and time delay (cost of migration) which will be paid by user. Migration cost is given by $q^{mig}_u = \delta D_u$.That is assumed to be related to task size, task type and distance between fog and UE. Therefore, cost of case 2 is given by:

$$q^{F,2}_{uf} = q^F_{uf}\left(c^F_{uf}\right) + q^{mig}_u, \ \forall u \in N_u, f \in N_f \tag{13}$$

Finally cost in both cases will be summarized as:

$$q^F_{uf} = \begin{cases} q^{F,1}_{uf}, P_{\tau_{uf}}(t^s_{uf} > T^F_{uf}). \\ q^{F,2}_{uf}, P_{\tau_{uf}}(t^s_{uf} \le T^F_{uf}). \end{cases} \forall u \in N_u, f \in N_f \tag{14}$$

Since we are dealing with random mobility and migration and so random cost calculation. Therefore, we will calculate expectation of total cost as following:

$$\overline{q}^F_{uf} = P_{\tau_{uf}}\left(t^s_{uf} > T^F_{uf}\right)q^{F,1}_{uf} + P_{\tau_{uf}}\left(t^s_{uf} \le T^F_{uf}\right)q^{F,2}_{uf}, \tag{15}$$

where

$$P_{\tau_{uf}}(t^s_{uf} \le T^F_{uf}) = \int_0^{T^F_{uf}} \frac{1}{\tau_{uf}} e^{\frac{-t^s_{uf}}{\tau_{uf}}} dt^s_{uf} = -e^{\frac{-t^s_{uf}}{\tau_{uf}}} \Big|_0^{T^F_{uf}} = -e^{\frac{-t^s_{uf}}{\tau_{uf}}} + 1$$

$$P_{\tau_{uf}}(t^s_{uf} > T^F_{uf}) = 1 - P_{\tau_{uf}}(t^s_{uf} \le T^F_{uf})$$

5 Problem Formulation

Revenue of UE is defined by time delay and energy consumption reduction. So, revenue equals cost of performing the task locally subtracted from cost of migration which is the reason of overhead. If this equation gives positive result, this means that offloading saves the cost and vice versa. The variable Q_u represents user revenue:

$$Q_u\left(s_u, a_{uf}, c_{uf}^F\right) = \begin{cases} q_u^{local} - \sum_{f \in N_f} a_{uf}\overline{q}_{uf}^F, & s_u = 1 \\ 0 & s_u = 0 \end{cases} \tag{16}$$

The problem will be formulated as follow:

$$P_1 : \max_{S,A,C_{off}} \sum\nolimits_{u \in N_u} Q_u(s_u, a_{uf}, c_{uf}^F) \tag{17}$$

Subjected to:

$$\begin{array}{ll} C1 : t_{uf}^{up} < T_{uf}^F \le T_u^{max}, \forall u \in N_u & C2 : \sum_{u \in N_u} c_{uf}^F \le c^f \\ C3 : a_{uf} = \{0, 1\}, \forall u \in N_u, f \in N_f & C4 : a_{uf} = \mathrm{I}\left(c_{uf}^F\right), \forall u \in N_u, f \in N_f \\ C5 : \sum_{u \in N_u} a_{uf} \le K, \forall f \in N_f & C6 : \sum_{f \in N_f} a_{uf} \le 1, \forall u \in N_u \\ C7 : s_u = \{0, 1\}, \forall u \in N_u & C8 : s_u = \sum_{f \in N_f} a_{uf}, \forall u \in N_u \end{array}$$

S is the offloading decision set and A is the FCN parameters including all users chosen to offload. C_{off} is the computation capacity. C1 means that maximum processing time on fog will not exceed maximum latency. C2 means that the amount of computing resources distributed to UE which will offload to fog f will not exceed total computation resources available for that fog f. C3 is the UE decision in case of offloading or performing locally. I in C4 is indicator function i.e. $I\,(.) = 1$ if "." > 0, otherwise $I\,(.) = 0$ indicates that no resources will be applied to UE will locally computing tasks. C5 indicates that the number of UE the fog will serve in the same time is at most K that depends on number of sub-channel. C6 indicates that each UE can access only one fog for offloading. C7 is a decision variable of UE in case of offloading equals 1, otherwise equals 0. C8 means that if UE decided to offload, it must access only one fog.

The constraints show that revenue of user depends on offloading decision, FCN selection and computation resources allocation. These factors highly affect mobility. In case of migration failure or migration to FCN with high latency, cost q_u^{mig} will increase and revenue of user will decrease. So, to improve the revenue and reduce probability of migration, we need to select the suitable FCN, allocate the sufficient resource based on sojourn time and choose to offload correctly. To do that, an algorithm is named Gini Coefficient algorithm is developed to optimize offloading decision with FCN selection properly. Another algorithm is required for resources allocation that is resource optimization algorithm based on Gini coefficient. Figure 3 depicts the performance of the model in terms of cost and different number of fogs. More fogs can process more tasks and gives more resources and this makes network accommodates more UEs.

$$\text{For } s_u = 1, Q_u = q_u^{local} - \sum_{f \in N_f} a_{uf} \bar{q}_{uf}^F$$
$$= \lambda_u^E k f_u \left(c_u^{local}\right)^2 + \lambda_u^T \left(\frac{f_u}{c_u^{local}}\right) - \sum_{f \in N_f} a_{uf} \left[\delta D_u \left(e^{-\frac{T_{uf}^F}{\tau_{uf}}+1} \right) + \lambda_u^T T_{uf}^F + \lambda_u^E E_{uf}^{up} \right]. \quad (18)$$

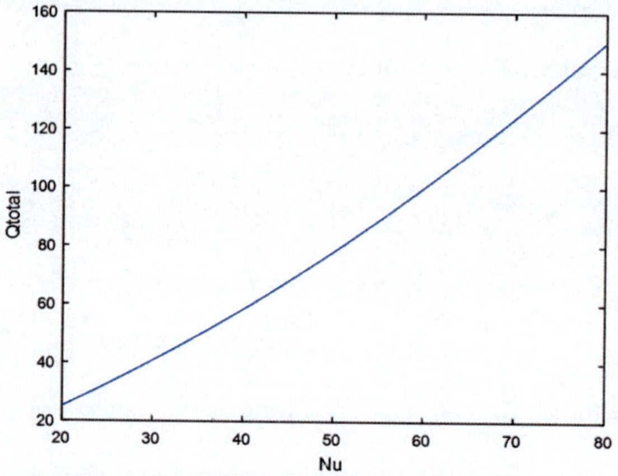

Fig. 3. Total cost performance with no. of fogs.

This paper represents the idea of cooperation between MEC servers instead of fogs. So in case of mobility and UE left the coverage area of the current base station, the user can communicate with the following MEC server rather than cloud. So, task will be transmitted to Back Haul network *BH*. Consequently, connection will be more rapid than cloud and can deal more efficiently to save latency. Figure 4 represents the expected performance of using cooperated MEC servers. So the final revenue of the UE is given by:

$$Q_u = q_u^{local} - \sum_{f \in N_f} a_{uf} \bar{q}_{uf}^F$$
$$= \lambda_u^E k f_u (c_u^{local})^2 + \lambda_u^T \left(\frac{f_u}{c_u^{local}}\right) - \sum_{f \in mec} a_{mec} \left[e^{\frac{T_{uf}^{mec}}{\tau_{uf}}} (\lambda_u^E E^{BH} + \lambda_u^T t^{BH}) + \lambda_u^T T_u^{mec} \right.$$
$$\left. + \lambda_u^E E_u^{up} \right] \quad (19)$$

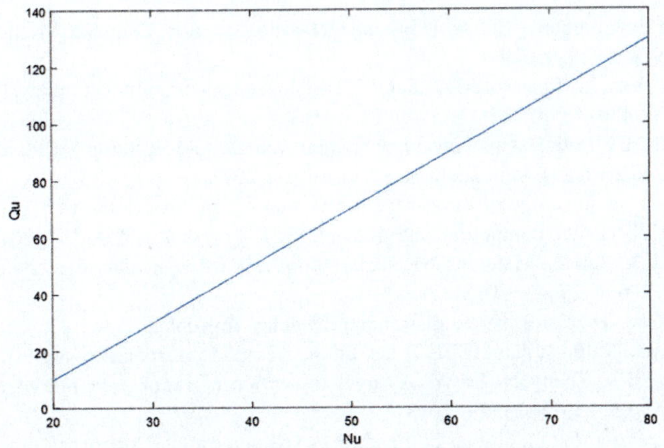

Fig. 4. Total cost performance with number of users.

6 Conclusion

The paper investigates two models; local computing model and fog computing model. Mobility of user is formulated by exponential function of sojourn time. The paper solves the problem of performance degradation due to service migration. Cooperated MEC servers are used instead of fogs because these servers can deal more closely and faster with moving users. The problem was formulated as mixed integer nonlinear programming and solved to investigate the performance. The proposed model is more efficient than using cloud in migration because it saves the user revenue which is a function of time delay and energy of sending task and processing it. In future work, machine learning will be suggested to model the users' mobility in order to be more realistic and represent real life networks.

References

1. Nashaat, H., Ahmed, E., Rizk, R.: IoT application placement algorithm based on multi-dimensional QoE prioritization model in fog computing environment. IEEE Access **8**, 111253–111264 (2020)
2. Abdel-Kader, R.F., El-Sayad, N.E., Rizk, R.Y.: Efficient energy and completion time for dependent task computation offloading algorithm in industry 4.0. PLoS ONE **16**(6), e0252756 (2021)
3. Yannuzzi, M., Milito, R., Serral-Gracia, R., Montero, D., Nemirovsky, M.: Key Ingredients in an IoT recipe: fog computing, cloud computing, and more fog computing. In: 2014 IEEE 19th International Workshop on Computer Aided Modeling and Design of Communication Links and Networks (CAMAD), pp. 325–329 (2014)
4. Mouradian, C., Naboulsi, D., Glitho, R.H., Morrow, M.J., Polakos P.A.: A comprehensive survey on fog computing: state-of-the-art and research challenges. IEEE Commun. Surv. Tutor. **20**, 416–464 (2017)
5. Bonomi, F., Milito, R., Natarajan, P., Zhu, J.: Fog computing: a platform for internet of things and analytics. In: Bessis, N., Dobre, C. (eds.) Big Data and Internet of Things: A Roadmap

for Smart Environments. SCI, vol. 546, pp. 169–186. Springer, Cham (2014). https://doi.org/10.1007/978-3-319-05029-4_7
6. Abbas, N., Yan, Z., Taherkordi, A., Skeie, T.: Mobile edge computing: a survey. IEEE Internet Things J. **5**, 450–465 (2017)
7. Uhlemann, E.: Initial steps toward a cellular vehicle-to-everything standard [connected vehicles]. IEEE Trans. Veh. Technol. **12**(1), 14–19 (2017)
8. ElSayed, H., et al.: Edge of things: the big picture on the integration of edge, IoT and the cloud in a distributed computing environment. IEEE Access **6**, 1706–1717 (2017)
9. Liu, P., Li, J., Sun Z.: Matching-based task offloading for vehicular edge computing. IEEE Access **7,** 27628–27640 (2019)
10. Du, L., Dao, H.: Information dissemination delay in vehicle-to-vehicle communication networks in a traffic stream. IEEE Trans. Intell. Transp. Syst. **16**(1), 66–80 (2015)
11. Wang, S., Xu, J., Zhang, N., Liu Y.: A survey on service migration in mobile edge computing. IEEE Access **6,** 23511–23528 (2018)
12. Tran, T.X., Pompili, D.: Joint task offloading and resource allocation for multi-server mobile-edge computing networks. IEEE Trans. Veh. Technol. **68**(1), 856–868 (2019)
13. Wang, Y., Tao, X., Zhang, X., Zhang, P., Hao, T.: Cooperative task offloading in three-tier mobile computing networks: an ADMM framework. IEEE Trans. Veh. Technol. **68**, 2763–2776 (2018)
14. Yang, C., Liu, Y., Chen, X., Zhong, W., Xie, S.: Efficient mobility-aware task offloading for vehicular edge computing networks. IEEE Access **7,** 26652–26664 (2019)
15. Zhan, W., Luo, C., Min, G., Wang, C., Zhu, Q., Duan, H.: Mobility aware multi-user offloading optimization for mobile edge computing. IEEE Trans. Veh. Technol. **69**(3), 3341–3356 (2020)
16. Sun, J., et al.: Joint optimization of computation offloading and task scheduling in vehicular edge computing networks. IEEE Access **8,** 10466–10477 (2020)
17. Zhao, J., Li, Q., Gong, Y., Zhang, K.: Computation offloading and resource allocation for cloud assisted mobile edge computing in vehicular networks. IEEE Trans. Veh. Technol. **68**(8), 7944–7956 (2019)
18. Yang, L., Zhang, H., Li, M., Guo, J., Ji, H.: Mobile edge computing empowered energy efficient task offloading in 5G. IEEE Trans. Veh. Technol. **67**, 6398–6409 (2018)
19. Li, J., et al.: Service migration in fog computing enabled cellular network to support real time vehicular communications. IEEE Access **7**, 13704–13714 (2019)
20. Wang, J., Feng, D., Zhang, S., Tang, J., Quek, T.Q.S.: Computation offloading for mobile edge computing enabled vehicular networks. IEEE Access **7**, 62624–62632 (2019)
21. Yang, C., Liu, Y., Chen, X., Zhong, W., Shengli, X.: Efficient mobility-aware task offloading for vehicular edge computing networks. IEEE Access **7,** 26652–26664 (2019)
22. Li, X., et al.: Energy-efficient computation offloading in vehicular edge cloud computing. IEEE Access **8,** 37632–37644 (2020)
23. Zhang, Y., Qin, X., Song, X.: Mobility-aware cooperative task offloading and resource allocation in vehicular edge computing. University of Birmingham, IEEE Explore (2020)
24. Wang, D., Liu, Z., Wang, X., Lan, Y.: Mobility aware task offloading and migration schemes in fog computing networks. IEEE Access **7,** 43356–43368 (2019)
25. Zhao, P., Tian, H., Qin, C., Nie, G.: Energy-saving offloading by jointly allocating radio and computational resources for mobile edge computing. IEEE Access **5,** 11255–11268 (2017)
26. Chen, X.: Decentralized computation offloading game for mobile cloud computing. IEEE Trans. Parallel Distrib. Syst. **26**(4), 974–983 (2015)

Comprehensive Study on Machine Learning-Based Container Scheduling in Cloud

Walid Moussa(✉), Mona Nashaat, Walaa Saber, and Rawya Rizk

Electrical Engineering Department, Port Said University, Port Said 42526, Egypt
eng.walidmoussa@gmail.com, {Monanashaat,walaa_saber, r.rizk}@eng.psu.edu.eg

Abstract. Containers are considered the best lightweight application in virtualization technology, and it is promising in enhancing cloud computing services quality. Due to cloud workload diversity, the scheduler module is considered the central part of the containers framework that optimizes resource utilization and reduces cost and energy consumption. Container scheduling algorithms can be classified into four main types: heuristic, metaheuristic, mathematical modeling, and machine learning. Machine Learning, with its high ability to analyze data and train the system to predict outputs based on previous data considered the best choice for predicting workloads and performance metrics. Such a vision allows schedulers to improve the quality of resource allocation with changing user requests rates in complicated work environments. This paper presents a comprehensive literature review for the current container orchestration machine learning-based algorithms. A detailed study is proposed for the main features, advantages, and disadvantages of existing algorithms.

Keywords: Machine learning · Container orchestration · Container engine · Virtual machine · Container cluster · Scheduling

1 Introduction

Recently Cloud and Fog computing became one of the most popular technologies that provide computing services for society, individuals, and economies worldwide [1]. Virtualization technology became the core of cloud and fog computing in the last few years because of its ease in deployment and migration from one server to another. Virtualization is an independent layer run on top of the host operating system [2]. Applications on standard cloud use Virtual Machines (VMs) hosted on top of data center physical servers sharing their resources. Scheduling algorithms allocate Physical resources to virtual machines based on application requirements executed by each VM.

Most recently, applications have migrated towards containers as lightweight virtualization technology because of their ease of deployment and less resource consumption [3] Containers are small (only megabytes) and have a short start time (just seconds to start) On the other hand, virtual machines have a large size of gigabytes and minutes as starting times. Containers have the same operating system; they have smaller management overhead, only manage and maintain a single operating system.

© The Author(s), under exclusive license to Springer Nature Switzerland AG 2022
A. E. Hassanien et al. (Eds.): AMLTA 2022, LNDECT 113, pp. 581–592, 2022.
https://doi.org/10.1007/978-3-031-03918-8_48

Container orchestration architecture consists of working nodes that run the containers and an orchestration platform representing the container orchestration backbone. Orchestration platform contains main functions such as scheduler, cluster monitoring, and working node management. The container's scheduler is responsible for allocating workload across available nodes in the container cluster and managing the container's life cycle. Many orchestration frameworks are proposed to manage applications, and schedule containers, the most important are Google Kubernetes [4], Docker Swarmkit [5], and Apache Mesos [6] schedulers.

Scheduling techniques can be categorized into four main categories: mathematical modeling, heuristic, metaheuristic, and machine learning [7–10]. Machine learning [11] is one of the most active research areas that has succeeded in many applications and fields such as image recognition, speech recognition, medical diagnosis, smart building, and container orchestration. Machine learning is very promising in the container cluster orchestration field. One of the critical factors of Machine learning success is the availability to train the model with big data. Due to the diverse nature of containers workload, machine learning, compared with other scheduling algorithms, have precedence in taking intelligent scheduling decisions based on workload and performance parameters prediction.

In the presence of widespread cloud container applications, there is a great need for improving resource utilization, quality of service, and optimizing energy consumption. With its high accuracy in prediction and classification, machine learning plays a vital role in container orchestration technology to deliver high-quality services with efficient load balancing, resource utilization, SLA assurance, and energy efficiency. Although there are many literature reviews in the field of container orchestration field [7, 12], the work that focuses on applying machine learning techniques to container orchestration is still limited. Still, it is proliferating due to the successes of machine learning achieved every day in this field. Therefore, it is necessary to present a literature review to summarize and evaluate the latest approaches and the new challenges in this field.

2 Background

2.1 Containers and Virtual Machines

Containers are self-contained units of software that contain their binaries, dependencies, and libraries in one package as an independent application isolated from the operating system level. Containers are similar to VMs as both represent virtualization techniques that allow multiple applications and processes to simultaneously share a single physical node. Besides the traditional cloud services: software as a service (SaaS), platform as a service (PaaS), and infrastructure as a service (IaaS), a new cloud service is introduced called container as a service (CaaS) and lie between the IaaS layer and PaaS layer and mix these two layers.

Containerized applications can be classified into three main categories based on their application architecture [13]: monolithic, microservices, and serverless. Monolithic is an example of an all-in-one architecture where the whole application is developed in one deployed unit like Microsoft azure [14]; they are easy to develop and deployable in small-scale applications but not designed for enterprise ones. In Micro-service architecture, the application is split into small, lightweight independent, self-contained units [15]. Micro-services interact with each other through simple communication methods like representational state transfer (REST). Serverless architecture defines the model for event-driven applications that handle stateless computational functions. Serverless applications are represented as workflows of individual functions chain.

The overall architecture of containers and virtual machines are presented in Fig. 1 and Fig. 2. As shown in Fig. 1, containers run on a single host sharing the same operating system kernel. Containers are managed through a container engine that runs, creates, and deploys containers on the host. On the other hand, virtual machines with different operating systems can run on the same physical server, as shown in Fig. 2. Virtual machines need Hypervisor as an intermediate layer between the guest operating system and node physical resources.

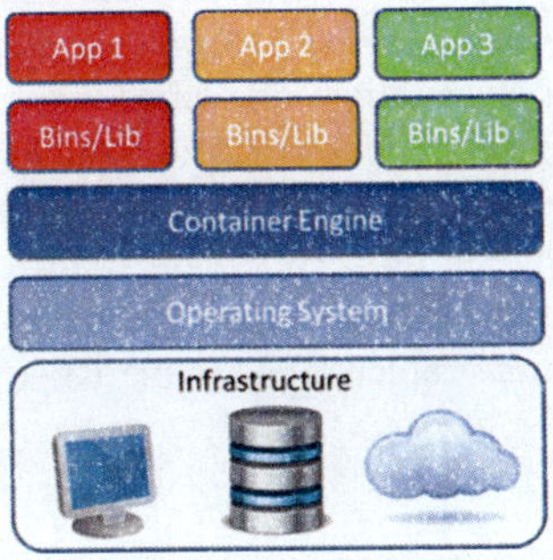

Fig. 1. Container architecture.

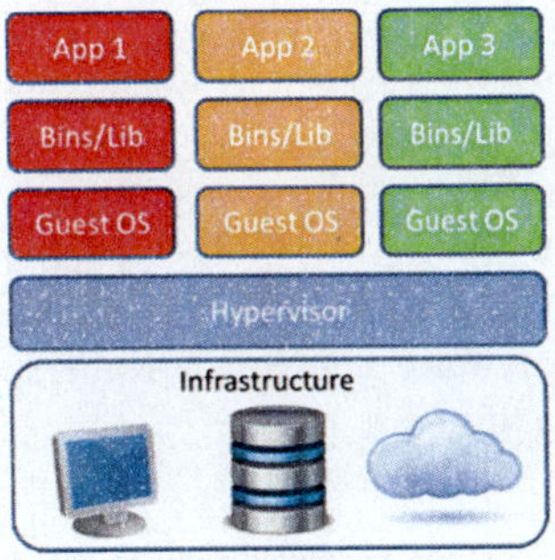

Fig. 2. Virtual machine architecture.

2.2 Container Engine

The container engine is responsible for managing containers on both states – rest and running, as shown in Fig. 3. In the rest status, a container is in the form of an image called a container image stored on a repository. When the container starts, the engine unpacks the image file from the repository or registry and runs it as a container process layer above the operating system kernel. The container engine can create container image files using some prepared configuration files. There are several competing container engines such as Docker, LXC, RKT, and Railcar. Docker is one of the most used container engines.

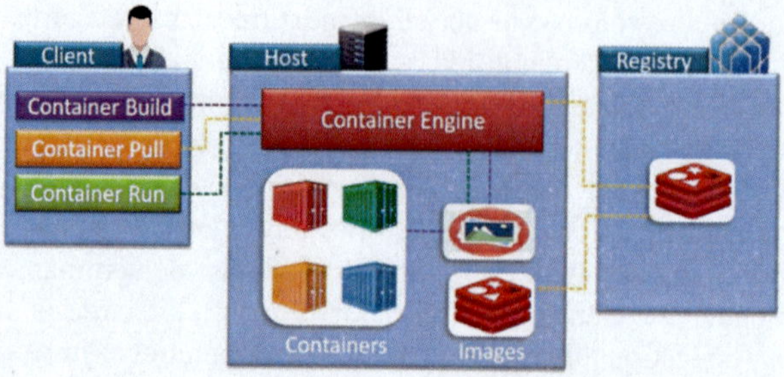

Fig. 3. Container engine architecture.

2.3 Container Cluster and Orchestration Architecture

Cloud data centers now rely on container cluster architecture. As shown in Fig. 4, the container cluster consists of a manager node and working nodes. Working nodes are responsible for running the containers submitted by clients using its local container engine. The manager node is responsible for managing the distribution and deployment of containers among the whole cluster. Manager node maintains status data about working nodes in the cluster. This process is called container cluster orchestration. The scheduler plays the most crucial role in container orchestration. The scheduler is part of the manager node that is responsible for distributing container workloads among working nodes. Container scheduler plays a crucial role in container orchestration due to the availability of cloud resources and the diverse nature of user workloads. Many orchestration frameworks are proposed for container cluster scheduling. Schedulers inside the Docker swarm and Google Kubernetes orchestration frameworks are the most widely used in this area.

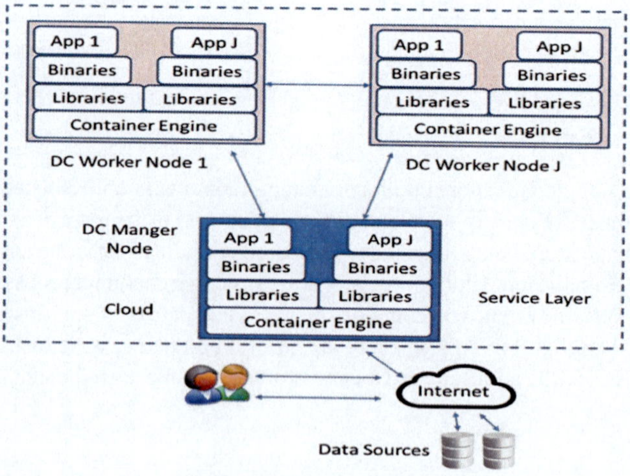

Fig. 4. Container cluster.

2.4 Machine Learning Types

Machine learning works in the same way as human learning; the machine learns from its previous experience by analyzing the data fed to its training model as input to classify and predict outputs. Machine learning aims to build a mathematical model that can predict the system's behavior by analyzing training data. Machine learning models used in container cluster orchestration scheduling can be divided into four types, as shown in Fig. 5.

- *Regression*: predict the output by finding a mathematical model to describe the relationship between multiple inputs and the output. Examples of regression algorithms include support vector regression, linear regression, random forest, and polynomial regression.
- *Classification*: classify the data based on the similarity of data input features. K-means and support vector machines (SVM) is one of the leading classification algorithms.
- *Decision-making*: simulate the process of decision-making and find the results with maximum cumulative rewards. Reinforcement learning is one of the most common techniques used in decision-making.
- *Time*-series: predict the future time series value of the studied variables based on analyzing previous values. Autoregressive integrated moving average (ARIMA) and recurrent neural network are standard time series algorithms that predict sequential data as cloud resource utilization and user requests arrival rate.

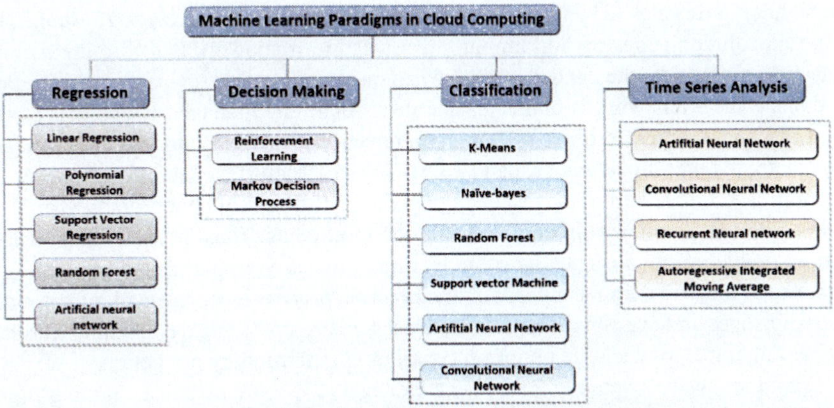

Fig. 5. Machine learning main types.

3 Related Machine Learning for Container Orchestration

In this section, a literature review of container orchestration based on machine learning techniques is introduced. After surveying existing work, Table 1 summarizes the studies.

As the table shows, for each study, the main objective is mentioned. Also, the table depicts the learning technique, the key features of the learning technique, and the working environment.

Container scheduling machine learning CSML algorithm [16] determines the number of containers based on current load pressure in real-time. Due to the diverse nature of container workload in real life. Machine learning algorithm used to predict the number of required containers for the next time window to serve the expected user workload. CSML uses a random forest regression model as a regression algorithm to find the relationship between input vector (total access, concurrency, average response time of users, error rate) and the number of containers required. Many studies find that these input parameters are the most critical factors that can accurately adjust the number of containers. Based on the prediction results, the number of containers of the next time window is adjusted at the end of the current time window. CSML saves 50% of the time required to serve user containers compared to other traditional algorithms. The author stated in his research that random forest regression is 10%–38% more accurate than the other machine learning algorithms. Experiments were conducted on three different workload pressure scenarios: the normal distribution, ladder-shaped, which represents the case where load pressure will not change quickly, and Zigzag-shape, in which the load pressure changes dramatically. CSML in all scenarios reacted more quickly while avoiding wasting containers and preserving load balancing.

On the other hand, a rescheduler learning algorithm [17] proposes a scheduling algorithm based on reinforcement learning as a decision-making algorithm. The reschedular receives the current cluster status as time-series information, allowing the scheduler to predict the following status based on the current and previous information received. The rescheduler in this article [17] is applied using the Docker containerization system. The proposed solution relies on building an infrastructure framework and forming a complete system that sets the needed work environment and performs cluster orchestration functions. The framework contains a separate module, the manager, which acts as the point of contact between other system components and working nodes. The manager can create, restore, move and suspend containers in the cluster. The manager receives scheduling commands from the rescheduler and redistributes containers among working nodes. The machine learning engine inside the rescheduler runs in two modes: learning and prediction mode. The algorithm learning process is implemented by running a sequence to form an iterative algorithm; the iteration process is represented by two main loops: episodes and attempts. This iterative mechanism represents the machine learning engine's attempts to reach the optimum container distribution by prediction.

Container-aware container placement strategy [18] uses machine learning algorithms to predict workload characteristics; this when containers co-located on the same node aggressively consume technique helps solve performance degradation problems resources. The algorithm proposed an answer to the question of how to share resources without degrading the node resources performance. The proposed scheduler engine uses the Docker swarm framework as the container cluster orchestration system. Docker swarm scheduler does not consider resource utilization but supports spread algorithm. The scheduler engine uses two machine learning clustering algorithms: k-mean++ and doubling, which take decisions based on the distance between data points, "distance" in these algorithms represent resource utilization. Workload characteristics can be predicted based on historical data; the k++ algorithm decreases the probability of assigning containers to the node with high resource contention by placing the container to the node with the smallest resource cluster distance. The doubling algorithm decreases cluster diameter and solves the k-cluster problem. The experiments normalized the throughput, which is measured against the number of nodes and tasks rate. The number of nodes is directly proportional to available resources and inverse proportional to server stress. In contrast, the number of tasks is directly proportional to server stress and inverse proportional to available resources. The experimental results showed that the proposed machine learning technique has improved applications performance by up to 14.5%, with an average of about 12%.

Using Bayesian optimization (EASY) [19], energy-aware service consolidation has been proposed to solve the data center container orchestration energy consumption problem. The main reason for high energy consumption on data centers is the improper scheduling of container workloads among servers, leading to high energy consumption due to high resource computational power. The problem of green containerized service consolidation is formulated as an NP optimization problem. The objective function is to distribute containers among servers efficiently to minimize the server's energy consumption and keep the workload QoE. BO uses Bayes Theorem to find the maximum or minimum solution of an objective function, so the EASY algorithm uses it to minimize container cluster energy consumption as an objective function. BO is the most practical algorithm for objective functions that are noisy, complicated, and expensive to evaluate. The cluster system architecture consists mainly of the manager node that runs the BO that runs the EASY algorithm to schedule containers among the cluster and worker nodes that run containers through its local container engine. CPU computational processes and IO operations data transmission is the primary source for cluster nodes' energy consumption. EASY, based on BO machine learning techniques, schedule the containers properly to minimize the CPU computational power and IO data transmissions, which lead to the total energy consumption would be minimal. Also, the EASY algorithm minimizes the number of active servers, which leads to decreased memory wastage. On the other side, the response time is a little more since fewer active servers make them more loaded.

Wattsapp, a software power-aware container mechanism [20] that applies six steps to decrease servers power consumption; authors [20] rely on neural networks to estimate power consumption for each container and distribute containers on servers based on power estimated such that the server total power consumption does not exceed the prespecified threshold set previously by the administrator. Authors [20] stated that a software-based algorithm did power capping with lower costs than hardware-based algorithms. The paper observed that containers with more server resources allocated consume more power than containers with the same workload but fewer resources allocated. The increase in power does not correspond to an increase in resources allocated. The resource performance starts to be degraded when power consumption violates the power threshold due to using power-saving techniques that lower the voltage and frequency of the CPU and other system resources. The scheduling technique overcomes this drawback by relying on software-based techniques to reduce servers power consumption. The wattsapp method uses neural network to predict container power consumption based on container resource utilization statistics. After the algorithm distributes containers among servers based on estimated power consumption, the experiments showed that neural network model prediction error does not exceed 10% over 90% of data samples. Testing on 10 Benchmark workloads, the approach achieved Mean Absolute Percentage Error (MAPE) less than 6% and minimal runtime scheduling overhead. The algorithm succeeded in decreasing power consumption without affecting resources performance. The workloads tested in this paper [20] are obtained from DC Bench and a collection of C/C++ -based scientific applications. However, it did not consider other workloads like Internet-of-Things, stream processing, and sensor-based applications.

A container consolidation scheme with usage prediction called CNCUP [21] reduces power consumption while complying with the service level agreement (SLA) of the services presented to users. The proposed algorithm uses the current and predicted CPU utilization in migrating containers between working nodes to avoid node overutilization and underutilization. Keeping the node running in health resource utilization conditions leads to minimal power consumption and SLA compliance. The author [21] built a CaaS simulation environment using Container Cloudsim toolkits. The paper introduces a linear regression algorithm for CPU usage prediction (LiRCUP) using previous utilization values as a training model. The main target of the algorithm is to predict whether the node will be overutilized or underutilized when hosting containers. Node is considered overutilized or underutilized for container migration if its current and predicted utilization are above or under the overutilized or underutilized threshold, respectively. CNCUP decides to assign a container to a node if the current and predicted node resources are sufficient to host it. The experiments showed that applying CNCUP in traditional scheduling algorithms such as LFHS, FFHS, MFHS, and RHS helps decrease power consumption, active nodes, SLA violation rate, and total container migration. CNCUP considers only CPU utilization while other factors as memory utilization and network communication are ignored.

Interference-aware container scheduler using machine learning (Co-scheML) [22] enhances the performance and increases resources utilization for GPU applications. The co-scheML algorithm introduces a solution for minimizing interference between GPU applications when sharing the same GPU resource. This algorithm uses a random forest regression model to predict the interference values among GPU applications based on profiling data. Application is offline profiled when submitted for the first time; this profiled information is stored in a repository as the application name and input metrics in a time series. The machine learning model uses the profiled information to predict the interference value, the ratio between executing the application alone, and the time of co-execution. When executed, the application progress is continuously monitored, and the profiling information is updated for further improvement for the prediction process. The experiments used 12 real-world applications: four HPC applications, five DL training jobs, and three DL inference jobs. The experiments showed that Co-scheML outperforms other traditional algorithms: Binpack, Load balance, and Mystics. Compared to traditional algorithms, average JCT improved by up to 30%, Makespan was improved by 26%, and resource utilization was improved by 24%.

Machine learning-based containerization autoscaling [23] introduces a machine learning algorithm for Docker containers auto-scaling with the workload dynamic changes. Long short-term memory (LSTM) Prediction model used to predict HTTP workloads to reduce or increase container numbers in the next time window. LSTM, a particular type of RNN neural network, overcomes some drawbacks, such as vanishing gradients, which make it efficient in detecting time-series sequences like workload over time. System architecture based on the Docker swarm framework consists of manager node and worker nodes. Manager node performs the autoscaling process and runs the machine learning model. In contrast, worker nodes run the containers, with one of them acting as a load balancer that receives HTTP requests and distributes them among worker nodes. Container auto-scaling in this study uses a controller that tunes the number of containers based on the MAPE (Monitor, Analyze, Plan, and Execute) control loop. The monitor collects different types of data needed by the prediction model to estimate future workloads like incoming HTTP requests, CPU utilization, and Memory utilization and store it time stamped in time series database. The analyzer uses LSTM neural network model to predict future workload based on historical time series stored data. Planner based on predicted workload decides to scale up or down several containers to face predicted future workload. Executer, the last phase, decreases or increases the number of containers based on planner phase results. The experiments evaluated the LSTM prediction model against ARIMA and ANN models. LSTM matches the accuracy of the ARIMA model with 600 times faster prediction time; on the other hand, it outperforms ANN in auto scalar metrics related to provisioning and elastic speedup. The study focuses only on horizontal scaling and will be extended in the future to include vertical scaling.

RLSched [24] Deep reinforcement learning-based scheduler predicts the best container placement decision based on the current cluster resources state. The algorithm observes the current cluster state, decides the best placement action, and observes the rewarding result from this action to train the model with an optimal policy that maps each state with each best action. The algorithm relies on three primary keys: State-space, action space, and Reward function. State-space represents current cluster resource usage and requests to be placed by the scheduler, Action space represents the PM number where the container will be placed, and reward function defines the algorithm's goal. In this algorithm, the main goal is to maximize the number of requests by reducing the number of wasted resources. RLSched utilizes a deep neural network as a function approximator, which takes a deep neural network as a function approximator that takes cluster current state as input and outputs the estimated reward for container placement to each physical machine. The experiments in this study are based on the placement of a single container and observe only memory and CPU resources as the current cluster state. For the algorithm to avoid falling into sub-optimal policies and reduce time convergence, an extra layer is added in the model that filters and removes invalid actions representing PM that can't satisfy container requirements. RLSched is compared against three Reinforcement learning methods used in cloud resource management: DQN, DDQN, and PPO. Experiments showed that RLSched converges quickly and finds the best possible container replacement than other RL algorithms. The algorithm can be expanded in the future to optimize network latency as part of the reward and schedule multiple containers.

Table 1. Summary of introduced Machine learning algorithms features

Ref.	Algorithm	Method	Objective	Metrics	Advantage	Limitations
[16]	CSML	Random forest regression model	Predict number of required containers	– Average – response time – error rate	Avoid wasting containers and maintain load balancing	Other container types rather than microservices not considered
[17]	Rescheduler learning	Reinforcement learning	Predict cluster next status	Reward	Load balancing	Not simple framework design
[18]	Contention aware container placement	K++, Doubling	Classify workload based on its characterics	Normalized throughput	Improve node resources performance	Single evaluation parameter
[19]	EASY	Bayesian optimization	Optimize energy consumption	Energy consumption – average response time – RAM wastage	Green container consolidation	More response time
[20]	Power aware container scheduling	Neural network	Predict container power consumption	Power estimation error – Power consumption	Decreasing power consumption without affecting resources performance	Didn't consider some important workloads
[21]	CNCUP	Linear regression	CPU usage prediction	Power consumption – SLA violation – active nodes – Number of container migrations	Improve power consumption and comply SLA	Memory utilization and network communication not considered
[22]	Co-scheML	Random forest regression model	Predict interference value among GPU applications	Job completion time – Makespan	Improve JCT, makespan and resource utilization	CPU utilization not considered
[23]	MAPE loop	LSTM neural network	Predict future HTTP workload	Mean square error – prediction speed	Enhance Horizontal container autoscaling	Didn't consider container vertical autoscaling
[24]	*RL*Sched	Deep reinforcement learning	Predict the best reward for input state	– Number of requests – Percentage of requests per episode	Maximize request number by reduce wasted resources	Single container scheduling

4 Conclusion

This work has reviewed how machine learning technology is used to solve the problem of container scheduling among cloud cluster nodes. The introduced techniques achieved many important objectives, including power consumption enhancement, resource utilization optimization, avoiding wasting containers, minimizing response time, and complying with SLA. Several studies have applied different machine learning types such as regression, decision making, classification, and time series analysis that achieved container scheduling objectives. Obviously, machine learning techniques take accurate decisions and provide better results than other traditional methods. Analyzing cloud cluster history to train the scheduling algorithm to predict the future workload and behavior is considered the powerful feature of machine learning models, especially with the diverse nature of cloud workloads. Research on machine learning-based container orchestration has been grown and proved that it is one of the most influential fields of research in container scheduling. Most of the introduced research has future work to enhance its performance by considering more environmental working conditions and increasing its enhanced parameters. Container scheduling need more researches to overcome some open challenges such as data-intensive serverless edge computing due to high energy consuming and high workload, resource utilization, resource allocation, auto scaling, collocated containers interference and network communication overhead.

References

1. Varghese, B., Buyya, R.: Next generation cloud computing: new trends and research directions. Future Gener. Comput. Syst. **79**(3), 849–861 (2018)
2. Saber, W., Moussa, W., Ghuniem, A.M., Rizk, R.Y.: Hybrid load balance based on genetic algorithm in cloud environment. Int. J. Electr. Comput. Eng. (IJECE) **11**(3), 2477–2489 (2020)
3. da Cunha, H.G.V.O., Moreira, R., de Oliveira, F.: A comparative study between containerization and full-virtualization of virtualized everything functions in edge computing. In: Advanced Information Networking and Applications, pp. 771–782. AINA (2021)
4. Shah, J., Dubaria, D.: Building modern clouds: using Docker, Kubernetes & Google cloud platform. In: 2019 IEEE 9th Annual Computing and Communication Workshop and Conference (CCWC), Las Vegas, pp. 184–189 (2019)
5. Menouer, T., Cérin, C., Leclercq, É.: New multi-objectives scheduling strategies in Docker SwarmKit. In: Vaidya, J., Li, J. (eds.) ICA3PP 2018. LNCS, vol. 11336, pp. 103–117. Springer, Cham (2018). https://doi.org/10.1007/978-3-030-05057-3_8
6. Kaushik, P., Raghavendra, S., Govindaraju, M., Tiwari, D.: Exploring the potential of using power as a first class parameter for resource allocation in apache mesos managed clouds. In: 2020 IEEE/ACM 13th International Conference on Utility and Cloud Computing (UCC), Leicester, pp. 216–226. IEEE (2020)
7. Ahmad, I., AlFailakawi, M.Gh., AlMutawa, A., Alsalman, L.: Container scheduling techniques: a survey and assessment. J. King Saud Univ. Comput. Inf. Sci. (2021)
8. Gamal, M., Rizk, R., Mahdi, H., Elnaghi, B.E.: Osmotic bio-inspired load balancing algorithm in cloud computing. IEEE Access **7**, 42735–42744 (2019)
9. Attia, R., Hassaan, A., Rizk, R.: Advanced greedy hybrid bio-inspired routing protocol to improve IoV. IEEE Access **9**, 131260–131272 (2021)

10. Mohamed, A., Saber, W., Elnahry, I., Hassanien, A.E.: Coyote optimization based on a fuzzy logic algorithm for energy-efficiency in wireless sensor networks. IEEE Access **8**, 185816–185829 (2020)
11. Pouyanfar, S., et al.: A survey on deep learning: algorithms, techniques, and applications. ACM Comput. Surv. (CSUR) **51**(5), 1–36 (2018)
12. Bentaleb, O., Belloum, A.S.Z., Sebaa, A.: Containerization technologies: taxonomies, applications and challenges. J. Supercomput. **78**, 1144–1181 (2021)
13. Goodarzy, S., Nazari, M., Han, R., Keller, E., Rozner, E.: Resource management in cloud computing using machine learning: a survey. In: 2020 19th IEEE International Conference on Machine Learning and Applications (ICMLA), Miami, pp. 811–816 (2020)
14. Bianchini, R., et al.: Toward ML-centric cloud platforms. Commun. ACM **63**(2), 50–59 (2020)
15. Kecskemeti, G., Marosi, A.C., Kertesz, A.: The ENTICE approach to decompose monolithic services into microservices. In: 2016 International Conference on High Performance Computing Simulation (HPCS), pp. 591–596, Innsbruck (2016)
16. Lv, J., Wei, M., Yu, Y.: A container scheduling strategy based on machine learning in microservice architecture. In: 2019 IEEE International Conference on Services Computing (SCC), Milan, pp. 65–71 (2019)
17. Rovnyagin, M.M., Dmitriev, S.O., Hrapov, A.S., Kozlov, V.K.: Algorithm of ML-based re-scheduler for container orchestration system. In: 2021 IEEE Conference of Russian Young Researchers in Electrical and Electronic Engineering (ElConRus), Moscow, pp. 613–617 (2021)
18. Chiang, R.C.: Contention-aware container placement strategy for docker swarm with machine learning based clustering algorithms. Clust. Comput. (2020)
19. Nath, S.B., Addya, S.K., Chakraborty, S., Ghosh, S.K.: Green containerized service consolidation in cloud. In: ICC 2020 - 2020 IEEE International Conference on Communications (ICC), Dublin , pp.1–6 (2020)
20. Mehta, H.K., Harvey, P., Rana, O., Buyya R., Varghese, B.: WattsApp: power-aware container scheduling. In: 2020 IEEE/ACM 13th International Conference on Utility and Cloud Computing (UCC), Leicester, pp. 79–90 (2020)
21. Liu, J., Wang, S., Zhou, A., Xu, J., Yang, F.: SLA-driven container consolidation with usage prediction for green cloud computing. Front. Comp. Sci. **14**(1), 42–52 (2019). https://doi.org/10.1007/s11704-018-7172-3
22. Kim, S., Kim, Y.: Co-scheML: interference-aware container co-scheduling scheme using machine learning application profiles for GPU clusters. In: 2020 IEEE International Conference on Cluster Computing (CLUSTER), Kobe, pp. 104–108 (2020)
23. Imdoukh, M., Ahmad, I., Alfailakawi, M.G.: Machine learning-based auto-scaling for containerized applications. Neural Comput. Appl. **32**, 9745–9760 (2020)
24. Lorido-Botran, T., Bhatti, M.K.: Adaptive container scheduling in cloud data centers: a deep reinforcement learning approach. In: Barolli, L., Woungang, I., Enokido, T. (eds.) AINA 2021. LNNS, vol. 227, pp. 572–581. Springer, Cham (2021). https://doi.org/10.1007/978-3-030-75078-7_57

Mobile Computation Offloading in Mobile Edge Computing Based on Artificial Intelligence Approach: A Review and Future Directions

Heba Saleh(✉), Walaa Saber, and Rawya Rizk

Electrical Engineering Department, Port Said University, Port Said 42526, Egypt
{Heba.saleh,walaa_saber,r.rizk}@eng.psu.edu.eg

Abstract. Mobile computation offloading (MCO) is one of the significant processes in mobile edge computing (MEC). MCO is a promising approach to contract with the restrictions in client devices by offloading resource-intensive tasks or at least a part of it to the nearby resource-rich servers in MEC. Since most of the MCO optimization models endeavor to solve an NP-hard problem, the approximation solutions with higher performance and lower complexity are proposed and evaluated in several studies. These solutions could be more optimized with the most recent developments in the Artificial Intelligence (AI) field such as machine learning (ML), and meta-learning (MTL). Lately, many AI techniques have been proposed to learn offloading policies through interacting with the MEC environment. This paper proposes a literature review for the recent mechanisms of ML-based and MTL-based MCO in MEC. A detailed study is proposed for the main issues, challenges, and future research direction in the field of MCO in MEC servers.

Keywords: Internet of Things · Mobile edge computing · Mobile computation offloading · Artificial Intelligence · Machine learning · Meta learning

1 Introduction

Internet of Things (IoT) is a computing system for connecting a set of terminals, objects, and sensors to deliver data over a network devoid of the need for human interaction [1]. With the popularity growth of IoT devices and applications, the requests for corresponding services have increased exponentially. This led to the classic cloud computing model has faced many challenges such as low performance, latency, bandwidth, and security. To defeat these challenges, new technologies and paradigms of communication can be structured to support IoT networks such as mobile edge computing (MEC) and mobile fog computing (MFC) [2]. MEC is used to extend cloud computing capabilities to MEC hosts at the edge of the network close to mobile devices, which can significantly improve network traffic and reduce latency. In addition, MFC is used to transport data and place it nearer to the user.

However, the network providers are not encouraged to expand continuously the network infrastructure because of heavy investments. Mobile computation offloading

© The Author(s), under exclusive license to Springer Nature Switzerland AG 2022
A. E. Hassanien et al. (Eds.): AMLTA 2022, LNDECT 113, pp. 593–603, 2022.
https://doi.org/10.1007/978-3-031-03918-8_49

(MCO) is one of the most hopeful keys used to deal with these problems with using different techniques to reduce the cost for the network providers and enhance the quality of service (QoS) and quality of experience (QoE) for the client. Optimal decision to choose whether the task should be offloaded to the edge server or cloud server is the most critical problem for edge offloading [3]. Numerous techniques were presented to solve this problem such as linear programming, nonlinear programming, game theory, queuing theory, Machine Learning (ML) and Meta-Learning (MTL). In general, linear programming is a single objective deterministic model, which is suitable for problems involving in minimization of costs or maximization of profits with clear defined constraints. Besides, the game theory treats with determining pure Nash equilibria as an efficient stochastic method. Since of uncertainties, conditions of diverse behavior, and assessment to measure the system usefulness, queuing theory pays attention to stochastic problems, and decision-making process. While ML methodology is used to solve the problem iteratively to reach the optimal policy in training data, MTL can adapt fast to new environments with a small number of samples and gradient updates.

This survey aims at reviewing the most recent studies and exploring various techniques in the MEC model, covering ML-based and meta-learning-based approaches comprehensively and systematically. Briefly, the main contributions of this survey are as follows:

- Reviewing some articles related to ML-based offloading mechanisms in MEC and presenting the advantages and the weaknesses for each one;
- Exploring the latest meta-learning-based approaches in the field of offloading in MEC;
- Providing a comprehensive systematic review of current approaches and proposing a comprehensive classification;
- Discussing future uncovered research challenges to improve MCO mechanisms in the MEC environment.

The rest of this paper is organized as follows: Sect. 2 provides the necessary background of offloading issues in MEC. In Sect. 3, related works are presented. The research methodology, paper selection mechanisms, the comparison and a discussion of the reviewed techniques are presented in this section. Finally, the conclusions and future directions are presented in Sect. 4.

2 Background

2.1 Multi-tier Computing Paradigm

Cloud computing, fog computing and edge computing are three technologies straight related to IoT. With the growing number of IoT devices and huge volume of the generated data, cloud computing has been proven inadequate to efficiently manipulate the corresponding loads, while meeting the client requirements in terms of latency and energy efficiency [4]. Consequently, edge computing represents the practice of processing data near the edge of the network. Whereas, fog computing is considered as a computing paradigm that implements IoT applications at the edge of the network. However, fog computing has limited resources and low computational capabilities for task execution.

MFC serves as an extension of the cloud computing paradigm to deliver more capabilities in processing, and storage. In MFC, tasks can be offloaded to the Fog devices, instead of a far cloud server for saving the time and energy. On the other side, MEC provides cloud computing services and capabilities at the edge of the network where the computation tasks are executed closer to the users. It realizes great benefits for IoT networks such as providing new services, increasing the efficiency, and reducing the communication latency by providing a way for data offloading.

2.2 Mobile Computation Offloading

MCO is the process for delegating computations from devices with low computational capacity and power to remote powerful servers in order to satisfy QoS requirements [5]. MCO improves the application performance by decreasing the execution time of the tasks, saving bandwidth, reducing the overall energy consumption, prolonging the battery life of the constrained mobile devices, increasing the service availability for the mobile devices, and minimizing the latency [6]. The main objective of data offloading is finding the best place for the optimal performance of the tasks, in the edge of the network, in the fog, or in the cloud servers. The architecture of MCO is illustrated in Fig. 1.

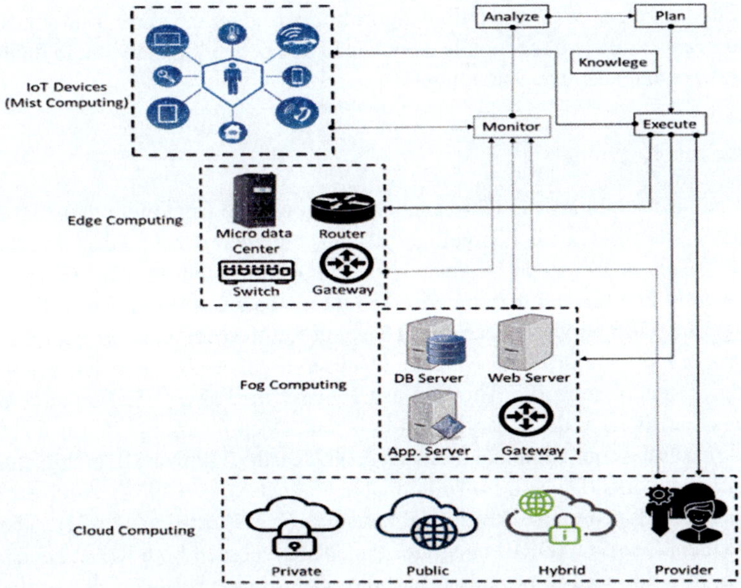

Fig. 1. Mobile computation offloading architecture.

To empower computation offloading, an offloading client has to be running on the host mobile device and an offloading server within the external resources [7]. They need to communicate for directing the offloading process. After the application begins executing, the task that has been distinguished as offloadable will be delivered to the offloading client. All needed data such as request parameters and resource files should be provided to the offloading server. The offloaded component begins executing and communicates straightforwardly with the mobile app for sending back the results of the remote execution. MCO has to be responsive to the changes in the environment and ensure a fast, accurate, user transparent, and low overhead techniques for any change detection. Since offloading migrates computation to a more powerful server, it involves deciding whether and what computation to migrate [8].

MCO Classification. In general, MCO can be classified into three categories, namely: (1) always offload; (2) all or nothing offloading, where either the entire data is offloaded or the entire data is processed locally, with the offloading decision typically to depend on energy thresholds; and (3) partial offloading, where some parts are offloaded with the remaining to be executed locally [9]. Partial offloading offers the greatest flexibility and potential for intelligence and optimization, based on both communication and computation environment awareness. Therefore, the proposed survey work focuses on the latter category.

The data offloading decision can be performed either during development (offline) or at runtime (online). The main difference between offline and online offloading algorithms is that offline algorithm assumes that the future predictions of the device load and network bandwidth are given, while the online algorithm considers the current progress of the application execution and the mobile device load and network bandwidth in the form of context-aware and adaptive offloading [10].

2.3 Machine Learning

ML is a multi-disciplinary model to extract the knowledge from input data by merging mathematics, statistics, artificial intelligence, and computer to make a decision automatically [11]. This decision can be made by learning from input to achieve a particular output without any manual manipulations in three general learning ways: Supervised Learning (SL), Unsupervised Learning (UL), and Reinforcement Learning (RL). Neural Networks (NNs) or artificial NNs are a subset of ML techniques, loosely inspired by biological neural networks. They are usually described as a collection of connected units, called artificial neurons, organized in layers. Deep Learning (DL) is a subset of NNs that makes the computational multi-layer NN feasible. Typical DL architectures are deep neural networks (DNNs), convolutional neural networks (CNNs), recurrent neural networks (RNNs), generative adversarial networks (GAN), and many more. The Deep Reinforcement Learning (DRL) integrates the effectiveness of both RL and deep learning. DRL does not need to provide the whole set of training data, such as supervised learning, but rather a few pieces that are sufficient for the model.

2.4 Meta-learning

MTL is the science of systematically observing how different ML approaches perform on a wide range of learning tasks, and then learning from this experience, or meta-data, to learn new tasks much faster than otherwise possible [12]. MTL leverages past experiences to ascertain a prior model's parameters and learning process. The benefits of introducing MTL are twofold: first is to avoid the need to re-train deep models from scratch; second is to increase reaction speed by adapting deep models to dynamic environment conditions. MTL is feasible in tackling situations where out-of-distribution task predictive performance is required [13].

MTL Stages. In general, one optimizes a meta-objective by using various tasks. This is done in three stages [14]:

- **First, in the meta-train stage:** the MTL algorithm is applied to the meta-train tasks.
- **Second, the meta-validation stage:** the meta-validation tasks can then be used to evaluate the performance on unseen tasks, which were not used for training. Effectually, this measures the meta-generalization ability of the trained network.
- **Third, the meta-test stage:** the meta-test tasks are used to provide a final performance estimate of the MTL technique.

MTL Techniques. MTL techniques can be categorized based on the type of meta-data they leverage into three categories [15]:

- **First, learn from model evaluations:** these techniques can be used to recommend generally useful configurations and configuration search spaces, as well as transfer knowledge from empirically similar tasks.
- **Second, characterize tasks:** these techniques can be used to more explicitly express task similarity and build meta-models that learn the relationships between data characteristics and learning performance.
- **Finally, transfer trained model parameters:** these techniques can be used to transfer parameters between tasks that are inherently similar by sharing the same input features, which enables transfer learning and few-shot learning among others.

MTL Models. Not all MTL models follow the same techniques. Types of meta-learning models include [16]:

- **Few Shots Meta-Learning:** the goal of few-shot MTL is to train a model that can quickly adapt to a new task using only a few data points and training iterations by train the model's initial parameters.
- **Optimizer MTL:** it is used to learn how to optimize a neural network NN to accomplish a task efficiently. There is one network (the meta-learner) which learns to update another network (the learner) so that the learner effectively learns the task.
- **Metric MTL:** this approach can be seen as a subset of few shots MTL in which a learned metric space to evaluate the quality of learning with a few examples can be used.
- **Recurrent Model MTL:** it uses the meta-learner algorithm to train a RNN model. It will process a dataset sequentially and then process new inputs from the task.
- **Initializations MTL:** it is for an initial representation that can be fine-tuned from a small number of examples effectively.

MTL Limitations. MTL cannot solve all ML problems for a variety of reasons [17]:

- The extracted meta-features should to be representative of their problem domain. Or else, an algorithm will fail to identify similar domains.
- MTL cannot be able to exploit past knowledge to improve prediction performance when a problem has not been seen before.
- Performance estimation may be unreliable because of the natural limitations of estimating the true performance of the dataset.

3 Related Work

3.1 Machine Learning in MCO

In recent years, the idea of employing ML in networking and in computation offloading in special has caught on, with many works addressing this research topic. Some of the earliest work focused on applying ML in wireless sensor networks in order to solve problems such as energy-aware communication, efficient sensor deployment, resource allocation, or task scheduling, while more recent uses of ML in networking include traffic prediction and classification, routing strategies, quality of experience optimization, congestion control, or performance prediction. Nowadays, the utility of ML methods in network offloading tends to focus on moving code from a mobile node to other devices, which can be mobile, but also edge, fog, or cloud nodes. Table 1 depicts a comparison between recent ML-based computation offloading mechanisms.

Erfan et al. [18] designed efficient offloading approaches that take into account offloading reservations such as user mobility and dynamic changes. The proposed algorithm addressed these challenges by minimizing the turnaround time of the applications, which is constituted by offloading latency, migration delay, and execution time. First, the proposed algorithm formulates this NP-hard problem as an integer programming model to obtain optimal offloading decisions. Second, it tackles the intractability by designing two novel offloading approaches, called S-OAMC and G-OAMC. S-OAMC is a sampling-based approximation dynamic programming approach that enhances scalability and obtains near-optimal solutions. Both approaches utilize ML-based predictions generated by the matrix completion method to obtain smart decisions on offloading in order to minimize the turnaround time of the offloaded applications. Experimental evaluations show that S-OAMC and G-OAMC lead to near-optimal turnaround time in a reasonable time, while they are highly scalable. Moreover, they reduce the number of migrations significantly compared to other non-optimal approaches.

Ali Shakaramiet al [19] designed and implemented an autonomous computation offloading framework based on Monitor-Analyze-Plan-Execute-Knowledge (MAPE-K) loop methodology to address some challenges related to time-intensive and resource-intensive applications. In addition, different simulations including DNNs, multiple linear regression, hybrid model, and Hidden Markov Model (HMM) as the planning module of the mentioned autonomous methodology have been conducted to cope with the large dimension of the offloading decision-making problem. Simulation results show that

Table 1. A comparison of recent ML-based computation offloading mechanisms

Ref	Utilized technique	Performance metric	Benefits	Weaknesses
[18]	Matrix completion method	Turnaround time	Leads to near-optimal turnaround time in a reasonable time Obtains low migration rates Reduces the number of migrations	Single evaluation parameter
[19]	Deep learning (DL)	Latency & energy	Fits the problem with near-optimal accuracy	The effects and side effects of handover between is neglected
[20]	Lagrange coded computing (LCC)	Resiliency & security	Fast and secure Shows the superior performance regardless of the numbers of stragglers and malicious workers	Not simple framework design
[21]	Deep reinforcement learning (DRL)	Latency	Ultra-reliable Has low latency Scalable than the baseline schemes	Single evaluation parameter

the proposed hybrid model can appropriately fit the problem with near-optimal accuracy regarding the offloading decision-making, the latency, and the energy consumption predictions in the proposed self-management framework.

Alia Asheralieva et al. [20] devised LCC-MEC, a novel auction-and-learning based encoding scheme. LCC-MEC framework is proposed based on Lagrange Coded Computing (LCC) for fast and secure offloading of computing tasks in the MEC network. The network is formed by multiple base stations acting as “masters” which offload their computations to edge devices acting as “workers”. LCC-MEC framework aimed to ensure efficient allocation of computing loads and bandwidths to workers, and provide them with proper incentives to finish their tasks by the specified deadlines. Thus, each master must decide on the amounts of allocated load and bandwidth, and a service fee paid to each worker. As such, masters compete for the best workers in a stochastic and observable environment. LCC-MEC represents the auction as a stochastic Bayesian game and develops ML algorithms to improve the auction solution. Compared to the state-of-the-art, LCC-MEC scheme shows the superior performance regardless of the numbers of stragglers and malicious workers.

Getenet Tefera et al. [21] explored decentralized adaptive resource-aware communication, computing, & caching for multi-access edge computing networks (DARMEC) framework based on DRL to handle the structural complexity and dynamic environment of MEC networks. Therefore, the framework can perform augmented decision-making policy, resource allocation, and scalability as well. Therefore, the problem is formulated using Non-cooperative game theory, which is non-deterministic polynomial. The proposed framework analyzed and showed that the game admits a Nash Equilibrium. Moreover, a decentralized cognitive scheduling algorithm is introduced by exploiting DRL technology to secure the optimal solution. As a result, numerical results and theoretical analysis reveal that the proposed DARMEC algorithm outperforms ultra-reliable low latency, and scalable than the baseline schemes.

3.2 Meta-learning in MCO

In recent applications, MTL is focused upon integration with other ML frameworks and forms meta-reinforcement learning and meta-imitation learning. A comparison of recent MTL-based computation offloading mechanisms is depicted in Table 2.

Jin Wang et al. [22] proposed a MRLCO method based on meta-reinforcement learning (MRL). More specifically, training for the meta-policy (outer loop) is run on the MEC host and training for the specific offloading policy (inner loop) is processed on user devices. Normally, the "inner loop" training only needs several training steps and a small amount of sampling data, thus the user devices with limited computation resources and data is able to complete the training process. The experimental results demonstrate that MRLCO can reduce the latency by up to 25% compared to three baselines while being able to adapt fast to new environments.

Md. Shirajum Munir et al. [23] proposed an effective energy dispatch mechanism for self-powered wireless networks with edge computing capabilities. First, a two-stage linear stochastic programming problem is formulated with the goal of minimizing the total energy consumption cost of the system while satisfying the energy demand. Second, a semi distributed data-driven solution is proposed by developing a novel multi-agent MRL (MAMRL) framework to solve the formulated problem. In particular, each base station plays the role of a local agent that explores a Markovian behavior for both energy consumption and generation while each base station transfers time-varying features to a meta-agent. Sequentially, the meta-agent optimizes the energy dispatch decision by accepting only the observations from each local agent with its own state information. Meanwhile, each base station agent estimates its own energy dispatch policy by applying the learned parameters from meta-agent. Experimental results show that the proposed MAMRL model can reduce up to 11% non-renewable energy usage and by 22.4% the energy cost with 95.8% prediction accuracy, compared to other baseline methods.

Liang Huang, et al. [24] proposed a MTL based computation Offloading (MELO) algorithm for dynamic computation tasks in MEC networks. MELO trains a general DNN for different MEC task scenarios and can quickly learn to adapt to a new one after a few training iterations. Numerical results show that the proposed algorithm can adapt to a new MEC task scenario and achieve 99% accuracy via 1-step fine-tuning using only 10 training samples.

Guanjin Qu et al. [25] proposed a Deep MRL-based Offloading (DMRO) algorithm, which combines multiple parallel DNNs with Q-learning to make fine-grained offloading decisions. DMRO aggregates the perceptive ability of DL, the decision-making ability of

RL, and the rapid environment learning ability of MTL. Simulation results demonstrate that the DMRO achieves obvious improvement and has strong portability in making real-time offloading decisions even in time-varying IoT environments.

Table 2. A comparison of recent MTL-based computation offloading mechanisms.

Ref.	Utilized technique	Performance metric	Benefits	Weaknesses
[22]	Meta-reinforcement learning (MRL)	Latency	Improves the training efficiency in learning new tasks More adaptive to the dynamic MEC environment Can efficiently run on the resource-constrained devices using its own data	Considers stable wireless channels, reliable mobile devices, sufficient resources, and no mobile devices as stragglers may drop out
[23]	Multi-agent meta-reinforcement learning (MAMRL)	Energy	High degree of reliability Overcomes a non-independent and identically distributed (non-i.i.d.) energy demand and generation	Single evaluation parameter
[24]	Meta deep learning (MDL)	Latency	Fasts a deployment of DL-based computation offloading algorithms in MEC networks	Single evaluation parameter
[25]	Deep meta reinforcement (DMR)	Delay & energy	Strong portability and rapid learning ability Highly scalable Can be easily extended with multiple edge servers and cloud servers	Only focuses on the scenario with one edge server and one cloud server

4 Conclusion

In recent years, MCO is becoming an adopted methodology to overcome the limitations of mobile devices. MCO algorithms decide to offload the tasks to be executed remotely, observing objective metrics. Because of the nondeterministic behavior in the MEC, ML methods are widely utilized in the literature. However, ML-based techniques have low sample efficiency and require full retraining to learn an updated policy for the new environment. MTL can speed up and improve the design of ML pipelines or neural architectures. However, research in this area is still in the early stage, a comprehensive survey of research progress in this field is provided.

In this survey, an extensive overview on AI-based MCO in MEC was provided hope to serve as handy references and valuable guidelines for further in-depth investigations of this area. First, the proposed survey presents a comprehensive overview of MEC from the perspective of service adoption and provision. Then, the survey explores different approaches associated with ML-based and MTL-based offloading mechanisms in the MEC. Next, the applied approaches were compared with each other based on the essential factors such as performance metrics, utilized techniques, and their advantages and weaknesses, as well. Finally, with consideration of the current gap among the researches in the literature, some critical research challenges as open issues in the ML-based and MTL-based offloading mechanisms are explained.

References

1. Laghari, A.A., Wu, K., Laghari, R.A., Ali, M., Khan, A.A.: A review and state of art of internet of things (IoT). In: Archives of Computational Methods in Engineering, Barcelona, Spain (2021)
2. Nashaat, H., Ahmed, E., Rizk, R.: IoT application placement algorithm based on multi-dimensional QoE prioritization model in fog computing environment. IEEE Access **8**, 111253–111264 (2020)
3. Abdel-Kader, R.F., El-Sayad, N.E., Rizk, R.Y.: Efficient energy and completion time for dependent task computation offloading algorithm in industry 4.0. PLoS ONE **16**(6), e0252756 (2021)
4. Saber, W., Moussa, W., Ghuniem, A.M., Rizk, R.Y.: Hybrid load balance based on genetic algorithm in cloud environment. Int. J. Electr. Comput. Eng. (IJECE) **11**(3) (2020)
5. Carvalho, G., Cabral, B., Pereira, V., Bernardino, J.: Computation offloading in edge computing environments using artificial intelligence techniques. Eng. Appl. Artif. Intell. **95**, 103840 (2020)
6. Hou, X., et al.: Reliable computation offloading for edge-computing-enabled software-defined IoV. IEEE Internet Things J. **7**(8), 7097–7111 (2020)
7. Meng, T., Wu, H., Shang, Z., Zhao, Y., Xu, C.Z.: CoOMO: cost-efficient computation outsourcing with multi-site offloading for mobile-edge services. In: IEEE 16th International Conference on Mobility, Sensing and Networking (MSN), pp. 113–120, December 2020
8. Farahbakhsh, F., Shahidinejad, A., Ghobaei-Arani, M.: Multiuser context-aware computation offloading in mobile edge computing based on Bayesian learning automata. Trans. Emerg. Telecommun. Technol. **32**(1), e4127 (2021)
9. Heidari, A., Jabraeil Jamali, M.A., Jafari Navimipour, N., Akbarpour, S.: Internet of things offloading: ongoing issues, opportunities, and future challenges. Int. J. Commun. Syst. **33**(14), e4474 (2020)
10. Lakhan, A., Ahmad, M., Bilal, M., Jolfaei, A., Mehmood, R.M.: Mobility aware blockchain enabled offloading and scheduling in vehicular fog cloud computing. IEEE Trans. Intell. Transp. Syst. (2021)

11. Alfarraj, O.: A machine learning-assisted data aggregation and offloading system for cloud–IoT communication. Peer Peer Netw. Appl. **14**(4), 2554–2564 (2021)
12. Monteiro, J.P., et al.: Meta-learning and the new challenges of machine learning. Int. J. Intell. Syst. **36**(11), 6240–6272 (2021)
13. Liu, Z., et al.: A joint optimization framework of the embedding model and classifier for meta-learning. Sci. Program. (2021)
14. Park, S., Yoo, J., Cho, D., Kim, J., Kim, T.H.: Fast adaptation to super-resolution networks via meta-learning. In: Vedaldi, A., Bischof, H., Brox, T., Frahm, J.-M. (eds.) ECCV 2020. LNCS, vol. 12372, pp. 754–769. Springer, Cham (2020). https://doi.org/10.1007/978-3-030-58583-9_45
15. Vanschoren, J.: Meta-learning. In: Hutter, F., Kotthoff, L., Vanschoren, J. (eds.) Automated Machine Learning. TSSCML, pp. 35–61. Springer, Cham (2019). https://doi.org/10.1007/978-3-030-05318-5_2
16. Aghapour, E., Ayanian, N.: Double meta-learning for data efficient policy optimization in non-stationary environments. In: IEEE International Conference on Robotics and Automation (ICRA), pp. 9935–9942, May 2021
17. Xu, H., Wang, J., Li, H., Ouyang, D., Shao, J.: Unsupervised meta-learning for few-shot learning. Pattern Recognit. **116**, 107951 (2021)
18. Maleki, E.F., Mashayekhy, L., Nabavinejad, S.M.: Mobility-aware computation offloading in edge computing using machine learning. IEEE Trans. Mob. Comput. (2021)
19. Shakarami, A., Shahidinejad, A., Ghobaei-Arani, M.: An autonomous computation offloading strategy in mobile edge computing: a deep learning-based hybrid approach. J. Netw. Comput. Appl. **178**, 102974 (2021)
20. Asheralieva, A., Niyato, D.: Fast and secure computational offloading with Lagrange coded mobile edge computing. IEEE Trans. Veh. Technol. **70**(5), 4924–4942 (2021)
21. Tefera, G., She, K., Shelke, M., Ahmed, A.: Decentralized adaptive resource-aware computation offloading & caching for multi-access edge computing networks. Sustain. Comput. Inform. Syst. **30,** 100555 (2021)
22. Wang, J., Hu, J., Min, G., Zomaya, A.Y., Georgalas, N.: Fast adaptive task offloading in edge computing based on meta reinforcement learning. IEEE Trans. Parallel Distrib. Syst. (2020)
23. Munir, M.S., Tran, N.H., Saad, W., Hong, C.S.: Multi-agent meta-reinforcement learning for self-powered and sustainable edge computing systems. IEEE Trans. Netw. Serv. Manag. (2020)
24. Huang, L., Zhang, L., Yang, S., Qian, L.P., Wu, Y.: Meta-learning based dynamic computation task offloading for mobile edge computing networks. IEEE Commun. Lett. **25**(5), 1568–1572 (2020)
25. Qu, G., Wu, H., Li, R., Jiao, P.: DMRO: a deep meta reinforcement learning-based task offloading framework for edge-cloud computing. IEEE Trans. Netw. Serv. Manag. **18**(3), 3448–3459 (2021)

Check for updates

Assessment of Driving Behavior on Edge Devices Using Machine Learning and Sensor Data

Amirabbas Hojjati(✉), Muhammad Saad Jahangir, and Ibrahim A. Hameed

Department of Information and Communication Technologies (ICT) and Natural Sciences, Faculty of Information Technology and Electrical Engineering, Norwegian University of Science and Technology (NTNU), Larsgårdsvegen 2, 6009 Ålesund, Norway
{amirabbh,muhamsj}@stud.ntnu.no, ibib@ntnu.no

Abstract. Road accidents and related casualties can be prevented by identifying dangerous driving styles and proactively taking action to make improvements. While different driver assistance systems and mobile app-based solutions partially cover some of the aspects like warning and profiling systems, we present a standalone comprehensive camera and sensor based IoT system that runs state of the art models on a low cost edge device and performs driving behavior analysis by keeping track of vehicles' acceleration, deceleration, turning, lane departure, and tailgating events as well as driver's drowsiness and attentiveness. The system is deployed on a NVIDIA Jetson Nano, running in two parallel pipelines with IMU smoothened sensor data updating at 4 Hz while the rest of the models run using TensorRT Engine at approximately 2 fps when using the lightweight classification model for secondary task detection and 1.2 fps when using the more accurate classification model. All the models can run seamlessly in the final solution near real-time which is required for the tasks to produce meaningful results. The raw data is processed internally and only the final results will be sent to the cloud database to ensure the user's privacy.

Keywords: Driving behavior analysis · Accident prevention · Internet of Things (IoT) · Edge computing · Transfer learning

1 Introduction

Road accidents are among major causes of death around the world. According to the World Health Organization's 2018 global status report on road safety, 1.35 million traffic-related deaths occur annually worldwide, in addition to this 20 to 50 million people suffer serious injuries, with many becoming disabled as a result. Injuries from road accidents are the most significant cause of death for individuals aged 5–29 years [1]. The United Nations General Assembly has aimed to reduce the number of deaths and injuries from road accidents to one-half by 2030 [2] under SGD target 3.6 (good health and well-being) and 11.2 (sustainable cities and communities) [3]. Over 90% of road accidents are partly caused by human errors or negligence [4]. Improving safety and driver assistance has been a major focus for vehicle manufactures this decade using collision or distraction warnings [5].

© The Author(s), under exclusive license to Springer Nature Switzerland AG 2022
A. E. Hassanien et al. (Eds.): AMLTA 2022, LNDECT 113, pp. 604–614, 2022.
https://doi.org/10.1007/978-3-031-03918-8_50

We have built a comprehensive driving behavior assessment system that relies on its sensor input (camera and IMU) and it takes advantage of advanced behavior detection systems. Building this IoT system as a standalone edge device also helps in its implementation in low-end vehicles to give them high-end features and at the same time gets rid of the dependency of using mobile phones while driving. Our focused tasks are acceleration, braking and turning events, driver's drowsiness, and secondary task detections, lane detection and car lateral position estimation, as well as vehicle detection and forward gap estimation using monocular images. Section 2 covers the prominent research works in literature of the related work. Section 3 explains the materials and methodology used for each task as well as integration and implementation on edge devices. Section 4 discusses the results and implementation techniques. Section 5 presents the final remarks and the possibilities for future improvements.

2 Related Works

Driving behavior analysis is a vast area of research since there are many aspects of driver, vehicle, and road that can be considered to give an overall view of a driver's behavior on the road. In this section, we have reviewed some of the research that has been conducted to help monitor each of these aspects and then reviewed some of the existing applications that are using these methods to assess the driver's behavior.

2.1 Methods

Liu et al. in [6] established a scoring model based on the on-board diagnostics data from cars. Using speed and accelerometer data, they constructed a model that would output a score based on the events of overspeeding, sudden acceleration and sudden braking. Similarly, in [7], Waltereit et al. presented a scoring model that used a car's CAN bus data to measure the aggressiveness of a driving session. In [8], Ouardini et al. presented an approach for driver profiling using v2x (vehicle-to-everything), gps and CAN bus numerical data to find the important features. In [9], Azadani et al. pointed out the growing interest and the importance of driving behavior analysis by mentioning the underlying risk factors such as drunk driving, aggressive driving, etc. In [10], Sarafraz et al. used convolutional neural networks and transfer learning to train a network that can be used to classify different distractive tasks during driving. In [11], Weng et al. collected a drowsiness detection dataset and built a temporal deep belief network based on it to detect drowsiness taking into account the movements of head, mouth and eyes. To measure the distance to another car in the front, Zhe et al. in [12] used Faster R-CNN neural network to estimate the 3D bounding box and dimensions of a vehicle and added a projection geometry model after that based on the area of the back of the car to estimate the distance to the other cars in the front. The proposed model does not have a competitive real-time performance. In [13], Qin et al. proposed a very fast lane detection method which works based on a row-based approach and detects global lane features based on height sample lines. The lightest model was claimed to reach almost 96% in accuracy and 322.5 fps in speed.

2.2 Existing Applications

One of the relatively early examples of driving behavior apps is CarSafe that was developed by You et al. in [14] for android smartphones. This app alerts users when they are involved in unsafe activities or interacting with dangerous situations using phone cameras and IMU sensors. On the other hand, DriveSafe is an iPhone-based app developed by Bergasa et al. in [15] that also detects unsafe behaviors and situations using the iphone's rear-facing camera, microphone, gps and inertial sensors. They measure two kinds of unsafe situations and generate scores based on them, drowsiness and distraction. Kashevnik et al. in [16] developed an android app called Drive Safely that is able to use the front-facing camera, IMU sensor, microphone and gps to determine the driver's distraction, drowsiness and aggressiveness. More recently, in [17], Marafie et al. build AutoCoach, a cloud-based android application that can receive data from inertial sensors such as accelerometers, gyroscopes and magnetometers in the phone to identify accelerations, turns, braking and lane-changing events.

The driving behavior assessments in these implementations were found to be either non-comprehensive in terms of types of behaviors, or not accurate enough based on today's standards and state of the art models. In our paper, we propose an IoT system implemented on an edge device, which extensively uses state of the art machine learning methods to address many aspects related to driving behavior. Compared with other attempts, our approach uses optimization techniques to implement heavy machine learning models on edge devices as well as analyzing sensor data, while ensuring the real-time performance of the whole system.

3 Materials and Methods

3.1 System Components

Our system is composed of the low cost NVIDIA Jetson Nano 4 GB, an IMU sensor to measure acceleration and angular velocity, two USB cameras, one for driver monitoring and one for road view with a 55° FOV (Logitech c270) and a 78° FOV (Logitech c920) camera respectively, a wifi adapter and a 20000 mAH power bank for supplying power to the components.

3.2 Dataset

Since our driving behavior assessment system consists of various tasks that require training of machine learning models we used publicly available datasets along with collecting and labeling our own data. We tried Tusimple [18] and CULane [19] datasets for training the lane detection model. For the vehicle detection model, we used the KITTI 3D bounding box dataset [20]. We also used the iBUG 300-W [21] dataset for drowsiness detection. The description of the datasets is given in Table 1.

We collected a custom dataset using a front-facing dashboard camera for secondary task classification with four classes. Table 2 shows the distribution of the classes.

Table 1. Distribution of dataset.

Task	Dataset	No. of images
Vehicle detection	KITTI	14999
Facial features	iBUG 300-W	+4000
Lane detection	TuSimple/CULane	6408/133,235
Secondary task	Custom Dataset	7629

Table 2. Distribution of secondary task dataset.

Class	No. of images	Training	Testing
Drinking	1513	1135	378
Talking on phone (both hands)	2865	2160	705
Texting	1283	963	320
Normal driving	1968	1476	492

3.3 Sensor Data

In order to record the harsh braking and acceleration we used an IMU sensor to record the acceleration or deceleration along the driving axis. While we used the gradient of rotational speed along its own axis to record events of high rotational acceleration. The sensor data is downsampled using average values of each 10 recordings. The threshold values are given in Table 3.

Table 3. Threshold values for sensor data.

Attribute	Threshold	Threshold time
Acceleration	0.2g	2 s
Deceleration	0.2g	2 s
Turning	0.2g	N/A

3.4 Secondary Task Identification

Secondary task identification is done by a classification network using transfer learning. We tried three pre-trained models, namely ResNet50 [22], Xception [23] and

MobileNetV2 [24], with ImageNet weights as feature extractors. These extracted features of the driver images are then fed to a classifier layer to classify each frame. If the number of secondary task events is more than the threshold in a time window such as two seconds, a distracting event is triggered.

3.5 Drowsiness Detection

Some facial gestures such as closed eyes and open mouth (yawning) can be used as an indicator for detecting asleep or drowsy drivers. We trained dlib's facial landmark detection model on modified iBug 330-W dataset with our own points of interest to detect relevant points for mouths and eyes on the faces. Then we detect if the eyes and mouth are open or closed. This is done by calculating the relative distance between the top and bottom points in the landmark points of eyes and mouth. If this distance is less or greater than their respective thresholds for the threshold time, then the events of yawning or closed eyes are detected.

3.6 Forward Gap Estimation

Our system relies only on monocular camera images to find the distance from the front vehicles, so it is essential to first detect the vehicles. We trained a YOLOv5 [25] object detection model on KITTI bounding box detection dataset. The main difference here is that we extracted 2D bounding boxes from the nearest side of the cars. We then find a homography matrix to map points in the image to real world coordinates. This can be done using a rectangular reference object and map four corners of it in image space to real world coordinates. Then, we can find the distance (dt) of the reference object from the bottom edge in the transformed image and the real distance of the object in the real world (dr). The distance of the point of interest (dm) and the base in the top view space is then converted to the real distance by multiplying it with proportion constant (dr/dt). Finally, the real world offset (do) from the image baseline to the location of our camera is added to get the actual distance of the point of interest from our position in the real world. If this forward gap is less than a threshold an event is recorded. Equation (1) shows the final equation to get the real distance:

$$gap = do + \left(\frac{dr}{dt}\right) \times dm \tag{1}$$

3.7 Lane Drifting

Detecting lane drifting requires lane detection first and then lateral position estimation of the car with respect to the lane center. We used the lane detection model developed in [13] which required low computational power to run at high speed; ideal for edge device implementation. The model also gives better performance at fast driving speeds and the scenarios in which the visual visibility is limited. The method uses predefined rows in the image instead of a local convolution. Once the lane detection is done, we need to find the lateral position of the car with respect to the lanes. This estimation can be made much more intuitive and easier if we shift from perspective view to top view. We use the same procedure described in 3.6 to generate the inverse perspective map of the image. After the inverse perspective mapping of the image, we can calculate the distance between the center of the car in the image and the position of the lanes, and trigger a lane drift even if the distance was smaller than a threshold.

3.8 Implementation in TensorRT

One of the main goals in this research is the implementation of the entire system in real time on edge devices. TensorFlow and PyTorch frameworks are ideal for prototyping phase (training and experimental purposes) and are not suitable for production deployment phase when running inference on low end computation devices. We addressed these challenges by using lighter models that required less computational power and using the TensorRT optimization (for NVIDIA's hardware) in the inference pipeline for deep learning models. The yolov5 model was converted to TensorRT using the method given in [26] while the rest of the models were first converted to Open Neural Network Exchange (ONNX) models and then to TensorRT.

3.9 Overall System Design

The overall system consists of each module packaged together. The system runs in two parallel pipelines, one of them is specifically specified for sensor data only, while the other is used for the rest of the tasks using front and back camera data. Figure 1 shows a diagram of the whole system that runs in two parts in the pipeline.

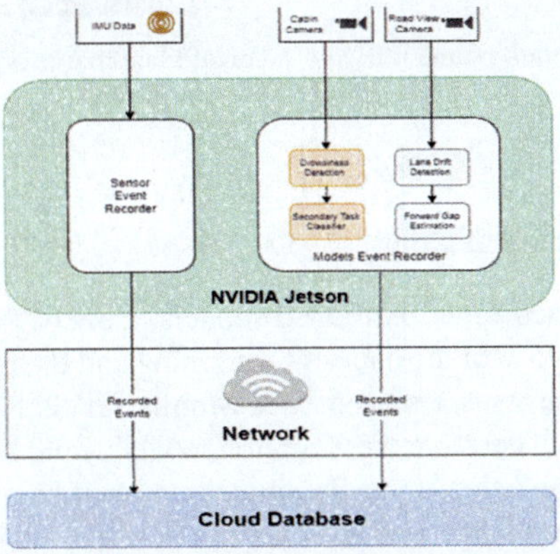

Fig. 1. Overall system design

The model runs entirely on the edge and only the detected events are sent to the database. We have used a MongoDB database in cloud to record our events. This makes our system more scalable and also eliminates possible privacy concerns by the users, as no data is sent or recorded.

4 Results and Discussion

4.1 Drowsiness Detection

We chose six points from the output of the facial landmark predictor for each eye, and six points on the mouth to use them for further processing. Figure 2 (left) shows all the points that can be found using the dataset. The points that we used include points 37 to 48, and points 48, 50, 52, 54, 56 and 58. Equation (2) shows the calculated ratio.

$$ratio = \frac{d_{p1,p2} + d_{p3,p4}}{2 \times d_{p5,p6}} \tag{2}$$

where p1 and p2 are the left vertical points, p3 and p4 are the right vertical points, and p5 and p6 are the horizontal points. This method was used to calculate the ratio for the left eye, the right eye and the mouth. The ratio used for the eyes which indicated closed eyes is 0.3 and the ratio used for the mouth which indicates yawning is 0.7. The system issues an alert if there is an instance of drowsiness, i.e., the violation of threshold values, for any time period more than 1 s. Figure 2 (right) shows the predicted landmarks on a real-time stream with red markers.

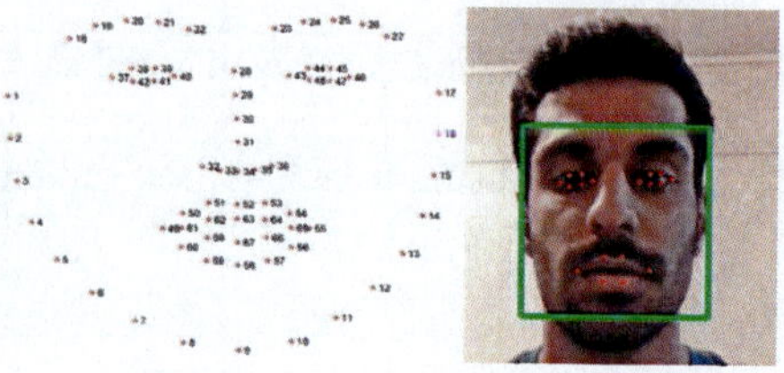

Fig. 2. Facial landmark points [20] and predicted landmarks on a real-time stream.

4.2 Secondary Task Classification

For classification, we used three pretrained models, namely ResNet50, Xception and MobileNetV2. We kept 75% of the dataset for training and the rest for testing. The classification accuracy on the test set was 65% for MobileNetV2, 87 percent for ResNet50, and 99.5% for Xception. Therefore the Xception model would be the best choice for the classification task. Figure 3 shows the classification accuracy and loss for the Xception model, and the confusion matrix across all classes.

To help the model generalize better, we augmented the dataset by modifying the brightness of the images and added a dropout layer with the probability of 0.2 after the feature extraction layer. We trained the models for 10 epochs using Adam optimizer with learning rate of 0.0001, as well as l1 and l2 regulizers with parameters of 0.0001 and 0.001 respectively, and categorical crossentropy as the loss function.

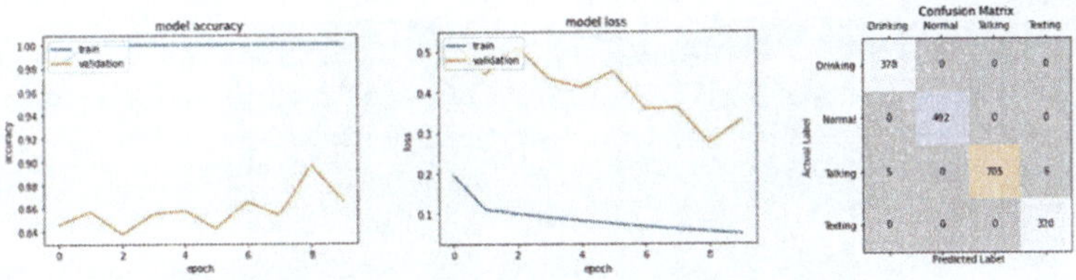

Fig. 3. Xception model's loss and accuracy over 10 epochs (left) and confusion matrix (right)

4.3 Car Distance Estimation

We fixed the 78° FOV camera at a desirable height and used a square target to obtain the homography matrix and the distance as explained in Sect. 3.6. The final calculation accounts for the lateral position of the cars as well and makes the method robust front. Figure 4 shows the original image and the mapped top-view in one of our tests.

Fig. 4. Original and mapped image. The midpoint of the lower center of the bounding box is marked with a yellow point. (Color figure online)

The actual distance was 6.28 m and the distance estimated using our method was 6.19 m with an error of 1.4%. Since we care mostly about near distances and collision avoidance, distances of more than 10 m in which the method can be less accurate are not considered.

4.4 Lane Drifting

The output of the ResNet-based model [13] trained on the TUSimple dataset are arrays of points in the image space where each array represents one lane. Figure 5 shows how the estimated points look like in a test image with green markers.

Once we find the lanes, we need to map them into a space where the points can be processed. We use the same technique of inverse perspective mapping explained in Sect. 3.7 and Sect. 4.3 to map the points from the original image forward view to a transformed top down view. In addition to distance estimation.

Fig. 5. An instance of lane estimation in a test image

4.5 Implementation in Jetson Nano

The system is powered via a usb power bank (20000 mAh) for upto 20 h with 4.8 Wh power consumption. The program is automatically initiated by a systemd service in linux. If it is connected to an internet network, it can also send the results to the MongoDB database in the cloud for later reporting and evaluation. We also compared the runtime of each model with a Nvidia GeForce GTX 1070 graphics adapter with 8 GB memory. When we switched between MobileNetV2 and Xception model, the system runs at approximately 2 and 1.2 frames per second respectively. Table 4 shows the comparison between two devices in terms of frames per second during runtime.

Table 4. Comparison of edge device and laptop performance.

Task	Jetson nano (FPS)	GeForce GTX 1070 (FPS)
Drowsiness detection	15	40
Lane detection	2.5	35
Car distance	9.5	142
Secondary task classification	2.5	15

5 Conclusions

We developed a low cost comprehensive IoT solution to perform driving behavior assessment for detecting careless driving patterns. The system uses simple IMU sensor inputs to observe acceleration, deceleration, and turning behaviour. We also developed a forward gap estimator to calculate the distance of the front vehicles using a monocular camera image with good accuracy for near range (<10 m) applications. The system also records events of lane drifting. It also uses the cabin camera to monitor drivers' attention and alertness by looking at the eyes and the mouth of the driver. The distraction events are detected by a classification model that record events of some secondary tasks like drinking, calling and texting with 99.5% accuracy.

We provide a comprehensive proof of concept as well as a deployable system. With some operational constraints and simplifications; the existing work serves as the baseline for further advancements such as driver profiling based on style, adding more modules, depth sensors for accurate depth calculations and implementation on better GPU kits like NVIDIA NX or AGX. Head pose estimation can be leveraged to augment the distraction detection method, and GPS data can be used to incorporate road context and environment into the assessment.

References

1. World Health Organization: Global status report on road safety 2018: summary (No. WHO/NMH/NVI/18.20). World Health Organization (2018)
2. Improving global road safety (A/RES/74/299), Agenda item 12, Resolution adopted by the General Assembly on 31 August 2020
3. Desa, U. N.: Transforming our world: The 2030 agenda for sustainable development (2016)
4. European Commission: Final report of the eSafety working group on road safety. Office for Official Publications of the European Communities (2003)
5. European Commission: Autonomous cars: A big opportunity for European industry (2017)
6. Liu, T., Yang, G., Shi, D.: Construction of driving behavior scoring model based on OBD terminal data analysis. In: 2020 5th International Conference on Information Science, Computer Technology and Transportation (ISCTT), pp. 24–27 (2020). https://doi.org/10.1109/ISCTT51595.2020.00012
7. Waltereit, M., Zdankin, P., Matkovic, V., Uphoff, M., Weis, T.: Online driving behavior scoring using wheel speeds. In: Proceedings of the 6th International Conference on Vehicle Technology and Intelligent Transport Systems - VEHITS, pp. 417–424 (2020)
8. Ouardini, A., El Ouazzany Ech-chahedy, I., Bouhoute, A., Berrada, I., El Kamili, M.: Analyzing driving behavior: towards dynamic driver profiling. In: Ad Hoc Networks, pp. 179–190 (2021)
9. Azadani, M.N., Boukerche, A.: Driving behavior analysis guidelines for intelligent transportation systems. IEEE Trans. Intell. Transp. Syst., 1–19 (2021)
10. Masood, S., Rai, A., Aggarwal, A., Doja, M.N., Ahmad, M.: Detecting distraction of drivers using convolutional neural network. Pattern Recognit. Lett. **139**, 79–85 (2020)
11. Weng, C.-H., Lai, Y.-H., Lai, S.-H.: Driver drowsiness detection via a hierarchical temporal deep belief network. In: Chen, C.-S., Lu, J., Ma, K.-K. (eds.) ACCV 2016. LNCS, vol. 10118, pp. 117–133. Springer, Cham (2017). https://doi.org/10.1007/978-3-319-54526-4_9
12. Zhe, T., et al.: Inter-vehicle distance estimation method based on monocular vision using 3D detection. IEEE Trans. Veh. Technol. **69**(5), pp. 4907–4919 (2020)
13. Qin, Z., Wang, H., Li, X.: Ultra fast structure-aware deep lane detection. arXiv [cs.CV] (2020)
14. You, C.-W., et al.: CarSafe: a driver safety app that detects dangerous driving behavior using dual-cameras on smartphones. In: Proceedings of the 2012 ACM Conference on Ubiquitous Computing, Pittsburgh, Pennsylvania, pp. 671–672 (2012)
15. Bergasa, L.M., Almería, D., Almazán, J., Yebes, J.J., Arroyo, R.: DriveSafe: an app for alerting inattentive drivers and scoring driving behaviors. In: 2014 IEEE Intelligent Vehicles Symposium Proceedings, pp. 240–245 (2014)
16. Kashevnik, A., Lashkov, I., Ponomarev, A., Teslya, N., Gurtov, A.: Cloud-based driver monitoring system using a smartphone. IEEE Sens. J. **20**(12), 6701–6715 (2020)
17. Marafie, Z., et al.: AutoCoach: an intelligent driver behavior feedback agent with personality-based driver models. Electronics **10**(11) (2021)

18. TuSimple: TuSimple benchmark. https://github.com/TuSimple/tusimplebenchmark
19. Pan, X., Shi, J., Luo, P., Wang, X., Tang, X.: Spatial as deep: spatial CNN for traffic scene understanding. In: Thirty-Second AAAI Conference on Artificial Intelligence, April 2018
20. Geiger, A., Lenz, P., Stiller, C., Urtasun, R.: Vision meets robotics: the KITTI dataset. Int. J. Robot. Res. **32**(11), 1231–1237 (2013)
21. Sagonas, C., Tzimiropoulos, G., Zafeiriou, S., Pantic, M.: 300 faces in-the-wild challenge: the first facial landmark localization challenge. In: Proceedings of IEEE International Conference on Computer Vision (ICCV-W), Workshop on 300 Faces in-the-Wild Challenge (300-W), Sydney, Australia, pp. 397–403 (2013)
22. He, K., Zhang, X., Ren, S., Sun, J.: Deep residual learning for image recognition. In: 2016 IEEE Conference on Computer Vision and Pattern Recognition (CVPR), pp. 770–778 (2016). https://doi.org/10.1109/CVPR.2016.90
23. Chollet, F.: Xception: deep learning with depthwise separable convolutions. In: 2017 IEEE Conference on Computer Vision and Pattern Recognition (CVPR), pp. 1800–1807 (2017). https://doi.org/10.1109/CVPR.2017.195
24. Sandler, M., Howard, A., Zhu, M., Zhmoginov, A., Chen, L.: MobileNetV2: inverted residuals and linear bottlenecks. In: 2018 IEEE/CVF Conference on Computer Vision and Pattern Recognition, pp. 4510–4520 (2018). https://doi.org/10.1109/CVPR.2018.00474
25. Jocher, G., et al.: ultralytics/yolov5: v5.0 - YOLOv5-P6 1280 models, AWS, Supervise.ly and YouTube integrations, April 2021
26. Xinyu, W.: TensorRTx, GitHub repository (2021). https://github.com/wang-xinyu/tensorrtx

Check for updates

Advanced Deep Reinforcement Learning Protocol to Improve Task Offloading for Edge and Cloud Computing

Walaa Hashem(✉), Radwa Attia, Heba Nashaat, and Rawya Rizk

Electrical Engineering Department, Port Said University, Port Fuad 42526, Egypt
{walaa.kasem,radwa_yousef,hebanashaat,r.rizk}@eng.psu.edu.eg

Abstract. Fog Computing (FC) is an extension of cloud computing, providing services closer to the users near the edge. In FC, the Mobile Devices (MDs) can offload heavy tasks to an edge or cloud server. The decision of whether a MD offloads tasks to a Mobile Edge Computing (MEC) or a Mobile Cloud Computing (MCC) servers should be carefully studied. In this paper, an Advanced Deep Reinforcement Learning (ADRL) protocol is proposed to improve task offloading. It can generate multi-class offload decisions for executing independent tasks locally, at the edge, or cloud while considering the current workload of MD, MEC, and MCC to maintain balance in the mobile system. The experiments show the efficiency of the proposed algorithm in terms of the computation time and convergence to the optimal solution.

Keywords: Fog computing · MEC · MCC · Machine learning · Task offloading

1 Introduction

Fog Computing (FC) is a decentralized computing infrastructure in which data, computation, storage, and applications exist between the data source and the cloud [1]. Edge-Cloud Computing (ECC) combines the benefits of both edge and cloud computing. It is one of the most promising ways to address the resource insufficiency of smart devices. Task offloading is an important challenge that should be addressed practically in ECC for deciding the place and the time for each task [2].

The main purpose of task offloading in ECC is to decide which tasks are offloaded from user devices and which edge or cloud an offloaded task is assigned. For edge/clouds, the task offloading problem is much more complex than the cloud or edge problem because it not only involves all the challenges of offloading tasks on both mobile clouds and edges but also introduces new challenges such as [1]; the heterogeneity between edges and clouds in terms of various resources, a more complex network, and the decision of which edge or cloud each offloaded task is assigned. Essential metrics in offloading are; energy, latency, Quality of Service (QoS), Quality of Experience (QoE), response time, and cost [3]. These metrics are considered in offloading problems disjointedly or jointly as multi-objective problems. However, due to the varying network conditions and

© The Author(s), under exclusive license to Springer Nature Switzerland AG 2022
A. E. Hassanien et al. (Eds.): AMLTA 2022, LNDECT 113, pp. 615–628, 2022.
https://doi.org/10.1007/978-3-031-03918-8_51

limited computation, the offloading actions taken by the mobile user may not achieve the lowest cost [2].

Achieving optimal offloading decisions usually requires accurate information about network resources and, in turn, increases the overhead and operating costs of Mobile Devices (MDs). Also, cloud environments don't need a dynamic approach all the time, as most of the traffic on the cloud can be predicted; Therefore, Machine Learning (ML) methods are incorporated into the task offloading algorithm to achieve the offloading decision based on workload information only rather than all information about the system [4]. ML is an application of artificial intelligence that provides systems the ability to automatically learn and improve from experience. Therefore, in the presence of high dynamic and time-critical behavior of intelligent systems, it is more practical, intelligent, and unavoidable to apply ML methods instead of classical methods. There are three types of ML; Supervised, unsupervised, and reinforcement learning.

- **Supervised Learning** [4]**:** Consists of a target/outcome dependent variable predicted from a given set of independent variables (predictors). Using these sets of variables, a function that maps inputs to desired outputs is generated. The training process continues until the model achieves the desired level of accuracy on the training data.
- **Unsupervised Learning** [4]**:** In this algorithm, there is no target or outcome variable to be predicted. It is used to cluster populations in different groups, which is widely used to segment customers in different groups for specific intervention.
- **Reinforcement learning (RL)** [4]**:** A type of learning that is based on interaction with the environment. It is rapidly growing and producing a huge variety of learning algorithms that can be used for various applications. In this technique, an agent learns from the environment by interaction. Then the decision is made through feedback. Deep Reinforcement Learning (DRL) [5, 6] is a subfield of ML that combines RL and deep learning, where machines can teach themselves based upon the results of their actions. Due to the efficiency and simplicity of DRL, as it doesn't require target data, it is used to generate the offload decision in this paper.

To minimize the computation time and convergence to the optimal solution while maintaining the QoS for MDs; an Advanced Deep Reinforcement Learning (ADRL) protocol is proposed. It improves the task offloading, generating the offload decision for executing independent tasks locally or cloud. Also, the current workloads of MD, Mobile Edge Computing (MEC), and Mobile Cloud Computing (MCC) are considered to maintain balance in the mobile system. The proposed algorithm is based on DRL, which generates and stores the offloading decisions with the workloads and system consumption together in a database and then trains and updates multiple parallel convolutional Neural Networks (CNNs) with a batch of labeled data.

The remainder of the paper is organized as follows: Sect. 2 introduces the related works of offloading approaches in MEC. Section 3 explains the system model and problem formulation. An explanation of the proposed task offloading algorithm in ECC is presented in Sect. 4. Simulation results using python programming are shown in Sect. 5. Finally, the conclusion is drawn out in Sect. 6.

2 Related Work

The decision of whether an MD offloads its computation task to a MEC/MCC server or not is presented in many works of literature. A weighted graph is constructed in [7] to optimize the execution time of a workflow in the ECC by dividing all tasks into two groups, the edge with the least weight is determined, finally, realize the mapping between offloaded tasks and edge/cloud layers. However, this work didn't concern task deadlines. Task offloading scheme is introduced in [8] for minimizing the total processing time of all tasks by making decisions dynamically. Nowadays, in the presence of high dynamic and time-critical behavior of smart systems, it is more beneficial to apply ML methods instead of classical methods. A comparison of the main ML-based offloading mechanisms in MEC is illustrated in [2]; they are categorized into three main classes, as seen in Fig. 1.

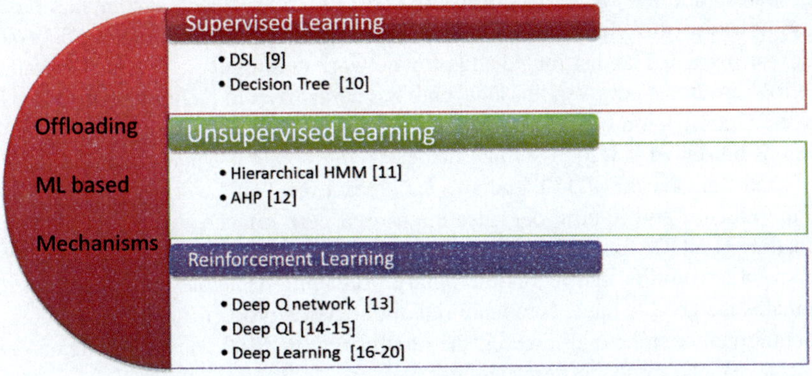

Fig. 1. Taxonomy of ML-based offloading mechanisms.

2.1 Supervised Learning-Based Offloading Mechanisms

A dynamically fine-grained offloading framework is presented in [9] to address the challenge of user mobility, considering the local overhead on the mobile terminal side and the limited communication and computation resources on the network side. However, the task offloading decision is used only for a single user and server. A Module Placement method by Classification and regression tree Algorithm (MPCA) is presented in [10], where the power consumption of MDs is checked; if the value is more significant than Wi-Fi's power consumption, offloading is done. However, this algorithm doesn't have a failure recovery policy.

2.2 Unsupervised Learning-Based Offloading Mechanisms

A hidden Markov model-based dynamic scheduling approach is presented in [11] to allow the system adaptation to the changing requirements with an improvement process towards multiple performance metrics. However, this algorithm is high state space. A computation offloading strategy is proposed in [12]. It is divided into three stages; the offloading decision-making stage, Analytic Hierarchy Process (AHP) stage, used to select the server that undertakes the offloading task, and, finally, the tasks scheduling stage. This strategy may achieve the global optimum; however, it is highly complex.

2.3 Reinforcement Learning-Based Offloading Mechanisms

For the MEC environment, a Deep Q-Network (DQN)-based tasks offloading and resource allocation algorithm is proposed [13]. A joint task offloading decision is designed to minimize the overall offloading cost in terms of energy, computation, and delay; however, this algorithm is centralized. In [14], a DQ-Learning (DQL) framework is proposed. The distributed edge/fog network controller scavenges the available edge/fog resources; however, this algorithm has complexity in reality with many mobile devices. Optimization of the vehicular energy and task execution time in a vehicular system is presented in [15], assuming that each edge server is equally shared among all connected vehicles and all of them have equal resources. This work assumes no departure of vehicles, and in turn, degrades the overall performance. A DRL-based Online Offloading (DROO) framework is proposed [16]. It implements a deep neural network as a scalable solution that learns the binary offloading decisions from experience. It eliminates the need to solve combinatorial optimization problems and thus reduces the computational complexity; however, the mobility of the MDs causes DROO harder to converge. Much more algorithms are presented in [17–20].

To minimize the response time of all tasks from different MDs, an ADRL algorithm is proposed, which considers the current workload of MD, MEC, and MCC.

3 System Model and Problem Formulation

The main objective of the ADRL is to generate a multi-class offload decision for executing an independent task locally or offloading them to the central cloud or edge; for execution; It utilizes multi-dimension CNN while considering multiple tasks with multiple features as input to CNN to generate offload decision while considering the previous workload in the MDs, edge, and cloud.

3.1 System Model

An overview of the ADRL system model is presented in Fig. 2. It consists of one edge, one central cloud, and multiple MDs. Each MD can offload its computation tasks either to the edge server or cloud server.

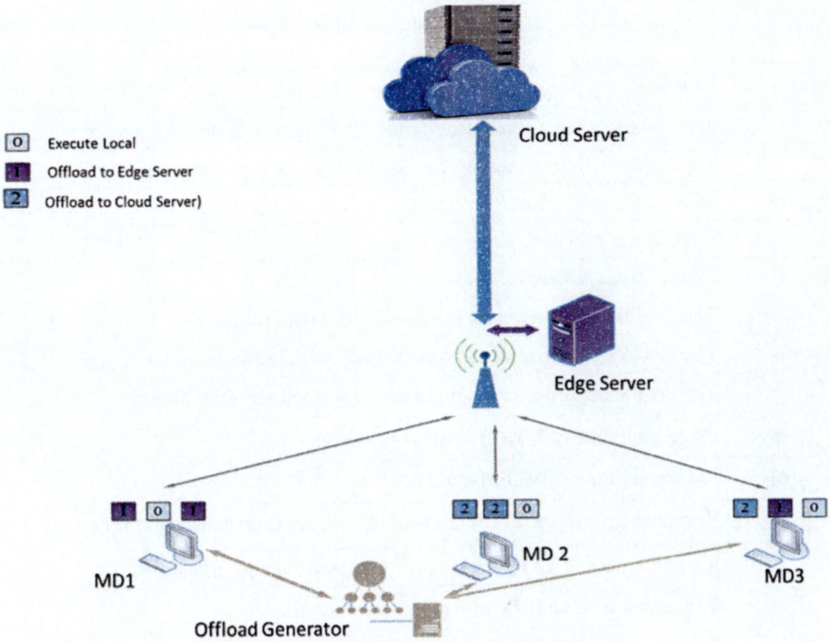

Fig. 2. The system model of task offloading.

The system model consists of multiple MDs, denoted by a set $N = \{1, 2, ..., n\}$ and some independent computational tasks, denoted by $M = \{1, 2, ..., m\}$, the length of the request of the mth task of the nth MD at time t is denoted as $\boldsymbol{REQ}_{nmt}$. Each MD can either execute their tasks locally or offload them to the edge server or center cloud server for further execution. The offloading decision is denoted by a numerical variable $a_{nmt} \in \{0, 1, 2\}$. The notations used in the ADRL are shown in Table 1.

In local computing, the total length of requests processing by MD(n), the processing time of MD(n), the response time of the mth request of nth MD at time t, and the response time of all requests of nth MD are obtained by (1), (2), (3) and (4); respectively.

$$TL_{md}(nt) = \sum\nolimits_{m=1}^{M} \mathrm{REQ}(nmt) \times \left[\frac{(1 - a_{nmt})(2 - a_{nmt})}{2}\right] \quad (1)$$

$$PT^{t}_{MD(n)} = PT^{t-1}_{md}(n) + \frac{TL^{t}_{d}}{PR_{md}(n)} \quad (2)$$

$$RT^{MD}_{req(m)}(nt) = PT^{t-1}_{md}(n) + \frac{\mathrm{REQ}(nmt)}{PR_{md}(n)} \quad (3)$$

$$RT^{MD}_{n} = \sum\nolimits_{m=1}^{M} RT^{MD}_{req(m)}(nt) \times \left[\frac{(1 - a_{nmt})(2 - a_{nmt})}{2}\right] \quad (4)$$

Table 1. List of notations

Meaning	Notation
ED	Transmission delay when user n offloads its task m to the edge server
CD	Transmission delay when user n offloads its task m to the cloud server
$PR_{md}(n)$	MD(n) processing rate
PR_E	Edge server processing rate
PR_C	Cloud server processing rate
$TL_{md}(nt)$	The total length of requests processing by MD(n) at time t
TL_E^t	The total length of requests processing by edge server at time t
TL_C^t	The total length of requests processing by cloud server at time t:
$PT_{md}^{t-1}(n)$	Processing time of MD(n) at the previous time
$PT_{md}^t(n)$	Processing time of MD(n) at time t
$RT_{req(m)}^{MD}(nt)$	Response time of the mth request of nth MD, executed locally at time t
RT_n^{MD}	Response time of all requests of nth MD, executed locally at time t
PT_E^{t-1}	Processing time of cloud server at a previous time
PT_E^t	Processing time of Cloud server at time t
$RT_{req}^E(nmt)$	Response time of the mth request of nth MD, executed in edge server at time t
RT_n^E	Response time of all requests of nth MD, executed in edge server at time t
PT_c^{t-1}	Processing time of cloud server at a previous time
PT_c^t	Processing time of Cloud server at time t
$RT_{req}^C(nmt)$	Response time of the mth request of nth MD if executed in Cloud server at time t
RT_n^C	Response time of all requests of nth MD if executed in Cloud server at time t
REQ_{nmt}	Length of the mth request of nth MD at time t
X_t	Channel at time t consists of 9 requests from 3 MD
D_{in}^t	An array of shapes (9,3) defines response time for each task in Xt
a_{nmt}	Offloading decision for the task (m) of MD(n)
K	Number of CNNs computed by trial and error

For edge computing, the total length of requests processing by edge server, the processing time of one edge server, the response time of the mth request of nth MD, executed in edge server at time t, and the response time of the all requests of nth MD executed in edge server at time t are executed by (5), (6), (7) and (8); respectively.

$$\mathrm{TL}_E^t = \sum\nolimits_{n=1}^{N} \sum\nolimits_{m=1}^{M} REQ(nmt) \times [(2 - a_{nmt}) \times a_{nmt}] \tag{5}$$

$$PT_E^t = PT_E^{t-1} + \frac{TL_E^t}{PR_E} \tag{6}$$

$$RT_{req}^{E}(nmt) = PT_{E}^{t-1} + ED + \frac{\text{REQ}(nmt)}{PR_{E}} \tag{7}$$

$$RT_{n}^{E} = \sum\nolimits_{m=1}^{M} RT_{req}^{E}(nmt) \times [(2 - a_{nmt}) \times a_{nmt}] \tag{8}$$

In cloud computing, the Total length of requests processed by the cloud server, the processing time of the cloud server, the response time of the mth request of nth MD executed in cloud server at time t, and the response time of all requests of nth MD executed in cloud server at time t is calculated by (9), (10), (11) and (12); respectively.

$$TL_{c}^{t} = \sum_{n=1}^{N}\sum_{m=1}^{M} REQ(nmt) \times \left[\frac{(a_{nmt} - 1)(a_{nmt})}{2}\right] \tag{9}$$

$$PT_{c}^{t} = PT_{c}^{t-1} + \frac{\text{TL}_{\text{c}}^{\text{t}}}{\text{PR}_{\text{C}}} \tag{10}$$

$$RT_{req}^{C}(nm) = PT_{c}^{t-1} + CD + \frac{\text{REQ}(nmt)}{PR_{C}} \tag{11}$$

$$RT_{n}^{C} = \sum_{m=1}^{M} RT_{req}^{C}(nmt) \times \left[\frac{(a_{nmt} - 1)(a_{nmt})}{2}\right] \tag{12}$$

3.2 Problem Formulation

In ADRL, the system input channel (X_t) consists of 9 input requests which are 3 tasks of different sizes from 3 MDs at time t. Given the previous processing time of MDs, edge server, and cloud server, the input Channel is changed from array of shape (9,1) to array (9,3), which is called D_t^{in} and contains response time of each task when executed in MD, edge server, or cloud server, The construction of D_t^{in} from X_t is shown in Fig. 3, the construction of each item in D_t^{in} is explained as follow:

Each REQ(nm), in the channel, become $[RT_{req(\text{m})}^{MD}(\text{n}), RT_{req(\text{nm})}^{E}, RT_{req(nm)}^{C}]$
Where n = {1, 2, 3}.
m = {1, 2, 3}.

The total response time required to execute all the channel tasks at time t can be given by (13).

$$\text{RT}(D_t^{in}, A_t) = \sum\nolimits_{n=1}^{N} \max\left(RT_n^{MD}, RT_n^{E}, RT_n^{C}\right) \tag{13}$$

To reduce the overall response time, the problem (P1) that defines the best value for a_{nmt} (0, or 1, or 2) is defined as in (14).

$$\text{(P1)}: \quad \text{minimize} \quad \text{RT}\ (D_{in}^{t}, A_t) \tag{14a}$$

$$\text{Subject to}: \quad a_{nmt} \in \{0, 1, 2\} \tag{14b}$$

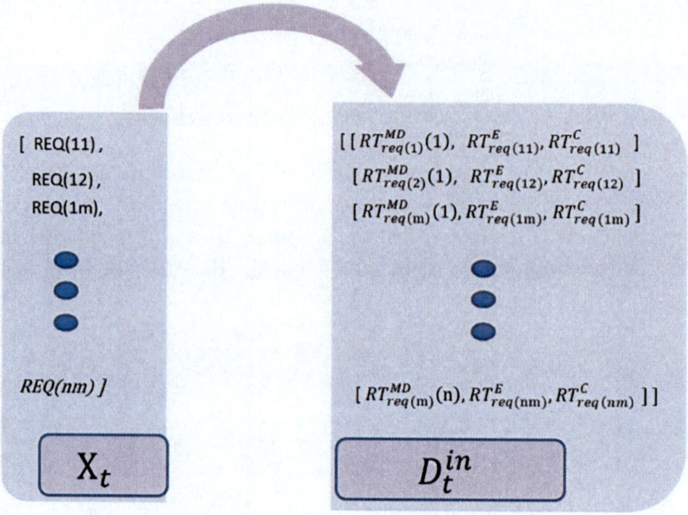

Fig. 3. The construction of D_t^{in} from X_t

P1 is a mixed-integer programming problem that is generally difficult to solve. An approximate algorithm based on deep learning is presented in the next section to solve P1 effectively.

4 ADRL Algorithm

ADRL task offloading algorithm is proposed for the MCC and MEC hybrid offloading; it is based on multiple parallel convolutional neural networks. The architecture of the ADRL algorithm is as depicted in Fig. 4. The proposal implements an offloading algorithm that generates offload decisions for executing independent tasks locally, at an edge, or a cloud while considering the current workload of MD, MEC, and MCC to maintain balance in a mobile system. To achieve this objective, DRL is used to make the optimal offloading decision by three main steps; offloading action generation, training, and testing.

The workloads D_{in}^t are regarded as the input to the neural networks and the offloading decisions A_t are regarded as output, they are stored into a database and used to train multiple parallel CNNs and generate new data to replace the old data in the database. Thus, the database can be updated, and therefore, the CNNs can be trained using the labeled database to solve the NP-hard problem (non-deterministic polynomial-time hardness) well.

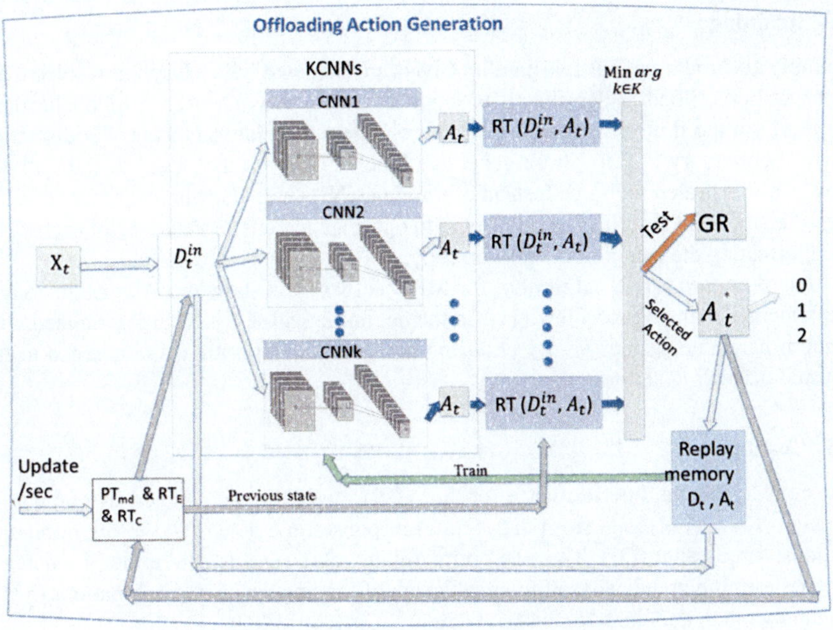

Fig. 4. The procedure of the ADRL algorithm.

4.1 Offloading Action Generation

The best offloading decision is generated by DRL, in which CNNs are applied and then storing the best offloading decisions with the workloads and system consumption together in a database. As depicted in Fig. 3, the generation process of the offloading decisions can be expressed as follows:

- When the inputs at time t D^t_{in} are given, the CNNs are used to get the outputs.
- The workloads D^t_{in} and the best offloading decisions A^*_t are stored in a database.
- Finally, the problem can be solved by training CNNs using the labeled database.

As shown in Fig. 3, for each input D^t_{in}, k offloading actors are used to generate k candidate offloading actions with one action per actor. Inside each offloading actor, a CNN is used to generate binary offloading action at time t, $A_t(k)$, a parametrized function can represent $f_{\theta k}$, as:

$$f_{\theta k} : D^t_{in} \longrightarrow A_t(k), \tag{15}$$

where θ_k denotes the parameters of the kth CNN. All those CNNs have the same structure but with different parameter values θ_k. The offloading action with the least $RT^*(D^t_{in}, A_t)$ is selected among all K candidates, as in (16).

$$A^*_t = \underset{k \in K}{\text{Minarg}} RT^*(D^t_{in}, A_t) \tag{16}$$

4.2 Training

Training and updating multiple parallel CNNs are executed with a batch of labeled data. Once the best offloading decision A_t^* is obtained, it is saved as a new entry of labeled data, (D_{in}^t, A_t^*) in the finite-size memory structure, where the oldest data entry is discarded when the memory is full. Those generated labeled data are then used to train all *K's* CNNs, as illustrated in Fig. 3. Instead of training CNNs with the whole memory of data, a gradient descent algorithm is performed to optimize parameter values θ_k of each CNN by minimizing the categorical cross-entropy.

The proposed ADRL algorithm for MEC networks is shown in Algorithm 1; it is implemented using TensorFlow [21]. In the beginning, all K CNNs are initialized with random parameter values θ_k. By choosing a $K \geq 2$, the algorithm is converged to the optimal offloading actions.

4.3 Testing

The convergence and performance for the ADRL algorithm are studied by computing the Gain Ratio (GR) gain from the optimal response time value (Q_2) to the minimum response time value (Q_1). The true optimal gain value (Q_2) is determined by using a time-consuming greedy algorithm, in which all offloading decision combinations (3^{NM}) are enumerated, and the ratio is calculated by (17).

$$\text{Gain}_{\text{Ratio}}(GR) = \frac{Q2}{Q1} \tag{17}$$

where $0 < GR < 1$ indicates how close the solution found by the ADRL algorithm to the true optimal solution achieved by the greedy algorithm.

Algorithm 1 The proposed ADRL for the MEC network

for each user(n) in the channel at time t **do**
    for each task (m) in the channel at time t **do**
        Compute D_{in}^t
    end for
end for
Initialize K CNNs with random parameters
Empty the memory structure
for i=1, 2, ……, N **do**
    Replicate different D_t^{in} to all CNN.
    Generate K-th offloading action Ak from the K-th CNN
    Select a best offloading decision A^*
    Compute GR (the convergence of ADRL algorithm)
    Store (D_t^{in}, A^*) into the memory structure
    if the database is more than 128 MB **do**
        Randomly Sample batches from the memory structure
        Train the CNNs and update θ_k
    end if
end for

5 Simulation Results

Simulation scenarios are carried out using Python programming with Google Colab [22]. It is a high-performance web framework to build APIs (Application Programming Interface) in Python, and it is a free cloud service hosted by Google.

In the simulation, the number of MDs (N) is assigned to 3, and each user had (M = 3) different tasks. In addition, the MD(n) processing rate is given by, $PR_{md}(n) = 1$, the processing rates of the edge cloud server is, $PR_E = 800$ MHz and the central cloud server is $PR_C = 1200$ MHz. The parameters satisfy: $PR_{md}(n) < PR_E < PR_C$, In all simulations, the request length of all tasks is randomly distributed between 10 MB and 30 MB. The offloading decision is denoted by a numerical variable $a_{nmt} \in \{0, 1, 2\}$, specifically, $a_{nmt} = 2$ denotes that user n decides to offload its task m to the cloud server, $a_{nmt} = 1$ denotes that user n decides to offload its task m to the edge server, and $a_{nmt} = 0$ means that user n decides to execute its task m locally, an experience replay memory of size 1024, a .00001 learning rate, and five CNNs, are selected to compensate for convergence and performance.

5.1 The Effect of Learning Rate on the Gain Ratio

The ADRL convergence is demonstrated in distinct situations, first, the convergence performance of ADRL is analyzed based on the learning rate. The Gain Ratio (GR) is plotted between ADRL and the greedy method for better comparison.

Figure 5 shows the performance of the relative optimality GR with different learning rates. The value of GR is at a minimum when the learning rate is 0. 1. However, when the learning rate is smaller, GR increases. The higher the learning rate, the faster the convergence speed of ADRL. However, when the learning rate reaches the .00001, the probability of obtaining an optimal local solution is increased rather than the optimal global one. Hence, the appropriate learning rate is carefully selected concerning specific situations.

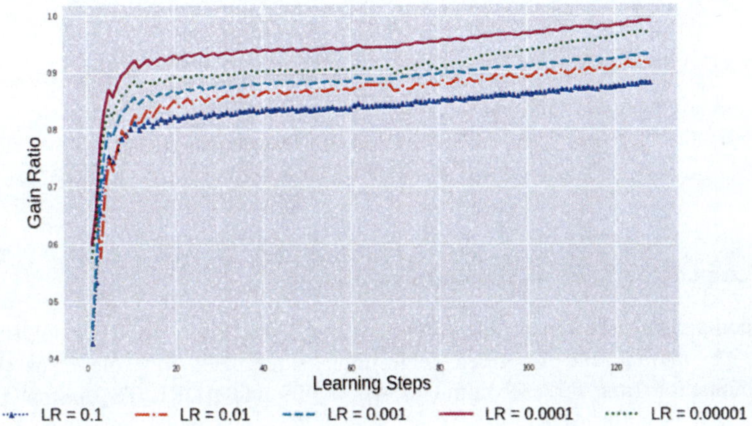

Fig. 5. Convergence performance under different learning rates.

5.2 Gain Ratio Comparison Analysis

To gain some insights and analyze the efficiency of the proposed ADRL algorithm, the following state-of-the-art offloading decision methods are implemented for comparisons:

- ADRL algorithm where the input of neural network is restricted to the request length of the data (ADRL_TS)
- ADRL algorithm where the input of neural network is restricted to the request length of the data and neural network is trained with the optimal offloading decision (ADRL_TS_optimal)

Figure 6 shows the Response time (RT) for ADRL, ADRL_TS, and ADRL_TS_optimal. The value of RT is at a minimum when ADRL is used to generate offload decisions. but, when ADRL_TS is used, RT increases. However, when the neural network is trained with the optimal offloading decision as in ADRL_TS_optimal The value of RT is decreased but still larger than ADRL because ADRL generates offload desition while considering previous processing time of both MD, edge server, and cloud server.

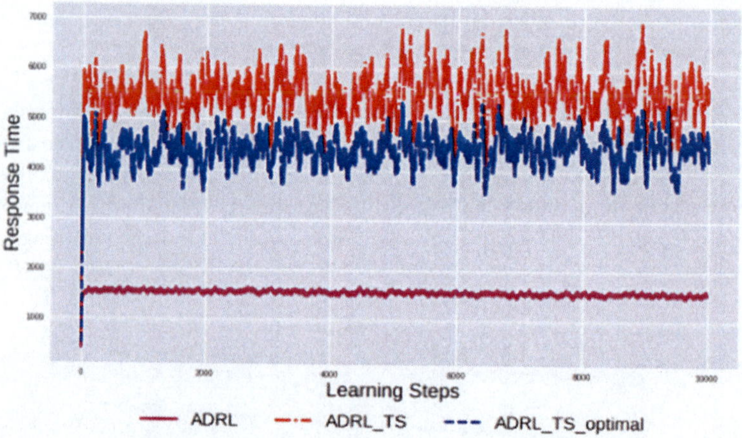

Fig. 6. Comparison with different offloading schemes

5.3 Computation Time and Comparison Analysis

The comparison between the computation time of ADRL, ADRL_TS, ADRL_TS_optimal, and the greedy algorithm, is presented in Table 2; as shown, the computation time for each input in ADRL_TS and ADRL_TS_optimal is less than ADRL because ADRL generates offload decisions while considering previous processing time of MDs, edge, and cloud. However, still, the time cost of the ADRL algorithm is less than the greedy algorithm with about 1227 s.

Table 2. Time cost of ADRL, ADRL_TS, and ADRL_TS_optimal versus greedy algorithm.

	Algorithm			
	ADRL_TS	ADRL_TS_optimal	ADRL	Greedy
Time cost (second)	0.0031	0.0059	0.5273	1228

6 Conclusion and Future Work

The offloading procedure for MDs to MEC, MCC, or locally at MDs is presented in this paper. An ADRL offloading algorithm is proposed, considering the previous processing time for MDs, edge, and cloud servers. It can also generate multi-class offloading decisions for executing independent tasks locally, at the edge, or in a cloud. The proposed algorithm is based on DRL, which works by generating and storing the offloading decisions with the workloads and system consumption together in a database and then training and updating multiple parallel CNNs with a batch of labeled data. The results demonstrate that the ADRL algorithm achieved a good gain ratio and computation time performance.

In ongoing work, more factors will be considered to improve the capability of the ADRL algorithm in handling realistic mobile offloading scenarios such as; the energy consumption and the bandwidth allocations for each task, transfer learning on a dataset that has a smaller shape.

References

1. Wang, B., Wang, C., Huang, W., Song, Y., Qin, X.: A survey and taxonomy on task offloading for edge-cloud computing. IEEE **8**, 186080–186101 (2020)
2. Abdel-Kader, R.F., El-Sayad, N.E., Rizk, R.Y.: Efficient energy and completion time for dependent task computation offloading algorithm in industry 4.0. PLoS ONE **16**(6), e0252756 (2021)
3. Nashaat, H., Ahmed, E., Rizk, R.: IoT application placement algorithm based on multi-dimensional QoE prioritization model in fog computing environment. IEEE Access **8**, 1–12 (2020)
4. Sarker, I.H.: Machine learning: algorithms, real-world applications and research directions. SN Comput. Sci. **2**(3), 160 (2021)
5. Lei, C.: Deep reinforcement learning. In: Deep Learning and Practice with MindSpore. Cognitive Intelligence and Robotics. Springer, Singapore, pp. 217–243 (2021)
6. Heuillet, A., Couthouis, F., Rodríguez, N.D.: Explainability in deep reinforcement learning. Knowl.-Based Syst. **214**, 106685 (2021)
7. Du, M., Wang, Y., Ye, K., Xu, C.: Algorithmics of cost-driven computation offloading in the edge-cloud environment. IEEE Trans. Comput. **69**(10), 1519–1532 (2020)
8. Kai, C., Zhou, H., Yi, Y., Huang, W.: Collaborative cloud-edge-end task offloading in mobile-edge computing networks with limited communication capability. IEEE Trans. Cogn. Commun. Netw. **7**(2), 624–634 (2021)

9. Yu, S., Wang, X., Langar, R.: Computation offloading for mobile edge computing: a deep learning approach. In: IEEE 28th Annual International Symposium on Personal, Indoor, and Mobile Radio Communications (PIMRC), IEEE, pp. 1–6, October 2017
10. Rahbari, D., Nickray, M.: Task offloading in mobile fog computing by classification and regression tree. Peer-to-Peer Peer Netw. Appl. **13**, 104–122 (2020)
11. Wang, X., Xu, W., Jin, Z.: A hidden Markov model-based dynamic scheduling approach for mobile cloud telemonitoring. In: 2017 IEEE EMBS International Conference on Biomedical & Health Informatics, BHI. IEEE, pp. 273–276, February 2017
12. Sheng, J., Hu, J., Teng, X., Wang, B., Pan, X.: Computation offloading strategy in mobile edge computing. Information **10**(6), 2 (2019)
13. Huang, L., Feng, X., Zhang, C., Qian, L., Wu, Y.: Deep reinforcement learning-based joint task offloading and bandwidth allocation for multi-user mobile edge computing. Digit. Commun. Netw. **5**(1), 10–17 (2019)
14. Alam, M.G.R., Hassan, M.M., Uddin, M.Z., Almogren, A., Fortino, G.: Autonomic computation offloading in mobile edge for IoT applications. Future Gener. Comput. Syst. **90**, 149–157 (2019)
15. Khayyat, M., Elgendy, I.A., Muthanna, A., Alshahrani, A.S., Alharbi, S., Koucheryavy, A.: Advanced deep-learning-based computational offloading for multilevel vehicular edge-cloud computing networks. IEEE Access 8, 137052–137062 (2020)
16. Huang, L., Bi, S., Zhang, Y.-J.A.: Deep reinforcement learning for online computation offloading in wireless powered mobile-edge computing networks. IEEE Trans. Mobile Comput. **19**(11), 2581–2593 (2020)
17. Huang, L., Feng, X., Feng, A., Huang, Y., Qian, P.: Distributed deep learning-based offloading for mobile edge computing networks. Mobile Netw. Appl. (2018)
18. Huang, L., Feng, X., Zhang, L., Qian, L., Wu, Y.: Multi-server multi-user multi-task computation offloading for mobile edge computing networks. Sensors **19**(6), 1446 (2019)
19. Liang, F., Yu, W., Liu, X., Griffith, D., Golmie, N.: Towards edge-based deep learning in industrial Internet of Things. IEEE Internet Things J. **7**(5), 4329-4341 (2020)
20. Wu, H., Zhang, Z., Guan, C., Wolter, K., Xu, M.: Collaborate edge and cloud computing with distributed deep learning for smart city Internet of Things. IEEE Internet Things J. **7**(9), 8099–8110 (2020)
21. Abadi, M., Agarwal, A., et al.: Tensorflow: large-scale machine learning on heterogeneous distributed systems. arXiv:1603.04467 (2016)
22. Carneiro, T., Medeiros Da NóBrega, R.V., Nepomuceno, T., Bian, G., De Albuquerque, V.H.C., Filho, P.P.R.: Performance analysis of google colaboratory as a tool for accelerating deep learning applications. IEEE Access **6**, 61677–61685 (2018)

Intelligent Optimization

Early Classification COVID-19 Based on Particle Swarm Optimization Algorithm Using CT-Images

Amira M. Hasan[1(✉)], Hala M. Abd El-Kader[2], and Aya Hossam[2]

[1] Communications Engineering Department, Alexandria Higher Institute of Engineering Technology (AIET), Alex, Egypt
amira.mohamed@aiet.edu.eg

[2] Electrical Engineering Department, Faculty of Engineering (Shoubra), Benha University, Cairo, Egypt
hala.mansour@feng.bu.edu.eg

Abstract. Diagnosis is a critical preventative step in Coronavirus researches. Because of the fast spread of this virus, it is necessary to present a computer-aided diagnostic (CAD) system which is very faster for radiologists. Feature Selection (FS) is a significant technique to obtain an accurate CAD system. This paper presented an effective FS model which based on wrapper approach as evaluator and Particle Swarm Optimization (PSO) as search method for classifying cases of COVID-19 using Computed Tomography (CT). This model was used PSO algorithm to identify the significant features subset within overall features set. Support Vector Machine (SVM), K-nearest neighbor (KNN) classifiers were used as evaluators with 10-fold cross validation and reached accuracy of 99.6% for SVM and 94.27% for KNN respectively. The results were shown that proposed PSO-FS model is an intelligent and outperforms other two traditional FS search methods, Genetic Algorithm (GA) and Greedy Stepwise (GS).

Keywords: COVID-19 · Feature selection (FS) · Particle swarm optimization (PSO) · Classification

1 Introduction

COVID-19 disease was occurred on 2019 at Wuhan region of China. This virus causes fever, cough, and fatigue to eventually leading to patient death [1]. Medical imaging was used to investigate human body. Image processing techniques are required to improve image analysis which provide semi-automated tissue detection. Computed Tomography (CT) is effective tool for detecting the cases of COVID-19. Physicians need a precise and reliable diagnosis procedure that allows to distinguish between cases of COVID-19 and reduce burden on clinician [2]. Therefore, it is necessary to present a diagnosis CAD system which based on Machine Learning (ML) to specify the result of CT images. CAD system contains of many stages as image collection, image enhancement, and image segmentation [3]. The dataset which used in [4] contains about thirty two features. The

© The Author(s), under exclusive license to Springer Nature Switzerland AG 2022
A. E. Hassanien et al. (Eds.): AMLTA 2022, LNDECT 113, pp. 631–642, 2022.
https://doi.org/10.1007/978-3-031-03918-8_52

challenges that face the CAD system is the great number of the features in dataset because not all these features are required. Large features appear great effects on management of dataset, decrease accuracy of classification model and cause over hitting. So, Researchers pay more attention for FS to improve performance of classification model [4]. FS is technique that reduces the set of possible input variables to decrease dimensionality of the data and facilitates learning algorithms to operate more quickly and efficiently. This paper proposed an effective FS model for detecting cases of COVID-19 with evaluation the optimized feature subset using PSO search method. We used the selected features as inputs to two classifiers such as, SVM and KNN. Symptoms of disease includes fever, cough, fatigue and mild to severe respiratory complications, eventually leading to the death of patient. Due to the limited availability of resources in hospitals and the delay of the results in the medical tests, so it is very necessary to build an automated model which based on ML approach for frontline clinical physicians to give appropriate medical.

2 Related Works

Recent works show that, CT images are mainly utilized to offer fast diagnostic methods to prevent the spread of COVID-19. Li et al. [5] proposed a methodology to recognize infection rate using the coronal and axial view of lung CT scans. The proposed work achieved AUC of 0.918, and sensitivity of 82.6%. Nour et al. [6] proposed a model that was trained using the CNN architecture model with five convolution layers that were used as deep feature extractor. The extracted deep features were fed into KNN and SVM models. The superiority of SVM classifier was demonstrated by its accuracy of 98.97%, sensitivity of 89.39%. Yang et al. [7] proposed scheme to evaluate lung CT scans and their scheme could achieve Specificity of 94%, AUC of 0.892%, and Sensitivity of 83.3% El-Kenawy et al. [8] compare proposed feature selection algorithm (SFS-Guided WOA) to other widely used optimization algorithms and find the proposed voting classifier (PSO-Guided-WOA) outperforms other voting classifiers.

3 Materials and Methods

3.1 Methodology

In medical diagnosis, CAD system involves many stages. First stage was select the images from the dataset to realize the highest level of quality and clarity. The second stage was employed several image enhancement techniques. The third stage was used the segmentation algorithms, fourth stage was extracted the general features from ROI, and the fifth stage was the feature selection. Finally the classification was the final stage to build the classification model, as shown in (Fig. 1).

3.2 Dataset

Data collection is considered the first step in CAD system. The authors have used a dataset which consists of 2,110 CT scans, which included 1,110 CT images of people infected by COVID-19 and 1,000 images of uninfected persons. The used dataset was

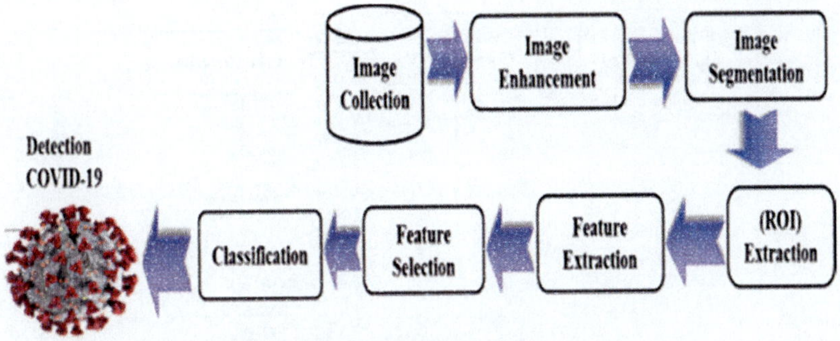

Fig. 1. Overview of detection system

collected from the open-source data repository GitHub and Kaggle [9, 10]. CT image has clearer information. It also offers better accuracy, because of its high sensitivity and low noise. So, we segmented ROI to extract all features of COVID-19, as shown in (Fig. 2).

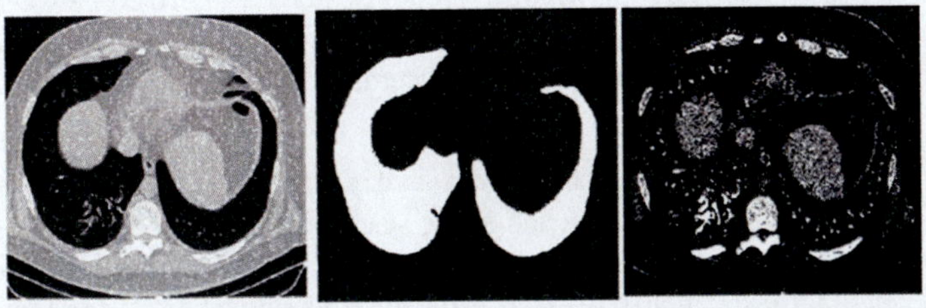

Fig. 2. (a) Original Image (b) Segmented Lung (c) ROI

3.3 Feature Selection (FS)

FS is crucial part of preprocessing steps in process of Knowledge Data Discovery. Feature Selection Algorithms (FSAs) are known as Attribute Selection and Data Selection. FS talked about three steps in image selection: screening, ranking, and selecting. During screening, it eliminates irrelevant and problematic predictors or cases, such as those with too many missing values or too little variation to be useful. Sorts and ranks the remaining predictors based on their significance. It denotes the subset of capabilities. It identifies subset of features by retaining only the important forecasters and filtering or excluding the others. The important forecasters are those that the FS screens, ranks, and selects. This process contains of Subset Generation, Subset Evaluation, a stopping criterion, and Result Validation, as shown in (Fig. 3).

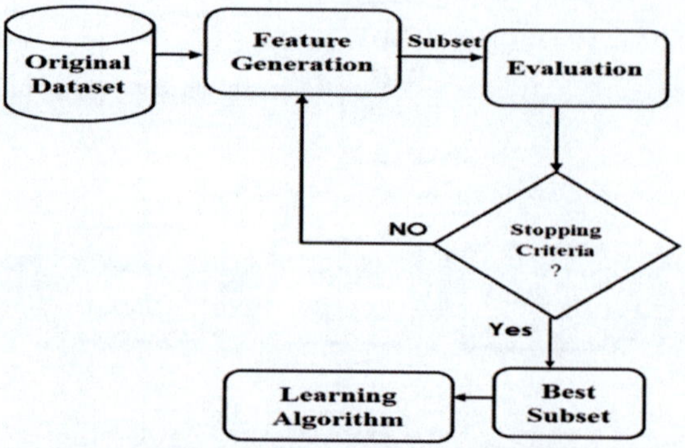

Fig. 3. The flowchart of feature selection process.

3.4 Feature Selection Approaches

There have been two categories of features selection approaches: filter and wrapper, as shown in (Fig. 4) Benefits of wrapper methods include interface between feature subset search and model selection. It is more accurate and computationally intensive. So, wrapper approach provides results more accurate than filter approach [11].

3.4.1 Filter Approach

In filter approach, evaluation of feature subsets depends on general characteristics of dataset. This approach uses ranking and space search methods on basis of strategy to select best subset of features. The disadvantage of the approach is the missing interactions between the features.it also disregards the possible effects on performance of ML algorithm of the selected feature subsets.

3.4.2 Wrapper Approach

Because of the disadvantages of the filter approach, wrapper approach was proposed in 1997. Here, prognostic model is used for scoring the feature subset. That is, no knowledge of the used algorithm is required; only the interface is required. Using the interface, feature subsets can be evaluated based on classifier accuracy to find best feature subset.

3.5 Particle Swarm Optimization Algorithm (PSO)

PSO is an algorithm which based on behavior of swarm of bees or flock of birds while searching for food. It maintains a population, named a swarm of particles, where each particle denotes potential solution to optimization problem. Each particle has a position in search space that is denoted by a vector $x_i = (x_{i1}.x_{i2}.\cdots\cdots.x_{iD})$, where D is the search space dimensionality. Particles move in search space searching for optimal solutions. So that, each particle has a velocity, which is represented by $v_i = (v_{i1}.v_{i2}.\cdots\cdots.v_{iD})$, the velocity value must be in a range defined by parameters v_{min} and v_{max}.During movement, each particle updates its position and velocity based on its own and its neighbors' experiences. The best previous position of particle denotes personal best position of particle, called local best (lbest) and the best position obtained by the population thus

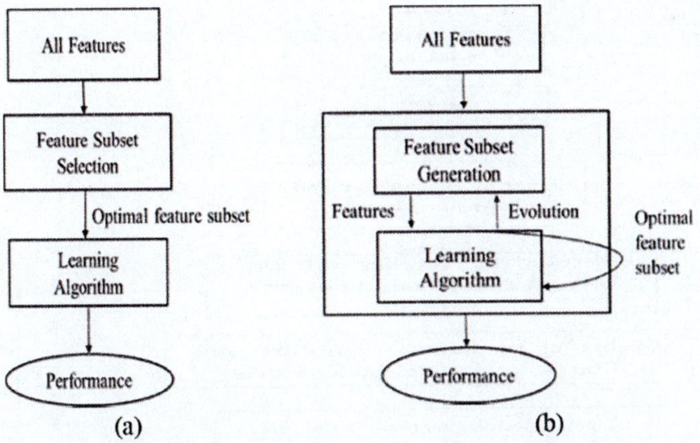

Fig. 4. The flowchart of features selection approaches (a) Filter; (b) wrapper.

far is called global best (gbest). According to lbest and gbest values, PSO searches for optimal solutions by updating velocity and position of each particle using the following equations:

$$x_{id}(t+1) = x_{id}(t) + v_{id}(t+1) \quad (1)$$

$$v_{id}(t+1) = w * v_{id}(t) + C_1 * r_{1i} * (P_{id} - x_{id}(t)) + C_2 * r_{2i} * \left(P_{gd} - x_{id}(t)\right) \quad (2)$$

where t denotes the tth iteration, d ϵ D refers to the dth dimension in search space, $\boldsymbol{w}$ is the inertia weight, c_1 and c_2 are the learning factors which called cognitive Parameter, and social parameter, respectively, r_{1i} and r_{2i} are random values uniformly distributed in [0, 1], and finally p_{id} and p_{gd} represent the elements of lbest and gbest in the dth dimension. The personal best position of particle i can be determined by the following equation:

$$y_i(t+1) = \begin{cases} y_i(t) if \ f(x_i(t+1)) \geq f(y_i(t) \\ x_i(t+1) if \ (f(x_i(t+1) < f(y_i(t) \end{cases} \quad (3)$$

The inertia weight is a new parameter that was added to original PSO algorithm to balance global and local search. It is calculated by the following equation:

$$w = \frac{w_{\max} - w_{min}}{t_{max}} * t_{current} \quad (4)$$

Where, t_{max}: the maximum number of iterations. $t_{current}$: The number of current iteration. The wrong value selection of these parameters will effect on speed of PSO algorithm convergence. So, the initial selection would be very important [12]. The flow chart of proposed PSO-FS algorithm is shown in (Fig. 5).

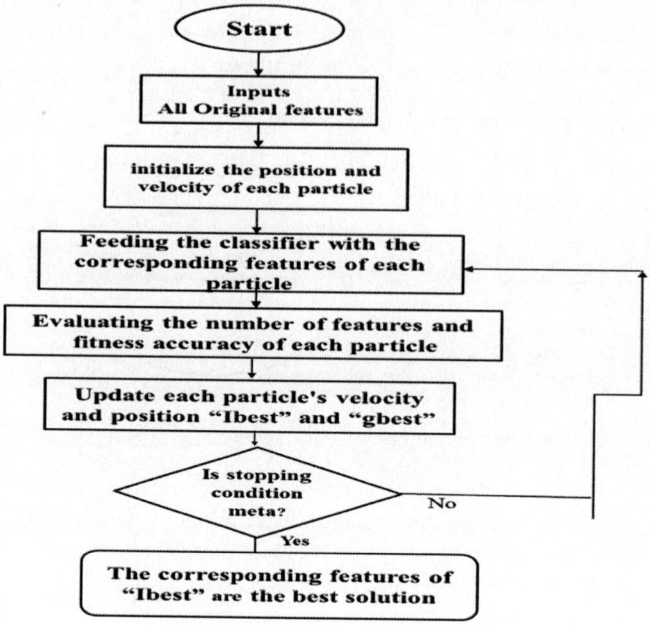

Fig. 5. The flow chart of PSO algorithm

3.6 Classification Methods

Classification is the final step in building COVID-19 classification model. In this paper, three features selection models were constructed using two classification algorithms which are SVM and KNN.

3.6.1 Support Vector Machine (SVM)

SVM is a powerful classification algorithm that performs classification tasks, by constructing hyper-planes in a space that splits the values of several class labels. In this classifier, it is easy to have a linear hyper-plane among two modules [13].

3.6.2 K-Nearest Neighbor (KNN)

The K-NN algorithm accepts similarity between new case/data and existing cases and assigns new case to the category with the highest similarity to the existing categories. The algorithm stores all available data and uses similarity to classify new data points. This means that using this algorithm, new data were classified into a suitable category [14].

4 The Proposed PSO-FS Model

This paper proposed an effective FS model based on wrapper approach as evaluator and PSO as search method for detecting the cases of COVID-19 using CT images. The model is more intelligent and accurate model by comparing to other two traditional FSA which are Greedy Stepwise (GS) and Genetic Search (GA). Support Vector Machine (SVM) and K-Nearest Neighbor (KNN) classifiers were used as evaluators to the optimized feature subset out from PSO search method, as shown in (Fig. 6).

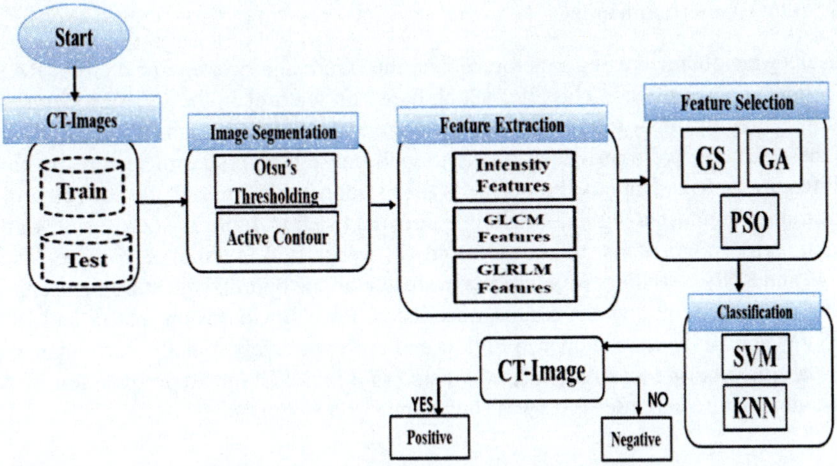

Fig. 6. The flow chart of proposed PSO-FS model

5 Experimental Results

5.1 Performance Metrics

By using WEKA, FS methods contain of two elements: Feature evaluator, the way by which feature subsets are evaluated and Search method the way by which space of feature subsets are searched. The dataset is separated to training set and testing set, in percentages of 70 and 30 with 10-fold cross validation, respectively. Performance is described in terms of accuracy, specificity and sensitivity by applying a confusion matrix. This matrix contains of four indices: true positive (TP), true negative (TN), false positive (FP), and false negative (FN).TP and TN represent the number of correctly predicted positive and negative samples, respectively, whereas FP and FN correspond to number of incorrectly predicted positive and negative samples, respectively. All metrics were listed in Table 1.

Table.1. Evaluation criteria of proposed model

Metrics	Formula
Accuracy	$Accuracy = \frac{TP+TN}{TP+TN+FP+FN}$
Specificity	$FP\text{-}Rate = \frac{TN}{TN+FP}$
Sensitivity	$TP-Rate = \frac{TP}{TP+FN}$

5.2 The Evaluation Model

Results were obtained using a laptop with an Intel Core i5 processor and 4 GB of RAM. The performance of each classifier which based on features in the COVID-19 dataset that prepared in [4]. A total of all features were extracted from each image, but not all of these features were provided to classifiers because the number of features is high. All features selected and their algorithms are listed in Table 2 the Firstly each classifier was trained and tested with 32 original features as listed in Table 3. Then, results of the traditional FS algorithms such as, GA and GS were listed in Tables 4, 5 respectively. SVM and KNN classifiers were used as evaluators to the optimized feature subset using PSO search method and the results were listed Table 6. Confusion matrix and ROC of SVM and KNN were shown in Figs. 7 and 8. Respectively. Finally Performance of all classifiers before and After FS was listed in Table 7.Graph Representation of All classifiers before and after PSO was shown in (Fig. 9).

Table 2. The models and features selected

Models	Parameters	Used method	Search method	No attributes
Model 1	Num execution slots is 1	Wrapper	Greedy Stepwise (GS)	4
Model 2	Max generation is 20, Crossover is 0.6	Wrapper	Genetic Algorithm (GA)	26
Proposed model	**Individual weight is 0.34, inertia weight is 0.33, iterations is 20 and population size is 20**	**Wrapper**	**Particle Swarm Optimization(PSO)**	**15**

Table 3. The performance of all classifiers using all original features

Classifier	TP rate	FP rate	Precision	Recall	F-measure	ROC area	Class
SVM	0.995	0.020	0.963	0.995	0.979	0.988	COVID-19
	0.980	0.005	0.997	0.980	0.989	0.988	Non COVID
	0.985	**0.010**	**0.986**	**0.985**	**0.985**	**0.988**	**Weighted Avg**
KNN	0.664	0.000	1.000	0.664	0.798	0.832	COVID-19
	1.000	0.336	0.849	1.000	0.918	0.832	Non COVID
	0.884	**0.220**	**0.901**	**0.884**	**0.877**	**0.832**	**Weighted Avg**

Table 4. The Performance of all classifiers using Genetic Algorithm (GA)

Classifier	TP rate	FP rate	Precision	Recall	F-measure	ROC area	Class
SVM+GA	0.995	0.015	0.972	0.995	0.984	0.990	COVID 19
	0.985	0.005	0.997	0.985	0.991	0.990	Non COVID
	0.989	**0.008**	**0.989**	**0.989**	**0.989**	**0.990**	**Weighted Avg**
KNN+GA	0.749	0.000	1.000	0.749	0.856	0.874	COVID 19
	1.000	0.251	0.883	1.000	0.938	0.874	Non COVID
	0.913	**0.164**	**0.923**	**0.913**	**0.910**	**0.874**	**Weighted Avg**

Table 5. The performance of all classifiers using Greedy Stepwise (GS)

Classifier	TP rate	FP rate	Precision	Recall	F-measure	ROC area	Class
SVM+GS	1.000	0.015	0.972	1.000	0.986	0.993	COVID-19
	0.985	0.000	1.000	0.985	0.992	0.993	Non COVID
	0.990	**0.005**	**0.990**	**0.990**	**0.990**	**0.987**	**Weighted Avg**
KNN+GS	0.782	0.000	1.000	0.782	0.878	0.891	COVID-19
	1.000	0.218	0.897	1.000	0.946	0.891	Non COVID
	0.925	**0.143**	**0.932**	**0.925**	**0.922**	**0.891**	**Weighted Avg**

Table 6. The performance of all classifiers using proposed model PSO-FS

Classifier	TP rate	FP rate	Precision	Recall	F-measure	ROC area	Class
SVM+PSO	1.000	0.005	0.991	1.000	0.995	0.998	COVID19
	0.995	0.000	1.000	0.995	0.997	0.998	Non COVID
	0.997	**0.002**	**0.997**	**0.997**	**0.997**	**0.998**	**Weighted Avg**
KNN+PSO	0.834	0.000	1.000	0.834	0.910	0.917	COVID19
	1.000	0.166	0.920	1.000	0.958	0.917	Non COVID
	0.943	**0.109**	**0.947**	**0.943**	**0.941**	**0.917**	**Weighted Avg**

Table 7. The performance of all classifiers before and after FS

Classifier	Correctly/incorrectly instances	Test time	RMSE	Sensitivity (%)	Specificity (%)	AUC (%)	Accuracy (%)
KNN	540/71	2.01 s	0.3407	66.3%	100%	83.6%	88.37%
SVM	602/9	2.48 s	0.1214	99.5%	98%	98.7%	98.52%
KNN+GA	558/53	1.99 s	0.2943	74.8%	100%	87.4%	91.32%
SVM+GA	604/7	2.17 s	0.107	99.5%	98.5%	99%	98.85%
KNN+GS	565/46	1.97 s	0.2742	78.5%	100%	89.1%	92.47%
SVM+GS	605/6	2.02 s	0.0991	100%	98.5%	99.2%	99.01%
KNN+PSO	**576/35**	**1.99 s**	**0.2392**	**83.4%**	**100%**	**91.7%**	**94.27%**
SVM+PSO	**609/2**	**2.04 s**	**0.0572**	**100%**	**99.5%**	**99.7%**	**99.67%**

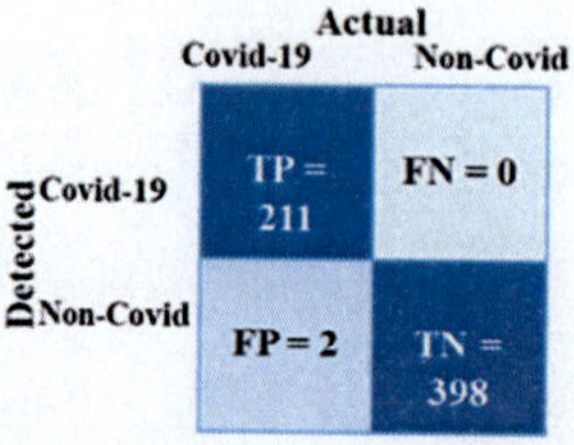

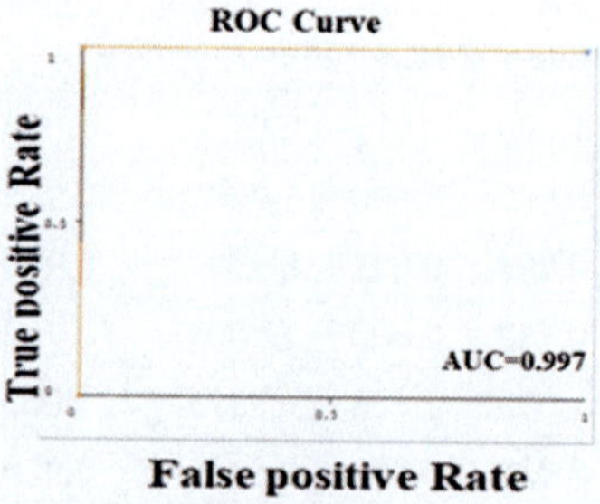

Fig. 7. Confusion matrix and ROC of SVM

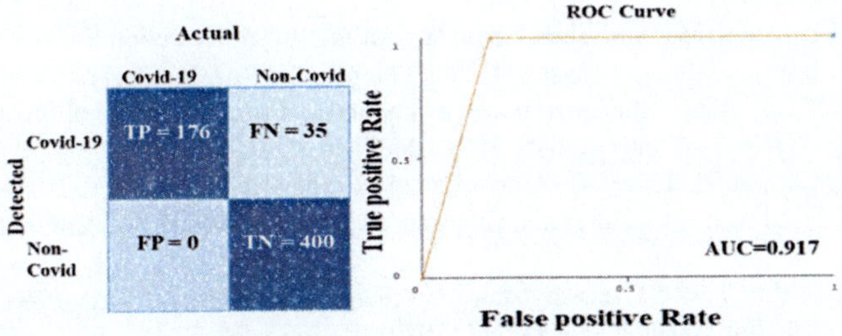

Fig. 8. Confusion matrix and ROC of KNN

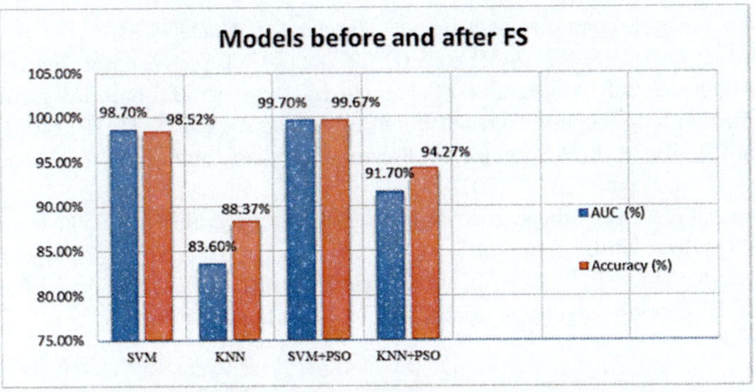

Fig. 9. Graph representation of all classifiers before and after PSO

6 Conclusion

This paper proposed an effective COVID-19 detection model which based on PSO Algorithm. The proposed PSO-FS algorithm was used to search for optimal set of features which achieve better classification performance than using the overall features. The experimental outcomes were showed the proposed PSO-FS algorithm outperformed other traditional FS algorithms as GS and GA. The classifiers SVM and KNN achieved accuracies reached to 99.67% and 94.27% respectively by using the proposed PSO-FS algorithm. The future works will be focused on using other classification algorithms and using X-ray images as dataset.

References

1. Huang, C., et al.: Clinical features of patients infected with 2019 novel coronavirus in Wuhan, China. Lancet **395**(10223), 497–506 (2020)
2. Chen, N., et al.: Epidemiological and clinical characteristics of 99 cases of 2019 novel coronavirus pneumonia in Wuhan, China: a descriptive study. The Lancet **395** (10223), 507–513 (2020)
3. Hui, D.S., et al.: The continuing 2019-nCoV epidemic threat of novel coronaviruses to global health—the latest 2019 novel coronavirus outbreak in Wuhan, China. Int. J. Inf. Dis. **91**, 264–266 (2020)

4. Amira, M.H., Hala, M., Abd El, K., Aya, H.: An Intelligent Detection System for Covid-19 Diagnosis Using CT-Images. Assiut. J. Eng. Sci. **49**(4), 476–508 (2021)
5. Li, K., et al.: CT image visual quantitative evaluation and clinical classification of coronavirus disease (COVID-19). Europ. Radiol. **30**(8), 4407–4416 (2020)
6. Nour, M., Cömert, Z., Polat, K.: A novel medical diagnosis model for COVID-19 infection detection based on deep features and Bayesian optimization. Appl. Soft Comput. **97**, 106580 (2020)
7. Yang, R., et al.: Chest CT severity score: an imaging tool for assessing severe COVID-19. Radiol. Cardiothor. Imag. **2**(2), e200047 (2020)
8. El-Kenawy, E.S.M., Ibrahim, A., Mirjalili, S., Eid, M.M., Hussein, S.E.: Novel feature selection and voting classifier algorithms for COVID-19 classification in CT images. IEEE Access **8**, 179317–179335 (2020)
9. https://www.kaggle.com/luisblanche/covidct?select=CT_NonCOVID
10. GitHub-UCSD-AI4H/COVID-CT: COVID-CT-Dataset:CT Scan Dataset about COVID-
11. Nouri-Moghaddam, B., Ghazanfari, M., Fathian, M.: A novel multi-objective forest optimization algorithm for wrapper feature selection. Exp. Syst. Appl. **175**, 114737 (2021)
12. Brezočnik, L., Fister, I., Podgorelec, V.: Swarm intelligence algorithms for feature selection: a review. Appl. Sci. **8**(9), 1521 (2018)
13. Thaiyalnayaki, K.: Classification of diabetes using deep learning and SVM techniques. Int. J. Curr. Res. Rev. **13**(01), 146 (2021)
14. Ariani, A., Samsuryadi, S.: Classification of kidney disease using genetic modified KNN and Artificial BEE colony algorithm. Sinergi **25**(2), 177–184 (2021)

MARL-FWC: Optimal Coordination of Freeway Traffic Control Measures

Ahmed Fares[1,2](✉), Mohamed A. Khamis[2,3], and Walid Gomaa[2,4]

[1] Faculty of Engineering (Shoubra), Benha University, Cairo 11689, Egypt
ahmed.fares@feng.bu.edu.eg

[2] Cyber-Physical Systems Lab, Egypt-Japan University of Science and Technology (E-JUST), New Borg El-Arab City, Alexandria 21934, Egypt
{mohamed.khamis,walid.gomaa}@ejust.edu.eg

[3] IME (Data Science), Ejada Systems Ltd., Alexandria, Egypt
maomar@ejada.com

[4] Faculty of Engineering, Alexandria University, Alexandria 21544, Egypt

Abstract. The objective of this paper is to optimize the overall traffic flow on freeways using multiple ramp metering controls plus its complementary Dynamic Speed Limits (DSLs). An optimal freeway operation can be reached when minimizing the difference between the freeway density and the critical ratio for maximum traffic flow. In this paper, a Multi-Agent Reinforcement Learning for Freeways Control (MARL-FWC) system for ramps metering and DSLs is proposed. MARL-FWC introduces a new microscopic framework at the network level based on collaborative Markov Decision Process modeling (Markov game) and an associated cooperative Q-learning algorithm. The technique incorporates payoff propagation (Max-Plus algorithm) under the coordination graphs framework, particularly suited for optimal control purposes. MARL-FWC provides three control designs: fully independent, fully distributed, and centralized; suited for different network architectures. MARL-FWC was extensively tested in order to assess the proposed model of the joint payoff, as well as the global payoff. Experiments are conducted with heavy traffic flow under the renowned VISSIM traffic simulator to evaluate MARL-FWC. The experimental results show a significant decrease in the total travel time and an increase in the average speed (when compared with the base case) while maintaining an optimal traffic flow.

Keywords: Multi-agent system · ITS · Adaptive traffic control · Freeway · Ramp metering · Max-plus · Q-learning · Sequential decision problem

1 Introduction

Generally, there are three freeway control strategies: on-ramp control, mainline control, and DSLs control. Two signals are only generated by a ramp metering

M.A. Khamis and W. Gomaa—Contributing authors.

© The Author(s), under exclusive license to Springer Nature Switzerland AG 2022
A. E. Hassanien et al. (Eds.): AMLTA 2022, LNDECT 113, pp. 643–656, 2022.
https://doi.org/10.1007/978-3-031-03918-8_53

device: red and green (no yellow), controlled by a smart or basic controller that regulates the flow of traffic entering the freeway according to the current traffic volume. We aim to control the number of vehicles entering the mainstream freeway from the ramp merging area. This optimizes the freeway density below the critical one. In order to achieve this flow, *optimal coordination* of the freeway traffic control *measures* over the network level is highly needed. In order to solve the problem of optimal metering rate on the network level; a *collaborative Multi-Agent System* (MAS) [1] based on reinforcement learning is appropriate. A collaborative MAS is defined as the system in which the agents are designed to act together in order to optimize the sequence of actions. The framework of the Coordination Graphs (CGs) [2] is utilized which is designed based on the dependencies between agents. It decomposes the global payoff function into a sum of local payoffs based on that dependency.

In this paper, a Multi-Agent Reinforcement Learning for Freeways Control (MARL-FWC) system for ramp metering and dynamic speed limit is proposed based on a Reinforcement Learning density Control Agent (RLCA). MARL-FWC introduces a new microscopic framework at the network level based on collaborative Markov Decision Process (Markov game) modeling and an associated cooperative Q-learning algorithm. The technique incorporates payoff propagation (Max-Plus algorithm) under the coordination graphs framework, particularly suited for optimal control purposes. A model for the local payoff, the joint payoff, as well as the global payoff is proposed. MARL-FWC provides three control designs: fully independent, fully distributed, and centralized; suited for different network architectures. Preliminary results of this work have been published in [3,4]. In this paper, a more detailed description and improvements on MARL-FWC are provided. An adaptive objective function plus DSLs are presented. The new approach for ramp metering plus DSLs have shown considerable improvement over the base case.

Several different techniques have been recently applied for ramp metering control including: intelligent control theory, fuzzy control [5,6], neural network control [7,8], and even some hybrid of them [9]. In this paper, a new framework of ramp metering in the network level is proposed based on modeling by *collaborative Markov Decision Process* (Markov Game) [1,10] and an associated *cooperative Q-learning* algorithm, which is based on *a payoff propagation* algorithm [11] under the CGs framework. MARL-FWC avoids not only the locality of all techniques mentioned above, but also the computational complexity and the risk of being trapped in local optimum of the Model Predictive Control (MPC) [12].

2 Single and Multi-agent Algorithms

A Modified Q-learning: Temporal Difference Algorithm: The RLCA is interacting with the traffic network. The agent receives the traffic state from the network detectors. This state consists of the number of vehicles N_t and the current density ρ_t. Based on this information the RLCA chooses an action a_t.

As a consequence of a_t; the agent receives a reward r_t that keeps the density of the network around a small neighborhood of the critical density. The aim of the agent is to learn a policy which maps an arbitrary state into an optimal action $\pi : S \longrightarrow A$. An optimal action is an action that optimizes the long-term cumulative discounted reward, thus the optimization is done over an infinite horizon.

As long as there is no *model* of the environment (unknown environment), so the infinite sum of the discount reward is no longer considered as a function of state s only as in Dynamic Programming (DP) value iteration [13], but as a function of the state action pair (s, a). That is why such a Q-function [14] is used. When the agent is in state s and the agent performs action a and its *long-term reward* is $Q(s, a)$; if each state and action pairs are visited infinitely often; then $Q(s, a)$ converges to $Q^*(s, a)$ [15]. Based on such long-term reward the optimal policy is calculated. Q-learning converges no matter how the actions during the learning are chosen, as long as every action is selected infinitely often (*fair action selection strategy*). A modification of the traditional Q-learning rule is in Eq. (1). This new scheme is called a Temporal Difference (TD) learning where α is the learning rate [15].

$$Q_t^{new}(s,a) = Q_{t-1}^{old}(s,a) + \underbrace{\alpha_t(s,a)}_{\text{learning rate}} \left[\overbrace{\underbrace{R_t(s,a)}_{\text{reward}} + \underbrace{\gamma}_{\text{dis. fac.}} \max_{a'} Q_{t-1}^{old}(s',a') - Q_{t-1}^{old}(s,a)}^{\text{TD-Error}} \right] \quad (1)$$

For alpha value, the easiest approach is to take a fixed value α, but even better to use a *variable* α. Equation (2) is one possibility where $b(s, a)$ is the number of times the action a has been chosen in the state s.

$$\alpha_t(s,a) = \frac{1}{1 + b_t(s,a)} \quad (2)$$

The state space, the action space and the reward function of the single control agent are described in [3] and Subsect. 3.1. The joint action space and the (joint and global) payoff function of collaborative multi-agent are described in detail in [4] and Subsect. 3.2.

Payoff Propagation with Max-Plus Algorithm: In this subsection, the problem of dynamics in MASs is discussed and the suggestion that the agents should take the behavior of other agents into account.

CGs [2] can be demonstrated as an undirected graph $G = (V,\ E)$, where each node $i \in V$ represents an agent, and an edge $e(i,\ j) \in E$ between agents $i,\ j$ means that a dependency exists between them ($j \in \Gamma(i)$ and $i \in \Gamma(j)$). The Maximum A Posteriori (MAP) configuration is then estimated by sending locally optimized messages between connected agents (nodes) in the CGs. Even though the message-passing algorithm is developed for estimating the MAP configuration in a probabilistic graph, it is applicable for estimating the optimal joint action of a group of agents in CGs. Because in both situations, the function that is being optimized is decomposed in local terms [16].

In collaborative multi-agent systems, the CGs framework [17] assumes the action of an agent i only depends on a subset of other agents ($j \in \Gamma(i)$) which may influence its state. The global payoff function $u(\mathbf{a})$ is then broken down into a linear combination of local payoff functions $f_i(\mathbf{a_i})$. The proposed design of the local and global payoff is found in Subsect. 3.2. The global Q-function $Q(\mathbf{s},\ \mathbf{a})$ is then decomposed into a sum of local functions given by:

$$Q(\mathbf{s},\ \mathbf{a}) = \sum_{i=1}^{n} Q_i(s_i,\ \mathbf{a_i}) \tag{3}$$

where: $\mathbf{a} = (a_1, ..., a_n)$ is the joint action resulting from: each agent i selects an action a_i from its action set A_i, $\mathbf{s} = (s_1, ..., s_n) \in S_1 \times \cdots \times S_n$ is the joint state. The joint Q-function $Q_i(s_i, \mathbf{a_i})$ depends only on $\mathbf{a_i} \subseteq \mathbf{a}$, where $\mathbf{a_i} \in A_i \times \prod_{j \in \Gamma(\mathrm{i})} A_j$, $\Gamma(\mathrm{i})$ is the set of neighborhoods of agent i. The decomposition in Eq. (3) can be demonstrated in Fig. 1a (an example of a CGs of four agents where the dependence among agents is decomposed into pair-wise functions [18]); where f_{ij} is the joint payoff between agent i and agent j. The Max-Plus algorithm approximates the optimal joint action by iteratively sending locally optimized messages between connected agents (*neighbors*) in the coordination graph. The MAP for each agent is estimated using only the incoming messages from other agents. It has been shown that the message updates converge to a fixed point after a finite number of iterations for cycle free graphs. The message mainly depends on the decomposition given by Eq. (3) and can be defined as follows:

(a) CG involving 4 agents with only pair-wise dependencies.

(b) Representation of the different μ_{ij} sent between neighboring agents.

Fig. 1. Payoff propagation: Coordination graphs and Max plus.

$$\mu_{ij}(a_j) = \max_{a_i} \left\{ f_i\left(a_i\right) + f_{ij}(a_i, a_j) + \sum_{k \in \Gamma(\mathrm{i}) \backslash j} \mu_{ki}(a_i) \right\} \tag{4}$$

where: $k \in \Gamma(\mathrm{i}) \backslash j$ is the subset of all neighbors connected to i except j, and $\mu_{ij}(a_j)$ is the message from an agent i to agent j after agent j preforms the action a_j as an evaluation of the influence of this action on agent i state. The message approximates the maximum reward agent i can achieve for a committed action of agent j. This message is the max-sum of the local payoff $f_i(a_i)$, the joint payoff $f_{ij}(a_i, a_j)$ and all incoming messages agent i received except that from agent j. The design of the joint payoff is found in detail in Subsect. 3.2. Figure 1b shows a CG with three agents and the corresponding messages.

If the graph has no cycle, Max-Plus always converges after a finite number of steps to a fixed point in which the messages are changed below a certain threshold [19]. At each time step, the value for an action a_i of an agent i can be determined as follows:

$$g_i(a_i) = f_i(a_i) + \sum_{j \in \Gamma(i)} \mu_{ji}(a_i) \tag{5}$$

When the convergence is held, optimal joint action $\mathbf{a}$ has the element a_i^* which can be computed as follows:

$$a_i^* = \arg \max_{a_i} g_i(a_i) \tag{6}$$

In the Max-Plus algorithm; at each iteration, an agent i sends a message $\mu_{ij}(a_j)$ to each neighbor $j \in \Gamma(\text{i})$ given the joint payoff f_{ij} and current incoming messages it has received and also computes its current optimal action given all the messages it has received so far that is why it is called *anytime algorithm* [16]. The process continues until all messages converge, or a deadline signal is received which means the agents must report their individual actions. In the anytime algorithm, the joint action Eq. 6 is only updated when the corresponding global payoff improves.

3 MARL-FWC Architecture

Figure 2 demonstrates MARL-FWC architecture. MARL-FWC consists of two layers; the intelligent control layer and the configuration layer. The configuration layer consists of three blocks:(i) ramp metering parameters: configure the ramp metering; (ii) DSLs parameters: configure the DSLs; and (iii) Agents' parameters: configure the control agent.

The main task of the configuration layer is to configure the control layer. The control layer consists of three interacting modules (blocks): (i) environment module (VISSIM simulator): models the traffic network; (ii) multi-agent reinforcement-learning module (Computing environment): implements different controls strategies; and (iii) interface module (VISSIM interface): facilitates the interaction between the Multi-agent reinforcement-learning module and the environment module.

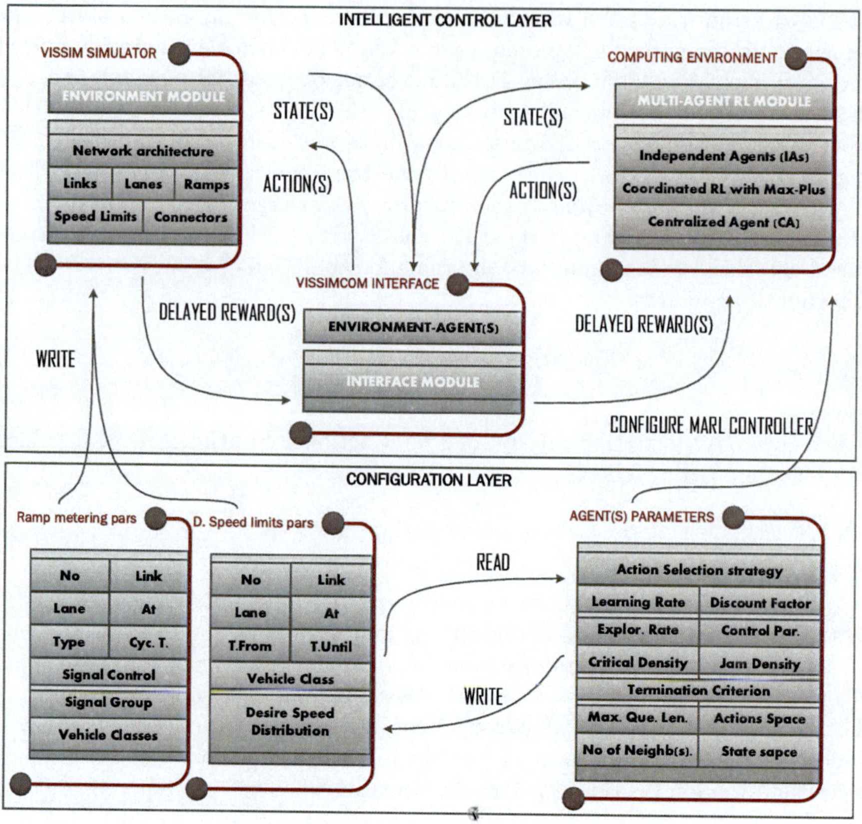

Fig. 2. MARL-FWC architecture.

3.1 MARL-FWC with Single Ramp and RLCA

State Space: The state space is three-dimensional where each state vector s_{t+1} at time $t+1$ consists of the following components: the number of vehicles in the freeway $N(t+1)$ (see [3]), the number of vehicles that entered the freeway from ramp $\Delta N(t+1)$, and the ramp traffic signal at the previous time step $Ts(t)$.

Action Space: By metering on the on-ramp, the red and green phases of the ramp metering change in order to control the flow entering the freeway from the merging point. So, the action space is modeled as consisting only of two actions: red and green.

Reward Function: Since the RLCA's goal is to keep the freeway density ρ around a small neighborhood of the critical density ρ_{cr}, the reward function is designed so as to depend on the current freeway density ρ and how much it deviates from the critical density ρ_{cr}. Hence, the reward of taking an action a in state s is designed as following:

$$Minimize\ O = |\rho - \rho_{cr}| : r(s, a) = -|\rho - \rho_{cr}| \tag{7}$$

As long as the Q-value Eq. (1) (as a function of the current state and action) is maximized, the difference between the freeway density ρ and the critical density ρ_{cr} is minimized. Therefore, the best utilization of the freeway is achieved without causing congestion.

3.2 MARL-FWC Based on Cooperative Q-learning

Independent Agents: The "independent learning" is the first proposed design to solve the freeway congestion problem (Fig. 3a). The Q-function is updated as follows:

$$\begin{aligned} Q_i^{new}(s_i, a_i) &= Q_i^{old}(s_i, a_i) + \\ &\alpha \left[R(s_i, a_i) + \gamma \max_{a'_i} Q_i^{old}(s'_i, a'_i) - Q_i^{old}(s_i, a_i) \right] \end{aligned} \tag{8}$$

where: $R(s_i, a_i) = -|\rho(s_i, a_i) - \rho_{cr}|$ from Eq. (7), with such conception of the objective function, the agents try to keep the freeway density within a small margin of the critical ratio; that ensures the maximum utilization of the freeway without entering in congestion. In this Q-function design, the agents are partially observing the environment. An agent observes its state which is defined as the number of vehicles in the areas of interest associated with that agent and chooses the local action either red or green, independent of all other agent actions. This cheap design is recommended when the distances between ramps are too long.

Coordinated Reinforcement Learning with Max-Plus: The second design is considered as a completely distributed design (Fig. 3b). It considers a MARL based design. This design works on the network level and is based on modeling by *collaborative Markov Decision Process (Markov Game)* and an associated *cooperative Q-learning* algorithm. This design incorporates *a payoff propagation* (Max-plus algorithm) under the *coordination graphs* framework, particularly suited for optimal control purposes. In this design, the control agent coordinates its actions with its neighboring agents (message passing algorithm). The agent updates the cooperative Q-function globally as follows:

$$\begin{aligned} Q_i^{new}(s_i, \mathbf{a_i}) &= Q_i^{old}(s_i, \mathbf{a_i}) + \\ &\alpha \left[R(\mathbf{s}, \mathbf{a}) + \gamma \max_{\mathbf{a'}} Q^{old}(\mathbf{s'}, \mathbf{a'}) - Q^{old}(\mathbf{s}, \mathbf{a}) \right] \end{aligned} \tag{9}$$

where: $Q(\mathbf{s}, \mathbf{a})$ can be computed using Eq. (3); α can be computed using Eq. (2); and R(s,**a**) can be figured out by the proposed conception of the global reward function using the *harmonic mean*. With such design, it is guaranteed the balance between the control agents payoffs, and hence optimal response to the freeway dynamics, as follows:

$$R(\mathbf{s}, \mathbf{a}) = \frac{1}{\frac{1}{n}\sum_{i=1}^{n} \frac{1}{R(s_i, \mathbf{a_i})}} \tag{10}$$

The maximum control action $\mathbf{a}'$ in state $\mathbf{s}'$ and its associated estimation of optimal future value $\max_{\mathbf{a}'} Q(\mathbf{s}', \mathbf{a}')$ can be computed using the Max-Plus algorithm. The function $f_{ij}(a_i, a_j)$ can be computed using the proposed design of the joint payoff between two neighboring control agents (with such design, it is guaranteed the balance between connected control agents), as follows:

$$f_{ij}(a_i, a_j) = \frac{2 * f_i(a_i) * f_j(a_j)}{f_i(a_i) + f_j(a_j)} \tag{11}$$

This moderate design is recommended when the distances between ramps are short and the traffic network has many ramps.

Centralized Agent: The third extreme design considers a centralized controller (Fig. 3c) using a collaborative MAS coordinated actions are orchestrated as a single action. The cooperative Q-function for the joint actions are updated using a single Q-function, as follows:

$$\begin{aligned} Q^{new}(\mathbf{s}, \mathbf{a}) &= Q^{old}(\mathbf{s}, \mathbf{a}) + \\ \alpha &\left[R(\mathbf{s}, \mathbf{a}) + \gamma \max_{\mathbf{a}'} Q^{old}(\mathbf{s}', \mathbf{a}') - Q^{old}(\mathbf{s}, \mathbf{a}) \right] \end{aligned} \tag{12}$$

where: R(**s**,**a**) can be figured out by Eq. (10). Nevertheless, this design leads to an optimal solution [14], it is not scalable as it suffers from the curse of dimensionality. This costly design is recommended when the distances between ramps are short and the traffic network has a fewer number of ramps.

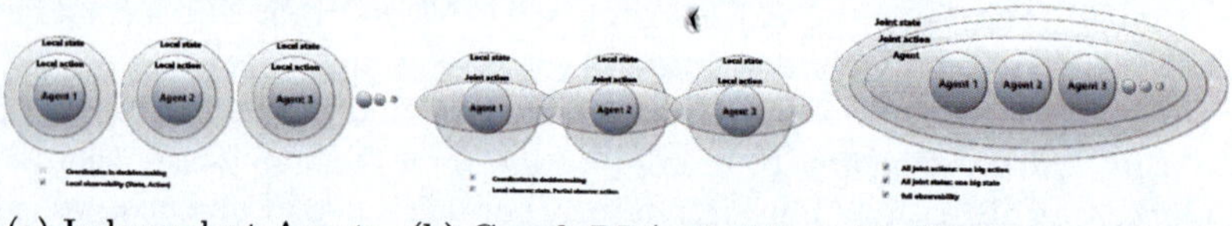

(a) Independent Agents. (b) Coord. RL/with M+. (c) Centralized Agent.

Fig. 3. MARL-FWC based on multi-agent and cooperative Q-learning.

3.3 MARL-FWC for Adaptive Ramp Metering Plus DSLs

This section illustrates the new conception of ramp metering objective function. This conception takes into account the ramp queue length. This section also demonstrates how DSLs can complement the ramp metering in order to mitigate the traffic congestion when the network is dense.

MARL-FWC Another Conception of Ramp Metering Objective Function: By considering the weighted sum of both freeway density and ramp, it is possible to minimize both the normalized difference from the critical density in the freeway O_h and the normalized queue length in the ramp O_r. The ramp metering objective function O is demonstrated as follows:

Minimize O:

$$
\begin{array}{lll}
O & = w_h O_h + w_r O_r : & \text{Collective Objective Function} \\
O_h & = 2\frac{|\rho-\rho_c|}{\rho_{max}} : & \text{Freeway Objective Function} \\
 & & \text{(normalized difference from critical density)} \\
O_r & = \frac{q}{q_{max}} : & \text{Ramp Objective Function} \\
 & & \text{(normalized queue length)} \\
w_h & = \frac{\rho\alpha}{\rho_{max}} : & \text{Adaptive Freeway Weight} \\
w_r & = \frac{q(1-\alpha)}{q_{max}} : & \text{Adaptive Ramp Weight} \\
\alpha & \in [0,1] : & \text{Control Parameter}
\end{array}
\tag{13}
$$

where: w_h is the adaptive freeway weight. It is adaptive because it depends on the current freeway state; as far as the density increases, the w_h also increases. w_r is the adaptive ramp weight. It is adaptive because it depends on the current ramp state; as far as the queue length increases, the w_r also increases. Finally, α is the control parameter. It is used to determine the importance of each weight (biasing factor). With such conception of ramp metering objective function, there is a balance between mitigating the traffic congestion at the freeway (capacity drop) and moving the problem to another road segment.

DSLs Objective Function: The DSLs analytical proof proves that changing the velocity affects the freeway density. This section illustrates the analytical proof. This proof is used to demonstrate how the change in velocity can affect both freeway density and flow. It shows that the slope between density and flow is the velocity, see Fig. 4.

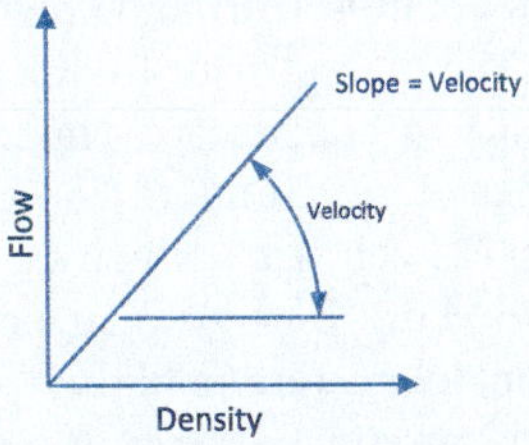

Fig. 4. ρ, q and v.

$$
\begin{array}{lll}
q & = \frac{N}{t} & \text{Freeway Flow (veh/h)} \\
\rho & = \frac{N}{A} = \frac{N}{(d*width(const.))} & \text{Freeway Density (veh/km/lane)} \\
v & = \frac{d}{t} = \frac{1/t}{1/d} & \text{Freeway Velocity (km/h)} \\
 & \text{Multiply both sides by } N & \\
v & = \frac{N/t}{N/d} & \\
v & \propto \frac{q}{\rho} &
\end{array}
\tag{14}
$$

where: N is the number of vehicles (veh) (see [3]) and t is the time (h). DSLs objective function: taking into account the queue length in designing the freeway objective function Eq. (13) raises a question; how to control the amount of vehicles that enter the merging area from the main stream freeway? Hence, the role of the DSLs can be considered as the main stream metering to complement the ramp metering function. When the ramp metering cannot mitigate the capacity drop at the merging area, the DSLs can complement by limiting the number of vehicles entering the merging area. That been said, the DSLs objective function is as follows:

$$O_{hDSL} = 2\frac{|\rho - \rho_c|}{\rho_{max}}, \text{Freeway DSLs Objective Function (normalized difference from critical density)} \quad (15)$$

4 Performance Evaluation

Multi-agent Reinforcement Learning: The studied network in Fig. 5 consists of a mainstream freeway with three lanes and three-metered on-ramp with one lane each. The network consists of four sources of inflow: O_1 for the mainstream flow, O_2, O_3, and O_4 for the three on-ramp flow. It has only one discharge point D_1 with unrestricted outflow. The mainstream freeway consists of three areas of interest: A_1, A_2, and A_3, plus three control agents (RLCA) which are located at the entrance points of each ramp. Each area of interest is associated with one control agent. Each area is about 1350 m as follows: 580 m before the ramp, 250 m as a merging area and 520 m after the ramp. There are four uncontrolled sections; three of them before A_1, A_2 (which is about 1500 m) and A_3 (which is about 100 m) and the fourth is after A_3. Table 1 represents the demand for both the mainstream freeway and ramps. That is because with such scenario the density ρ can exceed the critical density ρ_{cr} which is proven during the learning process. Random choice of action can lead to a scenario with density ρ equals 72 (Veh/km) which is higher than the critical one ρ_{cr} which is equal to 62 (Veh/km). The studied network is considered a dense network where the smart control system is highly needed.

Table 2 provides the results obtained from applying MARL-FWC with all the three approaches and compares it to the base case (no-metering). It can be seen that coordinated reinforcement learning with Max-Plus and centralized agent have shown considerable improvement over the base case. Centralized agent and cooperative Q-learning with Max-plus gave 6.5% and 6.9% in terms of total travel time and 6.74% and 7.5% in terms of average speed improvements over the base case, respectively.

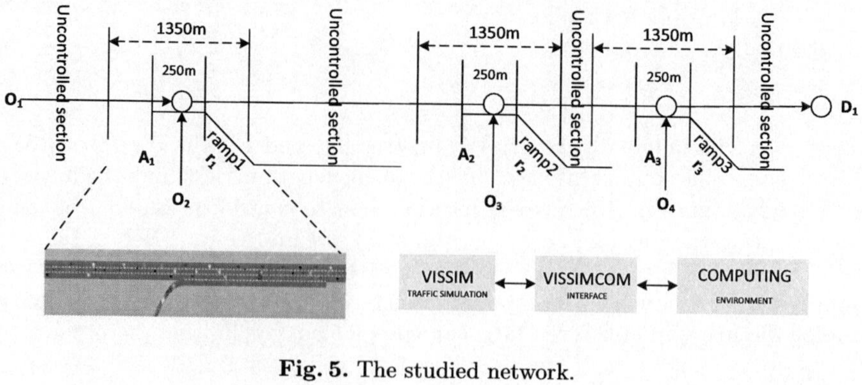

Fig. 5. The studied network.

Table 1. The demand scenario.

Simulation time (sec)	300	600	900	1200	1500	1800	2100
Freeway demand flow (veh/h)	4000	5500	7000	6500	6000	5500	4500
Ramp demand flow (veh/h)	1000	1000	1000	1000	1000	1000	1000

Table 2. MARL-FWC performance evaluation.

	No-metering	IAs	Max-plus	Centralized agent
Total travel time (h)	333	331.5	311.5	308.7
		0.5%	6.5%	6.9%
Average speed (km/h)	45.25	45.49	48.5	48.65
		0.5%	6.74%	7.5%

Adaptive Objective Function Plus DSLs: The studied network in Fig. 6 consists of a mainstream freeway with two lanes and an on-ramp with one lane. The network consists of two sources of inflow: O_1 for the mainstream flow and O_2 for the on-ramp flow and it has only one discharge point D_1 with unrestricted outflow. The mainstream freeway consists of two sections: S_1 and S_2. The first section S_1 lies before the ramp, it is 4 km long. The second section S_2 lies after the ramp, it is 2 km long. The mainstream freeway capacity is 4000 veh/h for both lanes, and the on-ramp capacity is 2000 veh/h. The network parameters are taken as follows: $V_{ff} = 60\,\text{km}^2/\text{h}$, $\rho(0) = 40$ veh/km/lane, $\rho_{jam} = 180$ veh/km/lane and $\rho_{cr} = 33.5$ veh/km/lane.

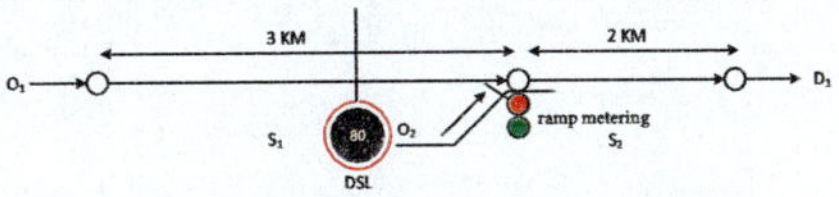

Fig. 6. Traffic network with DSLs and a metered on-ramp.

Table 3 represents the demand (for both the mainstream freeway and ramp) considered in the simulation experiment over a four hours period. This demand scenario gives the opportunity to study the effect of the on-ramp smart control and DSLs. In the freeway line, there was an increase in the demand flow over the first hour, then the demand flow remains steady at high level near the capacity of the freeway over two hours and half, and finally, the demand flow decreases in one hour to low level. In the ramp line, the demand flow went up over the first hour, then remains at high level near the capacity for half an hour, finally, it dropped to low value and remains steady for two hours and half. The studied network is considered a dense network where the smart control system is highly needed.

Table 3. Demand scenario for the simulation.

Simulation time (h)	0.5	1	1.5	2	2.5	3	3.5	4
Freeway flow (veh/h)	2500	3000	3500	3500	3500	3500	2500	1000
Ramp flow (veh/h)	500	1000	1500	500	500	500	500	500

Table 4 demonstrates the total time spent (TTS (veh.h.)) between three different strategies. There was a decline in the total time spent, when MARL-FWC for adaptive ramp metering and ramp metering plus DSLs is used as compared to the base case.

Table 4. MARL-FWC: adaptive ramp metering plus DSLs evaluation.

	No-metering	Ramp metering	Ramp metering plus DSL
TTS(veh.h.)	1240.0	1152.0	1055.6
		7%	14.8%

An important thing to be noticed here is that the new approach for ramp metering plus DSLs have shown considerable improvement over the base case. Adaptive ramp metering and adaptive ramp metering plus DSLs give 7% and 14.8% in terms of total time spent over the base case. Finally, in on-ramp queue constraint, the DSLs can complement the ramp metering by preventing traffic break down and maintaining the traffic flow at high level.

5 Conclusions

In this paper, a new system for controlling ramps metering and speed limits has been introduced based on a multi-agent reinforcement-learning framework. MARL-FWC comprises both a MDP modeling technique and an associated cooperative Q-learning technique, which is based on payoff propagation (Max-plus algorithm) under the coordination graphs framework. For MARL-FWC to be as optimal as possible, the approach is evaluated using different modes of operation depending on the network architecture. This framework has been tested on the state-of-the-art VISSIM traffic simulator in a dense practical scenarios. The experimental results have been thoroughly analyzed to study the performance of MARL-FWC using a concrete set of metrics, namely, total travel time, average speed, and total time spent while keeping freeway density at optimum level close to the critical ratio for maximum traffic flow. The findings have proved that MARL-FWC has achieved significant enhancement on three features as compared to the base case. It have been also shown that when the ramp metering could not mitigate the capacity drop at the merging area, DSLs could complement the ramp metering by adjusting the flow rate on the main street via changing the speed limits values. The advantages of applying MARL-FWC include saving car fuel, decreasing air pollution, mitigating capacity drop, and saving lives by reducing the chances of car accidents.

Acknowledgements. This work is mainly supported by the Ministry of Higher Education (MoHE) of Egypt through PhD fellowship awarded to Dr. Ahmed Fares. This work is supported in part by the Science and Technology Development Fund STDF (Egypt); Project id: 42519 - "Automatic Video Surveillance System for Crowd Scenes", and by E-JUST Research Fellowship awarded to Dr. Mohamed A. Khamis.

References

1. Vlassis, N., Elhorst, R., Kok, J.R.: Anytime algorithms for multiagent decision making using coordination graphs. In: 2004 IEEE International Conference on Systems, Man and Cybernetics, vol. 1, pp. 953–957. IEEE (2004)
2. Guestrin, C., Koller, D., Parr, R.: Multiagent planning with factored MDPs. In: NIPS, vol. 1, pp. 1523–1530 (2001)
3. Fares, A., Gomaa, W.: Freeway ramp-metering control based on reinforcement learning. In: 11th IEEE International Conference on Control Automation (ICCA), pp. 1226–1231 (2014)
4. Fares, A., Gomaa, W.: Multi-agent reinforcement learning control for ramp metering. In: Selvaraj, H., Zydek, D., Chmaj, G. (eds.) Progress in Systems Engineering. Advances in Intelligent Systems and Computing. Springer, Heidelberg, vol. 330, pp. 167–173 (2015)
5. Yu, X.F., Xu, W.L., Alam, F., Potgieter, J., Fang, C.F.: Genetic fuzzy logic approach to local ramp metering control using microscopic traffic simulation. In: 2012 19th International Conference on Mechatronics and Machine Vision in Practice (M2VIP), pp. 290–297 (2012)
6. Ghods, A.H., Kian, A., Tabibi, M.: Adaptive freeway ramp metering and variable speed limit control: a genetic-fuzzy approach. Intell. Transp. Syst. Mag. **1**(1), 27–36 (2009)
7. Liang, X., Li, J., Luo, N.: Single neuron based freeway traffic density control via ramp metering. In: 2010 2nd International Conference on Information Engineering and Computer Science (ICIECS), pp. 1–4 (2010)
8. Li, J., Liang, X.: Freeway ramp control based on single neuron. In: IEEE International Conference on Intelligent Computing and Intelligent Systems, ICIS 2009, vol. 2, pp. 122–125 (2009)
9. Feng, C., Yuanhua, J., Jian, L., Huixin, Y., Zhonghai, N.: Design of fuzzy neural network control method for ramp metering. In: 2011 Third International Conference on Measuring Technology and Mechatronics Automation (ICMTMA), vol. 1, pp. 966–969 (2011)
10. Guestrin, C.E.: Planning under uncertainty in complex structured environments. Ph.D. thesis, Stanford University (2003)
11. Pearl, J.: Probabilistic Reasoning in Intelligent Systems: Networks of Plausible Inference. Morgan Kaufmann (1988)
12. Ernst, D., Glavic, M., Capitanescu, F., Wehenkel, L.: Reinforcement learning versus model predictive control: a comparison on a power system problem. IEEE Trans. Syst. Man Cybern. Part B: Cybern. **39**(2), 517–529. IEEE (2009)

13. Bertsekas, D.P.: Dynamic Programming and Optimal Control, vol. 1. Athena Scientific Belmont, Massachusetts (1996)
14. Watkins, C.J., Dayan, P.: Q-learning. Mach. Learn. **8**(3-4), 279–292 (1992)
15. Kaelbling, L.P., Littman, M.L., Moore, A.W.: Reinforcement learning: a survey. J. Artif. Intell. Res. **4**, 237–285 (1996)
16. Kok, J.R., Vlassis, N.: Collaborative multiagent reinforcement learning by payoff propagation. J. Mach. Learn. Res. **7**, 1789–1828 (2006)
17. Papageorgiou, M., Hadj-Salem, H., Blosseville, J.-M.: Alinea: alocal feedback control law for on-ramp metering. Transp. Res. Rec. **1320**, 58–67 (1991)
18. Yedidia, J.S., Freeman, W.T., Weiss, Y.: Understanding belief propagation and its generalizations. Expl. Artif. Intell. New Millennium **8**, 236–239 (2003)
19. Wainwright, M., Jaakkola, T., Willsky, A.: Tree consistency and bounds on the performance of the max-product algorithm and its generalizations. Stat. Comput. **14**(2), 143–166 (2004)

Can Digital Finance Contribute to the Optimization of Industrial Structures: Empirical Evidence from Chinese 260 Cities

Jingliang Yue and Hongmei Wen(✉)

School of Finance, Harbin University of Commerce, Harbin, China
102242@hrbcu.edu.cn

Abstract. As a product of the deep combination of traditional finance and financial technology, digital finance has prominently increased the efficiency and widened the scope of financial services. To be an effective tool to regulate the use of funds in the digital era, digital finance has a significant impact on the optimization of industrial structure. In this paper, in order to analyze the effect of digital finance on the optimization of industrial structure with a fixed-effects model, the Digital Inclusive Finance Index, which was complied by Peking University and Ant Financial Services Group, and the panel data from 2011–2019 of 260 prefecture-level cities were selected. The empirical results indicate that digital finance can significantly contribute to the optimization of industrial structure, and the analysis from different dimensions reveals that the breadth of coverage of digital finance has the most vital contribution to the optimization of industrial structure. Finally, the paper puts forward relevant recommendations for the use of digital finance to optimize industrial structure.

Keywords: Digital finance · Industrial structure optimization · Economic growth

1 Introduction

With the rapid development of digital finance in recent years, the way and scale of industry development is being rewritten. With the convenience and low cost of the internet, digital finance has provided financial services to users who have long been excluded from traditional financial services. Digital finance has enabled more consumer demand to be released, more service models to be developed, and innovative methods of optimizing industrial structures. For the purpose of achieving advanced economic development, it is crucial to fully understand the relationship between digital finance and industrial structure optimization. In order to analyze the effect of digital finance on the optimization of industrial structure with a fixed-effects model, the Digital Inclusive Finance Index, which was compiled by Peking University and Ant Financial Services Group, and the panel data from 2011–2019 of 260 prefecture-level cities were selected. The remainder of the paper is organized as shown below: Part 2 combs through the relevant literature; Part 3 introduces the model and relevant data used in this paper; Part 4 reports the empirical results; and Part 5 provides conclusions and recommendations.

© The Author(s), under exclusive license to Springer Nature Switzerland AG 2022
A. E. Hassanien et al. (Eds.): AMLTA 2022, LNDECT 113, pp. 657–666, 2022.
https://doi.org/10.1007/978-3-031-03918-8_54

2 Literature Review

Digital finance first emerged in China in the form of internet-enabled traditional financial institutions, then fintech companies took advantage of their massive user base to make digital finance users grow significantly, and finally traditional finance significantly increased the depth of digital finance usage through deeper cooperation with fintech companies [1]. The starting point of digital finance in China can be counted from 2004 the year when the Alipay account system was launched, but the industry usually regards 2013, when Alipay appeared, as the first year of digital finance in China. In this year, the literature related to digital finance first appeared in China as well. In this paper, the author defined digital finance as a new pattern of financial business based on modern information technology such as the Internet and big data, and a new thing that emerged in the process of modern information enterprises going deeper into financial enterprises [2].

Scholars have paid more and more attention to digital finance from both macro and micro perspectives for the remarkable development of digital finance. From a macro perspective, digital finance can significantly contribute to economic growth, and it is growing more rapidly in economically disadvantaged areas, and can increase household income through increased entrepreneurial opportunities for households with low physical or social capital, thus contributing to economic growth in an inclusive manner [3]. At the micro level, scholars' research on digital finance has focused on enterprises and households. By analyzing the application of digital finance by SMEs, it can be found that digital finance can help enterprises improve their financing efficiency, alleviate financing difficulties, stabilize their financial situation and promote the increase of their technological innovation output [4]. For the convenience and low-cost, the household credit constrains has been eased and the impact on household was reflected in this way. The effect of digital finance on household consumption is more pronounced than that of non-recurring household expenditure [5].

According to the above literature, it can be found that digital finance can significantly increase the efficiency of finance and widen the scope of financial service. With the continuous development of financial technology, the digitalization of financial services has been upgraded, credit resource allocation and industrial financing efficiency have been significantly improved, and scholars have gradually paid attention to how digital finance affects the optimization of industry.

The current literature on industrial transformation and upgrading focuses on the direction of China's future industrial structure optimization and its influencing factors. Scholars believe that the optimization of industrial structure can effectively drive economic growth through a "structural dividend" [6], and that the ratio between productive services and manufacturing should be balanced to achieve high-quality economic development through industrial structure optimization [7]. Most of the literature on factors influencing industrial restructuring has focused on technological innovation and artificial intelligence. As an important outcome of technological innovation [8], the development of artificial intelligence will lead to the flow of production factors between different industrial sectors. The direction of the flow is related to the substitution of traditional production methods by artificial intelligence in different industrial sectors, thus promoting technological progress, improving labor endowment and catalyzing the growth of new industries [9].

In addition to the above studies, digital finance has also been identified as an important factor that can contribute to the transformation and upgrading of industries as the impact of the digital era on various industries gradually deepens [10]. Scholars explored how digital finance affect the optimization of industrial structure in detail. It can be found that the breadth of digital finance has a long-term and positive effect on the optimization of industrial structure, but the depth of use and the degree of digitalization have a non-linear effect on the optimization of industrial structure [11] such as a threshold effect [12]. However, in general, the contribution of breadth of coverage is greater than that of depth of use and digitalization [13]. When the different dimensions of industrial structure optimization are studied, it is easy to see that digital finance can have a positive impact comprehensively [14].

With the development of digital finance, the gradual maturity of financial technology and the improvement of data availability, scholars have gradually explored digital finance in depth, and their research has gone through several stages, including the effect of digital finance on the real economy, entrepreneurship and household consumption. The research methods have evolved from a single macro analysis or micro analysis to a combined macro and micro model. In general, digital finance can effectively lower the threshold of financial services and improve the quality of financial services, and scholars have given positive comments on the impact brought by digital finance. Digital finance has also been identified as an important factor that can contribute to the transformation and upgrading of industries as the impact of the digital era on various industries gradually deepens. Scholars explored how digital finance affect the optimization of industrial structure in detail. Drawing on the research methods of the existing literature and based on more detailed data, this paper empirically analyses how digital finance affects the optimization of industrial structure and analyses the underlying mechanisms involved.

3 Models and Data

In order to study how digital finance affects the industrial optimization, this paper sets up a model and describes the relevant data. In this Sect. 3.1 clarifies the idea of the model setting, Sect. 3.2 describes the variables and the data.

3.1 Model Setting

In order to test the impact of digital finance development on industrial structure optimization, the panel data from 2011–2019 of Chinese 260 cities was selected and first estimated using mixed regressions with the following expressions.

$$TS_{it} = a + \beta_1 df_{it} + \gamma X_{it} + \delta z_i + \varepsilon_{it} \tag{1}$$

where i represents the city, the t represents time, TS is the degree of optimization of industrial structure, df is the Digital Inclusive Finance Development Index, and X is a set of control variables, including *lnpgdp,popu*, *edu* and *lnfc*, denoting economic growth, population density, educational attainment and offshore investment, respectively. z_i are individual characteristics that do not vary over time, and a is a constant term, and ε_{it} is a random disturbance term. The mixed regression assumes that all individuals have exactly the same regression equation and that there are no individual effects, for which a statistical test must be conducted.

$$TS_{it} = \beta_2 df_{it} + \gamma X_{it} + \delta z_i + u_i + \varepsilon_{it} \tag{2}$$

The meaning of the variables in the expression is the same as in Eq. (1), but the perturbation terms are given by u_i and ε_{it} two components, where u_i is an unobservable intercept term that represents individual heterogeneity and is associated with an explanatory variable, and ε_{it} is a perturbation term that varies with individuals and time. If u_i is not correlated with all explanatory variables, it is called a random effects model and the expression is as follows.

$$TS_{it} = \beta_3 df_{it} + \gamma X_{it} + \delta z_i + u_i + \varepsilon_{it} \tag{3}$$

In order to select the appropriate regression model, this paper will verify the existence of individual effects and whether u_i is correlated with the explanatory variables through F-test, LM-test and Horsman test.

3.2 Data and Data Resources

(1) This paper measures the degree of optimization of industrial structure in terms of the ratio of tertiary to secondary production values (TS).
(2) Digital financial development. This paper selects the China Digital Financial Inclusion Index jointly compiled by the Digital Finance Research Centre of Peking University and Ant Financial Services Group to measure the extent of digital financial development in China. The index is compiled based on big data obtained from Ant Financial Services' transaction accounts and consists of breadth of coverage, depth of use and degree of digital support services: where breadth of coverage is described by the number of electronic accounts; depth of use is measured by actual use of Internet financial services; and the degree of digitalization is reflected by convenience and low cost. This paper uses this index 2011–2019 Chinese city-level digital financial inclusion index divided by 100 (df) as the core explanatory variable, with df1, df2 and df3 describing the breadth of coverage, depth of use and degree of digitalization of digital financial services respectively.
(3) Control variables. In this paper, the logarithm of gdp per capita (lnpgdp) is chosen to indicate economic growth, the logarithm of population per square kilometre (popu) to describe population density, the logarithm of the amount of foreign investment actually used in the year (lnfc) to indicate the extent of foreign investment, and the number of teachers in general higher education per 10,000 people to measure the level of education in the place (edu).

The panel data from 2011–2019 of 260 Chinese prefecture-level cities was selected. With the exception of the "Peking University Digital Financial Inclusion Index", all data are from the China City Statistical Yearbook. After removing some missing data, the observations are 2340. The Table 1 shows the selection and definition of variables, and the Table 2 shows the descriptive statistics of variables.

Table 1. Variable selection and definitions.

Variable name	Meaning of variables	Metrics
TS	Degree of optimization of industrial structure	Ratio of tertiary to secondary sector production values
df	Digital Financial Development	Digital Inclusive Finance Index/100
lnpgdp	Economic growth	Logarithmic value of gdp per capita
lnpopu	Population density	Logarithmic value of population per square kilometre
lnfc	Offshore investments	Logarithm of the amount of foreign a capital ctually used in the year
edu	Education level	Number of teachers in general higher education per 10,000 population

Table 2. Descriptive statistics for variables.

Variable	Obs	Mean	Std.Dev	Min	Max
TS	2340	.976	.531	.114	5.34
Df	2340	1.661	.653	.213	3.216
Lnpgdp	2340	10.734	.572	9.091	15.675
Popu	2340	5.794	.871	1.513	7.923
Lnfc	2340	10.006	1.887	0	14.544
edu	2340	10.581	14.03	0	83.467

4 Empirical Results

This section reports the empirical regression results. In Sect. 4.1, the effect of digital finance on the optimization of industrial structure is reported. In Sect. 4.2, the differences in the effects of digital finance on industrial structure optimization are empirically analyzed in different dimensions. Robustness tests are conducted in Sect. 4.3.

4.1 The Effect of Digital Finance on the Optimization of Industrial Structures

The results of the mixed regression, fixed effects regression and random effects regression are reported in the Table 3 respectively. Based on the test results of the F-test, LM-test

Table 3. Baseline regression results.

	(1) TS	(2) TS	(3) TS
Df	0.434*** (0.016)	0.378*** (0.012)	0.376*** (0.011)
lnpgdp	−0.338*** (0.021)	−0.160*** (0.028)	−0.190*** (0.025)
popu	−0.132*** (0.012)	−0.559*** (0.147)	−0.099*** (0.026)
lnfc	0.008 (0.006)	−0.040*** (0.006)	−0.032*** (0.006)
edu	0.016*** (0.001)	0.002 (0.003)	0.012*** (0.001)
_cons	4.391*** (0.211)	5.686*** (0.870)	3.151*** (0.281)
Obs	2340	2340	2340
FM test		17.62***	
LM test			3692.14***
Hausman test		54. 25***	

Standard errors are in parenthesis.
*** $p < 0.01$, ** $p < 0.05$, * $p < 0.1$

and Horsman test reported in the table, the fixed effects regression model should be selected as the baseline research model for this paper.

According to the regression results of model (2) in the Table 3, the coefficient of digital finance (df) is 0.378 and is significant at the 1% level, which empirically shows that digital finance development can promote the optimization of industrial structure. The reason for this is that digital finance can adjust the resource allocation of the industry from both the demand and supply sides. From the demand side, digital finance increases household recurrent expenditure and service consumption, thus increasing the demand of the tertiary industry. From the supply side, digital finance can ease the financing constraints of SMEs and promote innovation and entrepreneurship, thus increasing the supply of the tertiary industry. Ultimately, digital finance promoted the optimization of industrial structure prominently. The control variables of economic growth, population density and foreign investment are all prominently negative at the 1% level. That means that there may be a negative relationship between industrial structure optimization and the economy, population and foreign investment, the logical relationship of which is worth further exploration.

4.2 The Impact of Different Dimensions of Digital Finance on the Optimization of Industrial Structure

As the digital finance index consists of three dimensions: breadth of digital finance coverage (df1), depth of use (df2) and degree of digitization (df3), this paper uses each of these three dimensions of the digital finance index as the core explanatory variables and adopts a fixed-effects regression approach to analyze the impact of digital finance on the optimization of industrial structure from the three dimensions (Table 4).

Table 4. Fixed effects regression results for different dimensions of digital finance.

	(1) TS	(2) TS	(3) TS
df1	0.396*** (0.013)		
df2		0.380*** (0.012)	
df3			0.230*** (0.009)
Lnpgdp	−0.131*** (0.027)	−0.194*** (0.028)	−0.003 (0.028)
popu	−0.601*** (0.148)	−0.487*** (0.146)	−0.427*** (0.157)
lnfc	−0.043*** (0.006)	−0.034*** (0.006)	−0.053*** (0.007)
edu	0.001 (0.003)	0.004 (0.003)	0.005* (0.003)
_cons	5.658*** (0.876)	5.554*** (0.858)	3.494*** (0.919)
Obs	2340	2340	2340
R-squared	0.445	0.463	0.374

Standard errors are in parenthesis
*** $p < 0.01$, ** $p < 0.05$, * $p < 0.1$

The regression results in the table show that the breadth of digital financial coverage (df1), depth of use (df2) and degree of digitization (df3) are all significantly positive at the 1% level, indicating that digital finance can significantly contribute to the optimization of industrial structure in all three dimensions. According to the magnitude of the coefficients, breadth of digital finance coverage has the most significant impact on industrial structure optimization, while the degree of digitization has the weakest impact.

4.3 Robustness Tests

To ensure the reliability of the study, the following robustness tests were done: 1. Replacing the core explanatory variables, the logarithm of the digital finance index (df) is used to remeasure the development of digital finance, and the regression results are shown in column (1). 2. Replacing the explanatory variables, the logarithm of the degree of industrial structure optimization (TS) (lnTS) is used to remeasure the degree of industrial structure optimization, and the regression results are shown in column (2). 3. The regression was conducted after double scaling. In this paper, the core explanatory variables are double-tailed at the 1% level, and the regression results are shown in column (3). 4. Instrumental variables were selected to address possible endogeneity issues. In this paper, the ratio of the number of internet broadband access subscribers to year-end population in prefecture-level cities from 2002–2010 is chosen to portray local internet penetration (rate_inter) and used as an instrumental variable. Internet penetration as a prerequisite for digital financial applications satisfies the correlation condition as an instrumental variable, while the historical Internet penetration status cannot affect the current degree of industrial structure optimization, so the Internet penetration rate also satisfies the exclusivity requirement as an instrumental variable. The results of the one-stage regression of the instrumental variables are shown in column (4), and the results of the two-stage regression are shown in column (5) of the table. As can be seen, there is no change in the parameter estimates or significance of the main variables. That indicates that the findings of the study are robust and reliable (Table 5).

Tabel 5. Regression and robustness test results for instrument variables.

	(1) TS	(2) lnTS	(3) TS	(4) Df	(5) TS
df		0.395*** (0.009)	0.379*** (0.012)		0.328*** (0.048)
lnpgdp	−0.014 (0.028)	−0.157*** (0.021)	−0.160*** (0.028)	1.473*** (0.037)	−0.081 (0.078)
popu	−0.324** (0.157)	−0.806*** (0.109)	−0.529*** (0.147)	0.978*** (0.259)	−0.474*** (0.156)
lnfc	−0.056*** (0.007)	−0.024*** (0.005)	−0.040*** (0.006)	−0.092*** (0.011)	−0.045*** (0.008)
edu	0.006** (0.003)	−0.002 (0.002)	0.002 (0.003)	0.030*** (0.005)	0.004 (0.003)
lndf	0.381*** (0.015)				

(continued)

Tabel 5. (*continued*)

	(1) TS	(2) lnTS	(3) TS	(4) Df	(5) TS
rate_inter				1.915*** (0.162)	
_cons	1.595* (0.903)	5.830*** (0.643)	5.509*** (0.869)		
Obs	2340	2340	2340	2340	2340
R-squared	0.373	0.612	0.451		0.446
Anderson LM					131.757***
CDW					140.330***

Standard errors are in parenthesis.
*** $p < 0.01$, ** $p < 0.05$, * $p < 0.1$

5 Conclusions and Recommendations

This paper selects panel data from 2011–2019 of Chinese 260 prefecture-level cities and the Digital Inclusive Finance Index jointly compiled by Peking University and Ant Financial Services Group to empirically analyze the impact of digital finance on the optimization of industrial structure using a fixed-effects model. The empirical results show that digital finance can significantly contribute to the optimization of industrial structure. The analysis from different dimensions reveals that the breadth of coverage has the most significant contribution to industrial structure optimization, while the results pass the robustness test by selecting Internet penetration as an instrumental variable for regression.

The conclusion that digital finance can optimize the industrial structure is obvious. Therefore, the role of digital finance in promoting innovation and entrepreneurship among SMEs should continue to be exploited to promote the development of the tertiary sector from the supply side. In the mean time, the ability of digital finance to unleash the dynamism of household consumption should not be overlooked, as in this way the industrial structure can be adjusted from the demand side. Finally, there is value in investing more in the construction of digital finance in order to further realize the advantages of dividends from the optimization of the industrial structure.

References

1. Zhuo, H.: The Power of Digital Finance: Empowering The Real Economy, pp. 4–6. People's University of China Press (2018)
2. Kun, Z.: Focus on digital financial security and regulation - an overview of the workshop on digital finance and cybersecurity. Banker **11**, 114–117 (2013)
3. Haizhang, Q., Yunqing, T., Songwei, C., Yuyang, C.: Theoretical and empirical evidence on the development of digital finance and economic growth in China. Res. Quant. Econ. Technol. Econ. **37**(06), 26–46 (2020)

4. Song, T., Xuchuan, W., Jia, Z.: Digital finance and corporate technology innovation - structural characteristics, mechanism identification and differences in effects under financial regulation. Manag. World **36**(05), 52–66+9 (2020)
5. Jie, L., Yu, W., Jingjian, X.: The impacts of digital finance on household consumption: evidence from China. Econ. Model. **86**(317–326) (2020)
6. Shi, D., Li, P., Xu, M.: Industrial structure transformation and upgrading and high-quality economic development. Fujian Forum (Hum. Soc. Sci. Edn.) **09**, 108–118 (2020)
7. Yongheng, Z., Jiating, W.: Research on the direction of industrial transformation and upgrading in China under high quality development–a comparison based on data from China and the United States. Sci. Technol. Prog. Countermeas. **36**(23), 53–62 (2019)
8. Chuanzhong, D., Huawang, J., Wenhan, J.: Breakthrough technological innovation and industrial transformation and upgrading in China in the context of the new round of industrial revolution. Sci. Technol. Prog. Countermeas. **36**(24), 63–69 (2019)
9. Kaiming, G.: Artificial intelligence development, industrial structure transformation and upgrading and labor income share changes. Manag. World **35**(07), 60–77+202–203 (2019)
10. Lei, T., Degong, M.: Can digital finance promote high-quality development? Stat. Res. **37**(11), 80–92 (2020)
11. Wenjing, T., Shuang, L., Yuqing, T.: Digital inclusive finance development and industrial structure upgrading - empirical evidence from 283 individual cities. J. Guangdong Univ. Finance Econ. **34**(06), 35–49 (2019)
12. Rongjuan, T., Qiyuan, L.: Does digital inclusive finance promote the optimization and upgrading of industrial structure? Investment Res. **40**(09), 85–104 (2021)
13. Xiaolong, L., Guanghe, H.: Digital financial development, capital allocation efficiency and industrial structure upgrading. J. Southwest Univ. Natl. (Hum. Soc. Sci. Edn.) **42**(07), 152–162 (2021)
14. Jinmin, D., Shiwei, W., Wenyang, W.: Does digital inclusive finance promote the optimization of industrial structure? Comp. Econ. Soc. Syst. **06**, 38–49 (2020)

Optimization of Artificial Potential Field Parameters based on Enhanced Butterfly Algorithm

C. Chen[2], C. Y. Wu[1,2(✉)], F. J. Wang[2], M. T. Huang[2], and Z. L. Huang[2]

[1] College of Mathematics and Data Science, Minjiang University, Fuzhou 350121, China
wucy18259125950@foxmail.com

[2] Fujian Provincial Key Laboratory of Information Processing and Intelligent Control, College of Computer and Control Engineering, Minjiang University, Fuzhou 350121, China

Abstract. The artificial potential field algorithm in the global path planning of the robot is to calculate the combined force vector of the gravitational force of the destination and the repulsive force of the obstacle to determine the rotation angle and forward distance of the vehicle. It can effectively plan the path for the current robot position. Still, if the parameters in the algorithm are not appropriately set, it is easy to generate problems such as not being able to reach the specified location, a jagged planning path, or stagnant movement. In this paper, we propose an enhanced butterfly algorithm based on stochastic excellent community vectors for parameter optimization of the artificial potential field algorithm with the shortest path as the objective function under the condition of the finite number of steps. The butterfly algorithm converts the fitness values into impact factors for the update terms. It uses random number squares as coefficients, which has the disadvantages of requiring more iterations and slow convergence. Therefore, the concept of fitness value normalization and the stochastic excellent community vector mechanism is introduced to enhance the performance of the original algorithm. Finally, several robot path planning simulations show that the artificial potential field parameters based on the enhanced butterfly algorithm can effectively improve pathway planning.

Keywords: Global path planning · Butterfly algorithm · Artificial potential field · Parameter optimization · Stochastic excellent community vector

1 Introduction

Khatib proposed the concept of Artificial Potential Fields (APF) in 1985 to implement global path planning for robots in environments with obstacles [1]. Artificial Potential Fields is a reactive planning algorithm that calculates the following behavior pattern by the direct distance between the robot and the obstacle. In the process of path planning, the robot is subjected to the gravitational force of the target and the repulsive force of the obstacles to avoid all obstacles on the path to reach the target, but improper settings of the artificial potential field parameters can cause problems such as crooked

© The Author(s), under exclusive license to Springer Nature Switzerland AG 2022
A. E. Hassanien et al. (Eds.): AMLTA 2022, LNDECT 113, pp. 667–677, 2022.
https://doi.org/10.1007/978-3-031-03918-8_55

paths or unreachable targets. Many scholars have proposed improvements to the artificial potential field algorithm to address these shortcomings [2–4].

In general, such problems are usually evaluated by adjusting the parameters to meet the user's needs, which is very time-consuming and ineffective. With the development of swarm intelligence algorithms in recent years, it has also become a high-speed and effective solution to the parameter tuning problem [5–8]. The butterfly algorithm is a novel population intelligence algorithm proposed by Sankalap Arora in 2018 by observing the ecological patterns of butterflies. The algorithm's mechanism is inspired by the fact that butterflies are subjected to scent (fitness values) foraging or movement behavior. However, if the conversion relationship between the fitness value and the impact factor is not set correctly, it can easily produce calculation errors or affect the search performance.

In this paper, we will propose enhancement methods for the evolution mechanism of the original butterfly algorithm. (1) The fitness values are converted into impact factors of the update formula using exponential decreasing normalization; (2) the global exploration formula corrects random squares to double random values; (3) the random excellent community vector mechanism replaces the local search formula, and the above enhancement mechanisms can effectively avoid computational errors and improve the performance of the algorithm. Finally, the enhanced butterfly algorithm is applied to the optimization problem of artificial potential field parameters.

2 Original Butterfly Algorithm

The original butterfly algorithm proposes different mathematical models for butterfly ecology, global exploration and local exploitation. First, the algorithm converts the physical intensity of the butterfly stimulated by the odor into a mathematical formula for the following

$$P = cS^{\alpha} \tag{1}$$

where P is the degree of perception of smell, S is the stimulus intensity related to the fitness value, and c and α are set in the range [0,1].

For the global exploration needs, the individual of the algorithm determines the position of the next update of the individual by the position of the group optimum solution.

$$x_i^{t+1} = x_i^t + (r^2 \times G^* - x_i^t) \times P_i \tag{2}$$

where t is the number of iterations, x_i^t is the i-th individual at the t-th iteration, r is the random value of [0, 1], G^* is the group optimal solution, and P_i is the perceived degree of the i-th individual.

For local exploitation needs, two individuals from the cluster are arbitrarily selected as a reference for the update with the following formula

$$x_i^{t+1} = x_i^t + (r^2 \times x_j^t - x_k^t) \times P_i \tag{3}$$

where x_j^t and x_k^t are the j and k individuals at the t-th iteration. In addition, the original butterfly algorithm switches between two modes of global exploration and local exploitation by switching the parameter β.

3 Enhanced Butterfly Algorithm

There are several obvious problems with the original butterfly algorithm.

- Since the variable S is related to the fitness value, if the fitness value is large, it will cause the calculation results of Eq. 2 and Eq. 3 to fall on the boundary of the search range.
- Equation 2 and Eq. 3 use random value squared as the influence factor of the update, which will lead to an uneven update random range (range from 0 to 1 with the mean value of 0.25).
- Local exploitation using information from arbitrary two individuals to perform the update will reduce the search performance.

To overcome the above problems, the relevant mechanism to enhance the search performance is proposed based on the original butterfly algorithm.

Exponential Decreasing Normalization. First, an exponentially decreasing function is introduced to apply to the update step adjustment with the following formula,

$$\Omega(t) = \exp(-at/T)^b \tag{4}$$

From the figure below, we can observe that the parameters a and b affect the decreasing speed of the curve and the starting point of decreasing, respectively (Fig. 1).

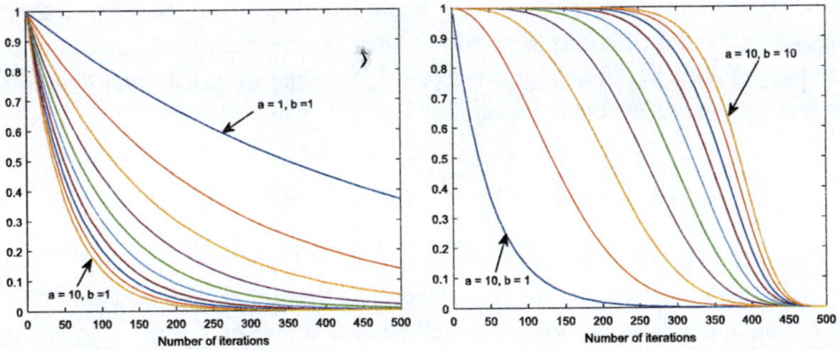

Fig. 1. Exponential decreasing curve (a) $a = 1, 2, \cdots, 10, \quad b = 1$; (b) $a = 10, \quad b = 1, 2, \cdots, 10$

Considering the relationship between the individual fitness value, the update step, and the evolutionary stage of the algorithm, Eq. 1 is changed to

$$\overline{P}_i = \Omega(t) \times \mathrm{N}_i^t \tag{5}$$

where N_i^t is the normalized value of the i-th individual of the t-th iteration.

Impact Factor Correction. To improve the performance of exploration and exploitation, double random values are used instead of the squared random values of the original formula to ensure that individuals can search over a sector centered on a reference point.

Stochastic Excellent Community Vector. Considering algorithmic heuristics and reducing the probability of falling in the local optimal solution. This paper proposes a concept of a random excellent community vector. Several individuals better than the current one are randomly selected as the update reference point, and their vector average is taken as the next move direction. The concept is illustrated below (Fig. 2).

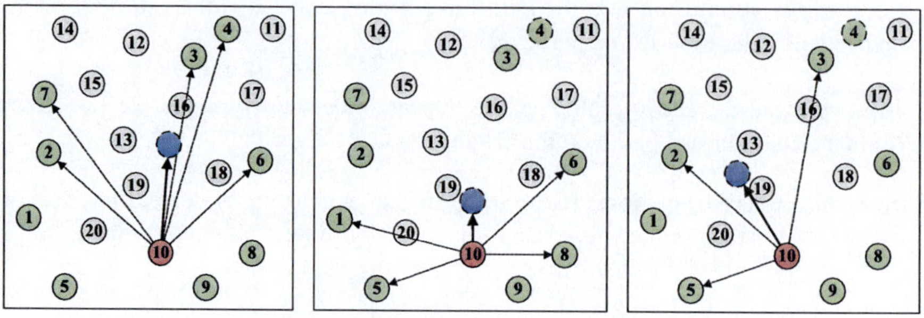

Fig. 2. Schematic diagram of the random excellent community vector

The numbers in the figure mean the fitness value ranking, the red circle is the current individual position, the green circle is the other individuals that are better than the current individual, and the blue circle is the position after the red circle is updated. From left to right are the possibilities of the algorithm position update when different numbers G_{num} of reference points are selected. The update formula is as follows.

$$x_i^{t+1} = (G_{num}(t))^{-1} \sum_{k=1}^{G_{num}(t)} (x_i^t + (r \times x_k^t - x_i^t) \times \overline{P}_i) \tag{6}$$

Linear Decreasing Attempts. In this paper, we introduce the concept of linearly decreasing attempts. The algorithm makes several attempts during each iteration to ensure that the updated fitness value is better than the current fitness value, and reduces the number of attempts as the fitness value converges. The flow chart of the enhanced butterfly algorithm is as follows (Fig. 3),

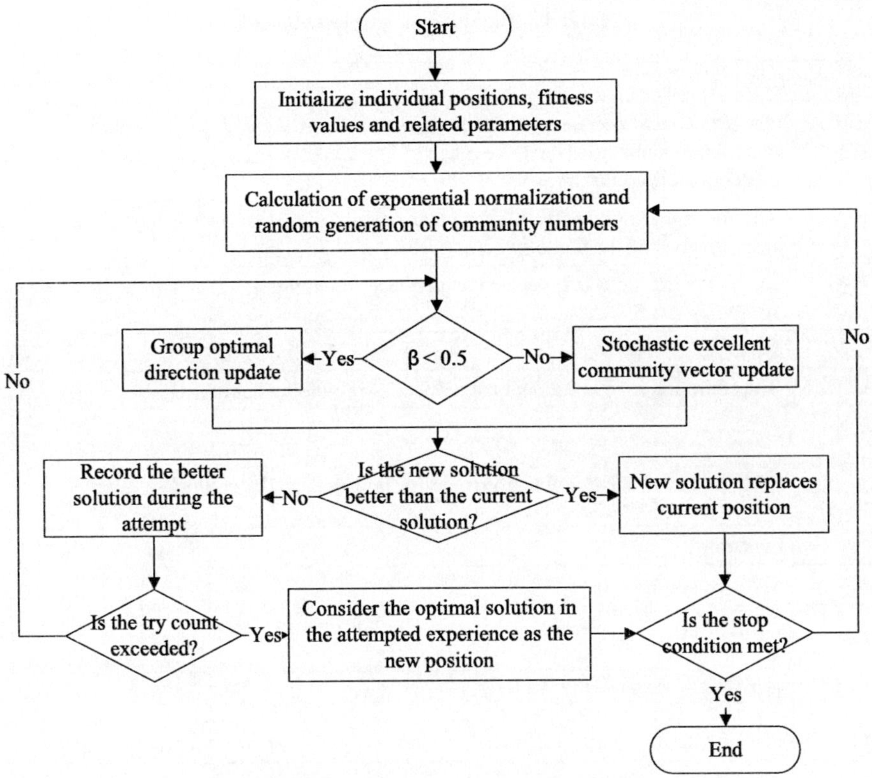

Fig. 3. Flow chart of the enhanced butterfly algorithm

4 Numerical Simulation

In this study, several mechanisms are proposed to enhance the performance of the exploration and exploitation of the algorithm in the solution space based on the above problem. To compare the performance of the enhanced butterfly algorithm(EBOA), the butterfly algorithm(BOA) [10], the particle swarm clustering algorithm(PSO) [11], the gray wolf algorithm(GWO) [12], and the lion swarm algorithm(LSO) [13] are implemented. The relevant algorithm parameters are set in Table 1.

The algorithm search performance was measured by the evaluation function in Table 2 under dimensional *n*, and independent tests are 30, 500 iterations, and populations size is 60. In addition, the population size was revised to 20 in order to reduce the number of calculations in the enhanced butterfly algorithm.

Table 1. Algorithm parameter setting

Parameter setting	
EBOA	The exponential decreasing coefficient [*a*, *b*] is [5, 10], the range of random community number is [1, 5], the range of linear decreasing is [1, 10], and the mode switching ratio β is 0.5
BOA	The perceptual weight *c* is 0.1, the index coefficient α is 0.1 increasing to 0.3, respectively, and the mode switching ratio β is 0.8
PSO	Inertial weighting is 0.7, personal experience weighting is 1.1, and group best weighting is 1.5
GWO	N/A
LSO	The adult lion ratio is 0.2, and the number of re-ranking iterations is 10

Table 2. The evaluation function, the search range, and the optimal solution

No	Function	Range	$f_{\min}$
f_1	$\sum_{i=1}^{n} \lvert x_i \sin(x_i) + 0.1x_i \rvert$	±10	0
f_2	$\left(\sum_{i=1}^{n} x_i^2\right)^2$	±100	0
f_3	$1 + \sum_{i=1}^{n} \frac{1}{4000} x_i^2 - \prod_{i=1}^{n} \cos(\frac{x_i}{\sqrt{i}})$	±600	0
f_4	$10n + \sum_{i=1}^{n} (x_i^2 - 10\cos(2\pi x_i))$	±5.12	0
f_5	$0.5 + \left(\sin^2(\sum_{i=1}^{n} x_i^2)^2 - 0.5\right)\left(1 + 0.001(\sum_{i=1}^{n} x_i^2)\right)^{-2}$	±100	0
f_6	$\sum_{i=1}^{n} \left(\sum_{j=1}^{i} x_i\right)^2$	±100	0
f_7	$\max_{i=1,\ldots,n} \lvert x_i \rvert$	±100	0
f_8	$\sum_{i=1}^{n} \lvert x_i \rvert + \prod_{i=1}^{n} \lvert x_i \rvert$	±100	0
f_9	$\sum_{i=1}^{n} i x_i^2$	±10	0
f_{10}	$\sum_{i=1}^{n} x_i^2 + (\sum_{i=1}^{n} 0.5ix_i)^2 + (\sum_{i=1}^{n} 0.5ix_i)^4$	±10	0

Table 3 shows the evaluation results of the above algorithms for different evaluation functions, and it can be observed that the enhanced butterfly algorithm can effectively search for the optimal solution in f_3 and f_4. The evaluation function f_8 significantly improves the search performance. Compared with the original butterfly algorithm, the search accuracy has been significantly improved, and it also outperforms other types of algorithms.

Table 3. Mean and standard deviation of the evaluation function for each type of algorithm.

	Alg.	Mean	Std		Alg.	Mean	Std
f_1	EBOA	**2.70e−33**	**2.53e−33**	f_6	EBOA	**6.93e−60**	**3.63e−59**
	BOA	2.85e−05	8.48e−07		BOA	1.41e−06	7.41e−08
	PSO	3.76e + 00	4.60e + 00		PSO	5.70e + 03	6.21e + 03
	GWO	5.66e−06	1.24e−05		GWO	1.84e−10	4.31e−10
	LSO	6.52e−24	2.35e−23		LSO	1.11e + 01	8.80e + 00
f_2	EBOA	**3.20e−100**	**8.06e−100**	f_7	EBOA	**4.98e−27**	**8.68e−27**
	BOA	2.93e−08	1.67e−09		BOA	4.29e−05	1.92e−06
	PSO	3.33e + 06	1.83e + 07		PSO	4.91e + 00	3.12e + 00
	GWO	1.23e−75	3.66e−75		GWO	4.86e−10	3.47e−10
	LSO	2.28e−04	5.32e−04		LSO	4.08e−01	2.24e−01
f_3	EBOA	**0**	**0**	f_8	EBOA	**2.59e−29**	**2.93e−29**
	BOA	2.37e−06	1.09e−07		BOA	1.31e + 42	2.95e + 42
	PSO	7.86e−02	1.29e−01		PSO	5.02e + 02	2.21e + 02
	GWO	8.02e−04	2.45e−03		GWO	2.02e−21	1.97e−21
	LSO	1.76e + 00	3.23e−01		LSO	4.83e + 01	1.08e + 02
f_4	EBOA	**0**	**0**	f_9	EBOA	**5.68e−58**	**1.14e−57**
	BOA	1.71e−06	9.51e−08		BOA	1.40e−06	5.30e−08
	PSO	1.15e + 02	3.16e + 01		PSO	2.00e + 02	3.45e + 02
	GWO	0	0		GWO	1.25e−39	1.31e−39
	LSO	0	0		LSO	2.58e−44	5.64e−44
f_5	EBOA	**2.22e−17**	**6.78e−17**	f_{10}	EBOA	**5.20e−50**	**2.35e−49**
	BOA	6.43e−07	2.57e−08		BOA	6.91e−07	8.32e−08
	PSO	5.87e−02	5.11e−02		PSO	2.46e + 01	3.92e + 01
	GWO	1.00e−03	8.91e−04		GWO	1.72e−11	3.24e−11
	LSO	7.44e−03	4.07e−03		LSO	7.93e−21	4.15e−20

Figure 4 shows the convergence curves of f_4 and f_5, where the convergence speed of EBOA is also better than the other algorithms.

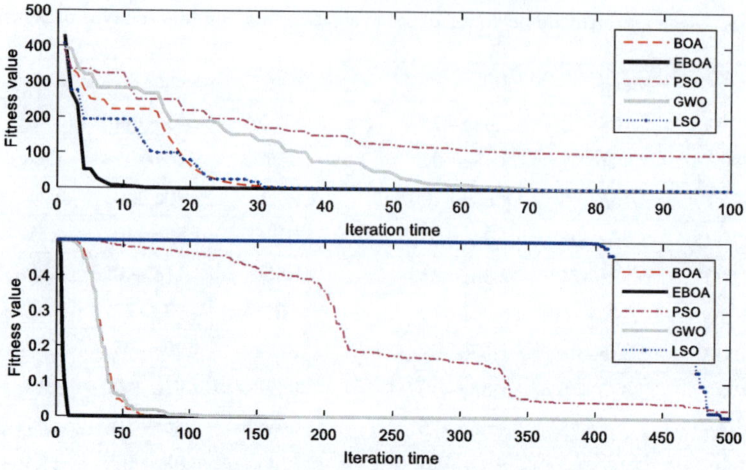

Fig. 4. Algorithm convergence diagram (a) f_4; (b) f_5

5 APF Parameters Optimization Problem

The artificial potential field is a method used for path planning that converts the distance between the robot and all obstacles in the environment into a repulsive potential field, which is then combined with the attractive field generated by the goal to determine the robot's travel direction and step length. It follows that the complexity of the obstacles in the environment affects the computational effort of path planning.

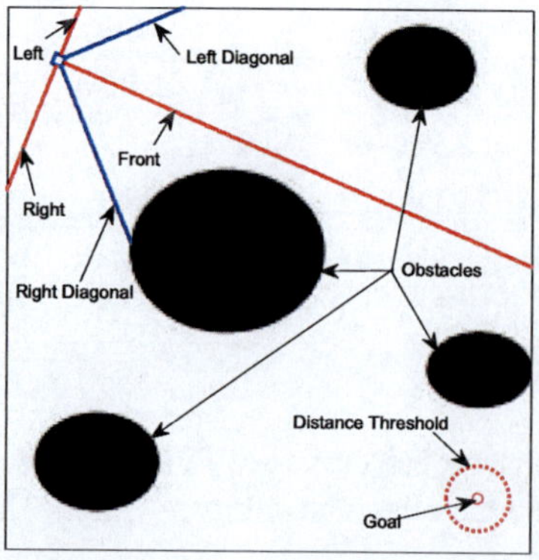

Fig. 5. Schematic diagram of simple artificial potential field

The literature [14] proposes to use the distances of the robot from five angles (front, left, right, left diagonal, and right diagonal) to the obstacles as the basis for implementing the simple artificial potential field concept. Figure 5 shows a schematic diagram of the simple artificial potential field implementation concept.

Therefore, to simplify the computational process and present the robot moving trajectory, the open-source code provided in this literature is cited. Adjustments of robot speed (*RS*), maximum robot speed (RS_{Max}), maximum speed change per unit time (Acc_{Max}), degree of calculating potential (κ), scaling factor for attractive potential (*APS*), scaling factor for repulsive potential (*RPS*) and minimum attractive potential at any point (AP_{min}) are performed.

In order to compare the route planning performance, the number of moving steps and the parameter values set in the literature [14] are considered the constraint and the search range maximum (not less than 1), respectively.

Then, five different path graphs were used to test the planning performance of the original APF and APF_{EBOA} under the step limit. Table 4 shows the original parameters and the artificial potential field parameters based on the enhanced butterfly algorithm for the different maps.

Table 4. Parameter setting based on APF_{EBOA} and original APF in different examples

	APF	APF_{EBOA}				
	Total	Map 1	Map 2	Map 3	Map 4	Map 5
RS	10	1.66	5.49	0.72	2.38	0.83
RS_{Max}	10	5.93	6.54	4.93	8.27	5.79
Acc_{Max}	10	2.12	1.67	4.82	0.94	4.20
κ	3	0.045	1.737	0.081	0.027	1.317
APS	3e + 06	6.39e + 05	8.22e + 05	5.47e + 05	1.64e + 05	1.85e + 05
RPS	3e + 06	9.20e + 05	1.17e + 06	2.35e + 05	2.25e + 06	6.07e + 05
AP_{min}	0.5	0.121	0.637	0.189	0.524	0.238

To present the obstacle avoidance characteristics of the artificial potential field, the robot size is assumed to be 10 × 10, and the simulation is represented by a square. The green and blue squares in Fig. 6 are the path trajectory maps of the original literature and the artificial potential field based on the enhanced butterfly algorithm, respectively.

Map1, Map2, and Map3 observe that the original parameters easily cause the robot to move substantially to avoid obstacles, and the optimized parameters based on the shortest path have a smoother route trajectory. Map4 is repelled by repulsive forces in different directions, and the simulation of the original parameters shows that it takes 974 steps to reach the destination. In comparison, the optimized parameters reach the target in 168 steps. The original parameters in Map5 are unable to reach the target because of the close distance between the target and the obstacle, and it can be observed from the simulation results that the proposed method in this study can overcome the above situation.

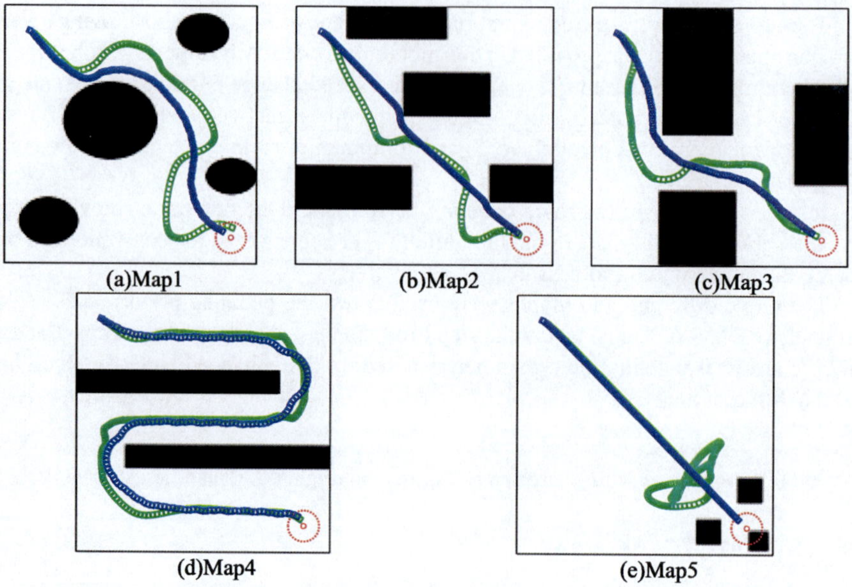

Fig. 6. Trajectory diagram based on the APF_{EBOA} and the original APF in different examples

6 Conclusion

In this paper, the evolution mechanism of the original butterfly algorithm is analyzed, and exponential decreasing normalization is proposed to construct the relationship between update term and fitness value; the performance of local exploitation is improved under the mechanism of random excellent cluster vector with the heuristic property; the number of linear decreasing attempts ensures the superiority of algorithm update. Finally, applied to the study of artificial potential field parameter optimization, the method is presented from the simulation trajectory graphs to effectively reduce the path length as well as the destination unreachability with the limited number of steps.

Acknowledgement. This work was partially supported by the Minjiang University under Grant MJY192026, 2021J011017, 103952021126 and 103952021128.

References

1. Khatib, O.: Real-time obstacle avoidance for manipulators and mobile robots. In: Proceedings 1985 IEEE International Conference on Robotics and Automation, pp. 500–505 (1985)
2. Ulises, O.-R., Oscar, M., Roberto, S.: Mobile robot path planning using membrane evolutionary artificial potential field. Appl. Soft Comput. **77**, 236–251 (2019)
3. Hu, H.Y., Zhang, C., Sheng, Y.H., Zhou, B., Gao, F.: An improved artificial potential field model considering vehicle velocity for autonomous driving. IFAC-PapersOnLine **51**(31), 863–867 (2018)

4. Wang, D.Y., Wang, P., Zhang, X.T., etc.: An obstacle avoidance strategy for the wave glider based on the improved artificial potential field and collision prediction model. Ocean Eng. **206**, 107356 (2020)
5. Lee, M.G., Yu, K.M.: Dynamic path planning based on an improved ant colony optimization with genetic algorithm. In: 2018 IEEE Asia-Pacific Conference on Antennas and Propagation (APCAP), pp. 1–2 (2018)
6. Zhang, X.Q., Wu, C.Y., Jang, L.Y., Shi, L., Guo, J.R.: Pole-placement control of hypothetical loop decoupling design based on improved grey wolf optimization algorithm. In: The 2nd International Conference on Artificial Intelligence and Computer Vision, AISC, vol. 1377, pp. 259–268 (2021)
7. Guo, Q., Zho, S. J., Chen, J.: Parameter tuning of linear active disturbance rejection controller based on chaotic quantum behaved particle swarm optimization. In: 2017 29th Chinese Control and Decision Conference, Chongqing: IEEE, pp. 7484–7489 (2017)
8. Zhang, X.Q., Wu, C.Y., Shi, L., Guo, J.R., Jang, L.Y.: Improved firefly algorithm based on community and migration strategy and its application of PID controller design. In: The 2nd International Conference on Artificial Intelligence and Computer Vision, AISC, vol. 1377, pp. 269–278 (2021)
9. Yin, Z., Du, C., Liu, J., et al.: Research on auto-disturbance-rejection control of induction motors based on an ant colony optimization algorithm. IEEE Trans. Industr. Electron. **65**(4), 3077–3094 (2018)
10. Arora, S., Singh, S.: Butterfly optimization algorithm: a novel approach for global optimization. Soft. Comput. **23**(3), 715–734 (2018). https://doi.org/10.1007/s00500-018-3102-4
11. Kennedy, J., Eberhart, R.: Particle swarm optimization. In: Proceedings of the IEEE International Conference on Neural Networks, pp. 1942–1948 (1995)
12. Mirjalili, S., Mirjalili, S.M., Lewis, A.: Grey wolf optimizer. Adv. Eng. Softw. **69**, 46–61 (2014)
13. Liu, S.J., Yang, Y., Zhou, Y.Q.: A swarm intelligence algorithm-lion swarm optimization. Pattern Recogn. Artif. Intell. **31**(5), 431–441 (2018). (Chinese)
14. Rahul, K.: Code for robot path planning using artificial potential fields. Robotics and Artificial Intelligence Laboratory, Indian Institute of Information Technology, India (2014)

Check for updates

A Discrete Grey Wolf Optimization Algorithm for Minimizing Penalties on a Single Machine Scheduling Problem

Riham Moharam[1(✉)], Ehab Morsy[1], Ahmed Fouad Ali[2], Mohamed Ali Ahmed[1], and Mostafa-Sami M. Mostafa[3]

[1] Department of Mathematics, Faculty of Science, Suez Canal University, Ismailia, Egypt
Riham_Moharam@science.suez.edu.eg

[2] Department of Computer Science, Faculty of Computers and Informatics, Suez Canal University, Ismailia, Egypt

[3] Professor of Computer Science, Faculty of Computers and Information, Helwan University, Cairo, Egypt

Abstract. This paper proposes a new penalty function for scheduling problems that aims to minimize tardiness penalties on a single machine in which each job has its own due date. The problem asks for finding a complete schedule of all jobs, minimizing the total penalties for tardiness jobs. According to this performance measure, a tardiness penalty will be applied to the job if it is finished after its due date. This problem is NP-hard problem. The problem has many real life applications especially in industries such as production systems and acceptance orders. A discrete grey wolf optimization algorithm (DGWO) is proposed for solving this problem. The problem under consideration is a discrete optimization problem. Therefore, we modified and improved the original grey wolf optimization algorithm (GWO) to adapt it with the problem. To ensure that the proposed DGWO is effective and efficient, the average and maximum relative errors between DGWO and optimal solutions are computed. According to our results, the proposed DGWO outputs near optimal solutions in most problem instances size.

Keywords: Scheduling problems · Tardiness penalties · Penalty function · Production systems · Grey Wolf Optimization Algorithm

1 Introduction

The scheduling problem is the problem of constructing a schedule for a certain set of jobs processed on a set of machines to optimize the value of a specified objective function. There are several well known variations of scheduling problems studied in the literatures. In this paper, we study a single machine scheduling problem with different due dates for all jobs with the goal of minimizing total penalties. The problem asks for finding a complete schedule of all jobs, minimizing the total penalties for tardiness jobs. Formally, a set $J = \{1, 2, \ldots, n\}$

© The Author(s), under exclusive license to Springer Nature Switzerland AG 2022
A. E. Hassanien et al. (Eds.): AMLTA 2022, LNDECT 113, pp. 678–687, 2022.
https://doi.org/10.1007/978-3-031-03918-8_56

of n jobs is assigned to a single machine for scheduling. Each job $j \in J$ has its own due date d_j and a processing time p_j. The completion time of job j is denoted by Cj. In case of a job j finished before its due date d_j, no penalty is assigned; and the if the job j finished after d_j, a tardiness weight w_j will be applied. We can formulate the penalty objective function in Equations (1) and (2), where Z_j represents the amount of penalty of job j. The underlying machine is always available from the beginning and can only process one job at a time. Furthermore, preemption is also prohibited.

$$Minimize\ Z = \sum_{j=1}^{n} Z_j \tag{1}$$

where

$$Z_j = \begin{cases} 0 & \text{if } C_j \leq d_j \\ w_j(C_j - d_j) & \text{if } C_j > d_j \end{cases} \tag{2}$$

The penalties scheduling problems has many real life applications especially in industries such as production systems and acceptance orders. In particular, it is applied in delivery contracts, which most of them have due dates for delivery. In case of the order is delivered early, then there is no penalty for it. If the order is not delivered by the due date, then a penalty will be assigned. Also, the problem is applicable in perishable goods and food industries [1,2].

In the literature, many penalty functions are introduced (See [2–12]). The proposed problem is NP-hard problem. This can be shown by simplifying the problem to a total weighted tardiness minimization problem with one due date for all jobs. A total weighed tardiness minimization problem with one due date for all jobs on a single machine is shown by Yaun [13] ordinary NP-hard.

This paper makes a significant contribution by proposing a new penalty function for the area of scheduling problems that allows each job to have its own due date. Due to the hardness of the problem, it is preferred to use meta-heuristic algorithm to get solutions that are close to optimal. We solved the proposed problem by using a discrete grey wolf optimization algorithm (DGWO). The original grey wolf optimization algorithm is modified to adapt it with the proposed discrete problem. We evaluated the proposed DGWO by calculating the difference between the solutions of DGWO and the corresponding optimal ones. The results of our experiments reveal that the proposed DGWO outputs near optimal solutions in most problem instances size.

The remainder of this paper is structured as follows. Section 2 reviews some of the results relating to the proposed problem. Section 3 illustrates the proposed DGWO algorithm. Section 4 evaluates the proposed DGWO by testing it on a set of problems generated at random. Section 5 comes to some conclusions.

2 Related Works

Results on related problems are reviewed in this section. Kianfar et al. [2] introduced the Tardy/Lost penalties scheduling problem, which aimed to minimize total penalties for tardy and lost jobs on a single machine with the same due dates for all jobs. They solved the problem by dynamic programming, branch-and-bound and heuristic algorithms. Weighted tardiness minimization function is a specific case of the Tardy/Lost penalty function. The objective of this problem is to design a schedule that minimizes $\sum_{j=1}^{n} w_j T_j$, where w_j and T_j are the weight and tardiness of job j, respectively. Cheng et al. [4] presented an $O(n^2)$ time approximation algorithm in a single machine for the broad total weighted tardiness problem. Because the overall total weighted tardiness minimization problem is NP-hard in the strict sense [14], they studied two special models of the problem. They first looked at a model in which work due dates are affine linear functions of processing times, while tardy weights are proportional to processing times. In this model, they classed the problem as NP-hard and designed an $O(n \log n)$ algorithm for it. The second is the equal slack (SLK) due date model, which assumes that all job due dates have the same amount of slack. In the conventional sense, they showed that this model is NP-hard. Also, they provided a pseudo-polynomial algorithm with an $O(n^2 P)$ time complexity, where P denotes the total processing times.

Minimizing the total late work scheduling problem on a single machine [15] is close to the Tardy/Lost problem. The purpose of the problem is to create a schedule that minimizes total late work, which is defined as the time a job takes to complete after the due date. In [15], Potts and Wassenhove designed a polynomial time algorithm for this problem. They then designed a branch-and-bound method to solve the same problem [16]. Kethley and Alidaee [17] gave the late work criterion a new definition by giving two due dates for each job, referred to as the due date and deadline. Recently, Chen et al. [18] investigated the issue of the late work criteria under the context of a deadline constraint. They also suggested a fully FPTAS and a pseudo-polynomial time approach for all jobs that have the same due dates.

3 The Proposed Algorithm

We selected the grey wolf optimization algorithm due to its capability of balancing between exploration and exploitation processes. In the following subsections, we highlight the standard grey wolf optimization algorithm (GWO) and the proposed DGWO.

3.1 The Standard Grey Wolf Optimization Algorithm (GWO)

Grey wolf optimization algorithm (GWO) [19] is a population based meta-heuristic algorithm which inspired by the behaviour of grey wolves in hunting strategy and leadership hierarchy. In GWO, each individual solution in the population represents a grey wolf. According to the social structure of wolves, the population is split into four groups of grey wolves (individual solutions): alpha, beta, delta, and omega. The fittest wolf is Alpha. In terms of mathematics, the best solution is alpha (α). Similarly, the second and third best solutions indicate the beta (β) and delta (δ), respectively. While omega (ω) is represented by the remaining individual solutions.

There are three main hunting steps for the prey, tracking the prey, encircling the prey, and attacking on prey. The encircling step of wolves around prey is obtained using the equations below.

$$D = |C.X_p(t) - X(t)| \tag{3}$$

$$X(t+1) = X_p(t) - A.D \tag{4}$$

where t stands for the current iteration and C and A stand for the vectors which are expressed as $C = 2.r_1$ and $A = 2a.r_2 - a$, such that r_1 and r_2 are random values from $[0, 1]$ and a decreases linearly from 2 to 0 over the iterations. The position vectors for the prey and grey wolf are Xp and X, respectively. Equation 4 is used to update the position of wolf X.

Grey wolves hunting can be mathematically modeled by assuming that the alpha, beta, and delta have superior knowledge of prey location. As a result, the first three best solutions are saved, while the remaining solutions (including omegas) are updated based on the best solutions. The following equations can be used to satisfy the position updating process.

$$X_1 = X_\alpha - A_1.D_\alpha,\; D_\alpha = |C_1.X_\alpha - X| \tag{5}$$

$$X_2 = X_\beta - A_2.D_\beta,\; D_\beta = |C_2.X_\beta - X| \tag{6}$$

$$X_3 = X_\delta - A_3.D_\delta,\; D_\delta = |C_3.X_\delta - X| \tag{7}$$

$$X(t+1) = \frac{X_1 + X_2 + X_3}{3} \tag{8}$$

where X_α, X_β and X_δ are the approximated positions of α, β and δ wolves, respectively. Equation 8 refers to the updated position of wolf X.

3.2 The Proposed DGWO

In this subsection, we propose our DGWO algorithm. At first, GWO was proposed to solve continuous optimization problems. The proposed problem is a discrete optimization problem, therefore we modified and improved the original GWO to adapt it with our proposed problem. Each grey wolf in the swarm represents a feasible schedule for the problem. Moreover, the prey represents the schedule with minimum total penalties for the problem. The steps of the DGWO are shown in Algorithm 1 and are described in details as follows.

- **Initial Population.** We generate the initial population by applying random initialization technique. In particular, we generate each schedule in the initial population by repeating the steps below as long as the set of unscheduled jobs is non-empty. Let J' denote the remaining (unscheduled) set of jobs to be scheduled (initially, $J' = J$). We first select random job j from J' and then schedule j after the last scheduled job. Next, we remove j from J' and repeat the above steps as long as $J' \neq \emptyset$.
  The initial population is constructed by repeatedly applying the above procedures with *pop_size* times, where *pop_size* is the population size.

Algorithm 1. Discrete Grey Wolf Optimization Algorithm (DGWO)

Input: A problem size n.
1. $pop\_size \leftarrow n$.
2. $maxgnr \leftarrow 3n$.
3. Initialize the population I_0 by applying random initialization technique.
4. Evaluate the fitness function for each population schedule.
5. Choose the first three best schedules, X_α, X_β, and X_δ.
6. $gnr \leftarrow 1$.
7. **While** ($gnr \leq maxgnr$) **do**
8.   **For** $i = 1$ to pop_size **do**
9.   Update the position of current schedule using Equations 9 and 10.
10.   **Endfor**
11.   Evaluate the fitness function for all new schedules.
12.   Update X_α, X_β, and X_δ.
13. $gnr \leftarrow gnr + 1$.
14. **Endwhile**
15. Output a schedule X_{gnr} with minimum fitness value.

- **Fitness Function.** We define the fitness value that evaluates the quality of the schedule. Recall that, $Z(X) = \sum_{j=1}^{n} Z_j$ is the objective value of the corresponding schedule X.
  Note that the proposed problem is a minimization problem, therefore schedules with the minimum objective values should be selected. For that purpose, the fitness value $fv(X)$ of schedule X is defined as the reverse of its objective value $Z(X)$ during the implementation of the proposed DGWO, i.e., $fv(X) = 1/Z(X)$.
- **Update Process.** As the parameters and arithmetic operators of original GWO cannot be used in discrete version in the same way, so we modified the original equations as shown in Eqs. 9 and 10.

$$D = C \otimes X_p(t) \ominus X(t) \tag{9}$$

$$X(t+1) = X_p(t) \oplus A \otimes D \tag{10}$$

  where C and A are random numbers between $[0, 1]$ and the operator ($\otimes$) means the probability of C that all jobs are selected from schedule X_p for swapping with jobs in the corresponding indices of schedule X. The operator ($\ominus$) means extracting a swap operators from X_p and X. We extract a sequence of swap operators D from Eq. 9. Then, the schedule X is updated according to Eq. 10 by applying the swap sequence D to X by a combined operator ($\oplus$) with probability A.
  Each schedule X in the current iteration is updated using Eqs. 9 and 10 by choosing random solution from X_α, X_β and X_δ as the best solution (X_p).

4 Experimental Results

The proposed DGWO is tested in this section by using it to solve the proposed scheduling problem with random instance sizes. All of the results in this section were carried out using MATLAB R2016a on a Laptop with a core i7 CPU and 16 GB of RAM.

4.1 Parameter Setting

We apply our proposed DGWO on data set contains for each problem size $n = 30, 50, 75, 100, 150, 200$, and 250, a set of 20 random instances. Processing times p_j and tardiness weight w_j are selected randomly from uniform distribution $[1, 100]$. We suppose that the due dates for all jobs d_j in each instance size are started from minimum processing time and does not exceed the total processing times.

For each instance size, the DGWO algorithm is implemented with a population size of $pop-size = n$ and a maximum number of iterations of $maxgen = n$.

Table 1. Problem parameters values.

Parameter	Value
n	$30, 50, 75, 100, 150, 200, 250$
p_j	Uniform $[1, 100]$
w_j	Uniform $[1, 100]$
d_{min}	$min(p_j)$
d_{max}	$\sum_{j=1}^{n} p_j$
d_j	Uniform $[d_{min}, d_{max}]$

4.2 Performance of DGWO

To select the best parameter values of the DGWO algorithm, we implement it with population sizes, $n/3, 2n/3, n, 4n/3, 5n/3$, and $2n$, where n is number of jobs. Table 2 shows the total penalties (objective value) at different population sizes with $n = 100$, $n = 150$ and $n = 200$. The obtained results show that n is the best population size value. Also, the maximum number of iterations $maxgnr$ is applied for that underlying instances $n/3, 2n/3, n, 4n/3, 5n/3$, and $2n$ (See Fig. 1). Figure 1 shows that at $maxgnr = n$, the algorithm can reach to near optimal solution.

The performance of the proposed DGWO is tested by calculating the average and maximum relative error between the DGWO solutions and the corresponding optimal ones. Also, the average running times for the DGWO algorithm are estimated for all problem sizes. Table 3 shows that the proposed DGWO algorithm outputs near optimal solutions in an acceptable time. The convergence of the DGWO algorithm to the problem with $n = 30, 50, 75, 100, 150, 200$, and $n = 250$ are explained in Fig. 2.

Table 2. The objective values at different population sizes.

pop_size	Z		
	$n = 100$	$n = 150$	$n = 200$
$n/3$	242359	1196531	713759
$2n/3$	**230633**	1186531	713759
n	**230633**	**1186531**	**713160**
$4n/3$	**230633**	**1186531**	**713160**
$5n/3$	**230633**	**1186531**	**713160**
$2n$	**230633**	**1186531**	**713160**

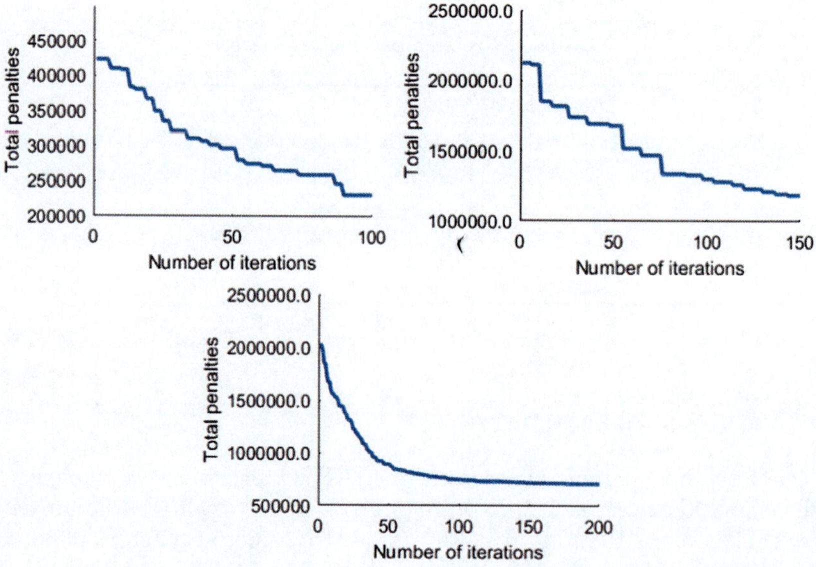

Fig. 1. The influence of maximum number of iterations on the total penalties (objective value) with (a) $n = 100$, (b) $n = 150$ and (c) $n = 200$.

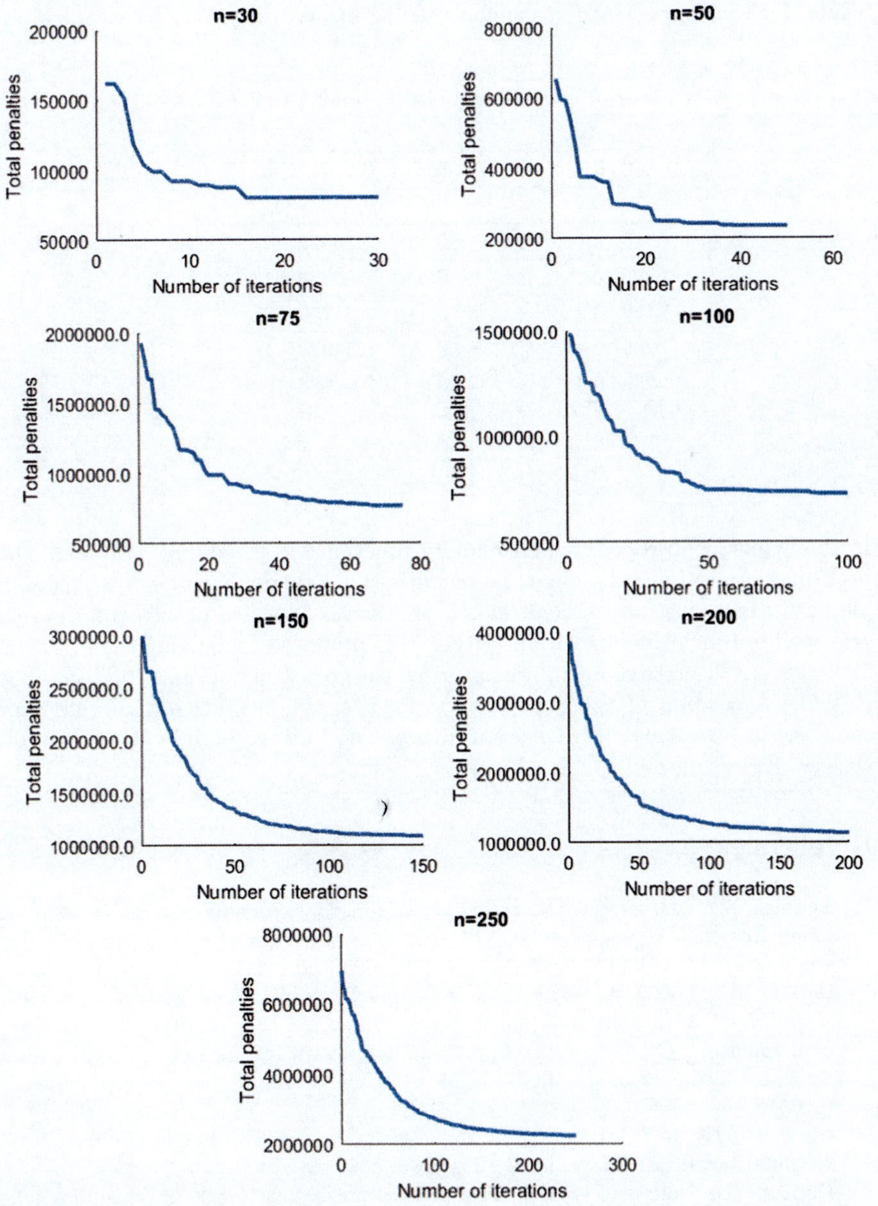

Fig. 2. The convergence of DGWO to the total penalties (objective value) for all problem sizes.

Table 3. The average and maximum relative errors of DGWO on random test instances.

n	Relative error	Max	Running time (s)
	Avg		Avg
30	0	0	0.20111175
50	0	0	0.46018691
75	0.1	0.2	1.18322232
100	0.3	0.5	2.80616031
150	0.2	0.4	4.217508225
200	0.4	0.5	9.84974171
250	0.4	0.6	12.35886176

5 Conclusion

In this paper, we propose a new penalty function for scheduling problem. The problem aims to minimize the total penalties on a single machine in which each job has its own due date. The proposed problem is NP-hard problem. A discrete grey wolf optimization algorithm (DGWO) is proposed for solving the problem. To tackle our discrete problem, we improved the grey wolf optimization technique (GWO). According to the experimental results, the DGWO outputs superior solutions in a reasonable time. As a future work, we may study the version of the problem using multiple machines.

References

1. Slotnick, S.A.: Order acceptance and scheduling: a taxonomy and review. Eur. J. Oper. Res. **212**(1), 1–11 (2011)
2. Kianfar, K., Moslehi, G., Nookabadi, A.S.: Exact and heuristic algorithms for minimizing Tardy/Lost penalties on a single-machine scheduling problem. Comput. Appl. Math. **37**(2), 867–895 (2016). https://doi.org/10.1007/s40314-016-0370-4
3. Kolliopoulos, S.G., Steiner, G.: Approximation algorithms for minimizing the total weighted tardiness on a single machine. Theor. Comput. Sci. **355**(3), 261–273 (2006)
4. Cheng, T., Ng, C.T., Yuan, J., Liu, Z.: Single machine scheduling to minimize total weighted tardiness. Eur. J. Oper. Res. **165**, 423–443 (2005)
5. Detienne, B., Dauzère-Pérès, S., Yugma, C.: An exact approach for scheduling jobs with regular step cost functions on a single machine. Comput. Oper. Res. **39**(5), 1033–1043 (2005)
6. Fathi, Y., Nuttle, H.: Heuristics for the common due date weighted tardiness problem IIE. Transactions **22**(3), 215–225 (1990)
7. Karakostas, G., Kolliopoulos, S.G., Wang, J.: An FPTAS for the minimum total weighted tardiness problem with a fixed number of distinct due dates. In: Ngo, H.Q. (ed.) COCOON 2009. LNCS, vol. 5609, pp. 238–248. Springer, Heidelberg (2009). https://doi.org/10.1007/978-3-642-02882-3_24

8. Kellerer, H., Strusevich, V.A.: A fully polynomial approximation scheme for the single machine weighted total tardiness problem with a common due date. Theor. Comput. Sci. **369**, 230–238 (2006)
9. Koulamas, C.: The single-machine total tardiness scheduling problem: review and extensions. Eur. J. Oper. Res. **202**(1), 1–7 (2010)
10. Lawler, E.l., Moore, J.M.: A functional equation and its application to resource allocation and sequencing problems. Manag. Sci. **16**, 77–84 (1969)
11. Shabtay, D.: Due date assignment and scheduling a single machine with a general earliness/tardiness cost function. Comput. Oper. Res. **35**, 1539–1545 (2008)
12. Sterna, M.: A survey of scheduling problems with late work criteria. Omega **39**, 120–129 (2011)
13. Yuan, J.: The NP-hardness of the single machine common due date weighted tardiness problem. Syst. Sci. Math. Sci. **5**, 328–333 (1992)
14. 2 Lenstra, J.K., Rinnoy Kan, A.H.G., Brucker, P.: Complexity of machine scheduling problems. Ann. Discrete Math. **1**, 343–362 (1977)
15. Potts, C.N., Wassenhove, V.: Single machine scheduling to minimize total late work. Oper. Res. **40**(3), 586–595 (1992)
16. Potts, C.N., Wassenhove, V.: Approximation algorithms for scheduling a single machine to minimize total late work. Oper. Res. Lett. **11**(5), 261–266 (1992)
17. Kethley, R.B., Alidaee, B.: Single machine scheduling to minimize total weighted late work: a comparison of scheduling rules and search algorithms. Comput. Ind. Eng. **43**, 509–528 (2002)
18. Chen, R., et al.: Single-machine scheduling with deadlines to minimize the total weighted late work. Naval Res. Logist. (NRL) **66**(7), 582–595 (2019)
19. Mirjalili, S., Mirjalili, S.M., Lewis, A.: Grey wolf optimizer. Adv. Eng. Softw. **69**, 46–61 (2014)

Chaos-Based Applications of Computing Dynamical Systems at Finite Resolution

Islam ElShaarawy[1,2](✉), Mohamed A. Khamis[2,3], and Walid Gomaa[2,4]

[1] Faculty of Engineering (at Shoubra), Benha University, Cairo 11689, Egypt
islam.elshaarawy@feng.bu.edu.eg

[2] Cyber-Physical Systems Lab, Egypt-Japan University of Science and Technology (E-JUST), New Borg El-Arab City, Alexandria 21934, Egypt
{mohamed.khamis,walid.gomaa}@ejust.edu.eg

[3] IME (Data Science), Ejada Systems Ltd., Alexandria, Egypt
maomar@ejada.com

[4] Faculty of Engineering, Alexandria University, Alexandria 21544, Egypt

Abstract. In this paper, a novel Pseudo Random Number Generator (PRNG) is introduced as an application of computing discrete time dynamical systems at finite resolution in chaos-based cryptography and chaotic optimization. It is based on constructing a combinatorial representation of a given chaotic map, expressing it as a graph, and walking randomly on the resulting graph. The limitations of current methods that use floating-point numbers in finite time compared to the proposed method that uses rational numbers at finite resolution are demonstrated.

Keywords: Dynamical systems · Finite resolution · Rigorous simulation · Pseudo Random Number Generator · Chaos-based cryptography · Chaotic optimization

1 Introduction

Roughly speaking, a dynamical system represents the *time evolution* of any physical or engineered system. In general, studying dynamical systems is very challenging due to many reasons. Some reasons are related to the system itself such as the difficulty to handle it analytically because of its mathematical complexity or even the difficulty to handle it computationally (numerically) because of its chaotic behavior.

There are three main reasons for which computing dynamical systems can be challenging. The first reason is related to state representation. System state usually comes from a continuous space and must be represented using finite number of bits. This results in round off errors. Therefore, the system state $x_0, x_1, \cdots$ are approximately represented by $\hat{x}_0, \hat{x}_1, \cdots$. The second reason is related to the map evaluation. Usually, the system map cannot be evaluated

M.A. Khamis and W. Gomaa—Contributing authors.

© The Author(s), under exclusive license to Springer Nature Switzerland AG 2022
A. E. Hassanien et al. (Eds.): AMLTA 2022, LNDECT 113, pp. 688–696, 2022.
https://doi.org/10.1007/978-3-031-03918-8_57

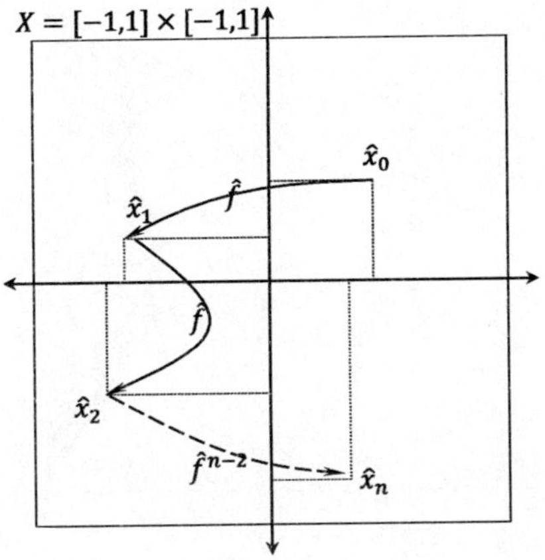

Fig. 1. No-rigorous simulation.

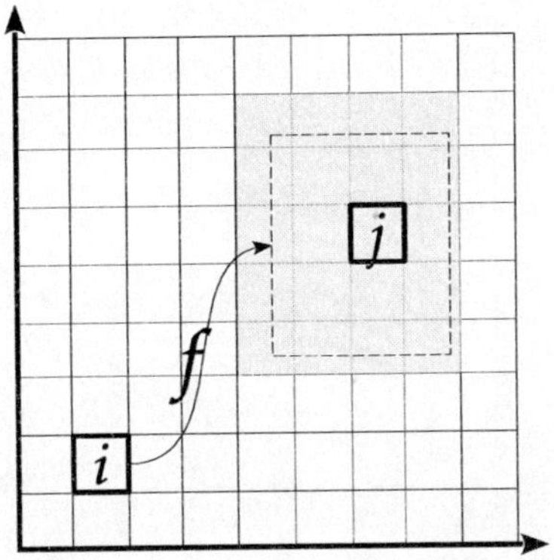

Fig. 2. Discretized phase space (the dotted rectangle is the exact image and the gray area is the ideal representation of $f(P_i)$).

exactly and an approximation method like talking the most significant terms of the corresponding Taylor series must be used. This results in truncation errors. Therefore, the map f is approximately represented by $\hat{f}$. The third reason is related to simulation time. As the time allowed for simulation is limited, only finite number n of evolution steps can be simulated. This is known as "finite or limited time simulation". Figure 1 demonstrates the three challenges. Those three reasons render the following questions hard to answer:

1. How $\hat{x}_n$ is related to x_n?
2. How the simulated orbit starting from x_0 is related to the real orbit?
3. How $\hat{x}_n$ for considerably large n is related to the steady state of the system?

When simple, well-behaved systems show complex behavior, the term chaos is used to describe this behavior. Although chaotic behavior is completely deterministic given the initial conditions, it looks erratic and almost random. Deterministic means that precise knowledge of the initial conditions of a system allows the prediction of the system future behavior [1].

The *logistic map* described by Eq. (1) is a typical example. The logistic system is widely used to model population of insects and animals in a closed environment with constant food supply represented by the parameter r. For different value of the parameter r, different behavior of the logistic map can be observed. The population will exponentially become distinct regardless of the initial value for $r \in [0, 1)$. For $r \in [1, 3)$ the population will eventually stabilize at the fixed-point $x = (r-1)/r$. For $r \in [3, 1+\sqrt{6})$ the population will oscillate between two fixed-points of the second order map f^2. The complete bifurcation diagram of the logistic map is shown in Fig. 3.

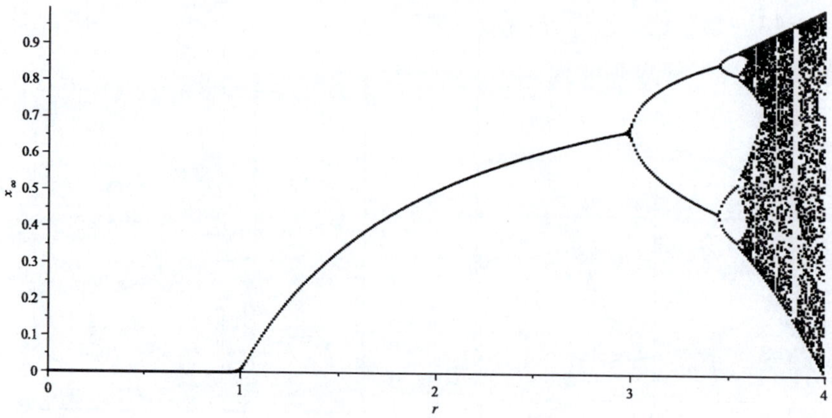

Fig. 3. Bifurcation diagram of the logistic map.

$$x_{n+1} = \mathcal{L}_r(x_n) = rx_n(1 - x_n), \quad x \in [0, 1] \tag{1}$$

The chaotic behavior of such dynamical systems has inspired many applications in cryptography and optimization. Chaos based cryptography [2–9] and chaotic optimization [10–12] have recently been proved to be successful. Such applications use a chaotic map as a source of noise or randomness.

Up to the authors knowledge, all published methods in this area are -in a way or another- based on non-rigorously simulating a "pseudo" orbit of the map starting at an initial point in the phase space for a finite time using floating-point numbers. Such methods are susceptible to being deceived by hidden periodicity [8,13] and suffer from the following four major limitations that will be illustrated in Sect. 4:

1. Initial condition(s) must be defined in order to simulate a trajectory.
2. Due to round off errors, the resulting trajectory is far from reality. What makes the problem worse is that there is no mean to find how far it is.
3. Only finite number of steps can be computed.
4. The computed value will depend on the selected initial conditions, the finite time, and the precision of floating-point numbers.

On the other hand, the proposed methods are based on finite resolution. Therefore, the phase space is being partitioned rather than sampled allowing capturing all the system dynamics and resulting in a better approximation. Moreover, such methods are stable, independent from initial conditions, and have nothing to do with finite time.

The paper is organized as follows. Section 2 covers related work. In Sect. 3, we introduce the proposed methods. In Sect. 4, we present experimental results. Finally, in Sect. 5, we draw conclusions and suggest future works.

2 Related Work

In 1995, Phatak demonstrated the possibility of using the logistic map in the chaotic regime as a pseudo random number generator. They have performed necessary statistical tests on obtained streams and they have reported that the map passed these tests satisfactorily [2].

In 2003, Alioto et al. suitability of digitally implemented maps for PRNG. They evaluated the statistical properties of the resulting streams. The performed analysis showed that digitally implemented maps are suitable for a PRNG only at the cost of circuit complexity. They have used the logistic map at parameter $r = 4.0$ [4].

In 2003, Caponetto et al. examined how chaotic sequence derived from the logistic, sinusoidal, Gauss, Tent, and Lozi maps can improve the performance of evolutionary algorithms. They were able to demonstrate that some chaotic sequences are always able to increase the value of some indexes of measured algorithm performance compared to random sequences [10]. In [12], a chaotic system for *particle swarm optimization* based on the *gradient model with perturbations* model was proposed. In this system, each particle searches locally for a solution intensively around its personal best and the global best without being trapped at any local minimum thanks to the chaotic behavior.

In 2006, Alvarez and Li provided a common framework of basic guidelines for basic cryptographic requirements for chaos-based cryptosystems addressing three main issues: implementation, key management, and security analysis [5]. In 2007, Amigó et al. proposed a conceptual framework for chaotic cryptography and they have illustrated it with different private and public key cryptography examples. In addition, they elaborated on possible limits of chaotic cryptography [3].

In 2001, Galias [14] investigated the possibility of using interval arithmetic for the rigorous computation of periodic orbits in chaotic discrete-time dynamical systems. Luzzatto et al. [15] introduced the idea of using open cover as a structure more closely related to the topology of the phase space. Their work is the most well formulated and the concepts they introduced are well defined.

In [16], we introduced a computational framework for studying dynamical systems. This framework can be used to prove the existence of certain behavior in a given dynamical system at any finite (limited) resolution automatically. In [17], we demonstrated the current limitations of quantifying chaos using floating-point numbers in finite time. Alternatively, we introduce a stable method for quantifying chaos of a given dynamical system using rational numbers at finite resolution.

3 Proposed Method

3.1 Finite Combinatorial Representation

In order to be able to simulate dynamical systems rigorously, the continuous phase space must be discretized first so as to suit the discrete nature of the computer. Finite family $\mathscr{P}$ of disjoint subsets of X such that $X = \bigcup_{P \in \mathscr{P}} P$ is said to be a *partition* of X. The elements of $\mathscr{P}$ will be used as a finite approximation of the topology on X (typically the Euclidean topology over $\mathbb{R}^m$ is assumed). In general, the smaller the partition elements are, the more accurate the approximation. Such partition is a *combinatorial representation* of the underlying phase space.

In addition, the continuous map of the dynamical system must be replaced with a combinatorial version. A *combinatorial map* over a partition $\mathscr{P}$ is a multivalued map $\mathcal{F} : \mathscr{P} \multimap \mathscr{P}$. Such map is said to be a finite representation of a map $f : X \to X$ if $\mathscr{P}$ is a finite partition of X and for every $P \in \mathscr{P}, \mathcal{F}(P) \supseteq \{W \in \mathscr{P} : W \cap f(P) \neq \emptyset\}$. Figure 2 illustrates the discretization process.

In order to build a transparent partition that enables constructing an ideal combinatorial representation of a given dynamical system, dyadic rationals $a/2^b, a \in \mathbb{Z}, b \in \mathbb{N}$ instead of floating-point numbers and partition elements with disjoint interiors instead of overlapping elements are employed [16]. Obviously, the finer the partition resolution $R = 2^{-b}$, the more accurate the representation will be.

The resulting combinatorial map can be represented as a directed graph $(\mathcal{V}, \mathcal{E})$ where every partition element P_i is represented by a vertex $v_i \in \mathcal{V}$ and there is an edge $(v_i, v_j) \in \mathcal{E}$ iff $P_j \in \mathcal{F}(P_i)$.

3.2 Combinatorial Pseudo Random Number Generator

Two variations of Combinatorial Pseudo Random Number Generator (CPRNG) can be constructed based on random walk on the graph [18] resulting from the combinatorial representation of the logistic map:

- CPRNG with memory in which a counter is associated with every graph node to remember which transition was made in the last visit as illustrated in Fig. 4, and
- memoryless CPRNG in which another PRNG is used to select a random transition at each node as illustrated in Fig. 5.

Both CPRNG gave promising results as shown in Sect. 4.

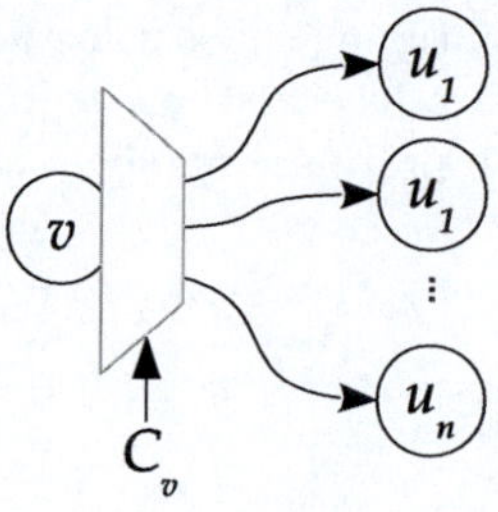

Fig. 4. Combinatorial random number generator with memory.

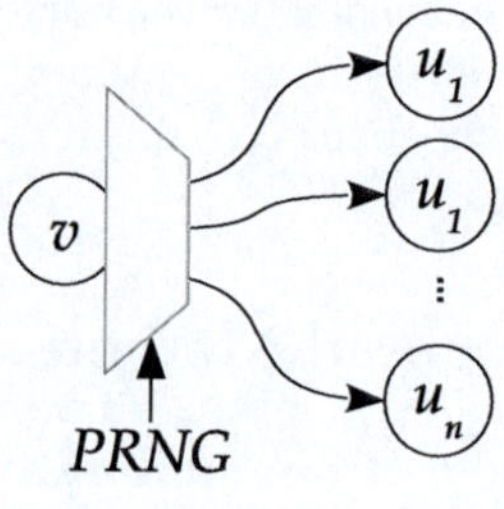

Fig. 5. Memoryless combinatorial random number generator.

4 Experimental Results

In this section, two experiments are performed to evaluate the proposed finite resolution methods compared to finite time methods. The first experiment is to compute the steady state probability distribution of the logistic map for $r \in [0, 4]$. The second experiment computes the PDF of the trajectories simulated using memoryless CPRNG and CPRNG with memory.

4.1 Steady State Probability of the Logistic Map

In the following experiment, interference of non-rigorous methods, that are based on floating-point numbers and finite time simulation, with the original behavior of the given map is illustrated. Figure 6 shows that using 3-bytes floating-point numbers reflects a completely wrong image about the logistic map. Moreover, dependence on the supplied initial condition is obvious from the reported probability density function (PDF).

On the other hand, Fig. 7 shows fast and stable convergence of the computed PDF using small size dyadic numbers. Independence from initial conditions, fast convergence, and stability of the proposed method make it far better than existing non-rigorous finite time methods in approximating the map behavior at the steady state.

The reason of the poor approximation of non-rigorous methods is that the phase space is sampled only at the floating-point numbers and it seems that 3-bytes representation results in too large spacing between the floating-point numbers to capture the dominant dynamics of the logistic map. Another side effect of the too large spacing is a bigger chance that whenever two points on the simulated trajectory are close, they may be represented by the same floating-point number resulting in being trapped in a fake periodic orbit. What makes the

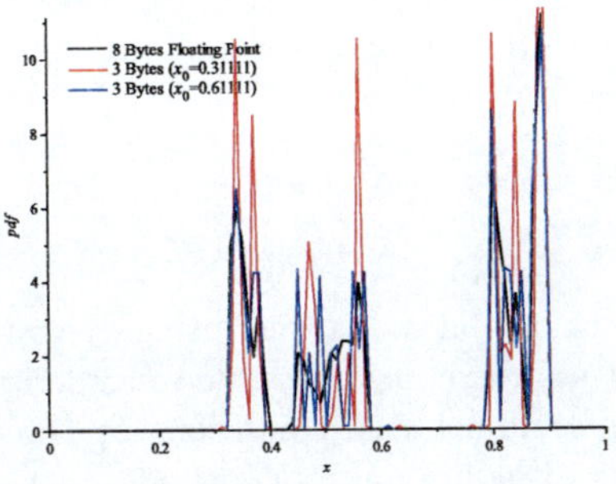

Fig. 6. Computing the PDF of $\mathcal{L}_{3.58}$ using floating-point numbers 19^5 in finite time $n \in [0, 10^4]$.

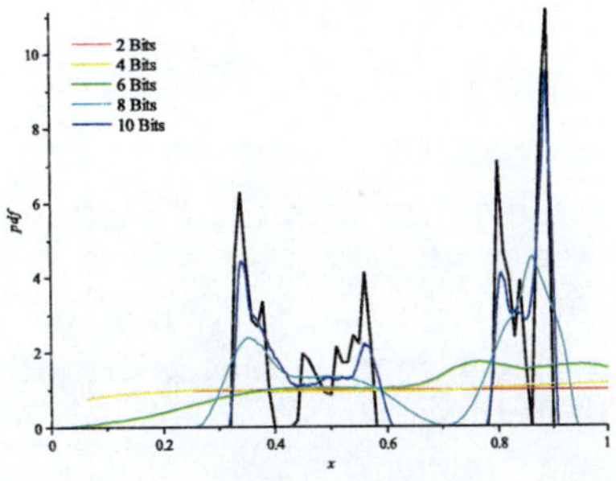

Fig. 7. Computing the PDF of $\mathcal{L}_{3.58}$ using dyadic numbers at finite resolution: $R = 2^{-2}, 2^{-4}, 2^{-6}, 2^{-8}$, and 2^{-10}.

problem even worse is the non-uniform distribution of the floating-point numbers resulting in larger spacing near 1 where the attractor of the logistic map tends to be. Although increasing the precision of the floating-point numbers seems to be a reasonable solution, dependence on the supplied initial condition will persist.

On the other hand, the proposed method uses dyadic partition. Therefore, the phase space is partitioned uniformly rather than sampled allowing capturing all the system dynamics and resulting in a better approximation. More details about this experiment can be found in [17].

4.2 Combinatorial Pseudo Random Number Generator

Two pseudo random stream of length 10^5 were generated using memoryless CPRNG and CPRNG with memory. A combinatorial representation of $\mathcal{L}_{3.58}$ was generated then two random walks were performed on the resulting graph. The steady state probability is computed from both trajectories by calculating the corresponding histogram. It was found that the two resulting PDF complies with each other as shown in Fig. 8 and with the one computed from the graph directly shown in Fig. 7.

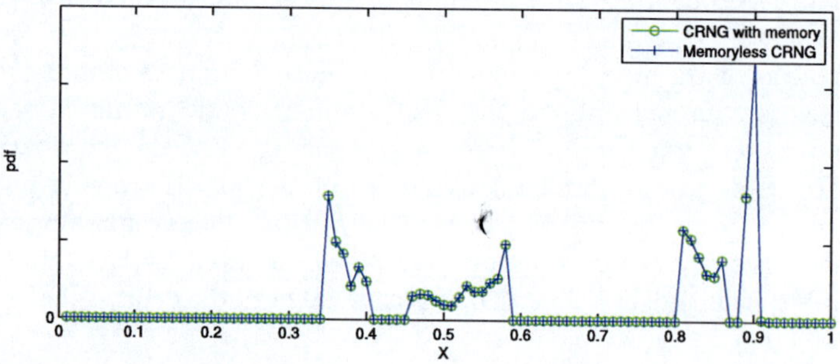

Fig. 8. Computed PDF of memoryless CPRNG and CPRNG with memory using $\mathcal{L}_{3.58}$.

5 Conclusions and Future Works

In this paper, it was shown that, unlike applications based on computing dynamical systems in finite time with floating-point numbers, it is possible to come up with methods that are stable and independent from initial conditions by computing dynamical systems at finite resolution with rational numbers.

For instance, this paper introduced a promising chaotic map based combinatorial pseudo random number generator. It was demonstrated that trajectories simulated using CPRNG are related to the used chaotic map more than that trajectories resulting from non-rigorous finite time simulation. It can be concluded that, using finite resolution instead of finite time allows observing the whole phase space rather than sampling it resulting in a more accurate, more transparent, and fast converging approximation.

We are working on performing further analysis for CPRNG and applying it as a source of unpredictable camera movement in smart surveillance systems.

Acknowledgment. This work is mainly supported by the Ministry of Higher Education (MoHE) of Egypt through PhD fellowship awarded to Dr. Islam ElShaarawy. This work is supported in part by the Science and Technology Development Fund STDF (Egypt); Project id: 42519 - "Automatic Video Surveillance System for Crowd Scenes", and by E-JUST Research Fellowship awarded to Dr. Mohamed A. Khamis.

References

1. Hilborn, R.C.: Chaos and Nonlinear Dynamics: An Introduction for Scientists and Engineers. Oxford University Press, New York (2000)
2. Phatak, S.C., Rao, S.S.: Logistic map: a possible random-number generator. Phys. Rev. E **51**(4), 3670–3678 (1995). https://doi.org/10.1103/PhysRevE.51.3670
3. Amigó, J.M., Kocarev, L., Szczepanski, J.: Theory and practice of chaotic cryptography. Phys. Lett. A **366**(3), 211–216 (2007). https://doi.org/10.1016/j.physleta.2007.02.021
4. Alioto, M., Bernardi, S., Fort, A., Rocchi, S., Vignoli, V.: On the suitability of digital maps for integrated pseudo-RNGs, pp. 349–352 (2003)
5. Alvarez, G., Li, S.: Some basic cryptographic requirements for chaos-based cryptosystems. Int. J. Bifurc. Chaos **16**, 2129–2151 (2006)
6. Arroyo, D., Alvarez, G., Fernandez, V.: On the inadequacy of the logistic map for cryptographic applications. arXiv:0805.4355 [nlin] (2008)
7. Pellicer-Lostao, C., Lopez-Ruiz, R.: Notions of chaotic cryptography: Sketch of a chaos based cryptosystem. arXiv:1203.4134 [cs, nlin] (2012)
8. Persohn, K.J., Povinelli, R.J.: Analyzing logistic map pseudorandom number generators for periodicity induced by finite precision floating-point representation. Chaos Solitons Fract. **45**(3), 238–245 (2012). https://doi.org/10.1016/j.chaos.2011.12.006
9. Tuncer, T., Celik, V.: Hybrid PRNG based on logistic map. In: 2013 21st Signal Processing and Communications Applications Conference (SIU), pp. 1–4 (2013). https://doi.org/10.1109/SIU.2013.6531517
10. Caponetto, R., Fortuna, L., Fazzino, S., Xibilia, M.G.: Chaotic sequences to improve the performance of evolutionary algorithms. IEEE Trans. Evol. Comput. **7**(3), 289–304 (2003). https://doi.org/10.1109/TEVC.2003.810069
11. Yang, D., Li, G., Cheng, G.: On the efficiency of chaos optimization algorithms for global optimization. Chaos Solitons Fract. **34**(4), 1366–1375 (2007). https://doi.org/10.1016/j.chaos.2006.04.057
12. Tatsumi, K., Ibuki, T., Tanino, T.: A chaotic particle swarm optimization exploiting a virtual quartic objective function based on the personal and global best solutions. Appl. Math. Comput. **219**(17), 8991–9011 (2013). https://doi.org/10.1016/j.amc.2013.03.029
13. Zelinka, I., Chadli, M., Davendra, D., Senkerik, R., Pluhacek, M., Lampinen, J.: Hidden periodicity - chaos dependance on numerical precision. In: Zelinka, I., Chen, G., Rössler, O.E., Snasel, V., Abraham, A. (eds.) Nostradamus 2013: Prediction, Modeling and Analysis of Complex Systems. AISC, vol. 210, pp. 47–59. Springer, Hidelberg (2013). https://doi.org/10.1007/978-3-319-00542-3_7

14. Galias, Z.: Interval methods for rigorous investigations of periodic orbits. Int. J. Bifurc. Chaos **11**(09), 2427–2450 (2001). https://doi.org/10.1142/S0218127401003516
15. Luzzatto, S., Pilarczyk, P.: Finite resolution dynamics. Found. Comput. Math. **11**(2), 211–239 (2011). https://doi.org/10.1007/s10208-010-9083-z
16. ElShaarawy, I., Gomaa, W.: An efficient computational framework for studying dynamical systems. In: 2013 15th International Symposium on Symbolic and Numeric Algorithms for Scientific Computing (SYNASC), Romania, pp. 138–145 (2013)
17. ElShaarawy, I., Gomaa, W.: Ideal quantification of chaos at finite resolution. In: Murgante, B., et al. (eds.) ICCSA 2014, Part I. LNCS, vol. 8579, pp. 162–175. Springer, Cham (2014). https://doi.org/10.1007/978-3-319-09144-0_12
18. Lovász, L.: Random walks on graphs: A survey. Combinatorics Paul Erdõs Eighty **2**, 1–46 (1993)

Author Index

© The Editor(s) (if applicable) and The Author(s), under exclusive license
to Springer Nature Switzerland AG 2022
A. E. Hassanien et al. (Eds.): AMLTA 2022, LNDECT 113, pp. 697–699, 2022.
https://doi.org/10.1007/978-3-031-03918-8